# TRANSPORTATION ENGINEERING

## *Planning and Design*

**THIRD EDITION**

**Paul H. Wright**
*Professor*
*Georgia Institute of Technology*

**Norman J. Ashford**
*Professor*
*University of Technology, Loughborough*

**John Wiley & Sons**
*New York • Chichester • Brisbane • Toronto • Singapore*

***Library of Congress Cataloging in Publication Data:***

Wright, Paul H.
Transportation engineering : planning and design / Paul H. Wright, Norman J. Ashford.—3rd ed.
p. cm.

Rev. ed. of: Transportation engineering / Radnor J. Paquette, 2nd ed. 1982.
Includes index.
ISBN 0-471-83874-8
1. Ashford, Norman. II. Paquette, Radnor Joseph. Transportation engineering. II. Title.
TA1145.W75 1989
629.04—dc19

86-26090
CIP

Printed in the United States of America

10 9 8 7 6

Printed and bound by Courier Companies, Inc.

# PREFACE

This book is the successor to *Transportation Engineering: Planning and Design*, which first appeared in 1972 and as a second edition in 1982. It is a basic textbook for transportation courses in civil engineering and city planning. It should also prove valuable as a professional reference for engineering consultants faced with transportation problems.

The book is divided into six parts. The first three parts—approximately half of the book—deal with planning and other nonengineering aspects of transportation. The material in these chapters is presented on a multimodal or functional basis.

Part One introduces the transportation system of the United States; the laws that have influenced its development; and administration organizations that continue to influence its development, finance, and use. Part Two deals with the operation and control of the vehicles that use the physical transport systems. It includes chapters on human factors, operational characteristics, and traffic control devices and procedures. Part Three extensively describes transportation planning including a general overview of the planning process, data collection procedures, mathematical and computer modeling, and procedures for the development and evaluation of transportation plans. This part also contains a new chapter on traffic impact analysis as well as a treatment of TSM and other short-term planning techniques.

Parts Four, Five, and Six, comprising 11 chapters, deal with the design of land, air, and water transportation facilities, respectively. In Part Four, a unified discussion of the design of streets and highways, railways, and guideway systems is given. Chapters are also included on the design of land transportation terminals and pipeline transportation. Specific planning procedures and design criteria for air transportation facilities are given in Part Five. Airport planning, site selection, layout, and design are discussed, along with such topics as terminal layout, runway orientation, airport configuration, runway and taxiway dimensioning, and airport lighting. Similar criteria for water transportation facilities, including up-to-date material on the design for the coastal environment, the design of harbors, and the design of various types of port facilities are discussed in Part Six. For the reader who wishes to study a transportation problem in greater detail, an excellent list of references is given at the end of each chapter.

This is a multimodal work, but we as authors recognize the realities of modal separation. Although planners and engineers in one mode have a real need to understand the workings of other modes, seldom do they engage professionally in other modes. Multimodal planning and design are intellectually appealing, but they are seldom achieved in the real world of engineering. Each mode has separate standards of planning, design, operation, management, and regulation. Financing is oriented modally and is largely destined to remain so, and careers become focused modally. Therefore the approach adopted in this work instills practical realism rather than idealism into transportation engineering and planning.

Necessarily, several areas have been either only lightly touched upon or have been omitted. The history of transport, especially in other civilizations, is one of them. This material, while of general interest, is largely irrelevant to the needs of today's professionals; furthermore, only a superficial and therefore necessarily misleading treatment could be given to a subject that would require a whole text for adequate coverage.

We are grateful to many organizations and individuals for supplying information, photographs, and sketches for this book. Direct credit is given in the text to these sources.

We particularly thank the individuals who have helped in the preparation of this work. They include Ian Galer and Brian Ratcliffe of Loughborough University, who contributed to Chapters 6 and 9, respectively, Vivien Grove, who provided invaluable assistance in the logistics of preparing the revised manuscript, and Virginia Tiernan, Jim Anderson, and Jon Wright, who assisted with research and revision. Thanks must also go to Jeff Griffith of the Federal Aviation Administration and to John Hassell formerly of the Georgia Department of Transportation for their helpful advice in updating the book.

PAUL H. WRIGHT
NORMAN J. ASHFORD

# ABOUT THE AUTHORS

**PAUL H. WRIGHT** is a Professor of Civil Engineering at Georgia Institute of Technology, where he has taught since 1963. He is a registered Professional Engineer in Georgia and a member of many professional societies, including the American Society of Civil Engineers, the Institute of Transportation Engineers (ITE), and the American Society of Engineering Education. Dr. Wright received a Sigma Xi Faculty Award and ITE's Technical Paper Award (Southern Section) in 1976. He has been active in continuing education programs and has served as a consultant to transportation agencies in the United States and Central America. The author of numerous papers, he is coauthor of three other books: *Highway Engineering, Fifth Edition* (1987), *Airport Engineering, Second Edition* (1984), and *Introduction to Engineering* (1989). He earned his Ph.D. degree in Civil Engineering (transportation planning) from Georgia Tech.

**NORMAN ASHFORD** is Professor of Transportation Planning at the Loughborough University of Technology in England. He is a registered Professional Engineer in the United States, Canada, and the United Kingdom. Dr. Ashford lived in North America for 15 years, during which he served as a civil engineering consultant in Canada, did postgraduate work at Georgia Institute of Technology, and began his teaching career at Florida State University. He is a Fellow of the Institution of Civil Engineers and of the Chartered Institute of Transport as well as a member of the American Society of Civil Engineers and of the Association of Professional Engineers of Ontario. He has published many papers and is the coauthor of *Airport Engineering* (1984), *Airport Operations* (1984), and the author of *Mobility and Transport for Elderly and Handicapped Persons* (1982).

# CONTENTS

# *PART 1*

# *INTRODUCTION TO TRANSPORTATION SYSTEMS*

*Part One describes the importance, nature, and overall size of the transportation system of the United States; the users of the system; the government agencies that subsidize and regulate its use; and the sources of income for its continued development, maintenance, and operation.*

# *Chapter 1*

# *The Transportation System of the United States*

An inherent need of highly developed and industrialized nations is a sophisticated and widespread transportation system. In developed nations there must be an easy mobility of persons and goods. As the degree of industrialization of an economy increases, there is a change in preponderance from *basic* production industries, sometimes called primary industries, to the service industries of a *secondary*, *tertiary*, and *quarternary* character. Primary industries have a great need for freight transportation. Industries in the service area display a need for both freight and extensive personal mobility. The United States, whose economy depends greatly on nonprimary industries, has found a need to provide the world's most sophisticated and widespread transportation system, which is expanding under the constant pressure of increasing demand. Because of the unprecedented need for nationwide mobility, even in the previously remote areas of the country, there is a requirement not only for various modes of transport but also for increasingly sophisticated interfaces between the modes of subsystems.

## 1-1. SUBSYSTEMS AND MODES

The overall transportation system is required to provide for both passenger movements and freight movements. Both subsystems of movement are able to supply various levels of service (measured by such criteria as convenience, speed, safety, and availability), at varying levels of price. The user of transportation service is seldom in the position of having only one service available. Monopoly conditions, prevalent in the days of early railroads, were broken by the introduction of motor vehicle and air transportation.

Modern technology has evolved a great variety of transporting devices either in the design or development stage. Five modes, however, account for all but a fractional percentage of all ton mileage of freight and passenger mileage of person travel.

1. Motor vehicles
2. Railroads
3. Air transport
4. Water transport
5. Pipelines

## 1-2. HIGHWAY SYSTEM CLASSIFICATION

The highway system of the United States is publicly owned. The elements of the system are customarily classified according to the jurisdiction or level of governmental unit that has the overriding influence in the construction and operation of the system. In Sections 1-7 to 1-13 the various subsystems of highways are discussed in detail. These subsystems are [1]:

1. The national system of interstate and defense highways.
2. The federal aid system—primary, secondary, and urban.
3. The state primary and secondary systems.
4. Local and county roads.
5. City streets.
6. Toll roads.

Roads are also classified by *function*. Used mainly for planning purposes, the rural and urban classifications are slightly different.

- Rural
  - Arterials
    - Principal—interstate
    - Principal—other
    - Minor
  - Collectors
    - Major
    - Minor
  - Local roads
- Urban
  - Arterials
    - Principal—interstate
    - Principal—other freeway and expressway
    - Principal—other
    - Minor
  - Collector streets
  - Local streets

## 1-3. RAILROAD SYSTEM CLASSIFICATION

Railroad systems are designed customarily by two separate system classifications, *revenues* and *use*.

*Revenue Classification.* As set down by the Interstate Commerce Commission, this divides the railroads into: Class I railroads; Class II railroads; Class III railroads; switching and terminal companies.

Before 1956, Class I railroads consisted of those carriers with annual operating revenues of $1 million or more. On January 1, 1982 the Interstate Commerce Commission adopted a procedure to adjust the threshold for inflation by restating current revenues in constant 1978 dollars when the Class I railroads were those companies whose average annual operating revenues were $50 million or more. Class II railroads were those companies with operating revenues in the range of $50 to $10 million and Class III railroads had revenues of less than $10 million. By 1984, the basis for Class I railroads was $87.3 million. Switching and terminal companies provide no line-haul service. Their facilities connect to Class I, II, and III facilities to provide switching and terminal service only, as the name implies. In urban areas, these facilities eliminate a duplication of switching facilities and also enable railroad service to be provided at one terminal location. In the United States in the mid-1980s, excluding Amtrack, there were approximately 25 Class I railroads, 26 Class II, and 375 Class III, and 172 switching and terminal railroads. The 25 Class I railroads accounted for 95 percent of all rail freight movement.

Even among the Class I companies there is a domination of traffic by a few large roads. Some observers have felt for years that the future development of the railroad system will lead to widespread mergers resulting in all rail service being provided by a few giant companies with a regional monopoly. In 1968 two of the six largest companies (which at that time accounted for over 40 percent of all freight ton mileage) merged with Supreme Court approval. The New York Central and Pennsylvania Railroads formed the Penn Central. By 1970 the Penn Central had filed bankruptcy proceedings. The federal government established the U. S. Railway Association to look into consolidating the six major railroads of the Northeast. The result was the formation of the giant Conrail which commenced operations on April 1, 1976, serving 16 states, Washington, D.C., and the provinces of Ontario and Quebec.

*Use Classification.* Railroads may also be classified by use. The various classes are *line-haul, switching, belt-line,* and *terminal* companies. Line-haul companies concentrate essentially on providing intercity movement of freight and passengers. Belt-line and switching companies connect to intercity facilities to provide special access to warehousing, wharves, and industrial areas. Terminal companies operate terminals and approach trackage; in this way, more than one line-haul company has use of terminal facilities, avoiding a duplication of plant.

## 1-4. THE AIRPORT SYSTEM CLASSIFICATION

There are more than 16,000 airports in the United States, including approximately 6000 open to public use. See Table 1-1. All airports that contribute significantly to the national transportation system are included in the National Plan of Integrated Airport Systems (NPIAS) [2]. Four categories of airport comprise the airport classification system.

1. Commercial service—primary
2. Commercial service—other
3. Reliever airports
4. General aviation

A *commercial service airport* is one which is served by at least one scheduled passenger service carrier at which there are a minimum of 2500 annual enplanements. *Primary airports* are defined as commercial service airports having 0.01 percent or more of total U. S. enplanements. This amounted to 30,915 enplanements when the system was set up in 1982. *Reliever airports* are general aviation airports in metropolitan areas which are intended to reduce congestion at large air carrier airports by providing alternative general avia-

*Table 1-1*
**Number of Airports by Ownership and Public Use as of January 1, 1984**

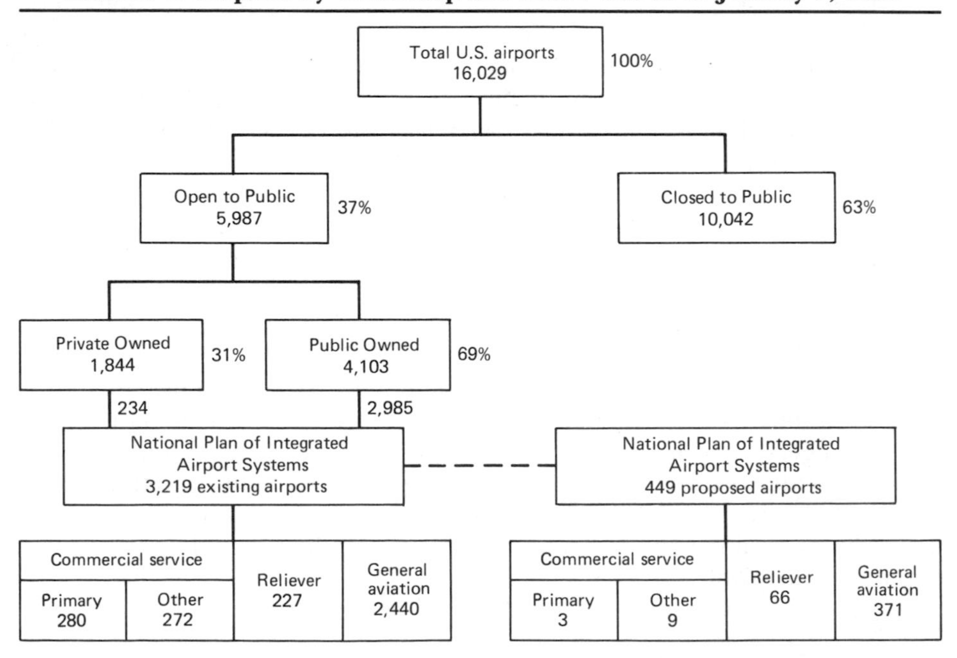

tion facilities. In the early 1960s Congress provided priority funding in special legislation to develop these airports. When originally set up the classification required that the reliever airport should have an activity level of at least 50 based aircraft, 2500 itinerant operations, or 35,000 local operations. The relieved airport should be operational at 60 percent of its capacity, at least, and should either serve a standard metropolitan statistical area (SMSA) of 500,000 population or should have 250,000 annual enplanements. A general aviation airport is normally included in the NPIAS if it satisfies any one of a number of criteria: significant local, regional, or national interest; receiving U.S. mail service; significant military activity; status of a general aviation heliport with more than 400 itinerant operations by air taxi, or have more than 810 itinerant operations, or have four based aircraft. The FAA has established general aviation airport categories based on aircraft design. In the NPIAS, some general aviation airports may have a very small amount of scheduled commercial service, less than 2500 annual enplanements. For convenience these are classified as general aviation.

A *basic utility* (BU) general aviation airport accommodates most single and many of the smaller twin engine aircraft, about 95 percent of the general aviation fleet.

A *general utility* (GU) airport accommodates virtually all general aviation aircraft with maximum takeoff weights of 12,500 pounds or less.

Other general aviation airports are based on transport type aircraft or business jets. These *transport airports* are designed to accommodate aircraft with approach speeds of more than 120 knots and are usually capable of handling turbo jet powered aircraft.

## 1-5. PIPELINE SYSTEM CLASSIFICATION

Pipelines are most commonly classified according to function and to type of commodity carried. The two principal classes of commodity carried are crude oil and its refined products and natural gas. A classification scheme for all pipelines would be of the following form:

- Oil Pipelines
  - Gathering lines
  - Trunk lines
    - crude oil
    - products
- Gas Pipelines
  - Field lines
  - Trunk lines

For further discussion of pipeline classification, the reader is referred to Section 1-17.

## 1-6. TRANSPORTATION DEMAND BY MODE [3]

The rising population and increasing industrialization of the United States have been accompanied by an increasing demand for transportation. As an industry, transportation amounts to a significant proportion of the economy. Economists estimate that the annual bill for transportation is close to 21 percent of the Gross National Product and the industry employs approximately 11 percent of the labor force. Approximately 13 percent of all federal taxes are collected from the transportation sphere. Figure 1-1 indicates the national trends in transportation demands since 1939. A clear relationship can be discerned between transportation demand and such indices as total population, industrial production, and the Gross National Product.

While overall demand for transportation, both passenger and freight, has been increasing rapidly with marked acceleration after World War II, the modal shares of this total demand have not been constant. Each mode has shown significant secular trends which tend to reflect governmental regulatory policy and technological development.

*Rails.* In the last three decades, rail intercity freight has accounted for a decreasing share of the total ton mileage, dropping from a peak of 71.3 percent in 1943 to 35.9 percent in 1985. Although the market share has declined, the absolute ton mileage has remained reasonably constant since 1970 (771 billion ton miles in 1970 *vs* 897 billion ton miles in 1985). The

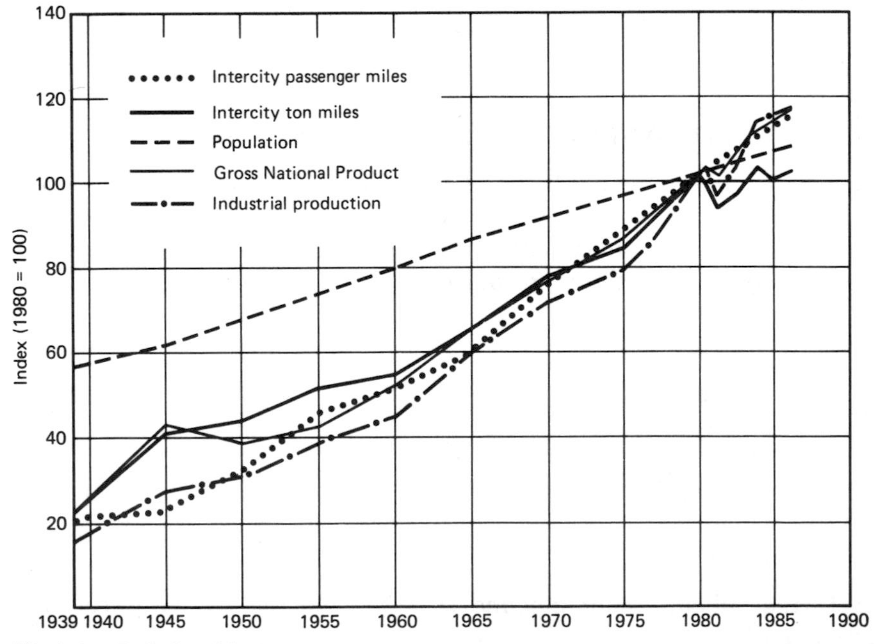


*Fig 1-1.* Relationship between transportation demand and growth indicators.

principal benefactor from rail's relative decline in the freight transportation market has been truck transport. The steadily increasing standards of the intercity road network has promoted intercity truck traffic from 53 billion ton miles in 1939 to 636 billion ton miles in 1985. This represents a 1100 percent increase, compared with an increase of only 16 percent for rail freight. Table 1-2 shows intercity freight ton mileage for all modes for the period 1939–1986.

Rail passenger traffic showed a long-term secular decline until the combined effects of better service with Amtrak and the large increases in fuel prices in 1973 had an impact. From the end of World War I, except for a brief period during gasoline rationing in World War II, intercity passenger mileage steadily decreased until 1972. In 1929, total intercity rail passenger mileage was 24.2 billion passenger miles. This decreased to a low of 8.7 billion passenger miles in 1972, reaching 11.9 billion passenger miles in 1985. A similar pattern is observed in rail commuting figures, which are 6.9 billion, 4.2 billion, and 5.1 billion passenger miles for the same years [4]. The degree to which air travel has provided competition to the railroads is indicated clearly by the dramatic rise of air passengers in the last 40 years. In 1939, air passengers amounted to 1.9 million persons or 0.5 percent of all intercity travelers. While railroads showed a significant long-term decline, air travelers reached 398 million in 1986 or approximately 38 percent of all intercity common carrier movements. Rail passenger traffic suffers stiff competition not only from air carriers but also from the auto mode. The greatest amount of intercity travel is not, however, made by any form of common carrier. Throughout the 1950s and first half of the 1960s private auto use accounted for over 89 percent of intercity passenger mileage. This figure remained

*Table 1-2*

**Intercity Freight by Modes (Billions of Ton Miles)**

| *Year* | *Rail* | *Truck* | *Oil Pipeline* | *Inland Waterways* | *Air* |
|---|---|---|---|---|---|
| 1939 | 334 | 53 | 56 | 96 | 0.01 |
| 1945 | 691 | 67 | 127 | 143 | 0.09 |
| 1950 | 597 | 173 | 129 | 164 | 0.30 |
| 1955 | 631 | 223 | 203 | 217 | 0.49 |
| 1960 | 579 | 285 | 229 | 220 | 0.89 |
| 1965 | 709 | 359 | 306 | 262 | 1.91 |
| 1970 | 771 | 412 | 431 | 319 | 3.30 |
| 1975 | 759 | 454 | 507 | 342 | 3.73 |
| 1980 | 932 | 555 | 588 | 407 | 4.84 |
| 1984 | 935 | 606 | 568 | 399 | 7.04 |
| 1985 | 895 | 610 | 564 | 382 | 6.08 |
| 1986 | 897 | 636 | 580 | 376 | 7.10 |

SOURCE: *Transportation in America*, 1987 ed., Transportation Association of America.

relatively constant over three decades, with the exception of the war years, but with gasoline shortages and rapidly rising gasoline costs there has been a significant decline in the percentage of intercity travel by auto. From a peak of 90.4 percent in 1960, the figure had dropped significantly to 81.3 percent in 1985.

*Highways.* The United States is the world's most mobile country. The right of personal mobility is accepted almost universally on a nationwide basis. This mobility is due mainly to the widespread provision of a superb road system combined with high per capita car ownership. In 1985 there was one motor vehicle for every 1.4 persons in the United States. Vehicle registrations followed the long-term increase shown in Fig. 1-2, reaching a figure of ap-

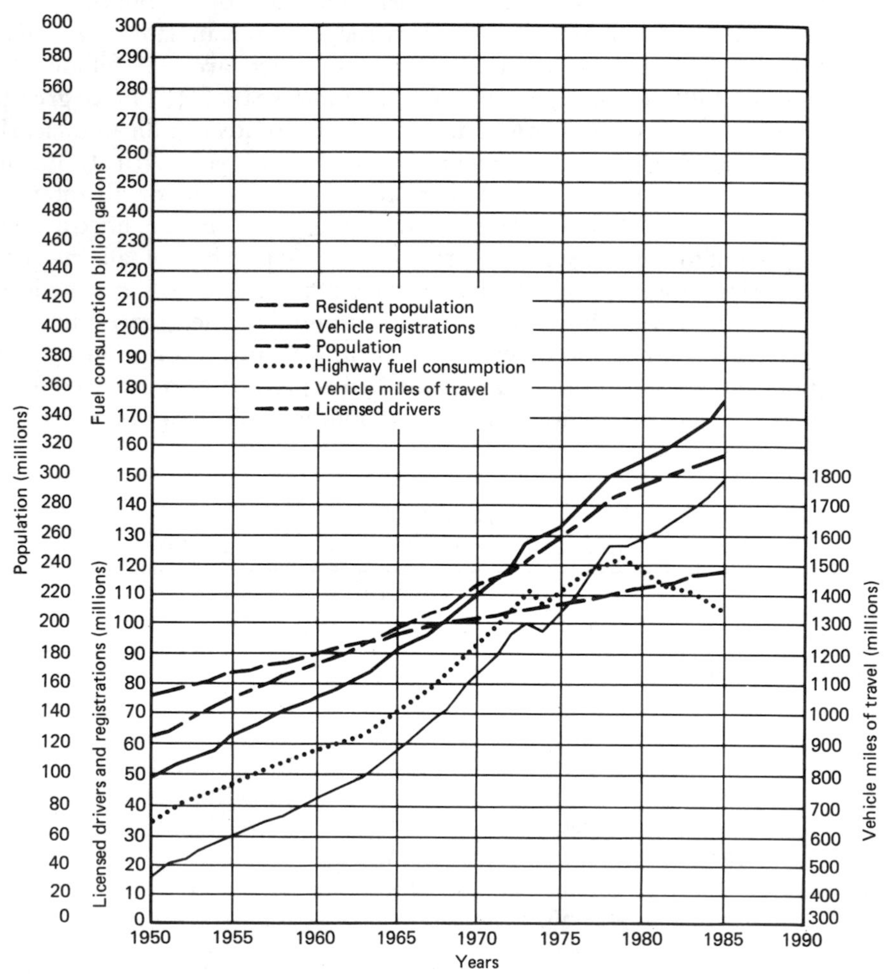


*Fig 1-2.* Resident population, vehicle registrations, fuel consumption, licensed drivers, and vehicle miles of travel.

proximately 171.7 million vehicles made up of 132 million autos, 39.0 million trucks, and 593,000 buses. The total amount of vehicle travel in 1985 was 1774 billion vehicle miles, an average of 10,330 miles per vehicle. The degree of personal mobility is underscored by the fact that of a population of approximately 239 million persons, over 156 million are licensed drivers [1].

Fifty-eight percent of the mileage traveled each year is estimated to occur in the urban areas. The amount of urban travel carried out by highway systems varies from city to city depending on availability of rail transit systems. Where no rail transit exists, urban mobility is totally dependent upon the highway system. Where rail transit does exist there is considerable diversion of downtown-oriented trips but a much smaller effect on a citywide basis. For example, in Chicago, a study found that the rail systems accounted for only 4.0 percent of all trips in the urban area. In comparison, 30 percent of all trips into the downtown area were made by rail, the remaining 70 percent being made by bus transit or automobile.

Similarly, intercity passenger and freight movements are heavily dependent on highway facilities. Rural road travel amounted to 730 billion vehicle miles in 1985. Of all intercity passenger miles traveled, private auto constantly accounts for approximately 80.4 percent of the total, and intercity bus accounts for an additional 1.3 percent. The remaining amount of travel is divided between air and rail, with an increasing share being taken by the former [3].

The relative importance of highway trucking on intercity freight movements can be inferred from reference to Table 1-2, which shows that freight ton mileage moved on the highways accounts for approximately 25.5 percent of all intercity ton mileage.

*Inland Waterways.* Since water transportation is slow and the modal interface relatively inconvenient, water transportation is primarily a question of freight movement. The type of freight moved tends to be limited to those classes that are bulky and of low cost per unit volume. Bituminous coal and lignite alone amounted to approximately 25 percent of all tonnage moved by water in 1980. Eight classes of commodities accounted for 81 percent of net tonnage in the same period. A breakdown of principal commodities transported is given in Table 20-1. The market share of total ton mileage moved over the last three decades shows a very small long-term decline, falling from 18 percent in 1939 to 15.1 percent in 1986. The actual ton mileage shows a steady growth, amounting to a quadrupling of ton miles in this period, as shown in Table 1-2. Without a dramatic breakthrough in technology, the economics of water transportation are likely to remain at approximately the same level. Under these conditions, this mode can be expected to retain approximately the same share of the total freight market with an increasing annual total ton mileage closely following the increase in the Gross National Product.

*Air Transport.* Air transportation is the fastest growing mode, both in the areas of passenger and freight transport. Table 1-3 shows the rapidly increas-

*Table 1-3*
**Intercity Passenger Travel by Public Carrier Modes (Billions of Passenger Miles)**

| *Year* | *Air* | *Bus* | *Rail* | *Total* |
|---|---|---|---|---|
| 1939 | 0.8 | 9.5 | 23.7 | 34.0 |
| 1950 | 9.3 | 22.7 | 32.5 | 64.5 |
| 1960 | 31.7 | 19.3 | 21.6 | 72.6 |
| 1970 | 109.5 | 25.3 | 10.9 | 145.7 |
| 1980 | 204.4 | 27.4 | 11.0 | 242.8 |
| 1985 | 277.8 | 24.0 | 11.4 | 313.2 |
| 1986 | 307.6 | 23.1 | 11.9 | 342.6 |

SOURCE: *Transportation in America,* 1987 ed., Transportation Association of America.

ing air passenger mileage in comparison with relatively stable or declining figures for other passenger modes. The increase in air travel is expected to continue but not at the rates of expansion of earlier years, say the 1960s. From 1985 to 1988 scheduled U.S. air travel is expected to increase from 333.6 billion passenger miles to 642.9 billion passenger miles [5]. Aircraft manufacturers have a somewhat similar forecast of 604 billion passenger miles for the same year [6]. The rapid increase in air travel is caused by many underlying factors including increasing Gross National Product, increasing affluence of the middle-income groups, a greater preponderance of nonprimary industry in the economic system, and improved technology.

In addition to the important needs of the passenger carrier sector, there is increasing demand in the areas of *general aviation*. General aviation is the term used for all flying done by aircraft other than military aircraft or by passenger and cargo carriers. From 1958 to 1985, general aviation aircraft handlings by the FAA at air route traffic control centers increased almost fifteenfold, from 535,517 to about 8.7 million [5]. The level of general aviation activity is expected to increase at a continuing rate as businesses use private aircraft and more middle-income individuals use aircraft for recreational purposes. By 1996 general aviation handlings are expected to rise to 12.5 million.

Other areas of demand for air transports are *express and freight* and *mail.* The trunk and local passenger/cargo carriers carry most of this material. The remainder is carried by a relatively small number of all-cargo carriers. Air freight is still a transportation mode in its relative infancy. Improved technology has resulted in a steady lowering of costs which has encouraged a rapid growth rate. If mail and express are included, the amount of intercity air freight has increased from 0.01 billion ton miles in 1939 to 7.1 billion ton miles in 1986. While this rate of increase is greater than for any other mode, it must be remembered that in 1986 air freight amounted to only about 0.28 percent of total intercity freight. By 1999, the aircraft manufacturers estimate that U.S. domestic freight will have grown to 16 billion ton miles.

*Pipelines.* In the terminology of transportation, pipeline transportation normally refers only to the oil and oil-products lines, on the reasoning that the natural gas lines are not in competition with other carriers.

As a mode of freight transportation, the pipelines are often overlooked. The total intercity mileage of oil pipeline in 1986 was over 205,650 miles compared with 121,122 miles in 1939 [3]. During the same period the ton mileage of oil moved showed a steady increase from 56 billion ton miles in 1939, reaching a peak of 608 billion ton miles in 1979 since then it has dropped to 580 billion ton miles in 1985 due to increased oil costs. The 1983 figure reached a low of 556 billion ton miles; even so it represented approximately 24 percent of all intercity freight ton mileage. In that year the nation's pipelines moved more than the total ton mileage of freight carried on the Great Lakes and the Inland Waterways.

## 1-7. NATIONAL SYSTEM OF INTERSTATE AND DEFENSE HIGHWAYS

In 1956, the Congress approved the construction of a 41,000-mile National System of Interstate and Defense Highways. This system, extending from coast to coast, was to provide a high level of service to interstate travel by the provision of nationwide freeway facilities. When completed, all of the Interstate routes will be at least four-lane divided highways, growing to six and even twelve lanes in and near the large metropolitan areas.[1] The 1956 act provided for federal participation in construction funding (including right-of-way acquisition) on a 90-10 matching basis. All maintenance costs were to be borne by the states.

The extent of the proposed system is shown in Fig. 1-3. Initially, the target date for completion of the system was 1975, but this was abandoned at a fairly early stage. By 1984, the extent of the system had been increased to 43,291 miles. More than 98 percent of the planned system was complete and open to traffic. The original proposed system cost of $27 billion, of which $2.7 billion was to come from the individual states, had increased to well over $100 billion due to increased design standards and inflation of construction costs.

The function of the system is to provide high-level-of-service roads both within and between urban areas. Urban areas account for over 10,900 miles of the total system. The Interstate constitutes only 1.1 percent of the total road and street mileage, yet it carries over 22 percent of all vehicle mileage [1]. In comparison with other road systems, the Interstate will be remarkably efficient. The degree of coverage that this limited system provides on a nationwide basis can be seen from the fact that 42 state capitals and 90 percent of all cities with populations in excess of 50,000 are located on its routes.

[1]The current Interstate "traveled way" includes certain relatively low-standard segments of roadway which will be improved or replaced later. Approximately 86 percent of the current system is divided highway with four or more lanes and full control of access. Two-lane undivided sections account for approximately 7 percent of the mileage.

*Fig 1-3.* The national system of interstate and defense highways. (Source: Federal Highway Administration.)

Houlton
Bangor
Augusta
Portland
Portsmouth
Concord
Manchester
Boston
Providence
New Bedford
Plattsburgh
St. Johnsbury
Mont-pelier
White Riv. Jct.
Springfield
Hartford
New Haven
Utica
Albany
Syracuse
Rochester
Lewiston
Buffalo
Binghamton
Scranton
Wilkes-Barre
Newark
Asbury Park
Trenton
Philadelphia
Wilmington
Harrisburg
Reading
Hagerstown
Baltimore
Washington
Strasburg
Staunton
Lexington
Richmond
Petersburg
Newport News
Norfolk
Roanoke
Winston-Salem
Greensboro
Durham
Rocky Mount
Statesville
Asheville
Charlotte
Fayetteville
Spartanburg
Florence
Columbia
Charleston
Augusta
Savannah
Macon
Atlanta
La Grange
Columbus
Jacksonville
Lake City
Tallahassee
Daytona Beach
Tampa
St. Petersburg
Naples
Miami
Erie
Conneaut
Cleveland
Youngstown
Akron
Canton
Pittsburgh
Wheeling
Marietta
Charleston
Beckley
Columbus
Dayton
Cincinnati
Frankfort
Lexington
Louisville
Knoxville
Nashville
Chattanooga
Huntsville
Gadsden
Birmingham
Tuscaloosa
Montgomery
Mobile
Pensacola
Sault Sainte Marie
Mackinaw City
Muskegon
Grand Rapids
Flint
Port Huron
Lansing
Battle Creek
Detroit
Benton Harbor
Toledo
Duluth
Superior
Minneapolis
St. Paul
Green Bay
Tomah
Albert Lea
Madison
Milwaukee
Chicago
Waterloo
Sioux City
Des Moines
Davenport
Rock Island
Council Bluffs
Peoria
Bloomington
Champaign
Indianapolis
Springfield
Effingham
St. Joseph
Topeka
Kansas City
St. Louis
Mt. Vernon
Evansville
Springfield
Sikeston
Cairo
Tulsa
Ft. Smith
Little Rock
West Memphis
Memphis
Texarkana
Shreveport
Monroe
Vicksburg
Jackson
Meridian
Baton Rouge
New Orleans
Lake Charles
Houston
Galveston

The Interstate System represents a massive public involvement in transportation. In its time it constituted the world's most expensive public works program. Originally the financing was to be carried out on a "pay-as-you-go" basis. Revenues from federal gasoline and other user taxes were placed in an earmarked Highway Trust Fund from which federal expenditures on the system would be drawn. Because insufficient revenue was available for construction, this feature of the 1956 act was suspended in 1960.

## 1-8. THE FEDERAL-AID PRIMARY SYSTEM

The main highway system interconnecting the cities and towns of the United States is the federal-aid primary system. It is composed of 301,000 miles of traveled way, of which 42,474 miles are in the urban areas. The Interstate System is included within the primary system. All but approximately 150 miles of the federal-aid primary system is surfaced; approximately 88 percent of the rural mileage and 39 percent of urban mileage is comprised of two-lane facilities. Federal highway funds are available for construction and improvement of the federal primary system. For those sections not on the Interstate System, the federal share of the funding is 75 percent including right-of-way purchase. Maintenance of these facilities falls totally on state and local authorities.

## 1-9. THE FEDERAL-AID SECONDARY SYSTEM

This is comprised of the principal, secondary, and feeder roads linking farms, distribution outlets, and smaller communities with the federal-aid primary system. It is under state, county, or local authority control and comprises approximately 398,000 miles of road, of which less than 1 percent is unsurfaced. As with the federal-aid primary system, federal highway funds are available for construction and improvement on a 70-30 federal-state matching basis. Maintenance of the system is carried out by state and local authorities.

## 1-10. THE FEDERAL-AID URBAN HIGHWAY SYSTEM

Prior to 1944, only token amounts of federal or state funds were expended for road construction within urban areas. To relieve the urban area, 25 percent of federal funds was earmarked, starting in 1944, for use inside urban areas. Arterial routes within urban areas connecting to the federal primary and secondary systems became known as the urban extensions. The federal-aid urban system was first authorized by the Federal-Aid Highway Act of 1970 and its definition altered by the Federal-Aid Highway Act of 1973. It is comprised of 144,055 miles of road, of which approximately 1400 miles is unsurfaced. The system serves major centers of activity within urban areas.

## 1-11. STATE SYSTEMS

Within each state, state highway departments have designated state primary and state secondary Systems. These systems overlap the federal primary and secondary systems. Local and county authorities bear different fiscal responsibilities on these two systems, with less state contribution to the secondary system. In 1986, state highways accounted for approximately 800,000 miles of road. Total expenditures on state-administered highways in 1985 amounted to $33.3 billion. Federal expenditures by the Federal Highway Administration amounted to $13.3 billion.

## 1-12. LOCAL AND OTHER SYSTEMS

County, town, township, and other local systems comprise a total of 3.86 million miles of road, of which about 43 percent is under county control. In addition, slightly over 960,000 miles of road fall under city street classification. Maintenance, improvement, and construction of these facilities are the responsibility of the local authorities.

In addition, there are approximately 253,000 miles of road not overlapping state, county, or other local systems. Generally these roads are national and state park, forest, and reservation roads. A large proportion of this system is not under the federal-aid system.

## 1-13. TOLL ROADS

There are over 3000 miles of toll roads in operation in the United States. Over 2000 miles are now under the federal-aid systems. Construction of these facilities stretched from the late 1930s to the middle 1950s. Toll facilities were conceived to solve the traffic problems of special areas, both urban and rural. For this purpose, the states created special local or statewide authorities empowered to build and operate toll facilities. Financing was carried out by revenue bonds paid off from toll charges levied against the users. The New York State Thruway and the Sunshine State Parkway in Florida are examples of statewide toll facilities. Local bodies such as the Jacksonville Expressway Authority are responsible for the provision of extensive intraurban toll networks.

The construction of toll roads, especially interurban facilities, fell into disfavor with the passage of the 1956 Federal Highway Act which authorized construction of competitive free facilities. In some areas, toll facilities have been incorporated into the Interstate System. It is anticipated that these facilities will become toll free when the bonded indebtedness is repaid.

## 1-14. RAILROADS [4]

Virtually every community of reasonable size in the United States is either served by or within close reach of rail service. Figure 1-4 shows the railroad routes between principal cities of the United States. This represented a net capitalized investment of $45.4 billion in 1984 in plant and equipment. In 1984 Class I railroads operated over 165,000 miles of line, 24,500 locomotives, and approximately 1.5 million cars. With minor exceptions, operations are carried out on company-owned and -maintained rights-of-way.

In addition, in 1985 railroads owned and operated approximately 47,000 service vehicles (ballast cars, snow removing cars, boarding outfit cars, etc.), and such floating equipment as tugboats, car ferries, lighters, and barges. These amounted to over 900 vessels for all railroad companies. Stationary plants both owned and operated by the companies include tunnels, bridges, stations, and buildings, water and fuel stations, grain elevators, warehouses, wharves, communications systems, systems powerplants, transportation systems, shops, and yards. Operating revenues amounted in 1984 to approximately $29.4 billion with income payments to employees amounting to $11 billion. Employment in the Class I railroads was 341,500.

## 1-15. THE AIRWAYS [7]

The National Airspace System has been established by the federal government for the purpose of providing a safe environment for air travel through the airspace above the United States. En route guidance is provided by the *very high frequency* (VHF) multidirectional system (see Chapter 5) with an airway mileage of 182,182 miles direct, and 3306 miles alternate at low altitudes, and 142,658 miles at jet altitudes in the 48 contiguous states [6]. The low-altitude system operates from 1200 feet (in some areas 700 feet) up to 18,000 feet while the high-altitude jet route mileage extends from 18,000 to 45,000 feet. The VHF system, often referred to as the VOR/VORTAC system, was introduced in 1952, phasing out the older *low and medium frequency* (L/MF) four-course system which constituted the airways prior to that time. Currently the airspace is also under Air Positive Control (APC). All airspace from 24,000 to 60,000 feet is covered by APC. Operations within this airspace must be under the control of the FAA using Instrument Flight Rules. In the busy north and northeast sections of the United States, the ceiling has been lowered to 18,000 feet.

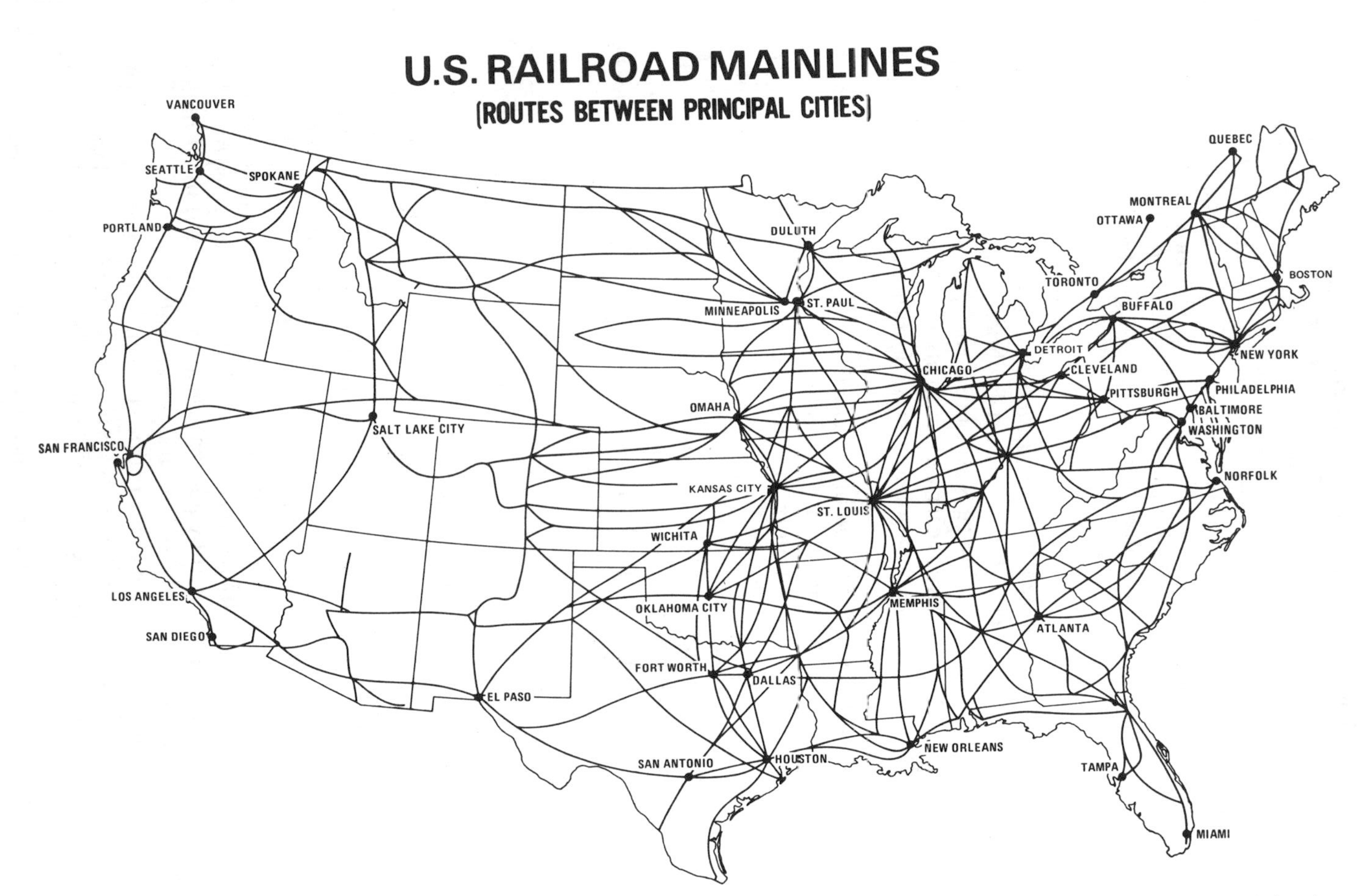


*Fig 1-4.* Railroad routes between principal cities in the United States. (Courtesy U.S. Department of Transportation.)

*Table 1-4*
**Hub Classification Related to Passenger Traffic**

| *Hub Classification* | *Percent of Total U.S. Emplaned Passengers* | *No. of Emplaned Passengers (1985)* |
|---|---|---|
| Large hub | 1.00 or more | 3,633,415 or more |
| Medium hub | 0.25 to 0.99 | 908,356 to 3,633,415 |
| Small hub | 0.05 to 0.24 | 181,671 to 908,353 |
| Nonhub | Less than 0.05 | Less than 181,671 |

Ground facilities represent an important element of the air transportation system. In 1985, there were 16,318 airport facilities in the United States, of which 3120 were heliports, 67 were STOLports, and 387 were seaplane bases only. Scheduled airline service is carried out in about 374 of these airports. It is apparent that small general aviation airports constitute the large majority of the facilities.

Airports can be assigned to the hub[2] classification shown in Table 1-4. In 1985 there were 124 hubs in the United States.

The degree of concentration of air passenger traffic to the hubs can be deduced from the fact that the 26 large hubs account for 72.8 percent of all enplaned traffic, the 37 medium hubs for a further 18.1 percent, and 61 small hubs for only 6.7 percent. Nonhubs accounted for only 2.4 percent of traffic.

Within the classification system under the National Integrated Airport System Plan there was a total of 3,219 airports in 1985. They could be assigned to the following categories:

| | |
|---|---|
| Commercial service airports: | |
| Primary | 280 |
| Other | 272 |
| Reliever airports | 227 |
| General aviation airports | 2440 |

The air vehicle, the remaining major element of the air transportation system is usually privately owned. While the air carrier fleet constituted in 1985 only 4678 airplanes, there were approximately 210,654 airplanes engaged in general aviation activities. At the same time there were 709,540 registered pilots of whom approximately 82,740 were certified for airline transport. The amount of flying time generated by general aviation amounted to 34 million hours, averaging approximately 161 hours per aircraft.

In addition to the provision of air traffic control in the interests of safety, federal aid to the air transportation system is available in the form of matching construction and planning grants for the airports themselves. To be eligible

[2]A hub is technically an area or metropolitan population center rather than a specific airport.

for federal aid an airport facility must be owned publicly and must be included in the National Plan of Integrated Airport Systems. User charges in part repay this federal assistance. Airlines contribute to user charges in the form of landing fees, and passengers are required to pay federal excise taxes. There is no doubt, however, that the air transportation system currently is subsidized heavily by government. The actual amount of subsidy is subject to dispute among the carriers of various modes.

## 1-16. THE INLAND WATERWAYS [7,8]

The extent of the inland waterway system of the United States is shown in Fig. 1-5. In total, there are approximately 25,500 miles of waterway on the system with approximately 15,000 miles having navigational depths in excess of 9 feet. As the map shows, the degree of coverage of the system is spread unevenly throughout the country with very few waterways west of the Mississippi River, where few rivers of navigable depth are available. The total system can be considered in three major divisions, the inland rivers and canals, the coastal waterways, and the Great Lakes system.

The inland rivers and canals are natural, improved, and artificially made waterways which thread deep into the eastern portion of the United States. The extent of the system is limited by the number of rivers that have navigable depths of 9 feet. Three rivers form the most important sections of the inland waterways: the Mississippi (which connects to the Great Lakes at Chicago by means of the Illinois Waterway), the Ohio, and the Tennessee. While limited in extent, the contribution of the inland waterways to freight transportation is considerable. The Mississippi River alone accounts for approximately 60 percent of the tonnage carried in domestic water transportation.

The coastal system includes the protected intracoastal waterways of the Atlantic and Gulf Coasts and the Pacific Coast Waterways in California, Washington, and Oregon. In addition to movements on these *intracoastal* waterways, there are also deep-water movements of oceangoing vessels on the Atlantic, Gulf, and Pacific Coasts. Because of the maritime nature of these vessels and their operation, they are controlled and regulated in interstate commerce by the United States Maritime Commission, rather than by the Interstate Commerce Commission (ICC).

The Great Lakes system includes Lake Superior, Lake Michigan, Lake Huron, Lake Erie, Lake Ontario, and Lake St. Clair, and connecting waterways such as the Soo Canal, the St. Mary's River, and the Welland Canal. Connecting systems such as the St. Lawrence Seaway system, the New York State Barge Canal, and the Hudson River are usually grouped within the Great Lakes System. Exclusive of Seaway traffic, the Great Lakes system handles approximately 20 percent of waterborne freight reported to the ICC.

The Great Lakes and Mississippi systems together provide an extensive network of navigable waters linking the eastern seaboard to states as far west

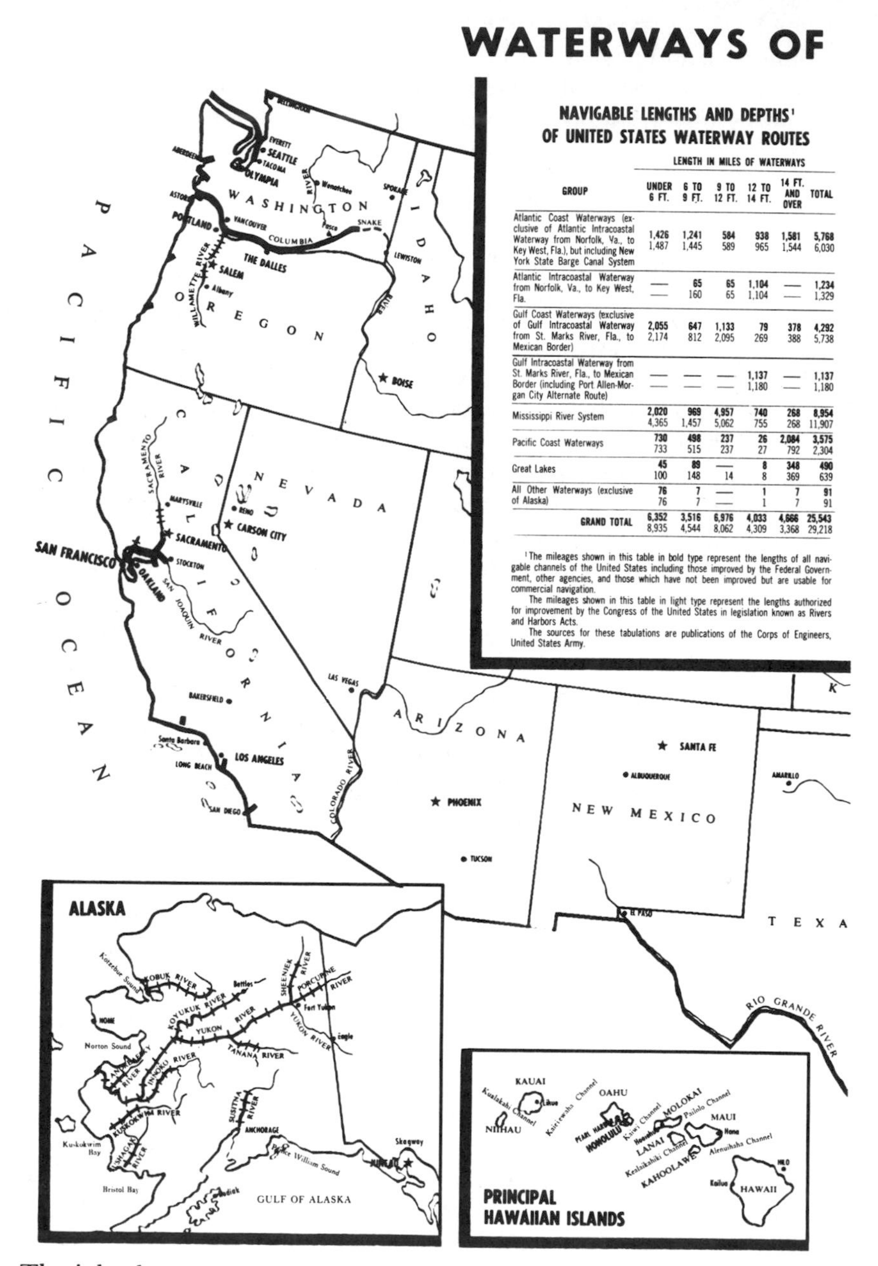


**NAVIGABLE LENGTHS AND DEPTHS[1] OF UNITED STATES WATERWAY ROUTES**

| GROUP | LENGTH IN MILES OF WATERWAYS | | | | | |
|---|---|---|---|---|---|---|
| | UNDER 6 FT. | 6 TO 9 FT. | 9 TO 12 FT. | 12 TO 14 FT. | 14 FT. AND OVER | TOTAL |
| Atlantic Coast Waterways (exclusive of Atlantic Intracoastal Waterway from Norfolk, Va., to Key West, Fla.), but including New York State Barge Canal System | **1,426**<br>1,487 | **1,241**<br>1,445 | **584**<br>589 | **938**<br>965 | **1,581**<br>1,544 | **5,768**<br>6,030 |
| Atlantic Intracoastal Waterway from Norfolk, Va., to Key West, Fla. | —<br>— | **65**<br>160 | **65**<br>65 | **1,104**<br>1,104 | —<br>— | **1,234**<br>1,329 |
| Gulf Coast Waterways (exclusive of Gulf Intracoastal Waterway from St. Marks River, Fla., to Mexican Border) | **2,055**<br>2,174 | **647**<br>812 | **1,133**<br>2,095 | **79**<br>269 | **378**<br>388 | **4,292**<br>5,738 |
| Gulf Intracoastal Waterway from St. Marks River, Fla., to Mexican Border (including Port Allen-Morgan City Alternate Route) | —<br>— | —<br>— | —<br>— | **1,137**<br>1,180 | —<br>— | **1,137**<br>1,180 |
| Mississippi River System | **2,020**<br>4,365 | **969**<br>1,457 | **4,957**<br>5,062 | **740**<br>755 | **268**<br>268 | **8,954**<br>11,907 |
| Pacific Coast Waterways | **730**<br>733 | **498**<br>515 | **237**<br>237 | **26**<br>27 | **2,084**<br>792 | **3,575**<br>2,304 |
| Great Lakes | **45**<br>100 | **89**<br>148 | —<br>14 | **8**<br>8 | **348**<br>369 | **490**<br>639 |
| All Other Waterways (exclusive of Alaska) | **76**<br>76 | **7**<br>7 | —<br>— | **1**<br>1 | **7**<br>7 | **91**<br>91 |
| **GRAND TOTAL** | **6,352**<br>8,935 | **3,516**<br>4,544 | **6,976**<br>8,062 | **4,033**<br>4,309 | **4,666**<br>3,368 | **25,543**<br>29,218 |

[1] The mileages shown in this table in bold type represent the lengths of all navigable channels of the United States including those improved by the Federal Government, other agencies, and those which have not been improved but are usable for commercial navigation.

The mileages shown in this table in light type represent the lengths authorized for improvement by the Congress of the United States in legislation known as Rivers and Harbors Acts.

The sources for these tabulations are publications of the Corps of Engineers, United States Army.

*Fig 1-5.* The inland waterways of the United States. (Source: American Waterways Operations Inc.)

# HE UNITED STATES

as Iowa and Nebraska, and also connecting the natural resource areas of the Great Lakes region with the Gulf Coast and the Port of New Orleans.

The St. Lawrence Seaway system, opened in 1959, was completed over a period of five years at a total cost of over $1 billion. A joint project of the Canadian and United States governments, the seaway permits oceangoing ships to travel 2430 miles into the heart of the continents to such ports as Chicago, Duluth, and Port Arthur. While usually not considered part of the inland waterways system because of the oceangoing nature of the vessels for which it was designed, the waterways of the Seaway are also used by domestic vessels.

There are approximately 42,000 vessels operating on the inland waterways of the United States, excluding the Great Lakes. In comparing the Mississippi River system to the Atlantic, Gulf, and Pacific Coasts, it can be seen that the river system has approximately two-thirds of all self-propelled towboats and tugs. The Mississippi, however, has a vast preponderance of non-self-propelled vessels, as indicated in Table 1-5. The high ratio of non-self-propelled vessels to self-propelled is indicative of the larger scale of tows which are common on the Mississippi system, as shown in Chapter 4, Fig. 4-21.

The waterways are constructed and maintained by the Corps of Engineers, because navigable waterways come under the jurisdiction of the federal government. Operation of the waterways is also supplied by the federal government since the United States Coast Guard assumes responsibility for the supply and operation of navigational aids.

## 1-17. PIPELINE SYSTEMS

Two principal pipeline systems serve the United States. These are the *oil* and *natural gas* pipelines.

*Oil* pipeline operations that engage in interstate common carriage are

*Table 1-5*

**Number of Towboats, Tugs, and Barges on Inland Waterways of the United States (1983)[a]**

| *Types of Vessels* | *Mississippi River System* | *Atlanta, Gulf, and Pacific Coasts* | *Great Lakes* | *Total* |
|---|---|---|---|---|
| Self-propelled: | | | | |
| Towboats and tugs | 3,289 | 1,556 | 148 | 4,993 |
| Non-self-propelled: | | | | |
| Dry cargo barges and scows | 24,712 | 4,760 | 258 | 29,730 |
| Tank barges | 3,404 | 669 | 41 | 4,114 |

[a]Figures exclude approximately 2900 self-propelled dry cargo, passenger, ferry and tanker vessels.

SOURCE: The American Waterways Operators, Inc.

controlled by the Federal Energy Regulatory Commission. The lines are classified as *gathering* lines, *crude oil* lines, and *products* lines. Crude oil and products lines are termed *trunk lines* to differentiate their line-haul function from that of the gathering lines whose purpose is to bring the crude oil in from the fields to the primary pumping station at the beginning of the trunk line. Products lines carry gasoline, fuel oils, and kerosene from the refineries. A description of the operation of oil pipeline systems is given in Chapter 15.

By the mid-1980s there were approximately 111,210 miles of crude and gathering lines and a further 94,442 miles of products lines. A very large proportion of pipelines originate in the oil-producing states of Illinois, Kansas, Ohio, Pennsylvania, Texas, and Wyoming. In the last reported breakdown by state, approximately 30 percent of all pipelines were in Texas. The orientation of pipelines is shown in Fig. 1-6, which displays the need to connect producing areas with the industrial areas of the north and northeast and with the Gulf ports.

*Gas* pipelines are under the control of the Department of Energy. In the mid-1980s, the total amount of transmission pipeline was approximately 280,000 miles. In addition there were about 95,000 miles of field and gathering lines and three-quarters of a million miles of distribution lines. This system conveyed about 17 trillion cu ft of gas to the consumer.

The total pipeline mileage of interstate and intrastate facilities for oil, gas, and oil products is approximately one-half million miles. It is interesting to note that pipeline transportation is the only mode that has no direct governmental aid.

## 1-18. THE MOTOR CARRIERS—PROPERTY

In comparison with other modes of transportation, the highway truck carriers offer the most varieties of forms of carriage. The most basic division of the carriers is between those carriers that are *for hire* and those that are *not for hire*, often called *private* carriers. Private carriers are not in the business of hauling the goods of others, but are concerned only with the moving of their own goods in their own trucks. For-hire carriers may be subdivided further into three categories: common carriers, contract carriers, and exempt carriers.

The *common carrier* is the one who presents a standing offer to carry the public's goods within the capability of the service and schedules that he publishes. Such carriers were previously subject to the extensive federal legislation when operating in interstate commerce, but were subject to the control of the state regulatory agencies when engaged in the intrastate commerce only.

Opposed to the concept of common carriage, available to the general public, is the concept of *contract* carriage, where the carrier hauls only under specific contract, and does not hold himself out to hire to the general public. Although the classification does not affect the type of goods that a carrier is capable of carrying, in general it may be stated that regulations have been

*Fig 1-6.* Oil pipeline systems of the United States.
(Source: *Oil and Gas Journal.* Petroleum Publishing Company.)

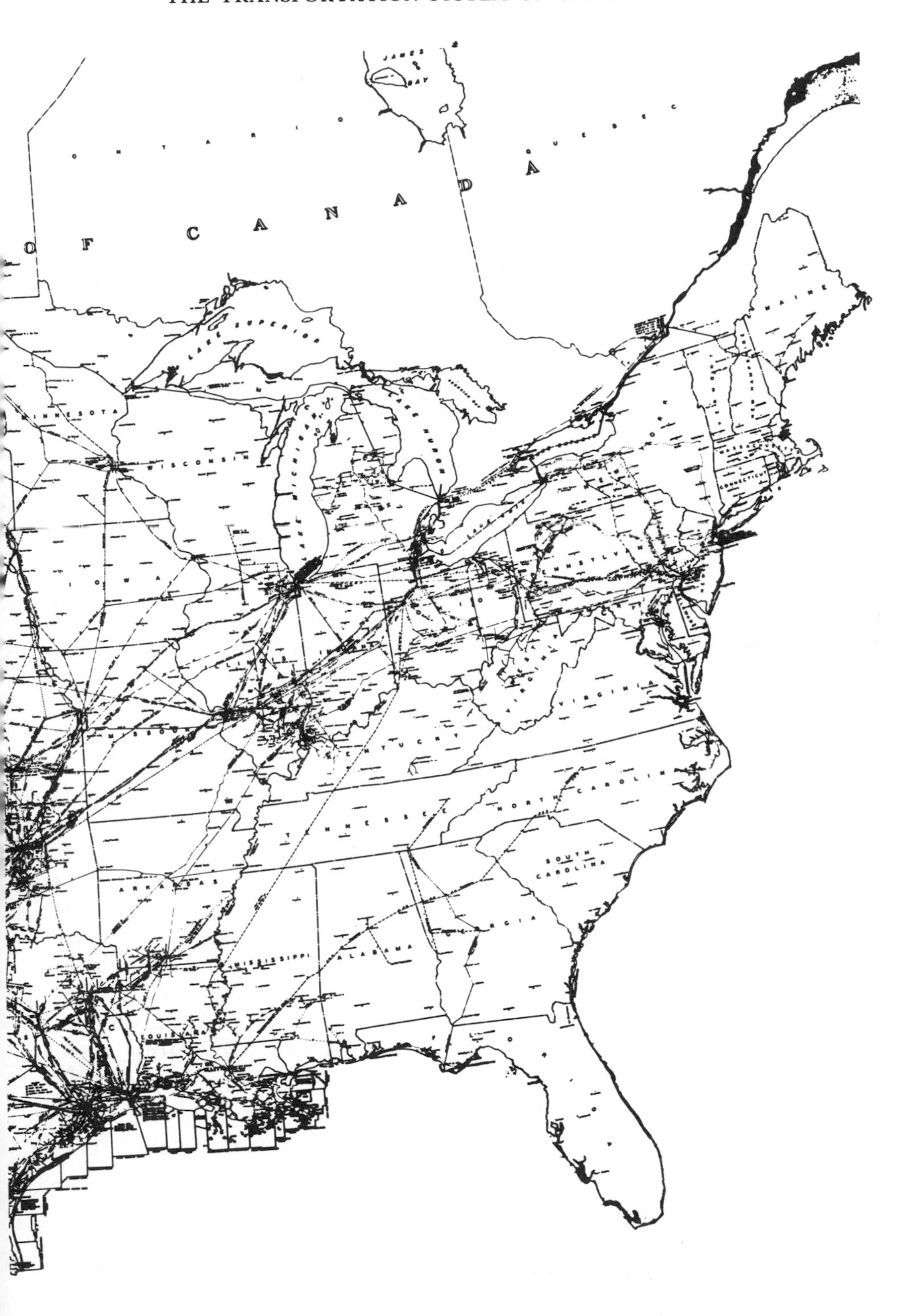
JAMES BAY
ONTARIO
QUEBEC
OF CANADA
LAKE SUPERIOR
LAKE MICHIGAN
LAKE HURON
MINNESOTA
WISCONSIN
IOWA
MAINE
CONNECTICUT
OHIO
KENTUCKY
TENNESSEE
NORTH CAROLINA
SOUTH CAROLINA
ARKANSAS
MISSISSIPPI
ALABAMA
LOUISIANA
VIRGINIA

considerably less stringent for contract carriers than for common carriers. In the latter instance, regulations were drawn up ostensibly for protection of the general public, whereas contract carriage is frequently negotiated between business firms. Since contract carriers were not subject to common carriage regulation, freight rates were not subject to regulation by state and federal agencies. Contract carriers carefully protected their classification, since failure to observe the requirements for contract carriage could result both in a loss of contract carriage classification and litigation against the carrier for engaging in illegal common carriage. With the advent of deregulation of the motor carrier industry, the former sharp division between common and contract carriage has largely disappeared.

Common carriers can now have dual authority, which means that they can engage in contract carriage. Method of entry into the market is easier and carriers do not have to file tariffs. Private carriers, on the other hand, can also apply for authority to enter into either contract or common carriage. In general it can be observed that in road transportation deregulation loosened rather than abolished the rules governing carriage. One important change in deregulation affected freight forwarders. Prior to deregulation, freight forwarders were considered as carriers; in 1987 they were relieved of all control with the exception of those engaged in the carriage of household goods.

*Exempt* carriers were exempt from governmental control either from the nature of the services they provide or from the nature of the goods carried. For example, private carriers were exempt from the nature of the service itself. Equally, carriers engaged even in interstate movement of unprocessed agricultural products could offer their services to the public without being classified and regulated as common carriers.

Under Interstate Commerce Commission regulations, common carriers are further classified by revenues. There are approximately 36,200 common carriers. Of these, 856 are Class I carriers with annual gross operating revenues in excess of $5 million, 1266 are Class II carriers with revenues in excess of $1 million and less than $5 million. The remaining 12,900 are Class III carriers with revenues below $1 million. The average motor common carrier obviously tends to be a much smaller operation than the average railroad counterpart. The trucking industry includes many very small operators, some ranging down to owners with single vehicle operations. The structuring of the transportation infrastructure, where rights-of-way and roadbed are government owned, enable small operations to flourish while the large capital investment required for railroad operation discourages the extremely small operator.

## 1-19. THE MOTOR CARRIERS—PASSENGER

Passenger carriers use the highways extensively. The form of operation differs greatly in type of service and in scale. *Intercity carriers* operate both on a regular service and special charter basis. Of the approximately $\frac{1}{2}$ million buses registered prior to deregulation, approximately 4 percent only were engaged in regular intercity service. The remainder operated in intercity charter service and, on a local level, in such categories as line-haul bus systems, local public and contract transit, and school buses. Intercity carriers that operate in interstate commerce are subject to regulation as common carriers by the Interstate Commerce Commission. The ICC classifies bus carriers according to operating revenues. Class I carriers are those with gross operating revenues in excess of $5 million, of which there were 44 in 1987.

Intercity operations that do not cross state lines are subject to state regulatory bodies only as are those operations, within most states, that cross municipal boundaries. Within most municipal areas, passenger transportation is regulated by municipal franchise without recourse to federal and state level of control.

Deregulation has also affected the passenger motor carriers. Substantial discounts in airline fares and declining gasoline prices have resulted in a continued decline in intercity bus ridership, a downward trend which started in the early 1960s. In the mid 1980s, the nation's 10 largest bus companies had revenues that were declining by approximately 6 percent per annum, whereas revenue passengers carried declined at over 7 percent annually. The ICC has adopted final rules of deregulation that allow bus companies to file new schedules of rates that become effective on one day's notice rather than up to 45 days as had been required previously. The Commission concluded that the current competitive environment requires that passenger carriers are able to initiate pricing changes quickly and to respond without delay to the pricing initiatives of their deregulated competitors, particularly those of other modes, for example, Amtrak and the airlines.

## 1-20. RAILROAD CARRIER CLASSIFICATION

With the exception of a relatively insignificant amount of *private* carriage, mainly by small operations affiliated with individual industries, the railroad system of the United States is operated by *common* carriers. A common carrier is one who holds himself out to provide transport to all on a nondiscriminatory basis. Service is provided at reasonable demand according to published schedules, rates, and charges.

Most commonly, railroad carriers are classified by revenues, as previously discussed in Section 1-3.

## 1-21. THE AIR CARRIERS

Prior to the deregulation of U.S. domestic air transport, the air carriers were subject to close regulation by the Civil Aeronautics Board. The market was largely divided into a small number of large *trunk* carriers, which operated between the major population areas of the United States, and *local service* carriers, which supplemented the trunk line carriers providing service for lower density movements. At the time of deregulation there were 11 trunk carriers, which accounted for 90 percent of all domestic passenger mileage. Additionally there were three *all-cargo* carriers, not permitted to operate passenger service, and a variety of *helicopter services*, intrastate carriers not subject to CAB regulation and supplemental carriers.

After deregulation, entry to the air transport market became open. The former structure of carriers became obsolete. Local carriers were able to compete over routes formerly considered part of the trunk network and trunk carriers were free to provide services formerly operated by local service carriers. It is, therefore, now more convenient to classify the air carriers in terms of an income breakdown as reported by the Federal Aviation Administration. In 1985, the FAA reported the following structure of U.S. scheduled airlines certificated by the Department of Transportation:

| *Class* | *Annual Revenue* | *Number* |
|---|---|---|
| Majors | $1 billion plus | 13 |
| Nationals | $1 billion to $75 million | 17 |
| Large Regionals | $75 million to $10 million | 42 |
| Medium Regionals | $10 million and less | 33 |

By the mid 1980s the U.S. air passenger carrier market had become very volatile with frequent entries to the market, mergers, takeovers, and company collapses.

During this period there were just over 20 *all-cargo* carriers certificated under Section 418 of the Federal Aviation Act.

The effect of deregulation has been much more noticeable in air transport than in any other mode. It is clear that the effect of the Deregulation Act is long term, and that the system had not settled down by 1988. Table 1-6 indicates the variation between the air carrier and general aviation fleets between 1973 and 1983. In 1978 there were 36 major and national carriers. By 1984 this figure had climbed to 123 but had declined to 78 by 1988 of which 37 operated entirely outside the contiguous 48 states. Carriers entered and left the scene with remarkable rapidity in the first decade of deregulation during which 210 new carriers were certificated and 168 either merged, were decertificated or were not in operation. Bankruptcy was the major reason for ceasing activity.

The most significant deregulatory change to the U.S. domestic air transportation system has been the growth of massive hub and spoke systems. Prior to deregulation the development of airline hubs was almost impossible. Since 1978 there has been freedom of access to certification and to the airports. By 1987, there were 48 airline hubs in the United States which were either developed or imminent. At 17 of these, a single carrier accounted for over 50 percent of the traffic and at 6 hubs over 75 percent.

*Table 1-6*
**Air Carrier and General Aviation Fleets**

| | *1973* | *1983* |
|---|---|---|
| U.S. Air Carriers | | |
| Total aircraft | 2,614 | 3,585 |
| Turbine | 2,463 | 3,043 |
| Piston | 138 | 530 |
| Rotorcraft | 13 | 12 |
| Percentage of total aircraft | 1.7 | 1.7 |
| General Aviation | | |
| Total aircraft | 153,540 | 207,000 |
| Turbine | 3,271 | 9,700 |
| Piston | 144,875 | 185,300 |
| Rotorcraft | 3,143 | 6,700 |
| Other | 2,251 | 5,300 |
| Percentage of total aircraft | 98.3 | 98.3 |

Another important change is the development of feeder agreements between large hub carriers and small feeder carriers. These did not occur prior to deregulation. By late 1986 these reached a peak of 66 and reduced to 50 by 1988. This reduction, however, does not indicate a reduced level of feeder activity; in fact, quite the contrary. Feeder activity is considered so important to the successful operation of a major hub operator that many independent operators have been acquired by the dominant airline.

Air traffic in route passenger miles almost doubled between 1978 and 1987, with an average growth rate of close to 6.8 percent. Load factors remained virtually constant at 62 percent. For the major and national airlines employee activity increased by 38 percent from 99 revenue ton miles per employee to 138 revenue ton miles. Airlines developed strong marketing initiatives using frequent flyer programs to retain customers and by investing heavily in powerful computerized reservation systems that competed in a fierce market. The large airlines also developed sophisticated yield management systems designed to maximize the carriers' revenues from expected demand. Surprisingly perhaps, fares, when adjusted for inflation, continued a secular decline which seemed unaffected by deregulation.

## 1-22. THE WATER CARRIERS

Up to 1978 domestic water carriers were classified by operating revenues into three divisions:

Class A: Annual gross operating revenues in excess of $500,000.

Class B: Annual gross operating revenues between $500,000 and $100,000.

Class C: Annual gross operating revenues less than $100,000.

Water carriers are much smaller in scale than railroads, their chief competitors. Approximately 1800 companies were engaged in commercial carriage on the inland waterways system. Only 112 of these were certificated by the Interstate Commerce Commission as regular route common carriers, and 31 companies were permitted to act as contract carriers. Because of the nature of their operation and the type of goods carried, more than 1200 companies were not under the jurisdiction of the ICC and, therefore, were referred to as *exempt* carriers. The remaining carriers, numbering about 400, were private carriers transporting their own products.

Considering both inland and coastal waterways, in 1978 there were 135 common carriers. Of these, 52 were Class A, 15 were Class B, and 68 were Class C. The Class A carriers alone accounted for 97 percent of operating revenues, and carried more than 98 percent of all freight on a tonnage basis. The remainder of waterborne traffic was shared, with approximately 2 percent being carried by Class B and Class C carriers. The small size of water carriers can be seen from statistics then reported to the ICC. The largest reported operating revenue was approximately $108,900,000, yet 10 of the 52 carriers accounted for 68 percent of all Class A revenues.

In 1979, in a general move toward deregulation of transportation water carriers were released from the necessity to report to the ICC. By 1987, the ICC had a count of only 315 inland water operators and six maritime carriers. However, the data collection system for water carriage had reached such disarray by the mid 1980s that the ICC recommended to Congress in 1986 that some formal system of carrier reporting be reinstated.

## 1-23. PIPELINE CARRIERS

Oil-pipeline operators engage in both private and common carriage. Those engaged in common carriage are required to file schedules of rates and charges with the Federal Energy Regulatory Commission in a manner similar to other common carriers.

## 1-24. SUMMARY

The transportation system of the United States is seen to be a complex combination of several modes which stretches into the most remote sections of the nation. Few areas are dependent on only one principal mode; the shipper, therefore, is guaranteed transportation at various levels of service and price. In the interests of economic development, federal, state, and local governments entered into the provision of transport systems by the provision of varying degrees of governmental aid and regulation. The result of this combination of private and public endeavor is an overall system that has provided American citizens and goods with an unprecedented degree of mobility.

Since the deregulation of transportation, which started in 1978, the transportation industry in the United States has gone through a remarkable transformation. In some modes, such as air transport, many new carriers have appeared in response to the deregulated environment. In air and in other modes many familiar names have disappeared. Other enterprises appeared and almost as quickly disappeared in the fiercely competitive climate. It is clear that even in the late 1980s the industry is still in a remarkable state of flux. It is hoped that in the long term the new competitive climate will create conditions that will continue to provide a high level of transport at a minimal cost.

## PROBLEMS

1. Describe the transportation system of some selected state. Show in the form of sketches the principal components of this system which account for the major intercity and interregional passenger and freight movements.
2. As the form of American society changes to an increasingly industrial society, what principal changes should be anticipated in the major model networks on a national basis?
3. The National System of Interstate and Defense Highways was constructed partially on the assumption that such a system would strengthen the national defense. Discuss the arguments both for and against this premise.
4. What has been the principal effect of deregulation on the structure of domestic air routes in the United States? What do you think will be the long-term effect?
5. As real fuel costs increase in the long term, what might be the effect on intercity passenger and freight transportation?

## REFERENCES

1. *Highway Statistics, 1985,* Federal Highway Administration, U.S. Department of Transportation, Washington D.C., 1986.
2. *National Plan of Integrated Airport Systems,* Federal Aviation Administration, U.S. Department of Transportation, Washington D.C., August 1985.
3. *Transportation in America,* Transportation Policy Associates, Washington D.C., March 1987.

4. *Yearbook of Railroad Facts, 1987 Edition,* American Associates of Railroads, Washington D.C., 1987.
5. *F.A.A. Aviation Forecasts, Fiscal Years 1987–1998,* Federal Aviation Administration, U.S. Department of Transportation, Washington D.C., February 1987.
6. *Current Market Outlook,* Boeing Commercial Airplane Company, Seattle, February 1987.
7. *FAA Statistical Handbook of Aviation, Calendar Year 1985,* Federal Aviation Administration, U.S. Department of Transportation, Washington D.C.
8. *Big Load Afloat,* American Waterway Operators Inc., Arlington, Va., (current edition).
9. *Waterborne Commerce of U.S.,* U.S. Corps of Engineers, Washington D.C., 1987.

# *Chapter* *2*

# *Governmental Activity in Transportation*

## 2-1. INTRODUCTION

Transportation has profound and enduring effects on a nation and its people. It is, in the words of a 1967 congressional committee report, "unquestionably the most important industry in the world."[1] The presence or lack of adequate transportation facilities shapes the boundaries of national, state, and local governments. Cities are located to take advantage of favorable transportation. Wars may be won or lost because of the mobility or lack of mobility of a nation's troops and weapons. The adequacy of the transportation system may determine a nation's foreign policy and the extent of its political control over its people. In the lives of people, transportation affects our economic well-being, our cultural development, and our social and recreational habits and customs.

In a free society, the interest and concern of the government in transportation matters, as in other matters, are a reflection of the moods and attitudes of the people. In turn, government interest and concern are manifest in legislation, court actions, and executive orders. Because of the vital importance of transportation to the well-being of the nation, it is not surprising that the federal and state governments through the years have seen fit to enact laws promoting the development of transportation facilities and the regulation of various carriers. In the paragraphs that follow, the dominant role the government has played in the development of the transportation system of the United States will be described.

## WATER TRANSPORTATION

## 2-2. CANAL BUILDING PRIOR TO THE CIVIL WAR

When studying the history of transportation in the United States, it is important to remember that formidable physical obstacles deterred the development of land transportation facilities. In addition to mountainous terrain, the early settlers were confronted with vast expanses of primeval forestland which was practically impenetrable. It is, therefore, understandable that the earliest settlers took advantage of existing lakes, streams, and rivers and that the first federal aid to transportation was for water transportation facilities.

Although federal aid was granted for improvement of harbors as early as 1789, practically all of the impetus for the development of water transportation prior to the Civil War came from the various states. Examples of canals constructed by states or with state aid are listed as follows:

1817–1825 The Erie Canal, the first large canal project undertaken in the nation, was constructed by the state of New York. This route was 364 miles long and cost about $7 million. Later, in 1918, at a cost of $176 million, the state built the New York Barge Canal by deepening and widening the main routes of the old Erie Canal.

1826–1834 The state of Pennsylvania constructed a canal-railroad route which extended from Philadelphia to Pittsburgh on the Ohio River. This project cost $10 million.

1828–1836 The state of Ohio built the Ohio Canal extending from Cleveland to Portsmouth on the Ohio River and the Miami and Erie Canal from Toledo to Cincinnati.

1828–1850 A joint project of Virginia and Maryland, the Chesapeake and Ohio Canal extended from Washington, D.C. to Cumberland, Maryland. This canal operated until 1924, and a portion of it has been restored as a historical park.

## 2-3. LATER GOVERNMENT INTEREST IN INLAND WATERWAYS

While the federal government began early to maintain river channels and to provide coastal surveys and lighthouses and other navigation aids, the amounts of federal aid to water transportation were relatively small until about 1890.

The Rivers and Harbors Act of 1902 created the Board of Engineers for Rivers and Harbors in the War Department, forerunner of the U.S. Army Corps of Engineers. This board was given the responsibility of investigating the feasibility of proposals for improvements of waterways.

In 1907, President Theodore Roosevelt appointed the Inland Waterways Commission to study the status of waterways and carriers on inland waterways. This commission recommended in 1908 that more suitable provision should be made for improving the inland waterways of the country.

Congress passed the Panama Canal Act in 1912 in anticipation of the completion of the Panama Canal. This act provided that without specific approval of the Interstate Commerce Commission, railroads could not own, operate, control, or have interest in common carriers using the Panama Canal. In cases where water lines were competitive with rail lines, the act made it illegal for other water carriers to be controlled by railroads unless specifically approved by the ICC. The increased activity in inland water transportation that subsequently occurred is attributed in part to the passage of this act.

The Transportation Act of 1920 stated the intent of Congress to promote, encourage, and develop water transportation. Four years later, Congress incorporated the Inland Waterways Corporation to demonstrate the service capabilities of water transportation. Historians generally date the beginnings of modern inland water carrier operations from this period [2].

The Transportation Act of 1940 established a system of regulation for domestic water carriers. It sought to control the route and service of water carriers and to prevent undesirable duplication of transportation services, undue competition, and the entry of incompetent carriers. This act established Part III of the Interstate Commerce Act extending its coverage to inland water transportation and charged the ICC with its implementation. Part III does not apply to private water carriers, and only about 10 percent of the for-hire carriers are regulated under this act.

In 1962, in his message to Congress on transportation, President Kennedy recommended that a user tax be applied to all fuels used in transportation on inland waterways. Similar recommendations were made by Presidents Johnson, Nixon, Ford, and Carter. In 1978, Congress passed P.L. 95-502 that authorized an inland waterway user fee in the form of a tax on diesel oil starting at 4 cents per gallon in 1980 and increasing in steps to 10 cents per gallon in 1985. That legislation also authorized $430 million for construction of a controversial new lock and dam at Alton, Illinois.

By 1978, a total of $17 billion in federal aid had been obligated for domestic water transportation, and annual obligations exceeded $90 million.

## 2-4. LEGISLATION RELATING TO OCEAN TRANSPORTATION

Events during the decade of World War I prompted increased interest in ocean transportation, and several acts were passed assisting the merchant marine. The Seamans Act of 1915 provided requirements of a minimum space per man, a separate berth for each seaman, and proper drainage, lighting, and ventilation of seamans' quarters.

The Shipping Board Act of 1916 created the U.S. Shipping Board, a five-member independent agency, the precursor of the Maritime Commission. The purpose of this act was to prohibit monopolistic, unjust, and discriminatory practices in ocean transportation. The act applied to common carriers operating over regular ocean and coastwise routes and on the Great Lakes.

In 1917, the Congress appropriated $500 million for a government shipbuilding program for World War I. Although the program was too late to aid the war effort significantly, vessels continued to be launched under the program until 1921.

The Congress passed the Merchant Marine Act of 1920 in an attempt to establish and to maintain temporarily ocean steamship lines operating over the major shipping routes from United States ports. The government intended to sell these lines, if necessary, at very low prices, to American companies for continued operation. The results of this act were disappointing.

Direct subsidies for ocean transportation were provided by the Merchant Marine Act of 1928. This act provided for a revolving construction loan fund of $250 million for ship building and permitted low-interest loans up to three-fourths of the cost of vessels to be built or reconditioned. Loans with interest rates as low as one-fourth of 1 percent were made under this act. Substantial mail subsidies were also provided by this law. The functions of the Shipping Board were transferred to the Shipping Board Bureau in the Department of Commerce in 1933. These functions subsequently were transferred to the U.S. Maritime Commission by the Merchant Marine Act of 1936.

The purpose of the 1936 act was to permit the transportation of a substantial percentage of American foreign commerce in American flagships. Two kinds of subsidies were provided by this legislation:

1. Construction subsidies for the difference between the cost of building a ship in a foreign yard and the cost of building it in the United States.
2. Operating subsidies to allow for the difference in wages between American and foreign flagships.

Increased concern for the U.S. Merchant Marine and the nation's shipyards was reflected in the passage of the Merchant Marine Act of 1970. This act, which is perhaps the most significant piece of legislation relating to water transportation since the 1936 act, was intended to end the nation's decline as a shipping power. The bill authorized $4 billion during the decade of the seventies in ocean transportation subsidies.

The Shipping Act of 1984 was enacted to provide greater stability in the shipping industry. The act had the effect of allowing ocean carriers greater freedom to make rate and service agreements with other carriers. Such agreements are made through conferences of liner companies that serve a particular trade route. The agreements usually involve collectively agreed upon rates, division of services, and shared capacity and services. The act gives the carriers broadened antitrust immunity and removes the power of the Federal Maritime Commission to disapprove rate and service agreements reached by the conferences [3].

## RAIL TRANSPORTATION

### 2-5. RAILROAD LEGISLATION PRIOR TO 1834

The first significant legislative act to provide assistance to railroads was the General Survey Bill of 1824. Haney [4] observes:

> In the early years of the century civil engineers were very scarce relatively to the demand for them. The settlement of our vast public domain and numerous roads and canals constantly required the service of surveyors and engineers, and on some of the larger work foreigners were employed. The period was one in which new lines of communication and transportation were rapidly projected—one of births and expansions. The work of plans, surveys and estimates was in great demand and of high relative importance at this stage, and it was natural that a government which was desirous of increasing the facilities for its commerce, and which maintained a corps of engineers in its employ, should have begun with aid to these initial steps toward the construction of such facilities. The first assistance ever granted by Congress for railway purposes was given in the shape of a survey by government engineers.
>
> The survey bill of 1824 authorized the President to cause the necessary surveys, plans and estimates to be made for the routes of "such roads and canals as he may deem of national importance, in a commercial or military point of view, or necessary for the transportation of the public mail"; and a limited number of civil engineers and officers of the corps of engineers was placed at his disposal. Thirty thousand dollars annually was the amount appropriated to cover expenditures.

Although railways were not mentioned specifically in this bill, by 1826 railways were being surveyed by government engineers under this bill. Before the repeal of this act in 1838, 61 railway surveys had been made.

Another law promoting railroad transportation was the Railroad Iron Bill of 1832. At this time the nation's iron industry was primitive and its output was relatively small. The railroads were commencing a period of vigorous growth, and the supply of rails had to be purchased from foreign sources. Earlier legislation had placed protective tariffs on imported iron to assist the growth and development of the iron industry in this country. After several years of debate, the Congress passed the Railroad Iron Bill of 1832, which removed the tariffs on iron imported for use as rails.

### 2-6. FEDERAL LAND GRANTS

The most significant legislative acts influencing the development of a national railroad system were the land grant acts. These acts generally provided grants of federal lands for rights-of-way and additional sections of land which the railroads could sell or use in some other way to raise capital for railroad construction. The purpose of these grants was to encourage the development of the western portion of the nation.

The Association of American Railroads [5] has reported that less than 8 percent—18,737 miles—of railroad mileage was involved in either federal or state land grants at the peak of expansion in the United States. Furthermore, the land grants imposed certain obligations on the railroads such as the transportation of mail and other federal property and the transportation of troops over the land grant portions without charge or at reduced rates. These obligations remained with the railroads until 1940, when the Transportation Act of 1940 canceled the rate concessions applicable to the transportation of mail and federal property. The concessions applicable to the transportation of troops were dropped in 1945.

The first railroad land grant was made to the Tallahassee Railroad in 1834. Several subsequent requests by railroads for land grants resulted in a number of grants being given in the years that immediately followed.

One of the largest land grants to railroads was made in 1850. This act provided alternate sections of land to the states of Illinois, Mississippi, and Alabama for railroads. Under this act, more than, 3,750,000 acres of land were granted for the Illinois Central, Mobile and Ohio, and Mobile and Chicago railroads.

By 1852, the demand for land grants was so great that a general law was passed providing for right-of-way through public lands. The act provided for a right-of-way of 100 feet in width for all rail and plank road or macadam turnpike companies then chartered or to be chartered within 10 years. These companies were also given the right to take construction materials from nearby public lands, and additional land was provided for watering places, workshops, and depots.

On July 1, 1862, enabling legislation for the construction of the first transcontinental railroad was signed by President Lincoln. The construction of this line, which culminated in the celebrated driving of the golden spike at Promontory, Utah, in 1869, played a prominent part in the development of the West (Fig. 2-1).

## 2-7. RAILROAD TAXATION AND REGULATION

During the Civil War the Revenue Bill of 1862 was passed, levying a tax on gross earnings from railroad passenger services and a tax on interest and dividend payments. Another wartime measure, the Revenue Bill of 1864, placed a tax on gross railway receipts including both passenger and freight earnings.

There had been earlier complaints about aid to the railroads for transportation of the mails and, by 1850, the mail service relation had ceased to be one

*Fig 2-1.* The driving of the golden spike at Promontory, Utah on May 10, 1869, signaling the completion of the first chain of railroads to span the American continent. (Courtesy Union Pacific Railroad and the Association of American Railroads.)

of aid and had become one of regulation. Opposition to government land grant policies, long present, grew stronger as railroads engaged in practices that were not in the public interest. There were frequent cases of overbuilding, and the economic struggles that followed often resulted in price discrimination against the shippers. Collusion among the carriers and other abusive practices were not uncommon. By about 1870, opposition to further land grants became so strong as to bring about the end of the land grant policy.

The first positive regulation of interstate commerce came in 1873 with the passage of a bill to prevent cruelty to animals in transportation. This act, which applied to the interstate transportation of livestock, provided that no railway should confine animals in cars, boats, or other vessels for a period longer than 28 consecutive hours without unloading them for at least five consecutive hours for rest, water, and feeding.

The pace of railroad construction increased and, by 1880, the railway system in the United States consisted of over 93,000 miles. The increase in railway mileage was accompanied by growing public fear of and hostility toward the railroad companies.

In 1887, the Congress passed the Interstate Commerce Act, a comprehensive measure which applied to railroads engaged in interstate or foreign commerce. This act:

1. Required "just and reasonable" rates.
2. Made it unlawful for a carrier to charge one person more than another person for like transportation services.
3. Prohibited "undue or unreasonable preference or advantage" to any person, place, or kind of traffic.
4. Made it unlawful for a carrier to charge more for transportation over a short distance than over a longer distance over the same line and in the same direction.
5. Prohibited the pooling of freights of different and competing railroads.
6. Required carriers to keep printed schedules of rates and fares for public inspection.
7. Established an Interstate Commerce Commission to administer the law.

In the years that followed, three laws were passed which strengthened the Interstate Commerce Act, and increased and broadened the powers of the Interstate Commerce Commission. The Elkins Act of 1903 strengthened the preferential treatment section of the Interstate Commerce Act by making both the carrier and shipper guilty of violations.

The Hepburn Act of 1906:

1. Extended the ICC's authority over related rail activities such as private car lines, sleeping car companies, and terminal facilities.
2. Required accounting reports as prescribed by the ICC.
3. Allowed ICC to prescribe *maximum rates*.
4. Strengthened ICC's enforcement procedures by providing for requests for court orders to comply.

## 2-8. RAILROAD LEGISLATION AFTER WORLD WAR I

At the beginning of World War I, the federal government took control of and operated the railroads during the war and until March 1920. The Transportation Act of 1920 returned the carriers to private operation.

The basic purpose of this act was to establish a stronger economy for the railroad industry by encouraging consolidation of rail carriers into a limited number of balanced systems under the supervision of the ICC. This act provided that carriers could earn a fair return on their investments. The commission was given permission to establish minimum rates in order to prevent rate wars and discrimination, and the power to prescribe exact transportation rates. The act also gave the ICC control over the issuance or purchase of securities by railroads.

During the years since 1920, the railroads have experienced relative declines in both passenger and freight traffic. In 1929, railroads accounted for 77 percent of the total ton-mileage of intercity freight transported and 75 percent of the common carrier passenger traffic. By the late 1950s, the railroads' share of freight had decreased to less than 50 percent of the total, and their portion of common carrier passenger traffic had declined to less than 30 percent [6].

The Transportation Act of 1958 sought to give relief to the railroads in two important ways:

1. It provided a more liberal policy in train service abandonment.
2. It provided that "the rates of a carrier shall not be held up to a particular level to protect the traffic of any other mode of transportation . . ."

During the 1950s and 1960s, the railroads' problems persisted and grew worse. The industry's share of the total transportation revenues declined from 56 percent in 1955 to 34 percent in 1970 [7].

## 2-9. RAILROAD LEGISLATION OF THE SEVENTIES AND EIGHTIES

During the 1970s, four important railroad laws were passed: (1) the Rail Passenger Service Act of 1970, (2) the Regional Rail Reorganization Act of 1973, (3) the Railroad Revitalization and Regulatory Reform Act of 1976, and (4) the Rail Transportation Improvement Act of 1976.

The purpose of the Rail Passenger Service Act of 1970 was to bring financial relief to railroads operating unprofitable passenger service. This legislation authorized the creation of the National Railroad Passenger Corporation, informally termed Amtrak. It was intended that this corporation take over the management of a national passenger train network in the hope that improved service would attract riders back to passenger trains.

The bill required that the new network and schedules of service be designated initially by the Secretary of Transportation. The law authorized the Amtrak corporation to contract with railroad companies to provide the right-of-way and service. Railroad employees would operate the trains under the management of Amtrak.

The legislation included a complex financial plan wherein each railroad that elected to participate would contribute cash or equipment to the new corporation in return for which the railroad would receive common stock in the company or a tax deduction. The law provided a direct grant of $40 million and loan guarantees of up to $100 million for the improvement of roadbeds and the purchase of new equipment.

By May 1, 1971, eighteen railroads had signed contracts with the National Railroad Passenger Corporation, and on that date the corporation began operating its intercity passenger network.

Since Amtrak came into being, the corporation has received federal grants exceeding $2.5 billion. By 1979, the system reported 21.5 million annual revenue passengers and an annual deficit of nearly $600 million [6].

Following bankruptcies of several railroad companies, the Regional Rail Reorganization Act of 1973 was signed into law. The legislation was designed to establish a viable rail system in the northeast quadrant of the United States. The law created the United States Railway Association (USRA), a government agency, to function as the planning and financial agency, and it created the Consolidated Rail Corporation (ConRail) to operate the restructured rail system as a private for-hire enterprise. The law provided $1.5 billion in federally guaranteed loans to be administered by USRA, interim loans of $150 million, and $558 million in direct grants.

The plan for the ConRail system developed by USRA consisted of the major portions of seven bankrupt lines: the Penn Central, Erie Lackawanna, Reading, Lehigh Valley, Jersey Central, Ann Arbor, and Lehigh and Hudson River. The plan recommended that 5700 miles of lightly used track either be abandoned or operated under a 70 percent federal and 30 percent local subsidy arrangement, and approximately 3000 miles of branch line was abandoned. ConRail took control of the restructured system of approximately 17,000 miles on April 1, 1976.

In the Railroad and Regulatory Reform Act of 1976, congress passed comprehensive landmark legislation that included a number of provisions that had been long sought by the railroad industry. Termed the 4-R Act, this measure provided a total of $6.5 billion in federal aid including $2.1 billion for ConRail, $2.4 billion for passenger services in the northeast corridor, $1.4 billion in loans and loan guarantees for solvent railroads, $400 million for five years for a national subsidy program for branch lines, and $200 million for electrification outside the northeast corridor.

The act also made extensive changes in federal regulatory procedures, requiring the Interstate Commerce Commission to expedite its deliberations over railroad rates, divisions, and mergers, and allowing the industry more freedom in raising and lowering railroad rates. Further, the law prohibited discriminatory state taxation of railroad property and provided several changes necessary to facilitate the restructuring of the northeastern rail system.

The Rail Transportation Improvement Act of 1976 amended several existing railroad laws. Informally titled the "Son of ConRail Act," this law provided $350 million in guaranteed loans for claims against the bankrupt railroads that formed ConRail. It also provided $900 million in operating subsidies for Amtrak and $260 for capital outlays. The measure included a formula that placed a ceiling on interest rates that the government could charge railroads using funds provided in the 4-R Act.

The Staggars Rail Act of 1980 removed some of the economic regulation of the railroad industry on the premise that competition would be a better safeguard of the public interest than regulatory discretion [8]. By 1987, the effects of that law were still being evaluated. Return on investment for railroads had improved during the 1980s, increasing the incentives for investment, innovation, and efficiency, and providing a climate for greater competition between railroads. However, some industry observers expressed concern that competition was being threatened by the growing number of mergers [8].

## HIGHWAY TRANSPORTATION

### 2-10. COLONIAL ROADS

Up until the Revolutionary War, the public roads consisted mainly of horse paths and were suitable for carts and wagons only near the large centers of population. Gradually, these paths, which often followed Indian trails or traces, were widened into wagon roads. Because of the formidable, heavily wooded terrain, the major emphasis in colonial times was on water transportation, and the roads were crude and localized.

In the colonies, the settlers instituted the statute labor system for the building and maintenance of roadways. Under this system, which had been used in England since 1555, several days were set aside each year in which the men were required to work on the roads. This practice was called "working out the road tax." It was supplemented by donations, bridge tolls, proceeds from public land sales, and fines from those who failed to perform the road work. Although the statute labor system was notoriously inefficient, it continued to be used in the United States until the beginning of the twentieth century.

### 2-11. ROAD BUILDING IN THE EARLY DAYS OF THE REPUBLIC

Prior to the Revolutionary War, the majority of long distance travel was either on foot or by horseback.

> Stage coaches had come into such general use, by 1785, that an urgent need arose for surfaced roads travelable at all seasons of the year. Impoverished by contributions of men, money and supplies during the War of Independence and by a ten-year postwar competition with British merchants, New England townships were unable to raise the necessary taxes for road improvement. The financial condition of the States was no better. The answer was found in turnpike companies chartered by the State and financed and operated by private citizens [9].

One of the most important of these roads was the Philadelphia and Lancaster Turnpike Road which was completed in 1795. This $62\frac{1}{4}$-mile stone and gravel road was built at a cost of $465,000, all from private sources. Other toll roads built during this period included:

1. The Northwestern Turnpike, which extended from Winchester, Virginia, to Parkersburg, Virginia. It was completed in 1838 at a cost of $400,000.
2. The Maysville Turnpike, a 64-mile section extending from Lexington, Kentucky, to Maysville, Kentucky, on the Ohio River. This stone-surfaced road was completed in 1835 at a cost of $426,000, one-half of which was provided by the state.
3. The Shenandoah Valley Turnpike, a 92-mile macadam road which extended from Staunton, Virginia, to Winchester, Virginia. The $385,000 road was completed in 1840.

Although the individual states frequently purchased stock in turnpike companies, no federal aid was provided for these roads. Indeed, the role of the federal government in roadbuilding during this period was rather limited, consisting principally of the construction of several primitive military roads and the National Pike.

## 2-12. THE NATIONAL PIKE

The National Pike, or Cumberland Road as it was also called, was the first important road to be built with federal funds. This road, which extended from Cumberland, Maryland, to Vandalia, Illinois, was built over a 33-year period ending in 1841. The legislation that authorized the construction of the National Pike provided for compacts or agreements between the federal government and Ohio, Indiana, and Illinois. Congress provided that a "two per cent fund" be established from the sale of public lands and that money from this fund would be used to reimburse the federal government for costs of the construction of the National Pike. This plan of reimbursement was never realized, and about $6.7 million was spent by the federal government in the construction of this road.

The steam railroads, which came into prominence in the 1840s, contributed to the financial failure of a large number of turnpike companies and decreased federal interest in road building. As a consequence, the responsibility for building and maintaining wagon roads and bridges was returned to the local governments for the remainder of the nineteenth century.

## 2-13. THE BICYCLE CRAZE

During the latter years of the nineteenth century, great interest was manifest in bicycles, first in "ordinary bicycles" which had large front wheels, and later in "safety bicycles" which resembled bicycles of the present day. Bicycle clubs were formed and, in 1880, a federation of local bicycle clubs was organized and named the League of American Wheelmen. Largely through the efforts of this organization, there grew a strong movement for the construction of good roads. This movement led to the establishment, in 1893, of the Office of Roads Inquiry in the U.S. Department of Agriculture, forerunner of the Bureau of Public Roads. Several state laws also were enacted at this time providing financial aid for local road improvements.

## 2-14. EARLY FEDERAL AID TO HIGHWAYS

Although the federal government has been involved in road construction since the early days of the republic, the current procedure whereby the states and the federal government share road building costs had its beginning in 1916, with the passage of the Federal Road Act. That act authorized the expenditure of $75 million over a period of five years for improvement of rural highways. The funds appropriated under this act were apportioned to the various states on the basis of area, population, and mileage of rural roads. The funds were divided on the basis of the ratio of the amount of each of these items within an individual state to similar totals for the country as a whole. Generally speaking, states were required to match the federal funds on a 50-50 basis.

The Federal-Aid Highway Act of 1921 strengthened the principle of federal aid in highway construction in two important respects. First, it required that federal funds be expended upon a designated system of interstate and intrastate routes, not to exceed 7 percent of the total rural mileage then existing within the state. Second, the act placed the responsibility for maintenance of these routes upon the individual states.

During the following years, appropriations for federal aid were increased steadily, while the basis of participation remained practically the same. Funds were provided for the improvement of roads in national parks and forests, Indian reservations, and other public lands.

In 1934, Congress authorized the expenditure of not more than 1.5 percent of the annual federal road funds received by the states in making highway planning surveys and other important investigations. Many planning surveys were conducted by state highway agencies during the period from 1935 to 1940.

World War II focused attention upon the role of highways in national defense. Although funds were provided for the construction of access roads to military establishments and for the performance of various other activities related to the war effort, normal highway development ceased during the war years.

The Federal-Aid Highway Act of 1944 provided $500 million for each of three years for postwar highway improvements including: $225 million for projects on the federal-aid system; $150 million for projects on the principal secondary roads; and $125 million for projects on the federal-aid system in urban areas. The act also required the designation of two new highway systems: the National System of Interstate Highways and the Federal-Aid Secondary System [9].

The Federal-Aid Highway Act of 1956 is regarded as one of the most important laws ever passed regarding transportation. The act provided nearly $25 billion[1] for the construction of the National System of Interstate and Defense Highways, the system envisioned by the 1944 highway act. The law specified that the funds for this system would be made available on a 90-10 matching basis, the larger share being provided by the federal government. This system connects all of the major cities of the United States with controlled-access freeways constructed to approved modern design standards. The act provided that the federal share of the costs of this system be paid from highway user taxes levied by the federal government and placed in a special trust fund called the Highway Trust Fund. The act also provided that several important studies be made, including a study of equitable tax allocations.

Prior to 1958, biennial highway appropriations were made by amending and supplementing the Federal Aid Road Act of 1916. On August 27, 1958, Congress passed a law that repealed all obsolete provisions of Federal-Aid Highway Acts and codified existing provisions into Title 23-Highways of the United States Code. Subsequent appropriations made by the Congress have been made "in accordance with Title 23 of the United States Code."

The growing problems of urban transportation were recognized by the Federal-Aid Highway Act of 1962. This act gave greater emphasis to planning and research and required cooperative and comprehensive planning of urban transportation facilities by state and local governments. Specifically the act:

1. Permitted more extensive use of federal aid secondary funds in urban areas.
2. Required that, after July 1, 1965, the Secretary of Commerce (Bureau of Public Roads) not approve any program for highway projects in an urban area of more than 50,000 population unless such projects are based on a continuing, comprehensive transportation planning process carried on cooperatively by the states and local communities.
3. Recognized the importance of urban mass transportation by requiring that urban highway systems be "an integral part of a soundly based, balanced transportation system for the area involved."
4. Required that the 1.5 percent planning and research funds be used only for these purposes, or otherwise lapse. (The Hayden-Cartwright Act, passed in 1934, was a permissive act. It authorized, but did not require, the use of up to 1.5 percent of the annual federal highway

[1]By 1988 the Interstate System was nearing completion, and it was estimated that its cost including the 10-percent state shares would exceed $100 billion.

funds for planning and research purposes.) The 1962 act provided that an additional one-half of one percent of federal and highway funds be available for expenditure for planning and research upon request by the state highway department.

5. Provided relocation assistance from construction funds for those required to move and relocate because of highway construction. Maximum relocation payments of $200 for individuals or families and $3,000 for business concerns and nonprofit organizations were authorized by this act.

## 2-15. HIGHWAY SAFETY LEGISLATION

Long a matter of national concern, the traffic safety problem worsened after World War II as gains in motor vehicle registrations were accompanied by similar increases in automobile crashes.

Highway engineers could take some satisfaction in the fact that thc traffic safety problem was not growing in proportion to the growth in travel. While the number of traffic fatalities had increased steadily over the years, the trend of the fatality rate in terms of deaths per vehicle mile had decreased. However, the troublesome highway safety problem reached epidemic proportions in the 1960s as the number of fatalities climbed higher than 50,000 per year. Reflecting an aroused public concern, the Congress passed two important safety measures: the National Traffic and Motor Vehicle Safety Act of 1966 and the Highway Safety Act of 1966. Administration of these acts was to be accomplished by a new federal agency, the National Traffic Safety Agency, later renamed and reorganized as the National Highway Traffic Safety Administration (NHTSA) in the Department of Transportation.

The National Traffic and Motor Vehicle Safety Act of 1966 provided for the establishment of federal motor vehicle safety standards and prohibited the manufacture or import of any motor vehicle that fails to conform to the established standards. Violators of these provisions may be penalized up to $1000 for each violation, up to a maximum of $400,000. This act required that motor vehicle manufacturers notify buyers and dealers of safety defects and provide for the correction of these defects. The act provided for research of the adequacy of safety standards and inspection requirements that apply to used motor vehicles, and for studies of the relationship between motor and vehicle performance characteristics, and accidents, injuries, and deaths. It was required that motor vehicle tires be labeled, giving the number of plies and maximum permissible load, and the new vehicles be equipped with tires that could support maximum loads safely.

The Highway Safety Act of 1966 required each state to have a highway safety program designed to reduce traffic accidents and deaths, injuries, and property damages resulting from these accidents. Administration of these programs is typically accomplished by a traffic safety coordinator on the governor's staff who oversees programs carried out by various state and local agencies. According to this act, the various programs were to be in accordance with uniform standards set forth by the federal government. The uniform standards promulgated in 1967 by the Secretary of Transportation provided for programs in 13 problem areas in traffic safety including programs in periodic motor vehicle inspection, motor vehicle registration, driver education, and emergency medical services. Five additional standards have since been established.

## 2-16. HIGHWAY LEGISLATION OF THE 1970s

The Federal-Aid Highway Acts of 1968 and 1970 demonstrated the government's increasing concern for transportation problems in urban areas and the social, economic, and environmental consequences of highway construction. The 1968 legislation contained three landmark provisions:

1. It authorized the use of federal aid urban funds for fringe parking projects to reduce urban congestion.
2. It established TOPICS, Traffic Operation Programs to Increase Capacity and Safety of urban facilities.
3. It provided for more liberal compensation to families and businesses displaced by highway projects.

The 1970 Federal-Aid Highway Act provided for the establishment of a federal-aid *urban* highway system and made $100 million available for expenditure on the new system during each of the fiscal years ending June 30, 1972, and June 30, 1973. The act specified that the urban system would serve major centers of activity, the highest traffic volume corridors, and the longest urban trips. Each route in the proposed system would connect with another route on a federal-aid system.

The 1970 act further emphasized the government's interest in encouraging the use of buses and other urban highway mass transportation systems. It provided that under certain conditions federal highway funds may be used to help finance the construction of exclusive or preferential bus lanes, traffic control devices, bus passenger loading areas and shelters, and fringe and transportation corridor parking facilities which serve mass transportation passengers.

The act required the Secretary of Transportation to promulgate guidelines designed to assure that possible adverse economic, social, and environmental effects are properly considered during the planning and development of federal-aid highway projects. (See Chapter 8.)

One of the sections of the 1970 legislation illustrates the important role of highways in carrying out national goals relating to land use and population distribution. This section authorized demonstration grants to states for the construction and improvement of *economic growth center development highways* on the federal-aid primary system. The purposes of these highways would be to revitalize the economy of rural areas and smaller communities, to enhance and disperse industrial growth, to encourage more balanced population patterns, and to check and possibly reverse the current migration from rural areas and smaller communities. The law provides payments of up to 100 percent of the cost of engineering and economic surveys and an increased federal share in the cost of construction of these facilities.

The act authorized the establishment of a National Highway Institute for the development and administration of training programs for federal, state, and local highway employees engaged in federal-aid highway work. The Secretary of Transportation was authorized to deduct up to $5 million per fiscal year from federal-aid appropriations for use in carrying out training programs. In addition, the act authorized the states (subject to the approval of the Secretary) to expend up to one-half of one percent of all federal-aid highway funds for payment of up to 70 percent of the cost of tuition and direct educational expenses of state and local highway department employees. The latter training could be carried out through grants and contracts with public and private agencies, institutions, and individuals.

In addition, the Federal-Aid Highway Act of 1970 provided that, beginning after June 30, 1973, the federal share of funds for the primary, secondary, and urban systems would increase from 50 to 70 percent.

The Federal-Aid Highway Act of 1973 greatly expanded the federal-aid urban system and provided for a major realignment of the federal road system based on anticipated functional usage in 1980. The 1973 act upgraded the economic growth center development highways to a standing program and expanded it to include all federal-aid highway systems except the Interstate.

The 1973 law made it possible for local officials in urban areas to decide not to use revenues from the Highway Trust Fund for road improvements, but instead to receive an equal amount from the general fund for mass transit. The legislation gave conditions under which trust funds could be used for mass transit, and placed a limit on the amount of such transfers.

Public Law 96-643, the Federal-Aid Highway Amendments of 1974, showed greater awareness of the need to conserve transportation energy. The law:

- Established a uniform national maximum speed limit on public highways.
- Established a demonstration program to encourage the development, improvement, and use of public mass transportation systems in rural and small urban areas.

## 2-17. HIGHWAY LEGISLATION OF THE EIGHTIES

With the passage of the Motor Carrier Act of 1980, major changes were enacted relative to the economic regulation of the trucking industry. The act removed certain restrictions on ratemaking and eased the regulations regarding entry of new carriers, the type of goods transported, and the routes served. The act was designed to eliminate over-regulation, and it was hoped that greater competition in the industry would save consumers $5 to $8 billion annually in decreased transportation costs.

The Surface Transportation Assistance Act of 1982 (STAA) was passed to address the problem of a deteriorating highway infrastructure. The act extended authorizations for the highway and transit programs from 1983 to 1986 and raised highway user fuel taxes from 4 to 9 cents per gallon. The act also instituted substantial increases in truck user fees. It specified that the additional funds were to be used for (1) accelerating the completion of the Interstate Highway System, (2) increasing the Interstate resurfacing, restoration, rehabilitation, and reconstruction (4R) program, (3) expanding the bridge replacement and rehabilitation program, and (4) providing greater funding for other primary and secondary projects.

The Surface Transportation Assistance Act allowed larger trucks to operate on the Interstate Highway System and certain other federal-aid highways. The law specified that states must allow twin-trailer combination trucks and may not establish overall length limits on tractor–semitrailer or tractor–twin-trailer combinations. The law established a width limitation of 102 inches and required that states allow

1. Semitrailers in a tractor–semitrailer configuration 48 feet in length.
2. Trailers in a tractor–semitrailer–trailer configuration 28 feet in length.

The law required that states allow the longer vehicles on interstate highways and on "those classes of qualifying Federal-aid Primary System highways as designated by the Secretary" (of Transportation). It provided that the Secretary designate as qualifying highways those Primary System highways that are capable of safely accommodating the longer vehicle lengths. The implementation of this part of the law has been accompanied by controversy, and by the late 1980s research was being performed to determine the classes of highways that could safely accommodate the larger trucks [10].

The Surface Transportation and Uniform Relocation Assistance Act of 1987 authorized $68.8 billion for federal-aid highway programs over a five-year period, fiscal years 1987–1991. Title II of the act authorized an additional $795 million for the National Highway Traffic Safety Administration's safety programs.

A major provision of the law allowed states to raise the 55 mile per hour speed limit on rural portions of the Interstate System to 65 miles per hour.

The higher speed limit is optional with the states and is allowed only along areas of the Interstate System that lie outside of designated areas of 50,000 or more.

## PUBLIC MASS TRANSPORTATION

### 2-18. THE GOVERNMENT AND MASS TRANSPORTATION

Until the early 1960s, there was practically no federal commitment to assist or promote mass transit systems.

One of the earliest legislative references to mass transportation was in the Federal-Aid Highway Act of 1962, which required that urban highway systems be "an integral part of a soundly based *balanced* transportation system." President Kennedy's transportation message of that same year to Congress reflected concern with mass transportation and requested a capital grant authorization of $500 million to be made available over a three-year period. Congress, by means of the Urban Mass Transportation Act of 1964, appropriated $375 million for mass transit assistance and gave the responsibility for the administration of these funds to the Housing and Home Finance Agency (later to become the Department of Housing and Urban Development). In 1968, this responsibility was shifted to the Department of Transportation when the Urban Mass Transportation Administration was formed within that department.

Despite increased government aid for mass transit, it became clear in the late 1960s that much more aid would be required to reverse the downward trends in mass transit usage. President Nixon reflected the deep concern of the federal government about public transportation in a message to Congress in 1969 in which he proposed that the government provide $10 billion out of the general fund over a 12-year period to help develop and improve public transportation in local communities.

The essence of the President's request was granted by Congress with the passage of the Urban Mass Transportation Assistance Act of 1970. This legislation, which amended the 1964 act, stated a federal intention to provide $10 billion for urban mass transportation during the period from 1970 to 1982. A total of $3.1 billion was authorized by the law, increasing amounts being specific for each of the fiscal years 1971 to 1975.

The act contained safeguards to protect the environment and required that public hearings be held on environmental as well as social and economic impacts of urban mass transportation projects. The act contained a provision that special consideration be given to the mass transportation needs of the elderly and the handicapped.

Although the act did not provide assistance for operating costs of mass transit systems, it required the Secretary of Transportation to conduct a study of the desirability of providing federal grants to assist local systems in meeting their operating costs.

The Mass Transport Assistance Act of 1974 provided $11.8 billion over a six-year period to aid mass passenger transport systems. The aid included $7.8 billion for capital outlays on an 80-20, federal-state matching basis, and $4.0 billion that could be used at local option for either capital improvements or operating subsidies. The latter amount was provided on a 50-50, federal-local matching basis.

The mass transit part of the Surface Transportation Assistance Act of 1978 authorized $13.5 billion over a four-year period for mass transportation. The authorizations included $5.9 billion in discretionary grants and $6.5 billion in "formula" grants to be apportioned on the basis of population and other considerations spelled out in the legislation. The 1978 law also authorized modest amounts for rural public transit assistance, intercity operating subsidies, and for bus terminal development.

Federal support of mass transit continued with the passage of the Surface Transportation Assistance Act of 1982 and the Surface Transportation and Uniform Relocation Assistance Act of 1987. Title III of the 1987 legislation authorized a total of $17.8 billion for mass transit over a five-year period, fiscal years 1987–1991.

## PIPELINE TRANSPORTATION

### 2-19. THE GOVERNMENT AND THE PIPELINE INDUSTRY

Pipeline companies are, for the most part, private corporations, and the federal government has given little direct assistance for the promotion of pipeline transportation. Indeed, the sole instance of direct assistance to pipeline transportation occurred during World War II, when the federal government helped finance the construction of two pipelines from Texas to New Jersey because of the threat of German submarines to coastal shipping. One of these pipelines was a 24-inch line, 1341 miles long, and the other was a 20-inch line, 1475 miles long.

On occasion, pipelines have benefited indirectly from government actions which adversely affected competitive transportation modes. For example, in 1890, the government aided the cause of pipeline transportation by outlawing the payment of rebates to the railroad industry.

Interstate oil pipelines were made common carriers subject to the Interstate Commerce Act by the Hepburn Amendment of June 29, 1906. Litigation followed and, in 1914, the U.S. Supreme Court ruled in favor of the ICC, except in the case of the Uncle Sam Oil Company which was simply transporting oil from its own wells to its own refinery. The decision in favor of the Uncle Sam Oil Company is the legal basis for present-day privately owned products pipeline systems that transport products from the owners' refineries to their own distribution terminals.

## AIR TRANSPORTATION

### 2-20. EARLY GOVERNMENT PROMOTION OF AIR TRANSPORTATION

There was incredible public apathy and governmental indifference to air transportation in the years immediately following the Wright brothers' first flight.[2] However, by 1908, the Army had expressed interest in the airplane and contracted with the Wrights to build an airplane for military use. The plane was delivered the following year.

Less than a decade after the Wrights' premiere flight, the government was becoming keenly interested in airplanes for the transportation of mail. In 1910, a law was passed to determine whether or not aerial navigation could be utilized for the safe and rapid transmission of the mails. The first official air mail flights were made the following year on Long Island, covering a distance of about 10 miles. Thirty-one additional air mail flights were made in 1912. The Post Office cooperated in these experimental flights, but did not share the expenses.

In 1916, Congress appropriated $50,000 for airmail service. However, because of World War I, nothing was accomplished with this appropriation. In 1918, a sum of $100,000 was appropriated for the establishment of an experimental air mail route, as well as the purchase, operation, and maintenance of airplanes. In May 1918, an experimental route was established between Washington, D.C., and New York City, with a stopover in Philadelphia. For about three months the air mail service was performed by six army pilots using army training planes. In August 1918, the Post Office bought four planes, hired six civilian pilots, and took charge of the air mail flights. The Post Office extended the air mail service to additional cities in 1919, and in 1920 transcontinental air mail service was achieved with the mail being transported by airplanes during daylight hours and by trains at night. During this period pilots often got lost and there were frequent crashes.

In 1925 the Kelly Act (Air Mail Act of 1925) was passed. This act provided for the transportation of mail by private airmail contractors and by the beginning of 1926, 12 contract air mail routes had been established. No subsidy was provided by this act as it stipulated that not more than 80 percent of the airmail postage revenue could be paid to the air carrier, with at least 20 percent going to the Post Office for ground handling expenses. Under this act

[2]On three occasions in 1905 the Wright brothers offered their invention and scientific knowledge to the War Department, which rejected their offers without even bothering to investigate [11].

payment was based on the number of letters, which meant that the letters had to be counted. The act was amended in 1926 to provide payments to the air carriers on the basis of weight and distance rather than number of letters. A second amendment to the Kelly Act, passed in 1928, opened the door for air mail subsidies to the carriers as it removed the provision in the original act that assured the government against losses.

## 2-21. THE AIR COMMERCE ACT OF 1926

In the 1920s, air transport was a risky business. The pilots were without radios and had neither navigational aids nor a system of aerial guideposts. In 1926, the Congress passed a law that marked the beginning of a recognition by the federal government of its leading role in the promotion and regulation of air transportation. Entitled the Air Commerce Act of 1926, this law gave the federal government the responsibility for the operation and maintenance of the airway system and other aids to air navigation, as well as the responsibility to provide safety in air commerce through regulation. The safety regulations, which were to be administered by the Department of Commerce, included registration and licensing of aircraft and the certification and medical examination of pilots. Thus, the scope of this act was much more comprehensive than the earlier airmail acts. Its purpose was to foster the growth and development of the air industry.

## 2-22. AIRMAIL ACTS OF THE THIRTIES

Two additional airmail acts were passed in the early 1930s: the Airmail Act of 1930 (McNary-Watres Act) and the Airmail Act of 1934 (Black-McKellar Act).

The McNary-Watres Act gave the Postmaster General unprecedented control over the airline industry, providing that airmail contracts could be granted without competitive bidding. The forceful administration of this act by Postmaster General Walter Brown contributed to the merger of several of the smaller airline companies and their placing greater emphasis on passenger transportation. This act also changed the method of compensating airlines for the transportation of mail, basing payments on space rather than weight.

The Black-McKellar Act placed the control of the air industry under three government agencies: the Post Office, the Interstate Commerce Commission, and the Department of Commerce. The Post Office retained its responsibility for awarding airmail contracts and enforcing airmail regulations. However, this act required that airmail contracts be made on the basis of competitive bidding. The Interstate Commerce Commission was given the responsibility of establishing and reviewing the rates of pay to the air carriers for the transportation of mail. The Department of Commerce was responsible for air safety, including the maintenance, development, and operation of the airway system.

The Airmail Act of 1934 also created a Federal Aviation Commission to study aviation policy. The findings of this commission and the difficulties encountered in regulating and promoting air transportation by three government agencies set the stage for the passage of the Civil Aeronautics Act of 1938.

## 2-23. THE CIVIL AERONAUTICS ACT OF 1938

This was a comprehensive act which provided for the development and regulation of air transportation. This act repealed or amended all major existing legislation having to do with aviation. It established three agencies for the regulation of air transportation. By amendment to the act in 1940, these agencies were reorganized and the various regulatory functions were placed under the control of:

1. The Civil Aeronautics Board, an independent organization, which was to exercise judicial and legislative authority over civil aviation and provide economic regulation of the industry. The CAB was also made responsible for the investigation of aircraft accidents.
2. The Civil Aeronautics Administration, a part of the Department of Commerce, which was given the responsibility for safety regulation and the operation of the airway system.

## 2-24. THE FEDERAL AIRPORT ACT OF 1946

This act must be regarded as one of the most important laws ever passed promoting air transportation. It provided $520 million to aid in the development of a comprehensive national system of airports. In addition, the act provided federal aid to airport sponsors for the construction of operational facilities (e.g., runways) on a 50-50 matching basis. Although the original bill provided funds for only a seven-year period, other bills have since been passed extending the federal-aid program.

## 2-25. THE FEDERAL AVIATION ACT OF 1958

The Federal Aviation Act of 1958 was largely a reenactment of the 1938 bill. Its principal effects were to remove the Civil Aeronautics Administration from the Department of Commerce and to give it increased authority in air safety. It was renamed the Federal Aviation Agency and under the new organizational structure, the Federal Aviation Administrator reported directly to the

President. The Federal Aviation Agency later became the Federal Aviation Administration, a part of the Department of Transportation, established in 1966.

In addition to outlining in general the organizations, powers, and responsibilities of the Federal Aviation Agency (FAA) and the Civil Aeronautics Board (CAB), this act:

1. Provided for the economic regulation of air carriers and the issuance of certificates of public convenience and necessity by the CAB. The act authorized the CAB to control the changing and publishing of fares and related charges and the consolidations and mergers of air carrier corporations and to fix rates of compensation to the carriers for transportation of mail.
2. Granted the FAA broad powers relating to safety regulation of civil aeronautics, including the issuance of licenses or certificates for aircraft and airmen.
3. Assigned the CAB the responsibility for investigation of aircraft accidents.
4. Authorized the Postmaster General to make rules and regulations, consistent with the provisions of the act, for the safe and expeditious transportation of mail by aircraft.

## 2-26. THE AIRPORT AND AIRWAY DEVELOPMENT ACT OF 1970

Recognizing the inadequacy of the nation's airport and airway system in meeting the current and projected growth in aviation, the Congress enacted Public Law 91-258. This law consisted of two parts: Title I, the Airport and Airway Development Act of 1970; and Title II, Airport and Airway Revenue Act of 1970.

Title I included the following provisions:

1. Authorized the expenditure of $280 million for each of the fiscal years 1971 through 1975 for airport development and an additional $250 million annually for acquiring, establishing, and improving air navigation facilities.[3] Generally, the airport development funds are to be provided on a 50-50 matching basis; however, the federal share of costs of certain lighting systems may be as much as 82 percent. The act does not allow federal aid for automobile parking lots or airport buildings except "such of those buildings or parts of buildings in-

[3]The act specified that $250 million would be for the purpose of developing airports served by air carriers and $30 million would be for general aviation airports.

tended to house facilities or activities directly related to the safety of persons at the airport."

2. Authorized an amount of up to $15 million annually for airport planning studies. Under this provision, two-thirds of the cost of these studies may be paid with federal funds.
3. Required that public hearings be held to consider the economic, social, and environmental effects of the airport location and its consistency with urban planning goals and objectives.
4. Directed the Secretary of Transportation to formulate and recommend to the Congress within one year after enactment of the law a national transportation policy considering the development of all modes of transportation. An annual report on the implementation of the national policy was also required.
5. Directed the Secretary to prepare and publish a national airport system plan for the development of public airports in the United States. The Secretary was directed to publish this plan within two years after enactment of the legislation and, thereafter, to review and revise it as necessary.
6. Established in Aviation Advisory Commission of nine members to formulate recommendations concerning the long-range needs of aviation.
7. Stated that "no airport development project involving the location of an airport, an airport runway, or a runway extension may be approved by the Secretary unless the public agency sponsoring the project certifies to the Secretary that there has been afforded the opportunity for public hearings for the purpose of considering the economic, social, and environmental effects of the airport location and its consistency with the goals and objectives of such urban planning as has been carried out by the community."

Title II of the 1970 legislation provided for taxes on aircraft fuel, air fares, and the transportation of property. This act also established the Airport and Airway Trust Fund and provided for the transfer from the general treasury to this fund amounts equivalent to the taxes imposed under the act. It was specified that the amounts in the trust fund were to be used to meet the obligations incurred under Title I of the act and under the Federal Aviation Act of 1958 as well as certain administrative expenses of the Department of Transportation.

## 2-27. RECENT AIRPORT LEGISLATION

The Airport Development Acceleration Act of 1973 increased federal funding for airport development from $280 million to $310 million annually and increased the federal share from 50 to 75 percent for airports with passenger enplanements less than 1 percent of the total national enplanements. The federal share for airport certification and security requirements was set at 82 percent. The 1973 law had a provision that prohibits the collection of "head taxes" at airports by state or local governments.

The 1976, Public Law 94-353 extended the airport development aid program for five years and increased the level of annual authorizations for airport development to $500 million in 1976, and increasing to $610 million in 1980. For airports enplaning less than 1 percent of the total national annual enplanements, the federal share of allowable project costs was changed to 80 percent beginning in 1979; for busier airports, the federal share was increased to 75 percent. The law also permitted the use of federal funds for certain non-revenue-producing parts of the passenger terminal.

The Airline Deregulation Act of 1978 was an air carrier regulatory reform law designed to deregulate domestic air passenger operations. The act stressed the goal of maximum reliance on competition and liberalized the criteria for entry into the air passenger market. It also made it easier for an air carrier to discontinue service and protected small communities from loss of service for 10 years by means of subsidized or replacement commuter service. The act decreased the authority of the Civil Aeronautics Board, removing its fare authority in 1981, its route authority in 1983, and calling for possible abolition of the board in 1985.

Significant changes to airport financing were made by the Airport and Airway Improvement Act of 1982. Major funding over a six-year period was authorized for airport improvement, ranging from $450 million in 1982 to $1017 million in 1987. The act authorized sums over the same period, ranging from $261 million to $1164 million for facilities and equipment associated with air traffic control and navigation, and a further $800 million to $1362 million for airspace system operation and maintenance [12].

## PROBLEMS

1. Prepare a report that outlines the various public statements regarding transportation which have been made by presidents of the United States.
2. Contrast the national transportation policy of the United States with that of one or more foreign countries.
3. Make a list of the various ways in which your state and federal government regulates:
   a. railroad transportation
   b. truck transportation
   c. water transportation
   d. air transportation
   e. pipelines

4. Prepare a report on the history of the Interstate Commerce Commission. Indicate how the powers of the ICC have changed since its establishment, and discuss the prospects for future changes in the Commission's regulatory powers and responsibilities.

## REFERENCES

1. *National Transportation Policy*, Committee on Interstate and Foreign Commerce (Doyle Report), January 1961.
2. Carr, Braxton B., "Inland Water Transportation Resources," *U. S. Transportation Resources, Performance, and Problems*, National Academy of Sciences, National Research Council, Publication 841-S, 1961.
3. Campbell, Thomas C., and Tai S. Shin, "Regulatory Reform, Ocean Shipping: The Shipping Act of 1984," *Transportation Quarterly*, Eno Foundation for Transportation, Westport, CT, April 1987.
4. Haney, Lewis H., *A Congressional History of Railways in the United States*, August Kelley, New York (1908, reprinted in 1968).
5. Association of American Railroads, "Land Grants, Public Service Role Put in Focus," News Release, September 17, 1969.
6. *Railroad Facts 1986*, Association of American Railroads, Washington, D.C., 1987.
7. *Transportation in America*, Fifth Edition, Transportation Policy Associates, Washington, D.C., March 1987.
8. Fitzsimmons, Edward L., "Can Railroads Compete with Each Other?" *Transportation Quarterly*, Eno Foundation for Transportation, Westport, CT, July 1987.
9. American Association of State Highway Officials, *Historic American Highways*, Washington, D.C., 1953.
10. Wright, Paul H., and Radnor J. Paquette, *Highway Engineering*, Fifth Edition, John Wiley, New York, 1987.
11. *The American Heritage History of Flight*, The American Heritage Publishing Company, New York, 1962.
12. Ashford, Norman and Paul H. Wright, *Airport Engineering*, Second Edition, John Wiley, New York, 1984.

# *Chapter* *3*

# *Administrative Structure and Finance*

Generally speaking, the administrative and regulatory machinery for transportation has evolved from a history of legislation enacted to cope with specific and sometimes transient problems. The result is an uneven patchwork of agencies, boards, and commissions at all levels of government that fund and oversee the development of the transport infrastructure and regulate, often unevenly, the carriers that use the system.

A variety of professional organizations, trade associations and unions, and lobbies of every description exist to influence transportation legislation, develop engineering standards and criteria, promote the use of products used to construct and maintain the physical system, and look after the interests of transport employees.

Chapter 1 described the physical transportation systems of the United States and generally described the users of those systems. Here, the administrative and financial structure for the development and operation of these systems will be described.

## FEDERAL TRANSPORTATION ORGANIZATIONS

Transportation is influenced to some extent by all of the three major divisions of government: legislative, judicial, and executive.

### 3-1. LEGISLATIVE COMMITTEES

At the federal level, transportation legislation mainly comes within the jurisdiction of five standing committees of Congress [1]:

Senate Committees
- Environment and Public Works Committee
- Commerce, Science, and Transportation Committee

House Committees
- Public Works and Transportation Committee
- Merchant Marine and Fisheries Committee
- Energy and Commerce Committee

In the Senate, the Environment and Public Works Committee is responsible for legislation relating to the improvement of water navigation along rivers and harbors and for the construction or maintenance of highways. The Commerce, Science, and Transportation Committee has jurisdiction over legislation affecting other modes, including measures affecting the regulation of interstate railroads, trucks, buses, oil pipelines, freight forwarders, and domestic water carriers, and both domestic and international air carriers. That committee also oversees measures dealing with the promotion of air transportation, including subsidies to the carriers and matching funds for airport construction.

In the House, the Public Works and Transportation Committee basically has the same jurisdiction as the Senate's Environment and Public Works Committee. In addition, it is concerned with laws affecting interstate trucks and buses, oil pipelines, freight forwarders, water carriers, urban mass transportation, and air carriers.

The Merchant Marine and Fisheries Committee has jurisdiction over international water transportation, including the approval of merchant marine programs. It is also concerned with laws affecting unregulated domestic ocean-going water transportation.

The Energy and Commerce Committee of the House has jurisdiction over railroad legislation, inland waterways, and the oil and gas pipeline functions of the Federal Energy Regulatory Commission.

Legislation that affects transportation may also be the concern of several other congressional committees, for example, the Appropriations Committees, which handle the actual appropriation of funds, the Judiciary Committees, which are concerned with the rules and procedures for the various regulatory agencies, and the Senate Banking, Housing, and Urban Affairs Committee which has jurisdiction over urban mass transportation matters.

## 3-2. FEDERAL TRANSPORTATION REGULATORY AGENCIES

Historically, in the United States, the railroads, airlines, trucking and bus companies, and other commercial carriers have been regulated by boards or commissions that were deemed to be independent of the executive branch of government. During the period 1977–1985, the regulatory powers resided in four agencies:

1. The Interstate Commerce Commission (ICC)

2. The Federal Maritime Commission (FMC)
3. The Federal Energy Regulatory Commission (FERC)
4. The Civil Aeronautics Board (CAB)

As arms of Congress these "independent" boards or commissions have regulated, in varying degrees, the right of carriers to operate over specified routes, the rates charged, and control over abandonments and mergers. These agencies are not courts, but they exercise quasijudicial powers, as well as quasilegislative powers. Members of the commissions are appointed by the President with the advice and consent of the Senate. The commissions are bipartisan. Federal laws require that not more than a majority of one can be from any political party.

The Interstate Commerce Commission has a complement of 11 members, and the Federal Maritime Commission and the Federal Energy Regulatory Commission have a membership of five each. Much of the routine work of these agencies is accomplished by a staff of specialists.

The oldest of the regulatory bodies, the Interstate Commerce Commission, is responsible for the regulation of railroads; motor carriers; water carriers that operate in coastwise, intercoastal, or inland waters of the United States; coal slurry pipelines; and freight forwarders.

The Federal Maritime Commission regulates U.S.-flag and foreign-flag vessels operating in the foreign commerce of the United States, as well as common carrier ships operating in domestic trade to points beyond the continental United States.

The youngest of the regulatory agencies, the Federal Energy Regulatory Commission was created in 1977 as part of the Department of Energy but with independent regulatory powers. It has jurisdiction over common carrier oil pipelines and natural gas pipelines.

Decisions of these regulatory commissions and boards may be, and often are, appealed in the courts.

Beginning in the mid-1970s and continuing into the 1980s, legislation was passed that deregulated large portions of the transportation industry. For example, the powers of the ICC over railroads, trucking firms, and bus lines were sharply reduced. Legislation passed in 1978 gradually phased out the authority of the CAB over civil aviation and disestablished the agency in 1985. The residual regulatory authority of the CAB was transferred to the U.S. Department of Transportation.

## 3-3. THE DEPARTMENT OF TRANSPORTATION

Within the executive branch of government, transportation matters at the federal level are administered by the U.S. Department of Transportation (DOT). The department is headed by the Secretary of Transportation, a member of the President's Cabinet. The Secretary is assisted by a deputy secretary, general counsel, assistant secretaries, and staff.

As Fig. 3-1 illustrates, the department has nine operating administrations:

1. U.S. Coast Guard
2. Federal Aviation Administration
3. Federal Highway Administration
4. Federal Railroad Administration
5. National Highway Traffic Safety Administration
6. Urban Mass Transportation Administration
7. St. Lawrence Seaway Development Corporation
8. Research and Special Programs Administration.
9. Maritime Administration

*The US. Coast Guard.* The U.S. Coast Guard is the nation's maritime law enforcement agency. It enforces federal laws regarding navigation, vessel inspection, port safety and security, and marine environmental protection. It maintains a network of rescue vessels, aircraft, and communications facilities to prevent maritime disasters and to provide effective search and rescue operations when offshore incidents occur. The Coast Guard conducts oceanographic research, furnishes icebreaking services, and develops, installs, maintains, and operates aids to maritime navigation. It enforces ship construction and safety standards and carries out a boating and safety education program for pleasure boats on domestic waters. An armed force, the Coast Guard polices the nation's coasts, and, in time of war supplements the operations of the U.S. Navy.

*The Federal Aviation Administration.* The primary responsibility of the Federal Aviation Administration (FAA) is ensuring the safety of air transportation. The administration issues and enforces air safety regulations, licenses airmen, inspects and certifies aircraft, and develops and operates the airways. Approximately one-half of the administration's employees are air controllers who are responsible for aiding pilots and monitoring and controlling aircraft movements. The FAA also conducts extensive research programs and provides financial assistance for airport construction and improvements through the Federal Airport and Airways Assistance Program.

*Federal Highway Administration.* The Federal Highway Administration (FHWA) administers the federal-aid highway program that provides matching funds to states for highway construction, reconstruction, and management. The administration develops and administers programs to promote highway safety and provides financial aid for several state-administered safety programs. FHWA cooperates with other federal agencies in the construction of roads in national parks and forests and on Indian reservations. It conducts and sponsors research in a wide range of subject areas and develops long-range transportation plans as well as short-range improvement programs. FHWA's

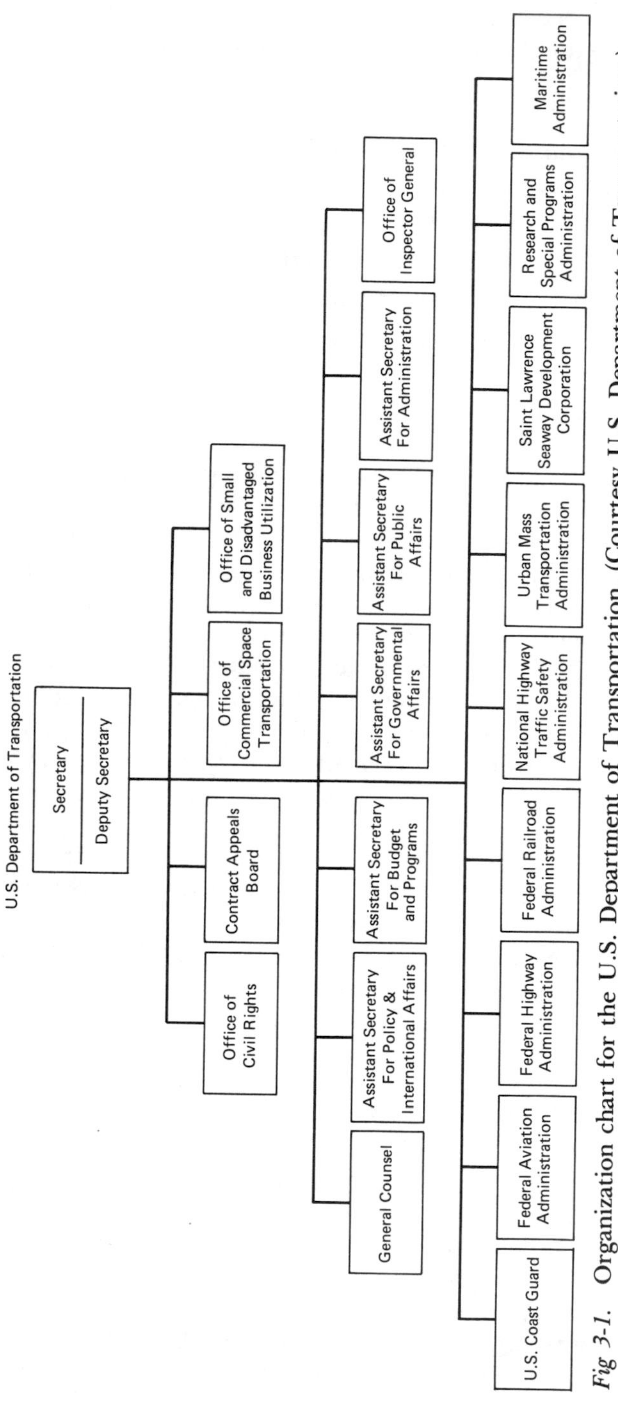


*Fig 3-1.* Organization chart for the U.S. Department of Transportation. (Courtesy U.S. Department of Transportation.)

Bureau of Motor Carrier Safety has jurisdiction over the safety performance of approximately 125,000 motor carriers engaged in interstate commerce. The bureau's employees make vehicle inspections, analyze truck and bus accidents, and enforce federal regulations regarding driver qualifications and their hours of service on the road.

*Federal Railroad Administration.* The Federal Railroad Administration (FRA) plans, develops, and administers programs to achieve safe operating and mechanical practices in the railroad industry. It enforces federal laws and regulations that promote the safety of railroads and administers financial-aid programs for railroad improvement. The FRA operates a test center at Pueblo, Colorado, where new railroad equipment is tested for safety and durability. It operates the U.S.-owned Alaska Railroad which has a 483-mile line that provides common carrier service between Seward and Fairbanks.

*National Highway Traffic Safety Administration.* The National Highway Traffic Safety Administration (NHTSA) establishes performance standards to improve the safety of motor vehicles and equipment and conducts test programs to evaluate and assure compliance with the standards. The administration provides financial assistance for safety programs carried out by the state and local agencies, including motor vehicle inspection, driver education, emergency medical services, and a variety of other activities that traditionally have been handled at the state or local level. NHTSA is also responsible for establishing fuel economy standards for automobiles and, in cooperation with the Environmental Protection Agency and Department of Energy, assessing penalties for violation of the standards.

*Urban Mass Transportation Administration.* The overall goal of the Urban Mass Transportation Administration (UMTA) is to improve urban mass transportation. Programs administered by UMTA provide matching funds to local agencies to build and rebuild fixed rail transit systems, modernize and replace equipment, and help offset operating deficits. For example, federal funds administered through UMTA pay 80 percent of the cost of new transit buses. The administration also provides aid for technical studies, planning, engineering, and designing transit systems.

*St. Lawrence Seaway Development Corporation.* The St. Lawrence Seaway Development Corporation was created in 1954 to construct the United States facilities for the St. Lawrence Seaway project. Since 1959, when the Seaway was opened to navigation by oceangoing ships, the Seaway Corporation has operated and maintained the part of the Seaway between Montreal and Lake Erie that is within the territorial limits of the United States. The corporation collects tolls, and earnings in excess of operating costs are used to pay the U.S. Treasury for a $131-million construction loan.

*Research and Special Programs Administration.* The youngest administration in the Department of Transportation, the Research and Special Programs Administration (RSPA), was created in 1977 to coordinate research and development activities and to conduct certain technical functions such as the regulation of hazardous materials shipments and the development of safety standards for oil and gas pipelines. The administration operates The Transportation Systems Center in Cambridge, Massachusetts, and is responsible for DOT's University Research Program. RSPA has ongoing programs designed to reduce maritime cargo losses due to theft and to simplify the distribution and processing of transport and trade documentation (docking papers, shipping orders, invoices).

*Maritime Administration.* Formerly a part of the Department of Commerce, the Maritime Administration (MARAD) was made a part of the Department of Transportation in 1980. MARAD promotes the merchant marine and maintains the National Defense Reserve Fleet. It sponsors research and development for ship design, propulsion, and operation. Through its Maritime Subsidy Board, the administration awards ship operating subsidies to liner carriers and formerly granted subsidies for ship construction in U.S. shipyards [1].

Table 3-1 shows the budget outlays and fulltime permanent positions for various administrations and offices of the Department of Transportation. In

*Table 3-1*

**Budget Outlays and Authorized Full-Time Permanent Positions in the U.S. Department of Transportation, Fiscal Year 1988**

| *Administration / Office* | *Budget Outlays $Millions (percent)* | *Full-Time Permanent Positions* |
|---|---|---|
| U.S. Coast Guard | $2,775 (11.1) | 5,662 civilian<br>38,606 military |
| Federal Aviation Administration | 5,251 (20.9) | 45,364 |
| Federal Highway Administration | 12,933 (51.6) | 3,482 |
| Federal Railroad Administration | −41[a] (−) | 634 |
| National Highway Traffic Safety Administration | 266 (0.9) | 645 |
| Urban Mass Transportation Administration | 3,369 (13.4) | 415 |
| Maritime Administration | 369 (1.5) | 1,028 |
| St. Lawrence Seaway Development Corporation | 1 (Nil.) | 194 |
| Research and Special Programs Administration | 23 (0.1) | 700 |
| Office of the Secretary | 72 (0.3) | 1,037 |
| Office of Inspector General | 30 (0.1) | 458 |
| ICC Rail Programs | 16 (0.1) | 249 |

[a]The negative number for the FRA outlays results from the proposed sale of approximately $200 million in rail preference shares to private sector lenders.

SOURCE: *Fiscal Year 1988 Budget in Brief,* Department of Transportation, January 1987.

fiscal year 1988, the department operated with nearly 100,000 employees and a budget of over $23 billion.

### 3-4. OTHER FEDERAL TRANSPORTATION AGENCIES

There are dozens of other federal agencies that directly or indirectly have responsibilities relating to transportation. Of special importance are the Corps of Engineers and the National Transportation Safety Board.

The U.S. Army *Corps of Engineers* is responsible for the construction and maintenance of river and harbor improvements. A part of the Department of Defense, the corps administers laws for the protection of navigable waterways. Through its Waterways Experiment Station at Vicksburg, Mississippi, and the Coastal Engineering Research Center at Ft. Belvoir, Virginia, the corps conducts research in coastal protection works, wave mechanics and phenomena, environmental protection of coastal areas, and other matters related to water transport services and facilities.

The *National Transportation Safety Board* (NTSB) is an autonomous five-person board appointed by the President. With the aid of its staff, NTSB promotes transportation safety by conducting independent investigations of accidents involving all transportation modes. The board is also authorized to regulate accident reporting procedures and review appeals of suspensions or denials of licenses issued by Department of Transportation agencies.

## STATE TRANSPORTATION ORGANIZATIONS

### 3-5. STATE REGULATORY AGENCIES

A wide variety of transportation organizations exist at the state level. The economic regulation of carriers engaged in intrastate transportation usually is performed by an autonomous board or commission whose members may be elected by the people or appointed by the governor. In the various states, regulatory agencies are given a variety of names. For example, in Georgia the regulatory body is called the Public Service Commission; in Illinois, it is the Commerce Commission; in Texas, it is the Railroad Commission.

Often the state regulatory agency is also charged with regulating utilities and a variety of other responsibilities unrelated to transportation. In certain instances, the transportation function comprises a relatively minor part of the board's responsibilities. Other regulatory activities such as the collection of revenue and the monitoring of truck weights are often handled by other departments of state government.

Routine decisions of these commissions usually are relegated to a staff of specialists.

## 3-6. STATE DEPARTMENTS OF TRANSPORTATION

Prior to the mid-1960s, the development of transportation facilities in most states was handled by several agencies. The activities often were separated by mode, and the developmental activities were fragmented and uncoordinated.

Since the establishment of the U.S. Department of Transportation, most states have established multimodal departments of transportation. In several states the administrative changes creating these departments have involved evolutionary modifications to the organizational structure of the state highway agency; smaller agencies for airports, mass transit, ports, and so on have been appended to the highway organization. In most states, including those in which major organizational restructuring has occurred, the highway division remains the dominant part of the state department of transportation.

State departments of transportation vary a great deal in their administrative organization. In terms of administrative control, there are three classes of departments—those with:

1. a single executive;
2. a single executive with a board or commission acting in an advisory or coordinate capacity;
3. a board or commission.

The single-executive organization usually has a director, secretary, or commissioner who reports directly to the governor. The executive may be elected but more commonly is appointed by the governor.

Many state DOTs are under the administrative control of elected or appointed boards or commissions. The board may have total administrative control and may involve itself with the day-to-day management and operations of the department. More commonly, it serves in an advisory or policy-making capacity. In a few instances, the routine management functions have been shared by a single executive and a board or commission.

Some state departments of transportation have been organized by transportation mode, as illustrated by Fig. 3-2*a*. Others have been organized by function, with major divisions provided for planning, design, construction, and so forth (see Fig. 3-2*b*). Still others have a mixed organization as Fig. 3-2*c* illustrates.

An organization chart for Georgia's Department of Transportation is shown in Fig. 3-3. The department is organized by function. It has seven district offices, one in each of the congressional districts of the state. Port development is under the jurisdiction of the Georgia Ports Authority. Vehicle registration and the collection of highway or user taxes is a function of the Revenue Department, and driver licensing and highway patrols are handled by the Department of Public Safety.

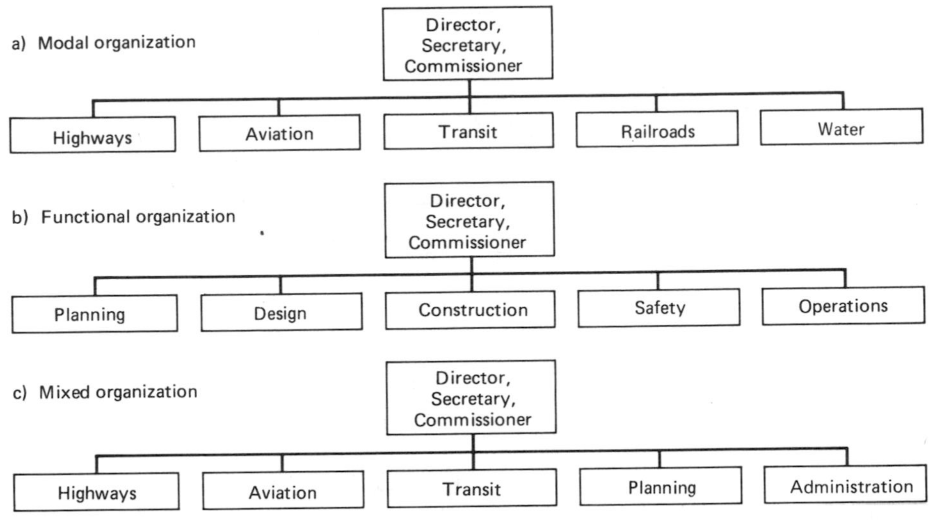


*Fig 3-2.* Basic organizational structures of state transportation departments. (Courtesy Transportation Research Board.)

## 3-7. LOCAL TRANSPORTATION AGENCIES

Ports, airports, and public mass transportation agencies[1] generally exist as an arm of local rather than state government. These facilities may be organized as public agencies, that is, as divisions of city government, or they may exist as quasigovernmental bodies such as authorities. Such bodies have some but not all of the powers of government. For example, authorities usually have the power to condemn land by eminent domain but do not have the power to levy taxes.

# TRANSPORTATION ASSOCIATIONS

In addition to the governmental agencies described, many other organizations contribute to the development of the transportation system. Some of the more prominent are described briefly in the following paragraphs [3].

*The American Association of State Highway and Transportation Officials* (AASHTO) is comprised of principal executive and engineering officers of the various state highway and transportation agencies and the U.S. Department of Transportation. The main purpose of AASHTO is to develop and improve

[1]The number of publicly owned transit systems in the United States has increased dramatically in recent years. In 1985, 43 percent of the transit systems were publicly owned. These systems carry about 89 percent of all transit passengers [2].

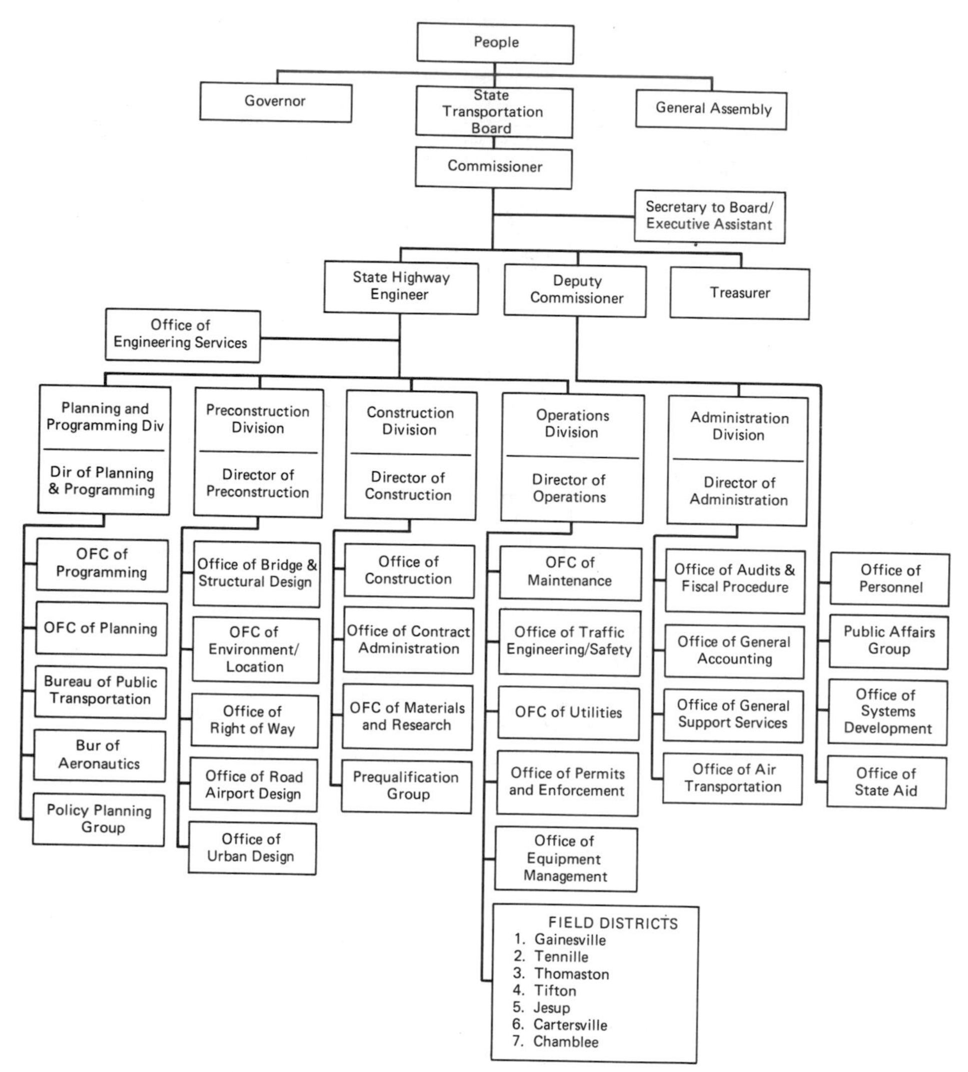


*Fig 3-3.* Georgia Department of Transportation organizational chart.

methods of administration, design, construction, operation, and maintenance of highways. AASHTO consults with Congress about highway legislation; develops technical, administrative, and operational standards and policies for highways; and cooperates with other agencies in the consideration and solution of highway problems [3].

*The Transportation Research Board* (TRB) is a national clearinghouse for transportation research. A division of the National Academy of Sciences, National Research Council, it is a private organization that encourages and correlates research efforts of educational institutions, governmental agencies, and industry. The TRB holds an annual research meeting in Washington,

D.C., has 270 technical committees, and issues a large number of technical publications on transportation subjects.

*The Institute of Transportation Engineers* (ITE) is a society of transportation professionals, with a membership of approximately 7700. Formerly called the Institute of Traffic Engineers, ITE's membership consists predominantly of traffic engineers and others interested in vehicular traffic control. The organization meets annually and publishes a number of handbooks and a monthly magazine called the *ITE Journal.*

*The Eno Foundation for Transportation* is a privately endowed organization that publishes and distributes technical literature on transportation subjects including *Traffic Quarterly* and various monographs.

*The American Railway Engineering Association* (AREA) is a professional organization of railway officers, engineers, and supervisors engaged in design, construction, and maintenance of railroad facilities. The AREA publishes a manual of railway engineering and various plans, specifications, and other technical materials relating to railroad engineering.

*The Association of American Railroads* (AAR) is the central coordinating and research agency of the American railroad industry. Its 148 members are the Class I railroad companies of the United States. The AAR has a staff of 800, and like most of the other associations, it is headquartered in Washington, D.C.

*The American Public Transit Association* (APTA) has as its members the rapid rail and motor bus transit systems in the United States, Canada, and Mexico, as well as manufacturers and suppliers of materials and services. It publishes *Passenger Transport* weekly and the *Transit Fact Book* annually.

*The American Trucking Associations* (ATA) is a federation of the trucking associations of the various states, the District of Columbia, plus 13 national conferences of truckers. With a staff of 250, the ATA promotes highway safety, supports highway research, and studies various technical and regulatory problems of the trucking industry.

*The Air Transport Association of America* (ATA) is comprised of domestic, international, territorial, intra-Hawaii, and intra-Alaska airlines of the United States. The association has a membership of 29 companies and a staff of 125. It publishes a *Quarterly Review of Airline Traffic and Financial Data* and an annual booklet entitled *Facts and Figures About Air Transportation.*

*The Aircraft Owners and Pilots Association* (AOPA) is an organization of 300,000 pilots and aircraft owners interested in making flying safer, more economical, and enjoyable. AOPA sponsors the Air Safety Foundation which conducts research, safety programs, and flight training clinics.

*The Airport Operators Council International* (AOCI) is an association of the various agencies, authorities, boards, and commissions that operate public airport facilities. The organization conducts research, publishes surveys and handbooks, and conducts an annual meeting of airport operators.

*The American Waterways Operators* (AWO) is comprised of the operators of towboats, tugboats, barges, and inland tankers and freighters that use the inland water system of the United States. The organization serves as a representative for its members on legislative matters and as liaison with the Coast Guard, the Corps of Engineers, and other public agencies.

*The American Association of Port Authorities* (AAPA) is an association of port administrative organizations of the United States, Canada, and Latin America. The association establishes standards for terminal design, operations, and cargo handling, fire prevention, maintenance, and administration.

Many other groups are engaged in activities relating to transportation, including organizations of materials producers, equipment manufacturers, colleges and universities, engineering associations, and labor unions. These organizations play an important role in defining and solving transport problems, promoting the development of the transportation infrastructure, and influencing the laws and regulations that control the use of those facilities.

## TRANSPORTATION FINANCE

The funding of transportation facilities in the United States is largely a public endeavor and is supported from tax funds by all levels of government: federal, state, and local. The financial support provided by governments has hardly been balanced, as Table 3-2 indicates. During the years 1980 to 1986, more

*Table 3-2*

**Federal, State, and Local Expenditures for Transport Facilities and Services (Millions of Dollars)**

| Year | | Mode | | | | | |
|---|---|---|---|---|---|---|---|
| | | *Airways* | *Airports* | *Highways* | *Rivers/Harbors* | *Railroad* | *Transit* |
| 1980 | Federal | 2,135 | 656 | 12,036 | 1,156 | 1,064 | 3,207 |
| | State and local | — | 2,501 | 27,152 | 1,168 | — | 3,806 |
| 1981 | Federal | 2,114 | 591 | 10,746 | 1,247 | 1,069 | 3,855 |
| | State and local | — | 2,743 | 28,115 | 1,508 | — | 4,488 |
| 1982 | Federal | 1,622 | 408 | 10,736 | 1,204 | 1,052 | 3,864 |
| | State and local | — | 2,818 | 30,545 | 1,422 | — | 4,672 |
| 1983 | Federal | 1,773 | 533 | 10,991 | 1,158 | 961 | 3,421 |
| | State and local | — | 3,014 | 32,986 | 1,458 | — | 6,325 |
| 1984 | Federal | 2,693 | 780 | 12,762 | 1,190 | 2,198 | 3,096 |
| | State and local | — | 3,585 | 35,090 | 1,329 | — | 6,325 |
| 1985 | Federal | 2,263 | 879 | 15,092 | 1,189 | 917 | 2,783 |
| | State and local | — | 3,744 | 40,623 | 1,495 | — | 6,945 |
| 1986 | Federal | 3,209 | 939 | 16,158 | 1,091 | 777 | 2,757 |
| | State and local | — | 3,900 | 45,310 | 1,400 | — | 7,200 |

SOURCE: *Transportation in America*, Fifth Edition, Transportation Policy Associates, March 1987.

than $148 billion was provided to transportation in federal financial aid. The distribution of these funds, by mode, in billions of dollars, was [1]

| | |
|---|---|
| Highway | 88.5 |
| Mass transit | 23.0 |
| Air | 20.6 |
| Rivers/harbors | 8.2 |
| Railroad | 8.0 |

Again, the differences among modes are striking. However, the funds for highways and air transport, and to a lesser extent, for the water mode, are primarily from taxes levied on the users of these facilities. Table 3-3 shows federal and state taxes and fees paid by users of various transportation modes during the years 1980 through 1985.

*Table 3-3*

**Federal and State Transport User Taxes and Fees (Millions of Dollars)**

| *Federal Taxes/Fees* | *1980* | *1981* | *1982* | *1983* | *1984* | *1985* |
|---|---|---|---|---|---|---|
| *Air Transportation* | 1,877 | 1,254 | 1,381 | 2,165 | 2,501 | 2,856 |
| Freight shipments | 92 | 13 | 0 | 118 | 134 | 134 |
| Passenger travel | 1,693 | 1,215 | 1,366 | 1,951 | 2,261 | 2,617 |
| Aircraft use/parts | 22 | 9 | 1 | 1 | — | — |
| Aviation fuel | 70 | 17 | 14 | 95 | 106 | 105 |
| *Motor Vehicles* | 6,419 | 6,247 | 6,744 | 7,777 | 10,546 | 11,810 |
| Gasoline | 3,970 | 3,761 | 4,120 | 5,640 | 7,699 | 7,588 |
| Diesel/special fuels | 518 | 570 | 594 | 890 | 1,470 | 2,225 |
| Large truck use tax | 260 | 271 | 333 | 236 | 180 | 379 |
| Lubricating oil | 77 | 76 | 76 | 9 | — | — |
| Automobiles | — | — | — | — | — | — |
| Trucks/trailers/buses | 736 | 688 | 725 | 338 | 865 | 1,396 |
| Tires and tubes | 634 | 643 | 672 | 616 | 332 | 222 |
| Parts and accessories | 224 | 238 | 224 | 48 | — | — |
| Inland waterways (Fuel) | 5 | 20 | 29 | 31 | 39 | 39 |
| Total—Federal taxes | 8,301 | 7,521 | 8,154 | 9,973 | 13,086 | 14,705 |
| State taxes/fees | 15,218 | 16,043 | 17,093 | 18,645 | 20,660 | 22,745 |
| Motor fuels | 9,486 | 9,972 | 10,484 | 11,508 | 12,846 | 13,683 |
| Registration/license/ mileage levies | 5,732 | 6,071 | 6,609 | 7,137 | 7,814 | 9,062 |
| Total—Federal/state | 23,519 | 23,564 | 25,564 | 28,618 | 33,746 | 37,450 |

SOURCE: *Transportation in America*, Fifth Edition, Transportation Policy Associates, March 1987.

## 3-8. HIGHWAY FINANCE

The funds necessary for the improvement of the nation's highway system are obtained from a variety of sources. The principal sources are (1) highway user taxes, (2) property taxes, and (3) tolls.

Highway user taxes comprise the dominant category of funds for highways at both the state and federal levels. Income from state highway user taxes totaled more than $22 billion in 1985, while federal user taxes amounted to approximately $12 billion.

Highway user taxes can be classified into three general types: fuel taxes, registration taxes and related fees, and special taxes on commercial vehicles.

Fuel taxes are by far the greatest revenue producer of all highway user taxes and, in most cases, this category is far and away the greatest single source of income for the operation and improvement of state highway systems. Fuel taxes constitute nearly two-thirds of the total obtained from state highway user taxes. State gasoline taxes range from 7 to 17 cents per gallon and average about 12 cents per gallon. Other fuels are taxed similarly.

First enacted as a regulatory measure, the registration of motor vehicles has long been recognized as a ready source of highway revenue. Registration fees are collected principally by state governments. Most states have graduated registration fees that increase with the weight or capacity of the vehicle. Several related fees are also levied by the states including fees for operators licenses, certificates of title, and title transfers. In many states, income from these latter sources is used for the operation (i.e., policing) of the highway system rather than for construction and maintenance.

In some states, special fees are levied against commercial vehicles. These taxes generally have taken the form of graduated fees computed on the basis of total mileage, ton miles or passenger miles of travel, or on gross receipts.

Property taxes are used primarily to finance highway improvements by local units of government. Two types of property taxes serve as a source of highway revenue: *ad valorem* taxes and special assessments. The first type is levied on property owners generally according to the value of the land and improvements, while the latter type applies only to the owners of property that lies on or adjacent to the highway or street improvement. Special assessments usually are based on a frontage basis rather than on the property value.

Tolls usually are levied by public agencies or authorities to repay costs incurred in the construction of a facility. Most commonly, tolls have been charged for particular structures, such as bridges, and tunnels. In 1985, toll receipts in the United States totaled more than $7 billion.

There are, of course, other sources of funds for highway improvements including parking meter receipts and funds from general revenue. These sources comprise a relatively small proportion of the total spent on the highway system.

## 3-9. RAILROAD FINANCE

Most of the railroads of the United States are privately owned and operated.[2] Prior to the 1970s, little government assistance was given to the railroads. Alarmed at the burdensome losses in passenger service, congress created the National Railroad Passenger Corporation (AMTRAK) in 1970 to operate most of the rail passenger service in the United States. Since that time the service has needed an average annual subsidy of about $350 million. The deficit for AMTRAK in 1986 was more than $700 million [4].

The Railroad Revitalization and Reform Act of 1976 provided a total of $2.1 billion for use by ConRail, the government-sponsored corporation created to restructure several financially troubled railroads in the northeastern and midwestern United States. The act also included arrangements for assisting other railroads in financing needed improvements, authorizing $1 billion in loan guarantees for plant and equipment and up to $600 million in lower-cost financing for rehabilitation and improvements.

The federal government provided approximately $917 million in financial aid to the railroads in 1985 [1]. During that year, the railroad industry spent $4.3 billion for maintenance of way and structures and $6.2 billion for maintenance of equipment [4]. No funds were requested for AMTRAK or other federal railway subsidy program in DOT's FY 1988 budget.

## 3-10. AIRPORT AND AIRWAY FINANCE

In the United States, most airports are owned by local governments which have limited capital resources. The funds for airport development therefore must come from a variety of sources including state and federal grants, local taxes, and borrowing.

In 1986, $3.9 billion was spent by state and local governments for airports, and $939 million was spent by the federal government. Much of state funding is derived from taxes on aviation fuel. Federal-aid funds for airports are generated by a series of user taxes first imposed by the Airport and Airways Development Act of 1970 including: an 8 percent tax on air fares, a 5 percent domestic cargo tax on tariffs, a 7 cent per gallon tax on noncommercial aviation fuel, and an airplane registration tax of $25 plus a levy based on the size and type of engine. These taxes are administered through the Airport and Airways Trust Fund.

Operating revenues for airports come from a variety of sources including landing fees, aircraft parking fees, fuel and oil sales, hangar rentals, and concessions. At large airports, income from concessions represents approxi-

[2] In 1988, there were 38 publicly owned railroads including the federally owned Alaska Railroad, plus 19 local line-haul railroads and 18 switching and terminal railroads owned by state or local governments.

mately 45 percent of the total operating revenue, while landing fees account for about 25 percent of the total. At small airports, hangar and building rentals tend to be the largest single source of operating revenue [5].

The development and operation of the airways in the United States is the responsibility of the Federal Aviation Administration. The development, maintenance and staffing of air traffic control systems are funded in large part by user taxes through the Airport and Airways Trust Fund. In 1986, the expenditures for the airways system exceeded $3.2 billion.

## 3-11. URBAN MASS TRANSPORTATION FINANCE

More than half of transit industry revenue comes from federal, state, and local government aid. In 1985, the breakdown of revenue for the transit industry was [2]:

| | |
|---|---|
| Passenger revenue | 36.9% |
| Federal operating assistance | 7.8% |
| State and local operating assistance | 49.6% |
| Other | 5.7% |

The average transit fare increased from 19 cents in 1963 to 38 cents in 1978, but in terms of real purchasing power (constant dollars), the average fare decreased. This decline reflected a shift from private to public ownership of the transit industry and a change in government policy toward transit [2].

In 1985, the transit industry received over $72 billion of financial assistance for operation and approximately $2.5 billion for capital improvements. The largest percentage of operating assistance came from state and local governments, but most of the capital assistance came from the federal government, which provides 80 percent of capital funding for purchases by transit systems.

Generally speaking, federal assistance for mass transportation is from general revenues. State and local assistance may be from general revenues, property taxes, gasoline taxes, sales taxes, bridge tolls, lottery proceeds, or some other source.

## 3-12. PIPELINE FINANCE

The construction and operation of a long-distance pipeline is an expensive and high-risk endeavor, posing problems in raising the necessary capital for development. Many pipelines have been constructed by specially formed corporations owned by oil companies that provide the necessary financial backing. Typically, the corporation issues debt instruments that are secured by the oil companies utilizing pledge of completion agreements or throughput and deficiency agreements. These debt instruments are usually privately placed with a consortium of lenders consisting of banks, insurance companies, pension trusts, and the like.

Another approach in financing multiple ownership pipelines is for each owner oil company to be responsible for completion of its share of the system. Each company then finances its share out of working capital, internal cash flow, or the issuance of securities.

In a few instances, pipeline ventures have been promoted or operated by owners who have had no other affiliation with the oil business. The most successful of these endeavors have been undertaken by companies of substantial size and borrowing capacity. Many of these ventures have failed because of lack of financing [6].

## 3-13. FINANCE OF WATER TRANSPORTATION FACILITIES

In the United States the funds for ports come from a variety of sources, and the financial structure varies among ports. Generally speaking, operating expenses are derived from cargo handling and storage fees, rentals of cranes and equipment, and for other services provided to ship owners and shippers. Some of these revenues are included as a part of the shipping tariff, and others are handled as rentals, lease agreements, or similar arrangements. Some port authorities serve as the terminal operator and deal directly with shippers and ship operators. Others lease the facilities to a terminal operator.

Funds for capital improvements for ports may be provided by state or city aid or from borrowing. Typically, capital is raised by revenue bonds or general obligation bonds with payback being made from operating revenue or with state or local government aid.

The federal government spent $1.09 billion for river and harbor improvements in 1986, while state and local governments spent $1.4 billion. Costs of improvements to the inland waterway system are borne in part by the users of the system by means of federal taxes on diesel fuel.

## PROBLEMS

1. Prepare a report on a selected transportation association. Describe its functions, administrative structure, sources of funding, and its impact on transportation.
2. Prepare a report on the procedures for initiating and enacting transportation laws in your state's legislature.
3. Prepare a report on the financial and administrative structure of a se-

lected bridge or road toll authority. Describe the authority's revenues, outlays, administrative structure, and its enabling legislation.

4. How is the airport in your area financed? Prepare a table detailing by type the annual receipts and expenditures for the airport agency.

5. Discuss the extent to which public transportation in your city is assisted by federal, state, and local governments, and how these sources of funds have increased or decreased during the past 20 years.

## REFERENCES

1. *Transportation in America,* Fifth Edition, Transportation Policy Associates, Washington, D.C., March 1987.
2. 1987 *Transit Fact Book,* American Public Transit Association, Washington, D.C., 1987.
3. *The Encyclopedia of Associations,* Gale Research Company, Detroit, Michigan.
4. *Railroad Facts 1986,* Association of American Railroads, Washington, D.C., 1987.
5. Ashford, Norman, and Paul H. Wright, *Airport Engineering,* Second Edition, Wiley Interscience, New York, 1984.
6. Wolbert, George S., Jr., *U.S. Oil Pipe Lines,* American Petroleum Institute, Washington, D.C., 1979.

## OTHER REFERENCES

*Fiscal Year 1988 Budget in Brief,* U.S. Department of Transportation, January 1987.

Wright, Paul H., and Radnor J. Paquette, *Highway Engineering,* Fifth Edition, John Wiley, New York, 1987.

# *PART 2*

# *OPERATION AND CONTROL OF TRANSPORTATION VEHICLES*

*The efficiency and safety of transportation facilities depend to a large degree on how well the physical facilities accommodate the variable and sometimes complex demands and needs of the vehicles and humans that use the system. Those who plan and design such facilities must therefore understand the operational and flow characteristics of the vehicles, the capabilities and limitations of the drivers, operators, and the pedestrians, and the devices and procedures used to control the use of the system. These matters are examined in the three chapters that comprise this part of the book.*

# Chapter 4

# Operational and Vehicular Characteristics

Although the civil engineer and planner have little professional involvement in the design of transport vehicles, the routeways and terminal facilities that must accommodate these vehicles and their traffic come within the domain of civil engineering design. The transport system is composed of a number of modes. Each mode has a variety of speeds and capacities depending on the type of vehicle being considered. From the viewpoint of understanding the impacts of vehicle demand on the geometric and structural design of the system a complete knowledge of operational characteristics and designs is desirable. This chapter attempts to provide a basic overview to some of the characteristics of the modes, while acknowledging that for actual design an in-depth study of system demands would be necessary.

## 4-1. SELECTION OF MODAL TECHNOLOGY

For both passenger and freight movement, there is a wide range of available transport technology. This technology depends on the type of movement and both the volume and speed at which it must be carried. It will later be shown that the demand *vs* distance relationship for transport is logarithmic in character. High-volume movements take place over short distances and long distance movements are of low volume. It is also true that speed is of real benefit only for long trips, and very high speed is viable only for very long trips given that it is expensive to design and operate very high-speed transportation modes. Because different classes of movement (e.g., intracity *vs* intercity) generate different volumes of movement and different trip lengths are suited to different speed ranges, the choice of mode can be seen to be a function in part of the two variables, volume and speed. This can be illustrated by examining the matrix of passenger modes shown in Fig. 4-1. The various available modes can be set in their speed-volume context within the triangular area below the demand limit line on the logarithmic graph. Three technologies define the corners of the triangular area:

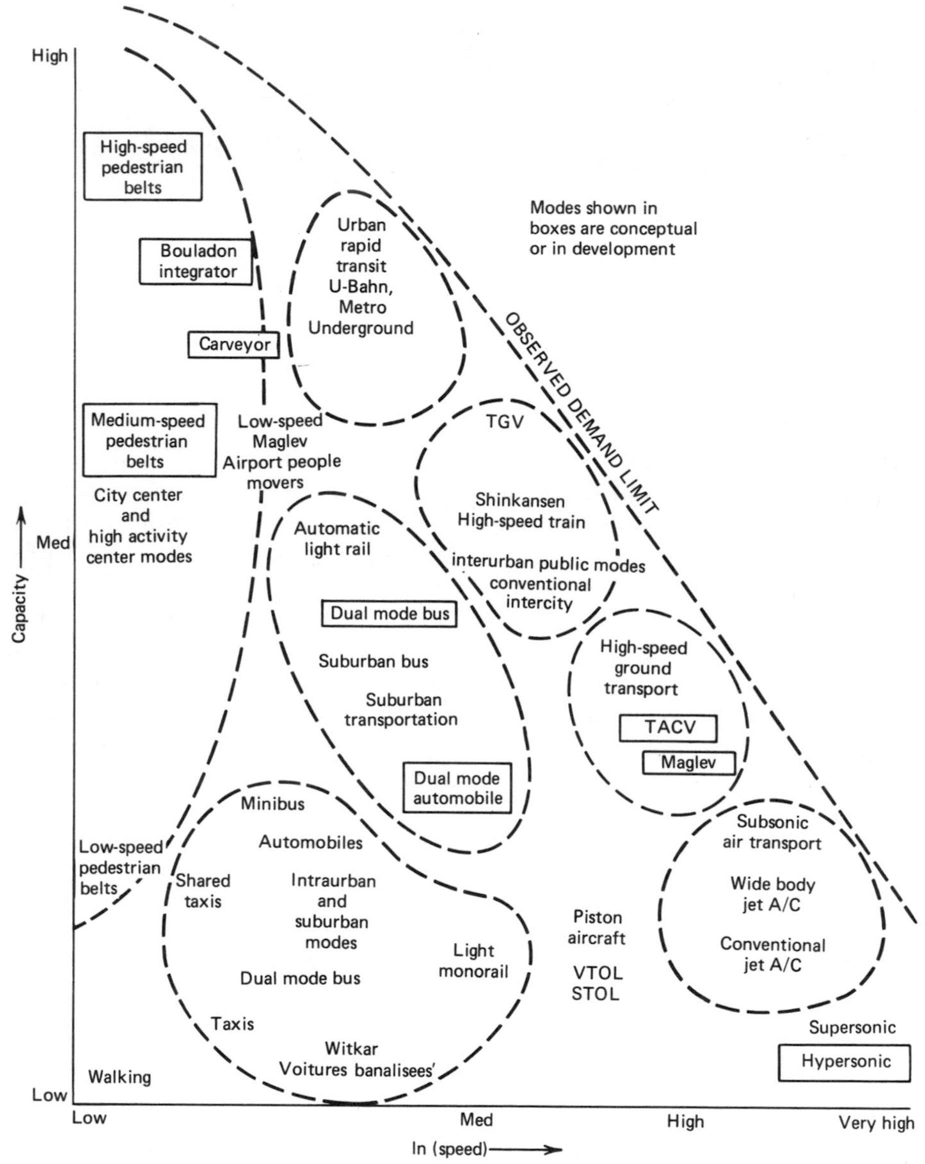


*Fig 4-1.* Location of modes in capacity speed spectrum.

| | |
|---|---|
| Low speed, low volume | —walking |
| Low speed, high volume | —high-speed pedestrian belts |
| Very high speed, very low volume | —supersonic and hypersonic transport |

Other modes are shown on the chart, grouped within general classifications.

## 4-2. ENERGY CONSUMPTION OF TRANSPORT [1]

The United States is an extremely mobile society. Its development and physical form reflect a need for high mobility. The overall economy and the structure of its cities were developed in an era of cheap energy; consequently transport energy consumption per capita is high by world standards even considering other developed countries. The United States with 5 percent of the world's population uses approximately 30 percent of total world energy. In comparison Western Europe with 9 percent of the world's population uses only 20 percent of all the energy consumed. Heavy urban reliance on the automobile and large distances between the United States markets has meant that a disproportionate share of the energy consumed has been used in the transport sector. Table 4-1 demonstrates that transport is the dominant sector of energy usage in the United States.

The same table indicates the very high level of personal mobility in the United States. The average American travels three times the amount traveled by the average Japanese and nearly twice as much as the average German. In doing so he uses 7.5 and 3.5 times as much energy as his respective Japanese and German counterparts.

The very large difference in energy usage per passenger mile is accounted for by the choice of passenger transport technology. The auto-dominant society of the United States is high in energy demand. Figure 4-2 shows the

*Table 4-1*
**Comparative Energy Consumption Statistics for Selected Western Economies**

| | *United States* | *Canada* | *United Kingdom* | *West Germany* | *Japan* |
|---|---|---|---|---|---|
| *Tons Oil Equivalent per Capita Annually* | | | | | |
| | 8.3 | 8.38 | 3.81 | 4.12 | 2.90 |
| *Energy End Use by Sector (percent)* | | | | | |
| Feedstocks | 8 | 6 | 9 | 10 | 9 |
| Fishing, agriculture, mining, and construction | 4 | 1 | — | 5 | 5 |
| Residential | 21 | 23 | 23 | 25 | 8 |
| Commerce and public | 8 | 17 | 12 | 10 | 8 |
| Transportation | 32 | 27 | 19 | 18 | 21 |
| Industry | 27 | 28 | 37 | 32 | 49 |
| *Passenger Transport Energy Consumption* | | | | | |
| Passenger miles per capita | 11,288 | 6,554 | 4,989 | 5,874 | 3,750 |
| Tons oil equivalent per capita | 1.227 | 0.820 | 0.280 | 0.336 | 0.164 |
| Tons oil equivalent per $ million GDP | 218 | 173 | 82 | 84 | 48 |

SOURCE: J. Darmstadter, J. Dunkerley, and J. Alterman, *How Industrial Societies Use Energy,* published for *Resources for the Future* by Johns Hopkins Press, 1977.

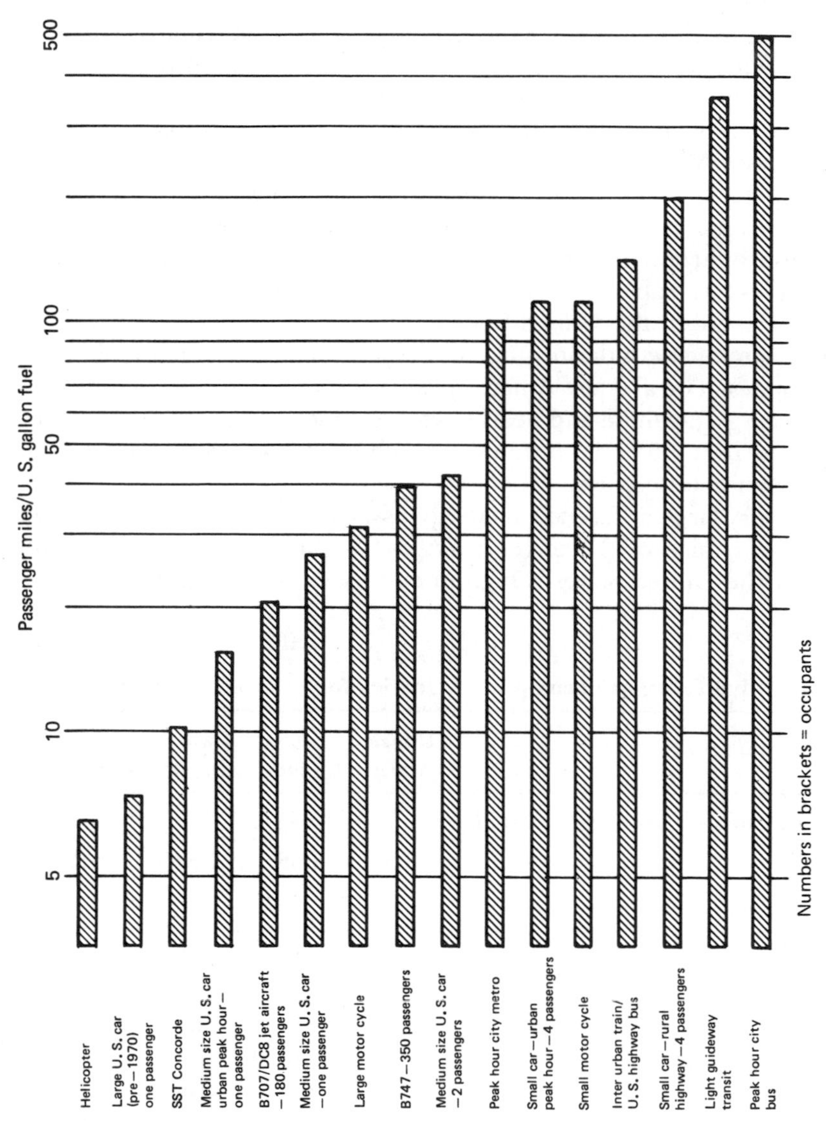


*Fig 4-2.* Net propulsive efficiency of various transport modes.

fuel consumption per passenger mile of a number of passenger transport vehicles under either assumed or normal levels of occupancy. It can be seen that most surface public transport modes are capable of attaining 40 passenger miles per gallon, a figure that is also approached by modern wide-bodied jet aircraft. Well-designed, fuel-efficient automobiles with more than single-person occupancy can be reasonably fuel efficient.

It is also worth examining, in energy terms, the price that must be paid to achieve high speeds in passenger movements. In general, it can be seen from Fig. 4-3 that there is a strong exponential relationship between speed and energy consumption per passenger mile. This has been especially critical in the economic viability of supersonic air transport operations. The Concorde, which was designed during a period of decreasing real oil prices, has had to operate in an adverse economic environment. The energy cost of supersonic transport is so high that its future has appeared questionable since the cancellation of the U.S. SST program in 1970. By the mid-1980s the collapse of world oil prices has engendered speculation that a large SST with a capacity of

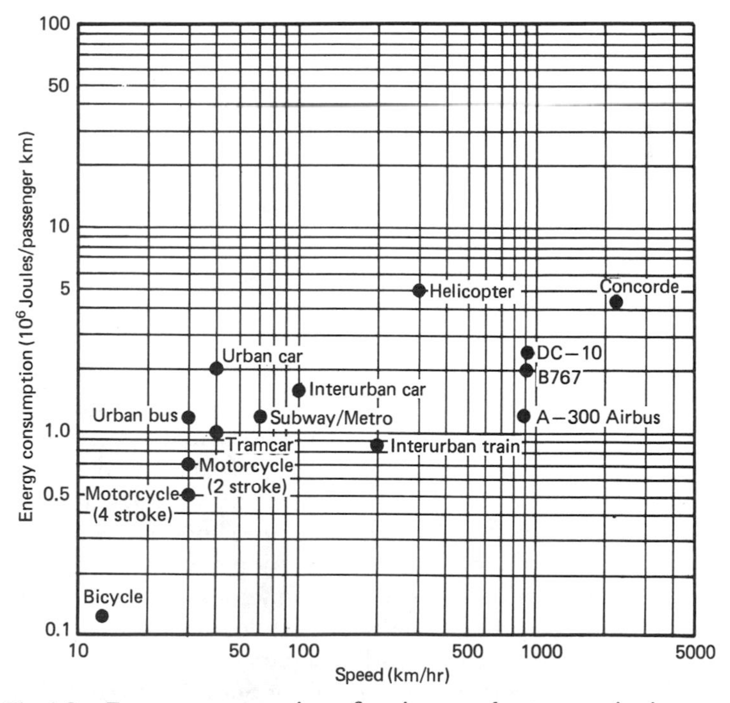


*Fig 4-3.* Energy consumption of various modes at capacity in terms of cruise speed.

300 passengers and a range of 6500 nm may be economically viable with the use of post-Concorde technology.

## 4-3. PASSENGER CAR DIMENSIONS AND WEIGHTS

The design of the passenger automobile reflects at any one point in time the status of available technology and consumer taste in styling. For passenger cars, styling and cost may be the two most important factors affecting the consumer's decision to select any particular model. Automobiles are more stylish, more luxurious, and less austerely designed than mass transit vehicles or trucks. To some degree energy conservation consideration has put some constraint on the consumer-oriented design philosophy. In the late 1970s the Congress imposed on the U.S. automobile manufacturers a requirement that by 1985 the average production fleet of each manufacturer should achieve a minimum of 27.5 mpg. This target was not reached but the legislation was effective in forcing the production of more energy efficient vehicles in the medium and small model ranges. Styling changes relate mostly to somewhat superficial body and finish innovations. The operating characteristics of the vehicles change far less than the striking modifications in style might at first indicate.

Certain physical limitations are placed on vehicles using the highway system, and the highways themselves are designed with the performance characteristics of the vehicles as a design constraint. The constraints and limitations apply only to the largest vehicles in the systems, that is, trucks; individual passenger car models can vary considerably without infringing upon these design constraints.

Figure 4-4 shows the variation in some of the principal dimensions of the American manufactured auto fleet since 1960 [3]. Average vehicle length had increased at a reasonably constant rate until World War II, after which length remained relatively constant until 1970. Since then there has been a significant decline in length, which will continue until U.S. government fuel consumption standards finally are met. The average width of cars increased from 67 in. in 1928 to 79 in. in 1959. There has been a decline since that time which also accelerated with the introduction of small model cars in 1970. In the long term the average width is likely to approach a figure in the region of 62 in. The most significant long-term trend of passenger automobiles was a continuous decrease in vehicle height up to 1960. After this date vehicle height has remained relatively constant. Along with a decrease in overall vehicle height, the position of the driver's line of sight has also dropped over time. Lowering the driver's eye height has influenced changes in the geometric design requirements of the road designer [4]. The trend of lower-slung autos prompted the American Association of State Highway Officials in 1984 to change its passing sight distance criteria from distances based on a driver's eye height to 3.5 ft [5].

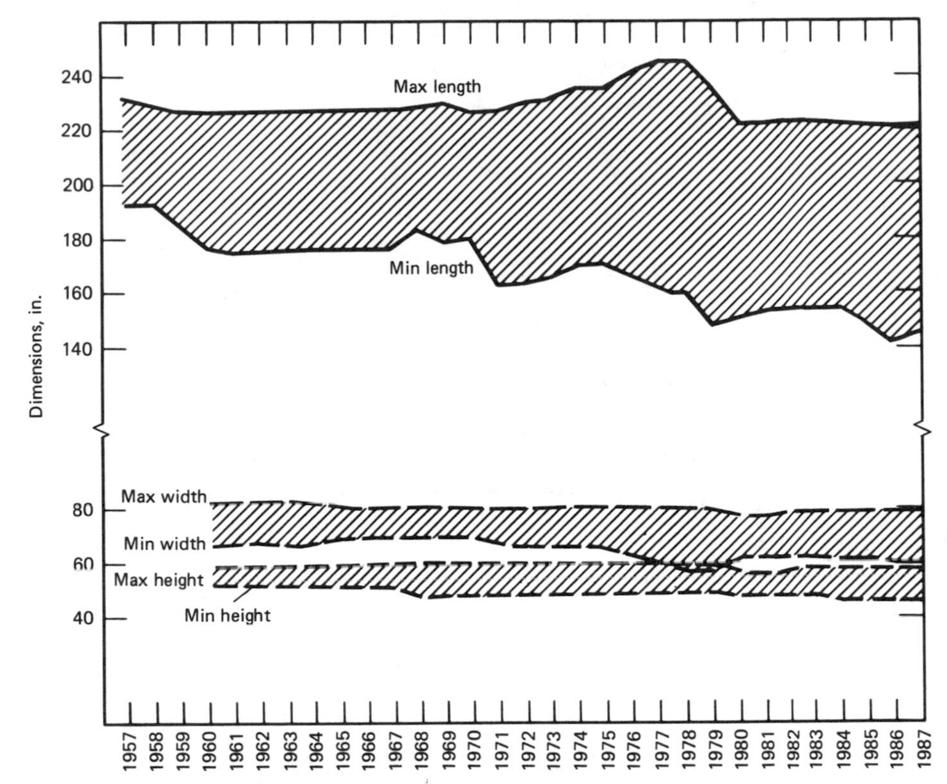


*Fig 4-4.* Trends in maximum and minimum dimensions of standard American passenger cars 1960–1981. (Source: Motor Vehicle Manufacturers Association of the U.S.)

## 4-4. HIGHWAY ACCELERATION RATES

The acceleration rates of the different vehicle types using roadways have a profound effect on both the design and operation of individual segments of highway. Two-lane highways can function well under low vehicular volumes where there is the ability to overtake slower moving vehicles. In multilane roads, the design of merging and weaving sections depends on the maneuverability and acceleration characteristics of the vehicular traffic. Figure 4-5 indicates the superior acceleration characteristics of passenger cars over commercial vehicles due to much higher power-to-weight ratios of private automobiles. In the presence of even medium grades, the lower performance of commercial vehicles is accentuated. Trucks are found to decelerate to very low operating speeds over long uphill grades. Figure 4-6 shows the effect of grades of different length on trucks from a previous operating speed of 55 mph. It can be seen that any grade appreciably greater than 2 percent will result in low operating speeds even over a relatively short distance. Grades in excess of 4 percent rapidly result in operation at crawl speeds and well below

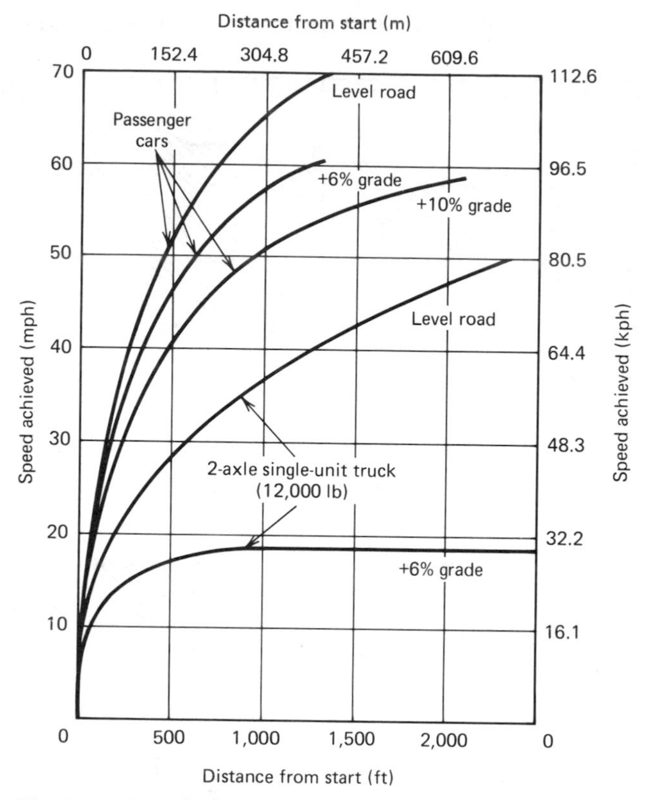


*Fig 4-5.* Speed–distance relationships observed during maximum rate accelerations. (Source: *Transportation and Traffic Engineering Handbook.*)

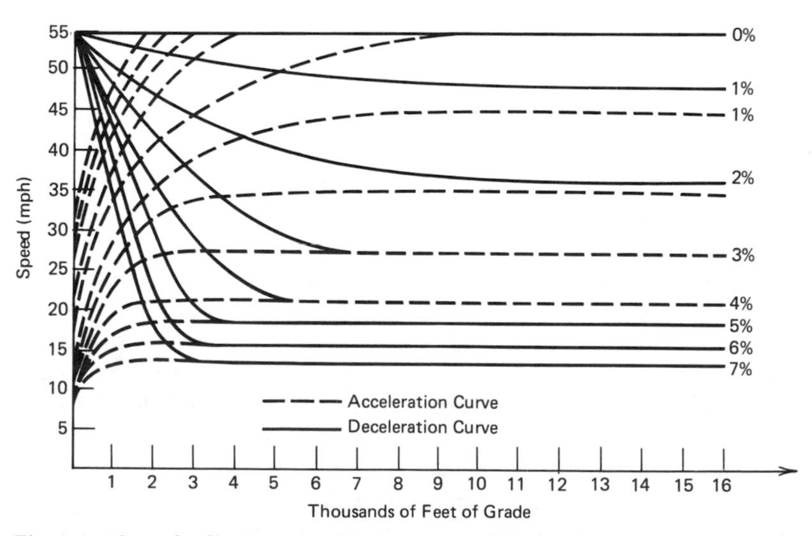


*Fig 4-6.* Speed–distance curves from road test of a typical heavy truck operating on various grades. (Source: Reference 7.)

tolerable conditions. Where extensive lengths of uphill grades are unavoidable, the presence of low performance trucks requires additional climbing lanes to allow higher performance vehicles to overtake others safely.

## 4-5. HIGHWAY VEHICLE DECELERATION RATES

The large range of highway vehicle types exhibits large variations in their abilities to decelerate. Empirical studies indicate that, in general, as the size of a vehicle increases from the ordinary passenger car to vehicular combinations such as trucks with full trailers with multiple axles, the deceleration capability decreases significantly. Figure 4-7 shows that while 94 percent of passenger vehicles can achieve a deceleration of 25 ft/sec$^2$ or better, only 13 percent of the class of trucks and full trailers were able to perform at the standard. The large range of deceleration characteristics in a heterogeneous traffic stream leads to hazardous conditions at high volumes and high speed. Under these conditions, it is not infrequent that the headway between vehicles falls below the amount required to compensate for the different deceleration characteristics, and rear-end collisions result.

## 4-6. DESIGN VEHICLES

Two general classes of vehicle use highways, namely passenger cars and trucks. Under the designation of passenger vehicles it is normal to refer to all light vehicles including light delivery trucks, vans, and pickups. Trucks are taken to

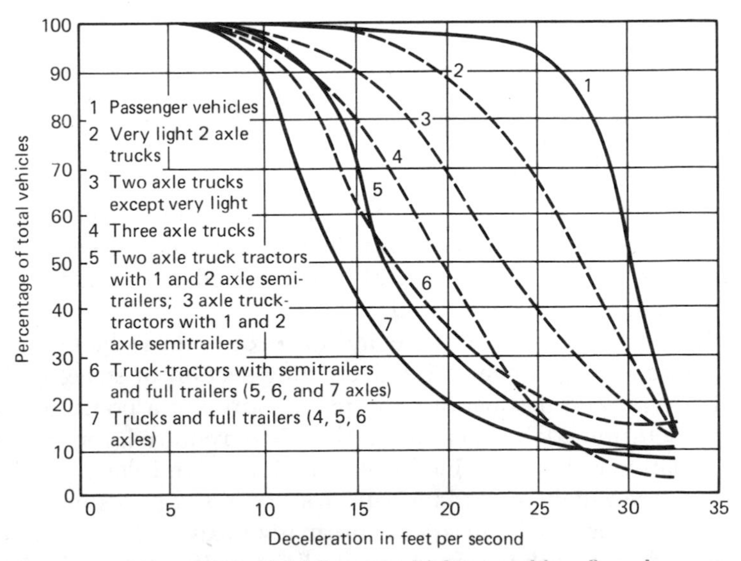


*Fig 4-7.* Percentage of highway vehicles capable of a given or greater deceleration. (Source: Reference 6.)

include single-unit trucks, recreation vehicles, buses, truck tractor–semitrailer combinations, and trucks or truck tractors with semitrailers in combination with full trailers.

Highways are designed to accommodate the types of vehicles that will use them. Prior to undertaking the geometric design of any facility, the proportions of travel that can be anticipated by any particular vehicle class are estimated and projected from available road use data. As a result the engineer can reasonably predict the type of usage any facility will receive prior to design. During the design phase the facility is proportioned and arranged to meet the needs of the class of vehicle that it will seek to accommodate. To facilitate the design process the American Association of State Highway and Transportation officials (AASHTO) has determined the dimensions and geometric characteristics of a number of "design vehicles." The largest of all the several design vehicles are usually accommodated in the design of freeways.

A design vehicle is defined as a selected motor vehicle whose weight, dimensions, and operating characteristics are used to establish highway design controls to accommodate vehicles of a designated type. To cover the large range of motor vehicles using highways, AASHTO has designated 10 design vehicles: passenger car (P), single-unit truck (SU), single-unit bus (BUS), articulated bus (A-BUS), semitrailer intermediate (WB-40), semitrailer combination large (WB-50), semitrailer full-trailer combination (WB-60), motor home (MH), passenger car with travel trailer (P/T), and passenger car with boat and trailer (P/B) [7]. The dimensions and characteristics of the design vehicles are chosen so that virtually all vehicles within these classes are found to be covered in the design case. Also in developing the dimensions of the design vehicles, trends of motor vehicle size were analyzed to make reasonable provision for future vehicles. Figures 4-8*a*–*c* show examples of the AASHTO vehicles.

## 4-7. HIGHWAY TRAFFIC STREAM CHARACTERISTICS [8]

The flow of vehicles along a highway are characterized by three fundamental traffic parameters: speed, volume, and density. Basically, speed is the total distance traveled divided by the time of travel. Speed is commonly expressed in miles per hour or feet per second.

Vehicle speeds vary both in time and in space and can be measured at a point or can be averaged over a relatively long section of street or highway between an origin and destination. Speeds measured at a point, termed spot speeds, are determined by measuring the time required for a vehicle to traverse a relatively short specified distance. Radar and various other electronic devices are widely used to measure spot speeds. Speeds measured over long sections of streets or highways are normally measured by means of a test vehicle that is driven at the average speed of the traffic stream. By recording

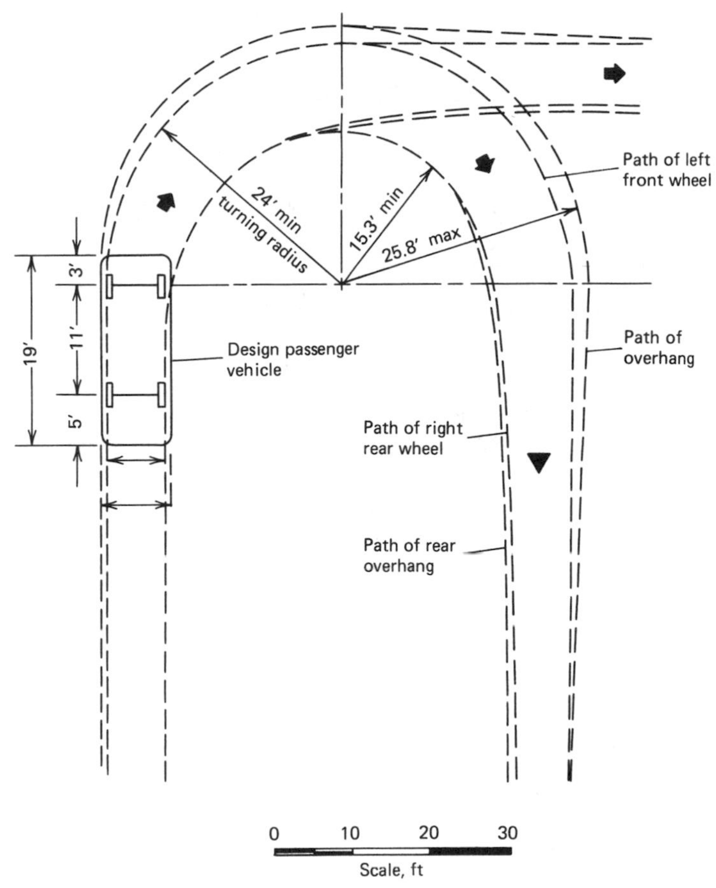


*Fig 4-8a.* Minimum turning path for P design vehicle. (Source: Reference 7.)

times and distances of travel, it is possible to calculate stream speeds along the route and identify sources of delay.

Speeds vary with environmental and traffic conditions, being generally higher along expressways and other well designed facilities and during times when traffic congestion is not a factor. At a given time and location, speeds are widely dispersed and can generally be represented by a normal probability distribution. Both the average and range of spot speeds decrease with an increase in traffic demand.

Traffic volume is defined as the number of vehicles that pass a point along a roadway or traffic lane per unit of time. A measure of the quantity of traffic flow, volume is commonly measured in units of vehicles per day, vehicles per hour, vehicles per minute, and so forth.

On a given roadway, the volume of traffic fluctuates widely with time. The variations that occur tend to be cyclical and to some extent predict-

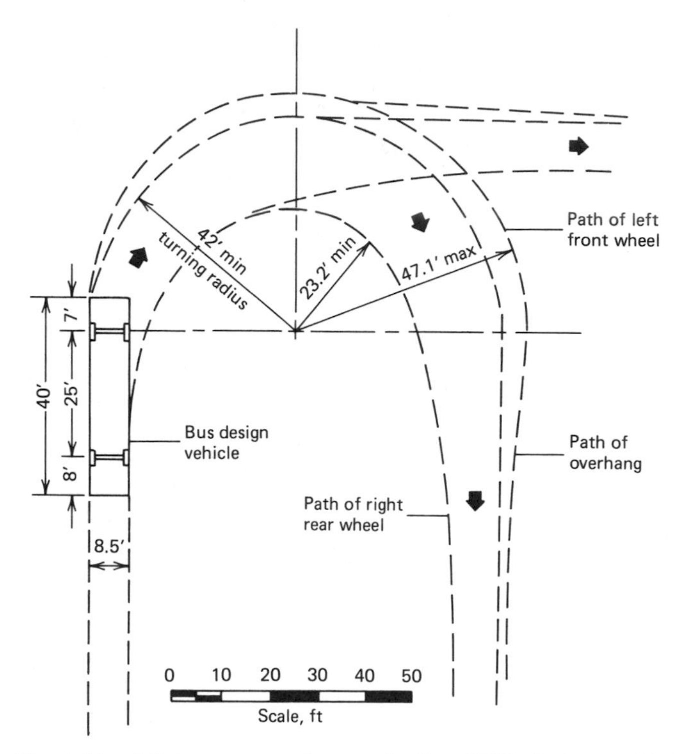


*Fig 4-8b.* Minimum turning path for BUS design vehicle. (Source: Reference 7.)

able. The nature of the pattern of variation depends on the type of highway facility. Urban arterial flow is characterized by pronounced peaks during the early morning and late afternoon hours, due primarily to commuter traffic. The peaking pattern is not generally evident on weekends, and such facilities experience lowest flows on Sundays. Rural highways tend to experience less pronounced daily peaks, but they may accommodate heaviest traffic flows on weekends and holidays because of recreational travel. Highway facilities generally must accommodate heaviest flows during the summer months. Peaks typically occur during July or August. As might be expected, the seasonal fluctuations are most pronounced for rural recreational routes.

The term *rate of flow* accounts for the variability or the peaking that may occur during periods of less than 1 hr. The term is used to express an equivalent hourly rate for vehicles passing a point along a roadway or for traffic during an interval less than 1 hr, usually 15 min.

The distinction between volume and rate of flow may be illustrated by an example. Suppose the following traffic counts were made during a study period of 1 hr:

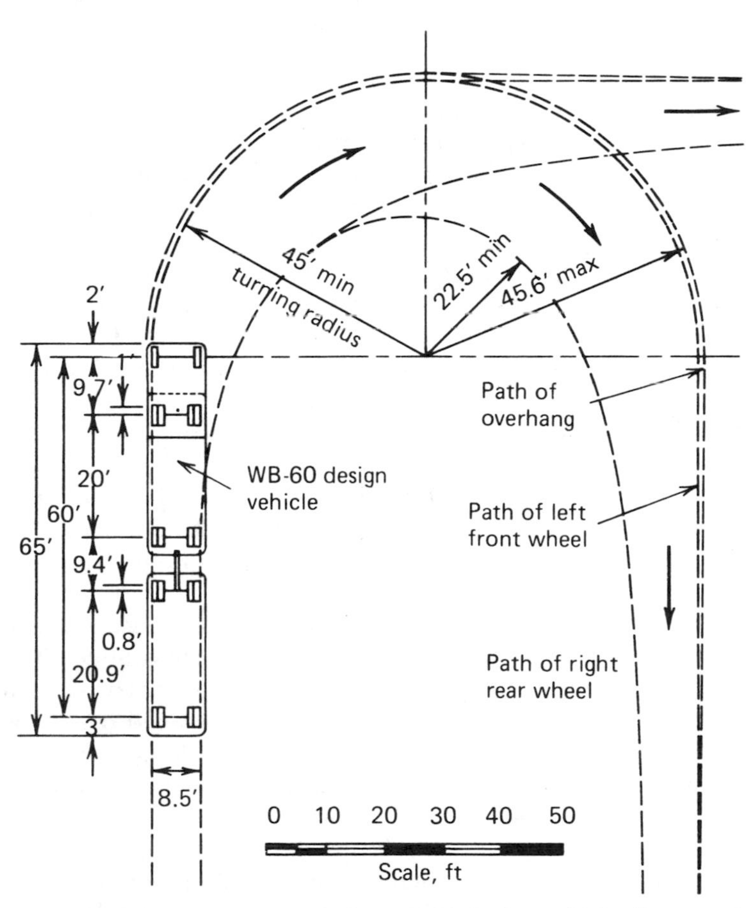


*Fig 4-8c.* Minimum turning path for WB-60 design vehicle. (Source: Reference 7.)

| *Time Period* | *Number of Vehicles* | *Rate of Flow (veh/hr)* |
|---|---|---|
| 6:00–6:15 | 500 | 2000 |
| 6:15–6:30 | 575 | 2300 |
| 6:30–6:45 | 500 | 2000 |
| 6:45–7:00 | 425 | 1700 |
| Total | 2000 | |

The total volume is the sum of these counts or 2000 veh/hr. The rate of flow varies for each 15-min period and during the peak period is 2300 veh/hr. Note that 2300 vehicles did not actually pass the observation point during the study hour, but they did pass *at that rate* for one 15-min period.

Consideration of peak rates of flow is of extreme importance in highway capacity analyses.[1] Suppose the example roadway section is capable of handling a maximum rate of only 2100 veh/hr. In other words, its capacity is 2100 veh/hr. Since the peak rate of flow is 2300 veh/hr, an extended breakdown in the flow would likely occur even though the volume, averaged over the full hour, is less than the capacity.

The *Highway Capacity Manual* [8] uses a *peak hour factor* to relate peak rates of flow to hourly volume. The peak hour factor is defined as the ratio of total hourly volume to the maximum rate of flow within the hour. If there was no variability in flow rate during the hour, the peak hour factor would be 1.00. Typical peak hour factors for two-lane roadways range from about 0.83 to 0.96.

Two measures of traffic volume are of special significance to highway engineers: average daily traffic (ADT) and design hourly volume (DHV). The average daily traffic is the number of vehicles that pass a particular point on a roadway during a period of 24 consecutive hours averaged over a period of 365 days. It is not feasible to make continuous counts along every section of a highway section. Average daily traffic values for many road sections are therefore based on a statistical sampling procedure.

The design hourly volume is a future hourly volume that is used for design. It is usually the thirtieth highest hourly volume of the design year. Traffic volumes are much heavier during certain hours of the day or year, and it is for these peak hours that highways are designed. The design hourly volume is discussed further in Section 12-6.

Traffic density is defined as the average number of vehicles occupying a unit length of roadway at a given instant. It is generally expressed in units of vehicles per mile. Density has not been extensively employed in the past by highway and traffic engineers to describe traffic flow; however, it is now recommended as the basic parameter for describing the quality of flow along

[1]Highway capacity analyses are discussed in more detail in Section 8-12.

freeways and other multilane highways. It has also been the focus of a number of theoretical studies.

In a stream of moving highway vehicles the relationship between speed, volume, and density is given by:

$$q = k\bar{u}_s$$

where

$q$ = the average volume or flow (veh/hr)
$k$ = the average density (veh/mi)
$\bar{u}_s$ = the space mean speed (mph)

Although a number of theoretical and analytical speed density relationships have been published, the exact shape of the $k$-$\bar{u}_s$ curve has not been established conclusively. One model assumes a linear relationship between speed and density. From this, a parabolic volume-density model can be deduced. Figures 4-9*a*, *b*, *c* illustrate the corresponding relationships between speed, volume, and density for a hypothetical situation with a maximum (or mean-free) speed of 50 mph and a maximum density of 175 veh/mi.

Empirical research indicates that speed in fact decreases exponentially with increasing density. Figure 4-9*d* shows empirical data from Lincoln Tunnel traffic fitted into a logarithmic flow-concentration curve of the form:

$$q = u_{\mathrm{m}} \ln \frac{k_{\mathrm{j}}}{k}$$

where

$u_{\mathrm{m}}$ = speed at maximum flow
$k_{\mathrm{j}}$ = maximum density

The fitted values for the figure shown are $u_{\mathrm{m}} = 17.2$ mph, $k_{\mathrm{j}} = 228$ veh/mi.

An examination of Fig. 4-9*b* indicates that the volume of traffic on a road reaches a maximum at fairly low speeds due to the exponential decrease in traffic density with increasing speeds. Another factor limiting the number of vehicles passing a point is interference between vehicles in the traffic stream. With low traffic volumes, the driver has a wide latitude in the selection of travel speed. As traffic volume increases, the speed of each vehicle is influenced by other vehicles, especially the slower ones. The capacity of a highway is its ability to accommodate traffic, and is usually expressed as the maximum number of vehicles that can pass a given point on the highway. By examining Fig. 4-9*b*, it can be seen that this maximum figure occurs in conditions in which the driver would feel some degree of operational constraint. To denote any of an infinite number of differing combinations of operating conditions on a given lane or roadway, the term *level of service* is used.

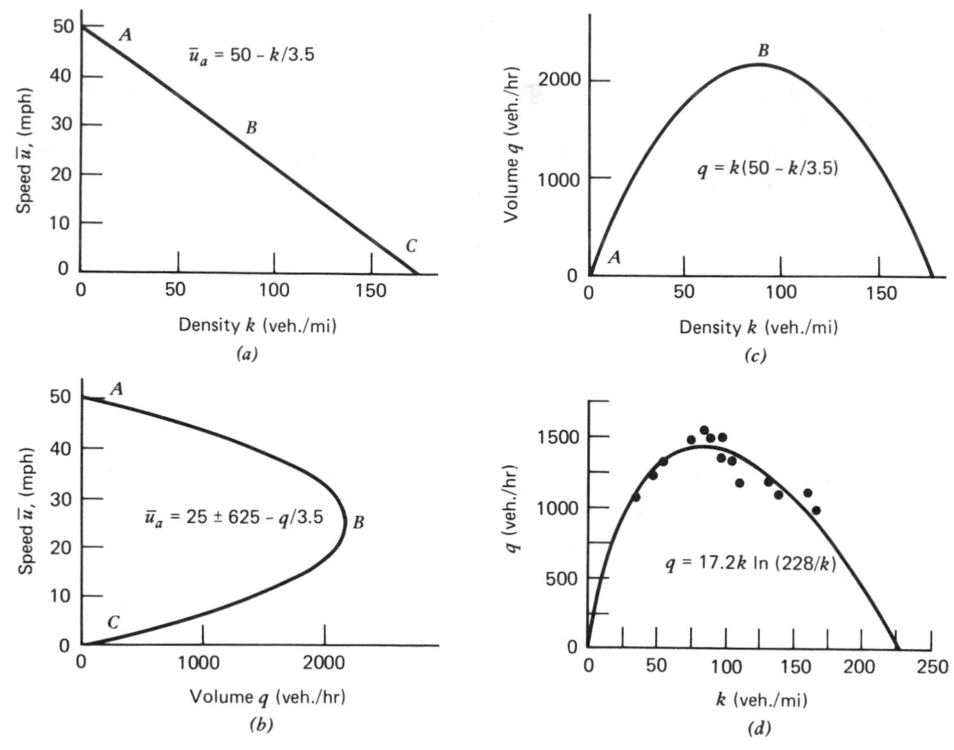


*Fig 4-9.* Highway stream characteristics.

Six levels of services, levels A through F, define the full range of driving conditions from best to worst, in that order. They are based on relationships of operating speed to volume/capacity ratio as shown in Fig. 4-10. Level A represents free flow, at low densities with no restrictions due to traffic conditions. Level B, the lower level of which is often used for the design of rural

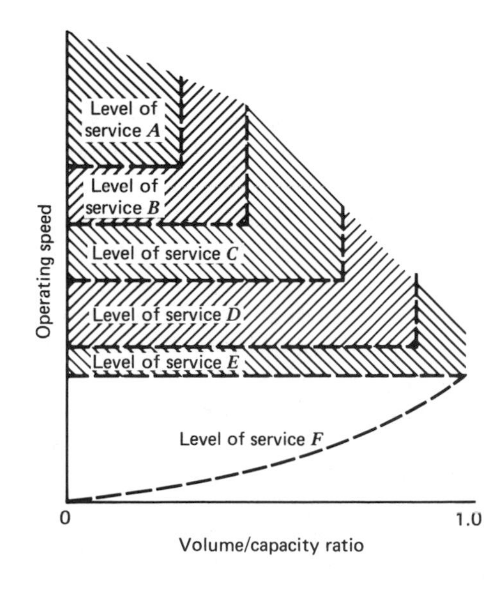


*Fig 4-10.* General concept of relationship of level of service to operating speed and volume/capacity ratio (not to scale). (Courtesy Transportation Research Board.)

highways, is the zone of stable flow with some slight restriction of driver freedom. Level C denotes the zone of stable flow with more marked driver constraint. Level D reflects little freedom for driver maneuverability, and while operating speeds are still tolerable, the condition of unstable flow is being approached. Low operating speed and volumes at or near capacity occur in level E, the area of unstable flow. Level F is the familiar traffic jam, with frequent interruptions to flow, low operating speeds, and volumes well below capacity.

## AIR VEHICLES

### 4-8. AIRCRAFT CHARACTERISTICS

Although the civil engineer and planner is not engaged in the design of aircraft, the characteristics of aircraft have very significant influences upon the design of airports. Satisfactory design of these facilities depends not only on a knowledge of carrier aircraft characteristics, but also on an understanding of the trends in these characteristics. The rapidly changing technology of aviation has made many airports technically obsolete long before the end of their economic lives. While aircraft weight and, specifically, wheel loadings largely determine whether or not an aircraft can use a facility, the size, capacity, and range also significantly affect the details of airport planning [9].

*Weight* affects the design of pavement thickness for runways, taxiways, aprons, and hardstands.

*Size*, as demonstrated in fuselage length, wingspan, deck height, and height of empennage, affects the design of aprons and parking areas, runway and taxiway widths, turning radii, and hangars and maintenance sheds.

*Capacity* in terms of passenger and cargo capacity, together with fuel requirements, determines the size and capacity of ground services that must be supplied to minimize the turn-around time of aircraft. Facilities affected by aircraft capacity include terminal size, baggage handling facilities, departure lounges and gate positions, off-loading facilities for cargo and freight, and fuel storage.

*Range* has an impact on frequency of operations, mix of type, and size of aircraft to be serviced by the airport with consequent effects on runway and gate capacities.

Tables 4-2 and 4-3 show selected characteristics of various classes of aircraft currently in use or in the stages of immediate development [9, 10].

### 4-9. WEIGHT TRENDS OF AIR CARRIER VEHICLES

The introduction of the jet engine permitted a continuing growth in gross weights of transport airplanes which otherwise would have been limited by a

*Table 4-2*

**Characteristics of Selected U.S. Transport and General Aviation Aircraft**

| *Manufacturer and Aircraft* | *Wing Span (ft)* | *Maximum Length (ft)* | *Maximum Height (ft)* | *Maximum Passengers* | *Maximum Takeoff Weight (lb × 1000)* | *Maximum Range (nm)* | *Cruising Speed (Kt)* | *FAR Field Length (ft)* | |
|---|---|---|---|---|---|---|---|---|---|
| | | | | | | | | *Takeoff* | *Landing* |
| *Transport Aircraft* | | | | | | | | | |
| Boeing: 747-200B | 195.7 | 231.3 | 63.4 | 500 | 770 | 7,140 | 506 | 10,500 | 6,150 |
| Douglas: DC-10-30 | 165.3 | 182.2 | 58.1 | 380 | 555 | 6,660 | 499 | 10,490 | 5,960 |
| British Aerospace/ Aerospatiale: Concorde | 83.8 | 203.8 | 37.9 | 128 | 400 | 4,480 | 1176 | 10,280 | 8,000 |
| Boeing: 707-320B | 145.8 | 152.8 | 42.5 | 219 | 333.6 | 6,500 | 478 | 10,000 | 6,250 |
| McDonnell Douglas: L-1011 Tristar | 155.3 | 178.6 | 55.3 | 400 | 430 | 4,845 | 495 | 7,750 | 5,700 |
| Boeing 767 | 156.3 | 159.2 | 52.0 | 255 | 300 | 2,220 | 506 | 5,650 | 4,700 |
| Boeing 757 | 124.5 | 155.3 | 44.5 | 233 | 220 | 1,200 | 494 | 6,180 | 4,820 |
| Airbus Industrie: A300 B4 | 147.1 | 175.8 | 54.3 | 345 | 330.7 | 3,400 | 481 | 8,740 | 5,950 |

| | | | | | | | | | |
|---|---|---|---|---|---|---|---|---|---|
| Boeing: 727-200 | 108.0 | 133.2 | 34.0 | 189 | 207.5 | 3,190 | 495 | 10,080 | 4,800 |
| Boeing: 737-200 | 93.0 | 100.0 | 37.0 | 130 | 115.5 | 2,740 | 460 | 6,550 | 4,290 |
| Douglas: DC-9-50 | 93.3 | 125.6 | 28.0 | 139 | 120 | 2,420 | 465 | 7,880 | 4,680 |
| Hawker Siddeley: | | | | | | | | | |
| HS-748-2A | 98.5 | 67.0 | 24.8 | 60 | 46.5 | 1,740 | 242 | 5,380 | 3,370 |
| Fokker: F27-500 | 95.2 | 82.3 | 28.6 | 56 | 45 | 2,180 | 248 | 5,470 | 3,290 |
| *General Aviation Aircraft* | | | | | | | | | |
| Gulfstream: Gulfstream 2 | 68.9 | 79.9 | 24.5 | 19 | 62.0 | — | 480 | 5,000 | 3,190 |
| Hawker Siddeley: | | | | | | | | | |
| HS-125-600 | 47.0 | 50.5 | 17.2 | 14 | 25.0 | — | 441 | 5,350 | 2,550 |
| Lear: Learjet 36 | 38.1 | 48.7 | 12.3 | 8 | 17.0 | — | 460 | 3,500 | 3,690 |
| Beech: Beech 99 | 45.8 | 44.6 | 14.4 | 15 | 10.9 | — | 248 | 3,100 | 2,220 |
| Cessna: Cessna 421 | 41.9 | 36.1 | 11.6 | 8 | 7.45 | — | 235 | 2,507 | 2,178 |
| Piper: PA-31P | 40.8 | 34.6 | 13.1 | 8 | 7.80 | — | 244 | 2,200 | 2,700 |
| Cessna: Cessna 337 | 38.0 | 29.9 | 9.1 | 6 | 4.63 | — | 170 | 1,675 | 1.650 |
| Piper: Cherokee 180 | 32.0 | 23.3 | 7.3 | 4 | 2.15 | — | 115 | 1,700 | 1,075 |
| Cessna: Cessna 150 | 32.7 | 23.8 | 8.1 | 2 | 1.60 | — | 102 | 1,385 | 1,075 |

SOURCE: Reference 9.

*Table 4-3*

**Characteristics of Selected North American Short Takeoff and Landing (STOL) Aircraft**

| *Manufacturer* | *Model* | *Passenger Capacity* | *Max. Payload* | *Wingspan (ft, in.)* | *Length (ft, in.)* | *Height (ft, in.)* | *Max. Takeoff Weight (lb)* | *Cruise Speed (mph)* | *Landing Speed (kt)* | *STOL Takeoff Distance to 50 ft at Best Speed* | *STOL Landing Distance from 50 ft at Best Speed* | *Range (mi)* |
|---|---|---|---|---|---|---|---|---|---|---|---|---|
| DeHavilland of Canada | DHC-5 Buffalo | 53 | 18,000 | 96-0 | 77-4 | 28-7 | 49,000 | 290 | 70 | 1,265 | 1,170 | 1,900 |
| DeHavilland of Canada | DHC-6 Twin Otter | 20 | 5,300 | 65-0 | 51-9 | 18-7 | 12,500 | 204 | 63 | 1,200 | 1,050 | 118 |
| DeHavilland of Canada | DHC-4 Caribou | 30 | 8,750 | 95-7½ | 72-7 | 31-9 | 28,500 | 170 | 65 | 1,185 | 1,235 | 236 |
| Fairchild Hiller | Heliporter | 7 | 3,502 | 49-10 | 35-9 | 10-6 | 4,850 | 140 | — | 560 | 560 | 545 |
| Helio Aircraft Div of Gen. Aircraft Corp. | Helio Super Courier H295 | 5 | — | 39-0 | 31-0 | 8-10 | 3,400 | 150 | — | 635 | 515 | 615 |

power ceiling on conventional internal combustion engines. Figure 4-11 indicates past and future possible trends in gross weight from 1930 to 1995. Aircraft manufacturers anticipate that gross weights in the region of one million pounds are possible by 2000 or even prior to this time if market demand is sufficient to warrant the production of such large vehicles. Although gross weights of the aircraft are continuing to rise, there is no similar increase in required thicknesses of airfield pavements. By means of elaborate landing gears, pavement thicknesses required to support the Boeing 747 are less than those for the 707, in spite of the fact that the jumbo jet is in excess of twice the weight of its predecessor.

## 4-10. FUSELAGE LENGTHS

Figure 4-12 shows trends in fuselage lengths since 1945. After the introduction of jet transports, aircraft fuselage lengths grew steadily. Most significantly, in 1970 aircraft lengths took a stepped increase with the introduction of the B747-100. The first supersonic transport to go into service, the Concorde, countered the tendency of increasing length. However, this has not proved to be an economic design and in 1978 production ceased on all new aircraft. Subsonic aircraft most likely will approach a ceiling on body length as designs with multiple decks are introduced.

## 4-11. WING SPAN

Due to more efficient wing design and greatly augmented power systems, wing span of transport aircraft has increased only slightly in comparison with growths in gross weight and payload. This is fortunate because apron design is affected especially severely by increases in wing span, and requires considerable expansion to handle turning movements and parked aircraft when span dimensions increase. In addition, runway shoulders have experienced unexpected erosion due to jet blast from the outermost engines of the larger airplanes. Figure 4-13 shows wing span trends for long and medium range transports from 1930 to 1980.

## 4-12. RUNWAY LENGTH REQUIREMENTS

Until the late 1950s, runways that had been designed for conventional piston aircraft were up to about 8000 ft long. With the introduction of the large jet transports such as the Boeing 707 and the Douglas DC-8 it became necessary to design and provide runways of over 12,000 ft. Longer runways were required by the new jet aircraft because of low thrust characteristics at low speeds and the introduction of swept wing aircraft with high wing loadings.

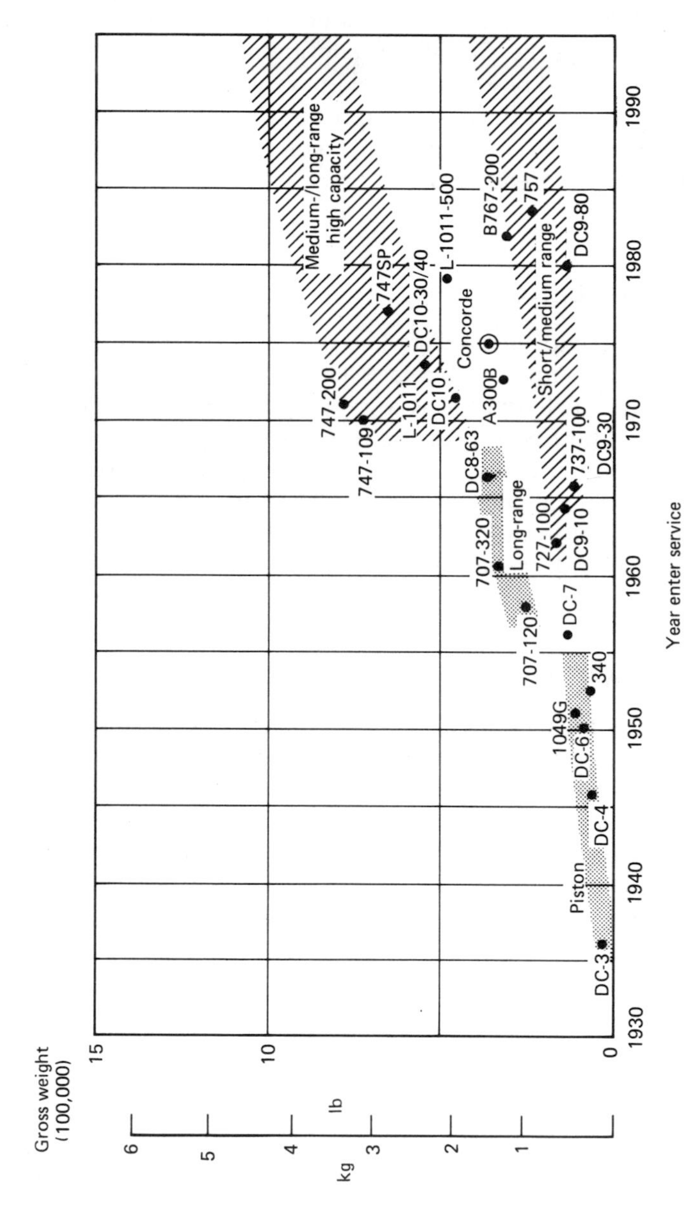


*Fig 4-11.* Trends in gross weight growth of air transport aircraft. (Source: Aerospace Industries Association of America.)

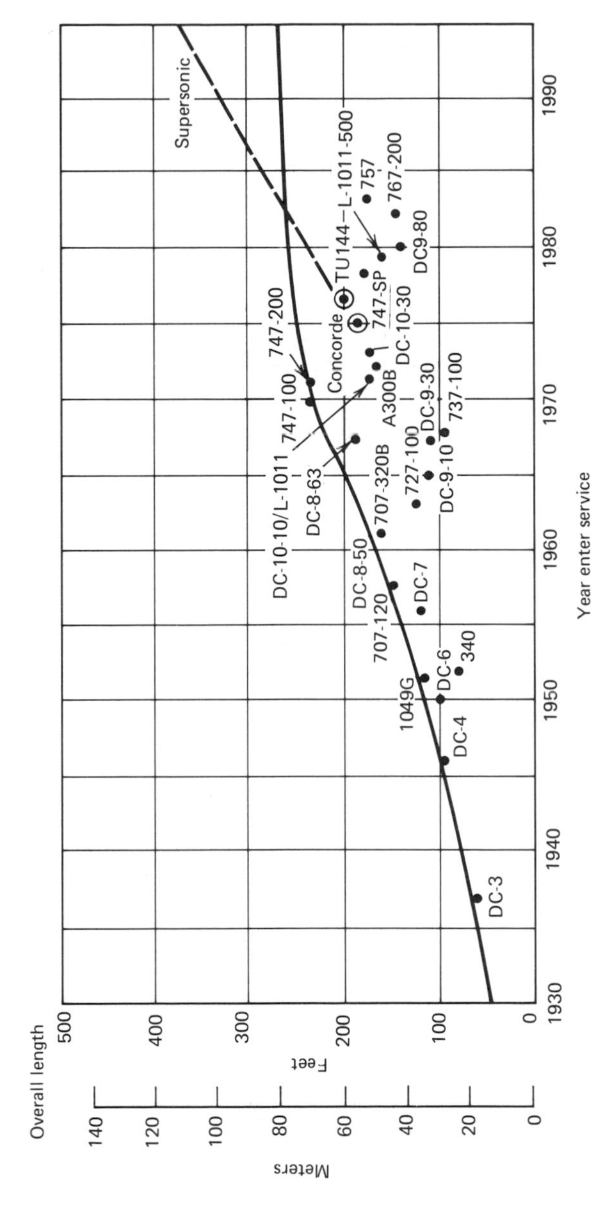


*Fig 4-12.* Trends in overall length growth of air transport aircraft. (Source: Aerospace Industries Association of America.)

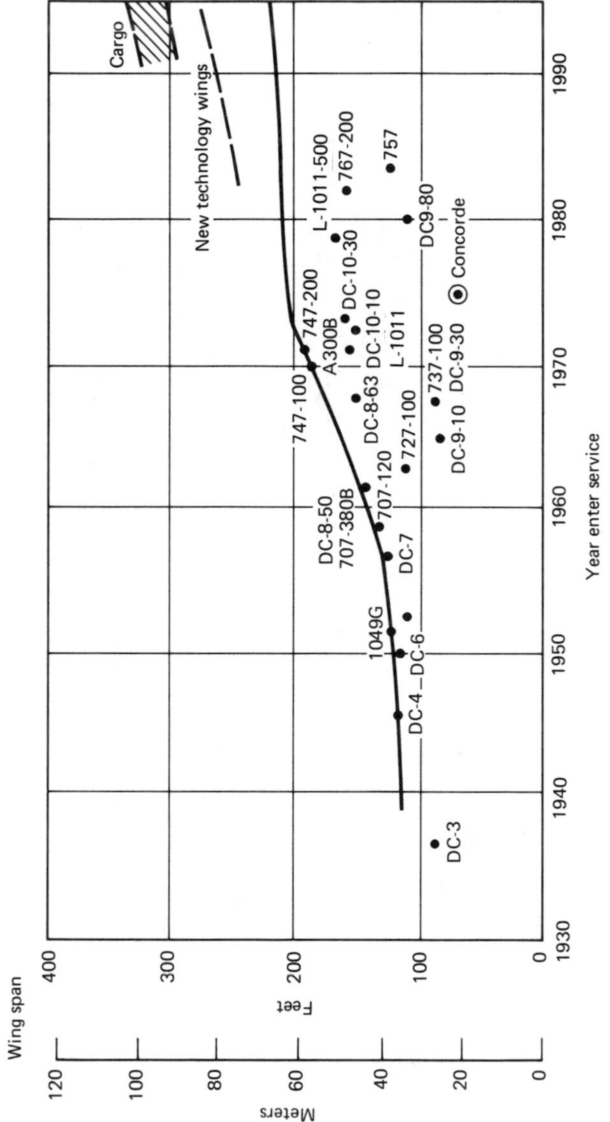


*Fig 4-13.* Trends in wing span growth of air transport aircraft. (Source: Aerospace Industries Association of America.)

The trend of increasing runway lengths, however, was ended by the introduction of turbofan engines, as shown in Fig. 4-14. With the provision of a fan either before or after the turbojet, there is a significant increase in thrust. As a result, the aircraft has a much improved climb-out capability with a subsequent decrease in runway requirement. An added advantage of turbofan engines can be the decrease of perceived noise due to take off operation because of the steep climb out. While more airports can be expected to construct longer runways to handle existing aircraft, no continued trend of increased runway lengths is expected. Supersonic aircraft are able to operate on existing runways at major air centers.

## 4-13. PASSENGER CAPACITY

Figure 4-15 shows the long-term trend in aircraft passenger capacity. Air vehicle designers predict continued increases in the capacities of air transports at a rate of about 2½ percent per year. By 2000, 800 passenger aircraft almost certainly will be in operation.

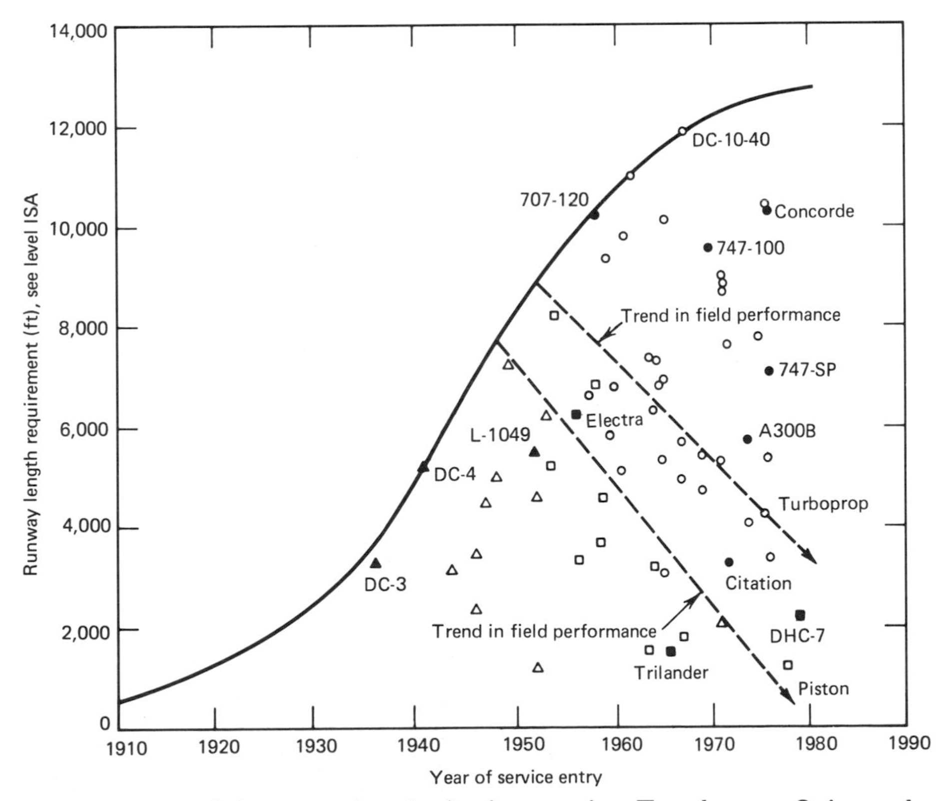


*Fig 4-14.* Trends in runway length. Δ. piston engine; □. turboprop; O. jets and fans.

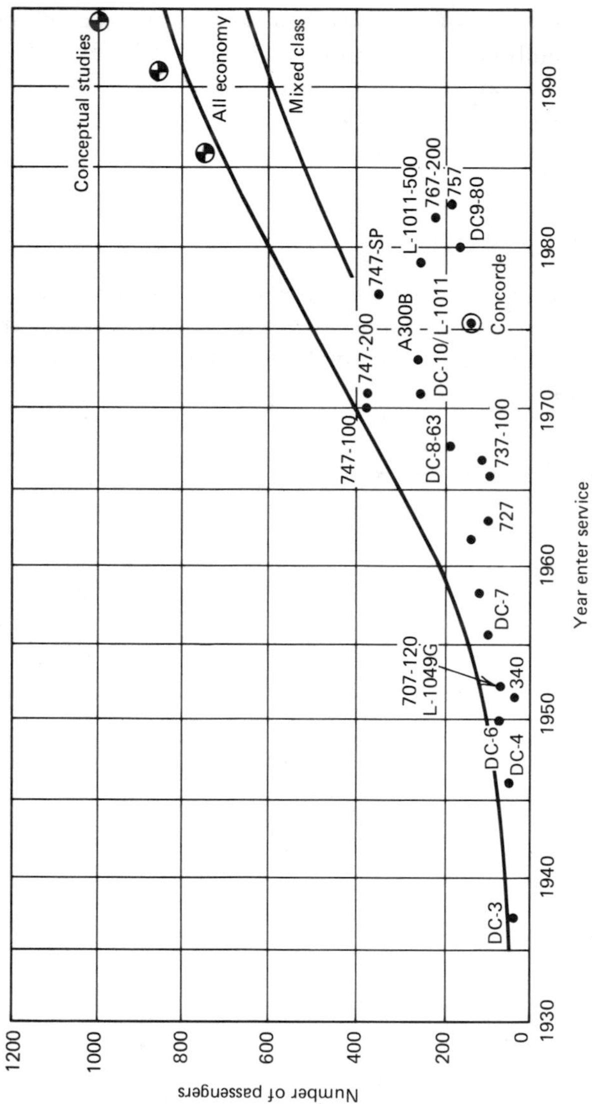


*Fig 4-15.* **Passenger aircraft capacity growth trend. (Source: Aerospace Industries Association of America.)**

The steady increase in passenger capacity has led to significant problems in air passenger terminals. Although the use of high capacity aircraft tends to ease airside capacity constraints, the landside has found itself inadequately designed for the handling of wide-bodied aircraft, which first went into operation in 1970. Departure and arrival lounges, baggage handling, and access and egress procedures have not kept pace with the rapidly changing capacities of air vehicles. The economic life of airport terminals greatly exceeds that of the air vehicle. Consequently, many air terminals that were not designed for the impact of wide-bodied operation will be required to handle these very high capacity transports for many years.

The drive toward higher capacity aircraft has been motivated by the ability of such vehicles to reduce operating costs from both a labor and fuel viewpoint. The trend to more efficient vehicles and increased load factors is expected to continue.

## 4-14. CRUISING SPEED

Since the beginning of commercial aviation there has been a long-term trend toward higher cruising speeds for air transports, as shown in Fig. 4-16. The first significant step in the general trend took place with the introduction of the large monoplanes (for example the DC 3) in the early 1930s. Speeds increased from an upper limit of approximately 190 mph in 1934 to 350 mph with increasingly advanced piston-powered aircraft. With the introduction in 1952 of the Comet I, the first commercial jet transport, upper limits of speed increased to 500 mph overnight. Improvements in jet transports have raised

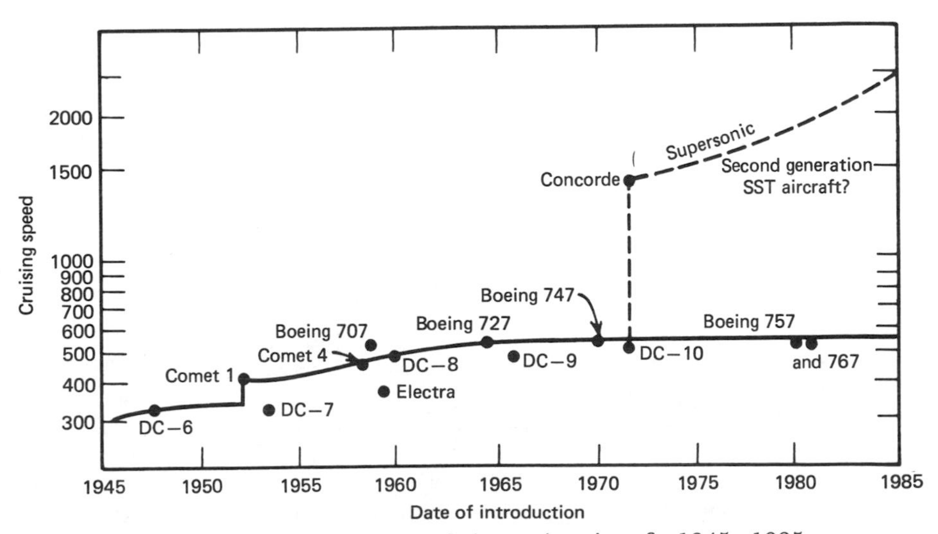


*Fig 4-16.* Cruising speeds of air carrier aircraft, 1945–1985.

cruising speeds to a point just below the speed of sound. The Boeing 707, 727, and 747 all have cruising speeds of about 550 mph. With the introduction of the Concorde in 1972, the upper limit of speed took another significant step to 1450 mph. SST designs proposed by U.S. manufacturers indicate another stepped increase to a design cruising velocity of 1800 mph. In the distant future, there is discussion of the hypersonic transport with speeds in excess of 3000 mph.

In spite of past long-term trends the logic of further increasing cruising speed is open to serious question. Increased speed gives steadily decreasing marginal returns in savings of travel time. On domestic flights, supersonic flight is uneconomic with the exception of transcontinental runs. Even on these very long trips the actual time saved would be relatively small. The use of supersonic transports over the continental United States has generated considerable unfavorable reaction, similar to protests that have caused the banning of such flights over many other countries. Development of economic high-speed aircraft operation hinges on the creation of an adequate intercontinental demand for very fast flight. It would appear that at the price that is necessary this scale of demand is not likely to be generated in the near future. In fact, the rapid fuel cost increases of the 1970s and the 1980s have led to lower subsonic cruising speeds for more fuel-efficient operations.

## 4-15. FUTURE TRENDS IN AIR TRANSPORT CHARACTERISTICS

The long-term trends in the characteristics of air transports are not clear. Figure 4-17 shows how the effect of increases in two dominant variables—capacity and speed—have affected the development of air transports. The

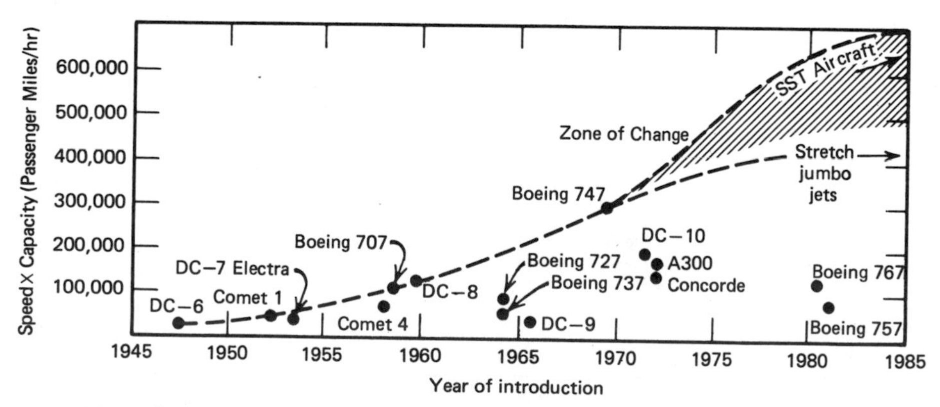


*Fig 4-17.* Trends in passenger miles per hour for air carrier aircraft.

graph shows the relationship between passenger miles per hour and year of introduction for various selected aircraft. The upper limit of the relationship defines the trend for long-distance transports; it would appear that this trend could continue. Both factors, however, seem to be reaching limiting values. Increased speeds appear to have marginal economies. Increased capacities offer the possibility of major air disasters with a shocking loss of life. Further changes in characteristics are likely to occur within the zone of change. Future improvements in technology are more likely to concentrate on decreased costs of operations, especially from the viewpoint of fuel efficient vehicles, than on increased speeds or giant capacities in excess of 1000 passengers.

## WATERBORNE VESSELS

### 4-16. THE DEVELOPMENT OF SHIP TYPES

During the twentieth century there has been a remarkable and relatively rapid development of a number of specialized ship types. Especially since 1960, highly specialized ship types have been introduced largely in the area of cargo. The trend has been toward highly mechanized special-purpose vessels. Such ships have low adaptability to uses outside those for which they are designed and they therefore risk rapid technological obsolescence or commercial redundancy should the market change. Their over-riding advantage is that being specially designed they are remarkably efficient and suitable for their design purpose. As ships become more specialized they tend to become more automated and more technically complicated, as shown in Fig. 4-18. This implies large capital outlays which are recouped over the life of the vessel by substantially reduced labor costs.

Parallel with the development of ship types, the shipping industry is now grouped into three principal classes of service: liner, nonliner, and tanker. *Liner service* refers to a definite, advertised schedule of relatively frequent sailings at regular intervals between specific U.S. ports and designated foreign ports. *Nonliner service* includes chartering or otherwise hiring for the carriage of goods on special voyage where sailing schedules are not predetermined or fixed. *Tanker service* refers to vessels primarily designed for the carriage of bulk liquid cargoes. The various subcategories within each type of service is shown in Fig. 4-19. Table 4-4 shows the tonnage of freight handled for these three categories of service through U.S. ports.

Container service is frequently expressed in 20-ft equivalent units (TEUs), a TEU being an $8 \times 8 \times 20$ ft container. In 1983, a total of 4.1 million containers (TEUs) were handled at U. S. ports, amounting to a tonnage of 38,999 thousands of long tons. Most of this cargo was shipped by liner service.

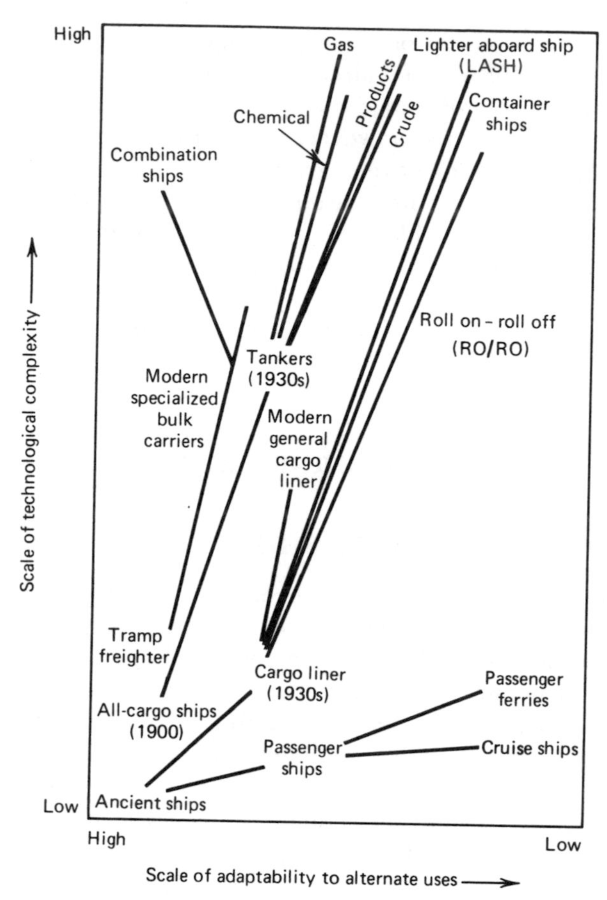


*Fig 4-18.* The development of specialized ship types.

## 4-17. PASSENGER CARRYING SHIPS

The U.S. Maritime Administration classifies any vessel that has accommodations for more than 12 persons as a passenger carrying vessel. Some ships that function principally as cargo vessels carry more than 12 passengers. For certain legal purposes, these ships are then recognized as passenger vessels and are subject to the more stringent regulations pertaining to passenger ships. It is common, therefore for cargo vessels to accommodate exactly 12 passengers to avoid such regulations.

Passenger liners are vessels, the primary function of which is to transport persons over substantial distances. Because of the level of accommodation provided they differ from passenger ferries. Few passenger liners now remain in service; they have disappeared under the pressure from the superior economics and performance of air transport aircraft. Those liners that remain in service can be considered as cruise ships offering not merely a transport

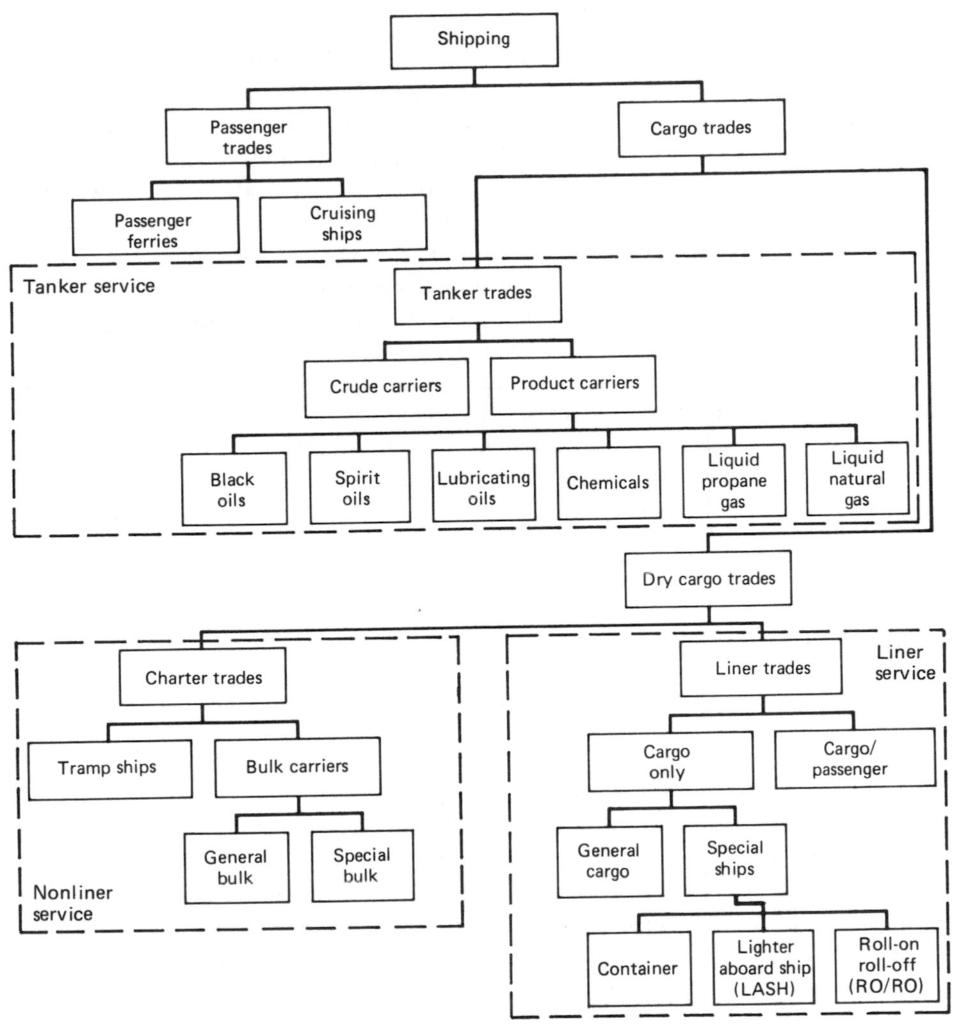


*Fig 4-19.* Breakdown of shipping types.

*Table 4-4*

**U.S. Oceanborne Freight Shipped in 1984 (Thousands of Long Tons)**

| | *Export* | *Import* |
|---|---|---|
| Liner service | 30,620 | 32,892 |
| Nonliner service | 250,366 | 95,982 |
| Tanker service | 29,236 | 237,681 |

SOURCE: *United States Oceanborne Foreign Trade Routes,* U. S. Department of Transportation, Maritime Administration, August 1986.

service but also a recreation service. The *S.S. Queen Elizabeth 2,* with a gross tonnage of 67,000 tons and a cruise speed of 28 Kt, is an example of a modern passenger liner cruise ship. Introduced into service in 1969, this strikingly modern ship carries 1970 passengers and a crew of 906. At the time of the design of this vessel it was realized that it could not compete directly with air transport. Consequently its design required a reduced draft, recognizing that during the winter it would not be used on North Atlantic runs, but would be required to use cruise port facilities in such areas as the Caribbean. During the summer the vessel travels the traditional New York to Southampton route long abandoned by the old prestige liners such as the *United States,* the *France,* and the *Queen Mary.* Table 4-5 gives an example of some of the characteristics of a few remaining passenger ships. [11]. In Europe and in some other parts of the world there are passenger ferries that have passenger accommodation and roll-on/roll-off facilities for both trucks and automobiles. A modern example of this is the Tor Scandanavia vessel of 15,500 tons which operates on a 24-hour run from Felixtowe, England, to Goteburg, Sweden, carrying up to 1400 passengers, 420 cars, or 70 trucks.

## 4-18. GENERAL CARGO CARRIERS

Figure 4-20 shows a modern high-speed general cargo carrier operating in transoceanic trade. The class of ship is one of the world's most modern designs; it has operating speeds in excess of 24 kt. Automated devices that have been incorporated into the design include such major functions as direct bridge control of the engine, automatic boiler control, constant tension winches, and hydraulic quick-acting hatch covers. Automation has been applied in less important areas also, such as temperature registers in reefer spaces, and air-conditioned crew and passenger accommodations. Typically, general cargo carriers have become more sophisticated since World War II. In addition to a great deal of automated control and operation, ships now have such innovations as reinforced main and 'tween decks for forked-lift truck operation, triple hatches, dehumidifying equipment, and modern ship's gear for rapid and efficient on-loading and off-loading of cargo compartments.

*Table 4-5*
**Operating Characteristics of Passenger Ships**

| *Ship* | *Tonnage* | *Passengers* | *Length (ft)* | *Speed (kt)* | *Power (hp)* |
|---|---|---|---|---|---|
| S.S. Cunard Princess | 17,496 | 947 | 537 | 21 | 21,000 |
| S.S. Queen Elizabeth 2 | 67,140 | 1970 | 963 | 28 | 110,000 |
| S.S. Oriana | 40,350 | 1700 | 804 | 27 | 80,000 |
| S.S. Norway | 66,348 | 2033 | 1035 | 17 | 40,000 |
| S.S. Sea Princess | 27,670 | 840 | 660 | 21 | 25,200 |
| S.S. Canberra | 43,975 | 1702 | 818 | 28 | 88,200 |

*Fig 4-20.* Fast cargo liner, Mormacargo. (Source: British Shipbuilders)

Trends in general cargo ships indicate that modern freighters now exceed 500 ft in length compared with lengths of slightly under 400 ft in the 1930s. Operational speeds of modern carriers are much higher than they were 30 years ago. Most freighters at that time operated at speeds of 14 kt or less. Modern powerful cargo liners are capable of speeds in excess of 20 kt, with the faster vessels operating at 24 kt. Increased speed and length have been accompanied by increased vessel size of typical general cargo vessels. The C1-A Cargo Vessel, one of the original designs in the U.S Maritime Commission's long-range building program of the 1930s, had a dead-weight tonnage of 7343 tons. The *Liberty* ships of World War II had a comparable tonnage of 10,920 tons. By the early 1980s, dead-weight tonnages of typical modern general cargo vessels had grown to approximately 15,000 tons.

## 4-19. SPECIAL SHIPS

The *containership* is an innovation in freighter design, brought about by rapid growth of containerization since the first deep-sea container operations were introduced in 1966. Figure 22-1 shows a modern container ship capable of cruising at 23 knots, making round trips from Europe in just over two weeks with two ports of call. Fully loaded, such a fully automated ship can carry 2500 TEUs (20-ft equivalent units).

*Barge-carrying* ships such as the LASH vessels use, in effect, floating detachable cargo holds. Used initially to link New Orleans to Rotterdam, the system is capable of even quicker turn around of the transoceanic vessel than is afforded by containerships. Each LASH vessel carries up to 85 individual lighters, with a total capacity of approximately 1750 TEUs. Vessels are operating at 40,000 dead-weight tons. A principal advantage of the barge-carrying ship is that the vessel carries its own unloading equipment (portal crane or floodable compartments) and is not required to wait for vacant berth space in a dock. This system is especially useful in developing countries or for estuary work.

*Roll-on/roll-off* (RO/RO) ships accept automobiles and large, wheeled units that can be loaded easily under their own power by ramps in the prow or stern of the ship. Multiple-deck RO/RO ships are common often in combination with container hold facilities (a partial container vessel). RO/RO operation is especially suitable for short sea crossings and minimal customs delays; tractor-trailer units are left intact during the voyage. For longer crossings the trailer unit only is shipped.

## 4-20. TANKERS AND BULK CARRIERS

Tankers are vessels whose primary function is the carriage of liquid cargo, such as petroleum, oil, asphalt, bitumen, gasoline, molasses, and chemicals. Since World War II the change in characteristics of tankers has been astounding.

In 1945, the largest tanker could carry approximately 16,000 tons with drafts of 30 ft. By 1950, the largest tanker in the world was designated a "supertanker" and carried 30,155 tons with a draft of approximately 35 ft. In 1966, the scale of tankers had jumped to the order of "mammoth tankers," with the launching of the *Idemitsu Maru* having a carrying capacity of just over 200,000 tons, a length of 1122 ft, and a draft of 58 ft. Tankers continue upward in size as vessels with a capacity over 600,000 tons are being constructed in the early eighties, and there is discussion of 1 million ton vessels in the near future. Obviously, few ports have the capability of handling the extensive draft requirements and large lengths of the mammoth tankers. It is becoming more frequent for such vessels to moor at special dolphins some distance off-shore to take on and discharge cargo via pipeline to shore.

Bulk carriers are special purpose vessels that are designed to carry one cargo or one cargo type only. They include *ore carriers,* and *coal carriers.*

## 4-21. INLAND WATERWAY VESSELS

A variety of vessels ply the inland waterways: towboats, tugboats, barges of various types, carfloats, and scows. Origin-destination speeds on a systemwide basis average close to 6 mph. On the longer rivers and waterways, giant integrated tows composed of powerful towboats and numerous barges make speeds of 15 mph. Figure 4-21 shows a typical large tow composed of open hopper barges. Almost all towboats and tugboats are powered by diesel engines; few steamboats remain in service. With the introduction of powerful diesel units, large barges and integrated tows become possible. Many towboats now use the Kort nozzle, a funnel-shaped structure enclosing the propellers which can improve propeller thrust up to 25 percent in favorable conditions. One of the most important technological advances came during World War II with the introduction of reversing-reduction gears permitting the operation of the engines at the most favorable rpm. Prior to this time the high efficiency of diesel engines at high rpm was not matched to maximum propeller efficiency at low propeller rpm. Still unsolved is the problem of gaining maximum efficiency from diesel propulsion under large variations of load which can occur by different tow sizes. Techniques that are now being tested to overcome this problem are diesel electric power units, variable pitch propellers, and overload limiters [12].

Table 4-6 summarizes selected operating characteristics of inland waterway craft.

Table 4-7 shows summary statistics for the world's fleets by classification of freighters, bulk carriers, combination passenger and cargo ships, and tankers. It is interesting to note that though freighters, bulk carriers, and tankers show strong trends of increasing size with date of introduction into service, the largest passenger ships are relatively old [8].

*Fig 4-21.* An integrated tow. (Source: American Waterways Operators, Inc.)

*Table 4-6*

**Selected Operating Characteristics of Inland Waterway Craft**

| | *Length (ft)* | *Breadth (ft)* | *Draft (ft)* | | |
|---|---|---|---|---|---|
| Towboats | | | | Horsepower | |
| | 117 | 30 | 7.6 | 1000 to 2000 | |
| | 142 | 34 | 8 | 2000 to 4000 | |
| | 160 | 40 | 8.6 | 4000 to 6000 | |
| Tugboats | | | | Horsepower | |
| | 65 to 80 | 21 to 23 | 8 | 350 to 650 | |
| | 90 | 24 | 10 to 11 | 800 to 1200 | |
| | 95 to 105 | 25 to 30 | 12 to 14 | 1200 to 3500 | |
| | 125 to 150 | 30 to 34 | 14 to 15 | 2000 to 4500 | |
| Deck barges | | | | Capacity (tons) | |
| | 110 | 26 | 6 | 350 | |
| | 130 | 30 | 7 | 900 | |
| | 195 | 35 | 8 | 1200 | |
| Carfloats | | | | Capacity Railroad cars | |
| | 257 | 40 | 10 | 10 | |
| | 366 | 36 | 10 | 19 | |
| Scows | | | | Capacity (tons) | |
| | 90 | 30 | 9 | 350 | |
| | 120 | 38 | 11 | 1000 | |
| | 130 | 40 | 12 | 1350 | |
| Open hopper barges | | | | Capacity (tons) | |
| | 175 | 26 | 9 | 1000 | |
| | 195 | 35 | 9 | 1500 | |
| | 290 | 50 | 9 | 3000 | |
| Covered dry cargo barges | | | | Capacity (tons) | |
| | 175 | 26 | 9 | 1000 | |
| | 195 | 35 | 9 | 1500 | |
| Liquid cargo (tank) barges | | | | Capacity (tons) | Gallons[a] |
| | 175 | 26 | 9 | 1000 | 302,000 |
| | 195 | 35 | 9 | 1500 | 454,000 |
| | 290 | 50 | 9 | 3000 | 907,200 |

[a]Based on an average of 7.2 bbl/ton and 42 gal/bbl.

SOURCE: *Big Load Afloat,* American Waterways Operators, Inc.

*Table 4-7*

**Average Age Speed and Draft of World's Merchant Fleets**

| *Dead-Weight Tons* | *Average Age (years)* | *Average Speed (kt)* | *Average Draft (ft)* |
|---|---|---|---|
| *Merchant Type Freighters* | | | |
| Under 2000 | 20 | 12 | 15 |
| 2,000–3,999 | 15 | 13 | 18 |
| 4,000–6,999 | 12 | 14 | 22 |
| 7,000–8,999 | 12 | 16 | 25 |
| 9,000–9,999 | 16 | 17 | 27 |
| 10,000–10,999 | 22 | 16 | 28 |
| 11,000–12,999 | 16 | 17 | 29 |
| 13,000–14,999 | 13 | 17 | 30 |
| 15,000 and over | 9 | 18 | 33 |
| Overall average | 14 | 15 | 24 |
| *Merchant Type Bulk Carriers (Dry)* | | | |
| Under 10,000 | 14 | 12 | 19 |
| 10,000–19,999 | 13 | 15 | 30 |
| 20,000–29,999 | 11 | 15 | 33 |
| 30,000–39,999 | 9 | 15 | 36 |
| 40,000–49,999 | 10 | 15 | 38 |
| 50,000–59,999 | 10 | 15 | 41 |
| 60,000–79,999 | 8 | 15 | 43 |
| 80,000–99,999 | 11 | 15 | 47 |
| 100,000–124,999 | 9 | 15 | 53 |
| 125,000–149,999 | 6 | 15 | 55 |
| 150,000–199,999 | 10 | 16 | 58 |
| 200,000 and over | 9 | 16 | 67 |
| Overall average | 11 | 15 | 36 |
| *Combination Passenger and Cargo* | | | |
| Under 4,000 | 22 | 15 | 15 |
| 4,000–6,999 | 25 | 17 | 19 |
| 7,000–9,999 | 27 | 17 | 23 |
| 10,000–14,999 | 26 | 19 | 25 |
| 15,000–19,999 | 30 | 19 | 26 |
| 20,000–29,999 | 19 | 22 | 27 |
| 30,000–49,999 | 13 | 23 | 28 |
| 50,000 and over | 18 | 24 | 34 |
| Overall average | 24 | 18 | 21 |
| *Tankers* | | | |
| Under 20,000 | 14 | 13 | 21 |
| 20,000–39,999 | 13 | 16 | 35 |

*continued*

*Table 4-7*
**Average Age Speed and Draft of World's Merchant Fleets—Continued**

| *Dead-Weight Tons* | *Average Age (years)* | *Average Speed (kt)* | *Average Draft (ft)* |
|---|---|---|---|
| *Tankers, continued* | | | |
| 40,000–59,999 | 12 | 16 | 40 |
| 60,000–79,999 | 11 | 16 | 42 |
| 80,000–99,999 | 9 | 16 | 45 |
| 100,000–124,999 | 11 | 16 | 51 |
| 125,000–149,999 | 8 | 16 | 55 |
| 150,000–199,999 | 8 | 16 | 57 |
| 200,000–249,999 | 10 | 16 | 65 |
| 250,000–299,999 | 9 | 16 | 69 |
| 300,000–349,999 | 8 | 15 | 74 |
| 350,000 and over | 7 | 16 | 77 |
| Overall average | 12 | 15 | 35 |

SOURCE: *A Statistical Analysis of the World's Merchant Fleets,* as of January 1, 1984, U.S. Department of Transportation, Maritime Administration, 1985.

## RAILROAD LOCOMOTIVES[1]

The operational characteristics of railroad trains are dependent on the type of motive power supplied by the *power units* or *locomotives.* The most common types of power units used by the American railroads are:

1. Locomotives
   - Electric
   - Diesel-electric
   - Steam
   - Other types including gas turbine-electric, diesel-hydraulic
2. Railcars
   - Electric
   - Diesel-electric
   - Gas-electric

This section will limit its discussion to the operating characteristics of electric, diesel-electric, and steam locomotives which move almost entirely the total passenger and freight traffic on the North American continent.

[1]This section is drawn extensively from Part 3, Vol. II, *Manual of Recommended Practice,* American Railway Engineering Association [13].

In dealing with the different operating characteristic of locomotives several definitions are necessary to clarify the following discussion.

*Horsepower* for the various types of locomotive is defined somewhat differently. The rated horsepower of electric locomotives is the power available at the rims of the driving wheels. For diesel or turbine-electric locomotives, rated horsepower refers to the power available as input to the turbine or diesel engine. Tractive effort is more usually used in conjunction with steam locomotives.

*Tractive effort* is the tangential force applied at the rims of the driving wheels by the locomotive.

*Rail horsepower* is the power available at the rims of the driving wheels.

## 4-22. ELECTRIC LOCOMOTIVES

Power is supplied to electric train systems by two types of distribution systems: direct current (dc) and alternating current (ac) systems. The method of current pickup is either by collector shoes riding on a third rail or by means of a pantograph or trolley wheel passing under an overhead wire. Third-rail systems are required where heavy currents are involved; these are normally suitable only where lower voltages are used. High-voltage systems require overhead wires from the viewpoint of safety. Four general types of electric motor traction are:

1. DC power supply with dc traction motors.
2. Single-phase ac power supply, with single-phase ac traction motors.
3. Single-phase ac power supply, intermediate rectifiers, and dc traction motors.
4. Single-phase ac power supplying an intermediate dc generator driving a dc traction motor.

Electric locomotives can be used in single units or coupled as multiple units under one controller.

Electric motors rely on an external source of supply for their power. The capacity of the motor, therefore, is not limited internally but is governed either by the power that can be drawn from the supply system without slipping the driving wheels or by the temperature or commutation capacity. Because of the time delay involved, electric motors have a short-term overload capacity. This overload capacity, coupled with a high adhesion between rails and wheels with the nonpulsating form of torque supplied by an electric motor, makes this form of traction desirable under certain load conditions, Figures 4-22 to 4-24 show graphs relating tractive effort and horsepower to the various forms of electric motor traction already discussed.

Manufacturers supply rating curves which furnish the relationship of tractive effort to speed for the various locomotives under short-term or con-

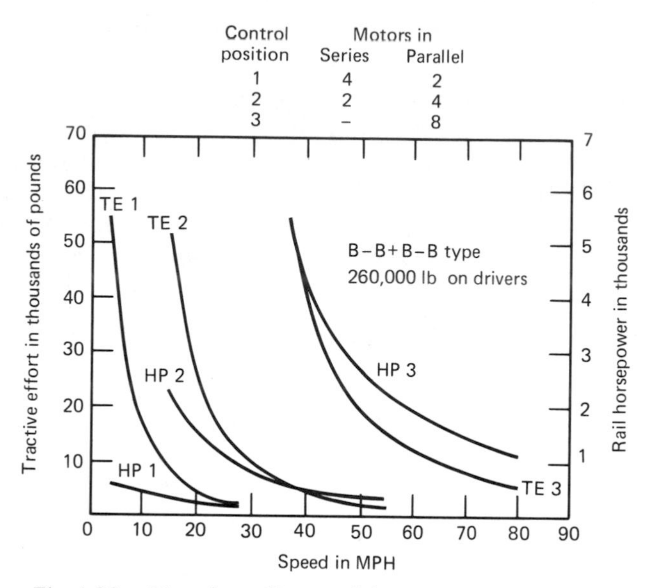


*Fig 4-22.* Tractive effort and horsepower curves for a 600-V dc locomotive, with series-parallel control. (Source: *AREA Manual of Recommended Practice.)*

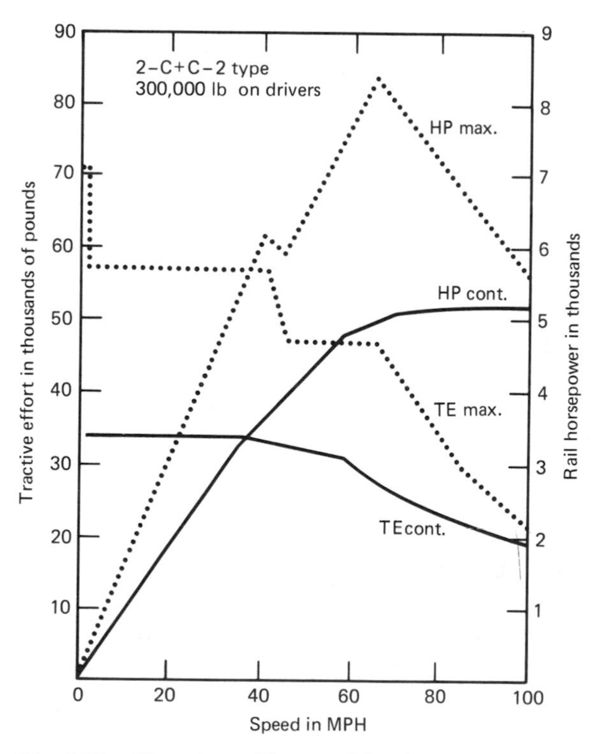


*Fig 4-23.* Tractive effort and horsepower curves for a single-phase ac locomotive. (Source: *AREA Manual of Recommended Practice.)*

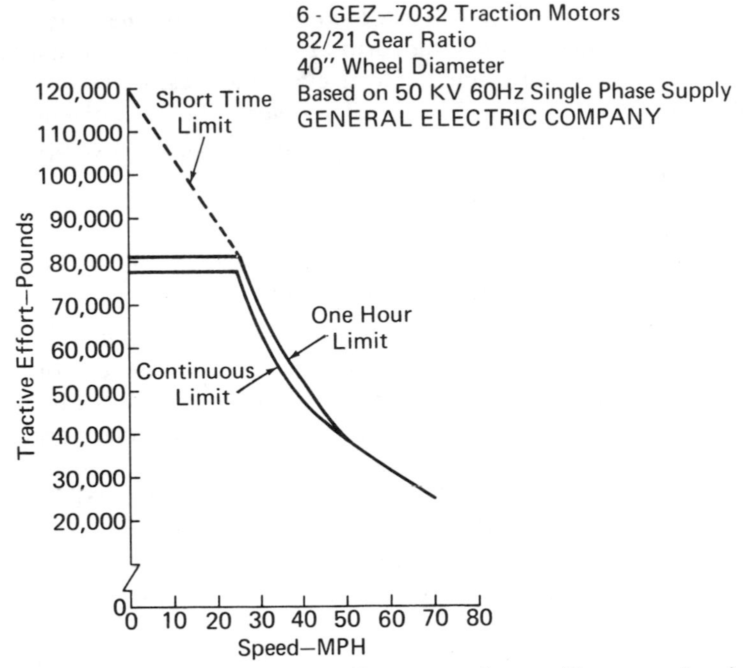


*Fig 4-24.* Speed *vs* tractive effort curve for rectifier-type electric locomotive. (Source: General Electric Co.)

tinuous-loading conditions. In the absence of these curves the following equations can be used to describe the relationships:

$$V = \frac{d \times p \times S}{336 \times G}$$

$$TE = \frac{T \times G \times 24 \times e}{d \times p}$$

where

$V$ = locomotive speed, mph
$TE$ = locomotive tractive effort, lb
$S$ = number of armature, rpm
$T$ = motor torque, lb-ft
$d$ = diameter of driving wheels, in.
$p$ = number of pinion teeth
$G$ = number of gear teeth
$e$ = gear efficiency (0.95 to 0.97)

The overload capacity of electric locomotives combined with the high accelerations at low speed makes this form of traction most desirable for short runs with frequent stops and starts. Electric locomotives, therefore, are used widely for suburban and mass transit rail systems. They are of equal use on short steep grades where other locomotives would require the operation of multiple units to avoid overload.

## 4-23. DIESEL-ELECTRIC LOCOMOTIVES

The diesel-electric locomotive is simply a complete power plant composed of a diesel engine prime mover connected directly to a dc generator. The dc generator itself feeds a series-wound dc traction motor. Each diesel-electric locomotive unit is a self-contained system with both power plant and traction motor. For relatively low load conditions such as in yards and at switches, diesel-electrics usually are operated singly; for line-haul operation it is common to use multiple units under unit control from one cab.

In essence, diesel-electric locomotives are constant maximum power units since their horsepower ratings are set according to the power delivered by the prime mover diesel to the main generator. This power is independent of the track speed of the locomotive. The capacity of the unit at low speeds is limited both by rail adhesion of the driving wheels and by overheating of the windings of the traction motor. At high speeds, there is again a decline in capacity due to overheating of the main generator in a similar manner. Because the actual losses are difficult to predict it is customary for manufacturers to supply curves which define the operating characteristics of their locomotives under short-term loading conditions. Figure 4-25 is a typical curve showing the short-term relationship between tractive effort and speed for a diesel-electric unit, while Fig. 4-26 shows the relationship between locomotive horsepower and speed under short-term loading conditions. Where the rated shaft horsepower is known, the tractive effort can be calculated for predictive purposes from:

$$TE = \frac{375HPe}{V}$$

where

$TE$ = tractive effort, lb
$HP$ = rated horsepower of the diesel prime mover (0.93 of gross horsepower if not otherwise known)
$V$ = track speed, mph
$e$ = efficiency of electromechanical drive system (0.82 – 0.83)

Because of their ability to take short-term overload, diesel-electric locomotives are used in situations similar to those where electric motors are most

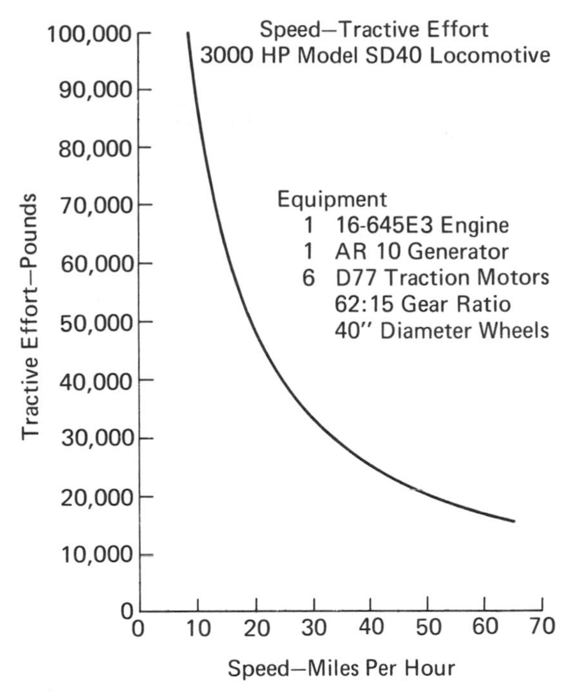


*Fig 4-25.* Performance curve for a diesel-electric locomotive. (Source: Locomotive Division, General Motors.)

useful. The reserve of power for diesel-electrics is somewhat less than all electric systems. The principal advantage of the diesel-electric locomotive is its ability to operate *without* the extensive power distribution systems required by electric units. These distribution systems are both expensive and create safety hazards under certain conditions.

## 4-24. STEAM LOCOMOTIVES

Steam engines are no longer widely used in railroad locomotives in North America or other developed countries. Because of the inherently low efficiency of the reciprocating steam engine, they have been widely replaced by diesel-electric units. Although the residual units represent an outmoded technology, the steam engine is worth discussion because it still forms the basic power unit in many parts of the world.

The advantages of the steam engine are relatively few but, in the case of capital-starved countries, may be decisive. The engine is a relatively simple unit and can be fabricated and maintained by a less advanced technology. However, the most compelling reason for continued use of the steam engine

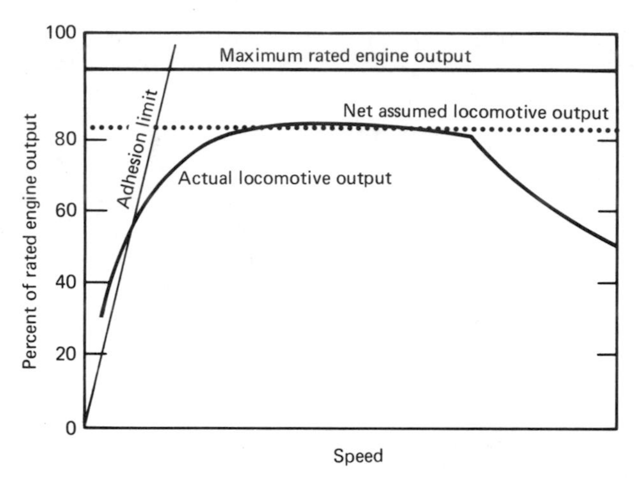


*Fig 4-26.* Typical horsepower curve, diesel-electric locomotive. (Source: *AREA Manual of Recommended Practice.)*

lies with low capital cost per unit of horsepower. In capital-scarce economies, the equipping of the railroad system with lower-cost steam locomotives serves to free capital for other uses. The disadvantages of the steam locomotive lie in its low thermal efficiency, its inability to be adapted easily for different types of service by simple gear changes, the need for an adequate supply of water throughout its line of operation, and, finally, the sensitivity of engine efficiency to changes in the type of fuel used. Steam engines customarily are rated by tractive effort rather than by horsepower. There are two types of tractive effort, *cylinder tractive effort* and *driver tractive effort.*

*Cylinder tractive effort* relates to the force exerted by the steam pressure on the pistons. It may be calculated from the following formula [14]:

$$\text{Cylinder } TE = \frac{KPd^2S}{D}$$

where

$K$ = the ratio of the mean effective pressure in the cylinders to that of the boiler (usually assumed at 0.85)
$P$ = boiler pressure in psi
$d$ = cylinder diameter, in.
$S$ = length of stroke, in.
$D$ = diameter of the drivers, in.

*Driver tractive effort* is less than cylinder tractive effort by the amount of mechanical losses in the cylinders, valve gear, and driving rods.

The *capacity* of the steam locomotive is limited by the capacity of the cylinders at low speeds and by steaming capacity at high speeds. Cylinder capacity is related to both the diameter of the cylinders and the length of available stroke. The steaming capacity of the boilers depends on the area of evaporating surface, the size of the firebox and grate, the fuel used, and the method of firing.

At low speeds, rail horsepower is limited by adhesion. The pulsating torque available from the reciprocating engine gives less favorable rail adhesion characteristics for steam engines than for other nonpulsating engines. Figure 4-27 shows the relation between track speed and both tractive effort and rail horsepower. For any particular locomotive it is usual for the manufacturer to supply such performance curves.

## 4-25. TRAIN RESISTANCE

The total resistance to movement of trains is composed of *inherent,* or *level tangent,* resistance and incidental resistances due mostly to curvature, grades, and winds. Level tangent resistance is due to a combination of factors including speed, cross-sectional area, axle load, type of journal, winds, temperature, and condition of track. Based on analysis of tests, W.J. Davis [15] recom-

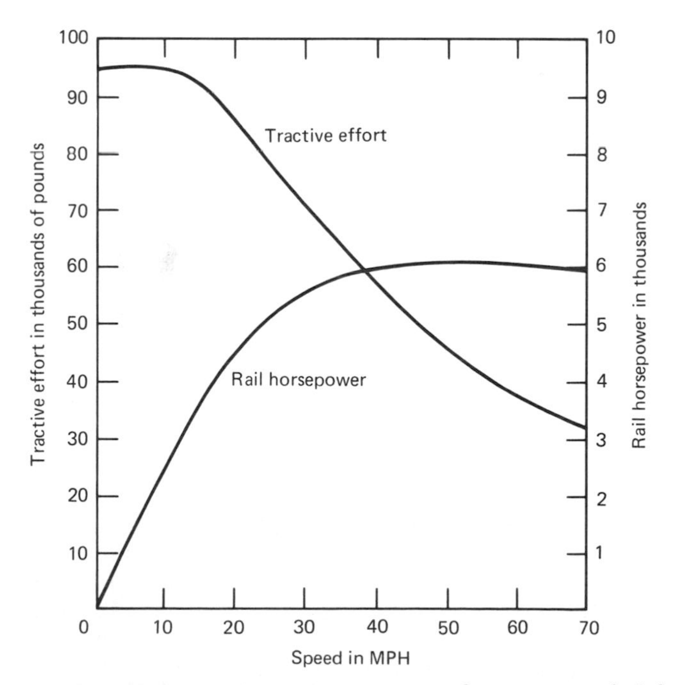


*Fig 4-27.* Tractive effort and horsepower curve for a steam freight locomotive. (Source: AREA.)

mended the following formula for computing level tangent resistance for axle loadings in excess of 5 tons:

$$R = 1.3 + \frac{29}{W} + bV + \frac{CAV^2}{WN}$$

where

$R$ = level tangent resistance in lb/ton
$W$ = average load per axle in tons
$N$ = number of axles
$WN$ = average weight of locomotive or car in tons
$A$ = gross section area, ft²
$b,C$ = coefficients which vary according to units considered
$V$ = track speed, mph
Values of $b,C,$ and $A$ are shown in Table 4-8.

While the Davis formula has been found to give satisfactory results within the speed range of 5 and 40 mph, modern freight operations with higher operating speeds, increased dimensions, and heavy loading and modern track structure necessitate use of a "modifield Davis formula":

$$R = 0.6 + \frac{20}{W} + 0.01V + \frac{KV^2}{WN}$$

where

$K$ = air resistance coefficient: 0.07 for conventional equipment; 0.16 for piggyback; 0.0935 for containers
$R$, $W$, $N$, and $V$ are as above.

Grade resistance usually is taken at 20 lb/ton for each percent of grade. Curve resistance is computed at 0.8 lb/ton for each degree of curvature. The computation of total train resistance is calculated from the relation:

*Table 4-8*
**Coefficient Values for Davis Level Tangent Resistance Formula**

| *Unit of Equipment* | *b* | *C*[a] | *A* |
|---|---|---|---|
| Locomotives | 0.03 | 0.0025 | 120 |
| Freight cars | 0.045 | 0.0005 | 85–90 |
| Passenger cars (vestibuled) | 0.03 | 0.00034 | 120 |
| Multiple unit trains, leading car (vestibuled) | 0.045 | 0.0024 | 100–110 |
| Multiple unit trains, trailing cars (vestibuled) | 0.045 | 0.00034 | 100–110 |
| Motor cars | 0.09 | 0.0024 | 80–100 |

[a]For streamlined equipment 0.0016 may be substituted for 0.0024 and 0.00027 for 0.00034.

Total train resistance = Level tangent resistance
+ grade resistance + curve resistance

## 4-26. CHARACTERISTICS OF URBAN RAIL TRANSIT

Urban rail transit vehicles are significantly different in operational characteristics from intercity passenger vehicles. Where intercity passenger trains hold sustained speeds for considerable periods of time, the urban rail transit car is required to accelerate and decelerate constantly between station stops. Therefore, it is not surprising that the choice of transit vehicle depends on several variables that strongly reflect this type of operation. In selecting transit cars the engineer must account for:

1. Maximum acceptable acceleration rates.
2. Desired station spacing.
3. Limit of speed capability of train.
4. Desired overall line-haul travel speeds, including stops.

It has been found in practice that maximum acceleration rates should be limited to 8 ft/sec$^2$ where all passengers are seated and 5 ft/sec$^2$ where standees are anticipated. The selection of any two of the remaining criteria will determine the last. The procedure used to determine line-haul characteristics of transit vehicles can best be described by an example.

EXAMPLE

A rapid transit vehicle has a top speed capability of 60 mph, a capacity of 63 persons seated, and a length of 60 ft, 3 in. The maximum passenger volume to be carried is 6000 persons in one direction. Manual control will be used, and the minimum headway will be 5 min. Station stops of $\frac{1}{2}$ min are to be used. What is the minimum station spacing if the full top speed capability of the vehicle is to be used; and what station lengths must be designed?

Number of trains per hour = 12 at 5-min headway

Required train capacity = Hourly passenger volume/no. of trains per hour
= 6000/12
= 500 passengers

Required number of cars per train = Train capacity/car capacity
= 500/63
= 8 cars

Platform length required = Individual car length × no. of cars per train
= $60.25 \times 8$
= 482 ft

Assuming a maximum acceleration and deceleration rate of 5 ft/sec² was used throughout the run, the distance-speed diagram would be as shown, attaining a maximum speed of 60 mph, or 88 ft/sec, at the halfway point.

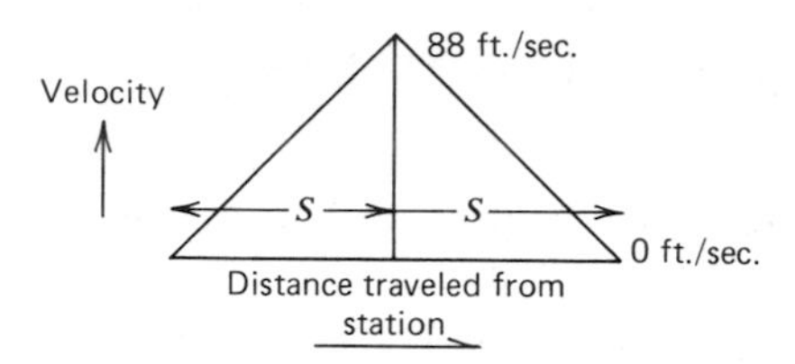


$$
\begin{aligned}
\text{Time to reach midpoint} &= \text{Velocity/acceleration} \\
&= 88/5 \\
&= 17.6 \text{ sec} \\
\text{Total running time} &= 17.6 \times 2 = 35.2 \text{ sec} \\
\text{Distance to midpoint} &= S = \tfrac{1}{2} \times \text{acceleration} \times \text{time}^2 \\
&= \tfrac{1}{2} \times 5 \times 17.6^2 \\
&= 774 \text{ ft}
\end{aligned}
$$

$$
\begin{aligned}
\text{Minimum station spacing} &= 774 \times 2 \\
&= 1548 \text{ ft} \\
&= 0.30 \text{ mi} \\
\text{Average running speed} &= 30 \text{ mph}
\end{aligned}
$$

$$
\begin{aligned}
\text{Time to travel between stations} &= \text{Acceleration time} + \text{deceleration time} \\
&= 17.6 \times 2 \\
&= 35.2 \text{ sec}
\end{aligned}
$$

$$
\begin{aligned}
\text{Total time/1548 ft of line} &= \text{Running time} + \text{station stop time} \\
&= 35.2 + 30.0 \text{ sec} \\
&= 65.2 \text{ sec}
\end{aligned}
$$

$$
\begin{aligned}
\text{Average overall travel speed} &= \text{Station spacing/overall travel time} \\
&= 1548/65.2 \\
&= 16.2 \text{ mph}
\end{aligned}
$$

It is worth noting that with stations set at 1548 ft spacing, there is no reason to use more powerful equipment, since overall travel speeds are limited by acceleration rates and station stop time. If high-speed equipment with limiting speeds of 90 mph were used for this example the overall travel speeds would be identical. It is interesting to note the effect of increasing the station spacing to $\frac{1}{2}$ mile. The additional 1092 ft are covered at top speed. The speed-distance diagram is now of the form:

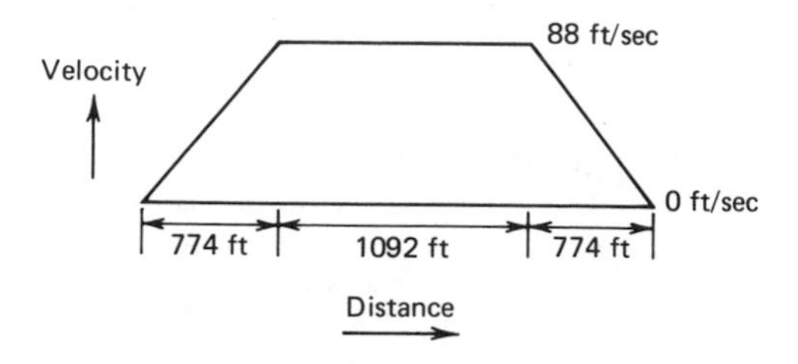

$$\text{Additional distance traveled in } 1092 \div 88 \text{ ft/sec} = 12.4 \text{ sec}$$

$$\text{Total travel time} = 65.2 + 12.4 = 77.6 \text{ sec}$$

$$\text{Overall velocity} = 2640/77.6 = 34.1 \text{ ft/sec or } 23.2 \text{ mph}$$

An increase of approximately 0.2 mi to the station spacing results in a virtual doubling of overall line-haul travel speeds. In practice, the calculations may become slightly more complicated by the nonuniform acceleration characteristics of the transit vehicles. Acceleration and tractive effort decrease with speed, as has been indicated in the discussion of electric locomotives. Speed-distance relationships must be computed from the characteristic speed-velocity relationships of the individual pieces of equipment. The overall approach is similar to the somewhat simplified example. It should be noted that the station spacings are normally greater in outlying areas than in the congested high-density central areas. Using constant acceleration rates throughout a system, overall travel speeds will vary significantly from CBD to suburban areas.

Figure 4-28 shows a rail rapid transit vehicle of conventional duo-rail design. The majority of vehicles in use in North America operate with one dc motor to each axle. Current is normally supplied to the dc motors from a collector shoe running over a "live" third rail. Alternatively, power can be supplied by a pantograph pickup as described in Section 4-22. The direct current voltage used by most systems is 600-V supply rectified and transformed from utility distribution voltage at special substations. For a fuller description of characteristics of the electric rail transit vehicles, the reader is referred to the section on electric traction locomotives.

*Fig 4-28.* A modern rail rapid transit vehicle. (Courtesy, Metropolitan Atlanta Rapid Transit Authority.)

## 4-27. SUMMARY

To summarize some of the major characteristics of the different modes of transportation, Table 4-9 and 4-10 show vehicle and route capacities of various types of vehicles and conveyances. These capacities can only be used as guideline figures since actual vehicle and route capacities in most cases will be limited by local conditions. The tables will serve, however, to show the immense variety of speeds and capacities that are currently available in the movement of passengers and freight.

## PROBLEMS

1. For many years the size and operating characteristics of American and European automobiles were very different. Discuss why the designs now are much more similar.

2. What would be the infrastructure effects of the widespread adoption of battery-powered automobiles in urban areas?

3. Discuss the problems involved in the fact that individual rapid transit systems have rolling stock peculiar to their own system, such as peculiar gage, power plant, and internal seating arrangements. Enumerate the advantages and disadvantages of standardization.

4. The capacities of systems often are quoted as measures of operational suitability. Discuss the pitfalls involved in restricting discussion to capacities as the only criterion of system selection. Relate your answer to urban transportation corridors in cities of varying sizes.

5. Compare the operations of STOL and short-haul airbus aircraft. Under what conditions will these aircraft compete with modern railroad passenger trains and where is each mode likely to dominate?

6. High-speed ground access modes have been suggested in many cities to link urban centers with airports. Given that most airports are less than 25 mi from the city center, discuss the likelihood of using technologies that have cruise speeds of 50, 100, 200, and 300 mph.

7. Using the performance curve shown in Fig. 4-24, compute the rated horsepower curve from the relationship:

$$RHP = \frac{\text{Tractive Effort} \times \text{mph}}{375}$$

*Table 4-9*
**Selected Vehicle Capacities and Operating Speeds**

| | *Freight (tons)* | *Passengers* | *Operating Speed (mph)* |
|---|---|---|---|
| *Highway Vehicles* | | | |
| Autos | | 2–5 | 30–55 |
| Minibus | | 6–20 | 25 |
| U.S. urban transit bus | | 55–85 | 25 |
| Streetcars, P.C.C. | | 125 | 20–30 |
| Streetcars, articulated Europe | | 200 | 20–30 |
| Trucks | 13–25 | | 20–55 |
| *Urban Rail Transit* | | | |
| New York PATH | | 140/car | 21 |
| Toronto duo-rail | | 327/car | 23 |
| San Francisco BART | | 216/car | 40 |
| Alweg monorail | | 200 | 50 |
| Morgantown (Boeing) | | 15 | 30 |
| *Interurban Rail* | | | |
| French TGV | | 386/train | 165 |
| British Rail High Speed Passenger Train | | 504/train and 72/car | 125 |
| Commuter railway-Budd Pioneer II | | 125 | 60 |
| U.S. passenger car | | 100+ | 60 |
| Japanese Tokaido car | | 1000/train | 130 |
| Open box car freight | 60/car | | 10–60 |
| *Airplanes* | | | |
| Transcontinental Boeing 707 | | 185 | 550 |
| Intercontinental Boeing 747 | | 500 | 582 |
| Concorde SST | | 128 | 1,354 |
| Beech 99 | | 15 | 285 |
| Piper Cherokee | | 4 | 132 |
| Boeing 707 Cargo | 45.5 | | 550 |
| Boeing 747 Cargo | 127 | | 582 |
| *Water Transport* | | | |
| Ocean liner, S.S. Queen Elizabeth 2 | | 1970 | 31 |
| Mammoth tanker | 600,000 | | 18 |
| Hovercraft, English Channel | | 450 | 50 |
| Mississippi barge | 1500–3000 | | 6 |

*Table 4-10*
**Route Capacities**

| | |
|---|---|
| *Highways (Uninterrupted Flow)* | |
| Multilane | 2000 passenger vehicles/hr/lane |
| Two lane-two way | 2000 passenger vehicles/hr total |
| Three lane-two way | 4000 passenger vehicles/hr total |
| *Railroads (Actual Capacity Passenger or Freight Trains)* | |
| A. Block signal system with train orders | |
| Single-track | 30 trains/day |
| Two or more tracks | 60 trains/day |
| B. Centralized traffic control | |
| Single-track | 65 trains/day |
| Two tracks | 120 trains/day |
| *Airports and Airways* | |
| Airway under instrument flying rules, 10-min separation | 6/hr |
| Airway under visual flying rules | Varies |
| Single-runway airport | Up to 110 operations/hr |
| Dual runways airport | Up to 220 operations/hr |
| Dual tandem runway airport | Up to 420 operations/hr |
| *Pipelines* | |
| 2 in. | 4–150 bbl/hr |
| 4 in. | 10–400 bbl/hr |
| 6 in. | 50–2000 bbl/hr |
| 8 in. | 100–4000 bbl/hr |
| 10 in. | 100–4000 bbl/hr |
| 12 in. | 400–5600 bbl/hr |
| *Belt Conveyors* | |
| 12-in. belt at 300 ft/min | 129 tons/hr |
| 34-in. belt at 450 ft/min | 843 tons/hr |
| 60-in. belt at 600 ft/min | 8100 tons/hr |
| *Heavy Rail Transit* | |
| New York, IND Queens Line (observed max. vol.) | 61,400 persons/hr |
| New York, 8th Ave. Express (observed max. vol.) | 62,030 persons/hr |
| Toronto, Yonge St. (observed max. vol.) | 35,166 persons/hr |
| *Busways* | |
| Individual vehicles, Runcorn | 6,000 persons/hr |
| Individual vehicles, New Orleans | 7,000 persons/hr (2500 observed) |
| 10-vehicle trains, 90–120 passengers/car | 36,000–48,000 passengers/hr |

| | |
|---|---|
| *Light Guideway Transit* | |
| Airtrains | 9,000 persons/hr |
| Morgantown | 10,000 persons/hr (ultimately) |
| Transit Expressway | 5,040 persons/hr |
| *Pedestrian Conveyor* | |
| 32-in. belt (1.5 to 2 mph) | 3,000 persons/hr |
| 48-in. belt (1.5 to 2 mph) | 10,000 persons/hr |

## REFERENCES


1. Over, J.A., (ed.), *Energy Conservation: Ways and Means,* Future Shape of Technology Publications, The Hague; 1974.
2. Darmstadter, J., J. Dunkerley, and J. Alterman, *How Industrial Societies Use Energy,* John Hopkins Press, Baltimore, 1977.
3. *Parking Dimensions 1960–1987 Model Passenger Cars,* Automobile Manufacturers Association, Detroit; 1960–1987.
4. Wright, Paul H., and R.J. Paquette, *Highway Engineering,* Fifth Edition, John Wiley, New York; 1987.
5. *A Policy on Geometric Design of Rural Highways 1965,* American Association of State Highway Officials, Washington, D.C.; 1966.
6. *Transportation and Traffic Engineering Handbook,* Second Edition, Prentice Hall, Englewood Cliffs; N.J., 1982.
7. *A Policy on Geometric Design of Highways and Streets,* American Association of State Highway and Transportation Officials, Washington, D.C.; 1984.
8. *Highway Capacity Manual, Special Report 209,* Transportation Research Board, National Academy of Sciences, Washington, D.C.; 1985.
9. Ashford, N., and P.H. Wright, *Airport Engineering,* Second Edition, Wiley Interscience, New York; 1984.
10. *CTOL Transport Aircraft Characteristics, Trends and Growth Projections,* Aerospace Industries Association of America, Washington, D.C.; January, 1979.
11. *Lloyds Register of Ships,* Lloyds, London; 1987.
12. *Big Load Afloat,* American Waterways Operators, Washington, D.C.; undated, current printing, 1987.
13. *Manual of Recommended Practice,* American Railway Engineering Association, Chicago (as amended to 1987).
14. *Manual of Recommended Practice,* American Railway Engineering Association, Chicago, Part 3, *Power,* (as amended).
15. Davis, W.J., Jr., *General Electric Review,* October 1926, pp. 685–707.

# *Chapter 5*

# *Traffic Control Devices*

Very light traffic, operating under ideal conditions, can move with ease and safety in any transportation network by following rudimentary rules of the road and strict adherence to schedules. As volumes and densities increase, conflicting movements bring both increased probability of collision and uneconomic operation through high average delays to vehicles in the system. The introduction of traffic control devices serves two principal purposes:

1. Increase of safety.
2. Reduction of system delays and increase of capacity.

The operating characteristics of the various modes and the variation in the degree of training required for the operators of each mode have resulted in a variety of control devices.

## HIGHWAY TRAFFIC CONTROL DEVICES[1]

### 5-1. PAVEMENT MARKINGS

The chief purposes of pavement markings are to regulate and provide guidance to traffic and to channel and separate movements with a minimum of physical equipment, relying on the psychological effect of the markings on behavior to bring about the desired driver reaction. The advantages of pavement markings lie in their lack of physical obstruction to the path of traffic, permitting their use on the traveled way itself, their low cost, and the ease of their removal when required. On the other hand, pavement markings are obliterated easily in snow conditions, are less effective under reduced visibility, can wear out rapidly under heavy traffic conditions, and provide no positive restraint against conflicting movements. Highway pavements in the

[1]The material contained in this section relies heavily on Chapter 23 of the *Transportation and Traffic Engineering Handbook* [1], and the *Manual on Uniform Traffic Control Devices for Streets and Highways* [2].

United States are marked in a uniform manner to provide nationwide standards [2]. The conformity to national standards by the individual states and cities provides the interstate driver with uniform markings and signalization throughout the United States.

*Center lines* are broken or unbroken yellow lines used to separate opposing traffic flows. On multilane roads with undivided pavements of four or more lanes, center lines may be composed of two unbroken yellow lines.

*Lane lines* are broken white lines, with the recommended standards being 15 ft of line segment with 25-ft gap. At intersections where weaving is undesirable, lane lines are frequently marked as solid lines for a short distance as in Fig. 5-1.

*No passing zone markings* indicate road sections on two- or three-lane roads where limited sight distance or other dangerous situations require the prohibition of passing movements. The zone is designated by a solid yellow line on the driver's side. No passing in either direction is indicated by a double yellow center line.

*Pavement edge lines* are solid white lines that delineate edges to provide guidance to the driver under lowered visibility conditions especially in the rain and at night. Most lines are white; those on the left edge of divided streets and highways or one-way roadways are yellow.

*Lane reduction transitions* are used to guide traffic at points where the pavement width changes to a lesser number of through lanes. The length of these yellow transition lines is computed from the design speed of the road to relate the visibility of the line to operating conditions.

*Approaches to obstructions* are marked in a manner that guides the traffic away from the obstruction. Depending on whether the obstruction is between

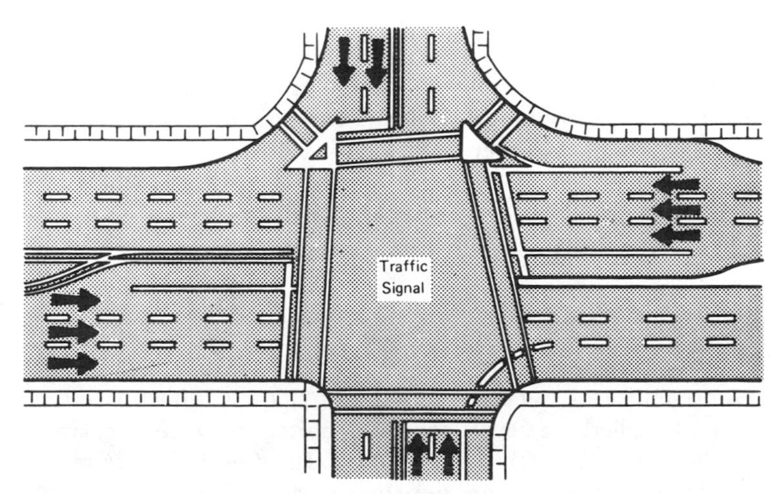


*Fig 5-1.* Typical applications of lane use control markings. (Source: *Manual of Uniform Traffic Control Devices for Streets and Highways,* U.S. Department of Transportation.)

opposing lanes of traffic or two lanes moving in the same direction the markings are yellow or white, respectively. The markings consist of a diagonal line or lines extending one or two feet from the center line of the obstruction, as shown in Fig. 5-2.

*Stop lines* extend across the full width of intersection approaches. Normally running parallel to the thin *white crosswalk lines,* the broad white stop line provides a well-defined stopping point for vehicles approaching intersections as shown in Fig. 5-1.

Other forms of pavement markings include *crosswalks and crosswalk lines, parking space markings, pavement word and symbol markings, preferential lane markings, speed measurement markings,* and *curb markings for parking restrictions.*

## 5-2. OBJECT MARKINGS

Obstructions within or close to the traveled way need to be marked clearly to reduce the traffic hazard that is presented by their physical proximity to moving vehicles. Typically, these obstructions may be structures associated with the highway such as bridge abutments, low bridges, and underpass piers, or they may be elements of the traffic control system such as signal and sign supports, traffic islands, and traffic beacons. Obstructions are marked by reflectorized hazard markers and, where possible, are painted in diagonal black and white stripes facing the flow of traffic. Where obstructions occur close to the line of traffic, guardrails should be used to deflect direct collisions and additional warnings should be given by means of signing and pavement markings.

## 5-3. DELINEATORS

To aid the driver under the reduced visibility of night driving conditions, light retroreflecting devices are mounted at the roadside to indicate roadway alignment. They are guidance rather than warning devices, which can be used on long continuous stretches of highway or in short stretches where changes in horizontal alignment may cause confusion. The color of delineators conforms to the colors of the roadway edgelines.

## 5-4. STUDS

While extensively used outside the United States, reflectorized pavement studs, sometimes called cat's eyes, have had limited use as center, lane line, and pavement edge markings within this country. In areas with little snow clearing problems, the use of studs can improve night driving conditions greatly, especially during poor weather conditions. The studs however, are quite susceptible to damage by snow plows.

a – Center of two-lane road.

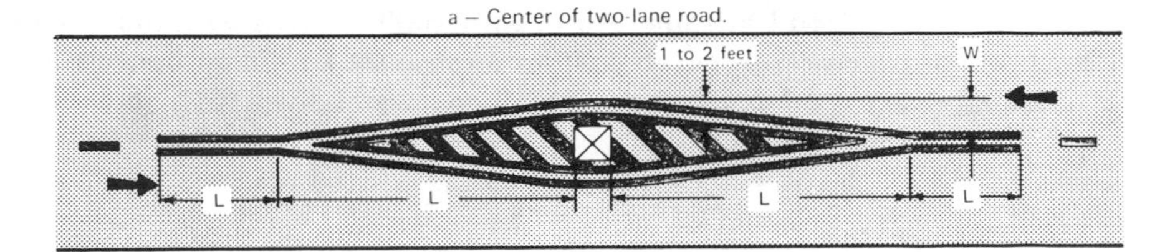


b – Center of four-lane road.

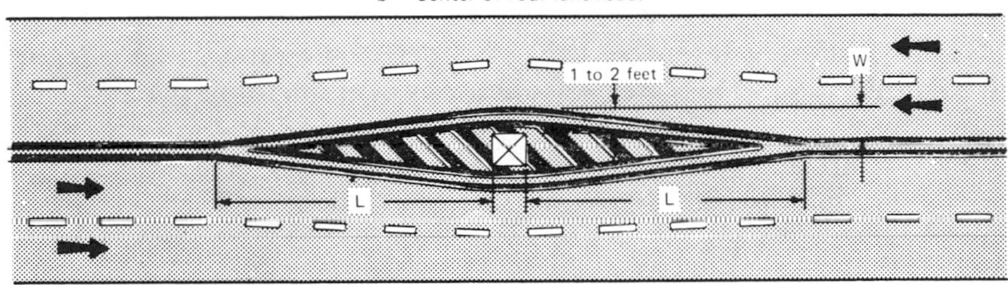


c – Traffic passing both sides of obstruction

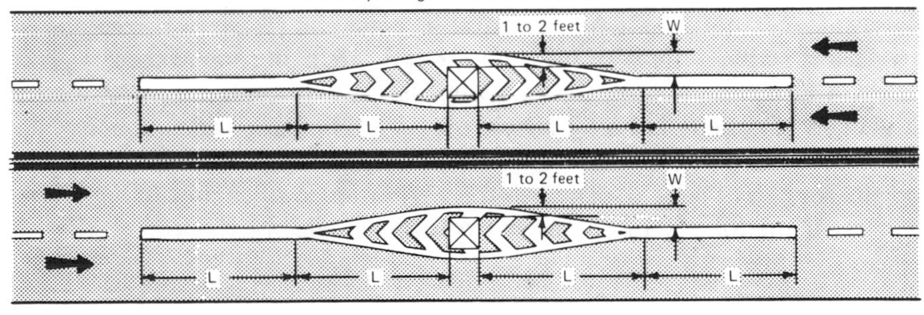


For speeds 45 or more $L = S \times W$ For speeds 40 or less $L = \frac{WS^2}{60}$

S = 85th percentile speed in miles per hour
W = Offset distance in feet

Minimum length of: L = 100 feet in urban areas
L = 200 feet in rural areas

Length "L" should be extended as required by sight distance conditions.

*Fig 5-2.* Typical approach markings for obstructions in the roadway. (Source: *Manual of Uniform Traffic Control Devices for Streets and Highways,* U.S. Department of Transportation.)

Another type of stud in some use in this country provides audible lane guidance when driven on by producing an unpleasant rumbling beneath the tire. These studs are used in urban areas to discourage weaving in high volume areas.

## 5-5. SIGNS

The most common and oldest method of traffic control is by means of the traffic sign. Signs function to control the movement of vehicles, to reduce the hazard of traffic operation, and to improve the quality of flow. These three functions are carried out by three different classifications of road sign, which are visually different to enable drivers to determine rapidly the category of any particular sign. The three classifications of traffic signs are

1. Regulatory signs.
2. Warning signs.
3. Guide signs.

## 5-6. REGULATORY SIGNS

Regulatory signs inform the driver of the applicability of specific laws and regulations along indicated sections of road. In the absence of such regulatory signs, enforcement would not be possible, since the restrictions are not recognized universally. Thus, it is necessary to indicate a one-way street by posting a "Do Not Enter" sign. It is not necessary, however, to inform the driver of the universally accepted law of driving on the right hand side of a street with two-way traffic.[2] Regulatory signs are posted where they can be seen clearly, and the section of road over which the regulation applies should be delineated clearly. Regulatory signs themselves can be subclassified into:

1. Right-of-way signs.
2. Speed signs.
3. Movement signs.
4. Parking signs.
5. Pedestrian signs.
6. Miscellaneous signs.

The form of all signs is defined clearly by the Federal Highway Administration [2]. Figure 5-3 shows typical signs from each subclassification.

[2]In only a few nations of the world does the driving-on-the-left rule prevail. In recent years several countries have changed to the driving-on-the-right rule.

TRAFFIC CONTROL DEVICES
RIGHT-OF-WAY SERIES
STOP
YIELD
SPEED SERIES
SPEED LIMIT 50
SPEED ZONE AHEAD
MINIMUM SPEED 40
MOVEMENT SERIES
ONLY
NO TURNS
PASS WITH CARE
PARKING SIGNS
NO PARKING ANY TIME
ONE HOUR PARKING 9AM-7PM
NO PARKING BUS STOP
2 HR PARKING 8:30 AM TO 5:30 PM
TOW-AWAY ZONE
PEDESTRIAN SIGNS
WALK ON LEFT FACING TRAFFIC
CROSS ON WALK SIGNAL ONLY
MISCELLANEOUS
ROAD CLOSED
ROAD CLOSED 10 MILES AHEAD LOCAL TRAFFIC ONLY
WEIGHT LIMIT 10 TONS

*Fig 5-3.* U.S. regulatory traffic signs. (Source: *Manual of Uniform Traffic Control Devices for Streets and Highways,* U.S. Department of Transportation.)

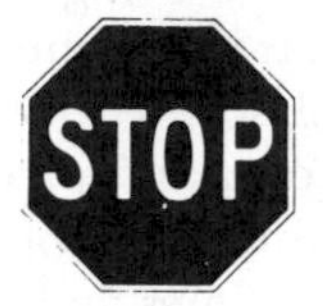

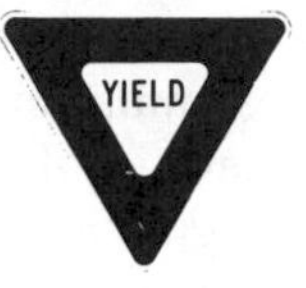

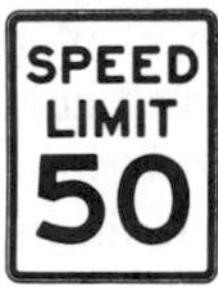

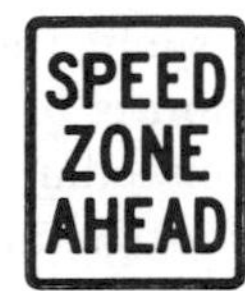

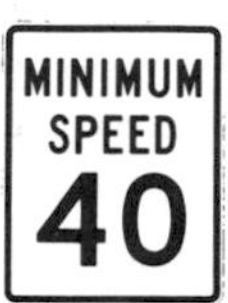

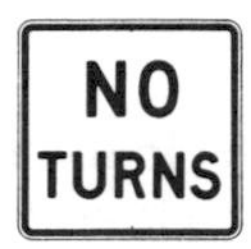

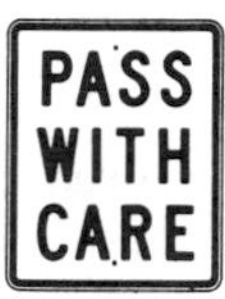

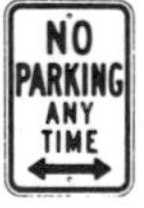

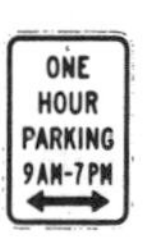

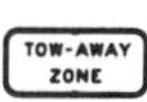

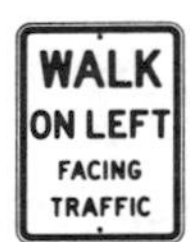

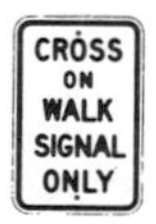

*Right-of-way signs.* Stop and Yield signs are placed at the junctions of traffic flows to indicate which stream of traffic has right-of-way and the type of movement minor flows must observe before moving into the intersection.

*Speed signs* indicate both daytime and nighttime speed limits and, in addition, delimit the beginning and end of speed zones.

*Movement signs* indicate legal, mandatory, and prohibited movements.

*Parking signs* state regulations governing the stopping and standing of vehicles. In general such signs may indicate times and days on which regulations are in force.

*Pedestrian signs* are signs directed principally at pedestrian movements both in urban and rural areas. As such, their method of placement differs from other signs whose chief function is to inform vehicular traffic.

*Miscellaneous signs* include a miscellany of regulations such as detour indications, information on road closings, and weight limits.

## 5-7. WARNING SIGNS

Where caution is required, including in some cases the reduction of speed or a special alertness to conditions on or close to the road, the driver is alerted by warning signs. Warning signs may be erected under the following conditions [2]:

1. To indicate changes in horizontal alignment.
2. To indicate an intersection.
3. To give warning that the driver should expect traffic control devices.
4. To warn converging traffic lanes.
5. To indicate narrow roadways.
6. To indicate changes in highway geometry such as the end of a divided highway.
7. To advise of unexpected or unusual grades.
8. To indicate sudden changes in surface condition, or poor pavement condition.
9. To advise of an at-grade rail crossing.
10. To indicate unexpected entrances and crossings.
11. Other signs such as advisory speed, clearance limits, animal crossings, and so on.

Warning signs are customarily yellow and diamond-shaped with black edging and lettering. Exceptions to this include the advised speed signs which are rectangular in shape.

## 5-8. GUIDE SIGNS

Guide signs are erected along highways to enable the traveler to find and follow routes in rural and urban areas, and to identify and locate items of need and interest. The class of sign normally is considered to be composed of three categories:

1. Route markers and auxiliary markers.
2. Destination and distance signs.
3. Information signs.

Examples of each type are shown in Fig. 5-4.

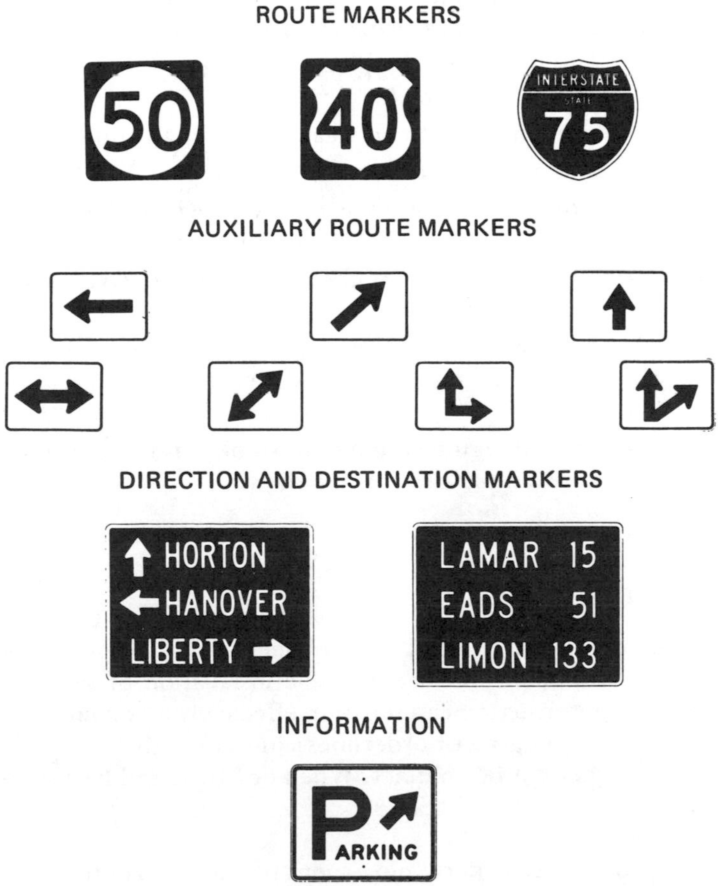

ROUTE MARKERS

50
40
INTERSTATE
75

AUXILIARY ROUTE MARKERS

DIRECTION AND DESTINATION MARKERS

HORTON
HANOVER
LIBERTY

LAMAR 15
EADS 51
LIMON 133

INFORMATION

PARKING

*Fig 5-4.* Guide signs on U.S. highways. (Source: *Manual of Uniform Traffic Control Devices for Streets and Highways,* U.S. Department of Transportation.

## 5-9. CHANNELIZATION

One of the most effective and efficient methods of highway traffic control is the adoption of high standards of geometric design at intersections. Islands and lane markings can be used to separate the intersecting, diverging, merging, weaving, and turning movements that can occur within any individual intersection. In determining the design of channelized intersections, it is considered that some of the following advantages can accrue from good design.

1. The paths of vehicles can be channelized so that not more than two paths cross at any one point.
2. The angle and location at which vehicles merge, diverge, or cross can be controlled.
3. The amount of paved areas can be reduced, thereby decreasing vehicle wander and narrowing the area of conflict between vehicles.
4. Clearer indications can be given of the proper manner in which movements are to be made.
5. The predominant movements can be given advantage.
6. Areas can be reserved for pedestrian refuge.
7. Separate storage lanes or areas can be provided that allow turning vehicles to wait, clear of the through traffic lanes.
8. Space can be provided for traffic control devices so that they can be perceived more readily.
9. Prohibited turns can be discouraged, if not prevented.
10. The speeds of vehicles can be controlled to some extent. [3]

Since channelization is a design technique this subject is treated more fully in Section 13-21, where example designs are shown.

## 5-10. TRAFFIC SIGNALS

Intersections that carry large vehicular volumes cannot be controlled safely and satisfactorily without traffic signals. The installation of power-operated traffic signals at an intersection can separate effectively all or most conflicting flows bringing about a degree of orderliness and safety that would otherwise be impossible at higher traffic volumes. When designed and located properly, traffic signals have several advantages.

1. They provide for orderly movement of traffic. Where proper physical layouts and control measures are used, they can increase the traffic-handling capacity of the intersection.
2. They reduce the frequency of certain types of accidents (particularly right-angle collisions).

3. Under conditions of favorable spacing, they can be coordinated to provide for continuous or nearly continuous movement of traffic at a definite speed along a given route.
4. They can be used to interrupt heavy traffic at intervals to permit other traffic, pedestrian and vehicular, to cross.
5. They represent a considerable economy, as compared with manual control, at intersections where the need for some definite means of assigning right-of-way first to one movement and then to another is indicated by the volumes of vehicular and pedestrian traffic, or by the occurrence of accidents [2].

Depending on their location and the type of traffic which they are designed to accommodate, individual traffic signals can be *pretimed, semiactuated, fully actuated,* or *volume-density* signals, according to the type of controller that is used in their operation.

*Pretimed* signals, sometimes called fixed-time, operate by means of an electric motor that drives an adjustable dial timing mechanism capable of providing cycles for 30 to 120 sec in length. Signals with only one timing dial provide the same cycle length and split on a 24-hr basis. More frequently, pretimed signal controllers operate with three dials, which are programmed to operate different cycle lengths and splits for the A.M. peak hour, the P.M. peak hour, and all off-peak times.

*Semi-traffic-actuated* signals operate on the basis of providing green time to the major artery until detectors on the minor roadway approaches actuate a demand for a signal. After providing green time for the minor roadway, the control returns to its normal position of permitting the main artery traffic to pass through the intersection. Traffic on the minor approaches receives variable lengths of green time up to some set maximum, dependent on the demand. Where all the traffic on the minor approach cannot clear the intersection during the maximum minor flow period, the green signal is returned to the major artery for some minimum predetermined time known as "minimum artery green," before returning again to the minor flow. In this manner the semi-traffic-actuated signal gives precedence to major arteries.

*Full-traffic-actuated* equipment can be useful where wide variations in traffic demand can occur on two or more conflicting approaches. Presence detectors are installed on all approaches. Maximum and minimum green phase times are selected. Arriving traffic triggers the detectors and the green phases adjust to reflect the arrival rates of the various approaches. Isolated intersections with large volume variations work well under full-traffic-actuated conditions.

*Volume-density* equipment is a complex fully actuated system, permitting variation of phase and cycle lengths according to input data. Sensing devices record data that gives information on arrivals, waiting times, and headways at all intersection approaches. Green times and cycle lengths are readjusted constantly, enabling the signal to provide a maximum degree of response to traffic flows and fluctuations.

## 5-11. SIGNAL SYSTEMS

Individually signalized intersections are frequently interconnected to form *signal systems.* The type of signal system adopted along the length of a route will influence the overall character of traffic flow through the area. In *progressive systems,* successive intersections along a street have identical cycle lengths. The cycle splits at individual intersections vary according to traffic demands of the cross streets. By off-setting the starting point of the cycles progressively along the artery it is possible for a car maintaining the design speed of the progression system to pass through all intersections without encountering a red phase at any time. Figure 5-5 shows a time-space diagram for a progressive system on a two-way street. While progressive systems work reasonably well

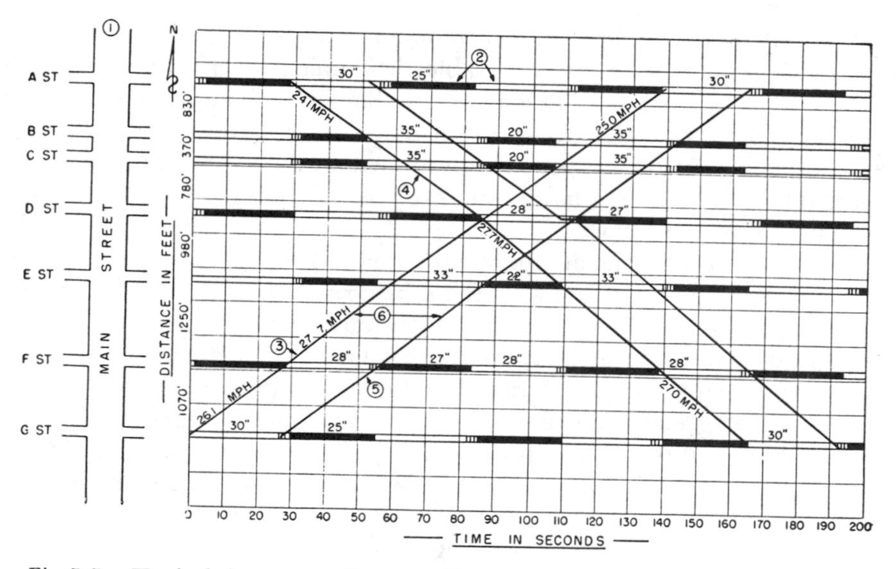


*Fig 5-5.* Typical time space diagram. (Source: *Transportation and Traffic Engineering Handbook,* Institute of Transportation Engineering Handbook, Institute of Transportation Engineers.)

(1) Plan of street drawn to scale.

(2) "Go" intervals for Main Street designated by open space; "Stop" intervals for Main Street designated by solid line; yellow" intervals for Main Street designated by crosshatching.

(3) Slope of this line indicates the speed and represents the first vehicle of a group or platoon moving progressively through the system from G Street to A Street.

(4) As in 3 above, this line represents the first vehicle moving from A Street to G Street in the southbound direction.

(5) This line is roughly parallel to line 3 and represents the last northbound vehicle in a group to go through all intersections.

(6) The space between lines 3 and 5 on the time scale is the width of the "through band" in seconds; this width may vary over different sections of street.

for light to medium-heavy traffic, they have been found to break down under very heavy loading.

*Simultaneous systems,* where all signals along one street are green or red at the same time, often are used under extreme loads, since all traffic tends to move at the same speed along the street and there is little speed variation between blocks. Obviously, in the operation of a simultaneous system all signal cycles along a street must be identical, and their starting points synchronized.

Under heavy loading conditions, *alternate systems* have also been known to work more effectively than progressive systems. In the alternate system, adjacent signals or signal groups show opposite color phases. Under favorable conditions, the system can lead to continuous movements of platoons of vehicles along the street.

The most sophisticated traffic signal systems are those where entire networks are *computer controlled.* Detectors throughout the system provide data on current traffic conditions to a central computer. The computer then can calculate the best sequence of signal changes throughout the whole system. In such a system, the computer has complete and direct control of all signals in the system. Constant readjustment of signal timing takes place, providing immediate response to system demand. In one area the adoption of a computer controlled system resulted in a 46 percent decrease in delay to vehicles and a 24 percent increase in overall speed [4].

## AIDS TO MARINE NAVIGATION

### 5-12. PURPOSE OF AIDS TO MARINE NAVIGATION

Because of the low volumes of vessel traffic that use the waterways and harbors, aids to navigation function more in the area of guidance devices than controls. Maneuvering controls are provided chiefly by rules of seamanship. Navigation aids, therefore, serve two main purposes.

1. To provide information that will permit overseas vessels to determine their exact position at landfall, and to allow precise positioning within the waterways and close to coastal waters.
2. To mark channels clearly and warn of hazards such as wrecks, sandbars, and other sudden changes in bottom topography.

### 5-13. BUOYS

Buoys are anchored floating markers which may be lighted or may give out some sound signal to indicate their position under conditions of limited visibility. A variety of buoy shapes is used as indicated by Fig. 5-6, which shows standardized buoy shapes and markings [5].

*Spar buoys* are in the shape of large logs, floating with one end out of the water. *Can buoys* are cylindrical markers built up from steel plates, while *nun buoys* are distinguished by their truncated conical top. *Bell, gong,* or *whistle buoys* are steel floats supporting small skeleton towers housing the signal-making equipment. Most buoys making sound signals are operated by the motion of the sea itself. Less commonly, compressed gas or electric power from batteries is used to produce the signals.

United States navigable waters are marked by buoys according to the *lateral system.* Buoys are numbered and marked in navigable channels from the point of entry to the head of navigation. Where channels do not lead from seaward, a direction is assumed for enumeration purposes. For example, directions of proceeding are southward on the Atlantic Coast and the Intracoastal Waterway, and northward and westward on the Gulf Coast.

In addition to the numbering and lettering by which buoys can be referenced to charts, coloring systems are used to denote both purpose and side on which they are to be passed. Black buoys mark port sides of channels or obstructions requiring passage to the starboard. Red buoys mark the starboard sides of channels and passage to the port. Figure 5-6 shows marking for preferred channels, quarantine and other anchorage, in addition to other markings of special significance.

Often buoys are equipped with reflectors. *Radar reflectors* improve the radar response of the buoy for easier location in limited visibility [6]. *Optical reflectors* of different colors are used to enable buoys to be picked up by ships' searchlights.

Lighted buoys may have red or green lights depending on the color of the buoy on which they are mounted. In addition, white lights are mounted on buoys to convey signals of special significance. For example, flashing lights indicate black, red, or special-purpose buoys and quick-flashing lights on black or red buoys indicate points at which special caution is required. Interrupted quick-flashing lights indicate junctions in channels or wrecks. Morse code flashes, groups of one short flash followed by one long flash, are placed only on buoys marking fairways and channel center lines that may be passed on either side.

## 5-14. LIGHTHOUSES

Lighthouses are used on prominent points and at areas of special hazard to aid ships in locating their position and to warn of possible danger. The function of a lighthouse is simply to support a light that is powerful enough to be seen at great distances. The type of structure used varies greatly according to location. This variation depends on the height at which the light must be supported, the supporting foundation conditions for the structure, the availability

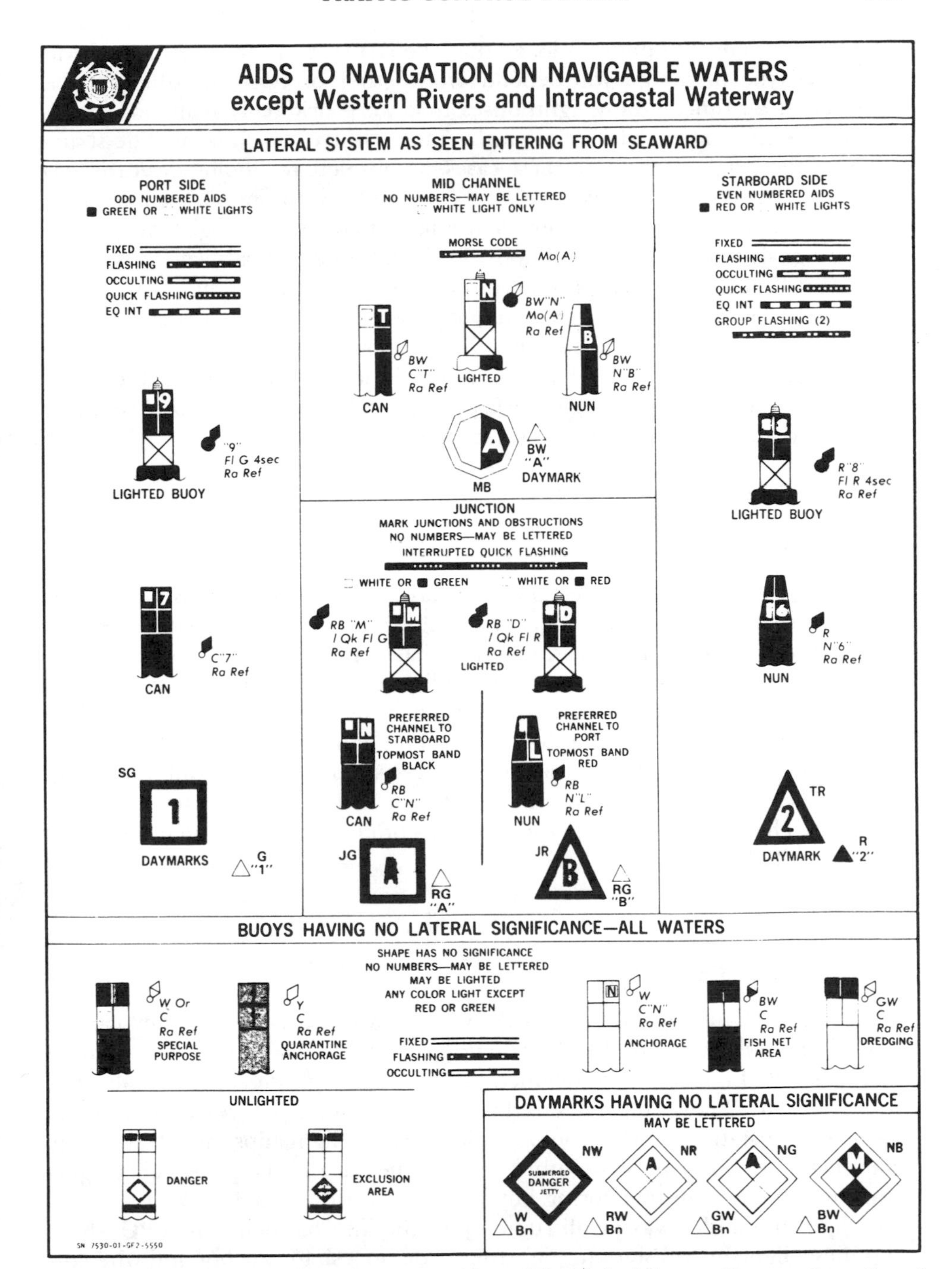


*Fig 5-6.* The lateral system of buoy marking of the United States. (Source: *Coast Guard Aids to Navigation,* 1977. U.S. Department of Transportation, United States Coast Guard.)

of materials, and the physical loads that the structure must sustain. In addition, the type of design will depend on whether or not the light is to be manned at all times. Many lighthouses that were originally manned continuously are now operated by electrical remote control machinery. The structures of these lights reflect, in these cases, an obsolete technology, yet there is no reason to replace such existing structures until it becomes uneconomic or undesirable to continue their use. Many light structures or lighthouses emit sound signals under foggy conditions. Each station emits patterns of blasts at prescribed intervals which permit vessels to identify the sending station.

Just as lighthouse fog signals uniquely identify the sending station within the area, the lights themselves vary in their periods of light and darknesses. Such "light characteristics" enable a vessel within any area to determine with certainty which signal is being observed. Light colors used are white, red, and green and can be either continuous in one color or alternating.

## 5-15. RANGE LIGHTS

Range lights are a simple yet effective navigation device consisting of two powerful lights some distance apart set along the axis of a channel, either within the channel or inland. Any vessel that aligns itself with the lights and steers a course that keeps the two lights aligned moves within a safe channel. Range lights are usually visible in one direction only and rapidly lose brilliance as the ship diverges from the range line. The lights can be any of the three standard colors and have characteristic phases which permit differentiation from other lights. When using range lights it is usually necessary to use, in addition, charts that designate the stretches of channel over which the range lights can be used safely.

## 5-16. LIGHTSHIPS

Lightships are functionally identical to lighthouses, except that they can be placed in positions where the construction of a lighthouse would be expensive and difficult. Lightships are manned vessels carrying lights, fog signals, and radio beacons. They are painted red with white superstructures, and carry the name of the station in white on both sides. When lightships must leave their station for servicing or some other reason, they are replaced by relief lightships which conform in coloring but carry the word "Relief" on either side. Any lightship under way or off station flies the international signal PC which signifies a lightship off station. Lightships ride on a single anchor and when on station exhibit only two lights, a brilliant masthead light, and a less brilliant forestay light. The position of the forestay light relative to the masthead light indicates the direction of current.

## 5-17. FOG SIGNALS

In periods of limited visibility, lights are supplemented by sound signals. Each station emits a signal characteristic to itself to help the mariner determine the exact position. When synchronized with lights, the time interval between seeing the light and hearing the signal will enable a vessel to determine its distance from the station. Fog signals vary in type. Commonly diaphragm horns, reed horns, sirens, diaphones, whistles, and bells are used at prescribed intervals. On buoys, bells and whistles operated by the motion of the sea operate at irregular intervals.

## 5-18. RADIO BEACONS

Electronic devices provide all-weather navigation aid both to coastal and deep-sea navigation. There are now a large number of installations of marine radio beacons operating in the 285 to 325 kilocycle range. In 1988, close to 200 stations were operated by the U.S. Coast Guard on the Atlantic, Gulf, and Pacific Coasts and on the Great Lakes. The chief advantage of the radio beacon system is its essential simplicity. All that is required is a relatively inexpensive receiver with a directional antenna. Signals from radio beacons are received on a short-to-medium range up to 100 mi. Most beacons transmit one out of six minutes continually while other stations transmit continuously 24 hr a day. Each station carries a characteristic signal permitting unique identification.

The LORAN system is a long-range radio system made up of transmitting pairs of stations which permit vessels to position themselves at great distances from shore. Location is carried out by determining the difference between time of reception of signals from pairs of stations transmitting a simultaneous signal. The locus of this time difference is known as a LORAN line. By receiving signals from two pairs of stations, intersecting loci can be plotted giving a firm position fix. Two LORAN systems are currently available, LORAN A, previously known as standard LORAN, operating at 1900 to 2000 kHz, and LORAN C which operates at a lower frequency of 90 to 110 kHz [5].

## 5-19. RADAR DEVICES

Since World War II, the field of electronics has supplied several powerful additional tools to maritime navigation devices. Most important of these is undoubtedly radar. Radar, originally a shortening of radio direction and ranging, makes use of the echo principle with electronic waves rather than the audible sound wave. Marine radar works on the basis of generating radio frequency oscillations of specific form and transmitting them into space in a narrow beam of radio waves. The beam is rotated continuously in azimuth about the sending aerial. Radio echoes from any suitable object return to a receiver and are displayed on a screen in such a way that the distance and direction of the surrounding objects can be directly determined. The essential equipment for radar, therefore, is a transmitter for generating radio waves, a rotating aerial or scanner to send out direct waves and pick up echoes, a receiver and a display for converting the received waves into directly readable information in the form of a radar pilot [7].

Radar is available both on board ship and at shore stations. Ships, therefore, are able to navigate close to shore under the worst visibility conditions and can determine the presence of other ships. On-shore radar enables close ship-to-shore cooperation at all times provided the ship has at least radio communication with land.

## RAILROAD SIGNALS AND TRAFFIC CONTROL

### 5-20. RAILROAD SIGNAL SYSTEMS

Railroads have the most extensive control system of any major transportation mode excluding pipelines. At all times, the locomotives remain under a high degree of control. This is made possible by the restricted freedom of movement of the vehicles and the operation of a self-contained system over an essentially exclusive right-of-way with a minimum of conflicts from other modes. Without the elaborate forms of traffic control currently used, the *modus operandi* of railroads would require extensive modifications from a safety viewpoint. With existing procedure, the railroads can operate high-speed, high-capacity freight and passenger service, often with opposing movements on single track, with a high degree of safety. Passenger trains can have capacities in the region of 1000 persons and operate at running speeds of well over 100 mph. The fatality rate per passenger mile of rail operations is approximately 1/25 of that for highway movement by private automobile. This remarkable safety record is built upon an elaborate signal and control system.

In areas where only light densities of traffic occur, operations are possible without signals by adherence to operating rules and time tables. (At low densities of air traffic, airports without radar also operate by using time separations between aircraft operations.) Under heavier densities signal systems help provide train control. The two types of signal used almost exclusively are *semaphore* and *light signals.* The Association of American Railroads has published the Standard Code of Operating Rules, Block Signal Rules, and Interlocking Rules which shows both the *aspect* and *indications* recommended for both semaphore and light signals.

The *aspect* of a signal is the position of the lights or semaphore blades and their relative arrangements. The *indication* of a signal is the message conveyed by the aspect of the signal according to the rules, indicating how the operator is to control the movement of the train. Figure 5-7 depicts some typical aspects of the AAR standard code showing the standard *indications* and *applications* (points where the aspects may be displayed). While the number of aspects is numerous, each signal light or semaphore has three basic positions.

Almost all semaphore signals currently in operation use the upper right-hand quadrant. The vertical position is used for *clear,* the 45-degree position for *approach* (requiring speed reduction), and the horizontal position for the most restrictive aspect, *stop* or *stop and proceed.*

Light signals are arrangements of electric lights and optical lenses which are capable of producing bright colored lights visible even in bright sunlight. Three colors of lamp are used. Green is used for *clear,* yellow for *approach,* and red for the most restrictive aspect.

## 5-21. AUTOMATIC BLOCK SIGNALING

The basic method of preventing rear-end collisions on double-track sections of railroad, and both head-on and rear-end collisions on single-track sections is the system of automatic block signaling. Through the use of track circuits a definite space interval can be provided between following or opposing trains. The term *automatic* is applicable because the control of the space interval is

| RULE | ASPECT | NAME | INDICATION | APPLICATION |
|---|---|---|---|---|
| 281 | 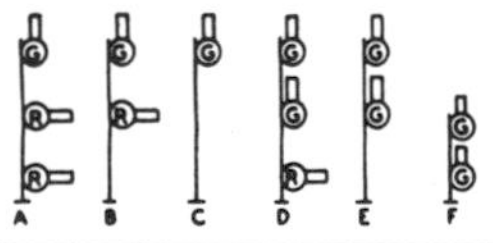 A B C D E F | Clear | Proceed | At entrance of normal speed route or block, to govern train movements at normal speed. |
| 281A | 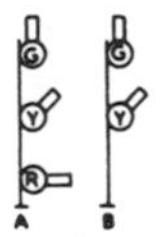 A B | Advance Approach Medium | Proceed approaching second signal at medium speed | At entrance of normal speed route or block, to govern the approach to approach-medium signal. |
| 281B |  | Approach Limited | Proceed approaching next signal at limited speed | At entrance of normal speed route or block, to govern the approach to limited-clear signal. |
| 281C |  | Limited-Clear | Proceed; limited speed within interlocking limits | At entrance of limited speed route or block, to govern train movements at not exceeding limited speed. |

*Fig 5-7.* Some typical aspects of the American standard code of signals. (Source: Association of American Railroads.)

provided by the presence of the trains themselves. Under this system, sections of track are divided longitudinally into smaller sections called *blocks*. By providing a signal system in which no two trains are permitted to occupy the same block at the same time, collision situations are avoided. Initially, block signal systems were controlled manually. The introduction of automatic block signals reduced the possibility of human error, affording a higher level of protection. In addition to the increased safety, additional benefits accrued from the increased capacity which results from reduced delays and lower time intervals between meets and passes.

Figure 5-8 shows in simplified form the track circuitry required for a two-position one-block signal system suitable for operation for one direction of movement. The circuit consists of both track rails, insulated from rails of adjacent blocks by insulated joints at the ends of the block. At one end of the block the rails are joined to a battery. At the other end the rails are connected to the coil of a high-voltage resistance relay. With no train in the block the current runs through the relay, keeping the contact in the signal circuit closed. With the contact closed, the signal at the beginning of the block shows the *clear* aspect. The presence of a train within the block provides a low-resistance shunt across the rails. The contact within the signal circuit is broken and the signal changes to *stop*.

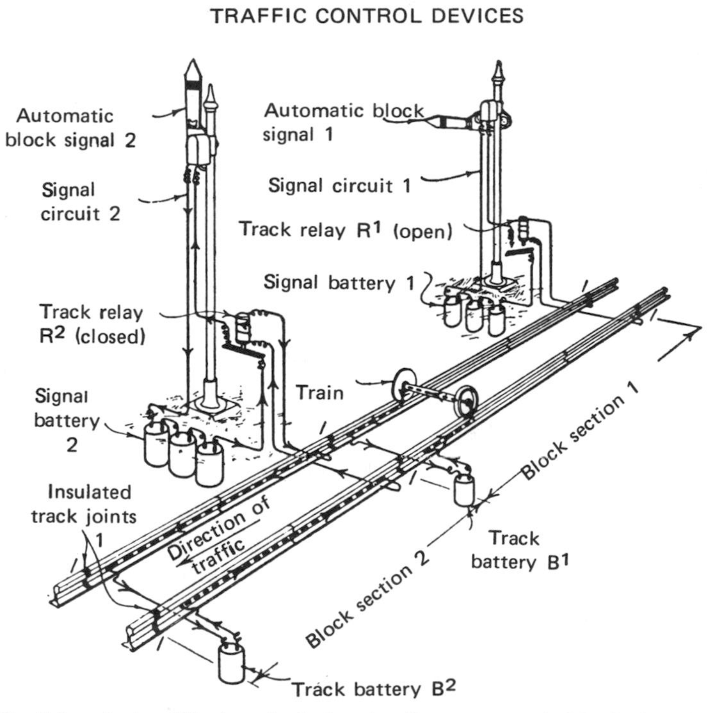

*Fig 5-8.* A simplified typical circuit of an automatic block signal system. (Source: Simmons Boardman Publishing Corp.)

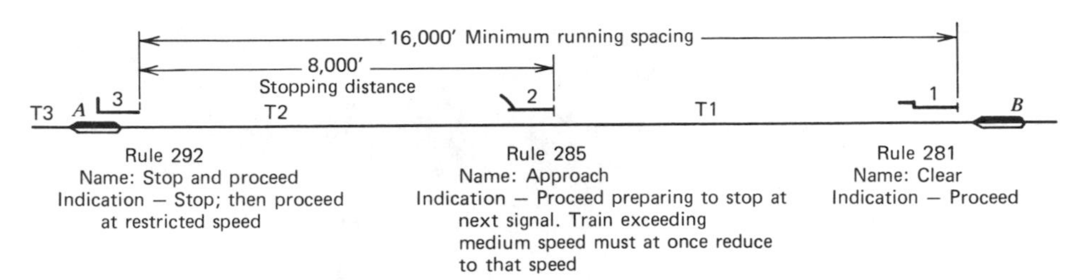


*Fig 5-9.* Three-aspect two-block signaling. (Source: Simmons Boardman Publishing Corp.)

Figure 5-9 indicates an arrangement that allows a higher capacity for single-direction track. Trains are informed under this system of the presence of trains two blocks ahead. Three-aspect two-block signaling permits a train to receive a *stop and proceed* signal if another train is in the block immediately ahead. An *approach* aspect is displayed if the nearest preceding train is two blocks ahead. If no trains occupy the next two blocks, then the *clear* aspect is displayed. This type of signaling permits the second train, B, to slow when approaching the first train, A, without coming to a complete stop unless the distance between them falls below a minimum stopping distance. With this type of block system, blocks can be shorter and unnecessary stopping minimized.

## 5-22. CENTRALIZED TRAFFIC CONTROL

By the use of centralized traffic control, a dispatcher at a central location controls the operation of switches and signals at key points such as the ends of sidings and junctions. The dispatcher can track the location of trains on each section by visual displays of lights on a track diagram. Using constantly changing displays, the operator can remain informed concerning the position of progress of each train in each territory.

Older forms of CTC operated with single wire connections from the control panel to the switches and signals. With the introduction of coded impulses to code-following relays, the amount of wiring could be significantly reduced and CTC systems were extended to control hundreds of miles of rail systems. The dispatcher supervises a large control board having direct operation of all switches and signals which remotely control the line switches and signals. Figure 5-10 shows the display board of a large central traffic control system for hundreds of miles of track. By judicious operation of the buttons and switches at the central board the dispatcher can produce the most desirable meets and passes. CTC is obviously applicable to multiple-track operation, but has found its prime application to be the increase of the capacity of single-track sections and lines.

Increase of capacity became important with the introduction of diesel-

*Fig 5-10.* Centralized traffic control. (Source: WABCO Signal and Communication Division.)

electric locomotives. With these more powerful engines, fewer trains could be run at higher speeds. By operating longer, faster trains at capacity, the number of tracks could be reduced. This has produced extensive saving in the area of maintenance in the face of mounting costs.

## 5-23. TOTAL OPERATIONS PROCESSING SYSTEM (TOPS) [9, 10]

From work that was carried out to improve and monitor the distribution of rolling stock, the railroads have developed an integrated information/operations systems which is computer based, known as the total operations processing system (TOPS). Developed in the late 1950s and 1960s by Southern Pacific and IBM, the system has been adopted and subsequently modified by railroads within the United States (e.g., Burlington Northern) and abroad (e.g., British Rail).

The Southern Pacific TOPS system, applied to over 5000 mi of railroad, became fully operational in 1972. Two large central computers in San Francisco controlled operations from Eugene, Oregon, to Houston, Texas. The system had 18 other linked computers which controlled regional stores and accounts and performed regional management functions. In addition, there were approximately 500 field terminal devices which varied from IBM 1050s down to teleprinters at small terminals. The mode of operation is simple in concept, but complex in the scale of data stored and processed. When a railroad car enters the TOPS system, details of the waybill are stored in a car file, which records both the freight and the physical characteristics of the car. Entry takes place either at an interchange point with another railroad or at a private siding. Any movement or change of status of the car produces computer output at the appropriate terminal. The terminal operator at this point re-enters the status or movement change into the computer system as an update. In this way, the TOPS system has a total record of location and status of its freight and rolling stock. The central operations control room is constantly attended to provide a backup to the field in periods of difficulty and to monitor the overall operating condition of the network. TOPS can provide the following useful functions:

1. Loaded cars are routed through the system according to waybill data, and trains are made up on instructions from TOPS.
2. Any inquiry on status from a shipper can be answered via a TOPS terminal device.
3. At all stages the central TOPS computer knows the location of each car and the makeup of each train even if on the move.
4. Empty cars can be directed according to standing instructions which can be overridden by the distribution. Allocation is possible by ownership, type, or class.
5. TOPS can monitor marshaling operations to ensure that there is no excessive delay at yards. Standard times can be set and exceptions reports are produced automatically.
6. There is an automatic check on train and car movement to ensure TOPS instructions have been followed.
7. Accounts and statistics are produced automatically.
8. Operational information produced includes train schedules, train files (motive power, crews, and consist), recalculation of schedules when delays occur, summaries of trains to arrive at a terminal or yard on request.
9. Motive power information includes such items as a complete maintenance schedule, and a motive power balance sheet of supply and demand of locomotives by terminal point for the next 48 hr.

For a number of years the United States railroads had attempted to bring in a nationwide freight car identification system which was known as *Automated Car Identification* (ACI). In concept the ACI system was simple. An optical scanning device at yards and other recording points scanned multicolored service stripes on the freight car which identified it uniquely. The device was to be capable of operating at 80 mph in the most inclement weather. The scanned information was transmitted to the railroad's own computer and then to a central AAR computer system in Washington. The interconnecting system, Tele Rail Automatic Identification Network (TRAIN), never functioned adequately. Similar technology already had been examined and rejected by European railroads which opted for a TOPS system. Three principal reasons were given for the 1978 decision to abandon the ACI system on which approximately $150 million had already been spent over a 10-year period.

1. The optical scanning system was only about 80 percent reliable.
2. It was an incomplete information system which stored redundant information that management should already know, that is, the number of each wagon in a particular train.
3. It has become redundant with the development of computer-based information systems such as TOPS which were available on an industry wide basis.

Although TOPS is a Southern Pacific system, it was subsequently used outside that particular organization. Other perpetual inventory systems have been developed which record car positions in a somewhat similar manner.

While not strictly within the purview of control as used elsewhere in this chapter, the perpetual inventory tracking computer systems have spawned operational methods. Using powerful mainframe computers, systems have been developed which not only locate railroad cars but also distribute cars throughout the rail network in such a way that the number of empty car miles is minimized by ensuring a supply of empty cars as close as possible to points of demand. On the Norfolk Southern Railroad this is achieved by the use of an enhanced linear programming model run on a monthly basis with flow rules that can be overridden on the basis of unusual daily demand.

A more sensitive model is required for the distribution of locomotives and cabooses. It is not unusual for one railroad to have the use of the locomotives of another system. Railways have agreements to balance out horsepower hours used and on a yearly basis this may be complicated by the highly seasonal nature of rail freight traffic. By the use of operational research techniques, such as the "out-of-kilter" algorithm, computerized models optimize locomotive and caboose distribution according to anticipated freight flows.

## AIR TRAFFIC CONTROL [11, 12]

### 5-24. PURPOSE OF AIR TRAFFIC CONTROL

Two basic factors in air traffic control (ATC) are safety and efficiency. Sufficient airspace must be provided to avoid collision, yet in order to maintain efficiency airspace cannot be wasted. ATC must give consideration to all users of airspace: the military, the commercial carriers, and general aviation.

Air traffic control measures have been instituted only after the need was demonstrated by human failures. As air traffic activity grows, increasing traffic problems will probably require further air traffic regulation and system development to provide for the expeditious movement of all aircraft. Greater attention will need to be given to the allotment of airspace and the compatibility of equipment between air carriers and general aviation as well as types of aircraft such as STOL/VTOL, SST, and new types of conventional aircraft. Safety, technology, regulation, and financing are four important factors that affect the development of a responsive air traffic control system.

Formal federal government involvement in air traffic came into existence with the Air Commerce Act of 1926 which provided for the establishment, maintenance, and operation of lighted civil airways using beacon lights. Today the Federal Aviation Administration (FAA) provides control and navigation assistance for the movement of air traffic. The federal governing authority in air traffic control exists by virtue of the fact that implications of air travel are nationwide and do not respect state boundaries.

## 5-25. CONTROLLED AND UNCONTROLLED AIRSPACE

The federal government designates airspace either as controlled or uncontrolled. In the case of the latter, flights may be conducted without reference to air traffic control although pilots are expected to conform to certain general rules of the air, and to some rules that are dependent on prevailing conditions of visibility and cloud coverage. Controlled airspace has various dimensions depending on whether it is an airway, a terminal control area, or a control zone, and pilots must communicate with the appropriate air traffic control facility in order to receive authorization to fly through their controlled airspace.

Extending upward from 1200 ft AGL (above ground level) and, in a few areas from 700 ft AGL, controlled airspace exists in almost all areas of the contiguous United States due to the designation of *control areas* and *transition areas.* In addition, controlled airspace extends upward from the ground in areas immediately surrounding an airport in *control zones.* The demand for airspace exhibits the need for shared airspace. Increased communications have helped and will become more important as they link computers and automatic air traffic control systems. In order to achieve greater airspace utilization and safety, an area above 14,500-ft altitude MSL (mean sea level) has been designated as a *continental control area.* Aircraft flying above this altitude are higher performance aircraft, usually jet powered. In *positive control areas,* above 18,000-ft altitude MSL, all aircraft are controlled by continuous surveillance and are required to have certain equipment to permit the higher aircraft densities of the higher performance aircraft.

Three different types of control in terminal areas are in current use in the United States:

Terminal control area (TCA)
Terminal radar service area (TRSA)
Airport radar service area (ARSA)

A *terminal control area,* such as those shown in Figs. 5-11 and 5-12, consists of controlled airspace extending upward from the surface or higher to specified altitudes, within which all aircraft are subject to operating rules and where there are requirements on pilot qualification and aircraft equipment. Terminal control areas are designated around major aviation hubs and they include at least one primary airport around which the TCA is located. In 1987, there were nine Group I TCAs and 14 Group II TCAs in the United States. The requirements for the Group I areas are more stringent than those for Group II.

In order to facilitate the flow of air traffic in terminal areas, the FAA initiated the concept of *terminal radar service areas,* and originally there were over 200 U.S. airports where approaching aircraft could use this advisory service. Within the airspace surrounding the designated TRSA airports, air traffic control provides radar vectoring, sequencing, and separation for all IFR and participating VFR aircraft. Pilot participation is urged but is not mandatory. Figure 15-13 shows the location and size of the TRSA at Augusta, Georgia in 1987.

Increasingly *airport radar service areas* are replacing terminal radar service areas. ARSAs are being established in busy smaller airports to protect aircraft which are landing or taking off. Aircraft wishing to enter an ARSA must establish two-way radio communication with air traffic control. The ARSA airspace is defined as 0-4000 ft height above airport within an inner circle of 5 nm and between 1200 and 4000 ft height above airport between 5 and 10 nm from the airport. The services provided by the ARSA on establishing two-way radio and radar contact are: sequencing arrivals, IFR/IFR standard separation, IFR/VFR traffic advisories and conflict resolution, and VFR/VFR traffic advisories. Figure 5-14 shows the location of the Daytona Beach ARSA.

Although airspace may be designated as controlled it does not necessarily mean in the United States that ATC instructions are mandatory in all circumstances. Pilots may elect to fly according to visual flight rules on a "see and avoid" basis. There is increasing pressure, however, to remove this option for traffic operating inside the busiest terminal areas. It is interesting to note that this option has not been available for many years in major European countries.

*Fig 5-11.* The New York TCA. (Source: Federal Aviation Administration.)

**ATLANTA, GEORGIA**

**The William B. Hartsfield**
**Atlanta International Airport**

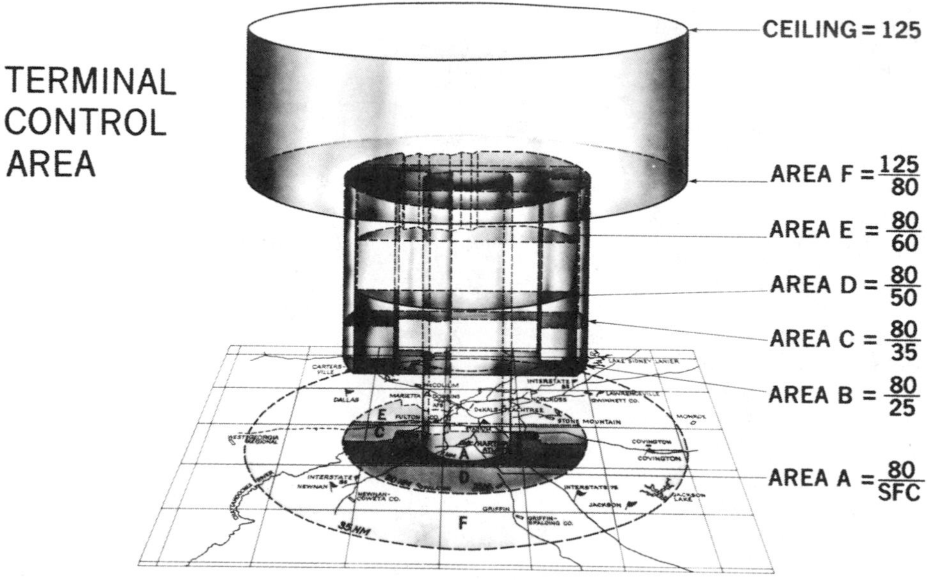


*Fig 5-12.* The Atlanta terminal control area. (Source: Federal Aviation Administration.)

## 5-26. VISUAL FLYING RULES AND INSTRUMENT FLYING RULES

The federal government prescribes two basic types of flight rules for air traffic which depend on weather conditions as well as on the location and altitudes of flight paths. These are known as visual flight rules (VFR) and instrument flight rules (IFR). In general, VFR means that the weather conditions are good enough for the aircraft to be operated by visual reference to the ground. On the other hand, IFR conditions prevail when the visibility or the ceiling (height of clouds above ground level) falls below those prescribed for VFR flight. In VFR conditions, there is essentially no en route air traffic control except where prescribed; aircraft fly according to "rules of the road" using designated altitudes for certain headings and are responsible for maintaining their own separation. Positive traffic control is exercised always in IFR conditions and designated control areas. Essentially, these rules require the controlled assignment of specific altitudes and routes and minimum separation of aircraft flying in the same direction at common altitudes, as shown in Figure 5-15.

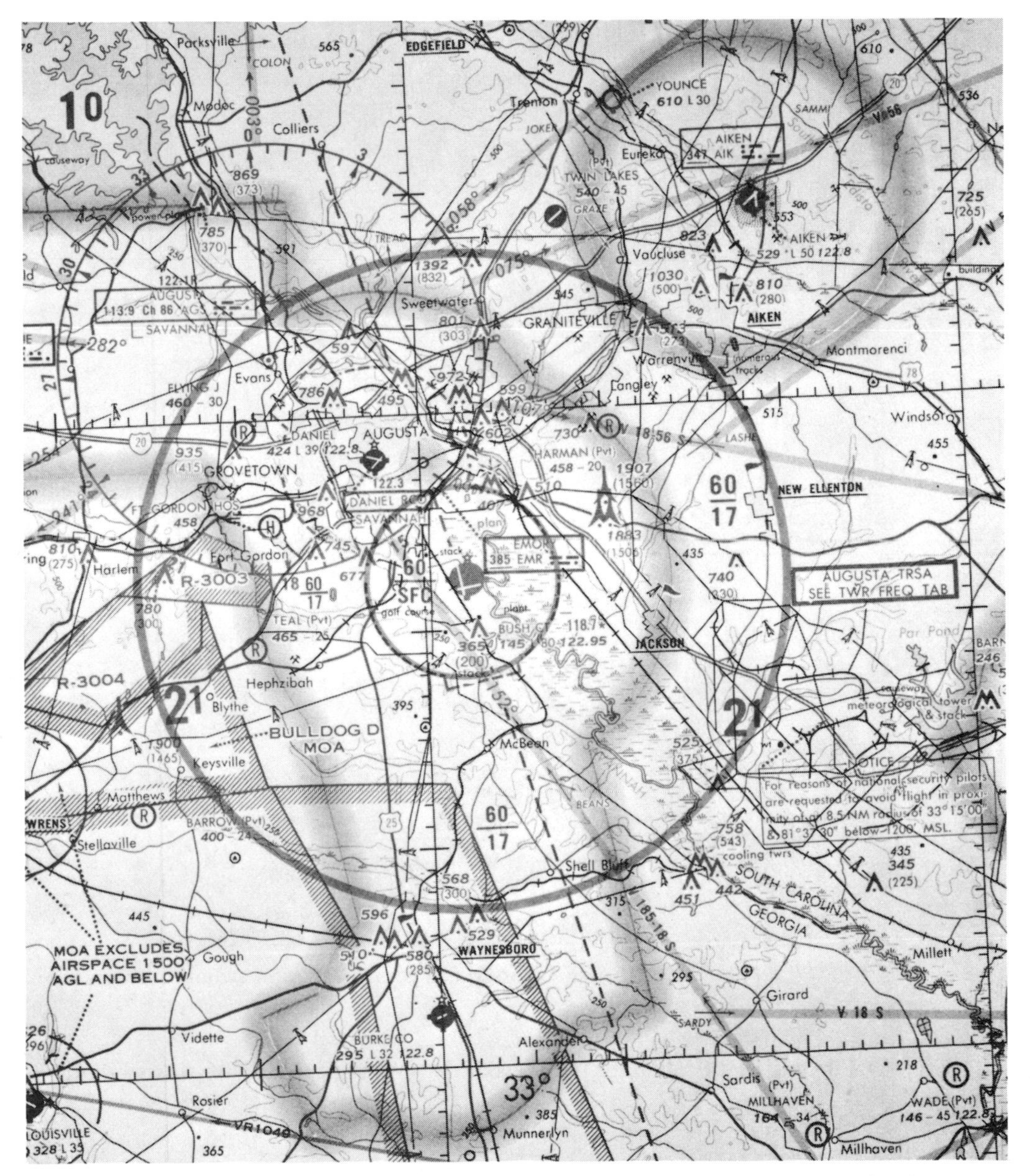


*Fig 5-13.* The Augusta TRSA. (Source: Federal Aviation Administration.)

## 5-27. THE U.S. AIRWAYS SYSTEM

Flights from one point in the United States to another are normally channeled along navigational routes that are as well identified as the surface highway system. Two separate route systems can be identified: (1) VOR and L/MF airways and (2) the jet route system. VOR airways are approximately 8 mi

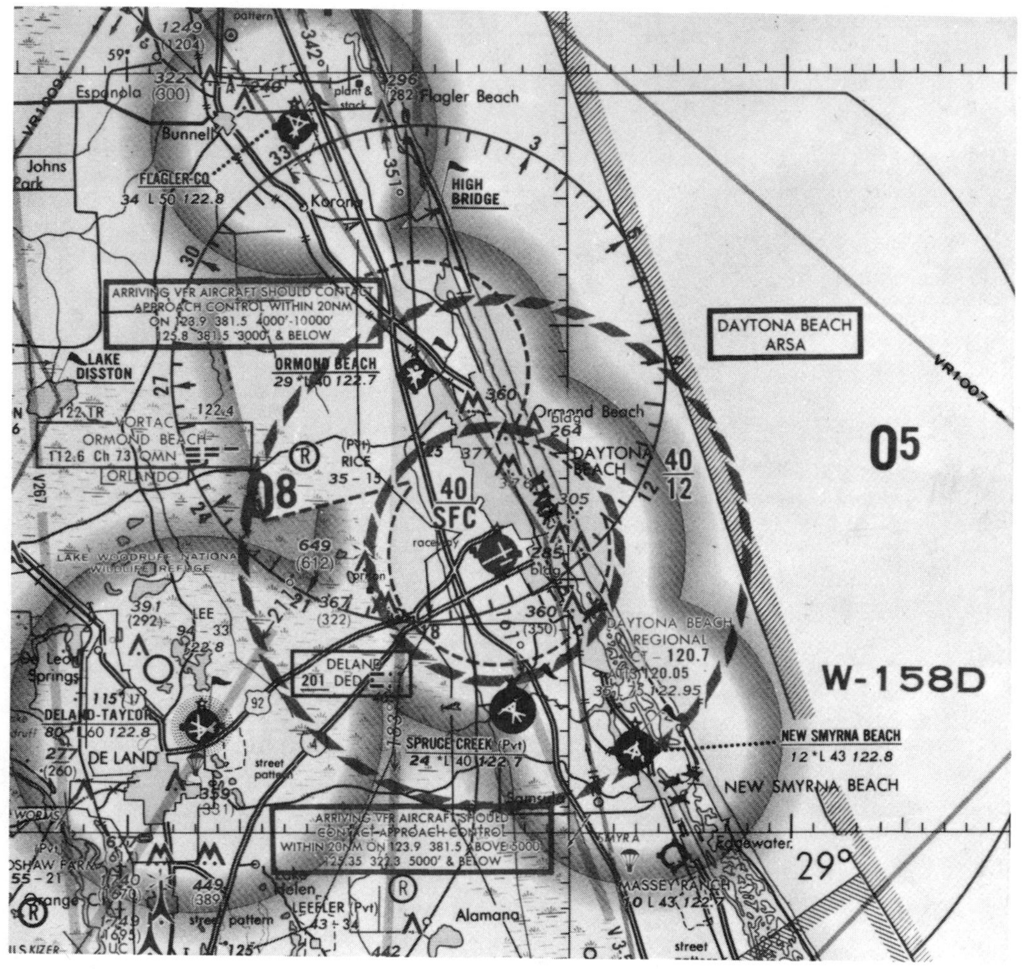


*Fig 5-14.* The Daytona Beach ARSA. (Source: Federal Aviation Administration.)

wide and between designated height bands. Jet routes are of unspecified width.

## 5-28. NAVIGATIONAL AIDS

The continued rapid growth in air traffic and the increasing congestion in airspace causes a continual reexamination of standards and procedures in air traffic control. Navigation technology is seen as a major possible contributor to the capacity of the already crowded airspace. Navigation aids and technology may be classified conveniently into en route navigation and terminal navigation. They are described as follows:

VFR ALTITUDES/FLIGHT LEVELS—CONTROLLED AND UNCONTROLLED AIRSPACE

Under visual flight rules (VFR) at
3,000 feet or more above the surface.

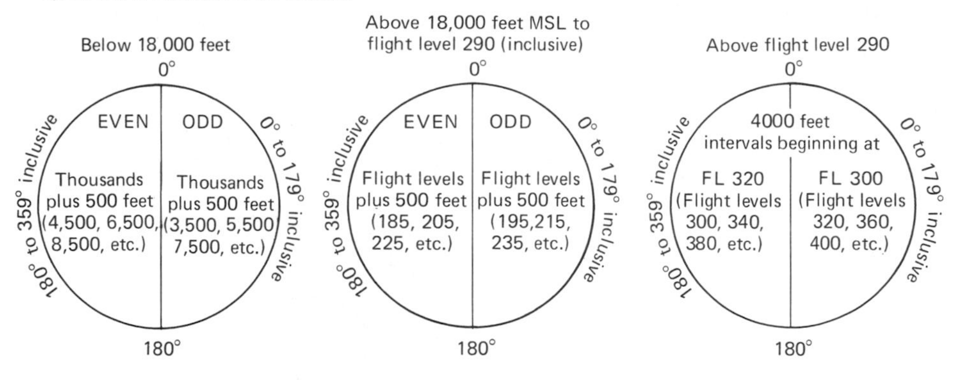


IFR ALTITUDES/FLIGHT LEVELS—UNCONTROLLED AIRSPACE

Outside controlled airspace.

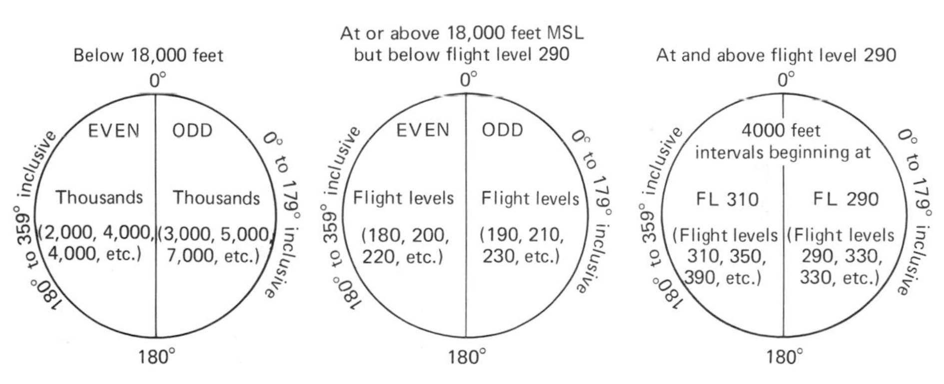


*Fig 5-15.* IFR and VRF altitudes/flight levels controlled and uncontrolled airspace. (Source: *Airman's Information Manual.*)

**En route navigation aids**

*Automatic direction finding (ADF),* sometimes known as nondirectional radio beacon (NDB), is a general-purpose low-frequency beacon or radio station on which aircraft can home. Unsophisticated and inexpensive, it is useful for long range (200 mi) but is subject to atmospheric noise and communications interference.

*Very-high-frequency omnidirectional range (VOR)* and the more modern development Doppler VOR (DVOR) is a very-high-frequency day-night, all-weather, static-free radio transmitter operating within the 108.0–117.95

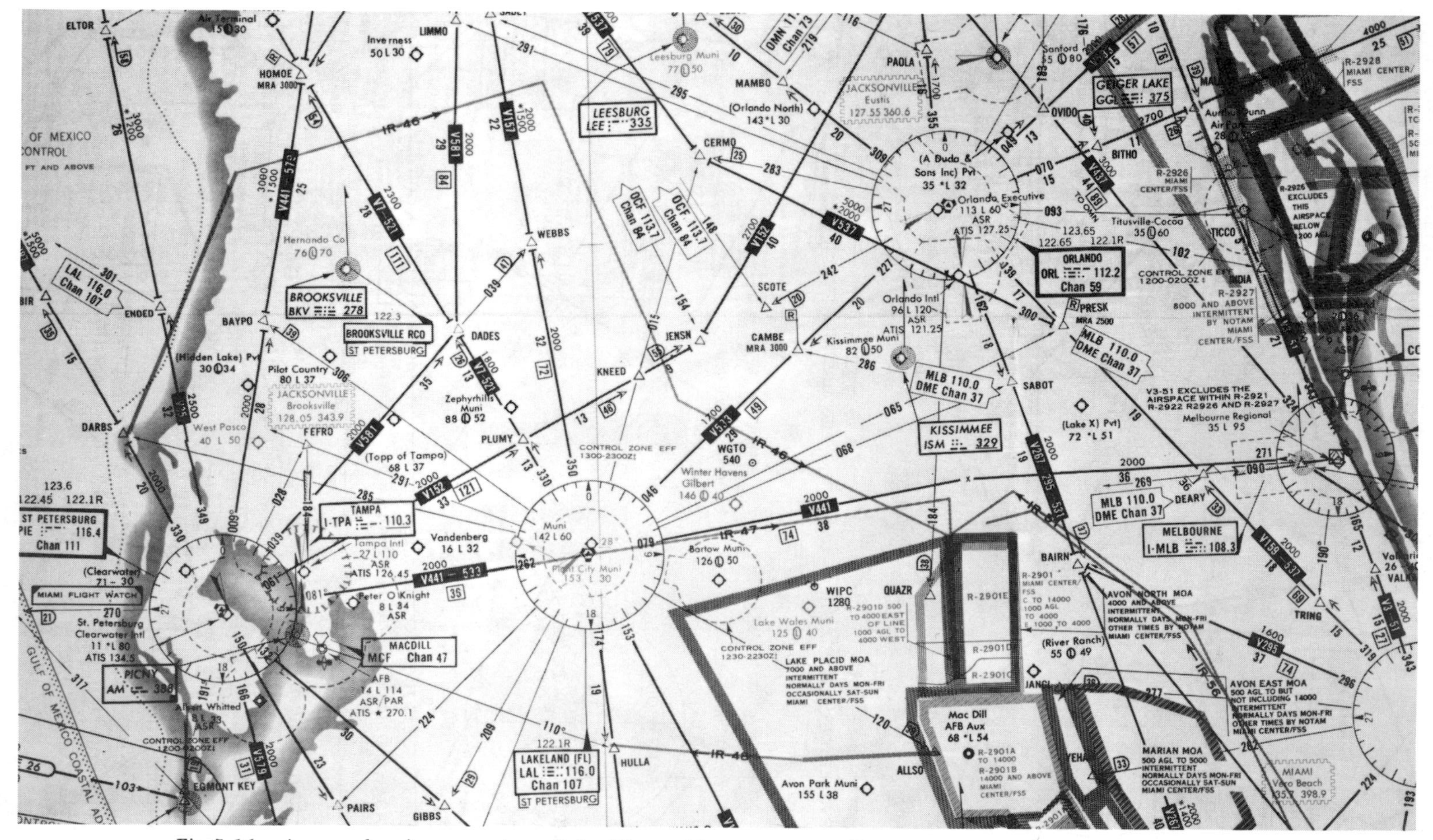


*Fig 5-16.* A map showing a section of the Victor Airways system. (Source: Federal Aviation Administration.)

MHz frequency band with a power output matched to the service area. Because units are limited to line-of-sight reception, the range is dependent upon aircraft altitude. High-altitude aircraft can suffer mutual VOR interference (multiple reception of facilities with like frequency assignments) because of the greatly increased horizon of the aircraft. The VORs form the basis of the present system of airways called "Victor Airways," a section of which is shown in Fig. 5-16. The numbering system for Victor Airways is even numbers for east and west and odd numbers for north and south.

*Distance measuring equipment (DME)* measures the slant range (maximum of 200 mi) to the DME facility located at the VOR site, and is subject to the same performance criteria as the VOR.

*Tactical air navigation (TACAN) and VHF omnidirectional range/tactical air navigation (VORTAC).* These navigational aids represent the incorporation of VOR and DME functions into a single channelized system utilizing frequencies in the ultrahigh-frequency range. Although the technical principles of operation of TACAN are quite different from those of VOR-DME, from the pilot's viewpoint, the outputs or information received are similar. Operating in conjunction with fixed or mobile ground transmitting equipment, the airborne unit translates a UHF pulse into a visual presentation of both azimuth and distance information. TACAN is independent of conventional VOR facilities but is constrained similarly to line-of-sight operation.

VORTAC is a combined facility composed of two different components, VOR and TACAN. It has a triple output: VOR azimuth, TACAN azimuth, and range. Although it consists of more than one component, operating at more than one frequency, VORTAC is considered to be an integrated navigational unit providing three simultaneous information outputs. The jet route high-altitude airways have been created for use with VORTAC stations separated by long distances.

*Marker beacons* are fan markers, small radio transmitters that send coded identification signals to identify a specific location on an airway or a given course using a 75-MHz directional signal to overflying aircraft. Useful only at low altitudes they are employed primarily in instrument approaches or departure procedures as holding fixes or position reporting points.

*Communications* are accomplished by radio receivers and transmitters located both in the aircraft and on the ground. Civilian aircraft primarily use VHF (very-high-frequency) radio bands; military aircraft, especially jet aircraft, utilize UHF (ultra high-frequency) radio bands. Air-ground communications are necessary to enable pilots to receive flight instructions as they progress along the airway or to their destinations if not on a flight plan, to obtain reports of weather ahead, and to alter their flight planning as required.

*Air route surveillance radar* is a long-range radar for tracking en route aircraft along an airway. These radars with a range of 200 mi are installed nationwide. They provide the controller in the air traffic control center with accurate information on the azimuth and distance position of each aircraft along the airway and thus reduce the frequency of communication between the pilot and the controller. Prior to radar, the controller had to rely upon the pilot's accuracy in reporting his position. The long-range radars permit the separation between aircraft flying at the same altitude to be reduced, thus increasing the capacity of the airways.

*Air traffic control radar beacon system* (ATCRBS) is a system with three main components: *interrogator, transponder,* and *radarscope.* Frequently the system is known as *secondary surveillance radar* (SSR). Whereas primary radar relies on the passive bouncing back of the transmitted radar signal, SSR is an active system in which the interrogation transmits, in synchrony with primary radar, discrete radio signals requesting all transponders on that mode to reply. The airborne transponder replies with a specific coded signal that is much stronger than the primary radar return. The radarscope displays the targets, differentiating between coded aircraft and ordinary radar primary targets, as shown in Fig. 5-17. Radarscopes are equipped to indicate aircraft identification and altitude on an alphanumeric display. The controller can thus differentiate between aircraft rapidly and with certainty. SSR is being refined further into a system known as discrete address beacon system (DABS) which provides for the semiautomatic exchange of information between the flight deck and air traffic control.

*LORAN C* determines aircraft positions using a velocity input as well as magnetic bearing from the LORAN station. The system uses a hyperbolic ground wave of low frequency for long-range utilization (approximately 2000 mi) and is extremely useful for over-water flights with position checking being accomplished by using cross bearings on LORAN stations. It is also used for ship navigation.

*Omega,* companion to LORAN C, is a network of eight transmitting stations located throughout the world to provide worldwide signal coverage. Because these stations transmit in the very-low-frequency (VLF) band, the signals have a range of thousands of miles. The Omega navigation network is capable of providing consistent fixing information to an accuracy of $\pm 2$ nm.

*Inertial navigation systems* (INS). Large modern air transports are fitted with INS which computes latitude and longitude from an onboard inertial device, which needs no ground equipment. This rapid navigational aid is especially useful for very long-range flights and long transoceanic sectors.

**Terminal area navigation and landing aids**

*Instrument landing system (ILS)* is an adaptation of the VOR and it is the most widely used system for instrument landings. It consists of two radio transmitters: the localizer transmitter situated at the end of the runway, and the glide slope transmitter situated near the side of the runway. The information in the cockpit can be incorporated with the VOR instrument display. In order to help a pilot further on an ILS approach, two low-power fan markers called ILS markers are installed so that the pilot may know how far along the approach to the runway he has progressed. The first is called the outer marker (LOM) and is located approximately 4.5 mi from the end of the runway, and the other, the middle marker (LMM), is located about two-thirds of a mile from the end of the runway. The ILS system is diagramed in Fig. 5-18.

**AIR NAVIGATION RADIO AIDS**

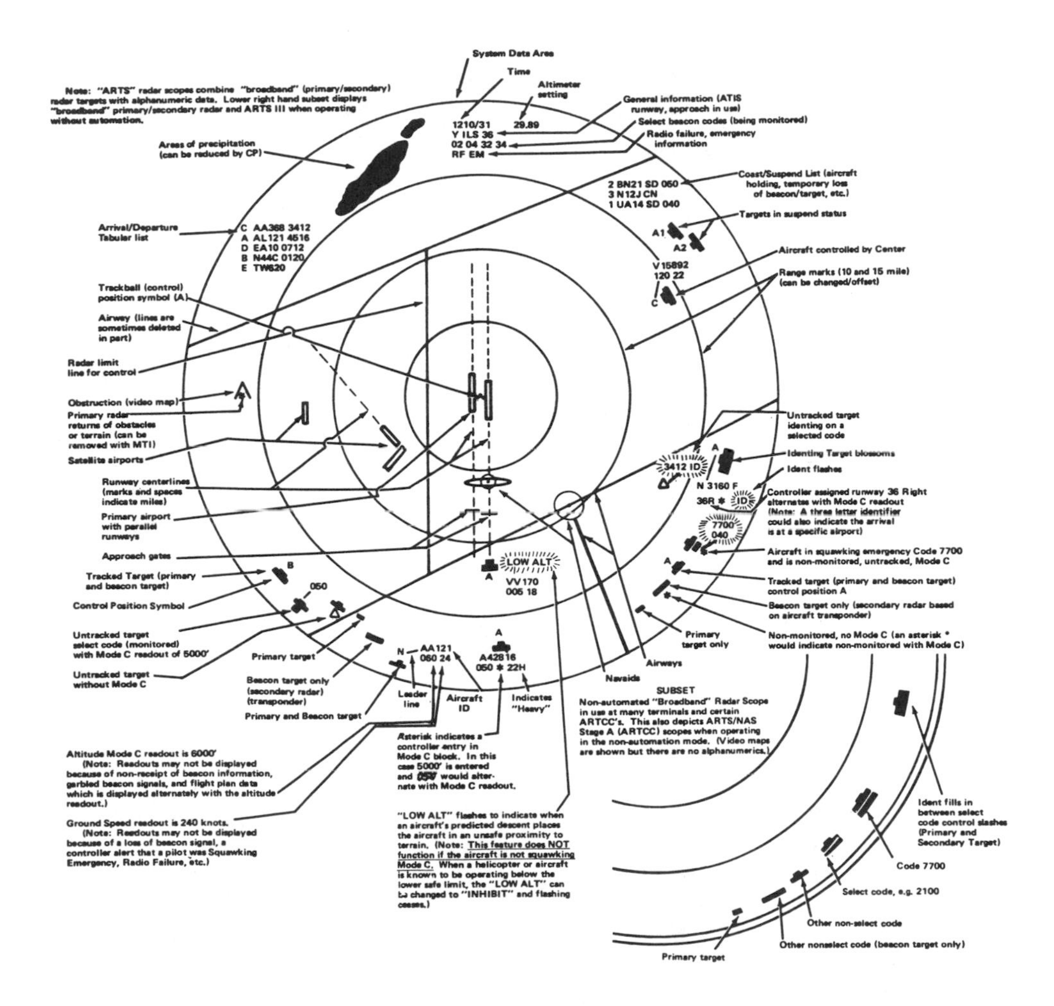


ARTS III Radar Scope with Alphanumeric Data. Note: A number of radar terminals do not have ARTS equipment. Those facilities and certain ARTCC's outside the contiguous US would have radar displays similar to the lower right hand subset. ARTS facilities and NAS Stage A ARTCC's, when operating in the non-automation mode would also have similar displays and certain services based on automation may not be available.

*Fig 5-17.* Controller's secondary surveillance radarscope display. (Source: Reference 12.)

*Microwave landing systems* (MLS) are under development and may be introduced in the 1990s. Until that time the use of ILS is safeguarded. Whereas ILS is affected by terrain and the presence of buildings and vehicles near the antennae, MLS is not subject to interference and can be installed using much smaller transmitting antennae. The system can provide not only direction but distance information and is ideal for providing continuous infor-

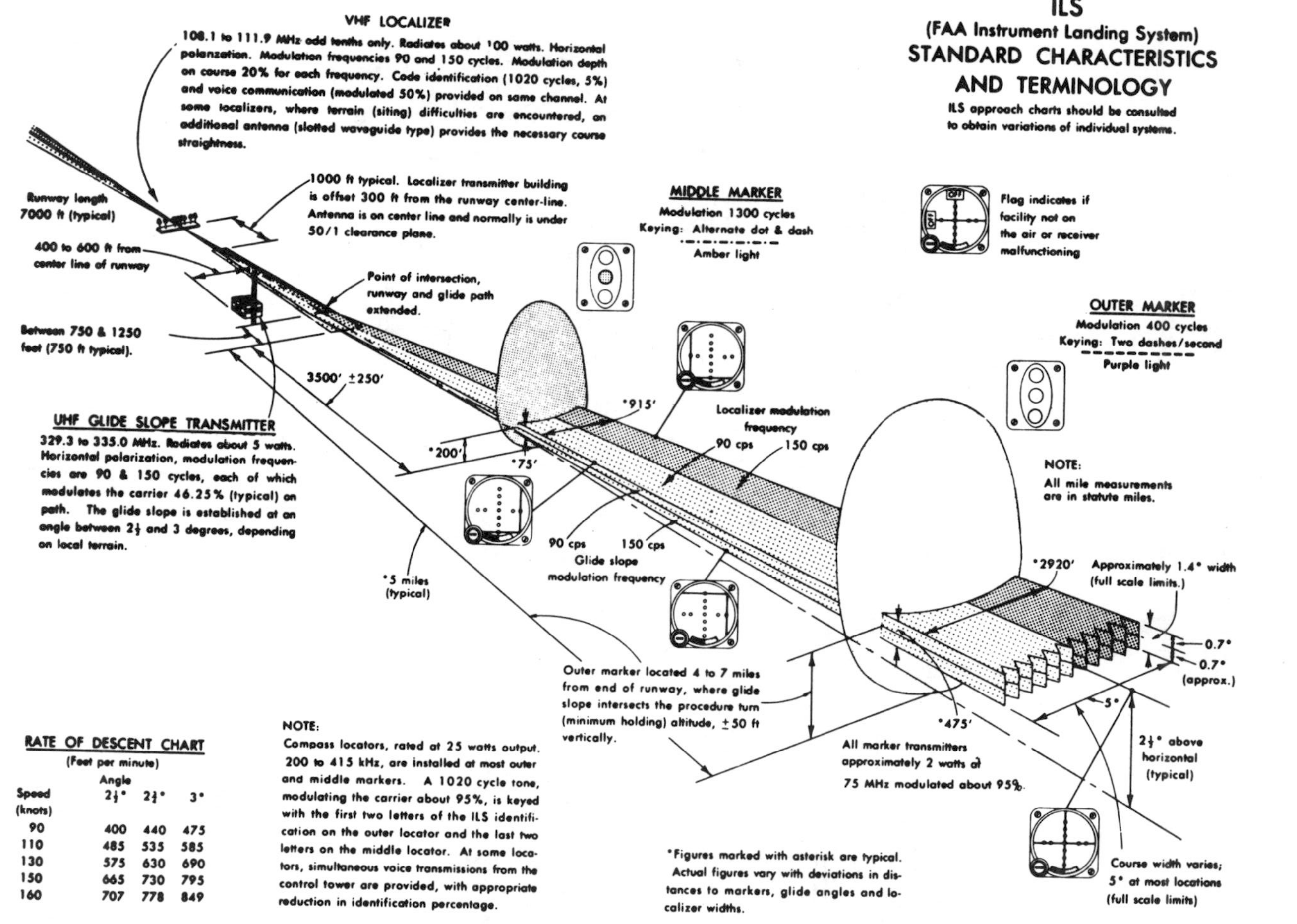


*Fig 5-18.* FAA instrument landing system. (Source: Federal Aviation Administration.)

mation needed for automatic "hands-off" landings. Curved approach paths are possible and even multipath approaches to the same threshold.

*Precision approach radar (PAR),* formerly called ground controlled approach (GCA), is not dependent on airborne navigation equipment, as the radar is located adjacent to the runway. The PAR radarscope gives the controller a picture of the descending aircraft in azimuth, distance, and elevation to enable determination of the aircraft's correct alignment and glide slope. Instructions for corrections are given by voice communications. At airports where both ILS and PAR are available, they are both used simultaneously with one being the backup for the other.

*Airport surveillance radar (ASR)* provides for terminal area air traffic control and aircraft location information to airport tower operators. For example, it will provide information for an aircraft transiting from an airway to a holding pattern to an ILS approach. The range of the ASR is 30 to 60 mi. It does not indicate the altitude of the aircraft.

*Approach lighting systems (ALS)* aid pilots in making the transition from instrument to visual conditions. Lights are installed on and in the runways as well as on the approaches to the runways. Various approach lighting systems are described in Chapter 19.

*Airport surface detection equipment (ASDE)* is a specially designed radar system for use at large, high-density airports where controllers have difficulty in regulating taxiing aircraft because they cannot see the aircraft in poor visibility conditions.

*Instrument approach procedures, standard terminal arrival routes, and standard instrument departures*—though these are not navigational aids, they provide the means for utilizing the en route navigational aids as well as the terminal area navigational aids. They are not only indispensable for IFR landing approaches but are also helpful to the VFR pilot landing at an unfamiliar airport. Instrument approach charts diagram every airport in the country where some kind of instrument landing aid is installed, that is, ADF, VOR, DME, TACAN, VORTAC, PAR/ASR, ILS, and so on. They depict prescribed instrument approach procedures from a distance of about 25 mi from the airport and present all related data such as airport elevation, obstructions, navigational aid locations, procedural turns, and so on. Each recommended procedure—and every airport has several—is designed for use with a specific type of electronic navigation aid. The pilot makes his choice depending upon instrumentation and prevailing weather conditions. To aid pilots on takeoff, *standard instrument departures* (SID) have been developed to facilitate the transition between takeoff and en route operation, thus alleviating the need for a great amount of oral communication between controllers and pilots.

## 5-29. AIR TRAFFIC CONTROL FACILITIES

Air traffic control facilities provide the basis for communication with aircraft and the relay and clearance of flight plans for air traffic. There are three basic types of manned facilities: the air route traffic control center, the airport traffic control tower, and the flight service station.

**Air route traffic control centers (ARTCC).** There are 22 domestic air route traffic control centers that control the movement of aircraft along the airways. Each center has control of a definite geographical area and is concerned primarily with the control of aircraft operating under instrument flight rules (IFR). At the boundary points marking the limits of the control area of the center, the aircraft is released either to an adjacent center or to an airport control tower. At present much of the aircraft separation is maintained by radar. With radar, off-airways vectors can be utilized maintaining positive control of the aircraft, and thus the ARTCCs can accommodate more aircraft.

Each ARTCC is broken down into sectors in order to increase the efficiency of personnel in the center. Sectors are smaller geographic areas and air traffic is monitored in each sector by remote radar units at the geographic location. It can be observed that an aircraft flight plan is transferred between sectors within an air route traffic control center and between air route traffic control centers when crossing the ARTCC boundaries.

**Central flow control facility (CFCF).** Formerly called the National Flow Control Center, the Central Flow Control Facility is established at the FAA Headquarters Office in Washington D.C. to maintain overall flow control of the airways and to coordinate the work of the ARTCCs in cases of disaster or severe weather conditions. Access to the CFCC in Washington is by telephone. Flow rates are set for each airport and aircraft are subject to a "wheels-up" or control departure time to ensure that the majority of delays in the system due to flow overload are on the ground before departure rather than in the air on arrival. This procedure reduces the inefficient practice of large numbers of stacked aircraft waiting to land.

**Airport traffic control tower.** Airport traffic control towers are the facilities that supervise, direct, and monitor the traffic within the airport area. The control tower provides a traffic control function for aircraft arriving at or departing from an airport for a 5- to 15-mi radius.

Some control towers have approach control facilities and associated airport surveillance radar (ASR) which guide aircraft to the airport from a number of specific positions called fixes within approximately 25 mi of the airport. The aircraft are brought to these positions by the air route traffic control centers (ARTCCs). It is often at these fixes that aircraft are held or "stacked" for landing during periods of heavy air traffic. The control towers without approach control facilities differ in that, under IFR conditions, the clearing of waiting aircraft for landing is done by the ARTCC and they are turned over to the control tower after they have started their landing approach.

**Flight service stations (FSS).** The flight service stations (FSSs), which are increasingly totally automated, are located along the airways and at airports. Their function may be described as follows:

1. Relay traffic control messages between en route aircraft and the air route traffic control centers.
2. Brief pilots, before flight and in flight, on weather, navigational aids, airports that are out of commission, and changes in procedures and new facilities.
3. Disseminate weather information.
4. Monitor navigational aids.

## PROBLEMS

**1.** Two urban four-lane undivided streets are to be designed to intersect at a rotary intersection. Draw a schematic intersection layout showing lane markings, channelization, and necessary signs and signals. Write a short report justifying the design.

**2.** The United States does not use international traffic sign designs. Present arguments both for and against U.S. adoption of these standards.

**3.** Under what conditions of traffic would you use:

a. traffic signals

b. stop signs

c. yield signs

(Refer to the *Traffic and Transportation Engineering Handbook* for the preparation of this answer).

**4.** It has now been agreed that microwave landing systems will be likely introduced widely in the 1990s. What will be the principal effects of their introduction?

**5.** Discuss the ramifications of the introduction of totally automatic railway locomotive control. In your answer deal with social problems (such as accident rates, reliability, and labor), economics, and technological feasibility.

**6.** What developments in marine navigation are foreseeable in the next 50 years?

**7.** Conduct a field survey of a half-mile length of a major urban street. Record what traffic control devices you find in the course of this inventory survey.

## REFERENCES

1. *Transportation and Traffic Engineering Handbook,* Second Edition, W.S. Homburger (ed.), Institute of Transportation Engineers, Prentice-Hall, Englewood Cliffs, N.J., 1982.
2. *Manual on Uniform Traffic Control Devices for Streets and Highways,* U.S. Department of Transportation, Federal Highway Administration, Washington, D.C., 1978 [as updated].
3. *A Policy on Geometric Design of Highways and Streets,* American Association of State Highway and Transportation Officials, Washington, D.C., 1984.
4. Parsonson, P.S., and J.M. Thomas, "A Case Study of the Effectiveness of a Traffic Responsive Computerized Traffic Control System," *Control in Transportation Systems,* Proceedings of the Third International Symposium of IFAC/IFIP/IFORS, International Federation of Automatic Control, Pittsburgh, 1976.
5. *Coast Guard Aids to Navigation,* CG-193, U.S. Department of Transportation, United States Coast Guard, Washington, D.C., 1987.
6. Quinn, A. de F., *Design and Construction of Ports and Marine Structures,* Second Edition, McGraw-Hill, New York, 1972.
7. Wylie, F.J., *The Use of Radar at Sea,* Elsevier, New York, 1968.
8. *Car and Locomotive Cyclopaedia of American Practices,* Third Edition, Simmons Boardman Publishing Corporation, New York, 1974.
9. *Modern Railways,* January 1976.
10. *Railway Gazette International,* November 1970.
11. Ashford, N., and P.H. Wright, *Airport Engineering,* Second Edition, Wiley Interscience, New York, 1984.
12. *Airman's Information Manual,* Federal Aviation Administration, 1987.

# Chapter 6

# Human Factors in Transportation

## 6-1. HUMAN FACTORS

Human factors is a term used to describe the study of man interacting with manmade objects and processes, and with the natural or manmade environment. It is also called ergonomics (Greek *ergon*-work; *nomos*-law) or human engineering. Human factors provides a key to the rational design of goods and services by providing a basic model of people in their working environments, and a set of measurement and evaluation methods appropriate to this model.

The basic model is shown in Fig. 6-1. It has four main components.

- man
- machine
- workspace
- environment

The term *machine* should be interpreted broadly as any object or process that the man uses in order to complete a task. Thus a machine may be the whole motor vehicle or just the components, an information display at an airport, a computer terminal, or a wrench for use on rail bolts. The model asserts that man and machine can interact in the completion of the task; thus the man can both receive information from the machine via displays, and can manipulate the operation of the machine by the use of controls. In many cases both these things can occur. For example, the driver of a vehicle gathers information from his or her view of the highway, from the instruments in the vehicle, from the sound it makes, and from the sensations of vehicle motion and vibration. In response to this the driver processes and integrates the information and uses the vehicle controls to vary the vehicle's position, velocity, and acceleration. In this example, man–machine interaction is continuous and two-directional as shown in Fig. 6-2. In other cases, however, this interaction may be neither continuous nor two-directional. The train traveler looking at a timetable will receive from it visual information about train times, will

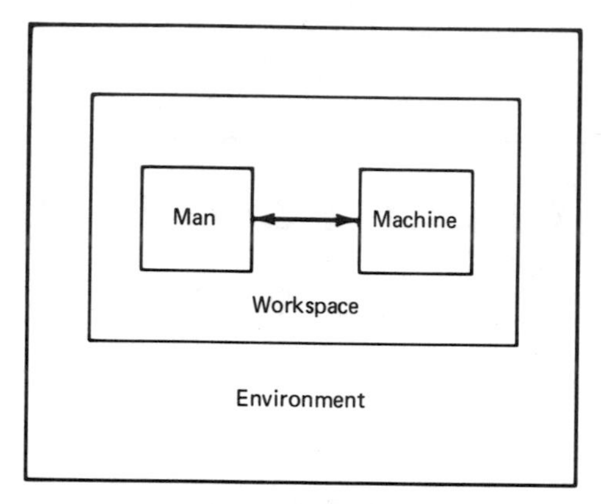


*Fig 6-1.* Basic model of man and his working environment.

process this information, and will take some action (such as buying a ticket) but this action will not affect the status of the timetable itself.

*Workspace.* The model further recognizes that man and machine will function within some workspace, and asserts that the form of that workspace will affect the efficiency of the man–machine interaction. The workspace usually is defined in terms of the physical dimensions of the individual's working position, in relation to the size of the body or of its components. There is ample evidence to show that the design of the workspace affects a person's performance; clearly, if the operator of a machine is unable to reach one of its controls because it is too far away, performance is impaired. If the control can be reached only with strain and effort, performance may be impaired, though to a lesser degree. In jobs that are dangerous or demanding, such as piloting an aircraft, the consequences of poor workspace design may be

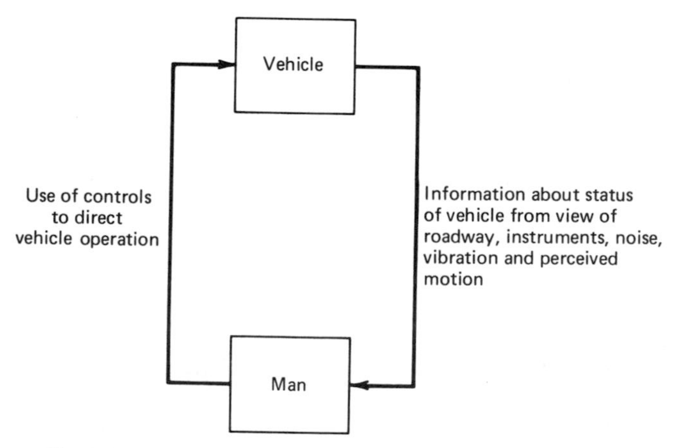


*Fig 6-2.* Man-machine interaction in driving a vehicle.

serious, expressing themselves as accidents or critical incidents. In less demanding jobs, poor workspace design may be less critical, although the worker, in adapting to this design, it likely to incur costs such as increased probability of errors and accidents, injuries and strains from bad working posture, and the loss of spare mental capacity to deal with emergencies. For example, anyone who has driven for some hours in an uncomfortable automobile seat will appreciate the importance of workspace design in the reduction of strain on the muscles in the back and shoulders.

*Environment.* The final aspect of the model is that the man–machine–workspace combination is located in some environment that affects the individual's performance. Traditionally, the working environment has been conceived and measured in terms of physical features: illumination, sound, vibration, and climate, as well as biological, chemical, and radiological pollutants. An airline ticketing clerk using a keyboard will need certain standards of illumination to locate the keys; high levels of vibration may impair the ability of a vehicle operator to make fine visual discriminations of an instrument; if a worker must communicate verbally with a colleague, the level and spectral characteristics of any background noise will affect the quality and comprehensibility of message reception; and the existence of pollutants in a transportation system may have short-term as well as long-term effects on the health and safety of operators and riders.

More recently, however, these well-established environmental factors have been supplemented (though in no sense replaced) by an additional group, summarized by the terms *psychological environment* or *psychosocial environment.* Human factors specialists increasingly have recognized that the psychological consequences of a task significantly affect performance at that task [1]. As an example, the organizational structure of a company or institution will affect the motivation, performance, and job satisfaction of the individual worker, as will the formal or informal contacts the worker has with colleagues. While in principle productivity may be maximized by giving each worker a single, simple task, it is becoming widely recognized that the effects of this on job satisfaction in practice may result in a lower level of productivity. Again, the psychological consequences of shiftwork, such as the stress and strain it can impose on the individual and on his family and social life, are receiving intensive study. In an age where productivity is the key to industrial and commercial success and where the humane design of jobs is recognized both as a vital component of productivity and as an end in itself, the psychosocial environment of a job cannot be overlooked.

## 6-2. MAN–MACHINE–ENVIRONMENT INTERACTION

At a number of points in the preceding discussion, the interaction between components of the human factors model has been presented. It is worthwhile to emphasize this interaction because this characterizes all human factors work. Although the four-component model contains six two-way interactions, we are mainly interested in only three: the interactions between man and machine, workspace, and environment, respectively. The nature of these interactions and their relationship to the physical, physiological, psychological, and social characteristics of the individual, will affect the volume and quality of the individual's output. It is the task of the designer to analyze, or synthesize, the interactions between man and machine, workspace, and environment in order to ensure that the fit between them is optimized both in terms of human characteristics and in terms of the output from the task carried out. Therefore the criteria by which human factors solutions must be assessed will include; system efficiency; personal well-being in both the short and long term, expressed in both physiological and psychological terms; and safety in emergencies and abuse of safety regulations. The last is of particular concern in transport systems, where the responsibility of carriers to their passengers, the wide range of ability and skill of these users, and the consequences of system failure mean that the relaxation of safety standards may have serious effects [1].

## 6-3. HUMAN VARIATION

If products and processes were individually tailored, that is, if they were designed for use by just one specified person, then human factors design would be relatively easy. However, this rarely happens. In the vast majority of cases the design must accommodate a large number of people. A seat on a train must allow for the range in body size of its users; information presented via the instruments of a vehicle must be comprehensible to those using it, accounting for their range of ability to process visual information; policy may dictate that a public bus system be accessible to those potential users who are usually classified as disabled. It is an obvious, but vital fact that human beings vary in every human characteristic, whether this is a description of the composition of the person (e.g., body weight or height) or some functional characteristic (e.g., mobility or visual acuity).

Any piece of equipment imposes some level of handicap on its users; the important thing is the point at which that handicap prevents use of the equipment. Mild levels of handicap result in inefficiency, inconvenience, and discomfort. Equipment that imposes a high level of handicap will admit the able-bodied and exclude the rest, while low-handicap equipment excludes only the severely disabled. Overall, it must be recognized that design is selective and may exclude just those people whose needs are greatest [2]. It must be remembered also that handicap may be due not only to permanent physical or mental disability, but also to temporary factors unrelated to human degeneration, such as the carrying of baggage, minor injury, pregnancy, and so on.

To maximize the application of a facility it is clearly of value to accommodate all of the population at which it is aimed. In other words, it must impose a low handicap on its users. A public bus system will be of more universal benefit if it admits all those from the very fit to the severely disabled. Such a system may well be more expensive in terms of both capital and operating costs, than one with narrower application. Providing special facilities for wheelchair-bound travelers may well involve high development and hardware resources and may reduce passenger capacity per vehicle. Designers recognize that the extremes of the distribution of human characteristics may present major problems to even the most innovative engineer. Conventionally, design solutions are attempted to accommodate only 90 percent of the whole population ranging from the fifth centile of the distribution to the ninety-fifth, see Fig. 6-3. This approach is particularly important since many human characteristics are known to be distributed normally. Extreme values occur, but with a very low frequency. It is known, for example, that the stature of a car driver is correlated to the seat position chosen. Whereas it is relatively simple to design a seat adjustment system that accommodates statures from the fifth to the ninety-fifth centile it would be significantly more difficult to include the whole adult range and to integrate this into the design of the rest of the vehicle. Those who fall below the fifth centile or above the ninety-fifth centile in any human characteristic are faced with three options: to press for specialized designs (such as the provision of special transport facilities for the severely disabled); to be excluded altogether from the facility or facilities being offered; or to attempt to use these facilities with varying degrees of handicap, discomfort, or inconvenience [3]. Clearly the designer would prefer the first option, although resource limitations sometimes prohibit this.

It is necessary to point out that although some human characteristics (such as stature or weight) are quantified readily and thus are directly amenable to the design approach already discussed, there are others with less reliable indices (such as mobility). In these cases the specification of centile distribution is difficult in quantitative terms, but the general policy and approach still can be similar.

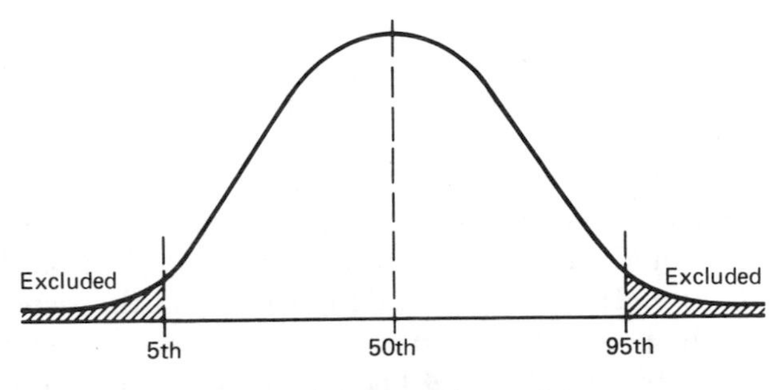


*Fig 6-3.* Range of values conventionally accommodated in human factors solutions.

## 6-4. INSTRUCTION AND ADAPTATION

Accepting that human characteristics vary, the designer is still left with a dilemma: "Fitting the individual to the job" or "fitting the job to the individual." It was perhaps this problem that led originally to the development of human factors; in recognizing that machines should be the servant of man and not the master, human factors argues that using a system inconsistent with his capacities and needs will be less efficient and may suffer inconvenience and a risk to personal safety. The costs of requiring a person to adapt to poor design may be greater than those incurred in developing a system that meets an individual's needs and accounts for his limitations. Yet is it clear that man is a highly adaptable creature and within limits can accommodate design deficiencies. Furthermore the effects of this adaptability may be enhanced by training and education in the use of the product or process [4]. By harnessing this adaptation, training can reduce the likelihood of an individual using inefficient task strategies and can guide a person in the actions most likely to cope with system failure. Systems, however, are constructed in the light of financial constraints as well as those of personal efficiency and well-being. The cost of developing a system with a transparently obvious function and method of operation may be so great that a more practical solution may be to provide a series of instructions (i.e., a short period of training) with a slightly more obscure but cheaper system. For example, the handbook for an automobile gives its owner information about the location and function of controls and display and the procedures for simple maintenance, and the demonstration of safety procedures in an aircraft alerts the passenger to actions that may prove to be necessary in an emergency.

These considerations may be illustrated schematically. Figure 6-4 shows that an individual's performance in relation to some product or process may be maintained at an acceptable level despite the fact that the design solution is not optimal for that person. This can be done partly because of man's adaptability and partly due to training or instruction. However, there will be a limit to this adaptation, and beyond a certain point the individual will not be able to adapt sufficiently to accommodate a large departure from the optimal solution. At this point efficiency will decrease, the individual will be more likely to make mistakes, and will suffer an unacceptable level in inconvenience in carrying out the task. In this context two things are apparent. First, the application of human factors principles to groups of people would be impossible unless the individual were adaptable; without this capacity, every person would need to have everything (both material and nonmaterial) specially designed. There would be no "graceful degradation" of performance as illustrated in Fig. 6-4. Human factors design requires the determination of both the shape of the performance curve and also the acceptable performance level. Second, the abscissa in Fig. 6-4 may be quantified in a number of ways and may be categorical rather than continuous. The departure from an optimal solution might be measured in the following ways.

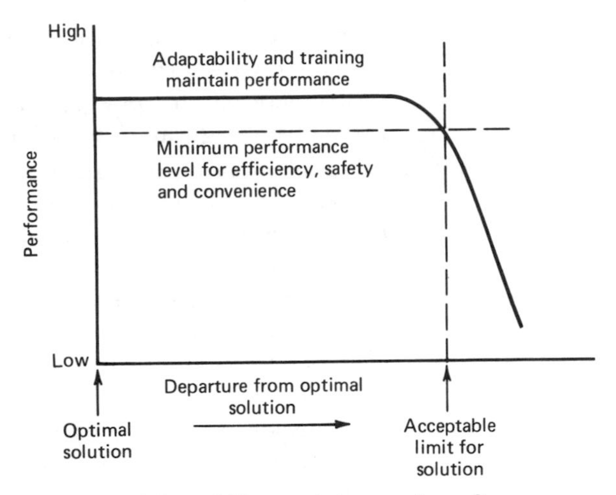


*Fig 6-4.* Adaptability, training, and performance.

1. A physical dimension—for example, the angle of a pedestrian ramp from the horizontal (the corresponding performance measure being the ability of people to negotiate it).
2. An operational procedure—for example, the ordered sequence of preflight checks an aircraft pilot makes before takeoff (the corresponding performance measure being the time taken to make these checks, or the number of errors and omissions made).
3. A temporal dimension–for example, the number of hours a truck driver may work in one shift (the corresponding performance measure being the number of accidents or critical incidents occurring, or any other of a number of indices of driver behavior).

## 6-5. HUMAN FACTORS IN A TRANSPORT SYSTEM

A transport system is a manmade phenomenon, incorporating both material and abstract components, designed and constructed to meet certain human requirements. From a human factors viewpoint these requirements may be considered in two categories, primary and secondary.

**Primary human requirements.** The *primary human factors requirements* of a transportation system are those that relate to the extension to human abilities. These requirements are:

| | |
|---|---|
| *Increase of speed:* | by enabling people to travel faster than is possible unaided (e.g., an aircraft). |

*Increase of range:* by permitting more efficient use of an individual's own energy (e.g., a bicycle) or by supplementing this energy with a powered mechanism (e.g., an automobile).

*Increase of carrying capacity:* by supplementing human strength with machines to permit the carriage of heavy loads for long distances (e.g., an ocean freighter).

**Secondary human requirements.** In addition to the primary requirements of the transportation system there is a set of secondary requirements which both supplement and qualify the primary set. These include:

*Safety:* Most transportation systems place man in an environment that is more hostile than his natural condition. Designs must therefore take safety into account.

*Comfort and Convenience:* Discomfort and inconvenience affect human efficiency and can compromise safety. Design solutions must accommodate this secondary factor. The range of consequences are large. Some are life threatening (e.g., the incorrect positioning of a car seat belt), while others are only slightly inconvenient and may have nonserious implications on system safety (e.g., the design of the work station and keyboard for the airline ticketing clerk). It should be noted that there is no general agreement on the meaning of concepts of comfort and convenience. Measuring the achievement of these criteria is difficult.

*Status:* Historically, the requirements of status have exerted a powerful influence on our transportation systems. Aircraft, ships, and trains provide different levels of service to passengers of different status. Some modes such as the automobile are seen as having higher status than others, such as the urban bus. Design solutions must recognize that the requirements of high-status users may differ substantially from those of the mass market.

## 6-6. HUMAN CONTACT WITH A TRANSPORT SYSTEM

Because a transport system is manmade and dedicated to meeting human requirements, it may be analyzed in much the same way as any other manmade object or process and can be evaluated in human factors terms by the same criteria. Human contact with the system will occur at many levels and will be governed by three main categories of factors operating alone or simultaneously, [5].

1. *Physical, physiological, and biomechanical factors.* These are concerned with the structure and mechanical operation of the human body. Some examples follow, with illustrations of how they are relevant to transport systems and components.

- *Body size and body component size.* Any space into which humans are placed must be designed to accommodate them. Thus body size will be reflected in the dimensions of seats, working surfaces, passageways, and workspaces. Furthermore, reference to published anthropometric data will show that, although it is expedient and customary to speak of "the fifth centile woman" or "the ninety-fifth centile man," these terms refer often to stature only. It does not follow that a person of ninety-fifth centile stature will have body components at the same centile level; such a person may by chance have particularly short legs or a particularly large waist circumference. Therefore, anthropometric data should be used with care.
- *Reach capacity* (functional anthropometry). Knowing the length of a person's arm will not allow us to specify accurately how far away a control should be placed. Not only can the effective arm length be extended by movement of the trunk, but that length may also be reduced by virtue of the task specified; grasping a control with the whole hand such as gear lever) will result in a shorter effective arm length than a task that requires only finger manipulation (such as pressing a button on a vehicle dashboard). Furthermore, the human body is not equipped with universal joints, and therefore functional anthropometry is also concerned with the directions in which the limbs and trunk can move, and the consequences of this for reaching and for task-performing ability.
- *Strength and biomechanics.* The body's ability to perform physical work is limited by the capacity of the muscles and their supporting systems. Thus freight-handling systems employing manual labor must be designed with muscular capacity in mind. Since muscular fatigue is, in part, temporally based, further distinctions must be made between tasks that require only short periods of manual work, and those that require continuous work; the ability of a driver to exert a force of 300 lb on a brake pedal under emergency conditions could not be maintained if that force is to be sustained for, say, 60 sec.
- *Body composition.* The human body comprises both bone and soft tissue. The bones' functions are to protect certain vital organs, to support the body structure, and to bear loads. If equipment is designed without

consideration of the location and strength of body components, discomfort and injury may result. For example a lap and diagonal car seat belt is designed to pass over the bone formations of the pelvis, chest, and shoulder, which are capable of sustaining the forces generated by rapid deceleration. A seat belt that passed over vulnerable areas such as the throat and the soft part of the abdomen would be very dangerous.

2. *Perceptual-motor factors.* Man receives information about the world via his senses, processes this information, and responds by making some physical or verbal action. The following factor categories apply.

- *Perception and sensation.* The primary modes of sensation in man are vision and hearing. The other modes, such as taste, smell, and touch are undeniably important but are used less often in the design of man-machine systems, either because these modes are less accurate and less consistent in their operation, or because the equipment necessary to use them is too complex and expensive in relation to the cost of designing for the visual and auditory channels. A great deal of attention is paid to the content and form of sensory input devices, since these lie squarely at the interface between man and machine. If man is unable to sense information adequately, the viability of the man-machine system is threatened immediately. It is important to ensure that information presented is in such a form that it can be perceived easily, and in a way that causes minimal effects on other tasks that the man is performing simultaneously. Thus the design of visual instruments for automobiles will be based on their readability and it will be necessary to assess the time taken to read them, as this is time that the individual would otherwise use in viewing the road ahead. Similarly, the design of auditory warning signals in aircraft flight decks must take into account a person's ability to sense different levels of sound frequency and amplitude, and must be able to operate efficiently in environments in which a variety of other sounds of varying frequency, amplitude, and temporal pattern may occur.
- *Information processing and decision making.* Man cannot process unlimited amounts of information. Furthermore, it is argued that an individual operates as a single-channel device, meaning that he is unable to process more than one item of information at a time. This places significant limitations on the range and pattern of tasks that can be performed. For example, even though it is technically simple to provide a car driver with a display that gives information about the acceleration of the car (rather than its speed), such information is unlikely to be usable by most drivers because man is poor at processing second-order differentiated data. Again, requiring a driver in heavy city traffic to attend to the roadway, to the sight and sound of other vehicles, to notice and interpret road signs and other information, to monitor the per-

formance and condition of the vehicle, and to control its progress, will probably overload information processing capacity. Consequently, some of these tasks may not be performed, or errors will be made in their performance.

- *Motor performance.* Man controls the machines he uses by the physical action of parts of the body, or by speech communication. This control will be made with varying degrees of skill, depending on the ability of the individual, the amount and distribution of experience with the task, and the nature of the task itself. For biomechanical, anatomical, and neurological reasons certain parts of the body are better than others at performing certain tasks. Thus the lower limbs are suited to the application of large forces, such as the use of brake pedals, by virtue of the musculature of the legs and back. For the precise setting of controls, the upper limbs are suitable, because of their agility, and the generous supply of touch sensors in hands and fingers. (To some extent the limitations of the body can be compensated by the appropriate gearing of controls; for example, vehicles adapted for use by the disabled may assign control of the braking system entirely to the arms and hands.)

3. *Cognitive and social factors.* It is well known that an individual's performance is mediated by motivation, which in turn is determined by view of the task, perception of the organization or system, the nature and extent of contact with other people, and a range of factors describing personality and attitudinal structure. In addition there are an indefinite number of other factors of a perhaps more temporary nature that affect these main factors, for example, recent experiences, diurnal variation, fatigue, and stress. To some extent these cognitive and social factors have been systematized, and formal means have been developed to measure and control them.

It is clear that in complexities as great as transportation systems, the implications of human factors considerations are very extensive. The following sections give examples of human factors design in principal areas:

Pedestrians
The transit and rail passenger
The automobile driver
Air transport passengers

## PEDESTRIANS

### 6-7. PEDESTRIANS IN THE CITY

Pedestrian movement is essential to the design of cities and other areas of human activity. The design of pedestrian space suitable for circulation and other activities has only relatively recently been recognized as a necessary design function for the engineer. Early cities which had, at their center, virtually all areas of commerce and industry were largely built around the provision of pedestrian space. Modern cities of the nineteenth and twentieth century have had to adapt to the difficult problem of accommodating both pedestrian and heavy vehicular traffic; in the twentieth century that traffic has become motorized in the form of the automobile placing increasing demand on urban space. In many of those U.S. cities which have largely developed since the invention of the automobile it is not unusual for more than one-half of the inner urban area to be taken up by streets and parking lots. Since the 1960s however, designers have paid considerably more attention to the need to provide for pedestrian movement along the ordinary city street, within pedestrianized areas and in specially designed pedestrian malls.

## 6-8. LEVEL OF SERVICE DESIGN

Human space requirements are behaviorally based; the human being does not inertly occupy space in the same way as freight fills a warehouse or as parked cars are stored in parking lots. Human beings use space for various activities, and react to that space in a manner that reflects their perception of its suitability to their comfort. In work which examined how individuals behaved in juxtaposition to others, Hall defined four different kinds of distance [5]. *Public distance* exists when there is no other person within 12 ft (4 m). An individual has only limited sensory involvement with others and communication requires a degree of exaggeration (loudness of speech, gesticulation) when the distance is greater than 25 ft (8 m). At lower public distances, the voice remains stylized, there is still some emphasis placed on general stance and gesture; facial expressions can be clearly seen. Between 12 and 4 ft (4 – 1.3 m), individuals are said to be within *social distance.* At the upper end of this space individuals can be in definite communication by gesture and normal voice levels, and at the lower end physical contact is just possible. *Personal distance* exists from 4 to 2.5 ft (1.3 to 0.75 m). This space conforms to a "circle of trust" [6]. Senses other than sight, such as smell and touch come into play within this distance. At the lower end of personal distance, bodily contact can be avoided but only with difficulty. Within a radius of $1\frac{1}{2}$ ft (0.4 m), an individual perceives what is termed *intimate distance.* Normally reserved for only the most intimate purposes, involuntary bodily contact is difficult to avoid. An individual is aware of personal scents, heat and sounds, and bodily movements. In crowded areas individuals are thrust into unwanted proximity which requires special behavior codes to prevent unacceptable intimacy [5]. Fruin developed levels of service standards with respect to the provision of pedestrian space. These range from level A in which the individual is completely unconstrained by the presence of others to level F at which the system has recognizably broken down into congestion and overcrowding. The categorization of space is similar in concept to the level of service structure shown in Fig. 4-10 which relates to highway traffic flows. Table 6-1 sets out space standards which have been in use for some years for the design of walkways, stairways, and queue areas [6]. The standards for queueing areas are graphically represented in Fig. 6-5, and the relationship between level of service and the pedestrian flow graph is shown in Fig. 6-6.

## 6-9. WALKING SPEEDS AND STREET CROSSING BEHAVIOR

Studies of pedestrians crossing streets in urban areas have shown that the average adult moves at 4.5 ft/sec (1.4 m/sec) [7]. The full range of speeds

*Table 6-1*
**Design Standards for Pedestrian Space**

| | *Level of Service* | | | | | |
|---|---|---|---|---|---|---|
| | *A* | *B* | *C* | *D* | *E* | *F* |
| *Walkways* | | | | | | |
| Average space/person (ft²) | >35 | 35–25 | 25–15 | 15–10 | 10–5 | <5 |
| Average flow (PFM*) | <7 | 7–10 | 10–15 | 15–20 | 20–25 | Variable up to 25 |
| | *A* | *B* | *C* | *D* | *E* | *F* |
| *Stairways* | | | | | | |
| Average space/person (ft²) | >20 | 20–15 | 15–10 | 10–7 | 7–4 | <4 |
| Average flow (PFM[a]) | <5 | 5–7 | 7–10 | 10–13 | 13–17 | Variable up to 17 |
| | *A* | *B* | *C* | *D* | *E* | *F* |
| *Queueing* | | | | | | |
| Average space/person (ft²) | >13 | 13–10 | 10–7 | 7–3 | 3–2 | <2 |
| | Free circulation | Restricted circulation | Personal comfort | No touch | Touch | Body ellipse |

[a]Pedestrians per foot width of walkway per minute.

SOURCE: John J. Fruin, *Pedestrian Planning and Design,* Metropolitan Association of Urban Designers and Environmental Planners, New York, 1971.

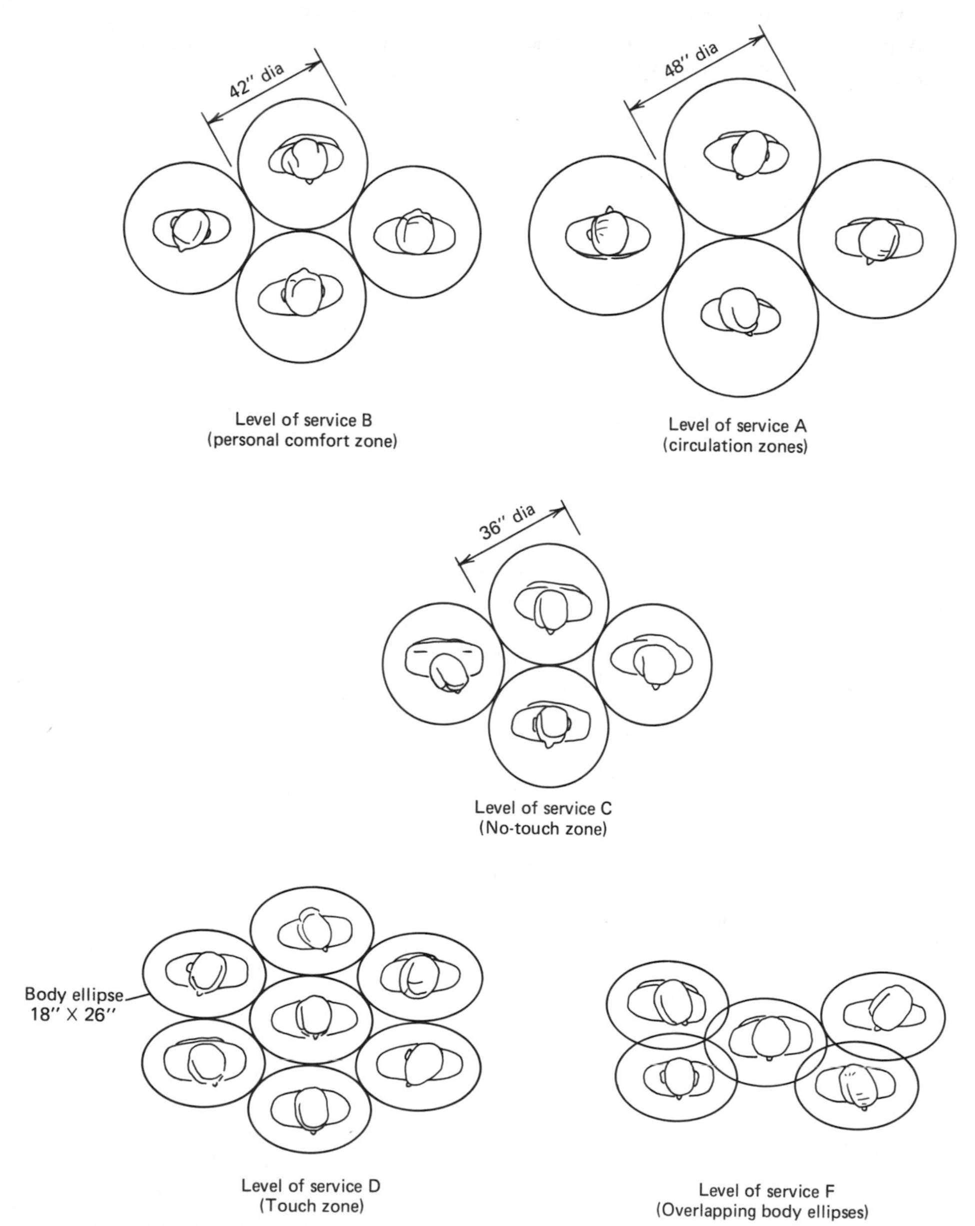


*Fig 6-5.* Relationship of space at varying levels of service with the body ellipse. (Adapted from Reference 6.)

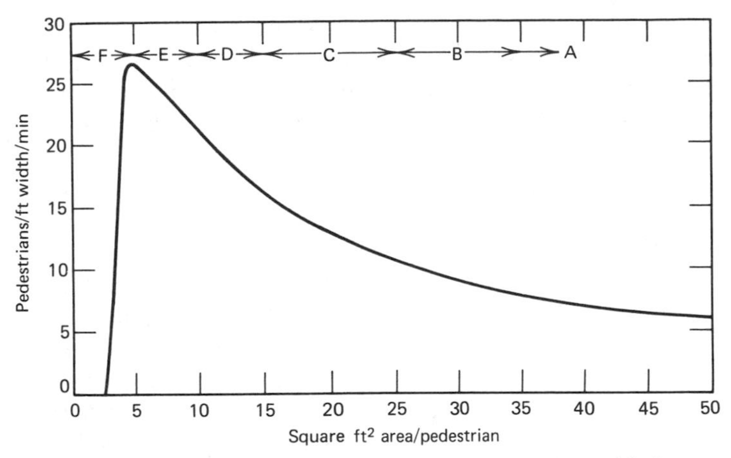


*Fig 6-6.* Relationship between pedestrian flow and provided area per pedestrian in walkways and sidewalks. Source: Reference 6.

observed varied fairly widely from this average figure. The average speed of the elderly was not significantly different from that of all adults but they were unable to attain the higher speeds of the upper percentile of adults. Children crossing speeds were, however, found to be significantly higher. These findings are indicated in Fig. 6-7. Somewhat lower figures have been found in U.S.

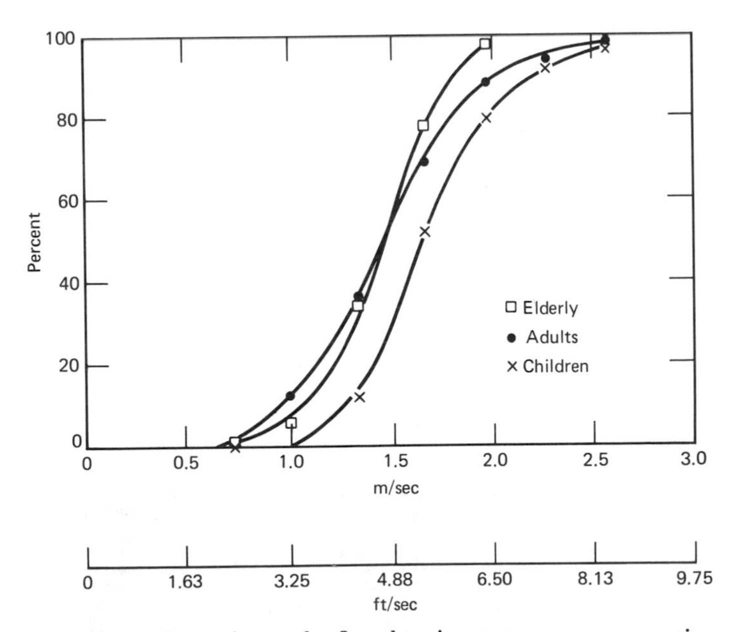


*Fig 6-7.* Typical speed of pedestrian movement at crossings. (After Sjostedt.)

studies. Men crossing streets alone were found to walk at 4.22 ft/sec and women alone at 3.70 ft/sec [8]. When walking with others, these speeds reduced, respectively, to 3.83 and 3.63 ft/sec. Many engineers assume crossing speeds of 4 ft/sec for design but some observers believe that a better design standard would be 3 to 3.25 ft/sec in order to safeguard the relatively slow walkers [9]. AASHTO recommends 3.0 ft/sec for urban road crossing design [10].

In addition to responsibility for designing space and circulation facilities which will comfortably accommodate pedestrian flow, the engineer has an interest in pedestrian behavior from the viewpoint of minimizing accidents. Pedestrian accidents occur largely when pedestrians cross roads and conflict with traffic flow. Pedestrian studies show large differences in crossing behavior. Figure 6-8 shows the very large variation in gap acceptance that has been observed [11]. Gap acceptance depends on a range of factors, but principally upon walking speed, ability to perceive oncoming traffic, level of attention, and training.

## 6-10. LOCATION AND TYPE OF PEDESTRIAN ACCIDENTS

The pedestrian in the roadway traffic environment has been likened to a small private plane flying on a "see and be seen" basis [9]. For safe crossing the pedestrian must be seen by the vehicle driver and the vehicle must be conspicuous to the pedestrian in order that a safe gap can be accepted. Table 6-2 shows the distribution of pedestrian accidents with reference to position on

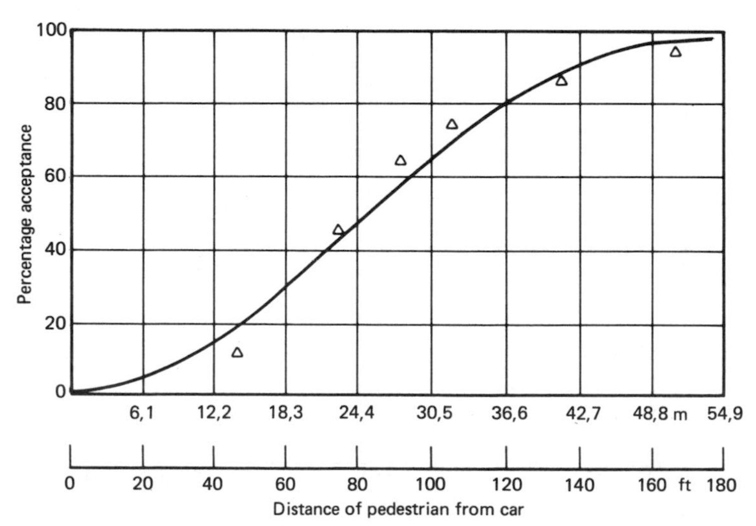


*Fig 6-8.* The percentage of pedestrians accepting gaps of given size in front of car.

*Table 6-2*
**Distribution of Pedestrian Accidents**

| | *Percent of Accidents Occurring* | | | |
|---|---|---|---|---|
| | *All Ages* | *Under 5* | *5–9* | *Over 65* |
| Crossing at intersection | 25.0 | 9.7 | 13.7 | 41.1 |
| Crossing between intersections | 27.5 | 38.0 | 51.8 | 25.0 |
| Walking with traffic in roadway | 13.0 | 22.2 | 1.4 | 10.1 |
| Walking against traffic in roadway | 10.1 | 9.1 | 8.5 | 6.9 |
| Other | 24.4 | 21.0 | 24.6 | 16.9 |

SOURCE: *Accident Facts,* National Safety Council, 1986.

the roadway [12]. It will be seen that the majority of accidents occur when pedestrians are crossing streets either at intersections or between intersections. When examined across all ages, the accident rates at intersections and between intersections are relatively similar. However, there are remarkable differences between the location of accidents for the elderly and the very young. For the elderly, most accidents occur at intersections. Studies have indicated that one of the problems is that conditions of intersections confuse the elderly who mistakenly step into the path of traffic which is maneuvering quite properly under traffic control. In other cases, the fault lies with drivers who fail to give priority to those less agile pedestrians who are unable to move quickly enough to avoid an accident.

Table 6-2 also indicates that children are most likely to be involved in an accident while trying to cross the road between intersections. Often this is due to undisciplined behavior of simply running out into oncoming traffic. Such thoughtlessness can be improved by child accident training schemes. In work carried out to compare the crossing behavior of adults and children it was found that there were significant differences [13]. Adults tend to assess the road situation before reaching the curb with the idea of minimizing curb delay; children tend to assess road conditions only at the curb itself. Adults are more likely to cross before the road is clear and to cross at an angle. They seldom run across the road and usually display a less "trained" behavior with respect to observing traffic properly before crossing. They are, of course, more conspicuous to car drivers than small children and move less rapidly from the curb.

## TRANSIT AND RAIL PASSENGERS

Human factors design considerations are taken into account in the design of both *terminals* and *vehicles* for transit and rail passengers [14]. In general, the civil engineer or planner is likely to be involved in the design of terminals only. However, transit system operation is now so closely linked to transport planning and transport systems management (see Chapter 7) that some knowledge of the factors influencing public transport vehicle design is essential in practice.

## 6-11. TRANSIT AND RAIL TERMINALS

The human factors requirements for the design of public transport passenger terminals have already largely been covered in Section 6-8. Bus and railway terminals are designed principally for pedestrians who are either waiting or walking on one level or using staircases. The space design standards shown in Fig. 6-5 and Table 6-1 can be used for such areas. These standards were in fact developed from observations made in the public transport passenger terminals of the former Port of New York Authority.

Another important element in the design of a terminal relates to the need for the passenger to absorb sufficient information to provide orientation and permit a smooth transit through the terminal. It is interesting to consider that whereas freight must be physically transferred through a freight terminal, passengers move themselves. The efficiency of the movement within any particular design is a function of how well the passenger is orientated within a terminal both by its intrinsic design and by information which is usually given by signs. Recently considerable research has been put into the question of orientation by design itself [15, 16]. By the use of *visibility indices,* the degree of intervisibility of the necessary node points in a terminal can be computed. Design options for the same terminal can then be compared by comparing the respective visibility indices.

Signing is of great importance in achieving smooth flows in terminals and in minimizing passenger disorientation. It is especially important where there is a significant proportion of passengers who are unfamiliar with the system or the terminal. The clearest and most precise form of signing is provided in the form of highly visible written messages such as "ENTRANCE" or "TO THE TRAINS." Such signs may be illuminated for maximum visibility or can be painted in highly contrasting colors. It has been found that black lettering on a yellow background gives maximum contrast. Written signs are of less use where there is a high proportion of the travelling public which is unfamiliar with the language of the country. This can present a problem at railroad stations and always poses a problem at international airports, see Section 6-18. In such circumstances, pictographs can be useful; they are also useful in countries with a high level of illiteracy. In the past, individual authorities have developed their own pictograph systems. Some have been excellent, others confusing. As a result international standards have been developed. Figure 6-9 shows a selection of the recommended international railroad pictographs. These can be compared with those shown in Figure 6-10 which indicates the signing recommended by the ICAO for international airports.

*Fig 6-9.* International railroad pictographs. (Courtesy: International Railway Federation.)

## 6-12. PUBLIC TRANSPORT VEHICLE DESIGN FACTORS

The internal design of public transport passenger vehicles is greatly influenced by passenger perception of comfort. The determinants of comfort are extremely complex and its definition is therefore difficult. Comfort would appear to be affected by three sets of inputs [17]:

Situational and social factors
Individual personal characteristics
Environmental factors

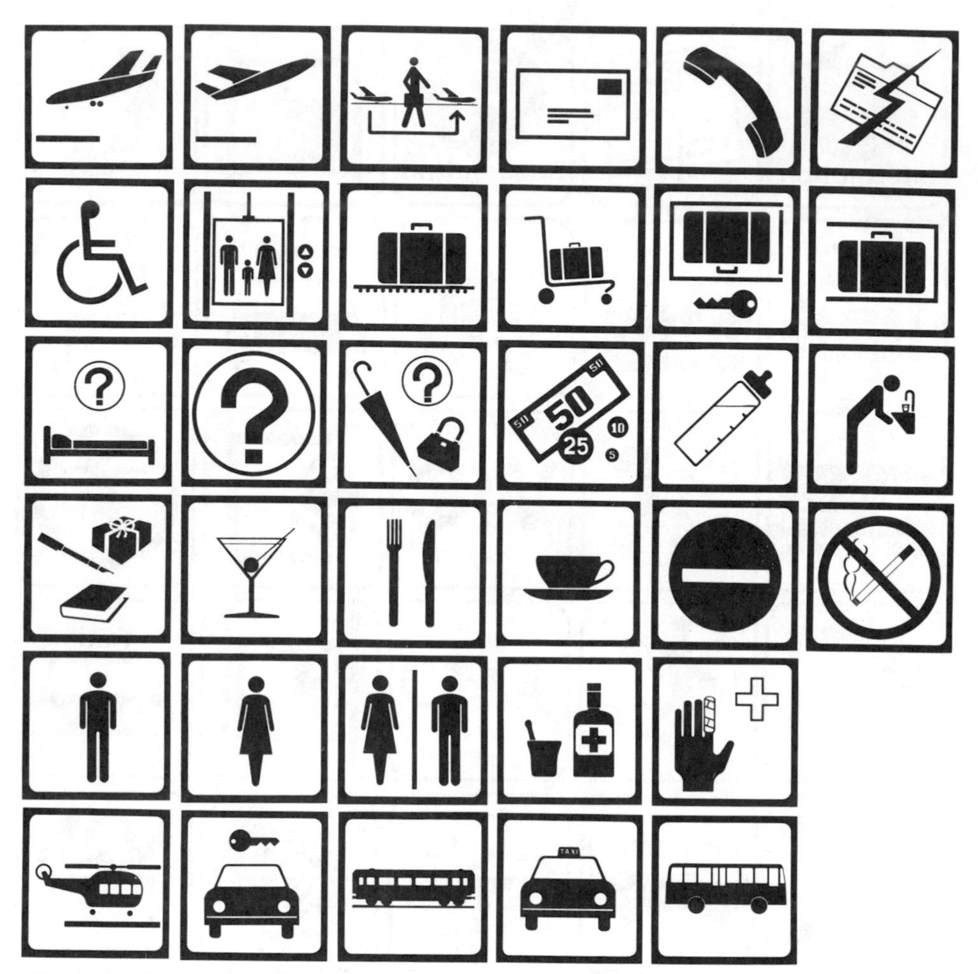


*Fig 6-10.* International signs to provide guidance to persons at airports, 1984. (Courtesy: ICAO.)

Social factors which affect perception of passenger comfort include level of crowding in stations and vehicles, presence of traveling companions, and familiarity with persons in adjacent seats.

Passenger characteristics include psychological, physical, and physiological factors such as previous experiences on this and similar vehicles, sex, size, proneness to motion sickness, physical and mental health, and so on.

Whereas the social and individual factor sets are beyond the control of the designer, environmental factors can be influenced. Table 6-3 shows the physical characteristics of vehicle environments which can be controlled to affect comfort. The importance of physical factors in comfort perception has been

*Table 6-3*
**Physical Characteristics of Vehicle Environments**

| *Dynamic Factors* | *Ambient Factors* | *Spatial Factors* |
|---|---|---|
| Vertical acceleration | Pressure | Workspace |
| Lateral acceleration | Temperature | Seat width |
| Longitudinal acceleration | Humidity | Leg room |
| Roll rate | Ventilation | Seat adjustment |
| Pitch rate | Smoke | Seat shape |
| Yaw rate | Odors | Seat firmness |
| Jolt and shock | Air quality | Standing space |
| Ascents, descents | Noise | |
| Change in speed | | |

SOURCE: Reference 17.

recognized for a long time. Some research has been carried out to develop fairly simple comfort models of ride quality based on vehicular motion. For example, the comparative ride qualities of luxury buses and trains have been evaluated from the following equations [18]:

$$C = 0.33 + 0.53w_p + 19.33a_v \text{ (Luxury bus)}$$

and

$$C = 1.14 + 37.33a_v \text{ (Railroad passenger car)}$$

where

$C$ = comfort index
$W_p$ = pitch rate (deg/sec)
$a_v$ = vertical acceleration ($g$)

For design such indices are insufficient to be of practical use. The designer must look at space provision and a variety of environmental factors.

*Space.* The amount of space required depends greatly upon the amount of time that space has to be occupied. Crowding which is acceptable for short periods of time becomes unacceptable over long periods. Equally important is the passenger's expectation of being able to sit down for longer journeys. The basic spatial requirements for standing and seated passengers is shown in Fig. 6-11 and the relationship between spatial requirements and journey times is shown in Fig. 6-12 [19, 20]:

*Environmental Considerations.* In addition to space, the transit vehicle designer must be concerned with the environmental factors shown in Table 6-3. Research has been carried out which classifies ranges of environmental conditions into "comfortable," "uncomfortable," and "unacceptable" categories. These are shown in Table 6-4.

| *Standing Passengers* | *Comments* | *Dimensions (meters)* ① | ② | ③ | ④ | ⑤ | *Projected Surface ($m^2$)* |
|---|---|---|---|---|---|---|---|
| 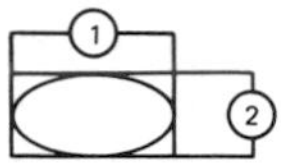 | Thorax | [0.49]<br>0.53 | [0.26]<br>0.31 | —<br>— | —<br>— | —<br>— | [0.13]<br>0.16 |
| 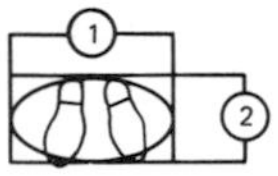 | Thorax with feet | [0.61]<br>0.66 | [0.29]<br>0.31 | —<br>— | —<br>— | —<br>— | [0.17]<br>0.21 |
| 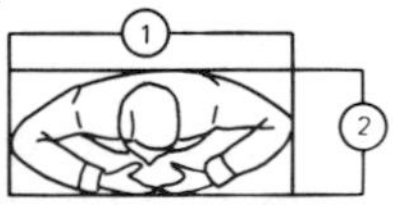 | Horizontal arms with tips of fingers touching each other | 0.95 | 0.43 | — | — | — | 0.40 |
| 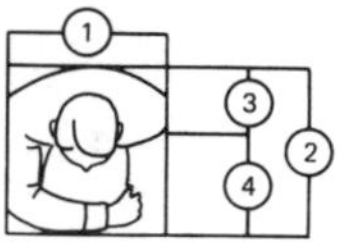 | Person carrying parcel | 0.53 | 0.57 | 0.16 | 0.41 | — | 0.30 |
| 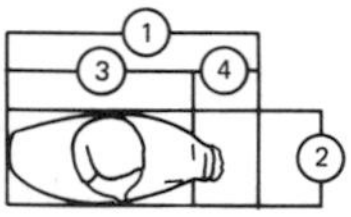 | Person carrying one small piece of luggage | 0.82 | 0.30 | 0.67 | 0.15 | — | 0.25 |
| 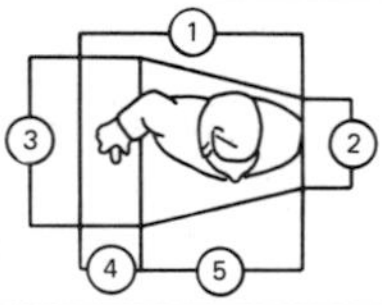 | Person carrying one large piece of luggage | 0.73 | 0.31 | 0.60 | 0.20 | 0.53 | 0.33 |
| 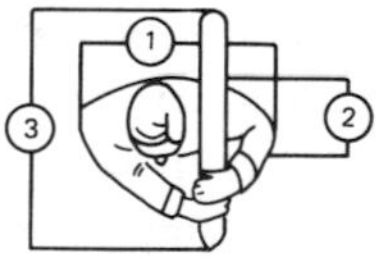 | Person carrying skis (for aerial tramway design) | —<br>0.70 | —<br>0.30 | —<br>0.75 | — | — | —<br>0.50 |
| 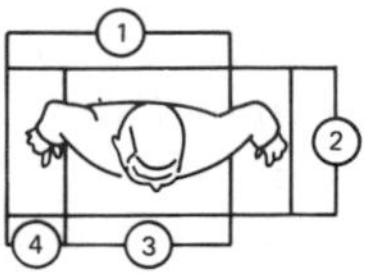 | Person carrying two large pieces of luggage | 0.93 | 0.60 | 0.53 | 0.20 | — | 0.56 |

*Fig 6-11.* Anthropometric measurements for seated and standing passengers. (Source: References 19 and 20.)

| *Standing Passengers* | *Comments* | *Dimensions (meters)* ① | ② | ③ | ④ | ⑤ | *Projected Surface ($m^2$)* |
|---|---|---|---|---|---|---|---|
| 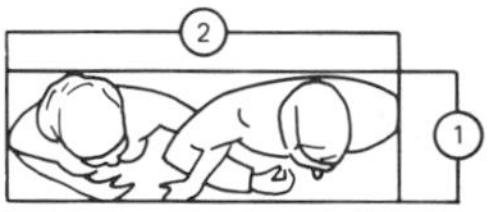 | Couple | 1.40 | 0.43 | — | — | — | 0.61 |
| 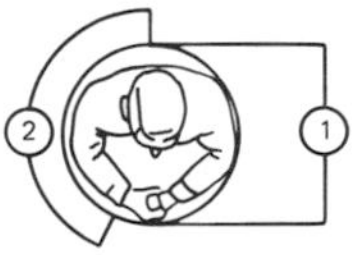 | Person holding on to stanchion | 0.58 | 0.905 | — | — | — | 0.26 |

| *Seated Passengers* | *Comments* | *Dimensions (meters)* ① | ② | | *Projected Surface ($m^2$)* |
|---|---|---|---|---|---|
| 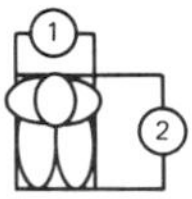 | Minimum space requirement | [0.40] 0.46 | [0.60] 0.66 | | [0.24] 0.30 |
|  | Normal, comfortable situation | 0.66 | 0.82 | | 0.54 |
| 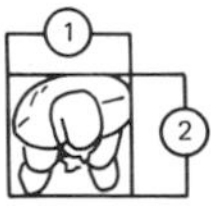 | Person leaning | 0.66 | 0.62 | | 0.41 |
| 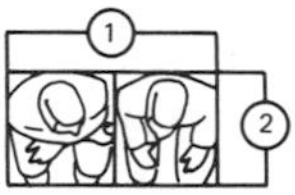 | Uncom-fortable arrangement | 1.08 | 0.66 | | 0.71 |
| 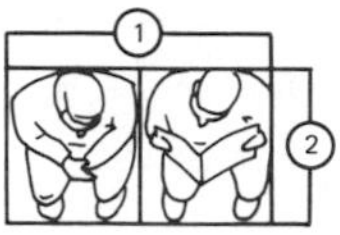 | Comfortable situation | 1.34 | 0.82 | | 1.10 |

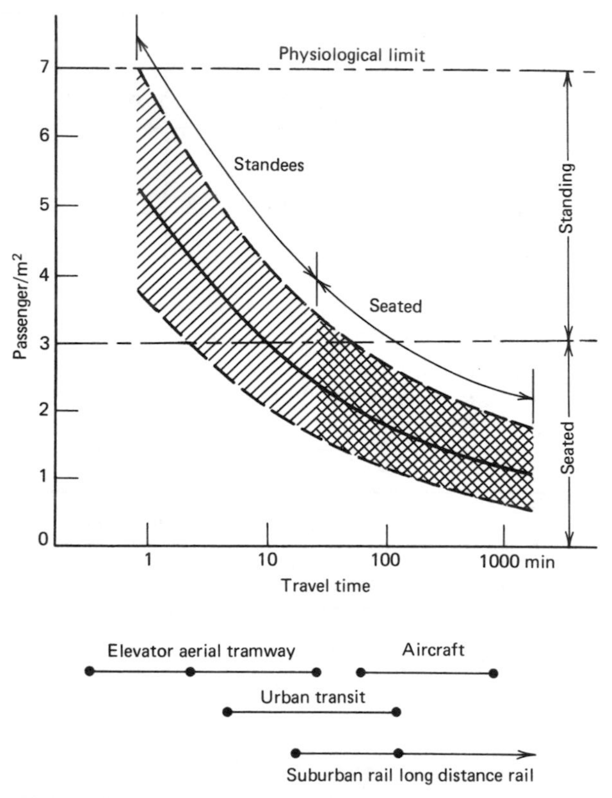


*Fig 6-12.* Passenger space requirements *vs* travel time. (Source: P. H. Bovy, *Amènagement du territoire et transports,* Vol. 2, EPF-ITEP, Lausanne, Switzerland, 1974.)

## THE AUTOMOBILE DRIVER [21]

### 6-13. THE HUMAN SENSORY PROCESS

A driver's decisions and actions depend principally on information received through the senses. This information comes to the driver through the eyes, ears, and the sensory nerve endings in the muscles, tendons, joints, skin, and organs. In general order of importance, the senses most used by drivers are

1. Visual (sight).
2. Kinesthetic (movement).
3. Vestibular (equilibrium).
4. Auditory (hearing).

*Table 6-4*
**Ranges of Environmental Conditions for Transit Vehicle Design**

| | *Comfortable* | *Uncomfortable* | *Unacceptable* |
|---|---|---|---|
| Temperature (°C) | 20–22 | 12–16 and 30–35 | <12 or >35 |
| Ventilation ($m^3$/hr/pass) | >35 | 8–20 | <8 |
| Humidity (%) | 30–70 | 30–20 and 70–80 | <20 or >80 |
| Vibration (mm/sec) | 0.2 | 1.2 | 3 |
| Lighting | Depends greatly on activity: Rough seeing tasks: hallways, corridors, inactive storage 30–200 lumens; Casual seeing tasks: reception, washrooms, service areas 40–300 lumens; Ordinary seeing tasks: switchboard, conference room, lounges 120–1000 lumens; Difficult and critical seeing tasks: offices, machine shops, detail work areas 250–5000 lumens | | |
| Noise (dBA) | <65 | 75–85 | >85 |
| Acceleration (m/$sec^2$) | 1 | 2 | 4 |
| Jerk (m/$sec^3$) | 0.6 | 1 | 1.5 |
| Floor slope (%) | 0–4 | 5–10 | >10 |
| Steps | 8 | 10 | >12 |

SOURCE: Adapted in part from Reference 20.

Visual information may come to a driver by means of foveal or peripheral vision. In foveal vision, images are concentrated in a small area of the eye near the center of the retina at which visual perception is most acute. At a given moment, a person's sharpest vision is concentrated within a cone with a central angle of about 3°. For most persons, visual acuity is reasonably sharp within a conical angle up to about 10°. Beyond this, a person's vision is less well defined. A person with normal vision can perceive peripheral objects within a cone having a central angle ranging up to about 160° [22].

A driver increases the amount of visual information received by movement of the head and eyes. The driver's eyes search and scan the field of view, moving the area with sharpest acuity. The direction of foveal vision repeatedly shifts and the eyes fixate, each such event requiring respectively 0.15 to 0.33 sec and 0.1 to 0.3 sec [22].

Mourant and Rockwell [23] found that experienced drivers scanned a wider range of horizontal fixation locations than novice drivers. The average horizontal fixation range for the experienced group varied from 30° to 48°, depending on the type of driving task. The novice drivers concentrated their eye fixations in a smaller area and looked closer in front of the vehicle and more to the right. By pursuit eye movements, the novice drivers tended to observe the guardrail and lane edges continuously for periods of about 1.0 sec to gain information on lateral position. Similar findings have been reported for drivers who have been deprived of sleep [23].

Visual acuity declines and the field of vision narrows with advancing age, especially when lighting conditions are poor [24]. Older drivers also tend to have more difficulty in judging distances and distinguishing colors than young drivers.

Drivers receive information by kinesthesia, the sensation of bodily position, presence, or movement that results primarily from stimulation of sensory receptors in muscles, tendons, and joints. Lateral or longitudinal accelerations exert forces on the driver that are transmitted by the seats, steering wheel, brake pedal, arm rest, and so forth. Proprioceptors in the muscles and joints are stimulated by the tendency of the body to move and shift as a result of the forces. This provides a feedback that may cause a driver to brake, slow down, or take some other action. This phenomenon has been termed "driving by the seat of the pants."

Drivers also receive messages from the vestibular nerve within the inner ear. Three fluid-filled semicircular canals in the inner ear enable a person to recognize the direction of movement and maintain balance or equilibrium. Movement of the head or body causes the fluid in the canals to move, affecting delicate fibrous cells located at one end of each of the canals. Impulses from the vestibular nerve are then transmitted to the brain, which, in turn, sends messages to the muscles, which take action to maintain balance.

The kinesthetic and vestibular senses provide important information to a driver about forces associated with change of direction, steering, braking, vibrations, and stability of the vehicle [25].

Sounds of horns or skidding tires may alert a driver to an impending collision, and engine noise aids the driver as well as pedestrians in judging vehicle speeds. When traveling along a curve or turning at an intersection, tire noises may indicate to the driver a need to slow down. Highway engineers employ sound in the form of special pavement textures, raised pavement markers, or "rumble strips" to signal to drivers to slow down or to avoid certain paved areas. Sounds within a vehicle such as loud conversation or radio music may mask out the sounds of sirens, bells, or other important warning devices.

Drivers may detect a fire or engine malfunction by their sense of smell. Hunger, thirst, or discomfort may cause drivers to change their trip pattern or stop. Drivers may also occasionally receive information through the sensory nerve endings in the head, skin, and internal organs and experience sensations of touch, coldness, warmness, fatigue, or pain.

## 6-14. DRIVER PERCEPTION AND REACTION

Driver perception-reaction time is defined as the interval between seeing, feeling, or hearing a traffic or highway situation and making initial response to what has been perceived. Take, for example, a simple braking situation. The "clock" for perception-reaction time begins when the object or condition first becomes visible and it stops when the driver's foot touches the brake.

Traditionally, perception-reaction time has been couched in terms of perception, intellection, emotion, and volition, termed PIEV time [22]. Perception is the process of forming a mental image of sensations received through the eyes, ears, or other parts of the body. It is recognizing and becoming aware of the information received by the senses. Intellection is another word for reasoning or using the intellect. Emotion is the affective and subjective aspect of a person's consciousness. It has to do with how a person feels about a situation. More often than not, emotion is detrimental to safe motor vehicle operation. Volition is the act of making a choice or decision. Perception-reaction times vary from less than 0.5 sec to 3 or more sec. Greenshields [26] has indicated that, in general:

1. The speed of all forms of reaction varies from one person to another and from time to time in the same person.
2. Reaction time changes gradually with age, very young and very old people being slower in their reactions.
3. People generally react more quickly to very strong stimuli than to weak ones.
4. Complicated situations take longer to react to than simple ones.
5. A person's physical condition affects his reactions. For example, fatigue tends to lengthen a person's reaction time.
6. Distractions increase the time of all reactions except the reflex.

Johansson and Rumar [27] measured brake-reaction times for a group of 321 drivers who had some degree of braking expectation. The median brake-reaction time was 0.66 sec and the range was from 0.3 to 2.0 sec. The frequency distribution of the brake-reaction times is shown in Figure 6-13. The researchers performed a second experiment to see how brake-reaction time in response to a completely unexpected signal compares with such time in response to a somewhat anticipated signal. They reported that unexpected brake-reaction times were larger by a factor of 1.35.

Elderly drivers have more difficulty than the young in perceiving traffic situations and tend to pause longer between successive acts. Data presented by Marsh [24] suggest that on the average, simple reaction times for persons aged 65 are about 16 percent higher than for those aged 20; complex reaction times are about 33 percent higher for the older subjects. However, because of the wide variability among drivers, "chronological age is not a sound criterion for appraisal of driving competence" [24].

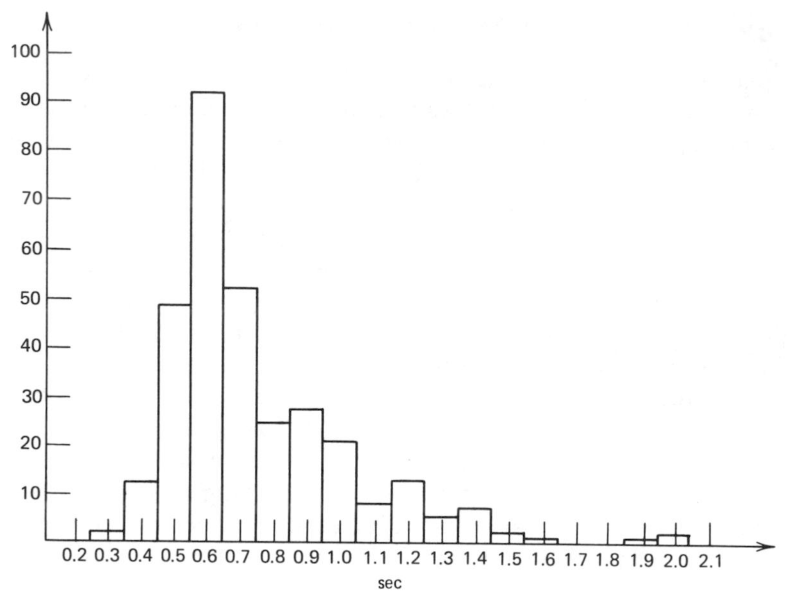


*Fig 6-13.* Frequency distribution of brake-reaction times. (Source: Reference 27.)

As Figure 6-14 illustrates, a person's response time depends on the sense stimulated. Generally, a person responds quickest to stimuli of touch and hearing and requires only slightly more time for response to visual stimuli. Response times for the kinesthetic and vestibular stimuli are considerably longer.

Figure 6-15 shows that response time increases with the complexity of a situation, as does the error rate.

Perception-reaction time tends to be detrimentally affected by fatigue, the use of alcohol or other drugs, and certain physiological and psychological conditions.

For the purpose of computing stopping sight distances, AASHTO recommends the use of a combined perception-reaction and brake-reaction time of 2.5 sec. [28].

## 6-15. DRIVER INFORMATION NEEDS

Given the ever increasing demands of the driving task, it is important that highway and traffic engineers understand the driver's need for information and the means of its transmission. Research was undertaken in the late 1960s by AIL, a division of Cutler-Hammer, to identify types of information needed by drivers and principal factors and interactions that affect the reception and use of this information. The results of one aspect of that work have been

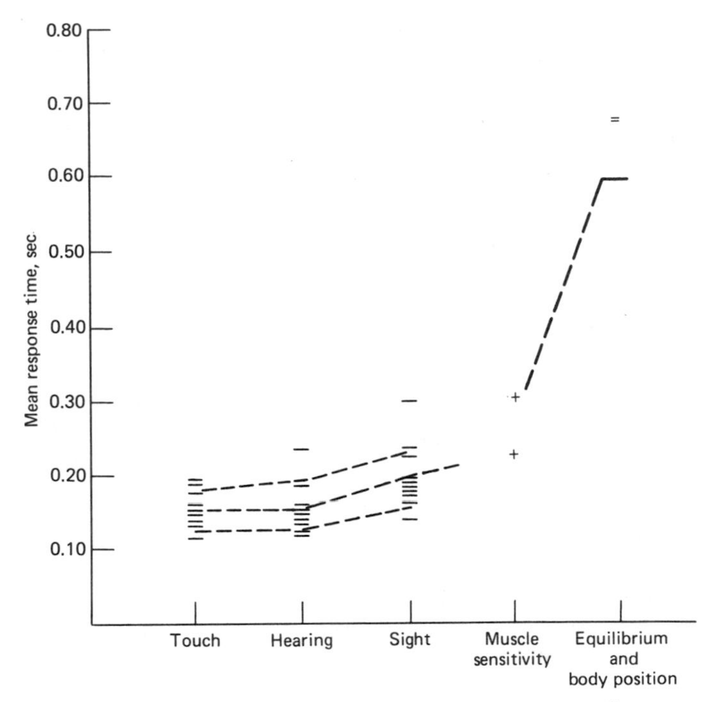


*Fig 6-14.* Mean response time due to various types of sensory stimulus. (Source: T. W. Forbes and N. S. Katz, Summary of Human Engineering Data and Principles Related to Highway Design and Traffic Engineering Problems; sketch courtesy of the Institute of Transportation Engineers.)

described in a paper by Allen, Lunenfeld, and Alexander [29], which forms the primary basis for the material presented in the following paragraphs.

In order to make sound decisions, a driver must receive reliable and understandable information to reduce uncertainty. A driver gathers information from a variety of sources, uses it as a basis for decision making, and then translates those decisions into actions to control the vehicle [30].

As Table 6-5 indicates, the various activities involved in driving a motor vehicle may be grouped into three categories or levels of performance: (1) the control level, (2) the guidance level, and (3) the navigation level.

The control level of driver performance is related to the physical operation of the vehicle. It includes lateral control that is maintained by steering and longitudinal control that is exercised by accelerating and decelerating. To maintain vehicle control, the driver requires information on the vehicle's position and orientation with respect to the road as well as feedback on the vehicle's response to braking, accelerating, and steering. Such information is received primarily through the visual, kinesthetic, and vestibular senses.

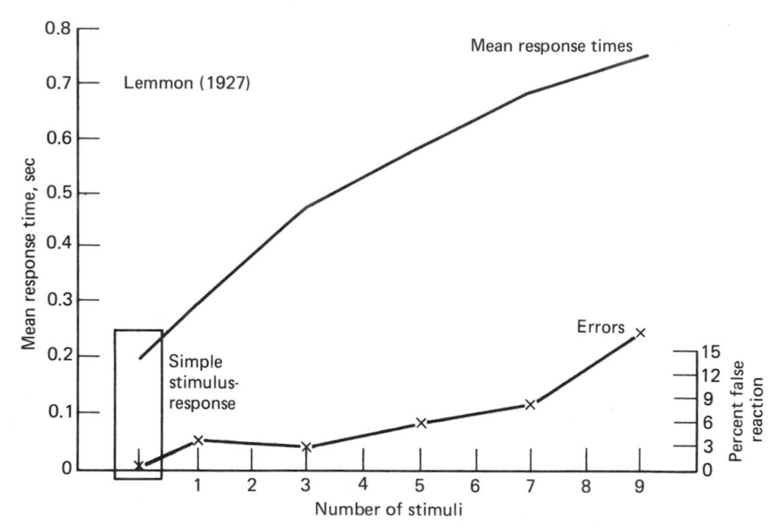


*Fig 6-15.* Increase in mean response time and errors with increasing number of stimuli. (Source: T. W. Forbes and N. S. Katz, Summary of Human Engineering Research Data and Principles Related to Highway Design and Traffic Engineering Problems; sketch courtesy of the Institute of Transportation Engineers.)

> Guidance level of driver performance refers to the driver's task of selecting a safe speed and path on the highway. While performance at the control level is overt action, performance at the guidance level is a decision process. The driver must evaluate the immediate situation, make appropriate speed and path decisions, and translate these decisions into control actions needed to survive in the traffic stream. Activities include decisions relating to lane positioning, overtaking, and passing. . . .
>
> The driver is constantly time-sharing activities at the guidance level with those at the control and navigation levels. Furthermore, two or more guidance level processing requirements may occur simultaneously or in close enough time proximity to require simultaneous action, which may not necessarily be compatible. This points to the importance of experience and prior knowledge throughout the driving task, since the driver may be required to make several decisions under heavy time pressure. [30]

The navigational level of driver performance refers to the driver's ability to plan and execute a trip. In includes trip preparation and planning and direction finding. The former activities may be performed before the trip begins, while the latter occurs en route. Information at the navigation level comes from maps, directional signs, landmarks, and from the driver's memory based on previous experience.

Allen et al. [29] have listed five basic principles for the systematic presentation of information required by drivers:

*Table 6-5*
**A Framework for Driving Task Conceptualization**

| *Subtask Category* | *Related to* | *Examples of Sources of Information* | *Importance of Information* | *Likely Consequence of Failure* |
|---|---|---|---|---|
| Control (Microperformance) | Physical operation of vehicle<br>Steering control<br>Speed control | Road edges<br>Lane divisions<br>Warning signs<br>Kinesthesia | Highest | Emergency situation or crash |
| Guidance (Situational performance) | Selecting and maintaining a safe speed and path | Road geometry<br>Obstacles<br>Traffic conditions<br>Weather conditons | Intermediate | Emergency situation or crash |
| Navigation (Macroperformance) | Route following<br>Direction finding<br>Trip planning | Experience<br>Directional signs<br>Maps<br>Touring service | Lowest | Delay, confusion, or inefficiencies |

1. First things first—primacy.
2. Do not overload—processing channel limitations.
3. Do it before they get on the road—*a priori* knowledge.
4. Keep them busy—spreading.
5. Do not surprise them—expectancy.

### Primacy

The concept of primacy is based on the realization that at any given instant certain driver information is more important than other information. In evaluating the primacy of information, the effect of failure to receive that information is considered. Lack of certain types of information may lead to an emergency situation or a crash; lack of other types of information may only cause momentary confusion or delay. Avoiding a car emerging from a driveway (a control subtask) is more important than deciding whether to pass a slow-moving vehicle (a guidance subtask). Either of these subtasks would take precedence over observance of directional signs (a navigation subtask).

> In the guidance category, where failures tend to be catastrophic, the question is one of degree. Generally, colliding with a guardrail is less serious than colliding with a bridge pier. Colliding with a vehicle moving in the same direction is less serious than a head-on collision. . . .
>
> A second aspect of primacy judgment is based on distance criterion. A hazard close by would have a higher primacy than a more distant hazard of the same or similar type. . . . Assessing the primacy (both distance and severity potential) of a single hazard is a task that drivers probably do well. Assessing the relative primacy of two competing hazards in a dynamic situation is a difficult task that is more prone to driver error. [30].

### Processing Limitations

As the complexity of a driving situation increases, so does the driver's response time. Obviously, if a driver is confronted with several complex situations within a short time span, a point may be reached when the driver is unable to receive and process the amount of information required for error-free response. Information needed by the driver should be presented systematically, requiring a series of simple choices rather than a smaller number of complex choices.

### A Priori Knowledge

Drivers should be given as much information as possible before they get on the road. To the extent feasible, *a priori* information should be communicated to drivers on distances and directions to destination points, average travel times, highway types and design features, and other such information that will facilitate the planning of trips.

**Spreading**

During the course of trips, drivers may be confronted with highly variable demands for their attention, too much information being provided in certain areas and too little in others. Efforts should be made to spread the information load in order to avoid boredom of drivers in certain areas and overloading of their processing capability in others. This is accomplished by performing a primacy analysis and transferring the less important information from the high-information "peaks" to the low-information "valleys."

**Expectancy**

A driver's readiness to respond to roadway conditions, traffic situations, or information systems is referred to as driver expectancy. Based on previous experience, drivers are conditioned to expect certain characteristics of road design and operation. For example, drivers generally expect exit and entrance ramps along freeways to be on the right-hand side. Drivers do not generally expect lane drops or other changes in the roadway cross section. At locations where driver expectancies are violated, drivers may require a longer response time and will have a greater tendency to make an inappropriate or hazardous response. Although expectancies occur at all levels of the driving task, their importance is probably greatest at the guidance level.

## 6-16. DRUGS AND ROAD SAFETY

*Alcohol* is one of the most widely used available drugs in human societies. The use of alcohol predates even biblical times; its long history of usage sometimes permits one to overlook the fact that it is a strong depressant drug which in itself is potentially lethal. Its effects when mixed with other drugs are often significantly enhanced. Up to 0.05 percent blood alcohol content (BAC), alcohol acts as a sedative inducing tranquility. For some individuals there may be a slight improvement in driving at low BAC levels due to the mild sedative effect. Between 0.05 percent and 0.15 percent BAC there is increasing lack of coordination; in most U.S. states 0.10 percent BAC is the lower legal limit of intoxication for drivers. Within 0.05 to 0.10 percent range, the individual appears to be stimulated rather than depressed. Outwardly the subject may become aggressive, talkative, or hyperactive in some other way. This is due to depression of brain functions which exercise control over these activities. Some researches have found that the basic driving task itself is not significantly affected by BAC levels up to 0.10 percent [31]. However, overall driving performance does become severely affected when a secondary task, such as a supplementary visual task is introduced in the form of detecting and interpreting road signs, detecting potentially hazardous conditions or avoiding potential collisions. Visual scanning rates decline even at BAC levels below legal intoxication; at 0.07 percent BAC level the visual scanning rate declines by 29 percent. Drivers within this range simply do not see hazardous situations such as other vehicles in potential conflict, pedestrians in danger, or hazardous road conditions. The depression of brain function also leads alcohol affected drivers to overestimate dangerously their driving performance. Above 0.15 percent BAC, the individual suffers from severe lack of mobility coordination, facility of speech, reasoning, and general brain function. Combined with fatigue, the effects of alcohol become more pronounced [32].

In conjunction with driving alcohol presents a huge problem [31]. If it can be assumed that 10 percent of all vehicle miles driven are driven by individuals who are under the influence of alcohol, a 50 percent saving of life could be achieved by banning these individuals from driving. Table 6-6 shows the increasing likelihood of accident involvement with increasing blood alcohol content; it is exponential in nature. The public and consequently legislative and judicial bodies are becoming more aware of this problem. Enforcement of driving while intoxicated (DWI) laws has been increased in the last 10 years. Consequently, the percentage of drivers who were killed in accidents being subsequently found to be intoxicated (BAC > 0.10 percent) decreased continuously from 50 percent in 1980 to 43 percent in 1984.

The involvement of intoxicated drivers in accidents is strongly affected by age and the time of day [31]. Table 6-7 shows that the worst offenders with respect to drunken driving are those within the 20–24 age group, although this group is only marginally worse than all adults aged 20–64. The social implications of alcohol become apparent when it is realized that during the weekday drunken drivers are involved in only 1 percent of accidents. On a weekend night, this figure leaps to 19 percent, and the worst affected age group is those between 20 and 24 years. The potential influence of alcohol on traffic fatalities can be inferred from Table 6-8 which shows the breakdown of traffic fatalities on a day-night/weekend-weekday basis.

The problems of the lack of coordination of the alcohol impaired driver become apparent also in the statistic that 59 percent of all alcohol related

*Table 6-6*
**Likelihood of Accident Involvement with Increasing Blood Alcohol Content (BAC)**

| *BAC Level (Percent)* | *Relative Likelihood of Accident* |
|---|---|
| 0.05 | 1 |
| 0.10 | 6.25 |
| 0.15 | 15.00 |
| 0.20 | 27.50 |

SOURCE: Reference 31.

*Table 6-7*
**Alcohol Related Car Driver Fatalities by Age**

| *Age* | *Percentage* |
|---|---|
| Under 15 | 23 |
| 14–19 | 51 |
| 20–24 | 68 |
| 25–64 | 60 |
| 65 and over | 22 |

SOURCE: Reference 31.

driver accidents are noncollision accidents, compared with 17.9 percent for nonimpaired drivers. In 1985, 22,680 persons died in accidents involving a drinking driver or a nonoccupant who was impaired in some manner [31].

Although not strictly related to the auto driver, it seems appropriate to introduce at this point facts relating to alcohol and pedestrian accidents. Approximately 25 percent of all pedestrians involved in fatal accidents are intoxicated (BAC $>$ 0.10 percent), and approximately 60 percent of these are heavily intoxicated (BAC $>$ 0.20 percent).

*Marijuana,* an illegal but widely used drug, contains THC ($\Delta^9$-tetrahydrocannabinol). The drug decreases human psychomotor performance producing a lack of motivation, an inability to concentrate, and a lack of ability to coordinate movement [33]. Taken in conjunction with alcohol, the effects of the drug are significantly increased. Tests on individuals under the influence of the drug show that there is a marked decrease in ability to cope with the driving task and a severe unsteadiness in balance. Other tests on groups of controlled users showed a marked detrimental effect on driving even at "social levels" of marijuana dosage. At very low THC levels some subjects slightly improved driving performance, possibly due to either overcompensation or to the sedative effect of the drug [34]. At higher levels of THC dosage, individuals overestimate their ability to cope with the overall driving task.

*Table 6-8*
**Temporal Breakdown of Traffic Accident Fatalities**

| *Weekday* | | *Weekend* | |
|---|---|---|---|
| Day | Night | Day | Night |
| 29.3 percent | 23.4 percent | 12.3 percent | 34.2 percent |

SOURCE: Reference 31.

## AIR TRANSPORT PASSENGERS

### 6-17. AIR TRAVELLER EXPECTATIONS

The air transport passenger is to a degree different from other transport users and the design standards of the mode tend to reflect perceived differences. First, the traveler is using what is still regarded as a premium mode and there is some residual expectation of high standards of design of equipment and facilities. Second, the average trip length is long and some trips are very long indeed, both in distance and time. Design must therefore acknowledge that the user will have to use the facilities and equipment for long periods. Conditions which would be tolerable in the short term are not necessarily acceptable for long periods or for individuals who are suffering journey fatigue. Third, trips are made infrequently and design must accommodate the unaccustomed traveler. Fourth, nearly all air passengers are heavily encumbered with luggage within the terminal and entering and egressing the aircraft. Finally, in almost all cases the air mode is a mode in competition, either with other modes or within the mode among airlines—one airline has the motto "We remember that you have a choice." For these and other reasons, human factors considerations weigh considerably in the design of airport terminals and the passenger cabins of air transport aircraft.

### 6-18. PASSENGER BEHAVIOR IN AIRPORT TERMINALS [14, 35]

The airport passenger terminal exists to permit a change of mode from the surface access mode to the aircraft and vice versa. As an example, passengers who arrive in a batch process (the aircraft) are distributed to a number of discrete-access vehicles (private cars, taxis, buses, trains, limousines) or transfer to other aircraft. For both the arriving and departing passenger, the terminal must cater to three basic functions: *processing* (e.g., check-in, enplanement, deplanement, immigration, customs, baggage claim), *waiting* (e.g., departures concourse, departure lounge, gate lounge, arrivals concourse), or *circulation* between the processing and waiting areas. Initially, it might appear that standards described elsewhere in this chapter (see Table 6-1) could be applied directly to airport terminal design. Processing requires some form of queueing behavior, circulation involves walking or carriage by some mechanical means, and waiting does not at first sight appear to be any different for the air passenger. In fact, because of the factors described in Section 6-17, design standards differ mainly because of the large amount of time spent in the terminal and the degree of encumbrance of the average passenger. A number of studies have been carried out to determine passenger behavior in airport terminals [35, 36] which in turn have led to space design standards specific to airports. Table 6-9 indicates the results of a survey to determine average

*Table 6-9*
**Typical Activity Times for Airport Terminal Passengers**

| *Arriving Passengers (International)* | *Minutes (average)* |
|---|---|
| Moving from aircraft to immigration | 4.0 |
| Immigration queue and processing | 4.6 |
| Baggage reclaim | 9.1 |
| Customs examination | 0.8 |
| Landside arrivals concourse | 5.9 |
| Waiting for surface transport | 2.6 |
| | 27.0 |

| *Departing Passengers (International)* | *Minutes (average)* |
|---|---|
| Check-in queue and processing | 5.6 |
| Waiting landside departures concourse | 18.4 |
| Passport control | 2.3 |
| Waiting airside departures lounge | 23.7 |
| Moving from lounge to gate lounge | 2.5 |
| Boarding aircraft including gate security check | 8.0 |
| | 60.5 |

SOURCE: After Deighton.

terminal dwell times for both arriving and departing passengers and the breakdown of the component times for the necessary activities. It is important to recognize that the times shown are average times only and the distribution of observations is highly skewed. Extremely high total terminal dwell times are observed for both arriving and departing passengers and for transfer passenger whose activities are not shown in the table. In response to perceived special needs of air passengers, IATA has published a set of design standards which relate space provision to specific areas of the terminal and perceived level of service, see Table 6-10 [37]. There is evidence, however, that this approach is oversimplified and that perceived level of service is dependent not only on space provision but on the time which the traveler must spend in that space; for example confined spaces in gate departure lounges may be acceptable for short periods of time but this provision for long-term waiting in the general departure lounge would be intolerable [38, 39].

For reasons discussed in Section 6-11, the satisfactory functioning of a

*Table 6-10*
**Air Terminal Building Space Standards (Square Meters per Occupant)**

| *Level of Service* | *A* | *B* | *C* | *D* | *E* | *F* |
|---|---|---|---|---|---|---|
| Check-in | 1.6 | 1.4 | 1.2 | 1.0 | 0.8 | |
| Wait/circulate | 2.7 | 2.3 | 1.9 | 1.5 | 1.0 | |
| Holdroom | 1.4 | 1.2 | 1.0 | 0.8 | 0.6 | |
| Bag claim area (without device) | 1.6 | 1.4 | 1.2 | 1.0 | 0.8 | |

| *Level of Service* | *Description* |
|---|---|
| *A* | Excellent level of service; condition of free flow; no delays; direct routes, excellent level of comfort. |
| *B* | High level of service; condition of stable flow, high level of comfort. |
| *C* | Good level of service; condition of stable flow; provides acceptable throughput; related subsystems in balance. |
| *D* | Adequate level of service; condition of unstable flow; delays for passengers; condition acceptable for short periods of time. |
| *E* | Unacceptable level of service, condition of unstable flow; subsystems not in balance; represents limiting capacity of the system. |
| *F* | System breakdown; unacceptable congestion and delays. |

SOURCE: IATA

passenger terminal depends greatly on preventing passenger disorientation. In airport terminals, signing is an extremely important method of providing the passenger with sufficient information to use terminal facilities or to pass through the terminal. Because a high proportion of passengers at airports with international flights may not understand the language of the country or may be illiterate, the ICAO has devised an internationally approved set of pictographs which can be used to guide passengers without using written words [40]. These are shown in Fig. 6-10.

## 6-19. DESIGN OF THE AIR TRANSPORT PASSENGER CABIN [41, 42]

During flight, the air passenger is entirely cocooned in the environment of the passenger cabin, often for long periods. The creation of the artificial environment of the passenger cabin of a commercial transport aircraft is a good example of an engineering design which relies heavily on human factors inputs. Cabin design requires the consideration of factors in four main areas:

1. *Spatial.* Cabin size, seating layout, circulation facilities, and provision of other personal comfort elements such as space for hand luggage and clothing.
2. *Physiological.* Air conditioning for temperature and odors, and lighting, noise, and vibration.
3. *Psychological.* Vista and space, safety, expectations, and impressions of service levels, efficiency of staff.
4. *Activity provision.* Minimize boredom by provisions for smoking, eating, drinking, conversation, reading, watching films, movement around cabin, sleeping.

*Spatial Factors.* In selecting aircraft cross sections, the designer must consider the number of seats across the width of the aircraft. Two-aisle wide-bodied aircraft have up to 10 seats, single-aisle aircraft have up to 6 seats. Minimum seat width is 15 in., but the center seat of three is normally increased to 17 in. to provide a more comfortable position for the center traveler. The position of the window seat offers a better view, but the curvature of the fuselage, especially in narrow-bodied aircraft can limit headroom and leg space for the window seat. Seating comfort depends greatly on seat design. Improved designs have done much to provide better support for the larger passenger and to provide more knee and foot space. Seats must also be designed to minimize injury to passengers in emergency conditions. Both comfort in flight and safety during evacuation are dependent upon seat pitch. Where seats are in pairs, seat pitch is usually 31 or 32 in.; seats in threes are normally more widely pitched at 33 to 34 in.

Aisles must be provided to permit circulation, evacuation, and cabin service. Minimum width is set at 15 in., but 19 in. is used where possible to minimize disturbance of aisle passengers. Other cabin accommodations are the provisions of lockers, wardrobes, and storage space to keep unstowed luggage to a minimum. It also must be remembered that evacuation should be possible in 90 sec in total darkness using only half the total number of emergency exits. Adequate toilet and galley space is necessary as is storage space for blankets, pillows, and serving equipment.

*Physiological Factors.* It is usual to control the *pressure* of cruising aircraft to 6000 ft at a time when ambient pressures may be anywhere up to 38,000 ft for subsonic aircraft. Pressurization is controlled by the flight engineer to limit pressure changes to a comfortable 0.16 lb/in.$^2$/min, equivalent to a climb or descent rate of 300 ft/min. Aircraft can in fact climb at 2000 ft/min and descend at 4000 ft/min.

*Temperature* is also controlled by the flight engineer at a comfortable 21 to 24°C. Ambient temperatures outside the aircraft can be anywhere in the range of −50 to +50°C. Heating is provided by 400°C compressed air from the engines. This must be reduced from its initial velocity of 5000 ft/min to 200 ft/min at the ventilators so that the passenger in his seat is heated by an air draft that is moving at 40 ft/min. Floor level extractors ensure good downward circulation and help to ensure a warm floor.

*Noise* in passenger cabins requires a design that recognizes that the speech interference level (SIL) is somewhere in the region of 60 to 65 dBA. Impairment of the hearing of cabin staff and frequent travelers can occur if they continually are exposed to noise at 80 dBA. Many jet-engine aircraft have noise levels in the 70 to 80 dBA range (c.f., 70 dBA for the interior of a modern intercity passenger railcar). To reduce interior aircraft cabin noise, attention is given at the design stage to reduction of transmitted engine vibrations and attenuation of aerodynamic skin noise by acoustical treatment.

Improvement in *vibrations* have been related to improvements in interior noise levels. Vibrations, especially those in excess of 10 Hz have been reduced significantly, with a consequential improvement in ride quality. This important factor has also improved with the widespread adoption of high-speed, high-wingload jets and turbojets, which has reduced vibrations of periods greater than 1 Hz. Vibration control is essential to passenger comfort. The 2 to 10 Hz range is critical for body resonance, and maximum sensitivity to motion sickness is found at 0.2 Hz. Maximum sensitivity to lateral vibration, that is, "tail-wagging," is in the 1 to 3 Hz range. Modern aircraft are designed to avoid vibrations in these ranges.

*Psychological Factors.* Designers of aircraft take account of psychological factors in a number of subtle ways. Safety devices for use in case of emergency must be evident, but not so intrusive as to cause unnecessary concern. Work spaces for cabin crews are designed to permit them to operate efficiently; a smooth-running cabin operation inspires confidence. A by-product of this last policy is that crews become less tired and are consequently less likely to show irritation and annoyance with passengers. Airlines also attempt to inspire confidence by imprinting a corporate image within the cabin which is aimed at impressing the passenger that he is in the care of a substantial and responsible organization.

*Activity-Related Factors.* While being transported the passenger is not entirely passive. In this way passengers differ significantly from freight. For a short journey all that a passenger requires is a seat and toilet facilities. For long journeys, the passenger is likely to be engaged in a number of tasks: sleeping, eating, drinking, writing, watching films, listening to music, engaging in conversation, and moving around to reduce travel fatigue. The seating and circulation space is designed to accommodate all these activities.

It can be seen that satisfactory design of an aircraft interior requires a procedure that is much broader than the conventional design procedure of the mechanical or civil engineer. Where there is a large element of consumer acceptance involved, human factors design is essential.

## 6-20. SUMMARY

In general, this chapter has emphasized the role of the human user of a transportation system, and has described in outline some of the ways in which this role may be accounted for and assessed. The individual interacts with a transportation system at the social, cognitive, perceptual-motor, and physiological levels, and the efficiency and acceptability of the system rests on a careful consideration of these factors very early in the process of system planning and design. The measurement rules and criteria applicable to human factors sometimes can be difficult to reconcile with the quantified and standardized measurement systems used in other areas; therefore, considerable care is needed in the interpretation of data. However, this process is vital in reflecting the very purpose of a transportation system in extending the abilities of the individual.

## PROBLEMS

1. Discuss the problems that would likely be encountered in attempting to improve traffic safety by identifying a small fraction of unsafe drivers and removing them from the driving population.
2. Prepare a report discussing the effect of human factors on:
   a. The length of approach lighting systems for general aviation and air carrier airports. (Refer to Chapter 19.)
   b. Passing and stopping sight distance for highways. (Refer to Chapter 12.)
3. Contrast the education and training programs employed to teach the necessary skills to:
   a. Ride a bicycle.
   b. Operate an automobile.
   c. Pilot a ship.
   d. Pilot a commercial airplane.
4. Prepare a report on the deceleration rate that humans can tolerate in a vehicular crash without sustaining injuries or death. Discuss the possible effects that the trend toward smaller automobiles could have on injury and fatality rates.
5. Discuss the design requirements that were incorporated in the Washington Metro System in order to make the system fully accessible to the elderly and handicapped.

## REFERENCES

1. Horodniceanu, M. *et al.*, "System Safety Techniques Useful for Transportation Safety," *Transportation Research Record No. 629*, Transportation Research Board, National Academy of Sciences, Washington, D.C., 1977.

2. *Travel Barriers,* U.S. Department of Transportation, Washington, D.C., 1970.
3. Ashford, N.J., and W.G. Bell (eds.), *Mobility for the Elderly and Handicapped,* Loughborough University of Technology, Loughborough, England, 1979.
Ashford, N.J., W.G. Bell, and T.A. Rich (eds), *Mobility and Transport for Elderly and Handicapped Persons,* Gordon and Breach, London, 1983.
Bell, W.G., and N.J. Ashford, Third International Conference on Mobility and Transport for Elderly and Handicapped Persons, DOT-1-85-07, U.S. Department of Transportation, Washington, D.C., 1985.
4. Rolfe, J.M., "Flight Simulator Research," *Applied Ergonomics,* Vol. 4, No. 2, 1973.
5. Hall, Edward T., *The Hidden Dimension,* Anchor Books, Garden City, N.Y., 1969.
6. Fruin, John J., *Pedestrian Planning and Design,* Metropolitan Association of Urban Designers and Environmental Planners, New York, 1971.
7. Sjosted, L., *Behavior of Pedestrians at Pedestrian Crossings,* National Swedish Road Research Institute, Stockholm, 1967.
8. Weiner, E.L., "The Elderly Pedestrian: Response to an Enforcement Campaign," *Traffic Safety Research Review,* Vol. 12, No. 4, 1968.
9. Sleight, R.B., "The Pedestrian," in *Human Factors in Highway Traffic Safety Research,* T.W. Forbes (ed), John Wiley, New York, 1972.
10. *A Policy on Geometric Design of Highways and Streets,* American Society of State Highways and Transportation Officials, Washington D.C., 1984.
11. Jacobs, G.D., *The Effect of Vehicle Lighting on Pedestrian Movement in Well Lighted Streets, RRL Report 214,* Road Research Laboratory, Crowthorne, England, 1968.
12. *Accident Facts, 1986 edition,* National Safety Council, Washington D.C., 1986.
13. Grayson, G.B., *Observations of Pedestrian Behaviour at Four Sites, TRRL Report 670,* Transport and Road Research Laboratory, Crowthorne, England, 1975.
14. Haight, Frank, and Roy Cresswell (eds.), *Design for Passenger Transport,* Pergamon Press, Oxford, 1979.
15. Modak, S.K., and V.N. Patkar, "Transport Terminal Design and Passenger Orientation," *Transportation Planning and Technology,* Vol. 9, No. 4, 1984.
16. Braaksma, J.P., and W.J. Cook, "Human Orientation in Transportation Terminals," *Transportation Journal of the American Society of Civil Engineers,* Vol. 106 (TE2), 1980.
17. Richards, L.G., "On the Psychology of Comfort," *Human Factors in Transport Research,* D.J. Oborne and J.A. Levis (eds.), Academic Press, London, 1980.
18. Jacobson, Ira D., Larry G. Richards, and A. Robert Kuhlthau, "Models of Human Comfort in Vehicle Environments," *Human Factors in Transport Research,* D.J. Oborne and J.A. Levis (eds.,) Academic Press, London, 1980.
19. Battelle Institute, *Recommendations en vue de l'amènagement d'une installation de transport compte tenu de données anthropomètriques et des limites physiologiques de l'homme,* Geneva, 1973.
20. Hulbert, Slade, "Human Factors in Transportation," in *Transportation and Traffic Engineering Handbook,* Second Edition, W. Homburger (ed.), Prentice Hall, Englewood Cliffs, 1982.
21. Wright P.H., and R.J. Paquette, *Highway Engineering,* Fifth Edition, John Wiley, New York, 1987.

**22.** Matson, Theodore M., Wilbur S. Smith, and Frederick W. Hurd, *Traffic Engineering,* McGraw-Hill, New York, 1955.

**23.** Mourant, Ronald R., and Thomas H. Rockwell, "Strategies of Visual Search by Novice and Experienced Drivers," *Human Factors,* Vol. 19, No. 4., pp 325–335, 1972.

**24.** Marsh, Burton W., "Aging and Driving," *Traffic Engineering,* Arlington, VA, Nov. 1960.

**25.** Greenshields, Bruce D., "The Driver," *Traffic Engineering Handbook,* John E. Baerwald (ed.), Institute of Traffic Engineers, Washington D.C., 1965.

**26.** Greenshields, Bruce D., "Reaction Time in Automobile Driving," *Journal of Applied Psychology,* Vol. 20, No. 3, p. 355, 1936.

**27.** Johansson, Gunnar, and Kare Rumar, "Drivers' Brake Reaction Times," *Human Factors,* Vol. 13, No. 1, pp. 23–27, 1971.

**28.** *A Policy on Design of Urban Highways and Arterial Streets,* American Association of State Highway and Transportation Officials, Washington D.C., 1973.

**29.** Allen, T.M., H. Lunenfeld, and G.J. Alexander, "Driver Information Needs," *Highway Research Record 366,* 1971.

**30.** Alexander, Gerson J., and Harold Lunenfeld, *Positive Guidance in Traffic Control, U.S. Department of Transportation, Washington, D.C., 1975.*

**31.** *Fatal Accident Reporting System 1986,* U.S. Department of Transportation, National Highway Traffic Safety Administration, National Center for Statistics and Analysis, Washington D.C., 1986.

**32.** Ryder, Joan M., Stacy A. Malin, and Craig H. Kinsley, *Effect of Fatigue and Alcohol on Highway Safety,* NHTSA Report DOT-805 854, Department of Transportation, Washington, D.C., March 1981.

**33.** Manno, Joseph E., Barbara R. Manno, Glen F. Kiplinger, and Robert F. Forney, "Motor and Mental Performance with Marijuana: Relationship to Administered Dose of THC and Its Interaction with Alcohol," in *Effects of Marijuana on Human Behavior,* Loren L. Miller (ed.), Academic Press, New York, 1974.

**34.** Klonoff, H., "Effects of Marijuana on Driving in a Restricted Area and on City Streets: Driving Performance and Physiological Changes," in *Effects of Marijuana on Human Behavior,* Loren L. Miller (ed.), Academic Press, New York, 1974.

**35.** Ashford, N., "Passenger Behavior and the Design of Airport Terminals," *Transportation Research Board Record No. 588,* Transportation Research Board, National Academy of Sciences, Washington, D.C., 1976.

**36.** Deighton, Richard, "Passenger Behavior and Expectations at an Airport," in *Design for Passenger Transport,* Frank Haight and Roy Cresswell (eds.), Pergamon Press, Oxford, 1979.

**37.** *Guidelines for Airport Capacity/Demand Management,* International Air Transport Association, Montreal, November 1981.

**38.** Mumayiz, S., and N. Ashford, "Methodology for Planning and Operations Management," *Transportation Research Record,* Transportation Research Board, National Academy of Sciences, Washington, D.C., 1987.

**39.** Ashford, N., "Level of Service Design Concept for Airport Passenger Terminals: An European View," *Transportation Planning and Technology,* Vol. 12, No. 1, 1988.

**40.** *International Signs to Provide Guidance to Persons at Airports,* Doc. 9430-C/1080, International Civil Aviation Organization, Montreal, 1984.

**41.** Bonney, K.V., and G.R. Allen, "Passenger and Crew Considerations, Designing from the Inside Out," Proceedings of the Royal Aeronautical Society Symposium, Royal Aeronautical Society, London, 1975.

**42.** Molony, G., "The Interior Design of Wide Bodied Aircraft," in *Design for Passenger Transport,* Frank Haight and Roy Cresswell (eds.), Pergamon Press, Oxford, 1979.

# *PART III*

# *TRANSPORTATION PLANNING*

*Before specific decisions can be made relating to the layout and design of transport facilities, a number of broad but important questions and issues must be addressed by transportation professionals on an urban, regional, or statewide basis. These include:*

- *What is the overall scale of the transportation problem, and what will be its magnitude at some specified future date?*
- *What mode of transport will best provide the needed service?*
- *How will changes to the system modify and possibly magnify the forecasted travel demands?*
- *How can the preferred transportation plan be implemented with minimum adverse impact on people and the environment?*
- *Given the scarcity of resources, what short-term solutions can be found and implemented pending the development of better but possibly more expensive solutions?*

*Such issues and the procedures for addressing them comprise the broad endeavor of transportation planning, the subject of Part Three.*

# Chapter 7

# An Introduction to Transportation Planning

## 7-1. RATIONALE

The provision and operation of a transportation system requires a continuous planning function to ensure that the mobility requirements of the community are supplied and maintained at a level found to be acceptable to its members at an economic, social, and environmental cost within its capabilities. The output of the planning function should be in terms of what needs to be done, what alternative approaches can be used, how well these alternatives match the community desires, and what steps ultimately need to be taken to implement plans satisfactorily.

The complexity of providing for transport needs defies the use of the simple design approach that can be applied to most engineering problems. Stated simply the usual engineering rationale for design is the determination of the demand to be placed upon a system and the provision of sufficient capacity to satisfy anticipated levels of demand. The system under design usually is isolated and independent. This enables the engineer to compute an optimal solution in a prescribed sequence of steps.

The formulation of a transportation plan is not approached as easily. First, the problem is not isolated and independent. For example, urban transportation solutions are an aggregation of a number of smaller transportation and traffic engineering solutions. Second, urban transportation systems are themselves a small part of the overall regional and national transportation infrastructure. Proper overall transportation planning requires an examination of problems at various levels, because policy decisions at any one level may have severe effects on proposed plans.

The most striking problem in plan design is not, however, the multiplicity of levels at which the solution must be considered. The chief difficulty lies in the fact that, unlike most engineering solutions, a transportation plan will affect its own environment when implemented. This change of environment will modify the demand on the system, possibly invalidating the criteria and input used in the initial formulation of the plan. The cyclic interaction of transportation facilities and land use is shown in Fig. 7-1. Land use has been

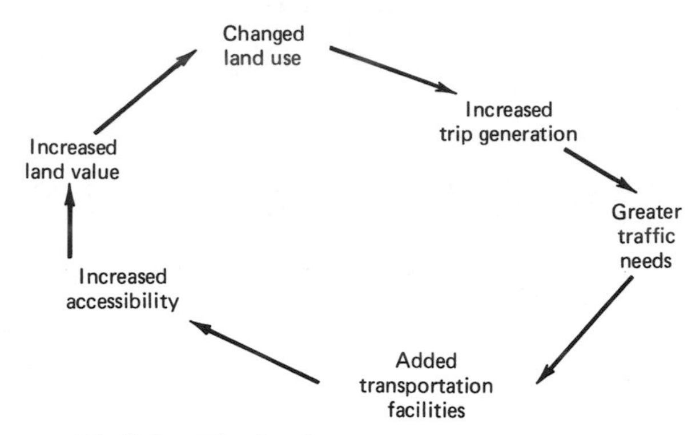


*Fig 7-1.* The land use transportation cycle.

found to be the prime determinant of trip generation activity. The level of trip generation activity and the orientation of trips within the study area will determine the need for facilities. Provision of these facilities alters the accessibility of the land itself, which in turn helps determine the value of land. Land value being the major determinant of land use, the planner is faced with a cycle in which alteration in any one element causes change both to all other elements and to itself.

## 7-2. TYPES OF PLANNING

Transportation planners find themselves involved with two different types of planning: *short- and medium-term* or *long-term (strategic)* planning.

Of these two, *short- and medium-term transportation planning* is usually far less complex. By its very nature it tends to place no great demands on construction activities and therefore has no large capital requirements. Largely it is concerned with obtaining maximum capacity or optimal operation from existing facilities. As such in the analysis and evaluation stages, the planner is dealing with a limited number of criteria. In proposing solutions, the number of options is likely to be limited and the planner usually is constrained to options that are totally contained within the budget allotted to transportation. Therefore, the scale of the problem usually is quite limited and the analysis and evaluation while being detailed are usually simple in structure. This type of problem comes under the generic term *transportation systems management* (TSM).

*Long-term, comprehensive,* or *strategic* transportation planning, on the other hand, can be and is likely to be a very complex problem indeed. It is probable that it will require huge financial expenditures and involve large and extensive construction programs which will affect the economic, social and natural environments. Furthermore, the desired solutions can be achieved only through carefully constructed policy making at the multiple levels of government and administration involved.

By the late 1960s it had become apparent that problems of this complexity could be solved best by the systems approach.

## 7-3. THE SYSTEMS APPROACH [1]

The systems approach is a decision-making process for complex problem solving composed of:

*Systems analysis.* A clear evaluation of the combination of all elements that structure the problem, and those forces and strategies needed for the achievement of an objective.

*Systems engineering.* Organizing and scheduling the complex strategies for problem solution, and the development of procedures for effecting alternate solutions.

1. The team tackling the problem is interdisciplinary in makeup. All facets of the problem are considered, not simply those facets that conveniently fall within one discipline.
2. Throughout the analysis the team uses scientific methods. This requires that a theory must be formulated to account for a set of observed facts. This theory is checked to determine whether or not it actually explains known facts and is also checked to determine its predictive validity.
3. The work is carried out according to a predetermined sequence.

The use of systems theory or scientific decision-making has one most important advantage. It is that in tackling even the most complex problem using a structured and repeatable methodology, different planners working independently would evolve very similar solutions. This inspires confidence from decision makers in planners' proposals.

The remainder of the chapter will deal firstly with strategic planning and secondly with transportation in systems management. It should be realized that whereas the two types of analysis can be separated theoretically, planning takes place in a time continuum, requiring in some cases the use of TSM in long-term planning. Equally, some of the techniques described under strategic planning will be used for short-term solutions.

# STRATEGIC TRANSPORTATION PLANNING

## 7-4. THE PHILOSOPHY OF LONG-TERM PLANNING

Stated succinctly, the aim of the comprehensive strategic transportation plan is to satisfy optimally the goals and objectives of the community with respect to mobility, subject to feasibility, resource, and impact constraints. A number of early comprehensive transportation studies failed or led to what were later to be seen as less than satisfactory solutions. It is clear now that these failures were due to an incomplete understanding of the interactive nature of transportation planning and comprehensive planning in the broadest sense. Figure 7-2 shows a systems analysis diagram of an approach to the development of a strategic transportation plan that is designed to meet the needs and aspirations of a community. The interactive processes can be divided into two separate areas: the philosophical and the operational elements of planning. In any successful plan these two areas must be in broad balance and the decisions taken in one area must take cognizance of those in the other.

*Philosophical Elements.* Largely neglected, these elements are essential to the development of an optimal working plan. Without adequate consideration of the philosophy of the planning process two major difficulties may arise. First, there is no way of defining with any degree of certainty what is an optimal plan or indeed in determining with any clarity the difference between a good plan and a bad plan. Second, a plan developed in isolation of community values will face substantial, if not overwhelming, obstacles to successful implementation.

In attempting to comprehend the philosophy of planning it is necessary to understand the hierarchy of the philosophical elements in Fig. 7-2. These have been defined in the following way [2]:

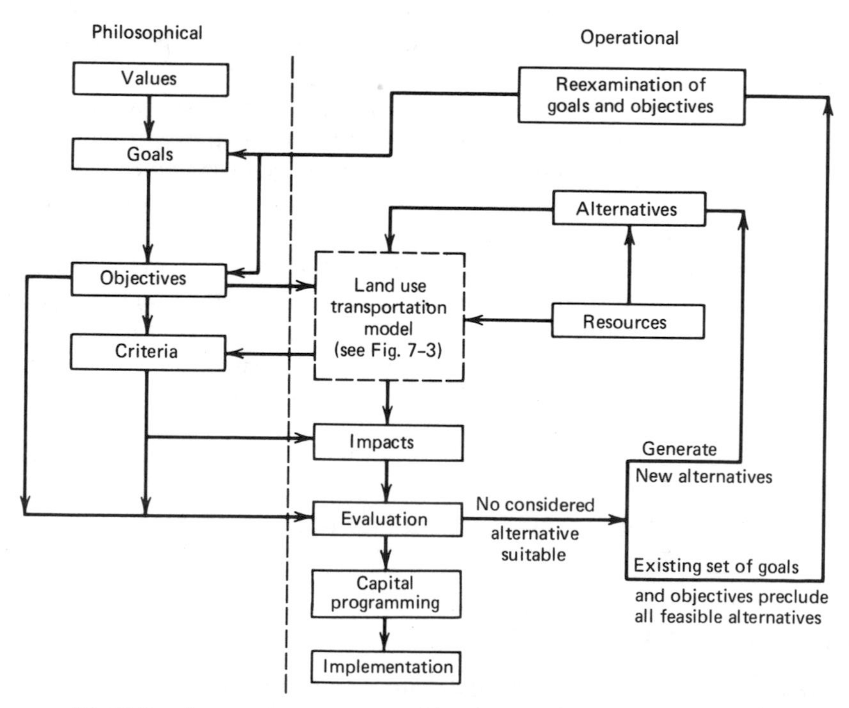


*Fig 7-2.* A systems structure of long-term or stategic planning.

*Values.* The underlying basic qualities upon which the ethics, morals, and preferences of societies, groups, and individuals are based. At least in the medium term these are irreducible and do not change.

*Goals.* The idealized desired end at which the planning process is aimed. Proposals within the plan are aimed at moving society toward these ideal goals.

*Objectives.* Measurable operational statements of individual goals, defined without reference to attainability in terms of budgetary or other resource constraints.

*Criteria.* Indices of measurement capable of defining the degree to which an objective or goal has been attained.

Figure 7-2 shows how the transportation planning process is a means of planning the use of resources to attain community goals and objectives. In this context, the term community is used in its broadest sense, ranging from the international or national community down to a local group of individuals such as a neighborhood. Goals and objectives of the transportation plan are set by the planner within the framework of the underlying and unchanging values of society. At the same time, criteria are established that can be used in the evaluation stage of operational planning. These criteria are dependent not only on the goals and objectives to be achieved, but on the form of the modeling process.

Once the philosophical side of the transportation planning process has been structured reasonably coherently, the planner is in a position to proceed with operational planning procedures. The one must precede the other to some degree if the evaluation phase of the modeling aspects are to have any validity. The results of a number of transportation studies have been questioned and their recommendations have been modified substantially because too much effort was expended on the mechanics of the modeling process at the expense of structuring a transportation plan that reflected the aspirations of the community.

Ideally, the modeling process is constructed taking into account community goals and objectives, resource constraints, and the range of available alternative solutions. The output of modeling is a number of definable system impacts which are examined with respect to the previously established criteria in the evaluation process. The process of system evaluation is of such vital importance to the production of a viable transportation plan that the authors consider that the topic requires separate and special discussion. The reader is referred to Chapter 10.

As a result of the evaluation of proposed systems either one system will be judged as optimal or all systems currently under evaluation will be rejected. In the first case, the planner will construct a capital program that will reflect the sequencing of capital investment in construction and equipment. In the other case, new system alternatives must be generated for a recycling through the modeling process. These are evaluated subsequently. It is possible that after several cycles through this process no satisfactory alternative has been found and no further alternatives can be generated realistically. This is the situation where the goals toward which the planner is working are beyond the constraints that resource availability puts on the system. The flowchart indicates that goals and objectives must be reassessed in this case in a downward direction so that at least one of the alternatives can satisfy the evaluation criteria being used. Although not shown in the chart, it is also possible that resources could be increased to achieve the same end.

## 7-5. REALISM IN THE SYSTEMS APPROACH

While the use of systems analysis can be of immense assistance in the solution of complex problems, this approach must be tempered in the light of the institutional framework within which planners must work and over which they have little control. For example, while the multimodal planning and operation of transportation is an end state which may seem to be ideal from many viewpoints, the realities of institutional structures would indicate that modes have been separated for many years and even in the long term are likely to remain so. The most that can be achieved in many cases may be multimodal coordination rather than multimodal integration. Traditionally the training, experience, and even the interests of planners, designers, and operators have been modally oriented. In many cases the modes are controlled or funded by separate arms or levels of government. Even different and well-established union structures and practices may be involved, which may make rationalization virtually impossible. In the late 1960s, under the Johnson administration, an attempt was made to introduce the systems approach to federal government operations by the use of program budgeting (PPBS). This approach failed in the face of programs that both overlapped the jurisdictions of a multiplicity of governmental departments and other organizations and failed to solve the problem of frequent administrative and operational reorganizations. These are needed to reflect the varied policies that government must adopt in the face of political, social, and economic change at the international and national level. An overall systems approach can be useful in transportation planning to bring about scientific decision making; the planners, however, must be careful to adapt the approach to the realities of the environment in which they are working.

## 7-6. ELEMENTS OF LONG-TERM TRANSPORTATION PLANNING

There are three basic elements that constitute any long-term transportation planning process [3]:

1. Forecast of demand for the system at the various levels of facility provision being considered.

2. Description of economic, social, and environmental changes that will accompany the development of the system at these same levels of facility provision.
3. An evaluation of the system in terms of benefits and disbenefits accruing from the various options considered.

In analyzing the system, transportation planners finds that they are dealing with three differing groups of individuals having very different viewpoints on the suitability of a system. These groups are the *operators,* the *users,* and the *nonusers.* The *operator* is concerned with such matters as capital costs, operating costs, operating revenues, and the viability of the plan from an institutional viewpoint such as union cooperation, or governmental funding and control. The *user,* whether a traveler or the customer of a freight transportation undertaking, is concerned with such factors as monetary cost (usually in terms of the fare or tariff), journey time from real origin to real destination, safety and security, reliability, and comfort and convenience. In addition to these directly involved parties, there remains frequently a large number of people, who while neither traveling themselves on a particular system nor causing goods or people to move, are affected by the proposals of the transportation planner. Such *nonusers* of particular transport facilities are affected by such factors as air, water, and solid waste pollution; noise; visual intrusion; safety; land use changes; and social disruption and economic effects (sometimes beneficial, sometimes not). A successful transportation plan is a plan that balances the needs of the operator and the user against the benefits and disbenefits accruing to the nonuser.

## 7-7. THE LAND USE TRANSPORTATION MODEL

The overall system diagram shown in Fig. 7-2 is, with the exception of the transportation specific modeling element, a chart that can be used to describe any long-term planning process. Figure 7-3 shows in detail the most common modeling procedure that has been used in comprehensive land use transportation studies. To be perceived in its correct context, Fig. 7-3 is a subsystem of the strategic planning system shown in the preceding diagram. The land use transportation model can be divided into two distinct phases: the *calibration phase,* in which, the models are built and tested using data from a base period; and the *projection phase,* where the developed models are used to determine future transport demand based on social economic projections for a design year. There are seven principal models used in the process of long-term planning. They are:

1. Population model.
2. Economic activity model.

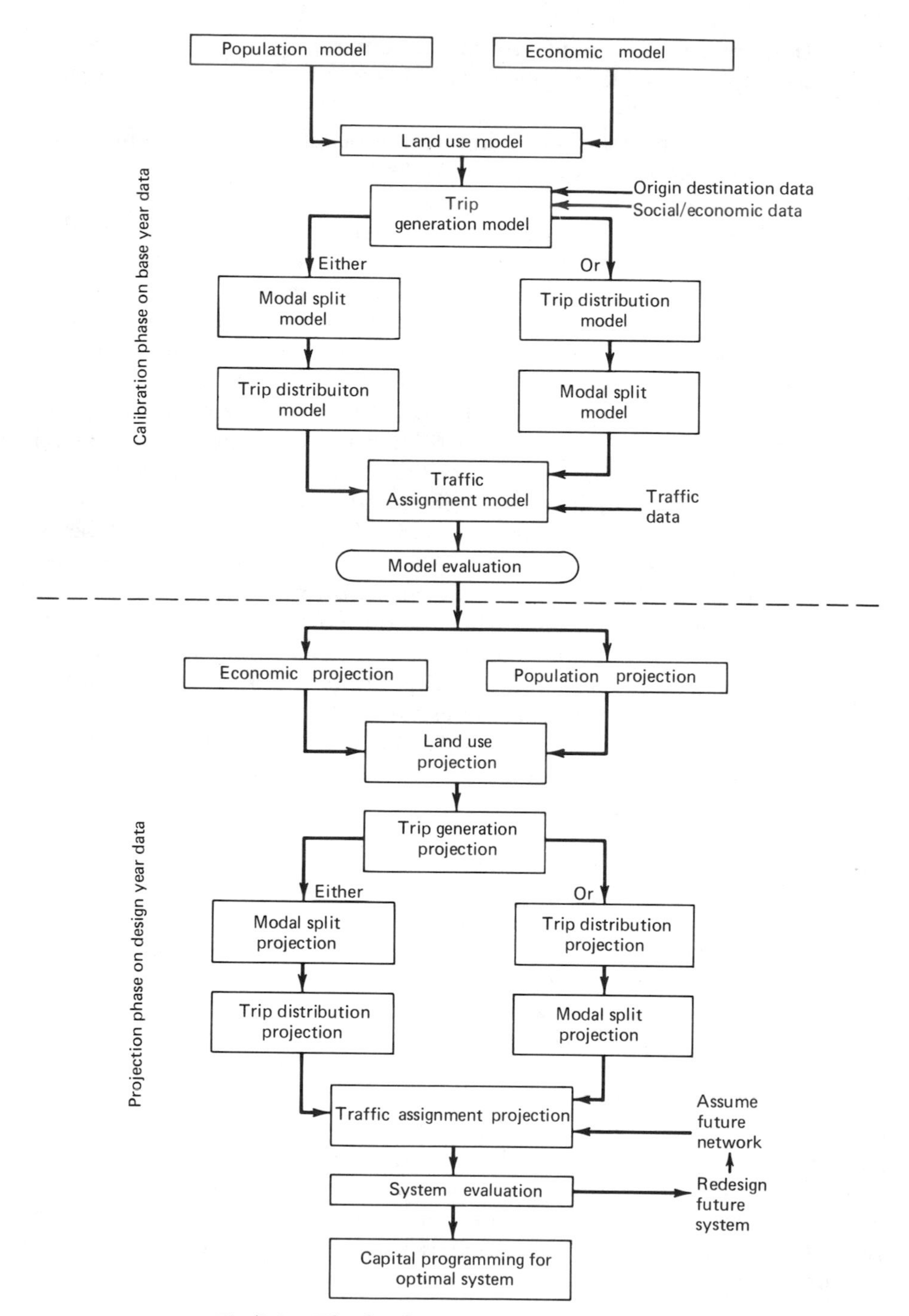


*Fig 7-3.* The land use transportation model.

3. Land use model.
4. Trip generation model.
5. Trip distribution model.
6. Modal split model.
7. Traffic assignment model.

*Population* and *economic* activity predictions generally are derived by specialist demographic and economic projections outside the scope of the transportation study. Because these areas usually would not require detailed examination by the transportation specialist in the preparation of the typical long-term transportation plan, they are omitted from this discussion.

There has been considerable debate on the third of these seven models, that for determining *land use* [4, 5]. A number of models have been developed that relate changes in land use to such independent variables as:

Accessibility to employment.
Percentage of available vacant land.
Land value.
Intensity of land use.
Measures of zone size.
Amount of land in different uses.
Net density of development in the base year.
Employment by land use type.
Time and distance to highest valued land of study area.
Degree of zoning protection expressed on a quantitative scale.
Transit accessibility.
Quality of water and sewer service.

Some of the dependent variables that have been predicted include:

Increase in residential units.
Increase in commercial land use.
Increase in industrial land use.
Increase in retail land use.

A number of models have been built [6], some of which have been extremely sophisticated and capable of calibration to remarkable accuracies [7]. These models are based on the interactive nature of the supply of development infrastructure and urban growth, as discussed in Section 7-1. There is, however, a very strong case against using land use models entirely for predicting urban growth. If used, the resultant land use plan will be self-fulfilling and the growth obtained is likely to be similar to that predicted by the model.

However, the juxtaposition of land uses achieved may not be desirable. Many planners feel that the layout of land use is vital to the proper development of any community and strongly resist any attempt to have this basic planning input modeled. In areas such as Europe where there is strong planning legislation, the land use pattern is set out by planners who design the development of communities with conventional design procedures. Land use modeling may be used, but only in the broadest sense, to ensure that adequate space has been supplied for each type of land use on a communitywide basis. Modeling of zonal changes would not be used in this case. On the assertion that land use models have not proved generally applicable in the land use transportation modeling process, it is considered that there is no need for further discussion of this topic. In summary, transportation planners generally are given land use plans where density and type of land use are set by design considerations rather than by modeling outputs.

The four remaining models are of considerable interest for strategic transportation planning: *generation, distribution, modal split,* and *assignment.* The planning area is divided into a number of relatively homogeneous traffic generating zones. The total traffic flow is modeled by treating the traffic as being generated by the center of gravity of each zone and moving between and within zones over the principal transportation network. Conventionally the models are developed sequentially. *Trip generation* models indicate how many trips are generated in each zone for a particular journey purpose. *Trip distribution* models describe how many trips originating in one particular zone end in each of the other zones. *Modal split* analysis models the proportion of trips that accrue to the various competing modes of transportation. Finally, the *traffic assignment* model indicates which individual routing will be taken by a trip between its origin and destination.

## 7-8. OTHER APPROACHES TO THE TRANSPORTATION MODEL

While the sequential approach of models in linked chains shown in Fig. 7-3 was used almost exclusively in the comprehensive urban and regional transportation studies carried out throughout the United States in the 1960s and 1970s, other approaches have been tried. It has been argued strongly with some justification that the decisions to make a trip from A to B are not made in the manner implied by a sequential model chain (e.g., "I wish to take a shopping trip, I will go to shopping center B, I will use the car, I will use the expressway route.") Rather, the decision is a composite one ("I will jump in the car and go on the expressway to shop at shopping center B."). Models have been developed that integrate distribution, modal split, and assignment [8], or even directly estimate traffic, integrating the model chain from generation to assignment into one step [9, 2]. Neither of these approaches has been used widely.

Much more common is the mode specific type of model, which is used fairly extensively in the special case of air transport planning where there may be very little modal competition in terms of surface transportation. Mode specific generation models are used to determine the level of air transport from a hub [10], or for air transport activity over a specific route or sector, such as the North Atlantic [11]. The techniques involved in mode specific models are in most ways entirely similar to the techniques described in Sections 8-1 to 8-7, under Trip Generation.

## 7-9. FUTURES ANALYSIS AND SCENARIO BUILDING

The modeling process that has been outlined in the previous sections and that is described in greater detail in Chapter 8, has a number of drawbacks when examined in the light of long-term planning during the 1980s and 1990s. The types of models that have been used traditionally in the strategic land use transportation model were developed in the United States and Western Europe during a period of what, in hindsight, was a period of relative economic, social, and political stability, from the mid-1950s to the mid-1970s. Consequently, the models deal with a relatively small range of variables which are expected to change in a not too dramatic manner over the planning period. When tested in their performance over time, there were strong indications that the models were temporally stable [12, 13, 14]. Typical variables that were used for predicting travel demand were travel cost, level of travel service in terms of waiting and walking times, relative travel times, relative income level of the trip-maker, and accessibility by various modes. The ranges over which these variables were considered were relatively restricted. Given that the western nations had enjoyed a sustained period of economic growth and continued prosperity the models looked adequately robust for the future.

The second half of the 1970s brought about changes that, if not unforeseeable, were at least unforeseen. A number of important factors that are intangibles in the traditional modeling process became apparent.

1. Very substantial fuel price changes that set the costs well above anticipated levels and the occurrence of fuel shortages induced by political actions outside domestic control.
2. The development of structural unemployment in the work force causing a hardening of management labor relations and increased resistance to changes both in work practices and the shedding of labor.
3. The rapid development of microchip automated technology having implications on the work place, the time spent at work, and the age of retirement.
4. A worldwide constraint on the rate of growth of energy demand and real energy shortages.
5. Unforeseen downward changes in population growth rate and a change in urbanization patterns in the United States. Large urban areas have not grown as rapidly as anticipated whereas there has been a

larger than expected growth around small- and medium-sized urban areas.

6. Whereas the environmental lobby had shown dramatic strength in the late 1960s and early 1970s, there were signs by the late 1970s of a retrenchment at least in the growth of environmental concerns in the face of severe economic pressures.
7. Political changes, both domestic and foreign, occurred much more rapidly than in the 1950s and 1960s. Abroad there was severe instability throughout the Far East and more importantly in the oil-dominant Middle East. Internally, the nation became more aware of the need for pluralistic policies and planning. Minority groups exerted a strong influence on transport policies which could no longer be decided on simple benefit-cost relationships.
8. Inflation, which had remained at tolerable levels between 1945 and 1975, became a prime economic problem for the governments of all Western nations and affected the public sector spending policies advocated under Keynesian economic theory and practice. The whole question of continued growth of the Western economies which underlay virtually all planning models both in the United States and elsewhere became suspect.

Faced with such massive levels of uncertainty, some planners have recognized that the conventional type of modeling shown in Fig. 7-3 is too dependent on the stable framework of the past to enable prediction of a future where serious disruptions to continuity have occurred. A technique that has been used to model futures of this nature is *scenario building.*

In this type of modeling a series of future end states are predicted. These end states include the intangibles, which are so difficult to model in more conventional analysis. The effect of intangibles on response variables of interest (such as transport demand) are estimated where precise calculations prove impossible. For example, for a particular national plan one end state might be predicted on the following assumptions:

- Low economic growth rates over the next 20 years.
- High environmental concern.
- Strong centralized government with little decentralization of governmental activity.
- Weak trade union activity.
- High levels of automation.
- Medium levels of population growth.

Other end states would be predicted using different assumptions about these same factors. Scenario building has a number of advantages.

1. The end states are not necessarily date specific and are therefore more realistic for really long-term plans.

2. The interrelationships between a number of intangibles can be estimated, as can their effects on response variables in the end states.
3. The models are robust in that they are capable of predicting across social, technological, and economic discontinuities.
4. They are truly a long-term tool even though they are of limited use in the short or medium term.

Figure 7-4 shows a systems chart of the scenario building analysis that was used in the mid-1970s for the strategic planning of the rail network in the United Kingdom [15].

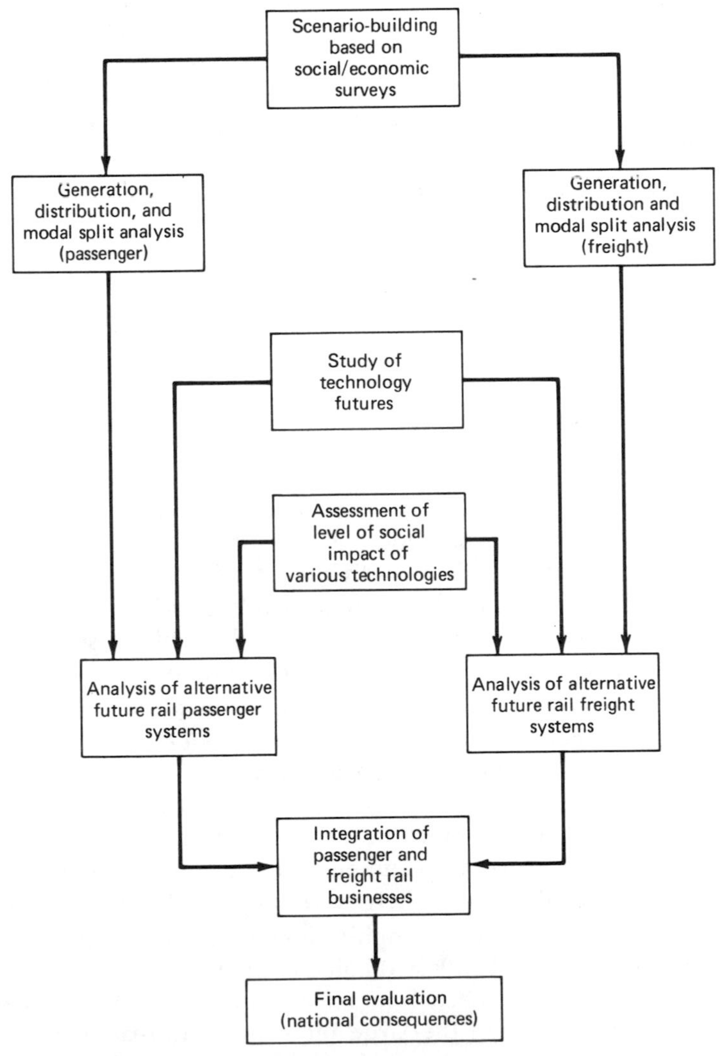


*Fig 7-4.* Schematic of the scenario analysis of the British Rail Stategic Plan.

## TRANSPORTATION SYSTEMS MANAGEMENT

### 7-10. NEED FOR TRANSPORTATION SYSTEMS MANAGEMENT (TSM)

Prior to the early 1970s, transportation planners had been almost entirely preoccupied with the long-term or strategic planning of transportation facilities. The requirements of the 1962 Federal Highway Act ensured that all cities with populations in excess of 50,000 had comprehensive long-term transportation plans. One of the requirements of the long-term plan was that it should be "continuing," that is, the process was set up to be modified continuously. For a number of reasons it became apparent by the mid 1970s that the almost total emphasis on long-term planning was no longer acceptable in the light of planning experience gained in the previous 20 years. In the fall of 1975 the Federal Highway Administration and the Urban Mass Transportation Administration jointly published guidelines which recognized the role of short- and intermediate-term planning within the context of the total transportation planning process [16]. These joint guidelines set the stage for a radical reorientation of thinking by United States transportation planners, and the widespread development of short- and intermediate-transportation plans that relied on TSM techniques for local and system improvements.

The "strategic" comprehensive planning programs of the 1960s were of a systemwide nature and concentrated on the construction of new long-term facilities in the area of auto and public transportation. A number of reasons have been put forward for what appeared to be a rather sudden deemphasis of long-term planning in favor of short- and medium-term solutions.

1. By its nature, long-term planning was normative and of a systemwide nature. Planning of this type had not been found to be sufficiently responsive to the needs of individual groups of people, such as the elderly, the handicapped, and the socially disadvantaged. There was also an inflexibility of response to special problems such as environmental impact, energy use and conservation, and sociological impact.
2. The public sector of the economy, which previously had shown continuous growth became subject to constraint. As a result, in the increasing competition for public funds, transportation was no longer seen to have the former level of priority. Coupled with this the escalation of construction costs at a much faster rate than the general inflation rate indicated that an overall planning strategy emphasizing the construction of new facilities was likely to be unworkable in the long term.
3. To some degree the demographic and land use developments that had been forecast in the 1960s had failed to take place. Population increase was lower than had been forecast and growth that had been anticipated to take place in the large metropolitan areas had to a degree been taken up by an unexpected growth around smaller urban centers.

4. The philosophy that new transportation facilities could be built to cope with exponential growth of traffic demand became untenable in the face of huge real increases in fuel costs that continued throughout the 1970s and through the early 1980s. The realization that the era of cheap energy to fuel transport was over was paralleled by a fierce resistance to the construction of surface infrastructure for transport in the cities. In this climate the public was making a clear demand that the existing supply of transportation should be made substantially more efficient before it should be increased to cope with expanding demand.
5. It was apparent that a broad reliance on the construction of new transportation facilities continued a trend to increasing inequity in society. New facilities often caused increasingly vocal sections of the community to bear financial and social costs that were unmatched by their transportation benefits. Existing inequities were more likely to be removed by service improvements rather than by construction-intensive projects.

The general aim of transportation system management techniques is to improve the efficiency of transportation systems by bringing the supply of facilities into better balance with demand by an approach toward the optimization of moving people and goods rather than vehicles.

## 7-11. DIFFERENCES BETWEEN STRATEGIC PLANNING AND TSM [17].

There are a number of fundamental differences between long-term transportation planning and short- to medium-term transportation systems management. While they must be seen as complementary methods of providing optimal solutions to the problems of moving goods and people, it is useful to make the following comparisons between the two approaches.

1. Long-term transportation studies have tended to be capital intensive, normative to the global community, of a comprehensive nature, and of long lead times. TSM schemes characteristically tend to be of a low-budget nature, and of short term both in the planning and implementation stages. Whereas long-term plans are based on secular trends in demand, the need for individual TSM schemes may vary over the long term.
2. Short- and medium-term improvements may be unattainable by the capital-intensive projects engendered by long-term planning due to long planning and implementation lead times.
3. Capital-intensive plans may require complementary TSM measures to ensure that the major projects are fully effective.

4. The monitoring and reevaluation of TSM measures must be virtually continuous. While it is recognized that even long-term planning is a continuous process with updating procedures throughout the life of the plan, a much more active and less formal monitoring system is required for short-term plans.
5. Because TSM measures are low budget they can be designed in a more flexible and experimental manner. The planner can use a more heuristic approach to solving transport problems.
6. TSM should not be confused with traffic operational improvements only. The latter form only a part of the former, which is the whole. Transportation systems management is a continuing, comprehensive monitoring and adjustment of the working of a transportation system to ensure a constant modification of both supply and demand by means of low-cost system management techniques.

## 7-12. THE TSM PLANNING PROCESS

Transportation system management programs are undertaken with a planning process that is considerably less well defined than that which has been set out for long-term or strategic planning. One major difference is that the planning is unlikely to be carried out by a specialized team assembled specifically for the purpose of TSM planning. More usually the responsibilities for the program projects are scattered in a number of departments within the transportation authority; in some cases the responsibilities will lie in different authorities. This less structured approach is a reflection of the range of medium-term transportation planning and short-term traffic management projects which are subsumed under the heading of transportation systems management.

Figure 7-5 indicates the individual technical steps that are integrated into a general planning philosophy and procedure. The principal steps involved are: program initiation, derivation of performance standards, evaluation of existing transportation system performance, enumeration of TSM options, determination of costs and impacts of individual options, selection of projects and project groupings, priority setting, and implementation program.

Initiation of TSM projects is likely to be a rather informal process in comparison with the initiation of the long-term strategic transportation plan. Project initiation may result from an apparent deterioration of transportation conditions, from social pressures resulting in changing goals and priorities, from routine or periodic planning and management decisions, or from the carrying over of projects from previous programs. Performance standards for the system and its elements may be set either by generally accepted norms such as standards set by manuals, handbooks, and codes or by warrants or may be internally designed standards used as system criteria that relate to the overall operating philosophy of the transportation agency. Design, cost, and revenue estimates normally are devolved to the appropriate operating authorities and agencies since these require a range of specialized knowledge unlikely to be

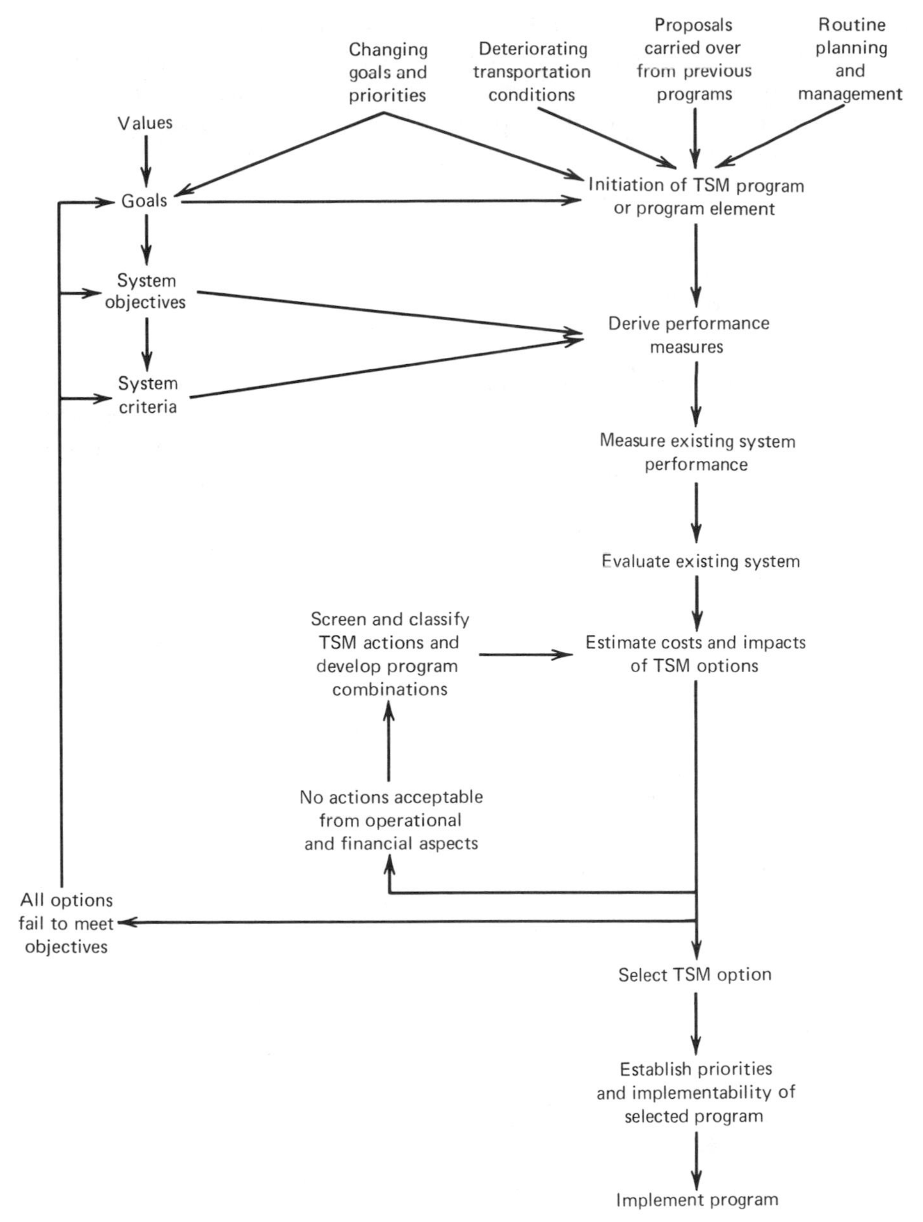


*Fig 7-5.* A general planning procedure for a TSM program. (Adapted from Reference 2.)

centralized within the planning agency. In selecting appropriate projects for inclusion within the TSM program, attention should be paid to the compatibility of various projects with program groupings. This is discussed at greater length in Section 7-12.

Finally, in setting out a capital program, it is necessary to examine the problems of coordinating the funding available from federal, state, and local sources. Budgetary constraints will be imposed by bodies such as the Federal Highway Administration, the Urban Mass Transportation Administration, local government, and possibly operating authorities such as metropolitan transit authorities.

## 7-13. FORMS OF TRANSPORTATION SYSTEM MANAGEMENT

Traditionally, transport planning has amounted to estimating the long-term demand for transportation and adjusting the supply to the requisite demand level using largely capital-intensive methods. TSM methods are best seen as actions or groups of actions that produce shifts in the supply-demand equilibrium of the transportation system. In a framework of supply-demand theory, equilibrium can be shown by plotting the supply and demand curves against axes of travel disutility and vehicle miles traveled, as in Fig. 7-6. The system functions at the equilibrium point where demand and supply curves intersect. Changes in equilibrium can be achieved by moving to another demand curve, by changing the supply curve or by changing both simultaneously. TSM measures can be conceptualized conveniently as a means of bringing about equilibrium changes by levels of transportation supply and demand [18]. It will be seen that transportation system management measures can be classified into four classes:

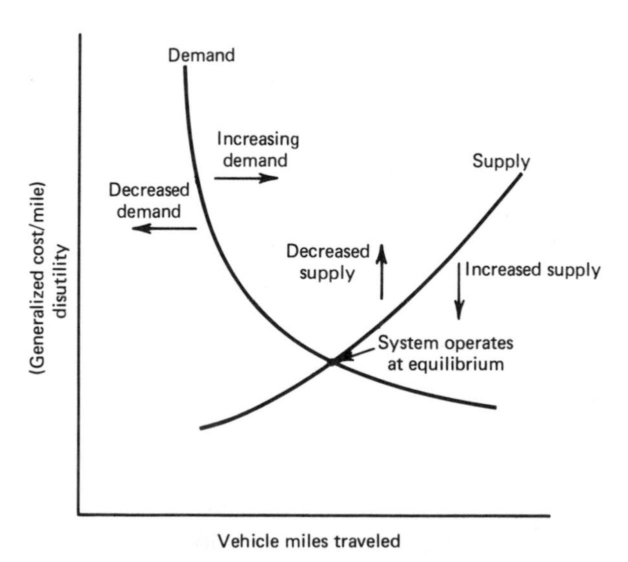


*Fig 7-6.* Supply and demand relationships for urban transportation.

1. Measures by which facility demand is reduced effectively.
2. Measures by which facility supply is increased effectively.
3. Measures by which facility demand is reduced effectively and facility supply is reduced effectively.
4. Measures by which facility demand is reduced effectively and facility supply is increased effectively.

## 7-14. MEASURES TO REDUCE DEMAND

A number of measures can be adopted that, without altering the level of transportation supply, could induce travelers to use existing vehicles at higher load factors, to switch to nonmotorized modes, or to reduce either trip-making frequency or average trip length. The result would be a downward shift in the equilibrium point by a downward shift in demand at a constant level of supply, as shown in Fig. 7-7. Among the measures that can be considered as effecting demand reduction are:

- *Ride sharing* including car pooling, van pooling, shared taxi systems.
- *Transit marketing.*
- *Improved bus services* including increased network density, increased frequency, express bus systems, improved vehicle design, and improved operational characteristics.

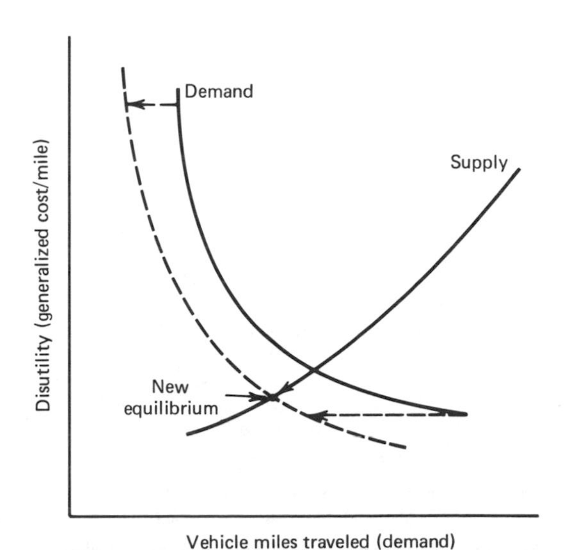


*Fig 7-7.* Change in equilibrium brought about by TSM measures that reduce demand.

- *Park-and-ride systems* including car pool fringe parking.
- *Paratransit systems* including dial-a-ride, subscription bus systems, jitneys, specialized volunteer systems.
- *Road pricing* including increased taxes and tolls for low-occupancy vehicles and decreased taxes, tolls, and fares for high-occupancy vehicles.
- *Improvements to bicycle and pedestrian facilities.*
- *Shortening work week*—by reduction of work week from five to four days the overall demand in terms of vehicle miles is reduced. Note that staggering of work hours does not fall into this category, but is discussed in Section 7-9.
- *Use of communications in lieu of transport* by promoting use of telephone or mobile facilities such as shops, libraries, clinics.

## 7-15. MEASURES TO INCREASE SUPPLY

It is usually possible to increase the supply of transportation (or in the terms of this conceptual construct to decrease the overall vehicle hours of travel at a given level of vehicle miles of travel) by adopting a number of low-cost traffic engineering and traffic control measures. The effect on the equilibrium point is shown in Fig. 7-8; there equilibrium is seen to occur at the intersection of

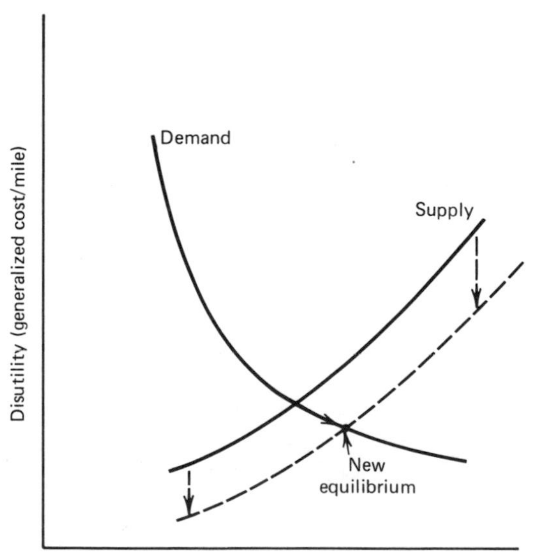


*Fig 7-8.* Change in equilibrium brought about by TSM measures that increase transportation supply.

the new supply line and the existing demand line. Actions in this category would include:

- *General street traffic engineering improvements* including minor design improvements, improved control devices, and regulations which aim to speed traffic flow.
- *Freeway traffic management* including surveillance ramp control, driver advisory information.
- *Truck restrictions* aimed at reducing conflict between trucks and autos and increasing general traffic flow.
- *Staggered work hours* to move vehicle trips into times when congestion is lower.

## 7-16. MEASURES THAT REDUCE DEMAND AND DEGRADE SUPPLY

By increasing the general travel time on the system, the level of transportation supply is seen to be degraded. If preference can at the same time be given to high-occupancy vehicles on the system, the vehicle miles of travel are decreased as more individuals switch from low- to high-occupancy vehicles. Measures that are designed to decrease demand and simultaneously degrade supply include:

- *Preferential treatment* of high-occupancy vehicles, for example, reserved freeway lanes for buses and other high-occupancy vehicles, priority bus ramps on freeways, freeway ramp metering and bypass positions, curb bus lanes, median bus lanes, bus streets.
- *Auto restricted zones* such as pedestrian malls, bus-only access areas.
- *Reduction of off-street parking supply.*

All of these measures remove some level of facilities from low-occupancy vehicles, and result in lower line haul travel speeds, more circulation time, or greater walking time after parking, while reducing overall demand in terms of vehicle miles traveled by inducing a switch to the higher-occupancy vehicles. The resulting change in equilibrium conditions is shown in Fig. 7-9.

## 7-17. MEASURES THAT INCREASE SUPPLY AND REDUCE DEMAND

These measures cause a switch of mode that by selectively increasing the level of transportation supply to high-occupancy vehicles, and by decreasing the overall travel times of these vehicles, thereby decreasing the overall demand in

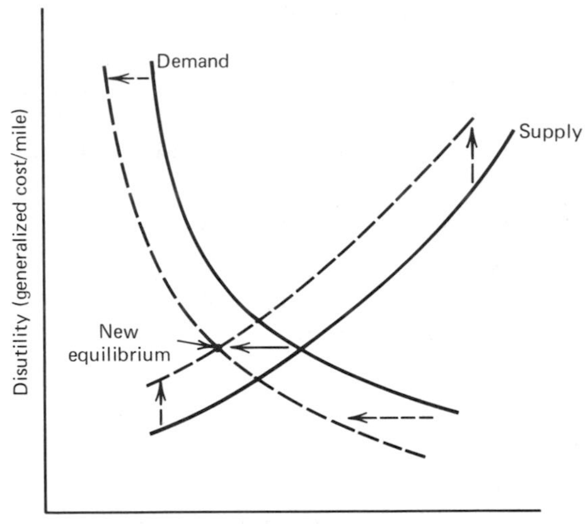


*Fig 7-9.* Change in equilibrium brought about by reduced demand and degraded transportation supply.

terms of vehicle miles traveled. The change in the equilibrium condition is shown in Fig. 7-10. Measures that come under this category include:

- *Additional preferential treatment lanes.*
- *Restriction of on-street parking* to speed bus movement in curb lanes.
- *Contraflow lanes* for high occupancy vehicles.

## 7-18. COMPATIBILITY OF TECHNIQUES

Any comprehensive program of TSM measures will be composed inherently of a number of the individual techniques discussed previously. These techniques may be applied to either the whole or part of the planning area. Theoretically, any mixture of techniques could therefore attempt to reduce demand, increase supply, reduce both demand and supply simultaneously, or reduce demand while increasing supply. In assembling TSM projects into packages or programs it is essential to examine the mutual compatibility of individual measures. Not all techniques are compatible. For example, a program of staggering working hours while reducing peak hour demand is not really compatible with a policy of encouraging car pooling. The latter requires close identification of trip ends and trip times. Similarly, a policy aimed at substituting communications for travel is likely to be incompatible with a transit marketing campaign that seeks to induce car drivers to switch to the high-occupancy public mode.

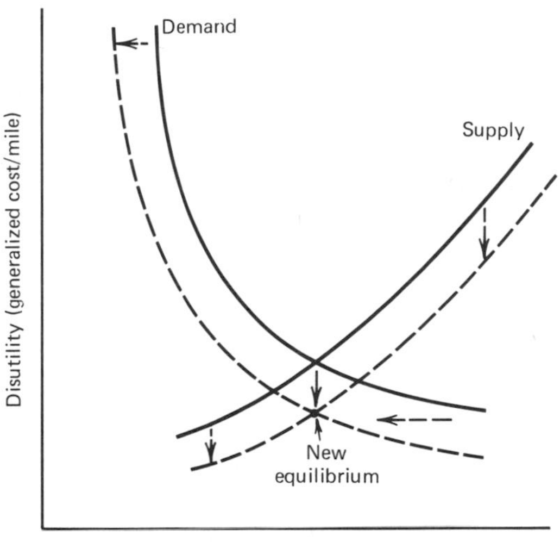


*Fig 7-10.* Change in equilibrium brought about by reduced demand and increased transportation supply.

Figure 7-11 shows a matrix indicating the relative compatibility of the various techniques. These techniques can be assembled into a number of program packages of internally consistent modules.

*Module 1.* *Work hour changes* including as component measures, staggered work hours, parking controls, road pricing, incentives to transit, and extended area transit.

*Module 2.* *Pricing techniques* including component measures of road pricing, parking controls, car pooling and other ride sharing, incentives to transit, priority transit for arterials and expressways, transit circulation, and extended area transit.

*Module 3.* *Restricting access* including creation of traffic cells, auto-free zones, maximizing use of existing facilities, and transit circulation.

*Module 4.* *Changing land uses* including component measures of new towns, planned neighborhoods, and revised zoning and building codes.

*Module 5.* *Prearranged ride sharing* including car pooling and ride sharing, road pricing, and parking controls.

*Module 6.* *Communications substitute for travel* including substitution of communications, road pricing, parking controls, new town design.

*Module 7.* *Traffic engineering techniques* including freeway surveillance and control measures and maximizing use of existing facilities.

*Module 8.* *Transit treatments* including incentives to transit use, priority transit, transit circulation and extended area transit [19].

| | Carpooling | Zoning and building codes | Planned neighborhoods | New towns | Communications | Traffic cells | Auto-free zones | Staggered work hours | Parking controls | Road pricing | Circulation | Transit priority: arterials | Transit priority: highways | Extended area services | Transit incentives | Maximum use of existing facilities | Freeway surveillance and control |
|---|---|---|---|---|---|---|---|---|---|---|---|---|---|---|---|---|---|
| Road pricing | + | 0 | 0 | 0 | + | + | 0 | + | + | + | + | + | + | + | + | - | - |
| Parking controls | + | + | 0 | 0 | + | + | 0 | + | + | + | + | + | + | + | + | - | - |
| Staggered work hours | - | 0 | 0 | 0 | 0 | 0 | 0 | + | + | + | + | + | + | + | + | - | - |
| Auto-free zones | 0 | + | + | 0 | 0 | + | + | 0 | 0 | 0 | + | + | + | + | + | 0 | 0 |
| Traffic cells | 0 | + | + | 0 | 0 | + | + | 0 | + | + | + | + | + | + | + | + | + |
| Maximum use of existing facilities | - | + | + | 0 | - | + | 0 | - | - | - | - | - | - | - | - | + | + |
| Freeway surveillance and control | - | 0 | 0 | 0 | 0 | + | 0 | - | - | - | - | - | - | - | - | + | + |
| Zoning and building codes | + | + | + | + | 0 | + | + | 0 | + | 0 | 0 | 0 | 0 | 0 | 0 | + | 0 |
| Planned neighborhoods | 0 | + | + | + | 0 | + | + | 0 | 0 | 0 | + | 0 | 0 | 0 | 0 | + | 0 |
| New towns | - | + | + | + | + | 0 | 0 | 0 | 0 | 0 | 0 | 0 | 0 | 0 | 0 | 0 | 0 |
| Communications | - | 0 | 0 | + | + | 0 | 0 | 0 | + | + | - | - | - | - | - | - | 0 |
| Carpooling | + | + | 0 | - | - | 0 | 0 | - | + | + | + | - | - | - | - | - | - |
| Transit incentives | - | 0 | 0 | 0 | - | + | + | + | + | + | + | + | + | + | + | - | - |
| Circulation system | + | + | + | 0 | - | + | + | + | + | + | + | + | + | + | + | - | - |
| Transit priority: arterials | - | 0 | 0 | 0 | - | + | + | + | + | + | + | + | + | + | + | - | - |
| Transit priority: expressways | - | 0 | 0 | 0 | - | + | + | + | + | + | + | + | + | + | + | - | - |
| Extended area services | - | 0 | 0 | 0 | - | + | + | + | + | + | + | + | + | + | + | - | - |

*Fig 7-11.* Matrix Showing Mutually-supportive Techniques Arrayed to Suggest Program Packages. (Source: Reference 19.)

Just as individual techniques have inherent incompatibilities, the modules themselves are not necessarily applicable to the same area simultaneously. However, incompatible modules can be used for different areas within the planning region. Figures 7-12*a* through 7-12*d* show operating examples of TSM measures in the four supply-demand categories.

## 7-19. DEVELOPING CRITERIA FOR EVALUATING TSM PROJECTS [20]

As with any proposal for the improvement of transportation, it is necessary to measure the strength of the proposal against a set of evaluation criteria. These criteria should be related to the area planning goals and objectives as previously shown in Fig. 7-5. Some authorities have indicated that the objectives

*Fig 7-12.* Transportation system management measures: a. vanpool, b. exclusive lane for car pool, c. contraflow bus lane, d. busway. (Photo a courtesy 3M Corporation, photos, b, c, d courtesy Federal Highway Administration and California Department of Transportation.)

should be stratified into *transportation-related* and *indirect* objectives. Because transportation is a derived demand, rather than an economic good in the true sense, this division would hardly seem necessary. Ideally the developed criteria should be:

1. *Quantifiable.* Where possible, quantification is desirable, because it eases the problem of evaluation, as discussed in Chapter 10. Where quantification is impossible, qualitative assessments may have to be made, even in the extreme subjective judgments.
2. *Related to objective.* The more closely the criteria is related to the objective, the easier the task of project evaluation becomes. Even if a closely related variable presents difficulty in measurement it is preferable to a more remote, more easily obtained measure.

3. *Flexible in scale.* Many TSM projects are local in character. System-wide criteria may not be suitable for monitoring schemes with only limited system effects. Therefore criteria must be selected with a view to the level at which they are to be used and their ability to be applied at the appropriate level.

Examples of TSM objectives and suitable evaluation criteria used in the United States are shown in Table 7-1.

## 7-20. SCOPE OF TRANSPORTATION SYSTEMS MANAGEMENT PROGRAMS

TSM programs should not be regarded simply as extended traffic engineering measures. Traffic engineering is a part of TSM, but only a part. Transportation systems management is the use of short-term non-capital-intensive measures to ensure that the existing infrastructure can be used to its utmost. For

*Table 7-1*
**Examples of TSM Criteria**

| *Objectives* | *Criteria* |
|---|---|
| To improve the performance of the urban transportation system | System-wide travel time<br>Average system-wide travel speeds<br>Average corridor speeds |
| To reduce the dependence of the urban dweller on the automobile | Transit ridership levels<br>Modal split percentages |
| To reduce the fuel consumption for the urban transportation system | Aggregate gasoline and diesel fuel consumption<br>Person miles of travel/gallon consumed |
| To improve safety | Number of persons injured and killed<br>Total number of accidents<br>Accident rate per million vehicle miles of travel |
| To provide mobility to the disadvantaged | Percent of population with access to specialized transport<br>Take-up rates on special fare concessions and passes |
| To improve the urban environment | Noise levels in $L_{eq}$<br>Pollutant levels of carbon monoxide, nitrogen oxides, and hydrocarbons<br>Particulate pollutant levels |
| To increase general levels of mobility while conserving both fuel and the environment | Trip-making rates<br>Person miles of travel<br>Person miles of travel per unit of fuel consumed and per unit of pollutant |

*Table 7-2*
**Examples of TSM Programs (Twin Cities and Portland)**

| | |
|---|---|
| *Twin Cities* | Bus on metered freeways |
| | Preferential transit lanes |
| | Computer-based traffic control system |
| | Pedestrian skyways to segregate and enclose pedestrian movements |
| | Car-pooling organization program including company-based van pools. |
| *Portland* | Freeway lanes for high-occupancy vehicles |
| | Bus-only streets in CBD |
| | Downtown parking policies to discourage commuter parking |
| | Provision of park-and-ride facilities |
| | Bicycle paths |
| | Express bus service for workers to industrial centers |
| | Contraflow arterial bus lanes |
| | Peak-hour signal preemption for bus |
| | Bus stop improvements for bus operations |
| | Transit fare policies to encourage commuter riders |
| | Bus shelter improvements |

example, our existing highway and transit systems are capable of carrying vastly increased numbers of person trips by moving from low-occupancy automobiles to high-occupancy vehicles. For instance, freeway bus lanes are capable of carrying 25,000 persons in peak hour flow. Volumes of 26,000 have been recorded on the I495 contraflow bus lanes in New Jersey. To achieve the goal of stretching the capacity of existing infrastructure, a range of physical and organizational improvements must be made to both the line-haul elements and to the access, terminal, and interchange facilities. Two early TSM programs were carried out in Twin Cities and Portland. Table 7-2 indicates the major items within the scope of these two rather typical programs.

Within the urban context, TSM can be applied through a variety of wide ranging measures which can conveniently be grouped as application in a number of general areas which include [21]:

### Freeway Corridors

- Ramp access control through ramp closure and ramp metering.
- Separate, reversible lanes for high-occupancy vehicles (HOVs).
- Freeway lanes for exclusive use by HOVs during peak periods, (either *add-a-lane* or *take-a-lane*).
- Contraflow lanes for HOVs during peak periods.
- Preferential access for HOVs.

### Arterial Corridors

- Improving intersection design by: correcting offsets, addition of left-turn and right-turn lanes, intersection widening, improved intersection geometrics, location of bus loading at midblock, turn prohibitions, traffic signal modernization, traffic signal coordination, optimization of signal timing.
- Improving roadway conditions: removal of on-street parking, restricting curb openings and property access, restricting median openings, left-turn prohibitions, bus turn outs and loading bays, provision of lanes for unbalanced flow, conversion to one-way pairs, restriction of truck traffic.
- Improving vehicle occupancy: better car pool, van pool, and transit marketing, separate HOV lane for buses, car pools, van pools, preferential HOV treatment at traffic control devices, intersection access control, and preferential bus/car pool access.

### Central Business Districts

- Designation of auto-restricted zones to encourage auto use and improve pedestrian movement.
- Transit malls to improve transit operations and encourage transit ridership.
- Fare-free transit within and to and from the CBD to improve the competitive business position of the central urban area.
- Parking programs either to provide more parking or to restrict undesirable parking.
- Flex-time to spread peak travel demand.
- Express transit to facilitate movement of person trips to CBD.
- Coordinated signal systems to improve vehicular flow and pedestrian accessibility.
- Programs to facilitate and control the delivery of goods and to reduce conflicts between freight traffic and other CBD users.

### Regional Operating Environment

- Ride sharing, (car pools and van pools), on a wide area basis in contrast to schemes implemented by an employer or neighborhood.
- Coordinated transit and paratransit services over a large area.
- Brokerage: the use of a coordinating agency to match demand to car and van pool supply and to arrange for public and private suppliers of transportation.

- Alternative work schedules: staggered working hours, flex-time, etc.
- Reduced transit fare and free fare transit over a wide area.

### Neighborhoods

- Parking constraints, permit parking, and other measures to limit all-day nonresident parking.
- Traffic management plans to restrain through traffic and to create artificially a hierarchical street system in older neighborhoods by means of traffic restraint devices such as speed ramps, stop signs, traffic circles, and median barriers.
- Ride sharing including car pool, van pool, and subscription bus.
- Bikeways for internal bicycle trips and trips to and out of neighborhood facilities such as work, shopping, and recreation.
- Improvement of arterial streets on edge of neighborhood streets.

### Major Employment Sites Outside the CBD

- Ride sharing, (van and car pools) with preferential parking for HOVs.
- Subscription bus or prearranged travel for either door to door or park-and-ride service.
- Express or nonstop bus service on a non-prearranged basis.
- Brokerage (as above).
- Alternative work schedules (as above).

### Outlying Commercial Centers

- Fixed route transit to the center from distant locations within the market area.
- Shoppers bus shuttle from adjacent neighborhoods.
- Dial-a-ride service within marketing area.
- Park-and-ride and express bus service terminals at the center's on-site parking.
- Special transportation (primarily for the elderly and handicapped) to the center on a charter or prearranged basis.
- Traditional traffic engineering measures—increased turn radii, protected left turns, adequate lane widths, adequate pavement markings and traffic signs, well-designed signal systems.

Major Activity Centers

- Parking constraints, parking fees, and fringe parking.
- Alternative work schedules (as above).
- Brokerage (as above).
- Ridersharing in car pools and van pools.
- Subscription bus and other prearranged travel.
- Event scheduling to minimize peaking and other travel conflicts to permit sequential rather than concurrent use of transportation and parking facilities.

Modal Transfer Points

- Good design of the facility to avoid conflicting flows and to provide adequate space.
- Timed transfer between feeder and line-haul facilities.
- Simplified fare collection and control procedures.
- Pedestrian islands for easier bus loading.
- Information systems which promote flow.

## 7-21. PUBLIC REACTION TO TSM MEASURES

Wherever the transportation authorities envisage changes either to the demand or the supply levels of transport, it is important that they have confidence that the proposed modifications will be both effective in the desired manner and acceptable to the public at large. There is now sufficient evidence that the strategies discussed in this chapter can be successful on both these counts [22]:

- *Car pool/Bus priority lanes.* Very low growth in vehicular traffic has been observed in corridors that have had substantial person-travel growth at the time that car pool and bus priority lane facilities have been used. Such priority facilities carry up to 35,000 bus passengers in peak morning hours and over 2000 car pool vehicles.
- *Variable work hours.* 15 to 25 percent reductions in the 15-min peak vehicular and passenger counts have been observed on radial bus routes and major terminal facilities where variable work hours programs have been adopted.
- *Car pooling.* Employer-based car-pooling encouragement schemes have proved much more effective than have area-wide matching programs.

There is little evidence that car pooling competes significantly with transit usage.

- *Bus pools and van pools.* In some cases over half the commuters living in areas poorly served by public transit have been attracted into bus and van pool programs.
- *Area restraints to autos.* Where adopted, restraint programs have had substantial effects in reducing traffic in localized areas. There is some evidence, however, of adverse impact outside the restraint area. With carefully designed schemes, traffic restraint can be attained with either no economic loss to the area or with up to 20 percent increases in trading activity.
- *Road pricing.* Where road pricing has been used in the form of tolls and parking fees, traffic has been found to have an elasticity of between -0.15 and -0.40. It would appear that the availability of alternative travel options has a strong influence on the degree of elasticity.
- *Transit schedule improvements.* There appears to be a wide variation in the elasticity of transit usage with respect to increases in schedule frequency. The mean level of observed elasticities for this relationship appears to be about 0.5.
- *Expansion of bus transit coverage.* Although the market takes between 1 and 3 years to adjust fully to the provision of the new bus routes, the expansion of bus transit is estimated to bring about an increase of 0.3 to 0.8 percent per percent increase in areawide bus miles of service ($e = 0.3$ to 0.8).
- *Transit fare changes.* The elasticity of transit to fare increases has for many years been judged to average between -0.3 and -0.05. Evaluation of a number of TSM projects indicates that ridership is more sensitive to service frequency changes than to fare changes.
- *Transit marketing.* Special campaigns providing information on routes and schedules have been found to be more effective than the general advertising of transit.

In work that evaluated the results of a large range of TSM projects, Roark determined that a number of characteristics were found to be common among most successful TSM experiences [21]:

1. The direction of a strong innovative personality keenly interested in urban transportation.
2. Coordinated teamwork among the various transportation organizations.
3. A clearly identifiable problem.
4. Adequate planning analysis to determine system impacts.
5. Proper packaging of TSM strategies and support measures.

The lack of the above attributes was not necessarily associated with unsuccessful TSM projects. However, among the unsuccessful experiences, Roark found a number of common problems:

1. The public perception of the withdrawal of traveling privileges which it had previously enjoyed (e.g., the take-a-lane concept for HOVs on freeways).
2. Lack of a perceived problem.
3. Imbalance between operating costs and level of service rendered.
4. Inadequate or no enforcement.

## 7-22. SUMMARY

It is apparent that the massive urban and rural highway construction programs of the 1960s and the 1970s are unlikely to be repeated. The need for strategic planning, however, continues and the techniques developed in the 1960s for comprehensive planning will continue to be used and updated. The development of powerful microcomputers and their attendant software packages has put the techniques of the strategic planner within the grasp of all transportation planners with access to relatively inexpensive desk top machines. These techniques are described and discussed in some detail in Chapters 9 and 11. At the same time as there is a continuing need for a strategic view of transportation plans, every effort is made to ensure that existing infrastructure is used in the short and medium term to its fullest capability, and it is envisaged that the still developing techniques of transportation system management will make that possible. The old attitude which divided strategic planning from TSM has largely disappeared since transportation planners have been able to use the full range of planning techniques for solving short- or long-term problems.

## PROBLEMS

1. Set out a hierarchy of goals, objectives, and criteria that might be used usefully for the transportation plan for a city of 100,000 population.

2. Develop five different scenarios for the year 2000 for a public transportation plan of a city of 2 million persons. Include among your scenarios the following factors: economic growth, fuel costs, population growth, environmental concern, level of automation, leisure habits. Discuss why you chose each combination of factors to make a scenario.

3. Examine any comprehensive transportation study that you have available. Describe the structure of the planning procedure, analyzing the objectives that the plan sought to achieve.

4. Explain why the following TSM techniques may be incompatible:
   a. Car pooling and staggered work hours.
   b. Transit priority and substitution of communications for transportation.
   c. Parking controls and improving freeway movement by surveillance control.
   d. Car pooling and incentives to use transit.
5. "Transportation Systems Management is a passing fashion. It can never replace solutions to transportation problems by massive capital investment." Discuss.
6. Explain why shortening the work week and staggering working hours fall into different categories of TSM.

## REFERENCES


1. *The Freeway in the City,* U.S. Department of Transportation, Washington, D.C., 1968.
2. Stopher, P.R., and A.H. Meyburg, *Urban Transportation Modeling and Planning,* Lexington Books, Lexington, MA, 1975.
3. Ashford, N.J., and J.M. Clark, "An Overview of Transport Technology Assessment," *Transportation Planning and Technology,* Vol. 3, No. 1, 1975.
4. Lupo, A., F. Colcord, and E.P. Fowler, *Rites of Way,* Little, Brown, Boston, 1971.
5. Steiner, H.M., *Conflict in Urban Transportation,* Lexington Books, Lexington, MA, 1979.
6. Swerdloff, C.N., and J.R. Stevens, "A Test of Some First Generation Residential Land Use Models," *Highway Research Record No. 126,* Washington, D.C., 1966.
7. Eastern Massachusetts Regional Planning Project, *Empiric Land Use Forecasting Model, Final Report,* Boston, February 1967.
8. Wilson, A.G., "The Use of Entropy Maximizing Models in the Theory of Trip Distribution, Mode Split, and Route Split," *Journal of Transport Economics and Policy,* Vol. 3, No. 1, 1969.
9. Schneider, M., "Direct Estimation of Traffic Volume at a Point," *Highway Research Record No. 165,* Highway Research Board, Washington, D.C., 1967.
10. U.S. Department of Transportation, Federal Aviation Administration, *FAA Aviation Forecasts San Francisco-Oakland,* Washington, D.C., June 1979.
11. Boeing Commercial Airplane Company, *Dimensions of Airline Growth,* Renton, Washington, March 1978.
12. Ashford, N. and F.M. Holloway, "Validity of Zonal Trip Production Models over Time," *Transportation Engineering Journal.* American Society of Civil Engineers, December 1972.

**13.** Smith, R.L., and D.E. Cleveland, "Time Stability Analysis of Trip Generation and Predistribution Modal Split Models," *Transportation Research Record No. 569,* Transportation Research Board, Washington, D.C., 1976.

**14.** Ashford, N., and D.O. Covault, "The Mathematical Form of Travel Time Factors," *Highway Research Record No. 283,* Highway Research Board, Washington, D.C., 1970.

**15.** Ashford, N., "Strategic Planning Studies within British Rail," *Transportation Research Record No. 687,* Transportation Research Board, Washington, D.C., 1978.

**16.** Federal Highway Administration and Urban Mass Transportation Administration, "Transportation Improvement Program, *Federal Register,* Vol. 40, No. 181, September 17, 1975.

**17.** *Transportation System Management, Special Report 172,* Transportation Research Board, Washington, D.C., 1977.

**18.** Wagner, Fred A., and Keith Gilbert, *Transportation Systems Management: An Assessment of Impacts,* prepared for the Urban Mass Transportation Administration by Alan M. Voorhees, Inc., Washington, D.C., November 1978.

**19.** Walton, C. Michael, and Sandra Rosenbloom, "Measures to Reduce Peak Period Congestion," *Urban Transportation Efficiency,* American Society of Civil Engineers, New York, 1977.

**20.** Lockwood, Stephen C., "TSM Planning—An Emerging Process, *Public Transportation: Planning Operations and Management,* George E. Gray and Lester A. Hoel (eds.), Prentice Hall, Englewood Cliffs, N.J., 1979.

**21.** *Experiences in Transportation System Management, Synthesis of Highway Practice No. 81,* National Cooperative Highway Program, Transportation Research Board, National Research Council, Washington, D.C., November 1981.

**22.** *Traveller Response to Transportation System Changes,* U.S. Department of Transportation, Washington, D.C., February 1977.

# Chapter 8

# Transportation Studies

## 8-1. PLANNING STRUCTURES

Ideally, transportation planning is a comprehensive procedure, carried out using the systems analysis techniques that were described in Chapter 7. Such techniques should recognize the influence of transportation across a continuum of both time and space, while taking into account physical, social, economic, and political constraints on the solutions of any plan. In practice, it must be recognized that transportation planning is carried out in a piecemeal manner, largely because of the realities of political fragmentation which bear little relationship to the dynamics of personal and freight movements. The failure of planning to deal with transportation across its full temporal and spatial continuum is a necessary result of the numerous governmental jurisdictions that have authority within a particular economic area.

Many of the comprehensive urban transportation studies carried out under the terms of the 1962 Federal Highway Act were truly less than comprehensive. Because of separate funding of the studies by the Federal Highway Administration and the Urban Mass Transportation Administration, transit planning frequently was considered in less depth than was highway planning. The movement of freight often was treated with benign neglect, and the air and rail intercity movement largely ignored.

Typically a comprehensive transport study is an examination of the supply and demand relationship of transport over a planning horizon. While every study differs in specific detail, the following procedures are general to all:

1. Setting up an administrative organization.
2. Data collection.
3. Analysis of the present and future status of transportation.
4. Development of a transportation plan and financial program.
5. Implementation.
6. Updating procedures.

Figure 8-1 shows the planning process used in one of the major regional transportation studies.

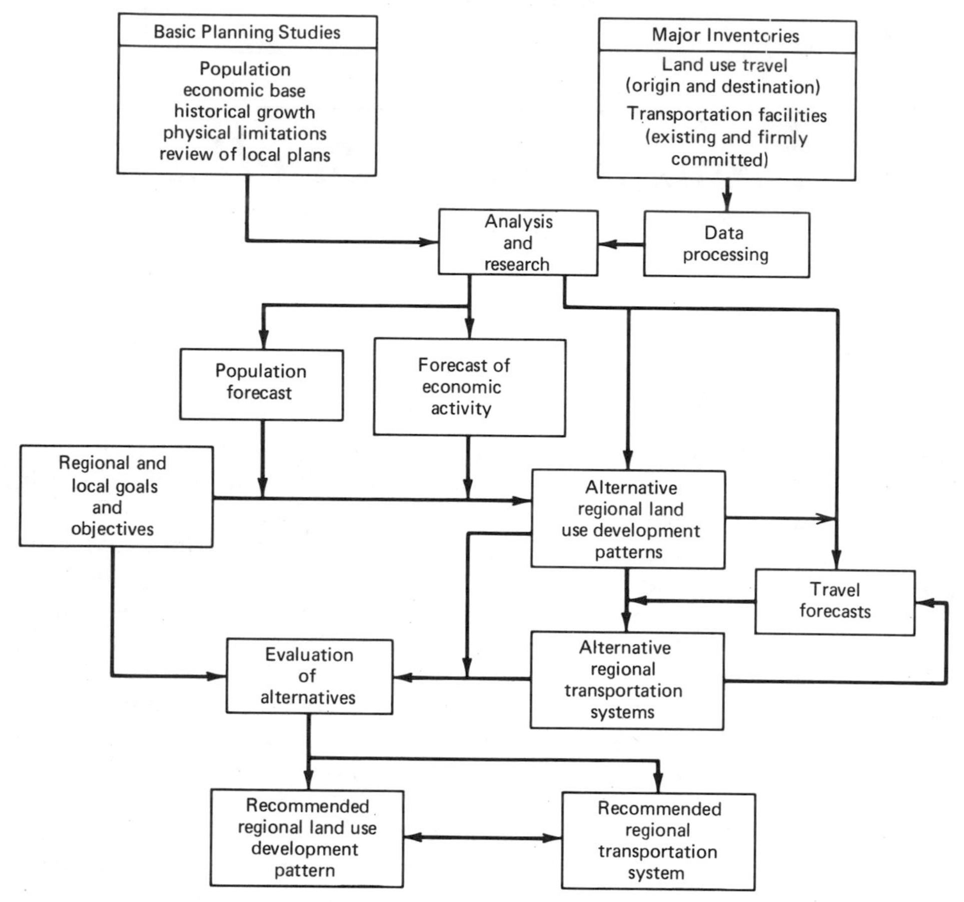


*Fig 8-1.* Puget Sound regional transportation planning process. (Source: *Puget Sound Regional Transportation Study.*)

With the widespread introduction of state departments of transportation, and the breakup of the constraints on use of the Highway Trust Fund monies, statewide transportation planning is now much more effective than it has been in the past. Policies and plans are developed with the broad involvement of other governmental agencies, organizations, groups, and individuals. Typically the planning process should include:

1. The state legislature.
2. The executive branch of state government.
3. Other state agencies such as commissions of industrial development, public service, recreation, travel, and so on.
4. Local and regional planning commissions.
5. Special interest groups.

6. Private and public transportation companies.
7. Multistate coordinating groups.
8. Citizens and citizen groups.

In the United States, planners of federally assisted projects must now coordinate planning activity through an area or state clearinghouse. In metropolitan areas, the clearinghouse is an areawide agency (such as a regional planning commission) that has been recognized by the federal government as an appropriate agency to evaluate, review, and coordinate federally assisted programs and projects. In nonmetropolitan areas, the clearinghouse is a comprehensive planning agency designated by the governor or by state law to review and coordinate such activities.

## 8-2. CITIZEN PARTICIPATION

Typically, coordination between public agencies, private groups, and individuals is accomplished by establishing two or more committees. For example, a *technical committee* may be formed representing the technically qualified members of involved governmental agencies and organizations and commissions. A lay group, sometimes called a *citizens advisory committee* or *citizens planning group* is also formed. Current regulations of the Department of Transportation require that citizens are given a full opportunity to participate in public hearings related to the construction of federally assisted transportation projects. Hearings of this nature permit a full opportunity for community comment at the planning and design stages of projects. Even where the project is not so significant that a public hearing is necessary, citizen participation may be necessary and desirable in terms of public information programs and public information meetings [1].

## 8-3. HIGHWAY AND MASS TRANSIT DATA

By means of field studies and the assembly of current data used by various agencies within the area, the planner assembles a complete inventory of existing highway and transit facilities and a cross-section of current travel behavior in the urban area. Together with the origin-destination survey and the land use studies, described in Sections 8-7 and 8-13, respectively, data are collected to give past economic and demographic trends.

*Street Classification and Street Use Map.* The objective of this study is the identification of all streets and highways with significant travel and their classification according to their present use.

All streets within the urban area are tentatively classified as expressways and freeways, other arterial streets, collectors, or local streets. Rural highways are classified by their place in the hierarchy of state and federal systems by functional class and by their design category as two-lane, three-lane, multilane, or freeway. Based on these tentative classifications, a street and highway use map is prepared.

*Existing Traffic Service Studies.* Those involved in traffic engineering in the area are required to provide an inventory of the traffic service provided on the existing facilities under current traffic demand. To permit an evaluation of current traffic service, the following studies normally are carried out:

1. *Traffic volume counts* [2]. Control counts are made at carefully selected stations, and sample counts on an areawide basis permit a reasonable estimation of traffic volumes throughout the urban area. A cordon-line count around the central business district can be coordinated with a parking study also carried out in this phase. The central business district parking study determines the mode of entry to the central area and hourly variations. At least one screen line count should be taken with the dual purpose of providing data on the annual trends of vehicular trips and providing verification of ground movements as predicted from the origin-destination survey samples.
2. *Travel time studies* [2]. Measurement of travel time on the major street system enables comparison of the levels of service provided by the various road sections comprising the existing network. Travel time studies are conducted at different periods of the day to permit this comparison under peak and nonpeak loading. This study usually is done by a driver and an observer using the "average speed" or "floating car" method.
3. *Street capacity.* Based on the geometry of the roadway, the type of traffic control, and the vehicular composition of the traffic stream, capacity calculations can be made for all sections of major streets and highways. These calculations usually are based on the techniques explained in the *Highway Capacity Manual* [3]. Under unusual conditions, special techniques may have to be developed by the traffic engineer in the field. In urbanized areas, capacity analysis is carried out on all intersections of two major streets. This normally amounts to approximately 5 percent of all intersections.
4. *Accident study.* Information on automobile and truck accidents is collected and assembled on a comprehensive basis from existing data sources, such as the files of local authorities and police records. Accident information gives a measure of the safety of the street and highway system. Safety is another criterion used to determine level of service which is generally considered in addition to congestion and convenience.
5. *Parking study.* The provision of adequate terminal parking facilities is an inherent part of any transportation plan that relies on auto and truck transportation for the movement of goods and people. A parking study therefore, is an important element in the inventory studies.

In the past, two principal techniques have been used, the *comprehensive parking study* and the *limited parking study.*

A comprehensive parking study is carried out where a complete analysis is required of the auto terminal problem in the central business district and other critical areas. The general scope of the comprehensive parking study includes [2]:

a. An inventory of existing parking facilities.
b. An examination of the adequacy of existing laws and ordinances.
c. An analysis of the limitations of administrative responsibility.
d. Usage patterns for existing parking facilities including duration of parking and walking distances.
e. Current demand patterns.
f. Possible methods of financing.
g. Traffic flow.
h. Parking characteristics.
i. Effect of large traffic generators.

A limited parking study is considerably narrower in scope and concerns itself chiefly with the examination of four elements only [2]:

a. Parking supply.
b. Parking usage.
c. Parking duration.
d. Parking meter revenues.

Conventionally, parking studies have been carried out using a combination of interview and tag techniques. Significant savings have been achieved by the use of colored aerial photography to determine parking use and duration [4]:

6. *Traffic control device study.* Because of the significant effect that control devices exert on the capacity of a street network, the planner carries out a comprehensive study that determines:

   a. Location, type, and functional characteristics of all major traffic control devices.
   b. Areawide parking regulations on a block-face basis.
   c. Transit routes and transit loading zones.

*Existing Transit Studies* [5]. In order to determine how well transit meets and stimulates passenger demand for service, it is essential that the planner have a comprehensive understanding of the existing level of transit service and demand. This can be achieved by carrying out the following studies.

1. *Routes and coverage study.* The existing route structure is inventoried to determine the relation of populated areas to areas within reasonable walking distance of service. Routes are examined further to determine whether service in general follows the desired lines of travel, and whether transit accommodates growing community needs.
2. *Transit route inventory.* A survey is made of the physical characteristics of each transit route.
3. *Service frequency and regularity, and transit running time studies.* From an evaluation of adherence of transit to schedule and the adequacy of the schedule itself to provide reasonable service, the analyst is able to compare existing service with recommended standards.
4. *Passenger load data.* Passenger load data can determine whether or not the service frequency is adequate to satisfy the existing demand and whether adequate standards of comfort are maintained. The feasibility of altered headways can be evaluated from this data.
5. *Transit speed and delay studies.* By means of speed and delay studies conducted on actual transit runs, identifiable causes and areas of delay can be delineated. Internal and external causes of delay are determined to assist future remedial measures.
6. *General operating data.* Six general yardsticks are used for comparative purposes.

   a. *Quality of service* in vehicle miles per capita.
   b. *Quality of service* in vehicle miles per revenue passenger.
   c. *Efficiency of service* in terminal-to-terminal time per vehicle mile.
   d. *Use of service* in revenue passengers per capita.
   e. *Route coverage* in route miles per capita.
   f. *Time convenience* as indicated by operating system speed.

   Other criteria may be examined relating to the cost, convenience, and overall comfort of the system.
7. *Passenger riding habits.* By sample survey methods, data are gathered concerning the riding habits of the passengers, their origins, destinations, their social and economic status. Where necessary, attitude surveys can be incorporated into this study.

*Physical Street System Inventory.* In order to evaluate properly the present and future traffic-carrying capacity of the street and highway system, it is necessary to conduct a physical inventory of the components including:

1. Street widths.
2. Block lengths.
3. Pavement conditions.

4. Geometric design.
5. Storm sewers and surface drainage.

## 8-4. AIR TRANSPORT DATA

The current status of air transportation can be analyzed from a series of studies that relate to existing physical plant, and existing demand and usage.

*Airport Classification.* An overall picture of air transport in an area can be gained from an examination of the existing FAA classification of all airport facilities within the area including reference to their spatial relationships among themselves and to the urbanized areas. Classification should also include the function and development plans of individual airports as indicated in the National Plan of Integrated Airport Systems [6].

*Surveys of Existing Traffic.* Existing traffic patterns and their relation and impact on existing airport facilities are determined by a variety of studies.

1. *Volume.* Air traffic volumes are available from FAA approach control facility records and individual airport records. In addition, FAA data on individual carriers can be surveyed to permit a breakdown of emplanements and deplanements.
2. *Travel time.* Airport-to-airport travel times for scheduled carriers can be determined from an examination of published schedules.
3. *Airport capacity.* The hourly and annual capacities of the airports are the respective hourly and annual movement rates that can be handled by the facility. Annual capacities have meaning only when the hourly flow rate is seen as a percentage of the total annual traffic in conjunction with peaking conditions over the day and week. Capacity can be calculated either on a maximum average acceptable delay basis (4 min) or on a flow basis [7].
4. *Peak loading.* The monthly, daily, and hourly variations of traffic demand and passenger movements are assembled from available airport data and from sample surveys where records are inadequate.

*Physical Inventory of Existing Airports.* A complete inventory of existing airports would record the physical characteristics of:

Runways, taxiways, turn-offs, lighting, drainage, aprons.

Aircraft parking areas, hangars, and maintenance shops.

Air traffic control facilities and navigational aids.

Terminal facilities for passengers, baggage, and freight.

General aviation support facilities.

Ground transportation facilities.

Vehicular parking and access.

Suitability of surrounding land use for expansion.

## 8-5. PORT AND HARBOR DATA

Analysis of the adequacy and functional performance of existing port and harbor facilities requires that the planner have an in-depth knowledge of the current level of traffic service and the physical characteristics of the port facilities.

*Traffic Service.* The evaluation of traffic service depends on the ability of the analyst to determine how well the existing facilities serve current demand and how well they can be expected to fill future needs. It is suggested that studies be carried out to permit the collection of adequate data in the following areas:

1. *Volume.* Passenger and freight volumes should be estimated from the port operators' records. Freight volumes should be classified by categories because of the different requirements for cargo loading, transfer, storage, and ground transport facilities.
2. *Seasonal variations.* Because of the seasonal nature of commerce, significant variations of traffic volumes occur throughout the year. Estimates of the seasonal variations of previous years can be gathered from the port operators' records.
3. *Berthing capacity.* Berthing capacity data require classification by the type of marine terminal; that is, public facilities operating on a first-come, first-served basis or private facilities under lease or private operation.
4. *Loading and unloading capacity.* The availability and general maneuverability of cargo handling equipment causes differentials of loading and unloading capacities from berth to berth. The presence of fixed cargo handling facilities limits certain unloading capacities to individual berths. These capacities must be determined.
5. *Storage capacity.* Storage capacity can be classified into short-term transit shed storage, long-term storage, and container storage. In addition, differentiation should be made between refrigerated and unrefrigerated space.
6. *Ground transportation interface.* The adequacy of the ground transportation interface should be examined. In the case of many older marine terminals, the facilities were designed for vertical freight movements rather than for the extensive horizontal movements which are handled more easily by modern materials handling devices. Many older terminals also suffer congestion problems in the transfer interface with trucks since they were designed at a time when virtually all land transportation was by rail. In such terminals there may be excess rail loading capacity along with a woeful shortage of trucking docks.
7. *Containerization.* Analysis should be made of the current containerization capacity, both from the viewpoint of space requirements and

the necessary unloading and handling requirements. This rapidly expanding area of waterborne commerce poses conversion problems at many older ports.

8. *Bulk cargo.* The growth of bulk carriers, both of dry cargo and tankers, requires a useful analysis of the level of supply of overall storage capacity in the port and individual berthing spaces.
9. *Specialized port services.* Around any port, special services develop that cater to the peculiar needs of the port itself. Since those services themselves serve as an attraction to waterborne commerce, it is essential that the planner include data on their availability in his analysis. Included within this category are such services as export packing, customs house brokerage, freight forwarding, dry docking, and specialized handling equipment.

*Physical Inventory.* An inventory of the port's facilities should include the physical characteristics of:

The harbor, channels, and turning basins.
Slips, wharves, and dolphins.
Navigational aids.
Aprons.
Transit and storage sheds.
Warehousing.
Rail and open storage.
Dry bulk, chemical, and tank storage.
Unloading devices.
Ground transportation interfaces.
Surrounding land and land use.

## 8-6. RAILROAD DATA

Under normal circumstances, the transportation planner within the United States will find himself only obliquely concerned with planning for railroads. By reason of the largely private nature of railroad ownership and operation, the degree of influence that can be exerted on this mode is slight. Where planning is required, however, the following studies should be carried out.

*Classification.* The rail facilities of the area can be classified in two ways: by ICC classification and by functional classification (see Chapter 1).

*Existing Traffic Service.* In establishing the current level of traffic service supplied, analysis must be made of several elements:

1. Inventory of freight and passenger routings available.
2. Service frequencies on individual routes.

3. Route capacities.
4. Capacities and service characteristics of classification yards.
5. Terminal facilities.
6. Passenger riding habits.
7. General operating data.
8. Route and classification yard volumes.
9. Traffic control devices.

*Physical Inventory.* In order to establish the physical condition of existing railroad plant, an inventory should be made of such items as rolling stock, power plant, bridges, structures, track, roadbed, traffic control devices, geometric design, right-of-way widths, and terminal facilities. Most of this information is available from the records of individual railroad companies and requires only a small amount of field work.

*Other Physical and Operational Patterns.* After the completion of inventories, the intermodal and multimodal relationships should be established. This would include an analysis of joint facility operations, reciprocal switching agreements, interchange routes, and overall handling of traffic volumes.

## SURVEYS OF CURRENT DEMAND

### 8-7. ORIGIN-DESTINATION STUDIES

Perhaps the most important, and certainly the most time-consuming and expensive planning study is the origin-destination survey. Properly designed, this study will identify (for the passenger movements) where and when trips begin and end, the socioeconomic characteristics of the trip-maker, the purpose of travel, the mode of travel, and the land use type at the beginning and end of the trip. For freight traffic, material type can be identified as well as many characteristics of the shipper and the shipping service.

The origin-destination study is a sample cross section of all movements on the average day within the area. This cross section is taken to be representative of the average travel demand on the system at the time the survey is carried out. After relating the demand to the characteristics of the area and to its inhabitants the planner can use the derived relationships to determine the future travel demand in conjunction with projections of future economic development and population growth. The origin-destination survey, therefore, is regarded as the prime data source of the transportation study. Figures 8-2 and 8-3 are examples of origin-destination questionnaires.

In the United States the most comprehensive origin-destination studies were done in conjunction with the highway and transit planning studies carried out under the terms of the 1962 Federal Highway Act which required comprehensive transportation studies for cities with a population in excess of

______ Urban Transportation Study
DWELLING UNIT SURVEY - DWELLING UNIT SUMMARY

General Instructions to Interviewer

- Assure respondent that all information will be held in strictest confidence
- Before the interview be sure questions numbered 1, 2 and 3 are completed
- During the interview complete items 4, 5, 6, 7 and 8
  After the interview complete items 9, 10 and 11

3 Interview Address:

Street Number ____ Street Name ____ Street Type (Ending: Rd., St., etc.) ____

City ____ State ____

Preceding Address:

Street Number ____ Street Name ____ Street Type ____

Succeeding Address:

Street Number ____ Street Name ____ Street Type ____

4. Structure Type: (Circle the best description)

Which Best Describes This Building?
(Include all apartments, flats, etc., even if vacant.)

A one-family house detached from any other house........1
A one-family house attached to one or more houses.......2
A building for 2 families..............................3
A building for 3 or 4 families.........................4
A building for 5 to 9 families.........................5
A building for 10 to 19 families.......................6
A building for 20 to 49 families.......................7
A building for 50 or more families.....................8
A mobile home or trailer...............................9
Other..................................................0

5. Household Data (Record information as given by respondent)

List each person (and his age) who lived here on the travel date, or was staying or visiting here and had no other home.

| Person = | Person I.D. | Age | Check if person interviewed |
|---|---|---|---|
| 1 | | | |
| 2 | | | |
| 3 | | | |
| 4 | | | |
| 5 | | | |

Total number of persons ____

6. Record the total number of persons listed above who are 5 years of age or older ____

7. How many passenger automobiles are owned or regularly used by members of this household? ____

8. How many other vehicles, such as pick-up trucks, are owned or regularly used by members of this household? ____

Describe ____

Respondent's Phone No. ____ (ask last)
Go to internal trip report

Administrative

Sheet ____ Of ____ Total

1. Sample Number ____

2. Travel Date ____

9. Total Person Trips Recorded ____

10. Interview Type: (Circle the best description)

Complete ____ 1
Partial Interview ____ 2
Refused ____ 3
Not at Home ____ 4
Date Past Due ____ 5
Trips Possible

Vacant ____ 6
Out of Area on Travel Date ____ 7
No Trips Possible

Demolished ____ 8
Non-residence ____ 9

11. Interviewer Identification and Certification:

I hereby certify that the information on this form has been obtained by me from the respondents and is accurate and complete.

Interviewer's Signature ____ Interviewer's Number ____

Contact with Respondents:

| Date | Time | By | Result |
|---|---|---|---|
| | | | |
| | | | |
| | | | |
| | | | |

Remarks: ____

Editing

Name ____

Date ____

Coding

| Station | Name | Date |
|---|---|---|
| | | |
| | | |

*Fig 8-2.* Dwelling unit survey form.

50,000. Guidelines for these and subsequent studies were set by the Department of Transportation [8, 9].

As the size of a city increases, the form of trip-making within and through the city changes in nature. Table 8-1 indicates the changes in the nature of traffic approaching the urban area with increasing city size. There is a dramatic decrease in the percentage of bypassable traffic, from almost 50 percent in areas with a population of less than 5000 to only 8 percent in cities over 500,000. This difference in patterns is reflected in the type of origin-destina-

______ Urban Transportation Study

DWELLING UNIT SURVEY INTERNAL TRIP REPORT

Sheet _____ of _____

(For trips which began on the travel date (4:00 a.m. to 4:00 a.m.)
by persons 5 years of age or older living in sampled dwelling)

Sample Number

| 12 Person No | 13 Trip No. | 14 Where Did This Trip Begin? | 15 Where Did This Trip End? | 16 How Did You Get There? | 17 When Did You Start? | 18 When Did You Arrive? | 19 Why Did You Go? Purpose (From / To) 20 | Remarks |
|---|---|---|---|---|---|---|---|---|
| | | Street No.<br>Street Name Ending<br>City<br>State | Street No.<br>Street Name/Ending<br>City<br>State | 1 Auto Driver<br>2 Auto Passenger<br>3 Transit Passanger<br>4 School Bus Passenger<br>5 Taxi Passenger<br>6 Truck Driver<br>7 Truck Passenger<br>8 Walked to Work<br>9 Worked at Home | ____ AM<br>PM | ____ AM<br>PM | 1 Work 1<br>2 Shop 2<br>3 Social 3<br>4 Recreation 4<br>5 School 5<br>6 Personal Affairs 6<br>7 Transfer to Another Means of Travel 7<br>8 Serve Passenger 8<br>9 Home 9 | |
| | | Street No<br>Street Name Ending<br>City<br>State | Street No.<br>Street Name/Ending<br>City<br>State | 1 Auto Driver<br>2 Auto Passenger<br>3 Transit Passanger<br>4 School Bus Passenger<br>5 Taxi Passenger<br>6 Truck Driver<br>7 Truck Passenger<br>8 Walked to Work<br>9 Worked at Home | ____ AM<br>PM | ____ AM<br>PM | 1 Work 1<br>2 Shop 2<br>3 Social 3<br>4 Recreation 4<br>5 School 5<br>6 Personal Affairs 6<br>7 Transfer to Another Means of Travel 7<br>8 Serve Passenger 8<br>9 Home 9 | |
| | | Street No.<br>Street Name/Ending<br>City<br>State | Street No.<br>Street Name/Ending<br>City<br>State | 1 Auto Driver<br>2 Auto Passenger<br>3 Transit Passanger<br>4 School Bus Passenger<br>5 Taxi Passenger<br>6 Truck Driver<br>7 Truck Passenger<br>8 Walked to Work<br>9 Worked at Home | ____ AM<br>PM | ____ AM<br>PM | 1 Work 1<br>2 Shop 2<br>3 Social 3<br>4 Recreation 4<br>5 School 5<br>6 Personal Affairs 6<br>7 Transfer to Another Means of Travel 7<br>8 Serve Passenger 8<br>9 Home 9 | |
| | | Street No.<br>Street Name/Ending<br>City<br>State | Street No.<br>Street Name/Ending<br>City<br>State | 1 Auto Driver<br>2 Auto Passenger<br>3 Transit Passanger<br>4 School Bus Passenger<br>5 Taxi Passenger<br>6 Truck Driver<br>7 Truck Passenger<br>8 Walked to Work<br>9 Worked at Home | ____ AM<br>PM | ____ AM<br>PM | 1 Work 1<br>2 Shop 2<br>3 Social 3<br>4 Recreation 4<br>5 School 5<br>6 Personal Affairs 6<br>7 Transfer to Another Means of Travel 7<br>8 Serve Passenger 8<br>9 Home 9 | |
| | | Street No.<br>Street Name/Ending<br>City<br>State | Street No.<br>Street Name/Ending<br>City<br>State | 1 Auto Driver<br>2 Auto Passenger<br>3 Transit Passanger<br>4 School Bus Passenger<br>5 Taxi Passenger<br>6 Truck Driver<br>7 Truck Passenger<br>8 Walked to Work<br>9 Worked at Home | ____ AM<br>PM | ____ AM<br>PM | 1 Work 1<br>2 Shop 2<br>3 Social 3<br>4 Recreation 4<br>5 School 5<br>6 Personal Affairs 6<br>7 Transfer to Another Means of Travel 7<br>8 Serve Passenger 8<br>9 Home 9 | |

Editing

Name ______ Date ______

Coding

| Station | Name | Date |
|---|---|---|
| | | |
| | | |
| | | |
| | | |

*Fig 8-3.* Internal trip report form.

tion survey that can be carried out to establish travel patterns satisfactorily. The following types of origin-destination surveys can be used.

1. *External cordon.* Normally carried out for cities under 5000 population where external traffic forms the major traffic patterns. Sample

*Table 8-1*
**Destination of Traffic Approaching Cities**

| | *Percentage of Total Approaching Traffic Destined to:* | | |
|---|---|---|---|
| *City Population Group (Thousands)* | *Central Business District* | *Other Points Within City* | *Points Beyond City* |
| Under 5 | 29 | 22 | 49 |
| 5 to 10 | 29 | 29 | 42 |
| 10 to 25 | 28 | 37 | 35 |
| 25 to 50 | 26 | 49 | 25 |
| 50 to 100 | 24 | 57 | 19 |
| 100 to 250 | 21 | 62 | 17 |
| 250 to 500 | 18 | 70 | 12 |
| 500 to 1000 | 15 | 77 | 8 |

SOURCE: AASHTO.

sizes range from 20 percent on high-volume roads to 100 percent on low-volume roads.

2. *Internal-external cordon.* Applicable for cities with populations of 5000 to 50,000. Direct interviews are carried out on two cordon lines. The external cordon is set at the limit of the urbanized area, and the internal cordon is set around the fringe of the Central Business District. Despite limitations, this type of survey can supply reasonably complete travel patterns. Typically, a 20 percent sample is used.
3. *External cordon-parkingsurvey.* Applicable for cities of 5000 to 50,000 with little transit and major problems in the central areas. The external cordon survey is carried out in the standard manner. The parking survey consists of three basic elements:

   a. Inventory of existing parking spaces
   b. Interview of all parkers at curb and off-street facilities to determine origin, destination, trip purpose, and length of time parked.
   c. A cordon volume count of traffic entering and leaving the central areas.

4. *External cordon-home interview survey.* This comprehensive type of study is applicable in areas of all sizes. The external cordon survey is carried out in the standard manner with a sample size of 20 percent of all vehicles passing through the cordon. In addition, within the cordon, household interviews are carried out on a sample basis. The sample size varies with the size of the population of the area, and to

some degree with the density of population. Recommended sample sizes vary from 20 percent of all households in areas less than 50,000 total population to 4 percent in areas over one million.

Sampling size is dependent on acceptable sample errors, the confidence level, and the trip volume [8, 9]. Each sample dwelling is visited by an interviewer, and questions are asked concerning how many persons live at the address, their ages, and occupations. The interviewer also determines the number of cars at the address and usually records information concerning the number of trips made by all individuals, their purposes, origins, destinations, modes of travel, and the times at the beginning and end of trips for the previous day. Figure 8-3 shows the recommended forms for the home interview.

When conducting a home interview study, it is also necessary to conduct *truck and taxi surveys;* otherwise, there will be no record of truck and taxi trips that do not cross the external cordon. It is customary to contact a proportion of truck and taxi owners for these surveys and to determine travel patterns from the daily reports. Sampling must be at a higher rate than for the home interview. The sample rate may be calculated from the total taxi population. It is common practice to sample truck owners at twice the rate of taxi owners.

5. *Transit surveys.* There are two chief types of study procedure used to define the patterns of transit usage.

   a. *Transit terminal passenger survey.* Passengers waiting to board a bus, streetcar, or tram are handed questionnaires which contain questions relating to the trip they are about to take. They are requested to mail these completed questionnaires which are on prepaid postcards. A variant of this system, in which the driver or a researcher hands out survey forms to boarding passengers, is also commonly used.
   b. *Transit route passenger survey.* Questionnaires are handed to passengers by two field interviewers who ride the bus. The questionnaires are collected as the passenger leaves the bus. As might be expected, the response rate for this type of survey is higher than for the mailed questionnaire. In most cases, however, the added expense may not be warranted. It should be noted that terminal passenger and route passenger surveys are used where the overall transit modal split is low. The incidence of reported trips in home interview surveys would, in this case, be very low and the transit data gathered would be very unreliable.

6. *Other passenger origin-destination surveys.* For rail, air, and water passenger modes, origin-destination data are best gathered by *terminal passenger questionnaires.* Figure 8-4 is an example of air terminal

passenger questionnaires used to gather a diversity of information on trip-makers and their travel habits. Some success has been attained using in-flight origin-destination surveys, but for large airports the number of different airlines involved makes coordination difficult. General aviation patterns are sampled best by personal interviews at the airport.

7. *Freight origin-destination studies.* The determination of origin and destination patterns of freight presents significant problems which have not been altogether solved, especially in the case of large-scale studies.

   For air and water carriage a sampling of waybills at the terminal points that these modes must use can give the desired information. With the cooperation of the railroads operating in the area, the same type of information is available but somewhat more difficult to obtain. Road motor freight, however, presents its own special problem. Because of the lack of localized terminals, there is no straightforward method of obtaining this information. Work in conjunction with descriptive sampling plans issued by the Interstate Commerce Commission indicates that continuous traffic study (CTS) data may be useful in the establishment of reliable commodity flow statistics on a regional basis. A number of successful studies have been carried out in this manner both in the United States and abroad.

## ANALYSIS OF PRESENT AND FUTURE STATUS OF TRANSPORTATION

### 8.8. FORMULATION OF CURRENT STATUS OF TRANSPORTATION

For the most part, much of the data assembled in the data collection phase of the study must be synthesized into numerous displays which more easily portray the state of the area and its transportation infrastructure. This, however, is not the sole use of the large amount of data collected. Data will be used later for modeling the development and transportation demand of the area and also in designing, evaluating, and programming a new system. At this stage, in order to indicate clearly the scope of the study, the analyst prepares a variety of displays and tabular summaries which enable a more rapid comprehension of existing conditions. Although from study to study the form of these summaries varies, a typical study would include at least some of the following:

1. *Population.* Several studies have shown current population along with the growth patterns over the historic period of the area. Figure 8-5 shows graphically the historical development of Chicago.

1. On which airline are you about to travel?

   (1) ☐ Eastern (3) ☐ Southern
   (2) ☐ National (4) ☐ United

2. Flight Number?________________

3. Date?________________

**FOR PASSENGERS NOT STARTING THEIR AIR TRAVEL AT THIS AIRPORT TODAY**

If you transferred to this flight from a different flight earlier today, or if you were on this flight when it arrived at this airport earlier today, please answer the questions in this box.

4. I was on this flight when it landed here today.
   (1) ☐ Yes (2) ☐ No

5. I was transferred to this flight at this airport from a connecting flight of:
   (1) ☐ This Airline
   (2) ☐ Another Airline________________
   Please Specify

6. If you are a transferring passenger, did you leave the airport between flights?
   (1) ☐ Yes (2) ☐ No

If you answered Question 6 **YES,** please complete all of the remaining Questions.

If you answered Question 6 **NO,** do not answer the remaining Questions, and please deposit this form in one of the boxes provided.

**ABOUT YOUR TRIP TO THIS AIRPORT TODAY**

7. From what location in this area did you leave for the airport?

   ________________
   No. and St. Address, Building Name, and/or nearest St. Intersection

   ________________
   City County State Zip

8. This location was:
   (1) ☐ Private Residence
   (2) ☐ Hotel/Motel
   (3) ☐ Your Place of Employment
   (4) ☐ Business you were visiting
   (5) ☐ Other________________
   Please Specify

9. **What time did you leave for the Airport?**
   ________ (1) ☐ A.M. (2) ☐ P.M.

10. **What time did you arrive at the Airport?**
    ________ (1) ☐ A.M. (2) ☐ P.M.

11. How did you travel to the Airport today?
    (1) ☐ Private Car (5) ☐ Taxicab
    (2) ☐ Rent-A-Car (6) ☐ Private Plane
    (3) ☐ Airport Limousine (7) ☐ Air Taxi
    (4) ☐ Bus (8) ☐ Hotel/Motel Courtesy Car
    (9) ☐ Other________________

12. If you arrived in a private car which was parked at the airport, about how long do you expect it will be parked there?
    (1) ☐ 0 to 4 hours (3) ☐ 10 to 24 hours
    (2) ☐ 4 to 9 hours (4) ☐ Over 24 hours

**ABOUT YOUR FLIGHT TODAY**

13. What is the **PRIMARY** nature of the trip you are taking today?
    (1) ☐ Business
    (2) ☐ Brief Pleasure (less than one week)
    (3) ☐ Vacation (more than one week)
    (4) ☐ Military orders or leave
    (5) ☐ Personal Matters
    (6) ☐ Other ____________
14. What is the total duration of this trip?

| | |
|---|---|
| (1) ☐ 1 day | (5) ☐ 5 days |
| (2) ☐ 2 days | (6) ☐ 6 days |
| (3) ☐ 3 days | (7) ☐ 7 days |
| (4) ☐ 4 days | (8) ☐ More than 7 days |

15. Other than yourself, how many persons accompanied you in the terminal? ____________
16. How many people who accompanied you will depart with you on this flight? ____________
17. At what city will you end your air travel today?
    ____________

**ABOUT YOURSELF**

PEARL RIVER COUNTY
HANCOCK COUNTY
HARRISON COUNTY
JACKSON COUNTY
MISS.
ALA
MOBILE COUNTY
BALDWIN COUNTY
ESCAMBIA COUNTY
ESCAMBIA COUNTY
SANTA ROSA COUNTY
OKALOOSA COUNTY
ALA
FLA

18. Do you reside within the area shown on the above map?
    1. ☐ Yes 2. ☐ No
19. About how many flights, including this one, have you taken from this Airport in the past 12 months?

| | |
|---|---|
| (1) ☐ 1 or 2 | (4) ☐ 7 or 8 |
| (2) ☐ 3 or 4 | (5) ☐ 9 or 10 |
| (3) ☐ 5 or 6 | (6) ☐ More than 10 |

**THANK YOU — PLEASE DEPOSIT THIS FORM IN THE BOX PROVIDED.**

UPPER GULF COAST
AIR TRANSPORTATION
ORIGIN SURVEY

conducted by

SOUTH ALABAMA REGIONAL PLANNING COMMISSION

ESCAMBIA-SANTA ROSA REGIONAL PLANNING COUNCIL OF FLORIDA

GULF REGIONAL PLANNING COUNCIL OF MISSISSIPPI

---

The information you provide in this survey is important in planning air transportation facilities in the coastal areas of Alabama, Mississippi, and Northwest Florida. In particular, this questionnaire requests information concerning:

1. The address of the place from which you began your trip to the airport,
2. The nature of your trip.

The information requested will take only a few minutes of your time. Your name or identification is not required.

When you have answered the questions, please deposit the questionnaire in box provided.

THANK YOU FOR YOUR COOPERATION.

---

**R. DIXON SPEAS ASSOCIATES**
*Aviation Consultants*

*Fig 8-4.* Air passenger terminal origin-destination questionnaire. (Source: R. Dickson Speas Associates.)

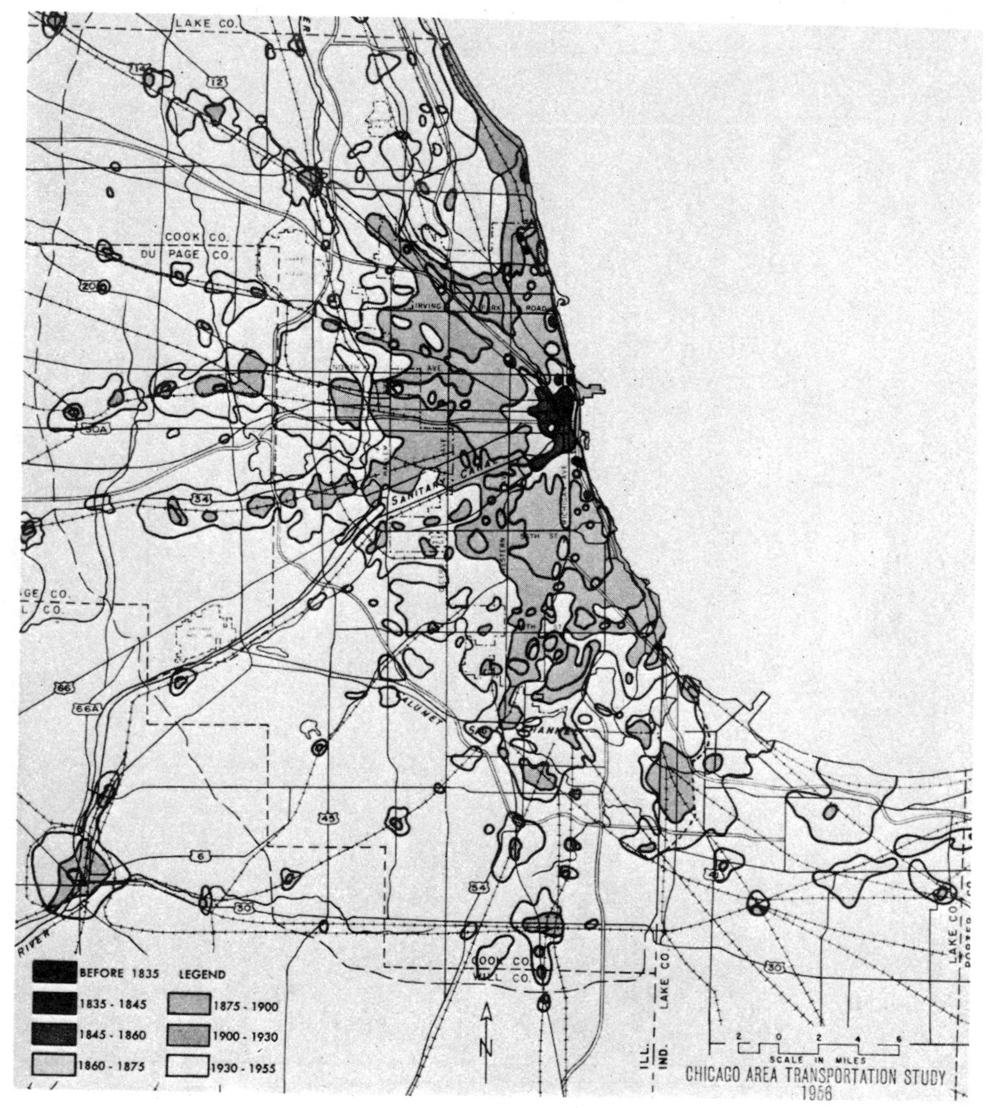


*Fig 8-5.* The historical development of Chicago.

2. *Intensity of land use.* Intensity of land use is as important as location of land use in an urban area. Land use intensity is frequently displayed in terms of population density. For a transportation study, a useful surrogate for intensity is the trip-end density, which is indicative of all land use intensity, rather than residential land use only.
3. *Land use.* The spatial relationships of the various types of land use normally are most easily displayed on a land use map compiled from

the dominant characteristics of the areas determined in the land use survey.

4. *Socioeconomic trends.* Under this category, it is customary to display those trends that are associated with increased demand for transportation facilities. Typically, in a comprehensive study some of the following would be displayed:

   a. Population growth over time.
   b. Average income over time.
   c. Growth of car ownership.
   d. Transit usage and transit revenues over time.
   e. Vehicle travel over time.
   f. Composition for employment.
   g. Growth of emplanements and deplanements in aviation.
   h. Growth of air traffic activity, both general aviation and carrier.
   i. Air cargo trends over time.
   j. Air carrier, air cargo, and general aviation vehicle sizes and requirements over time.
   k. Rail passenger volume and revenues within the area.
   l. Major commodities originating and terminating by rail.
   m. Major commodities originating and terminating by water.
   n. Revenue trends of ports and harbors.
   o. Trends in modal split of person miles traveled in the area.
   p. Trends in modal split of freight ton mileage in the area.
   q. Pertinent trends in other modes such as pipelines.

5. *Origin-destination information.* Origin-destination information can be displayed clearly by superimposing desire lines on areawide maps. While customarily used for the display of auto and transit person trips, their use can be legitimately expanded to all modes of transportation for both person and freight movements. Figure 8-6*a* shows how this technique is suitably used to display travel patterns for an urban area.

6. *System-wide facilities and functional characteristics.* The spatial arrangement of various systems over the study area are displayed by the use of systemwide maps. A concise exhibition of the various transportation systems permits the analyst to show clearly the relationships between land use, population, and transportation facilities.

   There is also usually a need to show a variety of system characteristics which can help to identify problem areas. Three typical maps of system characteristics are shown in Figs. 8-6*b*, *c*, and *d*, which display:

   a. Functional classification of a highway system.
   b. Base year traffic volumes.
   c. Travel time contours or isochrones.

**Internal and External Automobile and Truck Trips Combined, Volume Range of 250 or More in Alexandria, Louisiana**

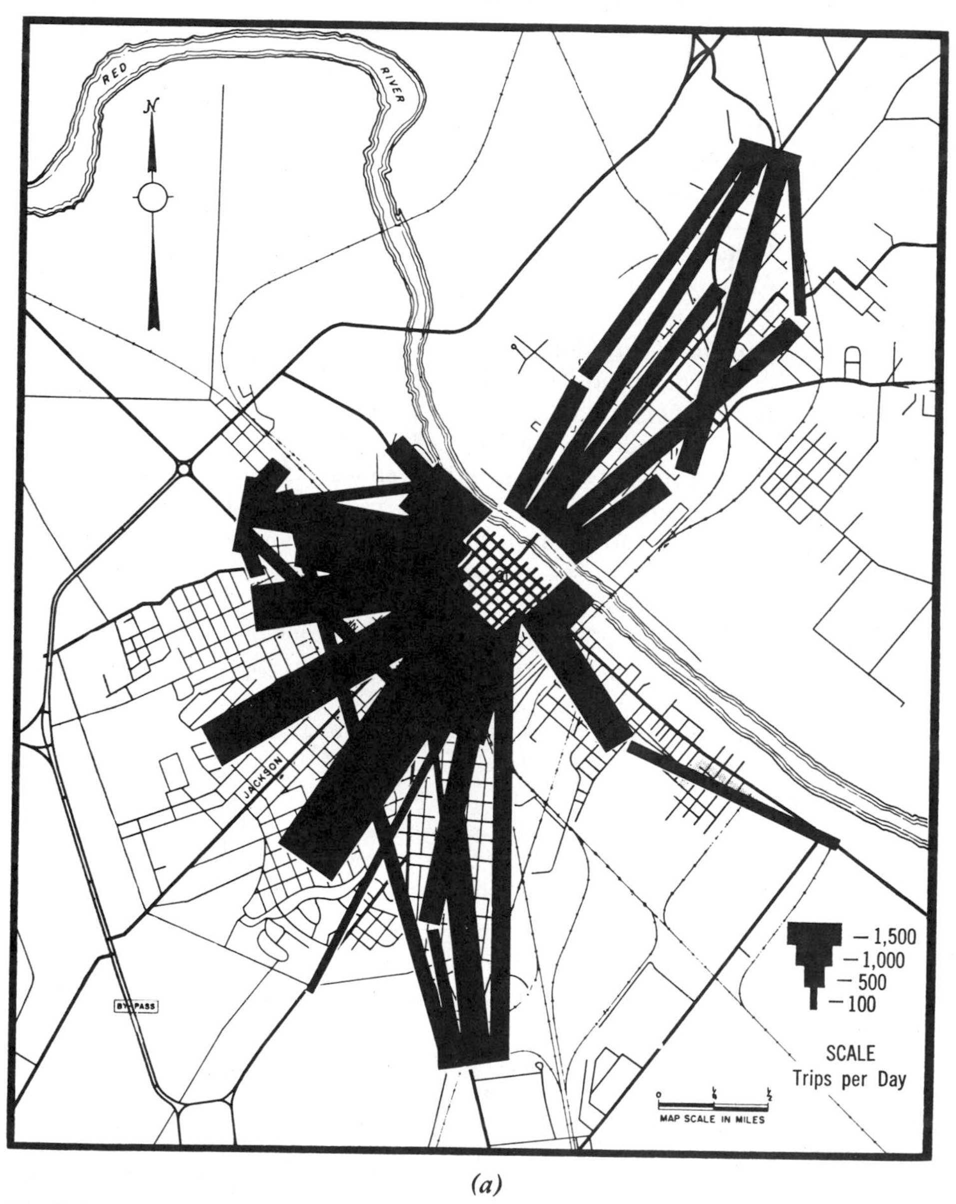


(a)

*Fig 8-6a.* Travel desire lines for an urban area. (Courtesy Public Administration Service.)

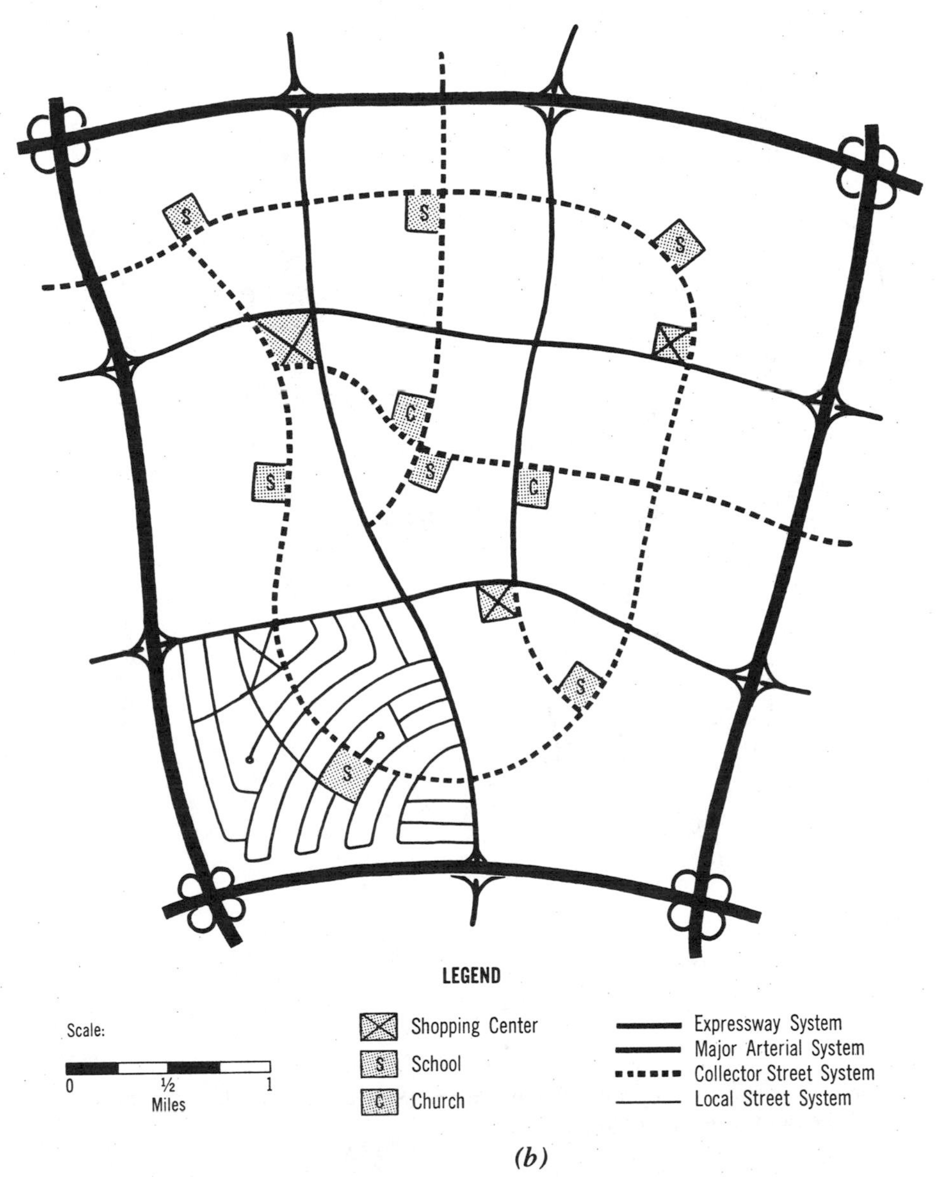


*(b)*

*Fig 8-6b.* A hypothetical functional classification map. (Courtesy Public Administration Service.)

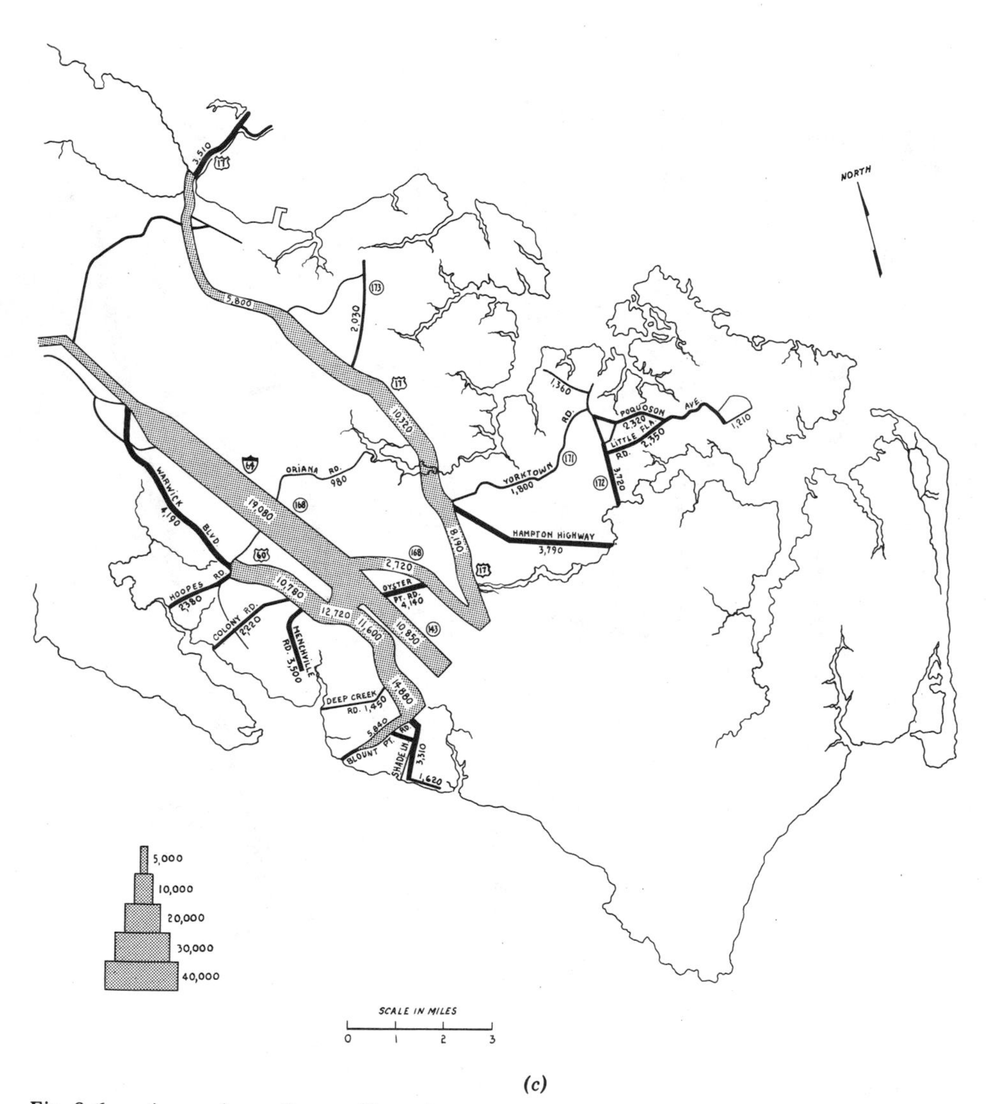


(c)

*Fig 8-6c.* A portion of a traffic volume map. (Source: *Peninsula Area Transportation Study,* DeLeuw, Cather and Co.)

## TRAVEL TIME CONTOURS
## OLD POINT COMFORT

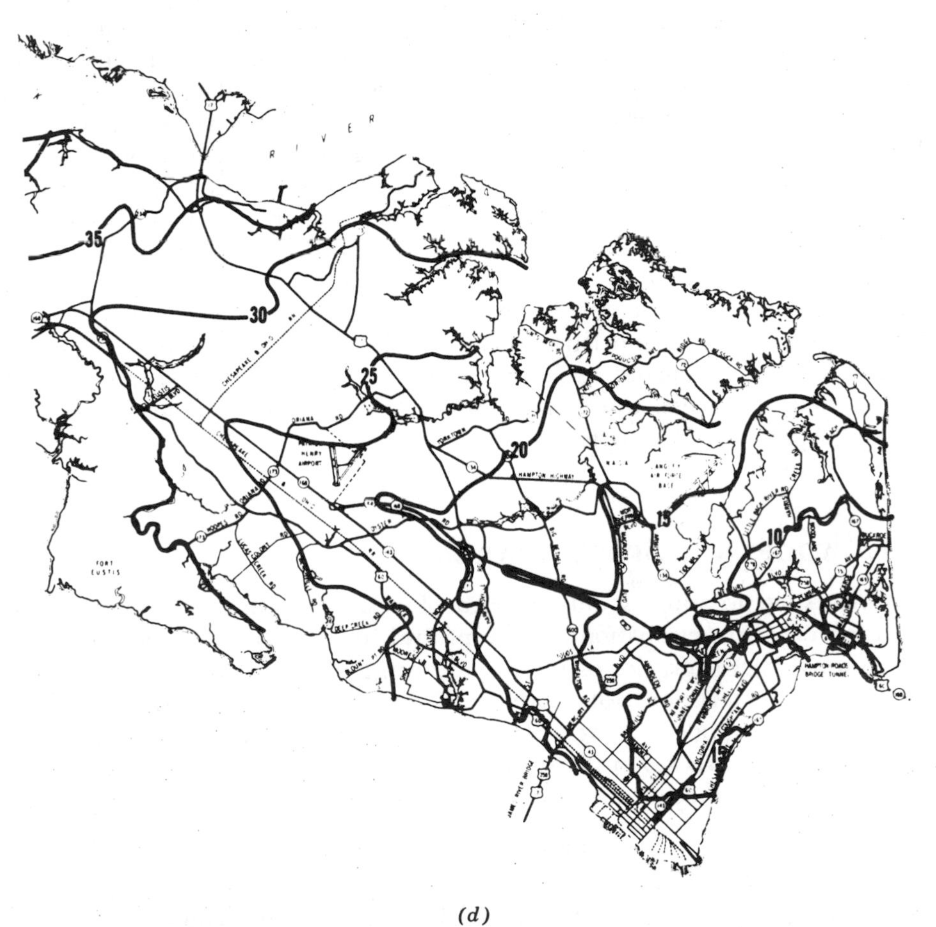


*(d)*

*Fig 8-6d.* A portion of a travel contour map. (Source: *Peninsula Area Transportation Study*, DeLeuw, Cather and Co.)

## 8-9. FUTURE STATUS OF TRANSPORTATION

*Land Use, Population, and Socioeconomic Projections.* After completing the analysis of data both collected and generated in the major inventories and the basic planning studies, the planner has reached the stage of projecting the future land use, population, and socioeconomic conditions in the study area. Population and economic projections are first made in conjunction with the goals and objectives of the region as articulated by local planning bodies. These forecasts are used to generate at least one future land use plan. Generally, in a large study, several land use configurations are generated and, after evaluation, the plan that best satisfied goals and objectives is selected for travel forecasts [10]. In areas of fairly loose planning control, valid alternatives can be produced by a variety of land use models [11]. In smaller studies with limited manpower and in areas with very strong planning legislation, land use plan options are more commonly based on more subjective goal-oriented techniques. For a complete discussion of population and economic projection techniques, the reader should refer to standard works in the fields of planning and economics [12, 13].

*Travel Forecasts.* On the completion of the projections of population, socioeconomic status, and land use, the planner is in a position to develop forecasts of travel demand, as shown by Fig. 8-1. Because of the complexity of the modeling process, the description of the forecasting procedure is described separately, in Chapter 9.

## 8-10. ADOPTION OF STANDARDS

It is necessary to adopt standards both of design and of levels of service to allow an objective evaluation of current and future system deficiencies. The operational characteristics of each mode must be matched to the changing demand characteristics to indicate clearly existing and developing problem areas. For each mode of travel there will be numerous minimum standards of each type. For purposes of illustration only, a few typical standards are given here for various modes and submodes.

Typical Design Standards

Highways and Transit
- Minimum average highway speed (AHS) on rural sections
- Minimum wheel-load capacities of pavements
- Minimum lane widths
- Recommended standards for bus lay-by areas
- Parking terminal standards
- Metro train capacity

Airports
- Minimum runway lengths
- Minimum apron sizes to accommodate anticipated aircraft dimensions

Terminal area space per 1000 emplaned passengers
Runway, taxiway, and apron bearing strengths

Waterways
Recommended channel depths
Loading-apron widths on wharves
Recommended areas of transit sheds per available berth
Container handling capacity

Railways
Maximum rail curvature
Recommended minimum overall operating speeds
Maximum acceleration and deceleration

Typical Level of Service Standards

Highways
Overall travel speeds
Level of service (as defined in *Highway Capacity Manual*)
Accident level per 100 million vehicle miles
Headway between transit vehicles
Percentage of urban population within walking distance of transit
Adequate parking supply/demand ratio

Airports
Maximum delay to an individual aircraft operation
Number of hours per year airport is closed due to atmospheric conditions
Load factors on aircraft

Waterways
Average delay to a ship arriving and waiting for a berth
Rate at which a container ship can be unloaded
Percentage of time that no storage will be available in port

Railways
Service intervals
Passenger demand to seat ratio
Percentage of cancellations of service
Age of rolling stock

## 8-11. DETERMINATION OF PRESENT AND FUTURE DEFICIENCIES

Based on the information gathered in the basic planning studies and the major inventories, the planner is in a position to identify those areas of the transportation network with deficiencies. Deficiencies occur where the transportation facilities provided do not conform to the standards that the study has adopted as minimally acceptable. Where inadequate facilities are provided, either with respect to design or service levels, the results are manifest in operating delays, congestion, low overall travel times, large and inconvenient service headways, unsafe operating conditions, lack of comfort and convenience, and, in the extreme, complete lack of service. In practice, the transportation analyst examines the existing transportation network and evaluates this with reference to existing deficiencies. Current inadequacies are displayed by deficiency maps and a variety of tabulations. In evaluating existing highway facilities, for example, it would be necessary to tabulate existing operating speeds by facility type, accident rates by facility type, and parking deficiencies by location. It would also be necessary to display graphically areas with inadequate transit service or coverage, isochrones from the Central Business District for auto and transit trips, and the capacity-demand relationships for individual facilities or corridors.

In addition to determining current deficiencies it is usual practice to project future travel demand and to apply this demand to the "committed" system. The committed system can be defined in various ways, but normally, it is regarded as that system of facilities that is so far into the process of implementation that there is little opportunity to introduce modifications. In some jurisdictions this is taken to be the point at which right-of-way purchase has started, in others it is taken as the beginning of the final contract design phase. By applying future traffic demand to the committed system, the planner becomes aware of the inadequacies that will arise if no further facilities are provided beyond those already in process. By including only the committed system, the options opened for planning the future system are at a maximum. Graphic displays of the areawide deficiencies are usually prepared to enable the planner to identify easily areas and corridors of insufficient facilities.

## SELECTED PLANNING STUDIES

A number of specialized studies must be carried out in the process of preparing an overall comprehensive transportation plan. In order to keep intact the thread of the planning process that has been described in the beginning part of this chapter, some of these studies are described in more detail in the following sections. The reader must understand, of course, that in practice these studies are integrated into those points of the study that attempt to establish the level of transportation supply and demand, covered in Sections 8-1 through 8-11.

### 8-12. HIGHWAY CAPACITY STUDIES

The *capacity* of a facility is defined as the maximum hourly rate at which persons or vehicles can reasonably be expected to traverse a point or uniform section of a lane or roadway during a given time period under prevailing roadway, traffic, and control conditions [3].

The time period used in most capacity analysis is 15 min, even though the hourly rate is quoted; this is because the lesser period is considered to be the shortest interval during which stable flow can be observed. The definition of capacity relates to prevailing roadway, traffic and control conditions, which should be reasonably uniform over a section taken for analysis. Furthermore, the definition assumes good weather and pavement conditions.

*Roadway conditions* refer to the geometrics of the street and highway such as type of facility, its development environment, number and direction of lanes, lane widths, shoulder widths, lateral clearances, design speed, and horizontal and vertical alignments.

*Traffic conditions* refer to the characteristic of the traffic stream and the stream components which use the facility such as the distribution of vehicle types comprising the traffic stream, the amount and distribution of traffic between available lanes, and directional distribution.

*Control conditions* refer to the type and specific design of control devices and traffic regulation on the facility including the location, type and timing of traffic signals, presence of STOP and YIELD signs, lane use and turn restrictions, and so on. While capacity denotes the maximum number of vehicles that a roadway can handle, there are lower volumes that a section can accommodate at higher operating speeds and decreased congestion levels. The different levels of operating speed and congestion are accounted for by the various levels of service at which a facility can operate. Graphically, this has been shown in Fig. 4-10.

*Level of service* denotes any of an infinite number of differing combinations of operating conditions that can occur on a given lane or roadway when it is accommodating various traffic volumes. The concept of levels of service is defined as a qualitative measure describing operational conditions within a traffic stream and their perception by motorists and/or passengers [3]. These conditions are generally described in terms of factors such as speed and travel time, freedom to maneuver, traffic interruptions, comfort, convenience, and safety. The various levels of service at which a facility can operate vary from level A, the free-flow condition, to level F, the congestion condition.

The concept is perhaps most easily described for the highway case. At high operating speeds and low volumes, the driver is free to maneuver. The facility is said to operate at the highest level of service—level A. As volume increases, operating speeds decrease. Level B is the zone of stable flow with some slight reduction in driver freedom. Rural highways are often designed to accommodate the service volume of level B. Level C is still within the zone of stable flow, but the driver is subject to more restrictions due to increasing volumes. As a result, he has less freedom to pass and to select his own operating speed. Level D approaches the zone of unstable flow. At this level of operation the speeds are still acceptable, but the driver has little freedom to maneuver within the traffic stream. Level E denotes volumes approaching capacity. The facility operates at low speeds and freedom to maneuver is eliminated almost totally. Small variations in volume can cause large speed changes as the facility operates in the range of unstable flow. Congestion and the familiar traffic jam are represented in level F, where densities increase rapidly as volume decreases and speeds drop from a crawl to a standstill. Table 8-2 indicates levels of service characteristics by highway type.

The *service flow rate* is the maximum hourly rate at which persons or vehicles can reasonably be expected to traverse a point or uniform section of lane or roadway during a given time period under prevailing roadway, traffic and control conditions while maintaining a designated level of service. Although quoted as hourly flows, service flow rates like capacity are usually taken over a 15-min period.

It will be seen that capacity and service flow rate analysis can be divided into two main areas of concern:

1. *Uninterrupted flow* where there are no fixed elements such as traffic signals, signs, or junctions external to the traffic stream which cause interruptions to the traffic flow (e.g., freeways, expressways, highways with access control, rural highways).
2. *Interrupted flow* where fixed elements cause periodic interruptions to flow (e.g., traffic signals and stop signs).

Levels of service can be defined by performance relative to one or more operational parameters for the facility. Depending on the facility, the best measure differs and while the concept of level of service comprises a wide range of parameters operating simultaneously, for practical purposes one parameter is selected as the most appropriate indicator. The parameters selected to define the level of service for an individual facility type are called *measures of effectiveness.* (See Tables 8-3a and 8-3b.)

It has been determined from extensive field observations of the behavior of traffic on operating facilities that capacities and levels of service provided by roadways under uninterrupted flow are dependent on two groups of factors.

1. *Roadway factors* which relate to the physical characteristics of the design of the facility. The following roadway factors are found to be pertinent to capacity analysis:

    a. Lane widths
    b. Lateral clearance from edge of pavement
    c. Width of shoulders
    d. Presence and nature of auxiliary lanes
        i. Parking lanes
        ii. Speed change lanes
        iii. Towing and storage lanes
        iv. Auxiliary lanes in weaving section
        v. Truck climbing lanes and passing bays
    e. Surface condition
    f. Alignment
    g. Grades

*Table 8-2*
**Level-of-Service Characteristics by Highway Type**

| *Level of Service* | *Controlled-Access Highways* | *Multilane Rural Without Access Control* | *Two Lanes* | *Urban and Suburban Arterials* |
|---|---|---|---|---|
| A | Free flow. Operating speeds at or greater than 60 mph. Service volume of 1400 passenger cars/hr on two lanes, one direction. Each additional lane serves volume of 1000 veh/hr lane. | Operating speed 60 mph or greater. Under ideal conditions, volume is limited to 600 passenger cars/lane/hr or 30 percent of capacity. Average speeds are likely to be influenced by speed limits. | Operating speeds of 60 mph or higher. 75 percent of passing maneuvers can be made with little or no delay. Under ideal conditions, a service volume of 400 passenger veh/hr, total two-way, can be achieved. | Average overall travel speed of 30 mph or more. Free flowing with volume/capacity ratio of 0.60. Load factor at intersections near the limit of the 0.0 range. Peak-hour factor at about 0.70. |
| B | Higher speed range of stable flow. Operating speed at greater than 55 mph. Service volume on two lanes in one direction not greater than 2000 passenger veh/hr. Each additional lane above two in one direction can serve 1500 veh/hr. | Beginning of stable flow area. Volume at which actions of preceding vehicle will have some influence on following vehicles. Volume will not exceed 50 percent of capacity of 1000 passenger vehicles/lane/hr at a 55 mph operating speed under ideal conditions | Operating speeds of 50 mph or higher. Volumes may reach 45 percent of capacity with continuous passing sight distance. Volumes of 900 passenger cars/hr, total two-way, can be carried under ideal conditions. | Average overall speeds drop due to intersection delay and intervehicular conflicts, but remain at 25 mph or above. Delay is not unreasonable. Volumes at 70 percent of capacity and peak-hour factor approximately 0.80. Load factor at intersections approximately 0.1. |

*continued*

*Table 8-2*
**Level-of-Service Characteristics by Highway Type *(Continued)***

| *Level of Service* | *Controlled-Access Highways* | *Multilane Rural Without Access Control* | *Two Lanes* | *Urban and Suburban Arterials* |
|---|---|---|---|---|
| C | Operation still stable, but becoming more critical. Operating speed of 50 mph. Service flow on two lanes in one direction at 75 percent of capacity or not more than 5-min flow rate of 3000 passenger cars/hr. | Stable flow to a volume not exceeding 75 percent of capacity or 1500 passenger cars/lane/hr, under ideal conditions, maintaining at least a 45-mph operating speed. | Flow still stable. Operating speeds of 40 mph or above with total volume under ideal conditions 70 percent of capacity with continuous passing sight distance, or 1400 passenger veh/hr total two-way. | Service volumes about 0.80 of capacity. Average overall travel speeds 20 mph. Load factor of 0.3 at most intersections. Peak-hour factor approximately 0.85. Traffic flow stable; acceptable delays. |
| D | Lower speed range of stable flow. Operation approaches instability and is susceptible to changing conditions. Operating speeds approx. 40 mph. Service flow rates at 90 percent of capacity. Peak 5-min flow under ideal conditions cannot exceed 3600 veh/hr for two-lanes, one direction; 1800 veh/hr for each added lane. | Approaching unstable flow at volume up to 90 percent of capacity or 1800 passenger cars/hr at an operating speed of about 35 mph under ideal conditions. | Approaching unstable flow. Operating speeds approximately 35 mph. Volumes, two-direction, at 85 percent of capacity with continuous passing opportunity, or 1700 passenger cars/hr total two-way under ideal conditions. | Beginning to tax capabilities of street section. Approaching unstable flow. Service volumes approach 90 percent of capacity. Average overall speeds down to 15 mph. Delays at intersections extensive with some cars waiting two or more cycles. Peak-hour factor approximately 0.90; load factor of 0.7. |

*Table 8-2*
**Level-of-Service Characteristics by Highway Type *(Continued)***

| *Level of Service* | *Controlled-Access Highways* | *Multilane Rural Without Access Control* | *Two Lanes* | *Urban and Suburban Arterials* |
|---|---|---|---|---|
| E | Unstable flow. Overall operating speeds of 30–35 mph. Volumes at capacity or 2000 veh/hr lane under ideal conditions. Traffic flow metered by design constrictions and bottle-necks, but long backups do not normally develop upstream. | Flow at 100 percent of capacity or 2000 passenger cars/lane/hr under ideal conditions. Operating speeds of about 30 mph or less. | Operating speeds in neighborhood of 30 mph but may vary considerably. Volumes under ideal conditions, total two-way, equal to 2000 passenger veh/hr. Level E may never be attained. Operation may go directly from Level D to Level F. | Service volumes at capacity. Average overall traffic variable but in area of 15 mph. Unstable flow. Continuous backup on approaches to intersections. Load factor at intersections in range 0.7 to 1.0. Peak-hour factor likely to be 0.95. |
| F | Forced flow. Freeway acts as a storage for vehicles backed up from downstream bottleneck. Operating speeds range from near 30 mph to stop-and-go operation. | Forced flow, congested condition with widely varying volume characteristics. Operating speeds of less than 30 mph. | Forced, congested flow with unpredictable characteristics. Operating speeds less than 30 mph. Volumes under 2000 passenger cars/hr, total two-way. | Forced flow. Average overall traffic speed below 15 mph. All intersections handling traffic in excess of capacity. Extensive vehicular backups. |

Adapted from *A Policy on Geometric Design of Highways and Streets.* American Association of State Highway and Transportation Officials, Washington, D.C. (1984)

*Table 8-3a*
**Measures of Effectiveness for Level of Service Definition**

| *Type of Facility* | *Measure of Effectiveness* |
|---|---|
| Freeways | |
| Basic freeway segments | Density (passenger car/mi/lane) |
| Weaving areas | Average travel speed (mph) |
| Ramp junctions | Flow rates (passenger car/hr) |
| Multilane highways | Density (passenger car/mi/lane) |
| Two-lane highways | Percent time delay<br>Average travel speed (mph) |
| Signalized intersections | Average individual stopped delay (sec/veh) |
| Unsignalized intersections | Reserve capacity (passenger car/hr |
| Arterials | Average travel speed (mph) |
| Transit | Load factor (persons/seat) |
| Pedestrians | Space (ft²/pedestrian) |

*Table 8-3b*
**Levels of Service (LOS) for Basic Freeway Sections**

| | | *70 mph Design Speed* | | | *60 mph Design Speed* | | |
|---|---|---|---|---|---|---|---|
| *LOS* | *Density (passenger cars/ mile per lane)* | *Speed*[a] *(mph)* | *v/c* | *MSF*[b] *(passenger cars/ hr per lane)* | *Speed*[a] *(mph)* | *v/c* | *MSF*[b] *(passenger cars/ hr per lane)* |
| A | ≤12 | ≥60 | 0.35 | 700 | — | — | — |
| B | ≤20 | ≥57 | 0.54 | 1,100 | ≥50 | 0.49 | 1,000 |
| C | ≤30 | ≥54 | 0.77 | 1,550 | ≥47 | 0.69 | 1,400 |
| D | ≤42 | ≥46 | 0.93 | 1,850 | ≥42 | 0.84 | 1,700 |
| E | ≤67 | ≥30 | 1.00 | 2,000 | ≥30 | 1.00 | 2,000 |
| F | >67 | <30 | c | c | <30 | c | c |

[a]Average travel speed.

[b]Maximum service flow rate per lane under ideal conditions. (Note: All values of *MSF* are rounded to the nearest 50 passenger cars/hr.)

[c]Highly variable, unstable.

SOURCE: *Highway Capacity Manual.* Transportation Research Board Special Report No. 209 (1985).

2. *Traffic factors* are those that relate to the composition of the traffic using the facility and the types of controls in use. The following have been found to exercise control over capacities and, therefore, are used in analyses:

   a. Percentage of traffic constituted by trucks
   b. Percentage of traffic constituted by buses

c. Lane distribution
d. Variations in traffic flow
e. Traffic interruptions
f. Type of driver (weekday/commuter or other)

At-grade intersections and urban streets operate under interrupted flow conditions. The following factors have been found to affect the capacities and service flow rates of such facilities:

1. Geometric or roadway factors:

   a. Area type (CBD or other)
   b. Number of lanes
   c. Lane widths
   d. Grades
   e. Provision of exclusive left- or right-turn lanes
   f. Length of storage bay at left- or right-turn lanes
   g. Parking conditions

2. Traffic factors:

   a. Volumes by movement
   b. Peak hour factor
   c. Percent heavy vehicles
   d. Conflicting pedestrian flow rate
   e. Number of local buses stopping in intersection
   f. Parking activity in maneuvers per hour
   g. Arriving type (platoon or random by position in signal cycle)

3. Signalization or control factors:

   a. Cycle length
   b. Green times
   c. Actuated versus pretimed operation
   d. Availability of pedestrian push button interrupt
   e. Minimum pedestrial green time
   f. Phase plan

The general approach to the procedures outlined in the *Highway Capacity Manual* is to give capacities and service flow rates in simple tables or graphs for specified standard conditions. The standard or ideal conditions must then be modified to account for the variations produced by differences in the roadway, traffic, or control conditions already discussed.

Ideal conditions for uninterrupted flow facilities include:

1. 12-ft lane widths
2. 6-ft clearance between the edge of travel lanes and the nearest obstructions of objects at the roadside or in the median
3. 70 mph design speed for multilane highways and 60 mph design speed for two-lane highways
4. All passenger cars in the traffic stream [3, 14]

The idealized signalized intersection approach has:

1. 12-ft lane widths
2. Level grades
3. No curb parking on the intersection approaches
4. All passenger cars in the traffic stream
5. No turning movements
6. Location in a noncentral business district location
7. Green signal available at all times [3, 14]

## EXAMPLE OF A SERVICE FLOW RATE ANALYSIS FOR A TWO-LANE RURAL HIGHWAY

For a full treatment of capacity and service flow rate analysis the reader is referred to Reference 3. However, for purposes of illustration an example is given of the analysis for a rural two-lane highway. Ideal conditions are defined in terms of no restrictions from geometric, traffic, or environmental conditions. As such the highway has:

1. A design speed greater than or equal to 60 mph.
2. Lane widths greater than or equal to 12 ft.
3. Clear shoulders wider than or equal to 6 ft.
4. No "no passing zones" on the highway.
5. All passenger cars in the passenger stream.
6. A 50/50 directional split of traffic.
7. No impediments to through traffic due to traffic control or turning vehicles.
8. Level terrain.

Under these ideal conditions the capacity of a two-lane rural highway is 2800 passenger cars per hour total in both directions. To examine the effect of various restrictions and assumption of a different service flow rate, a computation is made for the following problem:

Compute the service flow rate at level of service (LOS) B for a two-lane rural highway in rolling terrain with 11-ft-wide lanes and usable shoulder widths of 4 ft. The road has 20 percent no passing zones and a directional distribution split of 60/40. There are 10 percent trucks, 5 percent recreational vehicles, and 2 percent buses in the traffic.

The relationship that describes the effect of constraints on service flow rates for a two-lane road is [3]:

$$SF_i = 2800 \times (V/C)_i \times f_d \times f_w \times f_{HV}$$

where

$SF_i$ = total service flow rate (veh/hr) in both directions for prevailing roadway and traffic conditions for level of service $i$

$(V/C)_i$ = ratio of flow rate to ideal capacity for level of service $i$, see Table 8-4

$f_d$ = adjustment factor for directional distribution of traffic, see Table 8-5

$f_w$ = adjustment factor for narrow lanes and restricted shoulder width, see Table 8-6

$f_{HV}$ = adjustment factor for the presence of heavy vehicles in the traffic stream where:

$$f_v = 1/[1 + P_T (E_T - 1) + P_R (E_R - 1) + P_B (E_B - 1)]$$

$P_T$ = decimal proportion of trucks in traffic stream
$E_T$ = passenger car equivalent for trucks, see Table 8-7
$P_R$ = decimal proportion of recreation vehicles in traffic stream
$E_R$ = passenger car equivalent for recreation vehicles, see Table 8-7
$P_B$ = decimal proportion of buses in traffic stream
$E_B$ = passenger car equivalent for buses, see Table 8-7.

$$SF_B = 2800 \times (V/C)_B \times f_d \times f_w \times f_{HV}$$

1. For $LOS_B$, rolling terrain, 20 percent no passing zones, $(V/C)_B = 0.23$, see Table 8-4.
2. For directional distribution 60/40, $f_d = 0.94$, See Table 8-5.
3. For 11 ft lanes with usable shoulders of 4 ft, $fw = 0.85$, see Table 8-6.
4. $P_T = 10$ percent, $E_T = 5$, see Table 8-7
   $P_R = 5$ percent, $E_R = 3.9$, see Table 8-7
   $P_B = 2$ percent, $E_B = 3.4$, see Table 8-7

   Then

$$f_{HV} = 1/[1 + 0.1(5 - 1) + 0.05\ (3.9 - 1) + 0.02\ (3.4 - 1)]$$
$$= 0.63$$

$$SF_B = 2800 \times 0.23 \times 0.94 \times 0.85 \times 0.63$$
$$= 324 \text{ veh/hr}$$

The level B service flow rate under the prevailing roadway and traffic conditions is 324 veh/hr. The reader should check that the capacity under these conditions is 1409 veh/hr.

For worked examples of freeway sections and urban streets and intersections the reader is referred to Reference 14.

*Table 8-4*

**Level-of-Service Criteria for General Two-Lane Highway Segments**

| Level of Service | Percent Time Delay | V/C Ratio[a] | | | | | | | | | | | | | | | | | | | |
|---|---|---|---|---|---|---|---|---|---|---|---|---|---|---|---|---|---|---|---|---|---|
| | | Level Terrain | | | | | | | Rolling Terrain | | | | | | | Mountainous Terrain | | | | | |
| | | Avg[b] Speed | Percent No Passing Zones | | | | | | Avg[b] Speed | Percent No Passing Zones | | | | | | Avg[b] Speed | Percent No Passing Zones | | | | |
| | | | 0 | 20 | 40 | 60 | 80 | 100 | | 0 | 20 | 40 | 60 | 80 | 100 | | 0 | 20 | 40 | 60 | 80 | 100 |
| A | ≤30 | ≥58 | 0.15 | 0.12 | 0.09 | 0.07 | 0.05 | 0.04 | ≥57 | 0.15 | 0.10 | 0.07 | 0.05 | 0.04 | 0.03 | ≥56 | 0.14 | 0.09 | 0.07 | 0.04 | 0.02 | 0.01 |
| B | ≤45 | ≥55 | 0.27 | 0.24 | 0.21 | 0.19 | 0.17 | 0.16 | ≥54 | 0.26 | 0.23 | 0.19 | 0.17 | 0.15 | 0.13 | ≥54 | 0.25 | 0.20 | 0.16 | 0.13 | 0.12 | 0.10 |
| C | ≤60 | ≥52 | 0.43 | 0.39 | 0.36 | 0.34 | 0.33 | 0.32 | ≥51 | 0.42 | 0.39 | 0.35 | 0.32 | 0.30 | 0.28 | ≥49 | 0.39 | 0.33 | 0.28 | 0.23 | 0.20 | 0.16 |
| D | ≤75 | ≥50 | 0.64 | 0.62 | 0.60 | 0.59 | 0.58 | 0.57 | ≥49 | 0.62 | 0.57 | 0.52 | 0.48 | 0.46 | 0.43 | ≥45 | 0.58 | 0.50 | 0.45 | 0.40 | 0.37 | 0.33 |
| E | >75 | ≥45 | 1.00 | 1.00 | 1.00 | 1.00 | 1.00 | 1.00 | ≥40 | 0.97 | 0.94 | 0.92 | 0.91 | 0.90 | 0.90 | ≥35 | 0.91 | 0.87 | 0.84 | 0.82 | 0.80 | 0.78 |
| F | 100 | <45 | — | — | — | — | — | — | <40 | — | — | — | — | — | — | <35 | — | — | — | — | — | — |

[a]Ratio of flow rate to an ideal capacity of 2800 passengers cars/hr in both directions.

[b]Average travel speed of all vehicles (in mph) for highways with design speed ≥60 mph; for highways with lower design speeds, reduce speed by 4 mph for each 10-mph reduction in design speed below 60 mph; assumes that speed is not restricted to lower values by regulation.

*Table 8-5*

**Adjustment Factors for Directional Distribution on General Terrain Segments**

| *Directional Distribution* | 100/0 | 90/10 | 80/20 | 70/30 | 60/40 | 50/50 |
|---|---|---|---|---|---|---|
| *Adjustment factor,* $f_d$ | 0.71 | 0.75 | 0.83 | 0.89 | 0.94 | 1.00 |

*Table 8-6*

**Adjustment Factors for the Combined Effect of Narrow Lanes and Restricted Shoulder Width,** $f_w$

| *Usable[a] Shoulder Width (ft)* | *12-ft Lanes* | | *11-ft Lanes* | | *10-ft Lanes* | | *9-ft Lanes* | |
|---|---|---|---|---|---|---|---|---|
| | *Level of Service A–D* | *Level of Service E* | *Level of Service A–D* | *Level of Service E* | *Level of Service A–D* | *Level of Service E* | *Level of Service A–D* | *Level of Service E* |
| ≥6 | 1.00 | 1.00 | 0.93 | 0.94 | 0.84 | 0.87 | 0.70 | 0.76 |
| 4 | 0.92 | 0.97 | 0.85 | 0.92 | 0.77 | 0.85 | 0.65 | 0.74 |
| 2 | 0.81 | 0.93 | 0.75 | 0.88 | 0.68 | 0.81 | 0.57 | 0.70 |
| 0 | 0.70 | 0.88 | 0.65 | 0.82 | 0.58 | 0.75 | 0.49 | 0.66 |

[a]Where shoulder width is different on each side of the roadway, use the average shoulder width.
[b]Factor applies for all speeds less than 45 mph.

*Table 8-7*

**Average Passenger Car Equivalents for Trucks, RV's, and Buses on Two-Lane Highways Over General Terrain Segments**

| *Vehicle Type* | *Level of Service* | *Types of Terrain* | | |
|---|---|---|---|---|
| | | *Level* | *Rolling* | *Mountainous* |
| Trucks, $E_T$ | A | 2.0 | 4.0 | 7.0 |
| | B and C | 2.2 | 5.0 | 10.0 |
| | D and E | 2.0 | 5.0 | 12.0 |
| RV's $E_R$ | A | 2.2 | 3.2 | 5.0 |
| | B and C | 2.5 | 3.9 | 5.2 |
| | D and E | 1.6 | 3.3 | 5.2 |
| Buses, $E_B$ | A | 1.8 | 3.0 | 5.7 |
| | B and C | 2.0 | 3.4 | 6.0 |
| | D and E | 1.6 | 2.9 | 6.5 |

SOURCE: Reference 6.

## 8-13. LAND USE STUDIES

Early in the 1950s, it became clear from a variety of research studies that the characteristics of travel could be related to the intensity and spatial separations of land use [15].

For the purpose of developing traffic models and to understand clearly the physical impact of transportation facilities, the planner needs up-to-date land use data on an areawide basis. Typically, the planner will employ at least the following techniques:

1. Land use survey.
2. Land use classification.
3. Vacant land use study.
4. Presentation and storage of land use data.

*Land Use Survey.* Field studies are carried out to identify the type of use of all land parcels in the area. Two principle techniques are employed in the land use survey; these are *inspection* and *inspection-interview.*

The *inspection* method can be carried out on foot or, more commonly, "windshield inspection" is accomplished by a car with a driver and an observer. The use of individual parcels is noted directly on maps, or is recorded on schedules that are referenced in some manner to prepared parcel maps. Scheduled sheets are necessary where land use data are stored as computer records.

The *inspection-interview* method can be used in areas of more intense land use, where more accurate estimates are needed of actual floor areas by type of usage. The results are recorded on schedule sheets.

*Land Use Classification.* In the process of differentiating the spatial arrangements and activity patterns of the urban area, it is necessary to arrange land uses in some form of standard classification. There are several land use classification systems in current use that permit detailed differentiation of usage. It is recommended that in any study area of one of these standardized classification procedures be adopted. For the purpose of the transportation planner, only major categories of land use are classified. Therefore, it will be necessary in most cases to group the results of the land use survey for analysis and presentation. Table 8-8 indicates eight broad major categories of land use. Included in this table are the color codings suggested for standardized presentation.

*Vacant Land Use Study.* Prior to the design of alternate future land use plans, it is necessary to determine the development capability of existing vacant land in the area. This type of analysis is called *vacant land study* or sometimes *development capability analysis.*

Vacant land is classified by two chief criteria:

1. Capability as related to topographic and drainage standards.
2. Capability as related to utilities and improvements available.

*Table 8-8*
**Illustrative Major Urban Land Use Categories**

|  |  |
|---|---|
| Residence | |
| Low density | Yellow |
| Medium density | Orange |
| High density | Brown |
| Retail business | Red |
| Transportation, utilities, communications | Ultramarine |
| Industry and related uses | Indigo blue |
| Wholesale and related uses | Purple |
| Public building and open spaces | Green |
| Institutionalized building and areas | Gray |
| Vacant and nonurban land | Uncolored |

SOURCE: Chapin, F. S., *Urban Land Use Planning,* 3rd Ed., University of Illinois Press, Urbana.

Standard planning works suggest division into *prime* and *marginal* land.

Prime land is all land that can be drained adequately with topographic slopes considered economic for building in the area. Low-lying, steep, and derelict land is considered marginal because only with large capital investment is such land capable of conversion to the prime category. Prime land is often subcategorized to differentiate between different classes of slope. For example, land at slopes still satisfactory for residential areas, 10 to 15 percent, would be unsuitable for extensive one-story industrial plants. Thus, the basic categories are expanded to Class I Prime, Class II Prime, and Marginal.

Further subcategorization is undertaken with respect to the factor of available improvements. Subclassification A could indicate all improvements available, B water and sewer only, C power only, and so on. Based on the classification scheme of the planner it is possible to identify and summarize the amount of available land in each class and category by district over the whole area, for example, Class II Prime A.

*Presentation and Storage of Land Use Data.* The assembled general land use patterns are stored and presented using the following techniques:

1. Land use maps. These are prepared from aerial photographs with major land use categories colored as in Table 8-8.
2. Tabular summaries of land use studies which are readily made from data assembled in the various studies. Such summaries give the planner a clear idea of the nature of the area.
3. It is becoming widespread practice for cities to have computerized data banks with information on:

    a. Persons
    b. Parcels
    c. Street facilities

Computerized data banks where information is coded and adapted on a street address basis can be used in conjunction with computer mapping techniques. It is possible to produce by mechanical means graphic displays of the different characteristics of the urban area down to as fine a grain as the block face.

## 8-14. TRAFFIC VOLUME STUDIES

The automobile and truck mode is unique in the fact that while public agencies provide the roadbed they have little or no control over the amount of usage this roadbed receives, and unless special studies are carried out, the volume of traffic using the facilities is unknown. Motor vehicle volume studies have now been carried out for many years. The procedures used are well documented and, therefore, will be only outlined in the following section [2].

**Urban studies.** *Areawide Counts.* The comprehensive traffic volume study is carried out largely on a sample basis [2]. In order to develop factors to permit adjustment of sample counts to reflect the average daily traffic, it is necessary to conduct *control counts* throughout the system. There are three categories of control counts.

1. *Key count stations.* Selected control stations are used to determine hourly, daily, and annual traffic variations. It is recommended that one nondirectional seven-day count be carried out annually in conjunction with one nondirectional 24-hr count monthly. There will be at least one key station on each category of street classification.
2. *Major control stations.* Each major street will have a major control station which serves to estimate on a sample basis the traffic flows on the major street system. Therefore, each freeway, expressway, major arterial, and collector street is sample counted with a 24-hr nondirectional count every two years.
3. *Minor control stations.* These are set up to sample typical minor street traffic. In smaller cities it is recommended that three counts be taken on each of the three classes of minor street: industrial, commercial, and residential. A 24-hr nondirectional count is carried out every two years.

*Coverage counts* are the samples used to estimate average daily traffic (ADT) throughout the street system. On major streets, which would include freeways, expressways, arterials, and collectors, one nondirectional 24-hr count is made on each control section at a maximum interval of four years. Minor streets are sampled by coverage counts at the rate of one count for every mile of local street. In this case, the count frequency should be every four years.

**Rural studies.** Highways are generally classified by federal, state, and local system and also by function. *Control stations* are set up in a manner similar to urban major control stations, in such a way that seasonal and term variations can be analyzed. Traffic is counted continuously throughout the year at these points. At less important control stations the count may be for only one week per month. Coverage stations are located on all highway sections and are counted on a two- or four-day basis to provide estimates of annual average counts.

The sample counts are adjusted to give ADT for all segments of the system. Results of the volume survey normally are summarized by section and plotted for graphical presentation in a form of a traffic flow map. Figure 8-7 shows a portion of statewide traffic volume map which is updated annually.

**Cordon Counts.** Cordon counts can be used to measure the traffic activity of specific areas. They are often used in conjunction with traffic studies of central business districts of cities, large institutions such as universities, and in the case of external cordons. Information from such counts can indicate:

1. Number of persons entering and leaving the area.
2. Travel modes.
3. Hourly variations.
4. Accumulations within the cordon area of persons and vehicles.

Depending on the purpose of the cordon, it may be run on a 24-hr basis or for a shorter time, omitting the night hours during which there is little traffic movement. In general a cordon line is set that will minimize, as far as possible, the number of cordon crossing points, without loss of information. Figure 8-8 shows the graphical summaries from a central business district cordon.

**Screen Line Counts.** These are used most commonly to verify the volume of movements predicted by origin-destination sample surveys. Ground counts are compared with expanded sample data. The screen line that coincides with internal traffic zone boundaries can be used to determine the number of trips passing from one group of zones to another. Normally, in order to minimize the number of crossing points, the line incorporates such geographical and topographic barriers as rivers, lakes, railroad lines, and freeways. Figure 8-9 shows a graphical comparison of screen line ground counts and expanded survey data.

## 8-15. ATTITUDE SURVEYS

One of the most difficult areas in which transportation planners necessarily find themselves is the evaluation of community values prior to preparing a plan. It is not uncommon for a facility where assiduous attention has been paid to alignment, design speeds, and capacity to come under furious attack because the proposed facility violates what the community feels is essential to the

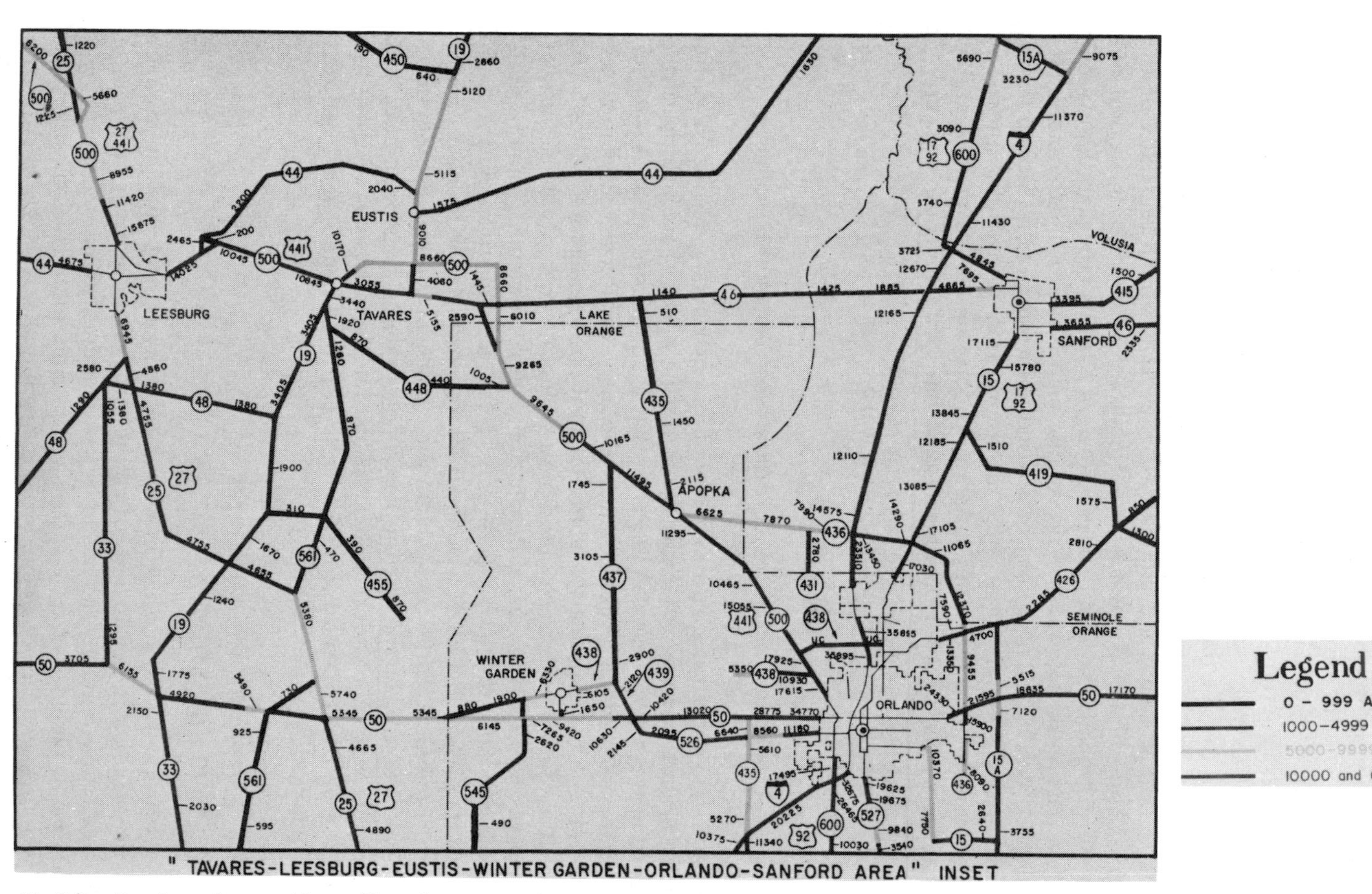


*Fig 8-7.* Portion of statewide traffic volume map. (Source: State of Florida Department of Transportation, Division of Transportation Planning, in Cooperation with U.S. Department of Transportation, Federal Highway Administration.)

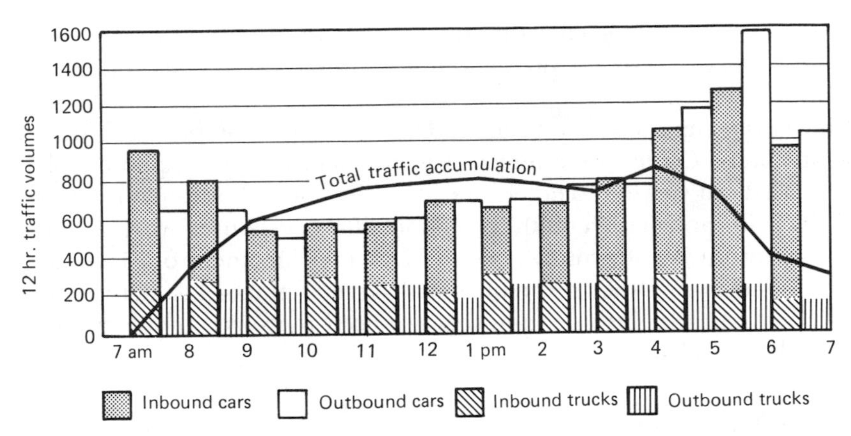


*Fig 8-8.* Graphical analysis of vehicle movements through a CBD cordon. (Source: *City of Galt, Traffic Planning Report,* Damas and Smith, Ltd.)

preservation of a proper environment. While federal law requires that the transportation plan take account of community values, planners find that the techniques for evaluation of attitudes are perhaps the least well defined of all planning techniques.

In determining community values it is necessary to determine the attitudes of those who live in the community. An attitude is a more basic feeling than an opinion, which can change over time when influenced by social pressure. Attitudes relate to basic value judgments. "An attitude is a learned predisposition to behave in a consistent manner in a given situation; as such,

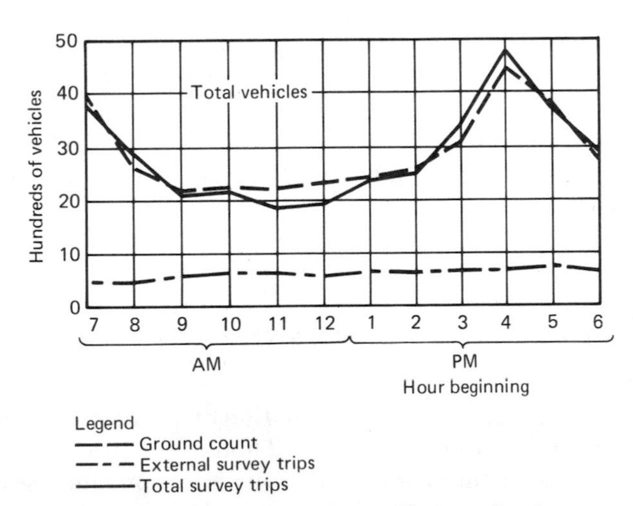


*Fig 8-9.* Screen line comparisons. (Source: *Southeastern Virginia Regional Transportation Study.*)

attitudes are much more enduring than opinions. Hence, attitude assessment is a more reliable basis for prediction of terminal action than opinion study [16]".

Various techniques have been used by transportation planners to evaluate basic attitudes. Using home interviews and mail return responses, planners have carried out surveys to determine attitudes toward housing, the town and state lived in, and leisure and recreation. The responses were analyzed to determine attitude patterns by persons of various income groups and living at different densities. Other techniques that have attempted to determine basic community attitudes include tools that are used more commonly in the social and psychological sciences.

1. Word association.
2. Sentence completion technique.
3. Semantic differential technique.

Other methods recommended for use in determining community values involve the use of groups and panels from specifically selected groups rather than random samples.

*Working review committees,* composed of elected officials, downtown leaders, business community spokesmen, and other community leaders, can evaluate preliminary proposals, raising objections that can be surmounted at an early date.

*Focus groups* are made up of people with common backgrounds and common interests. Under the direction of a discussion leader they articulate the viewpoints that the planner must consider in plans.

*Rating panels* are commonly constituted of professionals in the field of planning. They have been used to evaluate the merit of alternative development patterns with respect to criteria related to areawide goals and objectives.

*Delphi panels,* composed of academics, industrialists, and consultants, for example, are asked to respond to quantified questions on an iterative basis to obtain consensus opinions.

## 8-16. ENVIRONMENTAL IMPACTS OF TRANSPORTATION

The interrelationship between the construction of a transportation system and its environment has been mentioned in Chapter 7. It was shown there that transportation infrastructure has a very strong effect on land uses throughout its useful life. However, for many years the environmental impact of a transportation project has been recognized as having much broader impacts than simply those on land use. The construction of a facility can bring out sweeping environmental changes in an area that reaches far beyond the immediate environs of the project. The general neglect of environmental issues in the first rush of enthusiastic construction of Interstate and urban freeways in the 1950s and 1960s resulted in strong community-based environmental groups, bitterly opposed to freeway construction. These antihighway environmentalists soon were joined by anti-airport groups objecting mainly to the noise impact of rapidly increasing air transport demand in the newly introduced jet transport aircraft. As general concern for the obvious deterioration of the environment of modern cities spread, the U.S. Congress enacted a substantial body of environmental legislation, controlling the planning and construction of transportation facilities, as well as the design and operation of vehicles using those facilities. Early overemphasis of environmental considerations has been moderated and tempered to a large degree by the need to optimize energy usage, particularly oil fuels, since the formation of the OPEC cartel in the early 1970s. Massive increases in oil prices have depressed anticipated increases in traffic demand and, perhaps more importantly, have concentrated public opinion more on their own economic concerns rather than on general environmental problems.

Environmental impact analysis, however, is still of major concern to the transportation planner. Consequently, in the preparation of any plan it is necessary to examine some or all of the following areas:

1. Noise
2. Air pollution.
3. Long- and short-term land use and socioeconomic changes.
4. Water pollution and runoff changes.

## 8-17. NOISE POLLUTION [17]

Noise, which can be defined as unwanted sound, is a necessary by-product of the operation of transportation vehicles. Automobiles generate noise from such sources as the engine, the tires, and the gearbox. Aircraft produce noise from their engines and from the aerodynamic flow of air over the fuselage and wings. Railroad trains also generate noise aerodynamically in addition to the noises generated by rail wheel contact and traction motor noises. The scale of the problem of noise generation is illustrated by Fig. 8-10 which sets noise levels of transportation vehicles within the context of other everyday sounds.

Loudness which is the subjective magnitude of sound, is normally considered to double with an increase in sound intensity of 10 dB. The ear is particularly sensitive to frequencies within the A-range. Therefore sound normally is measured using A-weighted decibels (dBA) which reflect these sensitive ranges.

Another factor that affects the reactions of individuals and groups to

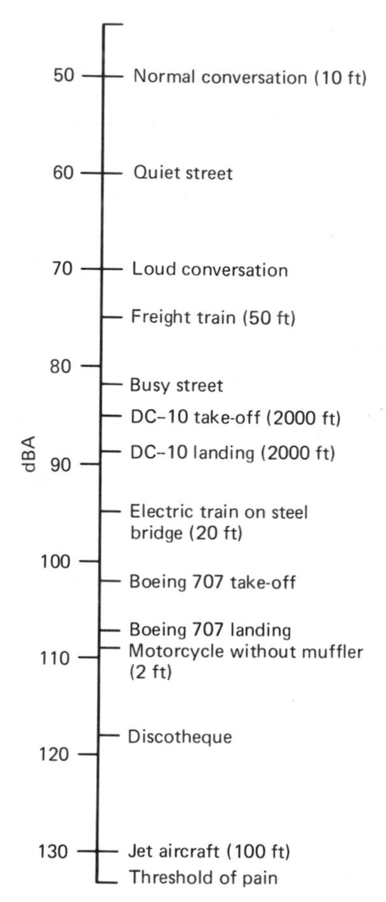


*Fig 8-10.* Sound levels in dBA. (Source: *Noise Pollution,* U.S. Environmental Protection Agency, Washington, D.C., 1972.)

noise is time. Sound is only rarely constant over time, and time is found to affect noise impacts in several ways:

1. The length of duration of the sound.
2. The number of times the sound is repeated.
3. The time of day at which the noise occurs.

## 8-18. AIRCRAFT NOISE

Aircraft noise can be described in terms of the level of noise produced by a single event. After the introduction of jet aircraft, research at Kennedy Airport indicated that measurement of jet noise by the dBA method was inadequate because the ear apparently registered noise in a more complicated way. As a result, a new single-event measure was produced, the perceived noise level (PNL), a D-weighted summation which is normally calculated by computer. Experience with aircraft noise problems has indicated that noise is best measured by indices that modify the equivalent levels of single events with factors reflecting duration, time, and repetition. The measure now in use in the United States is the *day-night average sound level,* or $L_{DN}$. This is a measure of the noise environment at a given location over a 24-hr period. In terms of sound energy, it is equivalent to the level of a continuous A-weighted sound level with nighttime noises increased by 10 dB to account for the undesirable effect of night noise disturbance. It is possible to measure directly the $L_{DN}$ by a sophisticated integrating noise measurement meter.

With the $L_{DN}$ method, the contribution of an aircraft operation is described in terms of the *sound exposure level* (SEL). The SEL is the A-weighted sound level integrated over the entire noise event and normalized to a reference duration of 1 sec. In other words, the SEL gives the level of a continuous 1-sec sound that contains the same amount of energy as the noise event.

Empirical graphs have been published [18] giving SEL values in dB for different classes of aircraft, modes of operation, and locations with respect to the flight path. The partial $L_{DN}$ value for those conditions can then be calculated by Eq. 8-1, which accounts for the number and time of day of such operations. For aircraft class $i$, and operation mode $j$,

$$L_{DN}(i, j) = \text{SEL}(i, j) + 10 \log (N_D + 10\, N_N) - 49.4 \tag{8-1}$$

where

$N_D$ = number of daytime operations for given conditions
$N_N$ = number of nighttime operations for given conditions

Daytime is taken as the period from 7:00 A.M. to 10 P.M., and nighttime is the remainder of the day.

After the partial $L_{DN}$ values have been calculated for each significant noise intrusion, they may be summed on an energy basis by Eq. 8-2 to obtain the total $L_{DN}$ due to all aircraft operations.

$$L_{DN} = 10 \log \sum_i \sum_j 10^{\frac{L_{DN}(i,j)}{10}} \tag{8-2}$$

Using charts such as Fig. 8-11, it is possible to compute the percentage of a community likely to be highly annoyed by aircraft noise. The FAA and HUD, on the basis of the findings of social surveys, have set guidelines of land use around airports, as shown in Fig. 8-12 [19]. Using land use guidance charts such as these, disturbance from aircraft noise impact can be minimized. One very important factor that must be remembered in the context of aircraft noise impact is that the aircraft introduced from the mid-1970s are very much quieter than are the early pure jet aircraft. Consequently, noise footprints,

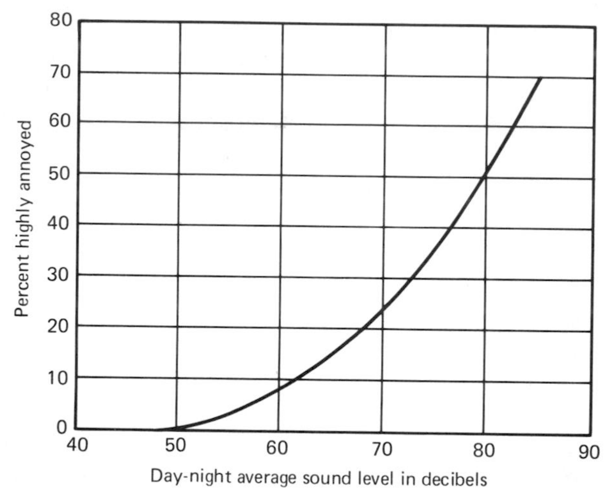


*Fig 8-11.* Synthesis of social survey results to indicate the degree of annoyance due to noise of all kinds. (Source: Shultz, T. J., *Journal of the Acoustical Society of America,* Vol. 64, August 1978.)

such as those shown in Fig. 8-12, will continue to shrink, probably through the 1990s.

## 8-19. HIGHWAY TRAFFIC NOISE [20]

A number of noise descriptors have been developed to characterize the nature of highway traffic noise. Ideally, any descriptor should be capable of reflecting frequency, sound pressure levels, and the fluctuation of these two variables over time. Figure 8-13 shows the variation of traffic noise over time for both high- and low-density traffic conditions. Also shown is the effect of monitoring distance from the highway. The noise pattern can be seen to be spikes of truck noise set against a variable pattern of lower level noise generated by automobiles. As traffic volumes increase, the individual peaks of automobiles coalesce and the general level of background noise from this source increases. Consequently, the spike effect of trucks passing is less pronounced. As the distance between the highway and the monitoring point increases, there is also a decrease in the magnitude of sound fluctuations as sound energy also results in a decrease in noise with distance to the monitoring point. Among noise descriptors currently in use are:

1. *Percentile exceeded noise levels, ($L_x$),* which defines the sound level that is exceeded x percent of the time. A common standard is the $L_{10}$ level, which is widely used in Europe.

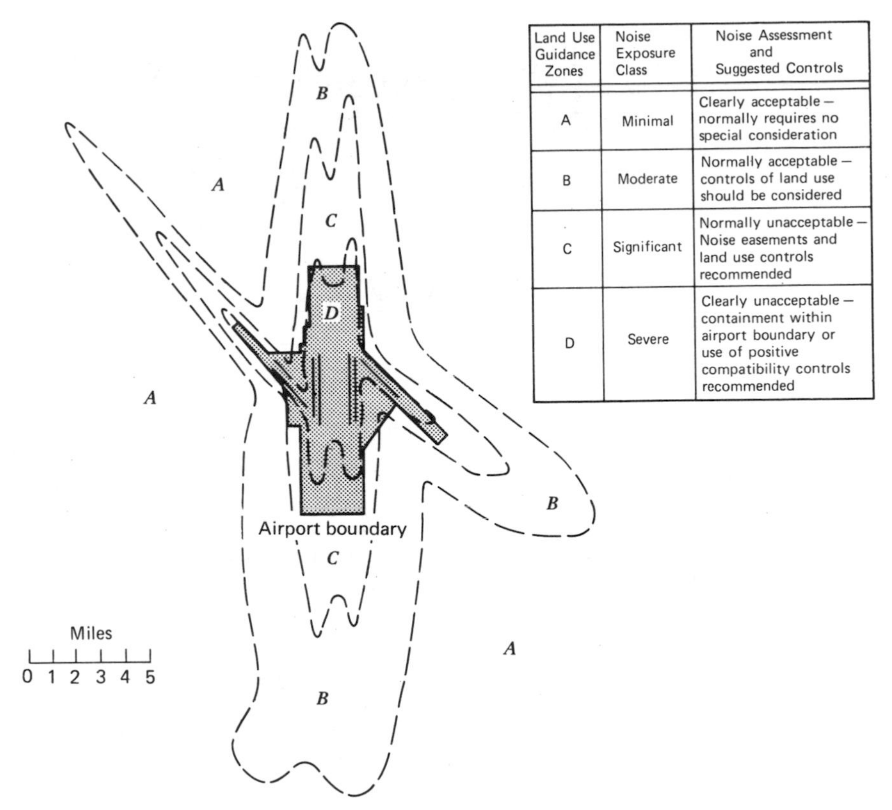


| Land Use Guidance Zones | Noise Exposure Class | Noise Assessment and Suggested Controls |
|---|---|---|
| A | Minimal | Clearly acceptable—normally requires no special consideration |
| B | Moderate | Normally acceptable—controls of land use should be considered |
| C | Significant | Normally unacceptable—Noise easements and land use controls recommended |
| D | Severe | Clearly unacceptable—containment within airport boundary or use of positive compatibility controls recommended |

*Fig 8-12.* Compatibility of land use around airport.

2. *Equivalent continuous (A-weighted) sound level ($L_{eq}$)*, which is the average sound level over a prescribed period of time. Common periods are 1 hr, 24 hr, daytime, or nighttime.
3. *Day-night average sound level, $L_{DN}$*, the one now widely used for airport noise disturbance (see Section 8-18). Care should be exercised, however, when combining in a single measure two different noise sources such as highway and aircraft noise.

## 8-20. AIR POLLUTION

Air pollution is the result of the discharge of unburnt and partially burnt engine fuels together with the by-products of complete combustion. The concentration and relative mixtures of these pollutants depends on the point of the operating cycle, that is, cruise, acceleration, deceleration, or idle, as shown in Fig. 8-14. The principal transportation cause of air pollution is the automobile. Although both the diesel engine of highway trucks and buses and the jet engines of transport aircraft give the appearance of producing massive

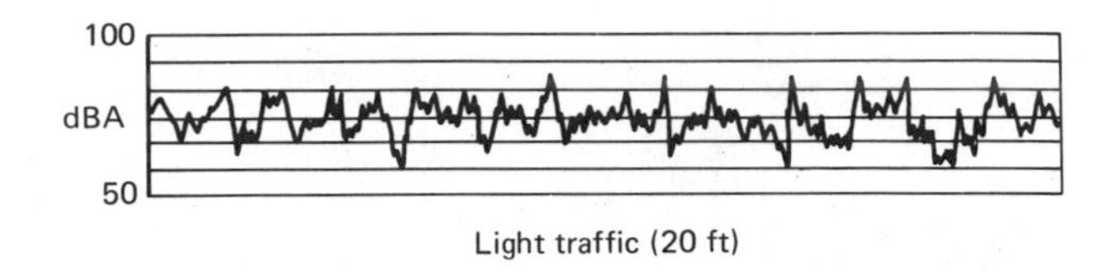


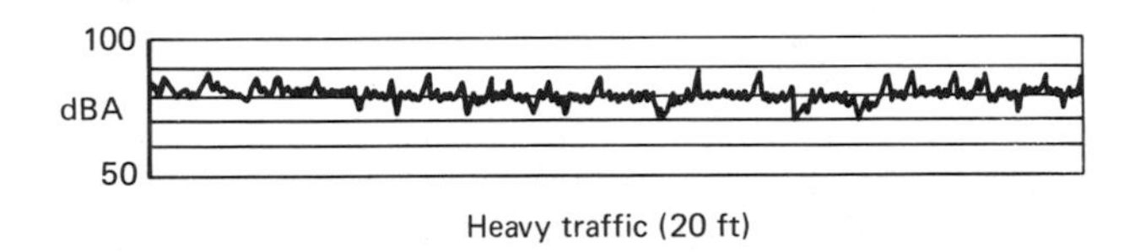


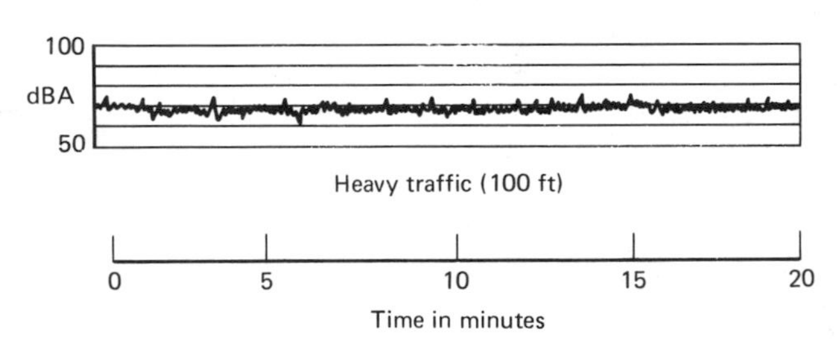


*Fig 8-13.* Variation of highway traffic noise with time, density, and distance to monitor.

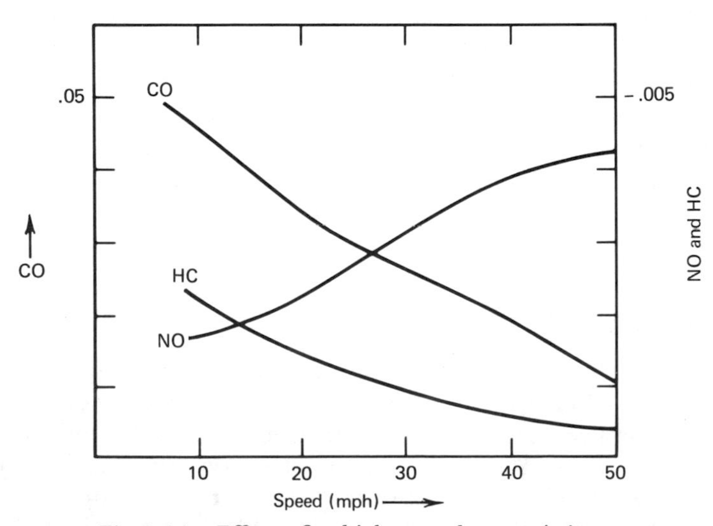


*Fig 8-14.* Effect of vehicle speed on emissions.

air pollution, these visible particulate elements of pollution are in fact of minimal importance in comparison with the much more substantial invisible outputs of the automobile. Indeed, air pollution studies around one major international airport have indicated that the level of air pollution in the area is less than the amount that would have occurred if the area had residential, commercial, and industrial uses similar to the rest of the metropolitan area [21]. The major components of air pollution due to transportation are:

1. *Carbon monoxide.* This is a principal by-product of the internal combustion engine of the automobile. It is produced in substantial quantities when the engine is idling or operating below normal cruise speeds. It is a poisonous gas that, even in low concentrations, can produce slow operating reactions, nausea, headaches, and dizziness. In large concentrations it can kill by reacting with blood hemoglobin.
2. *Oxides of nitrogen.* A by-product of combustion is the production of nitric and nitrogen dioxide. The latter is an irritant. This major component of smog causes very severe breathing problems at high concentrations.
3. *Oxides of sulfur.* Fuels containing sulfur form sulfur dioxide in the combustion process. When mixed with the normal water vapor chemical in the atmosphere it produces a highly irritant chemical that can cause severe discomfort to the eyes, nose, and throat; in high concentration it can produce pulmonary disease. The pollutant is corrosive and can damage structures of steel and stone.
4. *Hydrocarbons.* Hydrocarbons are largely the result of poor combustion. They result from unburnt fuels in the exhaust and from leakage around the crankcase. In the atmosphere they undergo photochemical change to produce the highly oxidizing chemicals of smog. There is also evidence that hydrocarbons are carcinogens.
5. *Particulates.* These highly visible products of poorly controlled combustion are mainly dust and carbon. The larger particles settle out of the air; finer material stays suspended and may form the basis of smog or fog. In themselves they are not toxic, but they may act to convey toxic pollutants to the lungs.

Because of the relative volumes of fuel burned, the highway mode is overwhelmingly the principal transportation source of pollution. For each gallon of fuel consumed, diesel engines produce approximately twice as much hydrocarbon pollution as do gasoline engines, and very nearly the same amount of the oxides of nitrogen. On the other hand, emissions from gasoline engines without catalytic converters contain 58 times as much carbon monoxide as do emissions from diesel engines. Federal and state emission controls have concentrated on cleaning up gaseous emissions from highway vehicles and on requiring airlines to refit their existing aircraft engines with kits that reduce particulate emissions.

## 8-21. WATER POLLUTION EFFECTS

In general the water pollution effects of transportation projects are small in comparison with the noise and air pollution implications. The exception can be airports where the vast paved areas can produce substantial increases in runoff into local streams. Equally, there may be local effects on aquifers and local groundwater level changes. In the 1960s, a proposal to develop an international airport to the west of Miami was abandoned largely due to the problem of the cutoff of ground water flows through the Everglades.

## 8-22. LONG- AND SHORT-TERM LAND USE EFFECTS

The construction of transportation facilities can influence the long-term use of land in a variety of ways. These land use changes are of mixed benefit to the community and therefore generate mixed reactions from those affected. The following are a few of the many long-term changes that have been associated with major transportation projects.

1. High-density corridor and station development following the construction of a rapid transit system. This was observed in Toronto but there appears to have been an insignificant corridor effect in Montreal, San Francisco, and Washington. Indeed, residents in suburban centers in the Bay area resisted high-density development in the area of BART stations.
2. Disruption of existing community structure with the need to provide huge park-and-ride parking lots at rapid transit locations.
3. Long-term "planning blight" as areas become classified as transition areas awaiting redevelopment to more intense uses.
4. The division of socially cohesive communities by rights-of-way for freeways or rapid transit.
5. Traffic "blight" caused by substantial increases in moving and parked traffic with concomitant increases in noise, fumes, dust, and general discomfort.
6. Economic segregation, caused by making suburban areas more accessible, with a consequential long-term abandonment of the urban center by the wealthier income groups and the secular in-migration of lower income groups.

Even in the short term there may be effects during the construction period that can have substantial effects on the existing land use and its occupants. These effects have been ignored too frequently in assessing the overall social costs associated with a project. They include:

1. Relocation of residents and businesses on both a permanent and temporary basis.
2. The economic consequences of the disruption of the construction period on business and industry.
3. The damage and disruption caused by construction traffic.

## 8-23. ENVIRONMENTAL IMPACT STUDIES

The creation of any new transportation facility or the significant expansion of an existing facility can have a very strong environmental impact. It is recognized that changes in transportation infrastructure are actions that can have among the strongest impacts of any developmental decision. Consequently, transportation-related proposals are placed under great scrutiny by all levels of government. Many countries have *ad hoc* or relatively vaguely defined and loosely structured procedures for the evaluation of environmental impact. In the United States, however, the procedure is well structured, clearly defined and exhaustive.

There is no single piece of legislation that defines what must be done to assess the environmental impact of transportation facilities. Among the more important are:

- Clean Air Act 1963
- National Environmental Policy Act of 1969
- Airport and Airways Development Act of 1970
- Department of Transportation Act of 1966
- Air Quality Act and Clean Air Act Amendments of 1970
- Uniform Relocation Assistance and Real Property Acquisition Policies Act of 1970
- National Historic Preservation Act of 1966
- Noise Control Act of 1972
- Endangered Species Act of 1973
- Coastal Zone Management Act of 1972
- Fish and Wildlife Coordination Act of 1972
- Water Bank Act
- Federal Highway Act of 1970

Environmental policy in the United States provides for several alternative courses of action with respect to environmental evaluations. Certain actions are so obviously innocuous that they categorically are excluded from the requirement to perform an environmental analysis. Examples include administrative procurement of supplies, promotions and hirings, and increases in costs to ongoing projects. The potential for significant environmental harm from other proposed federal actions may be in doubt, in which case a concise document called an *environmental assessment* must be prepared. That document provides a brief statement describing the environmental impacts of a proposed action and its alternatives and serves as a basis for determining whether an *environmental impact statement* (EIS) or a *finding of no significant impact* (FONSI) will need to be prepared. These reports must be submitted to appropriate government officials and made available to the public upon request.

Some general guidelines can be followed that would indicate to the planner the type of transportation impact that would require the preparation of an environmental impact statement. These include:

a. Impacts that are expected to be environmentally controversial.
b. Significant impacts on natural, cultural, ecological, or scenic resources, including effects on wetlands and endangered flora and fauna.
c. Significant effects on protected or historic sites.
d. Significant increases in surface traffic congestion.
e. Substantial impacts on local and regional planning and development.
f. Impacts that are inconsistent with federal, state, or local environmental statements.
g. Large relocation requirements or lack of suitable relocation housing.
h. Increases of noise impact in noise sensitive areas.
i. Significant deleterious impact on air quality.
j. Any other action with significant impact on the human environment.

The U.S. government has set out clear guidelines [22] on the form that an environmental impact statement should take. Assessments of impact must contain the following elements:

a. A description and justification of the proposed action and of the considered alternative actions.
b. A description of the planning process and structure of the affected area.
c. An estimate of the most likely impact that the environment will suffer from the proposed action.
d. Alternatives to the proposed action and their likely impacts.
e. A statement of the estimated likely adverse environmental effects that are unavoidable.
f. A discussion of the trade-off relationship between the short-term use of the environment and the benefits to long-term productivity that this use brings.
g. A statement of the irretrievable or irreversible commitment of resources.
h. A statement of impact on properties and sites of cultural significance.

i. A summary of the coordination of public involvement in the planning process.

The NEPA guidelines also specify the environmental elements that are to be studied. They include noise, air pollution, water pollution, solid waste pollution, effects on surrounding land use, ecological factors, resource usage, historical and cultural area effects, and economic impact.

As an example of the type of actions that would require an environmental impact study, the following may be cited in the context of airports:

a. The selection of a site for a new airport.
b. The development of a new airport.
c. A new runway or a major runway extension.
d. Any federally funded expansion of passenger facilities or access facilities including parking or any new construction associated with these areas.
e. Runway improvement permitting either the use of jet aircraft for the first time or the use of larger and noisier jet aircraft.
f. Land purchase for *a* to *e* above.
g. Land purchase causing relocation of homes and businesses.
h. Provision or relocation of navigational aids such as approach lights, ILS, runway end identification lights when federal funds are used.
i. Any airport development that affects endangered or protected species.
j. Any airport development affecting historical, architectural, archeological, or cultural land uses.

Additional details on the procedures for considering environmental impacts of transportation actions are given by DOT order 5610.1C [23].

## 8-24. SUMMARY

The preparation of transportation plans requires an extensive and expensive data-gathering program. These data serve as a basis for an analysis in the planning process itself. Even though the collection and analysis of base data require considerable expenditure in terms of time and manpower, the data-gathering exercise and its analysis must be given the highest priority. Strategies based on an inadequate or poorly analyzed data are of questionable value.

## PROBLEMS

**1.** Devise a questionnaire to compare bus riders' attitudes to bus service level changes.

2. It was stated in Chapter 7 that many land use planners use design techniques in preference to land use modeling based on data related to past trends in land use change. Discuss the advantages and disadvantages of both approaches.

3. Assume a highway by-pass is to be put around a town with which you are familiar that has a population of approximately 30,000. What issues should be raised in the environmental impact statement?

4. "Aircraft noise at airports will be of diminishing importance until the mid-1990s." Discuss this statement.

5. You need to obtain data relating to facility usage for the design of a major expansion to an airport terminal. The existing terminal, designed 15 years ago, is currently handling 9 million. The new design is for 30 million. What information would you need and how would you go about getting it?

## REFERENCES

1. *Citizen Participation in Airport Planning,* AC 150/5050-4, Federal Aviation Administration, Department of Transportation, Washington, D.C.
2. *Transportation and Traffic Engineering Handbook,* Second Edition, W.S. Homburger (ed.), Prentice-Hall, Englewood Cliffs, N.J., 1982.
3. *Highway Capacity Manual, Special Report 209,* Transportation Research Board, National Research Council, Washington, D.C., 1985.
4. Syrakis, T.A., and J.R. Platt, "Aerial Photographic Parking Study Techniques," *Highway Research Record 267,* Highway Research Board, Washington, D.C., 1969.
5. Gray, G.E., and L. Hoel, *Public Transportation: Planning, Operations and Management,* Prentice-Hall, Englewood Cliffs, 1979.
6. *National Plan of Integrated Airport Systems 1984-1993,* Department of Transportation, Federal Aviation Administration, Washington, D.C., 1985.
7. Ashford, N., and P.H. Wright, *Airport Engineering,* Second Edition, Wiley Interscience, New York, 1984.
8. *Urban Origin-Destination Studies,* U.S. Department of Transportation, Federal Highway Administration, Washington, D.C., 1973.
9. *Urban Mass Transportation Travel Surveys,* U.S. Department of Transportation, Washington, D.C., 1972.
10. Rajanikant, N. Joshi, and Fred Utevsky, *Alternative Patterns of Development—Puget Sound Region, Staff Report No. 5,* Puget Sound Regional Transportation Study, Seattle, 1964.
11. *Urban Development Models,* Special Report No. 97, Highway Research Board, Washington, D.C., 1968.
12. Chapin, F.S., and Edward J. Kaiser, *Urban Land Use Planning,* University of Illinois Press, Urbana, 1979.
13. Perloff, H., and L. Wingo (eds.), *Issues in Urban Economics,* Johns Hopkins Press, Baltimore, 1968.

14. Wright, P.H., and R.J. Paquette, *Highway Engineering,* Fifth Edition, John Wiley, New York, 1987.

15. Voorhees, A.M., "A General Theory of Traffic Movement," *Proceedings of the Institute of Traffic Engineers,* 1955.

16. Shaffer, M.T., "Attitudes, Community Values and Highway Planning," *Highway Research Record No. 187,* Highway Research Board, Washington, D.C., 1967.

17. Ashford, N.J., H.P.M. Stanton, and C.A. Moore, *Airport Operations,* Wiley Interscience, New York, 1984

18. *Calculation of Day-Night Levels ($L_{dn}$) Resulting from Civil Aircraft Operations,* Environment Protection Agency, Report No. EPA 550/9-77-450, National Technical Information Service No. PB 266165, January 1977.

19. *Airport Land Use Compatibility Planning,* AC 150/5050-6, Department of Transportation, Federal Aviation Administration, Washington, D.C., 1977.

20. Sharp, Ben H., and Paul R. Donovan, "Motor Vehicle Noise," in *Handbook of Noise Control,* Cyril M. Harris (ed.), McGraw Hill, New York, 1979.

21. Parker, J., "Air Pollution at Heathrow Airport London, April-Sept., 1970," *SAE Paper 710324,* Society of Automotive Engineers, Washington, D.C., February 1971.

22. *Regulations for Implementing the Procedural Provisions of the National Environmental Policy Act,* Council on Environmental Quality, Executive Office of the President, Nov. 29, 1978.

23. *DOT Order 5610.1C,* Including Change 2 U.S. Department of Transportation, July 30, 1985.

# *Chapter* *9*

# *Transportation Modeling*

Analytical models that form the core of the strategic planning process are considered in the following sections in the way in which they are developed in a conventional transportation plan: *trip generation, trip distribution, modal split,* and *traffic assignment.* The rationale and role these models play in the complete planning process has been the subject of much discussion in recent years. This has been considered in a systems context in Chapter 7, which should be read in conjunction with this chapter.

## 9-1. TRIP GENERATION MODELS

Trip generation is the analytical process that provides the relationship between urban activity and travel. The number of trips to and from activities in an area are related to land use and socioeconomic characteristics. Trips that originate or terminate within each zone are known as trip ends from origins or destinations, or in terms of the development of productions or attractions.

## 9-2. PRODUCTIONS AND ATTRACTIONS

When the terms *production* and *attraction* are used there is some implication of the land use at either the beginning or the end of the trip. All trips that either begin or end at the home of the trip-maker are classified as *home-based trips* while those that neither begin nor end at the trip-maker's place of residence are termed *non-home-based trips.* When a person makes two journeys in the day (e.g., to work and back home) the home is both the origin and destination, as is the place of work. In terms of productions and attractions, however, the place of residence generates two productions and the place of work generates two attractions. Only in the case of the non-home-based trip is the origin of the trip the production end and the destination the attraction end in all cases.

## 9-3. TRIP PURPOSE

All trips are not homogeneous in character and more reliable models can be derived if separate estimates are made for different trip purposes. In the same study area, for example, the length of work trips generally will have a different frequency distribution from shopping trips. The stratification of trip purpose at the trip generation stage depends on the distribution model to be used. Typical purposes encountered in a large study using the gravity model are:

1. Work
2. Personal business
3. School
4. Social
5. Recreational
6. Change travel mode
7. Shopping
8. Serve passenger
9. Non-home based

Many studies do not stratify trip purpose to this degree. A small study may be modeled with only three purposes: home-based work, home-based other than work, and non-home-based trips.

## 9-4. TRIP GENERATION ANALYSIS

The main aim of the trip generation model is to establish a functional relationship between travel, land use, and socioeconomic characteristics of an area. The rate of trip making within an area depends primarily on land use which, in conjunction with socioeconomic information concerning residential and working populations, is related to demands on the transportation system. Ultimately, the function of trip generation analysis is to establish meaningful relationships between land use and trip making activities so that changes in land use can be used to predict subsequent changes in transportation demand.

The three characteristics of land use that have been found to relate closely to trip generation are the *intensity, character,* and *location* of land use activities. Intensity of land use usually is expressed in such terms as dwelling units per acre, employees per acre, and employees per 1000 ft$^2$ of retail floor space. Character of land use has reference to the social and economic makeup of the users of the land and includes measures such as average family income and car ownership per capita. Location within an urban area has been found to be a variable that can express the combined effect of such variables as family size, stage in family life cycle, availability of parking, and index of street congestion.

Prediction of trip-making activity is possible by a variety of available methods. Two of the most frequently used techniques will be discussed at length in the following sections. These are:

1. Cross classifications analysis (category analysis)
2. Regression analysis

## 9-5. CROSS CLASSIFICATION ANALYSIS

Cross classification is a technique for trip generation in which the changes in one variable (trips) can be measured when the changes in other variables (land use, socioeconomic) are accounted for. In its use of independent prediction variables the method in some ways resembles multiple regression techniques. Cross classification is essentially nonparametric since no account is taken of the distribution of the individual values that compose the cells. The method has inherent advantages not found in the regression analysis method. "Families of curves" can be plotted showing the effect of changes of one independent variable at constant levels of the other independent variables. These plots are most useful to the planner enabling him to get a good feeling for the importance of the independent variables. This is sometimes difficult with multiple regression techniques, where the strength of the independent variable is described by significance levels and partial correlation coefficients (see Section 9.7). Another advantage of the technique comes from the fact that there is no assumption of linearity between the dependent and independent variables. The technique, therefore, is suitable for application where the effect of the independent variable is nonlinear and its actual form is not certain. Perhaps the greatest disadvantage of the method is the fact that the amount of the total variance explained by the independent variable is unknown and there is no examination of the underlying distribution of the individual values that make up the mean value entered in each cell. Where these distributions are highly skewed, the sample size to assure a meaningful cell entry may have to be large. One of the most serious weaknesses of the method, however, is the possibility that the "independent" variables that the analyst selects are not truly independent. The resultant relationships and predictions may well be spurious. An approach to forecasting residential trip generation based on cross classification analysis is used in the FHWA Trip Production Model [1]. The model consists of a sequence of four submodels which are developed using the origin-destination travel survey. The four submodels are as follows:

1. *Income submodel* reflects the distribution of households within various income categories (e.g., high, medium, and low).
2. *Auto ownership submodel* relates the household income to auto ownership.
3. *Trip production submodel* establishes the relationship between the trips

made by each household and the independent variables (see example in Fig. 9-2).

4. *Trip purpose submodel* relates the trip purpose to income in such a manner that the trip productions can be divided among various purposes.

The complete FHWA Trip Production Model procedure is shown in Fig. 9-1.

## 9-6. DISAGGREGATE APPROACH

A considerable amount of research and development has been directed into the area of disaggregate models for improved travel demand forecasting. The difference between aggregate and the disaggregate techniques is mainly in the data efficiency. Aggregate models usually are based upon home interview origin and destination data that have been aggregated into units (e.g., zones). Average values then are derived from model development. The disaggregate approach relies on samples over a range of household types and travel behavior and uses these data directly (without aggregation) for model calibration. Advantages of behavioral disaggregate models have been shown to include [2]:

- Savings in data required to calibrate models.
- Transferability to different situations, such as regional analysis and detailed corridor analysis.
- Tranferability between cities.
- Ability to express nonlinear relationships which are often lost in the case of aggregate analysis.
- Ability of more rapid data evaluation and analysis and development of relationships in a more timely fashion.
- Being understood more easily.
- More efficient monitoring and updating. Figure 9-2 shows an example of a disaggregate trip production model.

## 9-7. MULTIPLE REGRESSION ANALYSIS

Regression is a mathematically based procedure that has been programmed for most electronic computers. The technique is, therefore, readily available to the analyst. With reasoning similar to the cross classification method, the trip generation rate is treated as a *dependent* variable which is a function of one or more *independent* variables. The relationship assumed is linear, of the form:

$$Y = A_0 + A_1X_1 + A_2X_2 + \cdots + A_NX_N \tag{9-1}$$

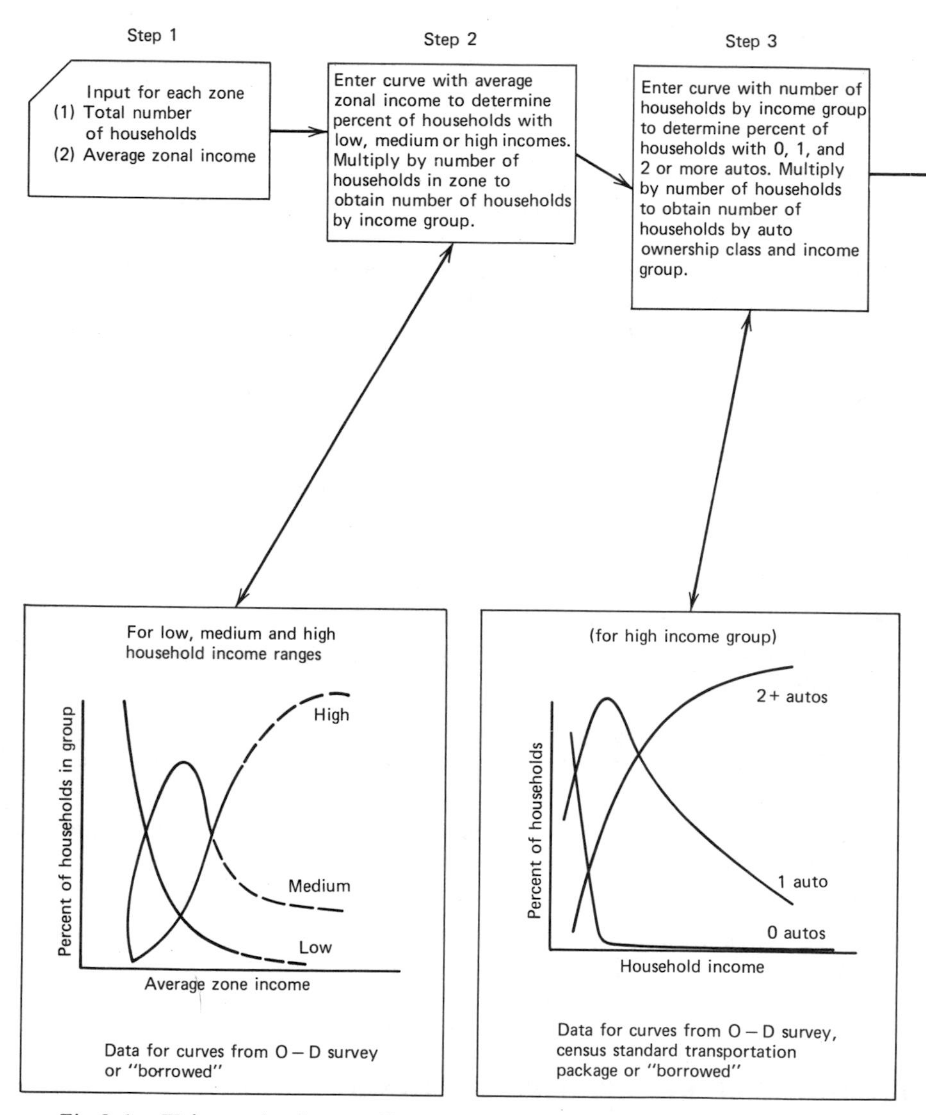


*Fig 9-1.* Trip production application procedure. (Source: *Computer Programs for Urban Transportation Planning,* U.S. Dept. Transportation/FHWA, April 1977.)

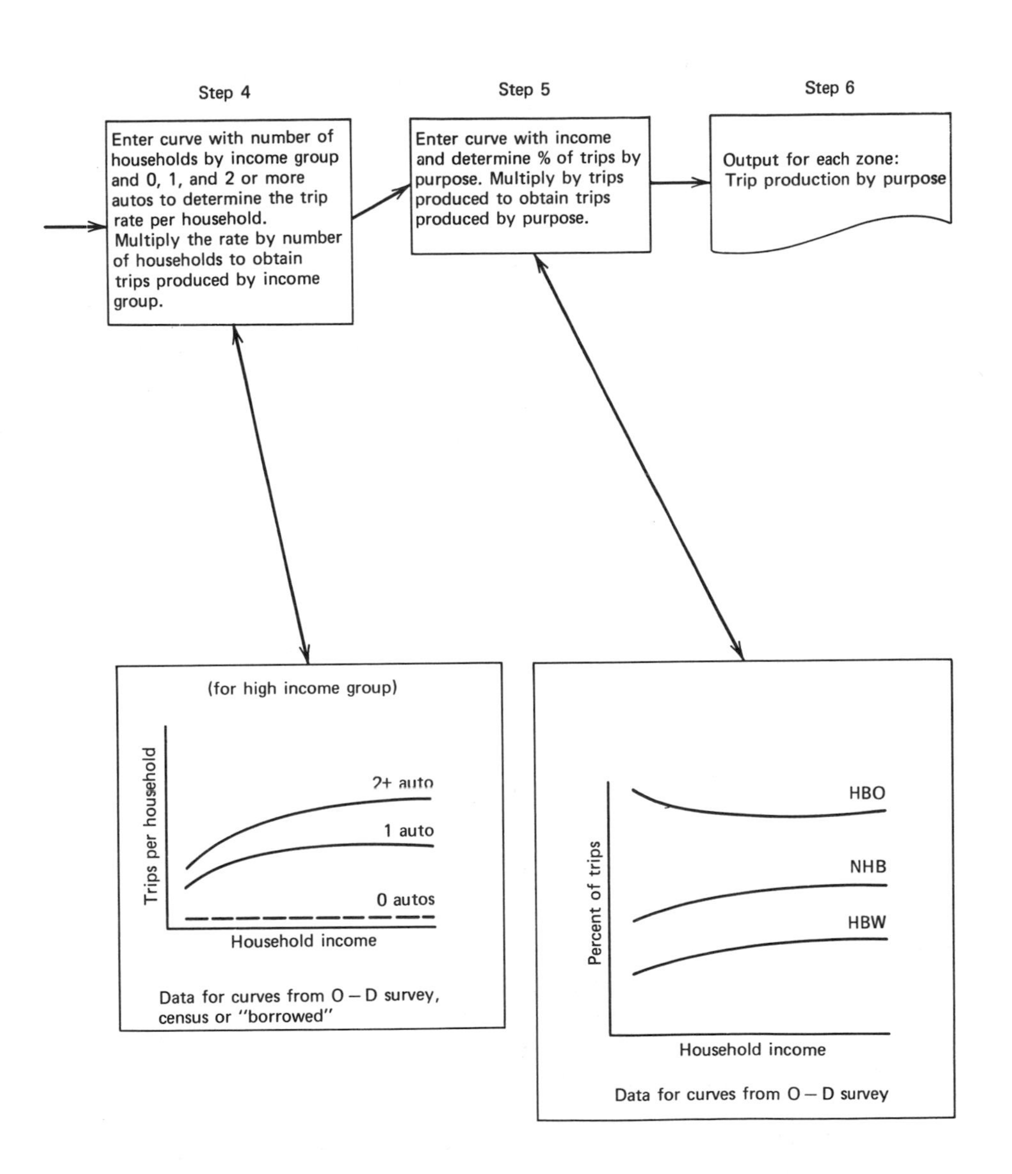

Step 4
Enter curve with number of households by income group and 0, 1, and 2 or more autos to determine the trip rate per household. Multiply the rate by number of households to obtain trips produced by income group.
Step 5
Enter curve with income and determine % of trips by purpose. Multiply by trips produced to obtain trips produced by purpose.
Step 6
Output for each zone: Trip production by purpose
(for high income group)
Trips per household
2+ auto
1 auto
0 autos
Household income
Data for curves from O – D survey, census or "borrowed"
Percent of trips
HBO
NHB
HBW
Household income
Data for curves from O – D survey

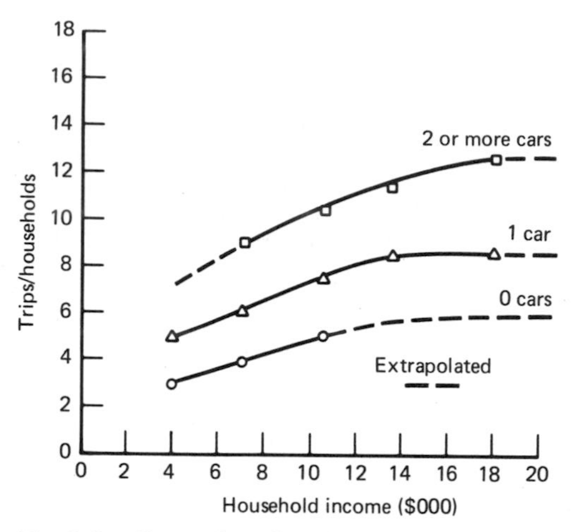


*Fig 9-2.* Example of Trip Production Submodel. (Source: *Computer Programs for Urban Transportation Planning,* U.S. Dept. Transportation/FHWA, April 1977.)

where

$$Y = \text{the trip rate}$$
$$X_1, \ldots, X_N = \text{independent variables}$$
$$A_0, \ldots, A_N = \text{constants}$$

A simple multiple regression equation relating shopping trips attracted to zones in a small urban area to various rates of employment in those zones exemplifies Eq. 9-1 [3]:

$$Y = 1.655X_2 + 4.082X_3 + 0.456X_5 - 3.004 \quad (9\text{-}2)$$

where

$Y$ = home-based shopping attractions
$X_2$ = wholesale and retail employment
$X_3$ = highway retail employment
$X_5$ = service employment

The regression approach is entirely mathematical; therefore, statistical tests of reliability of the derived relationships can be applied with ease. Two basic assumptions are made concerning the form of the independent variables. First, all variables are assumed to be random variables with a normal distribution. Second, the predictive variables are assumed to be independent of each other. The effect of violation of either of these assumptions is discussed later.

Possibly the greatest advantage of the regression approach is the ease with which the analyst can determine the degree of relationship between the dependent and independent variables, and can define the accuracy of the predictive equation itself. Some of the most common measures that are used in analysis need explanation.

1. *The coefficient of multiple determination* ($r^2$) is a measure of the amount of variance described by the model expressed as a decimal ratio of the total variance observed in the dependent variable. The value of this coefficient has an upper limit of 1.0, which would be the value for a perfect model.
2. *The standard error of estimate* ($S_y$) is a measure of the deviation of observed trip values from values predicted by the model. Where the model values agree exactly with observed values, the standard error reaches its lower limit, zero, the value for a perfect model.
3. *The partial correlation coefficient* ($r_j$) of an independent variable describes the relation between the dependent variable and the particular independent variable under consideration. Descriptively, it is a measure of this association, with the effect of all other independent variables eliminated. This measure is a particularly important indicator for it is possible for a regression equation to model observed values with suitable accuracy, with a low standard error, and a high value of $r^2$ and yet to contain an independent variable that does not have a close relationship to the dependent variable.

Multiple regression techniques have been packaged attractively in computer program routines to the point that very little effort is required of the analyst to produce equations that will fit adequately the zonal data. From this viewpoint, multiple regression is more convenient than are category analysis techniques. The method, however, has several pitfalls awaiting the unwary user that can produce erroneous relationships.

When the independent variables are not independent of each other, they are said to be *colinear*. Equations containing independent variables that demonstrate a high degree of colinearity must be avoided if spurious relationships are to be eliminated. The result of using two such variables is in effect to count the same factor twice. This can render the equation useless for projection purposes. Colinearity can be avoided by examining the correlation coefficients of the independent variables among themselves. If two independent variables in an equation have a high correlation coefficient, they may be presumed to be colinear; one variable should be eliminated. A distinct possibility, in the case where colinear variables are used, is the occurrence of an incorrect sign in the derived equation. A variable that apparently should contribute a trip generation and, therefore, should be associated with a positive coefficient will be found to have a negative coefficient. This frequently happens if care is not taken to assure independence of predictive variables. The need for wariness in variable selection should indicate to the analyst that hidden interactions may come into play if too many variables are used in the regression equation.

Usually, three variables are a limiting condition for most predictive equations. Introducing further variables, while marginally increasing the $r^2$ value, can result in a significant decrease in partial correlation coefficients.

Non-normality of data is a lesser problem. The analyst must examine the distribution of the variables used. Where a choice exists between a variable with a skewed distribution and one normal in character, the latter is preferable. The result of using non-normal data is to render the evaluation statistics inaccurate, but the form of relationship itself is not badly affected in most cases. In cases where the underlying distribution of an independent variable is very highly skewed, the planner is advised to avoid the use of such a variable in regression analysis.

Multiple linear regression is a process of fitting a regression plane to observed results with a minimization of the squares of the residuals. The process will force the results into the chosen linear model, whether or not the relationship is truly linear. Where the relationship is not linear, but *its form is known,* the method still can be applied by a transformation of the involved variable. For example, if the form is felt to be:

$$Y = A_0 + A_1X_1 + A_2X_2^2 \tag{9-3}$$

the variable $Z_2$ can be introduced, where

$$Z_2 = X_2^2 \tag{9-4}$$

and regression on the form:

$$Y = A_0 + A_1X_1 + A_2Z_2 \tag{9-5}$$

can be performed in a standard manner. Transformation of variables is useful in increasing the flexibility of the method. The difficulty lies, however, in the recognition that a nonlinear relationship exists. Its presence may be difficult to determine when mixed with the effects of other variables. In the case where the nonlinear effect is ignored, multiple regression will force the data into an incorrect linear relationship.

## 9-8. TRIP DISTRIBUTION MODELS

The trip generation phase of the planning process will have determined how many trips are generated from each zone and for what purpose they are made. Trip distribution models determine to what zones these trips are going by calculating trip interchanges in the base year. Most of the techniques used in the distribution process depend greatly upon the origin-destination study for *calibration* purposes, just as future trip generation was predicted from behavior patterns observed in base year generation.

## 9-9. THE GRAVITY MODEL

The gravity model is one of the most widely used trip distribution techniques in transportation planning. Early studies measured trip generation and attraction components in terms of zonal populations, and the resistance function was assumed to be related to an inverse function of distance. This gave similar relationship to Newton's theory of gravity, expressed in mathematical terms as:

$$I_{ij} = \frac{K \times P_i \times P_j}{D^n} \tag{9-6}$$

where
$I_{ij}$ = the interaction between $i$ and $j$
$P_i$ = the population at $i$
$P_j$ = the population at $j$
$D$ = the distance betweeen $i$ and $j$
$K$ = some constant
$n$ = some exponent

This work was followed by the development of more sophisticated models such as that by Voorhees [4] who suggested a constant exponent for the influence of distance, but recognized the need for a different exponent depending on trip purpose.

Gravity model formulations in current use are based on the hypothesis that the trips produced at an origin and attracted to a destination are directly proportional to:

- Total trip productions at the origin.
- The total attractions at the destination.
- A calibrating term.
- A socioeconomic adjustment factor.

The form of this relationship may be written:

$$T_{ij} = CP_iA_jF_{ij}K_{ij} \tag{9-7}$$

where
$T_{ij}$ = trips produced at $i$ and attracted at $j$
$C$ = a constant
$P_i$ = total trip production at $i$
$A_j$ = total trip attraction at $j$
$F_{ij}$ = a calibration term for interchange $ij$, (friction factor)
$K_{ij}$ = a socioeconomic adjustment factor for interchange $ij$
$i$ = an origin zone number
$n$ = number of zones

A value for $C$ for any origin zone $i$ ($C_i$) can be established when it is specified that the sum of all $T_{ij}$s for origin $i$ must be equal to $P_i$:

therefore

$$P_i = \sum_{j=1}^{n} T_{ij} = \sum_{j=1}^{n} (C_i P_i A_j F_{ij} K_{ij}) \tag{9-8}$$

$$= C_i P_i \sum_{j=1}^{n} (A_j F_{ij} K_{ij}) \tag{9-9}$$

therefore

$$C_i = \frac{1}{\sum_{j=1}^{n} (A_j F_{ij} K_{ij})} \tag{9-10}$$

and Eq. 9-7 becomes

$$T_{ij} = \frac{P_i A_j F_{ij} K_{ij}}{\sum_{j=1}^{n} (A_j F_{ij} K_{ij})} \tag{9-11}$$

Equation 9-11 is the standard form of the gravity model. The term $F_{ij}$ is the calibrating term and generally is found to be an inverse exponential function of impedance. In developing the model the output from Eq. 9-11 normally will show production (row) totals to be correct but attraction (column) totals will not necessarily match their desired values. In order to match the desired values an iterative procedure is employed.

## 9-10. CALIBRATING THE GRAVITY MODEL

The gravity model must be calibrated so as to establish a distribution parameter for each trip purpose, based on observed travel patterns. This distribution parameter is the travel time factor and calibration depends upon the repeated adjustment of a set of *friction factors*. Adjustment is continued until friction factors are obtained that result in a near enough approximation of base year data when the gravity model is applied to base year productions and attractions.

A procedure for calibrating the gravity model that was developed by the U.S. Department of Transportation is the computer program known as GMCAL, shown in Fig. 9-3. This calibration program can handle several trip purposes at the same time.

An example of a readjusted travel time factor curve for calibration runs

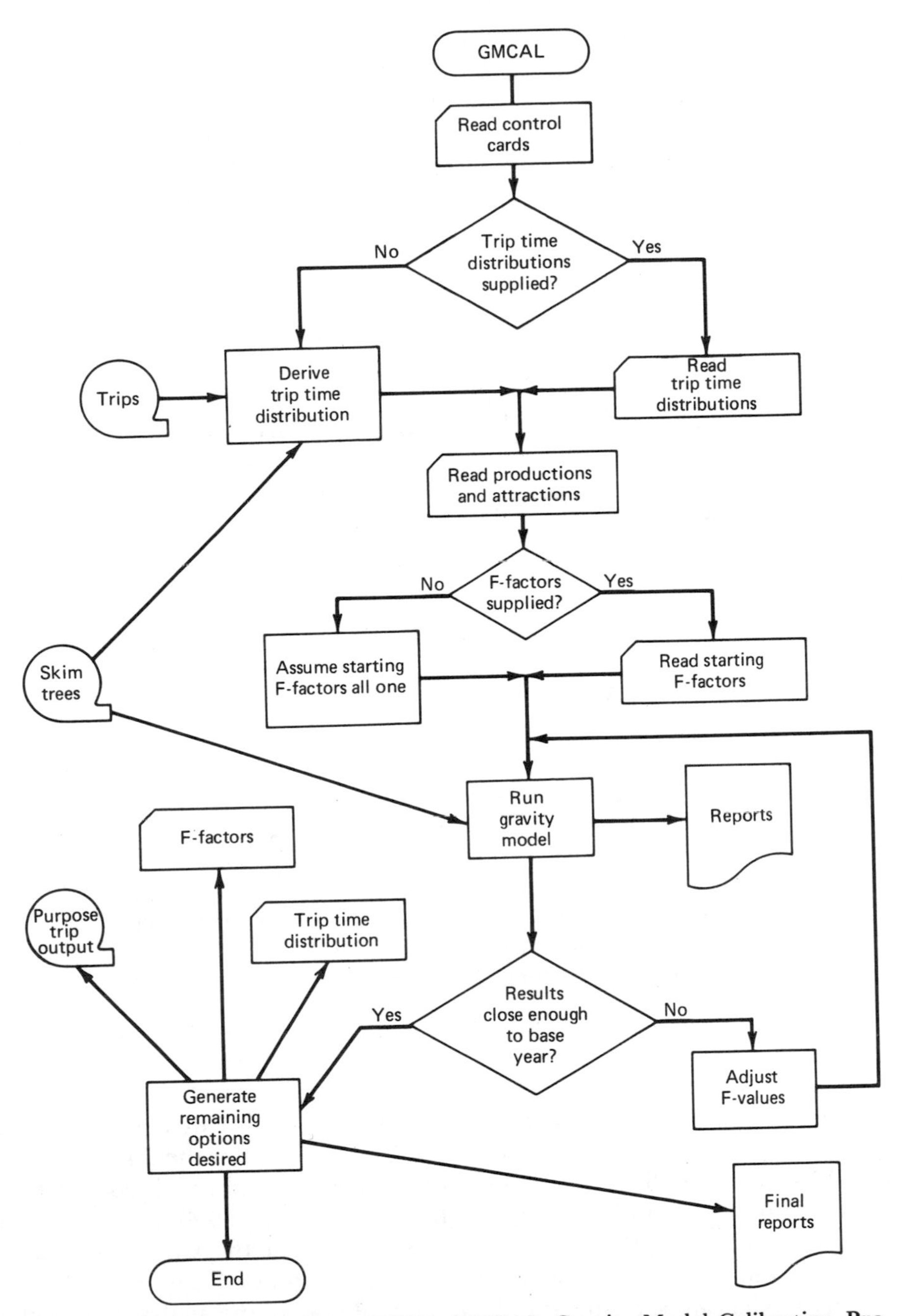


*Fig 9-3.* Flow diagram of the FHWA GMCAL Gravity Model Calibration Program. (Source: *Computer Programs for Urban Transportation Planning,* U.S. Dept. Transportation/FHWA, April 1977.)

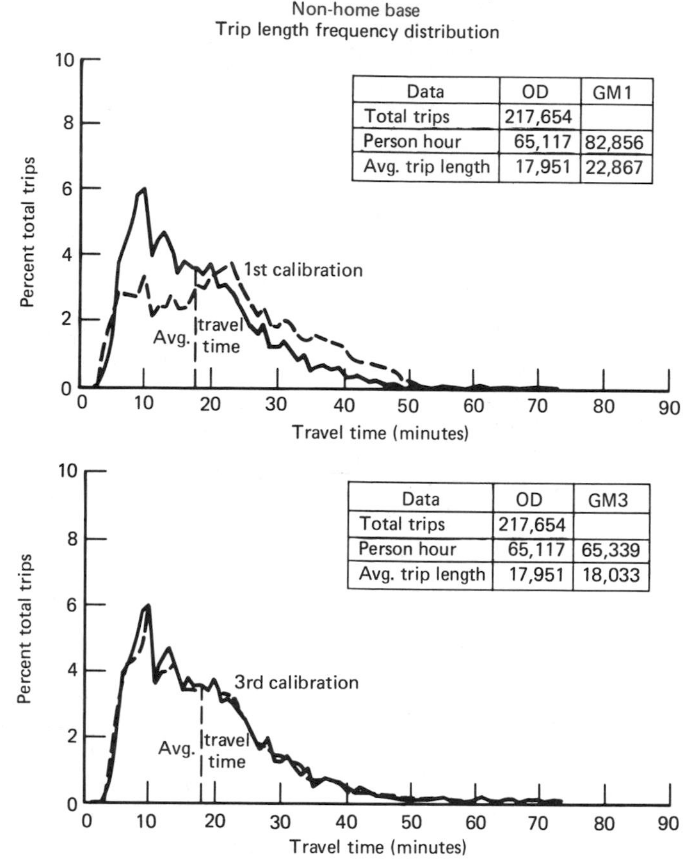


*Fig 9-4.* Comparison of O.D. and gravity model trip length frequency distributions. (Source: *Atlanta Area Transportation Study,* Georgia Department of Transportation.)

for non-home-based trips in the Atlanta study is shown in Fig. 9-4. These graphs show the results of the first calibration run and after the third iteration, by which time the difference between the observed and modified curve has shrunk to 0.34 percent for the person-hours of travel and 0.46 percent for the average trip length. With this degree of agreement the travel time factor curve of the last iteration is accepted as the calibrated curve. Before the gravity model can be assumed to be totally calibrated, the analyst must determine whether or not the socioeconomic adjustment factor is required to bring individual gravity model zonal interchanges into agreement with the base year zonal interchanges.

EXAMPLE 9-1

GRAVITY MODEL CALCULATION. Calculate the interzonal interchanges due to 100 productions at zone 1, with 250 attractions at zone 2,100 attractions at zone 3, and 600 attractions at zone 4. Assume that 1-2 is 5 min, 1-3 is 10 min, and 1-4 is 15 min. Assume all $K_{ij}$ factors are unity, and that $F_{ij}$ factors are as shown below.

| Zone | $A_j$ | $F_{ij}$ | $A_jF_{ij}$ | $\frac{A_iF_{ij}}{\Sigma_j A_j F_{ij}}$ | $P_l$ | $T_{ij}$ |
|---|---|---|---|---|---|---|
| 1 | 0 | 0 | 0 | 0 | 100 | $0 = T_{11}$ |
| 2 | 250 | 20 | 5000 | 0.732 | 100 | $73 = T_{12}$ |
| 3 | 100 | 5 | 500 | 0.073 | 100 | $7 = T_{13}$ |
| 4 | 600 | 2.22 | 1332 | 0.195 | 100 | $20 = T_{14}$ |
| | | | $\Sigma A_jF_{ij} = 6832$ | $\Sigma = 1.00$ | | $\Sigma = 100$ |

The reader is advised to verify that the exponent of time for the travel time factor is 2.

## 9-11. THE FRATAR METHOD

A technique for trip distribution utilizing *growth factors* was introduced by Thomas J. Fratar in 1954 [5]. While the technique is seldom used now as a study-wide distribution model it is considered by many to be a particularly useful way of dealing with external to external trips, that is, between external stations of the study area. The method is applied iteratively with the interchanges being computed according to the relative attractiveness of each interzonal movement from the point under consideration. The method involves the following steps:

1. Future traffic growth is estimated for each traffic zone and expressed as a "growth factor." (The growth factor is simply the ratio of expected future traffic to the existing traffic.)
2. Future traffic originating in (or destined to) a given zone is estimated by growth factors and existing traffic.
3. This traffic is distributed to other zones in proportion to existing interzonal travel and growth factors, for example,

$$(\text{Trips})_{AB} = (\text{Est. future total})_A \times \frac{(\text{present travel})_{AB}\,(\text{growth factor})_B}{\underset{\text{All destination zones}}{\Sigma}(\text{growth factor})\,(\text{present travel})} \tag{9-12}$$

4. This distribution will yield two values (different) for each movement, for example, $V_{AB}$ and $V_{BA}$. Average these two values.
5. The sum of these average values for a particular zone probably will be

different from the existing traffic to (or from) that zone multiplied by its growth factor (desired volume). Obtain new growth factors:

$$\text{New growth factors} = \frac{\text{(Desired volume)}}{\text{volume obtained from sum of movements}} \tag{9-13}$$

6. Make second approximation using these growth factors.
7. Repeat process until there is reasonable harmony between interzonal traffic sums and desired volume.

## EXAMPLE 9-2

1. For zones *A, B, C,* and *D* the present traffic volumes and patterns and growth factors indicated below.
   Determine: Future traffic volumes and patterns by the Fratar method.
2.

| | *Zone* | | | |
|---|---|---|---|---|
| | *A* | *B* | *C* | *D* |
| Present totals | 40 | 38 | 32 | 38 |
| Growth factors | 2.0 | 3.0 | 1.5 | 1.0 |
| Estimated future totals | 80 | 114 | 48 | 38 |

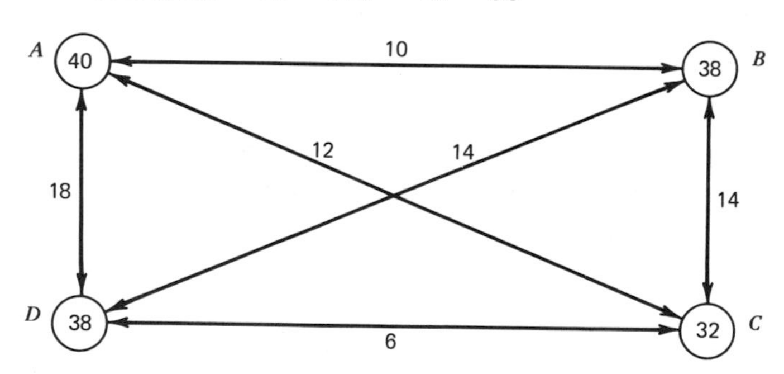


3. Distribute trips:

$$V_{AB} = \frac{(10 \times 3)}{(10 \times 3) + (12 \times 1.5) + (18 \times 1)} \times 80 = 36.4$$

$$V_{AC} = \frac{(12 \times 1.5)}{(10 \times 3) + (12 \times 1.5) + (18 \times 1)} \times 80 = 21.8$$

$$V_{AD} = \frac{(18 \times 1)}{(10 \times 3) + (12 \times 1.5) + (18 \times 1)} \times 80 = 21.8$$

Total 80.0

$$V_{BA} = \frac{(10 \times 2)}{(10 \times 2) + (14 \times 1) + 14 \times 1.5)} \times 114 = 41.5$$

4. Average the two values for each movement. Note, for example, that the traffic from $A$ to $B$ is not equal to the traffic from $B$ to $A$.

$$AB \text{ movement} = \frac{36.4 + 41.5}{2} = 39.0$$

Similarly,

| | |
|---|---|
| $AC$ movement = | 18.9 |
| $AD$ movement = | 18.8 |
| New total | 76.7 |
| Desired total | 80.0 |

5. Obtain new growth factors. For example, for zone A, new growth factor = 80.0/76.7 = 1.04.
6. Make second approximation; repeat until there is harmony between computed traffic sums and desired volumes.

First approximation results for all zones:

| | *A-B* | *A-C* | *A-D* | *B-C* | *B-D* | *C-D* |
|---|---|---|---|---|---|---|
| | 36.4 | 21.8 | 21.8 | 43.5 | 29.0 | 3.9 |
| | 41.5 | 16.0 | 15.8 | 28.0 | 18.3 | 4.0 |
| | 77.9 | 37.8 | 37.6 | 71.5 | 47.3 | 7.9 |
| First approximations | 39.0 | 18.9 | 18.8 | 35.7 | 23.6 | 4.0 |

| | *A* | *B* | *C* | *D* |
|---|---|---|---|---|
| | 39.0 | 39.0 | 18.9 | 18.8 |
| | 18.9 | 35.7 | 35.7 | 23.6 |
| | 18.8 | 23.6 | 4.0 | 4.0 |
| New totals | 76.7 | 98.3 | 58.6 | 46.4 |
| Desired totals | 80.0 | 114.0 | 48.0 | 38.0 |
| New growth factors | 1.04 | 1.16 | 0.82 | 0.82 |

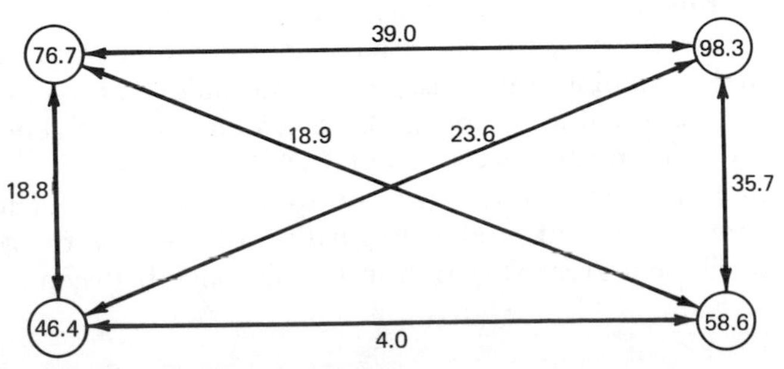

## 9-12. MODAL SPLIT

Modal split techniques are used by the planner to supply answers to the questions: "By what mode of transport will trips from this zone be made?" and "What is the vehicle occupancy that can be expected?" A knowledge of the variables affecting modal choice will enable the analyst to convert person trips into vehicle trips for the purpose of traffic assignment.

The analyst has a choice of position of the modal split model relative to the distribution process. In the case of a *predistribution split,* sometimes called a *trip end model,* the process follows directly after the generation of person trips on a zonal basis. These trips are then split into auto vehicle trips and transit person trips based on the criteria developed in the analysis of modal choice. Each mode is distributed separately using one of the trip distribution models discussed previously. A generalized model flow diagram for this method is shown in Fig. 9-5. This procedure usually is not suitable where there is minimal transit usage, as highly unstable distribution models can result from the small number of trip ends involved in their calibration.

Alternatively, the modal split analysis is applied after the distribution phase of modeling. This procedure is known as *postdistribution split* or *trip interchange model.* Figure 9-5 shows a flow diagram for this method. The distribution analysis is carried out to compute interzonal trip interchanges and these are split by the model into interzonal transit trips and, after an analysis of auto occupancy, the remaining trips are converted into interzonal vehicle trips.

## 9-13. FACTORS AFFECTING MODAL SPLIT

The decision of a trip-maker to take transit rather than go by auto is influenced chiefly by:

1. The type of trip.
2. The characteristics of the trip-maker.
3. The relative levels of service of the transportation system.

*Trip type* has been considered in studies by combining variables as trip purpose, length of trip, time of day, and orientation to the CBD.

The *characteristics of the trip-maker* have been determined to be of primary importance in determining transit usage. The variables that have been used to express this are personal (e.g., income), environmental (residential density), and related to location (distance to the CBD).

The *relative levels of service* of the transportation system ideally must be reflected in any meaningful model. Regardless of the type of trip and trip-maker, there will be a great variation in transit usage between areas of high

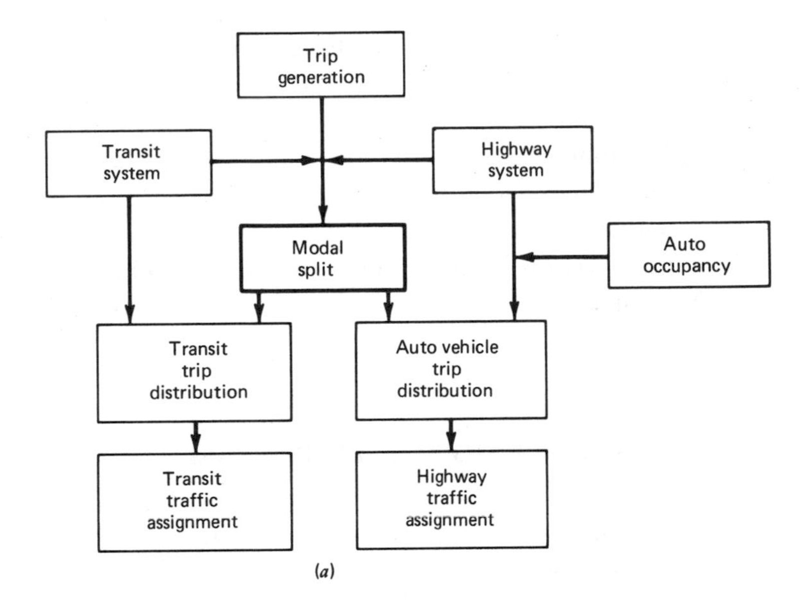


(*a*)

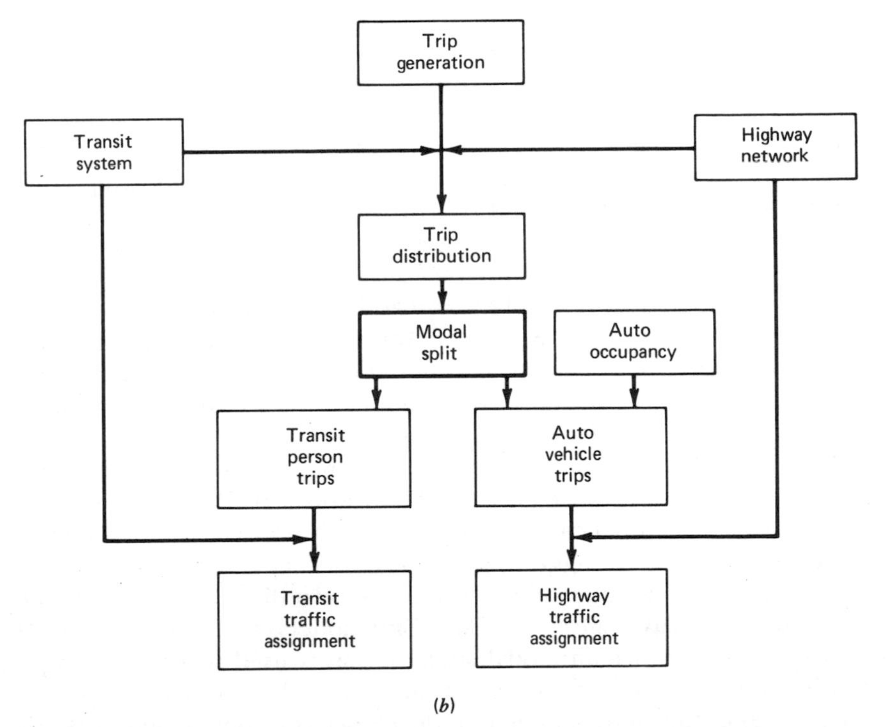


(*b*)

*Fig 9-5.* Flow charts showing the alternative location for modal split in the modeling process.

and low transit service. Variables that have been used as a measure of relative service include:

1. Travel time.
2. Travel cost.
3. Parking cost.
4. Extra travel time (time spent outside of the vehicle).
5. Accessibility (measured in terms of an accessibility index).

## 9-14. EXAMPLES OF MODAL SPLIT MODELS

Traditionally the trip end model has concentrated more on the trip type and trip-maker characteristics for modal analysis. Some early studies, for example, the 1960 Chicago study, ignored system characteristics, and transit usage was predicted from auto ownership with stratification by trip purpose and orientation to the CBD. Later trip end models tended to put their major emphasis on the characteristics of the transportation system itself. Postdistribution modal split models were initially very simple diversion-type curves that assigned a percentage of traffic to transit depending upon the ratio of travel times by each mode. This simple approach, even if accounting for different trip purposes by stratification, could not account for the third factor that is known to affect modal split—the characteristics of the trip-maker. An example of a model that attempted to recognize the three major factors affecting modal split is that used in the Washington, D.C. study. This was essentially a stratified diversion model using 160 diversion curves in all. The future modal split was determined by finding which of the 160 curves applied to the zone and trip purpose concerned. The transit split was read from the chart by entering the graph at the relative travel time ratio of transit to auto travel (see Fig. 9-6). Other long-range transit planning and modal choice models for large areas have been summarized in "A Review of Operational Urban Transportation Models" by the U.S. Department of Transportation [6]. More sophisticated methods of analysis have been attempted in recent studies such as the Multi-Modal Choice Models developed for the Minneapolis-St. Paul Urban Region [7]. These Twin City modal choice models are basically disaggregate multinomial logit formulations. Disaggregation being used here means that models are intended to describe the mode choice made by individuals rather than by large groups or zonal averages. The Twin City model is structured to estimate modal probabilities in a multimodal split. It also is used to estimate logical relationships between auto driver, auto passenger, and transit trips. Figure 9-7 shows examples of the mode split models developed for home-based work trips in this study [8].

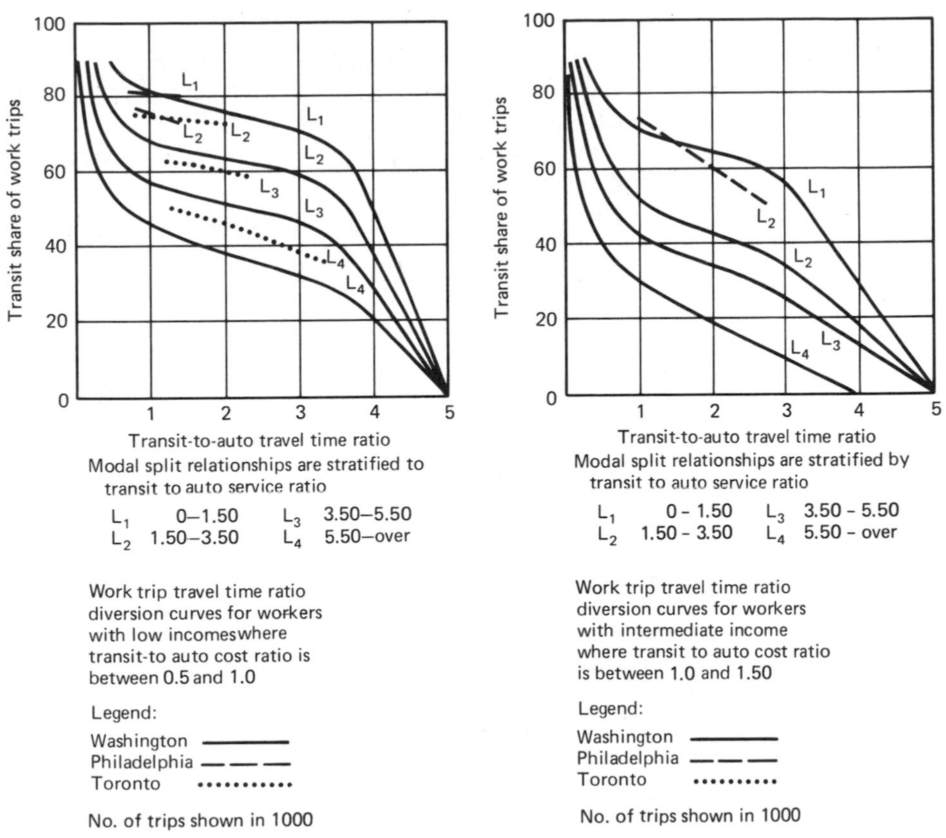


*Fig 9-6.* Modal split diversion curve. (Source: *Washington D.C. Study,* Peat, Marwick, Michell and Co.)

## 9-15. TRAFFIC ASSIGNMENT

The final step in the simulation of travel patterns in the planning area is the assignment of zonal trip interchanges to the individual transportation facilities. The basic resulting output is in the form of traffic volumes on each portion of the transportation system. As a result, the planner can determine for the base year how well the assignment process has simulated observed traffic volumes. When the traffic assignment is used with future trip interchanges, the model indicates how well the proposed facilities will serve the anticipated travel demand. As it has now evolved, the assignment process converts trip interchanges into traffic volumes by the application of specified criteria in an objective manner. By elimination of subjective decisions in route choice, the transportation planner can be more confident that the traffic volumes he predicts are reasonably likely to occur.

FIVE MODE MODEL

| | |
|---|---|
| TRN | = 0.0044 * (WALK + WAIT TWO) + 0.003 * WAIT ONE + 0.0014 * FARE + 0.0031 * (TRN RUN + AUTO ACC RUN) + 0.8659 * AUTO CONN − 0.0002 * TRN ACC (DEST) |
| ONE | = 0.0206 * HWY EXC + 0.0031 * HWY RUN1 + 0.0014 * HWY COST1 − 0.6063 * INCOME |
| TWO | = 0.0349 * HWY EXC + 0.0031 * HWY RUN2 + 0.0014 * HWY COST2 − 0.2392 * INCOME |
| THREE | = 0.0605 * HWY EXC + 0.0031 * HWY RUN3 + 0.0014 * HWY COST3 + 0.0778 * INCOME |
| FOUR+ | = 0.0605 * HWY EXC + 0.0031 * HWY RUN4 + 0.0014 * HWY COST 4 + 0.2818 * INCOME |

| *Mode Name* | *Definitions* |
|---|---|
| TRN | Bus trips |
| ONE | Auto trips with one occupant |
| TWO | Auto trips with two occupants |
| THREE | Auto trips with three occupants |
| FOUR+ | Auto trips with four or more occupants |

INDEPENDENT VARIABLES

| | |
|---|---|
| WALK | Walk time to and from the transit system |
| WAIT ONE | The waiting time to board the first transit vehicle |
| WAIT TWO | The waiting time to board the second and subsequent transit vehicles |
| TRN RUN | The time spend riding in the transit vehicle |
| AUTO ACC | The time spent riding in an automobile to the transit vehicle |
| FARE | The transit fare |
| AUTO CONN | A dummy variable signifying if an automobile was required to access the transit system (0 is no, 1 is yes) |
| HWY RUN (X) | The time spent riding in the automobile |
| HWY COST (X) | The out-of-pocket highway cost |
| HWY EXC | The time spent parking and unparking the automobile |
| INCOME | The four income quartiles |

*Fig 9-7.* Mode split models for home-based work trips.

## 9-16. DEVELOPMENT OF THE ASSIGNMENT PROCESS

With the introduction of major urban transportation studies in the 1950s it became apparent that a new approach was needed that would provide a substitute for "hand assignment" methods based on route selection by "experienced judgment." Machine-based methods became feasible with the widespread introduction of high-speed electronic computers in the late 1950s. At this time E. F. Moore published an algorithm for the calculation of minimum paths through networks [9]. This algorithm was well suited to computer use and was adopted rapidly by transportation planners for determining minimum time, minimum distance, or minimum cost routings between two points in an urban area. With the continued refinement and development of computer technology, programs became available that permitted assignment partially to the freeway and partially to the surface street system (*diversion*); other programs became available that took account of congestion on the transportation system (*capacity restraint*).

## 9-17. ASSIGNMENT NETWORK

In assignment techniques the center of activity of the study area zones are designated as zone centroids which are connected into a network representing all functional elements of the transportation system considered. For most assignment techniques the network comprises a series of *links* representing transportation facility segments. The ends of a link are denoted by *nodes,* which must appear at all link intersections. An example of a highway network is shown in Fig. 9-8 incorporating all important elements of the street system in a city [10]. All streets with signalization or carrying a significant amount of traffic are included.

## 9-18. ALL-OR-NOTHING ASSIGNMENT

Moore's algorithm can be used to calculate minimum paths from each zone centroid to all other zone centroids. These records of minimum time paths are known as trees, due to their graphical similarity to the plan view of a tree. All-or-nothing assignments, if used as the sole means of assigning traffic, usually will result in unrealistic traffic volumes. The only factor considered is usually travel time and there are no constraints placed upon the volume that any individual link can handle, nor is the operating speed related in any way to the ratio of the volume carried to the capacity. The method also neglects any preference that drivers show for arterial streets when travel time is close to that of an alternative freeway routing. These factors, however, are considered in either the *capacity restraint* or *diversion* techniques discussed in the following sections.

Assignment using the all-or-nothing method is a simple process. The minimum time path between two zone centroids is assigned the total volume of the trip interchange. The assigned volumes are accumulated for each link and the total volume is the accumulation of all individual link volumes for each trip interchange. The methodology of the all-or-nothing technique is illustrated in Example 9-3.

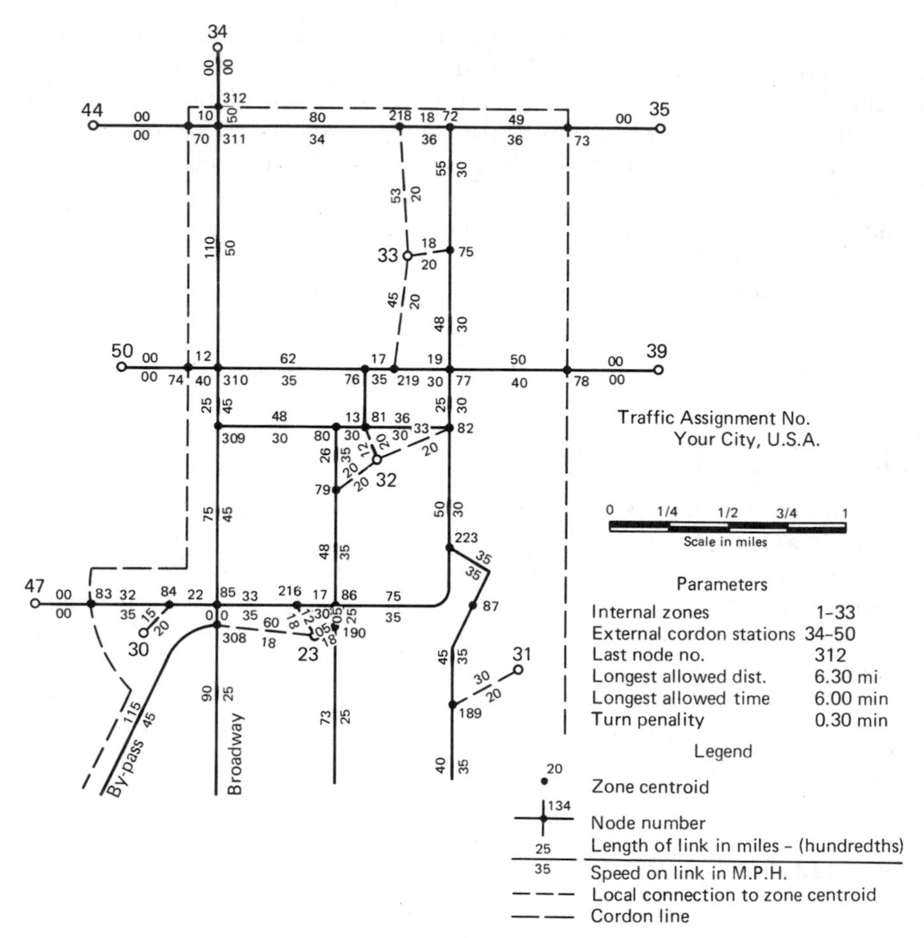


*Fig 9-8.* A portion of a typical network layout for a small city. (Source: *Traffic Assignment,* U.S. Department of Transportation, Federal Highway Administration, August 1973.)

## EXAMPLE 9-3

Assign the trip interchanges shown in Table 9-1 to the network of the four-zone area shown in Fig. 9-9. The times shown in this figure are the travel times along these links, expressed in minutes.

**Step 1.** *Compute minimum path trees (from Moore's algorithm):*

| *Tree No. 1* | *Time* | *Skim Tree* |
|---|---|---|
| 1-112-110-109-113-2 | 13 min | 1-2 13 min |
| 1-112-110-109-104-105-3 | 18 min | 1-3 18 min |
| 1-112-110-103-114-4 | 13 min | 1-4 13 min |

*Table 9-1*
**Trip Interchanges**

| *From Zone* / *To Zone* | *1* | *2* | *3* | *4* |
|---|---|---|---|---|
| 1 | — | 1000 | 5000 | 3000 |
| 2 | 4000 | — | 3000 | 4000 |
| 3 | 5000 | 6000 | — | 1000 |
| 4 | 2000 | 1000 | 3000 | — |

*Tree No. 2*

| | | |
|---|---|---|
| 2-113-109-110-112-1 | 13 min | 2-1 13 min |
| 2-113-109-104-105-3 | 13 min | 2-3 13 min |
| 2-113-109-110-103-114-4 | 18 min | 2-4 18 min |

*Tree No. 3*

| | | |
|---|---|---|
| 3-105-104-109-110-112-1 | 18 min | 3-1 18 min |
| 3-105-104-109-113-2 | 13 min | 3-2 13 min |
| 3-105-104-103-114-4 | 16 min | 3-4 16 min |

*Tree No. 4*

| | | |
|---|---|---|
| 4-114-103-110-112-1 | 13 min | 4-1 13 min |
| 4-114-103-110-109-113-2 | 18 min | 4-2 18 min |
| 4-114-103-104-105-3 | 16 min | 4-3 16 min |

**Step 2.** *A sample tree can be plotted using Tree No. 1.*

**Step 3.** *Assign zonal interchanges to each link comprising the minimum path for that interchange. The total volume for each link or the assignment is the summation of the individual zonal interchange link volumes.*

Table 9-2 shows a breakdown of the individual assignments for a part of a road network. The total assigned link volumes for the loaded network are displayed in Fig. 9-10.

## 9-19. DIVERSION

In the context of transportation models, diversion refers to the allocation of a trip interchange to two possible routes in a designated proportion that depends on some specified criterion. In most cases time is the criterion used, although some procedures use distance and others use generalized cost. Usually the two possible routes are the fastest all-surface route and the fastest route comprised partially or totally of expressway links. Experience has shown

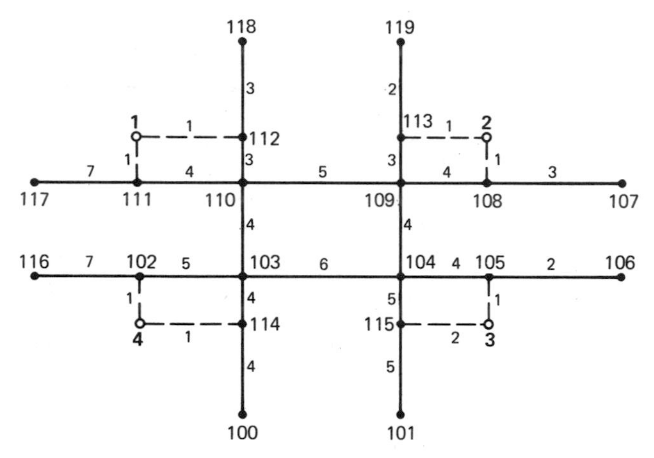


*Fig 9-9.* Network for a four-zone area.

that there are those who will travel a little longer both in time and distance in order to travel in the less congested atmosphere of a freeway. Equally, even though a freeway route offers time and distance savings, there is a proportion of drivers, often the elderly, who prefer to use surface routes.

One of the more widely used diversion techniques in early studies was that developed by Campbell [11]. This curve (Fig. 9-11) is based upon relative travel times between the new facility and the quickest alternative route.

The state of California method developed later [12] incorporated both time and distance savings in the diversion curves, shown in Fig. 9-12. The California curves consist of a family of hyperbolas. It can be seen that with equal time and distance on the freeway and on the best alternative arterial route, 50 percent of the trips are assigned to the freeway, using these curves. The most widely used diversion procedure in the 1960s was that available in the Bureau of Public Roads[1] series of traffic planning computer programs. This form of diversion is dependent on one parameter only, the ratio of travel times by the quickest combined arterial-freeway route to the quickest arterial-only route. With a one-parameter relationship, one single diversion curve (see Fig. 9-13) defines the relationship. The form of this S-shaped diversion curve is similar to those developed in the Detroit transportation study relating the percentage of freeway use to speed and distance ratios [13]. Diversion techniques were widely used in transportation studies carried out in the 1960s. The advantage in the spread of traffic between routes appeared to be more realistic than all-or-nothing assignments. The main disadvantage in the method was the considerable increase in computer cost and often unreasonable results. Although the method is not often used for current transportation planning it is useful for corridor studies. For most major studies more realistic results usually are obtained with *restraint* techniques.

[1]Former name of the Federal Highway Administration.

*Table 9-2*

**Assignment of Interchange Volumes by All-or-Nothing Assignment for a Portion of the Road Network**

| *Link* / *Interchange* | *1–112* | *113–2* | *117–111* | *111–110* | *110–109* | *109–108* | *108–107* | *116–102* | *102–103* | *103–104* | *104–105* | *103–114* |
|---|---|---|---|---|---|---|---|---|---|---|---|---|
| 1-2 | 1,000 | 1,000 | | | 1,000 | | | | | | | |
| 1-3 | 5,000 | | | | 5,000 | | | | | | 5,000 | |
| 1-4 | 3,000 | | | | | | | | | | | 3,000 |
| 2-1 | 4,000 | 4,000 | | | 4,000 | | | | | | | |
| 2-3 | | 3,000 | | | | | | | | | 3,000 | |
| 2-4 | | 4,000 | | | 4,000 | | | | | | | 4,000 |
| 3-1 | 5,000 | | | | 5,000 | | | | | | 5,000 | |
| 3-2 | | 6,000 | | | | | | | | | 6,000 | |
| 3-4 | | | | | | | | | | 1,000 | 1,000 | 1,000 |
| 4-1 | 2,000 | | | | | | | | | | | 2,000 |
| 4-2 | | 1,000 | | | 1,000 | | | | | | | 1,000 |
| 4-3 | | | | | | | | | | 3,000 | 3,000 | 3,000 |
| Total | 20,000 | 19,000 | | | 20,000 | | | | | 4,000 | 23,000 | 14,000 |

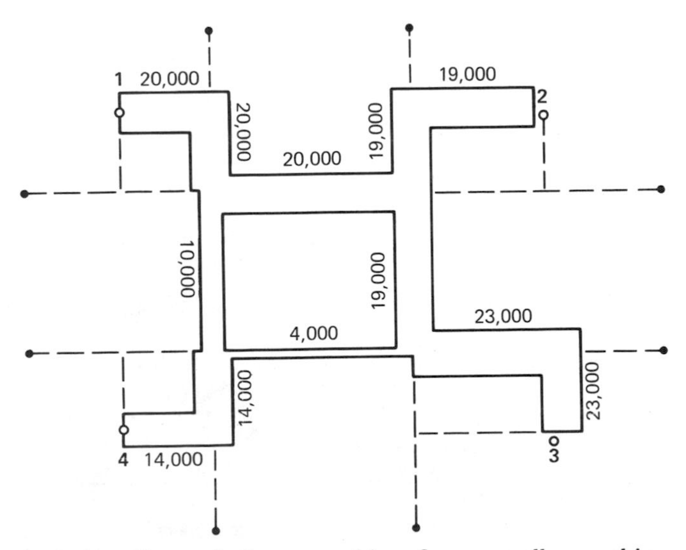


*Fig 9-10.* Network flows resulting from an all-or-nothing assignment.

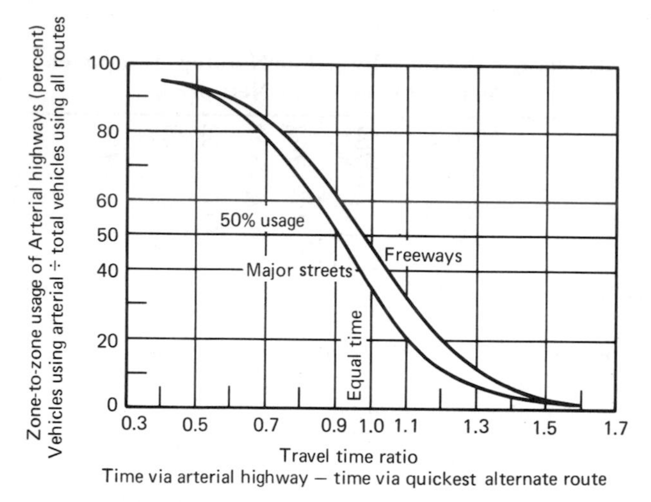


*Fig 9-11.* Time ratio diversion curves.

## 9-20. CAPACITY RESTRAINT

This technique is based on the fact that as traffic flow increases the speed of traffic decreases. At low volume/capacity ratios, traffic travels at a "free" speed on any facility. As volume increases, so does the volume/capacity ratio and an increasing restraint is placed on the drivers' freedom of action, with resulting decrease in speed (Fig. 9-14).

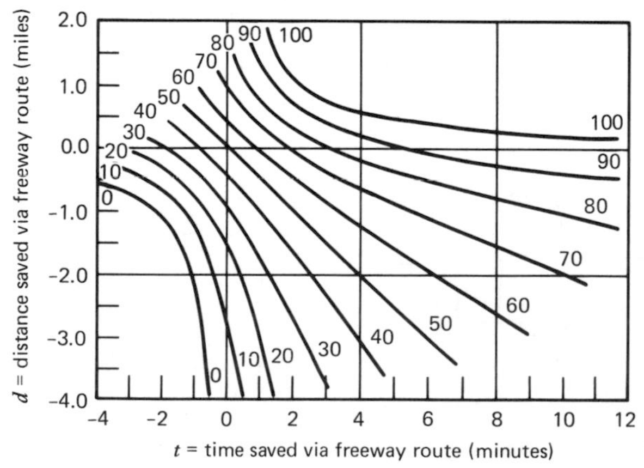


*Fig 9-12.* California time and distance savings diversion curves.

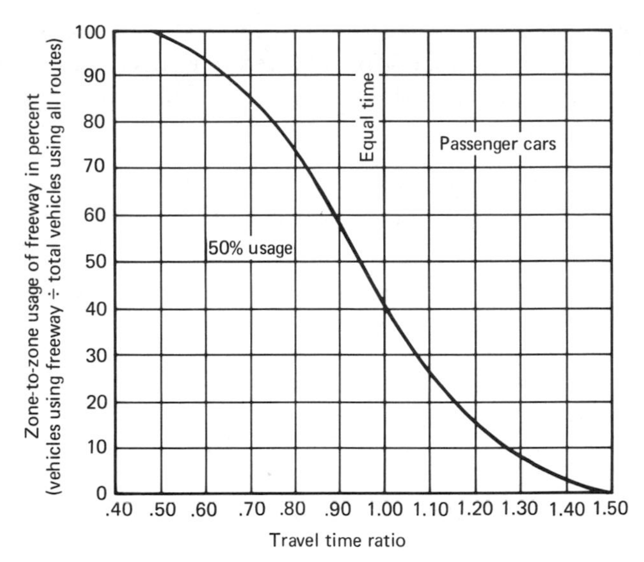


*Fig 9-13.* FHWA diversion curve. (Source: *Traffic Assignment Manual 1964,* U.S. Department of Commerce, Bureau of Public Roads.)

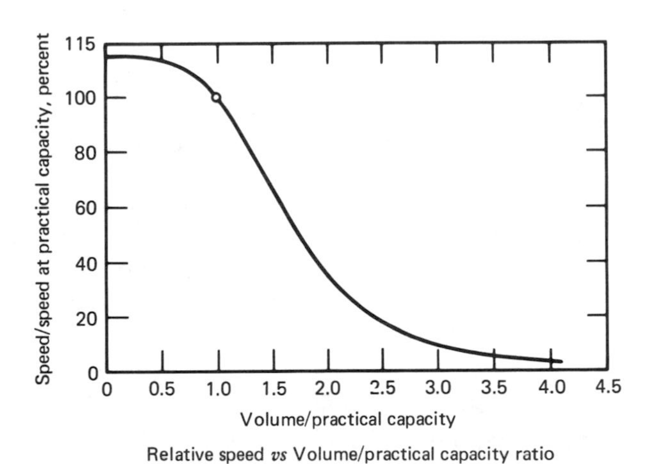


Relative speed *vs* Volume/practical capacity ratio
from equation:

$$T = T_0 \left[1 + 0.15 \left(\frac{\text{assigned volume}}{\text{practical capacity}}\right)^4\right]$$

*Fig 9-14.* Capacity restraint relationship. (Source: *CAPRES Capacity Restraint Pro-gram,* Federal Highway Administration.)

The capacity restraint model used in the FHWA computer program is applied in the iterative manner. The adjusted link speed and/or its associated travel impedence is computed by using the following capacity restraint function.

$$T = T_o[1 + 0.15(V/C)^4] \tag{9-14}$$

where

$T$ = balance travel time (at which traffic $V$ can travel on a highway segment)

$T_o$ = free flow travel time: observed travel time (at practical capacity) times 0.87

$V$ = assigned volume

$C$ = capacity

This is illustrated graphically in Fig. 9-14.

## 9-21. TRANSIT NETWORK ASSIGNMENT

Traffic assignment programs developed for highway networks have been applied to transit systems but this has often produced unrealistic results. These applications have been deficient mainly because of the following important differences between highway and transit analysis.

- The need to consider several lines of transit system operating on a single link.
- The need to handle special characteristics of a transit system such as passenger waiting time and transfer time.
- The need for summary statistics that are transit oriented.

An alternative approach to transit network assignment is to use a model that has special facilities to describe transit links. These facilities are available in the UMTA Transportation Planning System (UTPS) [16] which, in addition to link time and distance, includes facilities for several transit modes and many transit lines for each mode.

## 9-22. PRESENTATION OF ASSIGNED TRAFFIC FLOWS

The graphical presentation of traffic flows assigned to a network usually is shown by representing individual links flows by a scaled band width on a plan. With modern computer controlled plotting machines these network band plots can be drawn automatically to the required scale.

## 9-23. MICROCOMPUTERS AND TRANSPORTATION PLANNING

In the early years of transportation planning, (the 1960s), the activity was necessarily centralized because of the need to use mainframe computers, which in those days were of very limited capacity, typically 75K. The peripheral hardware for input and output was massive and expensive: high-speed card readers, tape drives, and hard copy on-line and off-line printers, all using the technology of batch processing.

The advent, first of the minicomputer and more recently, the microcomputer has made it possible for engineers and planners to have much easier access to relatively powerful machines. It is now common for an individual engineer to have immediate access at a very small cost to a personal computer (PC) with a capacity many times that of the mainframes of the 1960s. Input is by compact disk, keyboard, and interactive VDU screens and output can be either on screen display or on high-quality laser or dot matrix printers.

Widespread access to microcomputers has meant a decentralization of planning activities and the need for readily available user friendly software. Microcomputers are now available with a large range of software which is easy to use and which is well documented. The Department of Transportation, through the Urban Mass Transportation Administration (UMTA) and the Federal Highway Administration (FHWA) provides training and technical assistance on microcomputer use in a number of areas and maintains a record of available software and sources [15]. Extensive software is available in the following areas:

1. *Transit operations:* Scheduling and run cutting, maintenance, financial management, ridership reporting, ridership, and revenue estimation.
2. *Transportation planning:* Land use and trip generation, network-based highway and transit planning, site and subarea planning, mode choice, geographic data processing and display, network encoding and plotting, impact estimation, highway construction and maintenance.
3. *Traffic engineering:* General traffic operations, field data collection and analysis, signal timing simulation and optimization, traffic data management.
4. *Paratransit:* Planning and operations.
5. *Miscellaneous:* General support areas such as statistical analysis and specialist design programs.

## PROBLEMS

1. The ability of the multiple regression Eq. 9-2 to model observed shopping destinations was examined by considering the following statistics:

Standard error $(S_y) =$ 186

$$\begin{aligned} \text{Coefficient of multiple determination } (r^2) &= 0.902 \\ \text{Partial correlation coefficient of } X_{2_{(r_1)}} &= 0.882 \\ \text{Partial correlation coefficient of } X_{3_{(r_2)}} &= 0.764 \\ \text{Partial correlation coefficient of } X_{5_{(r_2)}} &= 0.720 \end{aligned}$$

Discuss these results in relation to the suitability of the model.

2. Rework Example 9.4 assuming the following socioeconomic adjustment factors.

$$\begin{aligned} K_{12} &= 0.85 \\ K_{13} &= 2.00 \\ K_{14} &= 1.00 \end{aligned}$$

Discuss the effect of these factors on the trip interchanges.

3. Complete the assignment of interchange volumes in Example 9.3, Table 9.2, for all links of the road network.

4. The three traffic zones *A, B,* and *C* shown here have interchanges $AB = 10$, $BC = 15$, and $CA = 20$ (nondirectionally). Growth factors of 2, 3, and 1 are forecast for zones *A, B,* and *C,* respectively. Using the Fratar method compute the zonal interchanges in the forecast year.

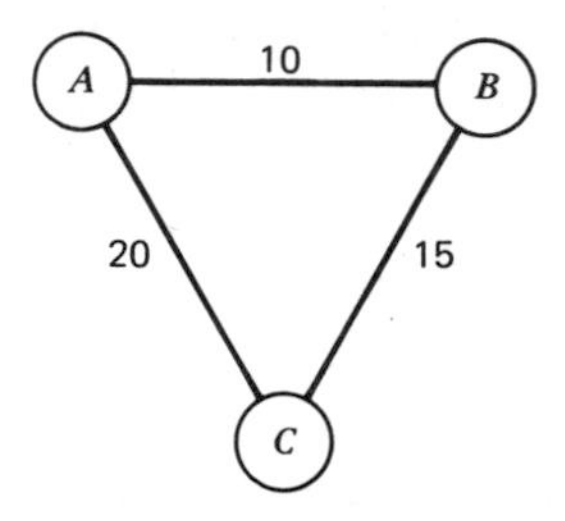


## REFERENCES

1. *Computer Programs for Urban Transportation Planning,* General Information, U.S. Dept. of Transportation/FHWA, April 1977.
2. *Trip Generation Analysis,* U.S. Dept. of Transportation/FHWA, Washington, D.C., August 1975.
3. Goode, L.R., "Evaluation of Estimated Vehicle-Trip Productions and Attractions in a Small Urban Area," *Transportation Research Record 638,* 1977.
4. Voorhees, A.M., *A General Theory of Traffic Movement,* Proceedings of the Institute of Traffic Engineers, 1955.
5. Fratar, T.J., *Vehicle Trip Distribution by Successive Approximation,* Traffic Quarterly, January 1954.
6. *A Review of Operational Urban Transportation Models,* U.S. Dept. of Transportation, Washington, D.C., April 1973.

7. *Mode Choice Application Manual for the Twin Cities Area,* R.H. Pratt Associates and DTM Inc., April 1976.
8. *Development and Calibration of Mode Choice Models for the Twin Cities Area,* U.S. Dept. of Transportation, February 1977.
9. Moore, E.F., *The Shortest Path Through a Maze,* International Symposium of the Theory of Switching Proceedings, Harvard University, 1957.
10. *Traffic Assignment, Methods, Applications, Products,* U.S. Dept. of Transportation/ FHWA, Washington, D.C. August 1973.
11. Campbell, M.E., *Assignment of Traffic to Expressways,* (as determined by diversion studies), Southwestern Association of State Highway Officials, Proceedings, 1951.
12. Moskowitz, K., *California Method of Assigning Diverted Traffic to Proposed Freeways,* Highway Research Bulletin 130, Highway Research Board, 1956.
13. Smock, R., *An Iterative Assignment Approach to Capacity Restraint on Arterial Networks,* Highway Research Board Bulletin 347, Highway Research Board, 1962.
14. *New System Requirements Analysis Program,* UMTA Transportation Planning (UTPS), Urban Mass Transportation Administration, Washington, D.C., October 1972.
15. *Microcomputers in Transportation: Software and Source Book,* UMTA Technical Assistance Program, Urban Mass Transportation Administration, Washington D.C., February 1985.

# *Chapter 10*

# *Development and Evaluation of Transportation Planning Options*

## 10-1. DEVELOPMENT OF PLANS FOR EVALUATION

There are quite obviously an infinite number of network and modal arrangements that will carry, with varying degrees of deficiency, the future transportation demands of a study area. These networks will themselves modify that demand since traffic generation, land use, and transportation facilities are highly interrelated. In attempting to develop an optimal system, however, the analyst is unable to generate but a few of the infinite variations that could be used. The question that naturally rises, therefore, centers on how to generate plans for evaluation.

Fortunately, perhaps, from the viewpoint of system generation, the transportation planner is not working in a vacuum. The existing and fully committed network provides a firm starting basis from which the planner is obliged to work. In general, it is unrealistic to abandon large amounts of existing capital investment and to abrogate plans to which substantial energy and capital have been devoted. The committed system, therefore, is generally used as a starting point for all future plans.

Even using the committed system as a basis, it is quite possible to generate poorly conceived plans. With proper evaluation techniques these solutions would be rejected. In the interest of efficiency, however, it is necessary to have a methodical and systematic way to develop study systems with a reasonable likelihood of satisfying the evaluation process. Since the evaluation techniques are predicated on comprehensive planning goals, it is essential to take note of these goals in system development. One comprehensive plan generated study systems that would meet the following objectives:

1. A "best" plan for a given regional land use plan.
2. An "optimum" transportation system to meet regional transportation needs.
3. An encouragement and a service to desirable patterns of land use, promoting desirable development [1].

Under normal circumstances, incremental plan development of this nature would follow the following pattern.

1. Future traffic demand is applied by traffic assignments to the committed system, and deficiencies of capacity with respect to demand are noted.
2. A second round of solutions is tried, based on the committed system, with facilities added to overcome deficiencies. The types of solution often vary substantially in their degrees of dependence on different modes. For example, in a study for a metropolitan urban area second-round solutions might vary according to the extent of the provision of bus or rapid transit systems on the public transport side, and, according to the degree of reliance on the automobile for the private transport side.
3. Deficiencies in the second-round solutions are corrected by changes in system design to provide third-round solutions, which, in turn, are tested.
4. This process is repeated until satisfactory systems are defined for comparative evaluation.

In the Atlanta Metropolitan Regional Transportation Study, system development was limited further by criteria that tested [2]:

Phase 1. The committed system.
Phase 2. The committed system plus maximum traffic engineering.
Phase 3. The committed system, maximum traffic engineering, plus improvements to existing alignments.
Phase 4. Phase 3 plus new facilities on new alignments.

## 10-2. FRAMEWORK OF EVALUATION

The process of selecting the best from a number of presented transportation plan options can be divided into four major parts [3]:

1. Structuring the options so that they may be tested and so that forecasts are possible.
2. Forecasting the consequences that can be expected to arise from option selection.
3. Obtaining the matrix of values of the individual consequences that are derived by both objective and subjective methods.
4. Developing the total value of all consequences from each option through the use of information on the relative value of the various types of consequences.

The actual method of evaluation can vary significantly in its particular technique. However, there are two basic methods of dealing with the matrix of information which describes the consequences of adopting the various plan options:

1. Weighting techniques, in which the information is collapsed into a single number.
2. Multidimensional techniques, in which the matrix is kept more or less intact as comparisons are made [4].

## TRADITIONAL ECONOMIC ANALYSIS

### 10-3. ENGINEERING ECONOMIC ANALYSIS[1]

By far the most common method of evaluation is engineering economic analysis, which attempts to collapse all impacts into a single monetary value or a single-evaluation parameter.

Four different methods of economic analysis are discussed:

1. Net present worth.
2. Equivalent uniform net annual return.
3. Benefit-cost ratio.
4. Internal rate of return.

Although upon preliminary examination, the various methods described appear to be quite different, in fact the underlying rationale is similar. The different forms of the economic analysis reflect the type of information required by different decision makers.

Benefit-cost ratio has been used extensively since its adoption by the U.S. Corps of Engineers as an evaluation procedure (under a congressional mandate) to show that the benefits accruing from public works projects exceed the costs of construction, maintenance, and operation. To some decision makers the relation of annual costs to annual benefits in the form of net annual benefits is more meaningful since it can be related to annual budgeting costs. The calculations that support net present value more clearly indicate the size of total costs and benefits over the life of the project. Rate of return analysis is more meaningful to some in that it makes direct comparisons between the earning power of invested capital. All methods will, however, lead to the selection of the same project when used for comparative purposes, and, therefore, are all appropriate criteria for selection.

[1] For underlying theory of interest and definition of interest factors the reader is referred to Reference 5.

*Net Present Worth (NPW).* The net present worth method of analysis permits comparison of costs and benefits throughout the life of a project. The periodic costs and benefits are treated as cash flows which are brought to equivalent worths at the zero time point. The discount rate used in the interest formulae is the minimum attractive rate of return over the period of project analysis. The net present worth is defined as the difference between net present benefits and net present costs. When the net present worth of a project is positive, the project is economically viable since it earns more than the minimum attractive rate of return. In the comparison of mutually exclusive alternatives, the greatest net present worth represents the most desired choice.

In the simplest case, a project could be shown as a single cash flow, $C$, at time zero bringing a benefit, $B$, $n$ years later, as shown here:

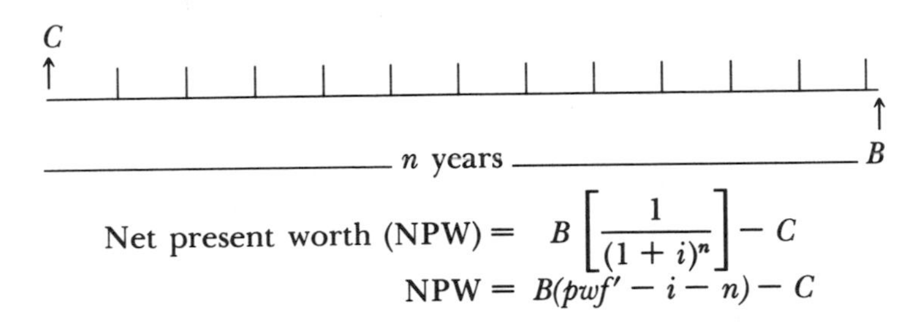


$$\text{Net present worth (NPW)} = B\left[\frac{1}{(1+i)^n}\right] - C$$

$$\text{NPW} = B(pwf' - i - n) - C$$

where

$C$ = cost
$B$ = benefit
$i$ = minimum attractive rate of return
$n$ = period of project analysis
$(pwf' - i - n)$ = present worth factor for a single benefit at discount rate $i$ for a period of $n$ years

Where a single project cost produces a uniform flow of annual benefits, the cash flow diagram is of the following form:

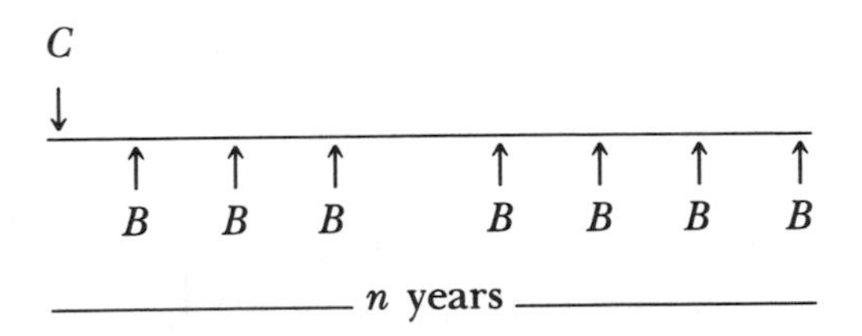


In this case the calculation of net present worth is made from the following equation:

$$\text{NPW} = B\left[\frac{(1+i)^n - 1}{i(1+i)^n}\right] - C$$

$$= B(pwf - i - n) - C$$

where

$B =$ annual benefit
$C =$ initial cost
$i =$ minimum attractive rate of return
$n =$ period of project analysis
$(pwf - i - n) =$ present worth factor for a series of uniform benefits at discount rate $i$ for a period of $n$ years

In the general case, both costs and benefits will be irregular cash flows over the whole time period of analysis. The general equation can be written for an irregular cash flow diagram.

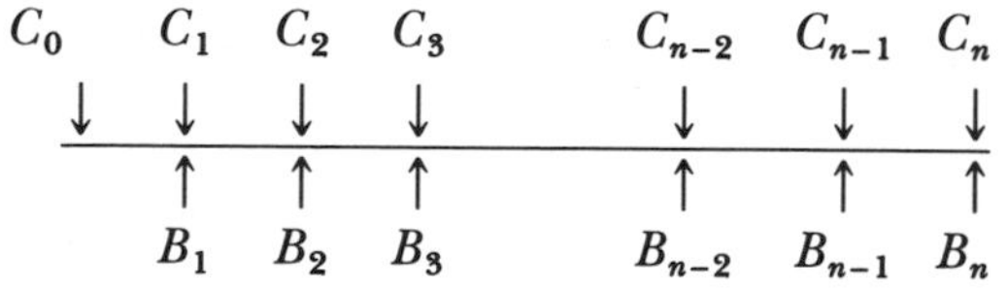

$$\begin{array}{ccccccc} C_0 & C_1 & C_2 & C_3 & C_{n-2} & C_{n-1} & C_n \\ \downarrow & \downarrow & \downarrow & \downarrow & \downarrow & \downarrow & \downarrow \\ \hline & \uparrow & \uparrow & \uparrow & \uparrow & \uparrow & \uparrow \\ & B_1 & B_2 & B_3 & B_{n-2} & B_{n-1} & B_n \end{array}$$

$$\text{NPW} = \sum_{k=0}^{n} B_k \left[\frac{1}{(1+i)^k}\right] - \sum_{k=0}^{n} C_k \left[\frac{1}{(1+i)^k}\right]$$

$$\text{NPW} = \sum_{k=0}^{n} B_k(pwf' - i - k) - \sum_{k=0}^{n} C_k\,(pwf' - i - k)$$

*Equivalent Uniform Net Annual Return (EUNAR).* The equivalent uniform net annual return can be calculated directly from the net present worth by multiplying the NPW by the capital recovery factor. This factor is the compound interest relationship which converts a cash flow at time zero into a series of uniform flows over a defined time period, in this case the analysis period. The relationship is shown by the following equations and cash flow diagram.

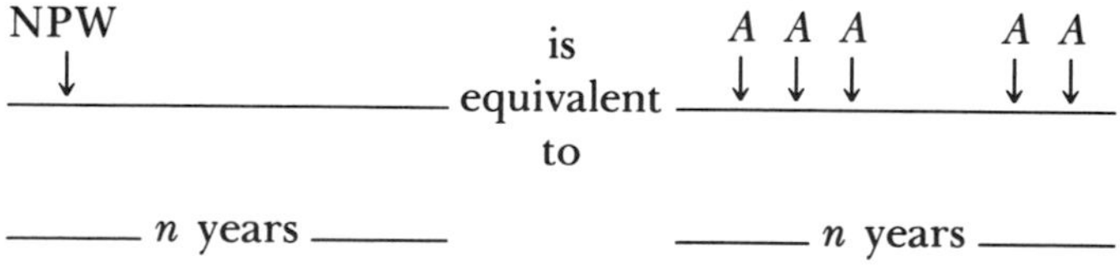

NPW ↓ ———— is equivalent to ———— A A A … A A (↓ ↓ ↓ … ↓ ↓)

—— $n$ years —— —— $n$ years ——

$$A = \text{NPW}\left[\frac{i(1+i)^n}{(1+i)^n - 1}\right]$$

or

$$A = \text{NPW}\ (crf - i - n)$$

where

$A$ = equivalent uniform net annual return
NPW = net present worth
$i$ = minimum attractive rate of return
$n$ = period of project analysis
$(crf - i - n)$ = capacity recovery factor for discount rate $i$ for a period of $n$ years

It can be seen that the equivalent uniform net annual return is the amount by which the uniform annual benefits exceed the uniform annual costs when discounted according to a minimum attractive rate of return. Provided that this difference is positive, the project is economically advisable since economic benefits outweigh costs at an attractive rate of return. Mutually exclusive alternate projects can be evaluated by comparison of the uniform annual net returns. The project with the higher net return is more desirable.

*Benefit-Cost Ratio (B/C).* One of the most widely used forms of economic analysis is the benefit-cost ratio. As the name implies, the method determines the ratio of benefits to costs after each has been discounted comparably with respect to time at the minimum attractive rate of return. Those projects with benefit-cost $(B/C)$ ratios greater than 1.0 are economically viable, while those with ratios below 1.0 are not. The ratio is computed from comparably discounted cash flows, and can be applied either to equivalent uniform annual costs and benefits or to present worth of costs and benefits. This equivalency is illustrated by the following cash flow diagrams and equations, where the redundancy of the capital recovery factor in the second form of the ratio is obvious.

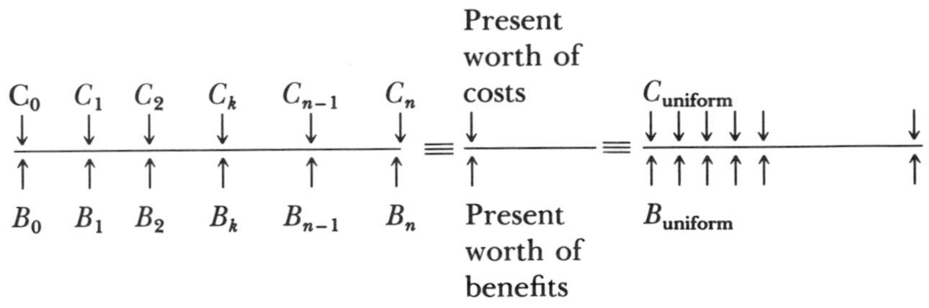

$C_0$ $C_1$ $C_2$ $C_k$ $C_{n-1}$ $C_n$ ↓ (costs) ≡ Present worth of costs ↓ ≡ $C_{\text{uniform}}$ ↓↓↓↓↓ … ↓

$B_0$ $B_1$ $B_2$ $B_k$ $B_{n-1}$ $B_n$ ↑ (benefits) ≡ Present worth of benefits ↑ ≡ $B_{\text{uniform}}$ ↑↑↑↑↑ … ↑

Using present worth values:

$$B/C = \text{Benefit-cost ratio} = \frac{\sum_{k=0}^{n} B_k \left[\frac{1}{(1+i)^k}\right]}{\sum_{k=0}^{n} C_k \left[\frac{1}{(1+i)^k}\right]} = \frac{\sum_{k=0}^{n} B_k(pwf' - i - k)}{\sum_{k=0}^{n} C_k(pwf' - i - k)}$$

or using equivalent uniform annual benefits and costs:

$$B/C = \frac{(crf - i - n) \sum_{k=0}^{n} B_k(pwf' - i - k)}{(crf - i - n) \sum_{k=0}^{n} C_k(pwf' - i - k)}$$

Direct benefit-cost analysis can be used to determine the economic feasibility of a project, since the ratio specifies the relationship between the benefits and costs associated with undertaking the project, as compared with no action at all. When attempting to evaluate the desirability of alternative projects with different associated costs and benefits, it is necessary to undertake incremental cost-benefit analysis. In this form of analysis the projects are examined in order of cost. The incremental benefit-cost ratio is the ratio between the incremental discounted benefits of the next least expensive alternative and its associated discounted costs. Where the incremental cost carries a benefit-cost ratio in excess of 1.0 the additional costs provide a return greater than the minimum attractive rate of return which has been assumed. Incremental benefit-cost ratios of less than 1.0 indicate that the additional investment is unwarranted economically.

*Internal Rate of Return.* One method of economic evaluation that makes no internal assumption concerning the minimum attractive rate of return on investment is called the internal rate of return method. In evaluating the economic feasibility of a scheme, the analyst computes, by a trial-and-error method, the discounting rate that exactly equalizes the discounted benefits and the discounted costs. While it is more usual to equalize the present worths of costs and benefits, the method also can be applied to equivalent uniform annual cash flows since these are directly calculable from present worths. It can be seen that the internal rate of return is that interest rate $r$ that satisfies the following equation of cash flows:

$$\begin{array}{cccccc} B_0 & B_1 & B_2 & B_k & B_{n-1} & B_n \\ \downarrow & \downarrow & \downarrow & \downarrow & \downarrow & \downarrow \\ \hline \uparrow & \uparrow & \uparrow & \uparrow & \uparrow & \uparrow \\ C_0 & C_1 & C_2 & C_k & C_{n-1} & C_n \end{array}$$

$$0 = \sum_{k=0}^{n} B_k \frac{1}{(1+r)^k} - \sum_{k=0}^{n} C_k \frac{1}{(1+r)^k}$$

Provided that the internal rate of return exceeds some external set minimum attractive rate of return, the scheme is judged to be economically feasible. The relative attractiveness of alternative schemes can be related directly to the size of their internal rates of return.

Internal rate of return analysis requires an incremental procedure similar to benefit-cost ratio analysis to determine the marginal rate of return on incremental investments. For a more extensive discussion of all methods of economic analysis, the reader is referred to Reference 6.

## 10-4. EXAMPLES OF ECONOMIC ANALYSIS PROCEDURES

In order to illustrate how economic analysis procedures can be used in practice, two examples are given:

### EXAMPLE 10-1

CONVERSION OF CAPITAL COSTS TO EQUIVALENT UNIFORM CAPITAL COSTS. A section of urban rapid transit line is to be constructed at a total cost of \$12,050,000. The breakdown of capital costs, the service lives of the individual portions of capital investment, and the assumed percentage of initial costs recoverable in salvage value are given below. The minimum attractive rate of return over the life of the project is assumed to be 5 percent.

| *Capital Investment Item* | *Initial Cost* | *Service Life (years)* | *Percentage of Initial Value Recoupable in Salvage at End of Service Life* |
|---|---|---|---|
| Land | \$5,000,000 | 100 | 100 |
| Earthwork | \$3,000,000 | 50 | 0 |
| Drainage | \$ 750,000 | 25 | 10 |
| Signalization and structures | \$3,000,000 | 40 | 10 |
| Roadbed | \$ 100,000 | 25 | 0 |
| Track | \$ 200,000 | 10 | 20 |

The total equivalent uniform annual capital cost can be calculated by summing the annual capital costs derived from the six separate items:

Land:

$$\text{EUAC}_l = 5{,}000{,}000\ (crf - 5\% - 100) - 5{,}000{,}000\ (sff - 5\% - 100)$$
$$\text{EUAC}_l = 5{,}000{,}000\ (0.05038) - 5{,}000{,}000\ (0.00038)$$
$$\text{EUAC}_l = \$250{,}000$$

Earthwork:

$$\text{EUAC}_e = \$3{,}000{,}000\ (crf - 5\% - 50)$$
$$\text{EUAC}_e = \$3{,}000{,}000\ (0.05478)$$
$$\text{EUAC}_e = \$164{,}340$$

Drainage:

$$\text{EUAC}_d = \$750{,}000\ (crf - 5\% - 25) - \$75{,}000\ (sff - 5\% - 25)$$
$$\text{EUAC}_d = \$750{,}000\ (0.07095) - 75{,}000\ (0.02095)$$
$$\text{EUAC}_d = \$17{,}484 - \$1571$$
$$\text{EUAC}_d = \$15{,}913$$

Signalization and structures:

$$\text{EUAC}_{s\text{-}s} = \$3{,}000{,}000\ (crf - 5\% - 40) - 300{,}000\ (sff - 5\% - 40)$$
$$\text{EUAC}_{s\text{-}s} = \$3{,}000{,}000\ (0.05828) - \$300{,}000\ (0.00828)$$
$$\text{EUAC}_{s\text{-}s} = \$174{,}840 - \$2484$$
$$\text{EUAC}_{s\text{-}s} - \$172{,}356$$

Roadbed:

$EUAC_r = \$100{,}000\ (crf - 5\% - 25)$
$EUAC_r = \$100{,}000\ (0.07095)$
$EUAC_r = \$7095$

Track:

$EUAC_t = \$200{,}000\ (crf - 5\% - 10) - \$40{,}000\ (sff - 5\% - 10)$
$EUAC_t = \$200{,}000\ (0.1295) - \$40{,}000\ (0.0795)$
$EUAC_t = \$25{,}900 - \$3180$
$EUAC_t = \$22{,}700$

$$\begin{aligned} EUAC_{total} &= EUAC_l + EUAC_e + EUAC_d + EUAC_{s\text{-}s} + EUAC_r + EUAC_t \\ &= \$250{,}000 + \$164{,}340 + \$15{,}913 + \$172{,}356 + \$7095 + \$22{,}700 \\ &= \$632{,}404 \end{aligned}$$

## EXAMPLE 10-2

**EXAMPLE OF AN ECONOMIC ANALYSIS USING BENEFIT-COST RATIO AND INTERNAL RATE OF RETURN ANALYSIS.** An engineer wishes to make an economic analysis of four mutually exclusive projects involving investments of $2 million, $4 million, $5 million, and $11 million. The following table indicates the capital, maintenance, and user costs for the proposed projects and no-investment, as-is situation. Each has been discounted to equivalent uniform annual costs based on a project life of 30 yr and minimum attractive rate of return of 7 percent.

*Benefit cost analysis:*

$$B/C_{1\text{-basic}} = \frac{R_{basic} - R_1}{H_1 - H_{basic}} = \frac{R_{basic} - R_1}{(C_1 + M_1) - M_{basic}} = \frac{2200 - 1860}{(161 + 30) - 60} = \frac{340}{131} = 2.60$$

$$B/C_{2\text{-basic}} = \frac{2200 - 1690}{(322 + 20) - 60} = \frac{510}{282} = 1.80$$

$$B/C_{3\text{-basic}} = \frac{2200 - 1580}{(403 + 30) - 60} = \frac{620}{373} = 1.66$$

| | | *Annual Items (×\$1000) i = 7% n = 30 years* | | | | |
|---|---|---|---|---|---|---|
| *Alternative* | *Investment (×\$1000)* | *Capital Cost* | *Maintenance Cost* | *User Cost* | *B/C Ratio* $\frac{R_{basic} - R_1}{H_1 - H_{basic}}$ | *Internal Rate of Return (percent)* |
| | *I* | *C* | *M* | *R* | | |
| *(Col. 1)* | *(Col. 2)* | *(Col. 3)* | *(Col. 4)* | *(Col. 5)* | *(Col. 6)* | *(Col. 7)* |
| Basic | — | — | 60 | 2,200 | — | — |
| 1 | 2,000 | 161 | 30 | 1,860 | 2.60 | 18.4 |
| 2 | 4,000 | 322 | 20 | 1,690 | 1.80 | 13.5 |
| 3 | 5,000 | 403 | 30 | 1,580 | 1.66 | 12.6 |
| 4 | 11,000 | 886 | 50 | 1,340 | 0.98 | 6.8 |

It can be seen from column 7 that alternatives 1, 2, and 3 are economically advantageous at the assumed minimum attractive rate of return. The $B/C$ ratio of 0.98 of alternative 4 would indicate that this alternative falls short of an attractive investment and can be disregarded. At this point, incremental cost-benefit analysis must be used to determine which of alternatives 1, 2, and 3 should be constructed. It would be incorrect to assume that the alternative with the highest benefit-cost ratio relative to the basic case is the most advantageous.

Alternative 1 is now regarded as the defender, and the incremental costs and benefits are compared for the increasingly expensive challengers.

$$B/C_{2\text{-}1} = \frac{\text{Incremental benefits}}{\text{incremental costs}} = \frac{R_1 - R_2}{H_2 - H_1} = \frac{1860 - 1690}{(322 + 20) - (161 + 30)}$$

$$= \frac{170}{151} = 1.13$$

Since the incremental costs associated with alternative 2 produce a rate of return that is greater than the minimum attractive rate of return, there is economic benefit in this incremental investment. Alternative 2, therefore, displaces alternative 1 and becomes the defender against the next most expensive challenger, alternative 3.

$$B/C_{3\text{-}2} = \frac{R_2 - R_3}{H_3 - \mathrm{H}_2} = \frac{1690 - 1580}{(403 + 30) - (322 + 20)} = \frac{110}{91} = 1.21$$

Alternative 3 displaces alternative 2 as the defender and alternative 4 is the new challenger.

$$B/C_{4\text{-}3} = \frac{R_3 - R_4}{H_4 - \mathrm{H}_3} = \frac{1580 - 1340}{(886 + 50) - (403 + 30)} = \frac{240}{503} = 0.477$$

Clearly, alternative 4 does not provide an economic investment of the incremental costs involved beyond the level of investment for alternative 3. Alternative 3, therefore, is accepted as the most appropriate level of investment.

*Rate of return analysis:* Similar investment decisions would be reached if, instead of benefit-cost analysis, the analyst were to use internal rate of return. This method also requires the use of incremental analysis. The engineer compares the internal rate of return on incremental investments with the minimum attractive rate of return.

$$(crf - r - 30)_{1\text{-basic}} = \frac{\text{Savings due to investment}}{\text{investment}}$$

$$(crf - r - 30)_{1\text{-basic}} = \frac{(M + R)_{\text{basic}} - (M + R)_1}{I_1 - I_{\text{basic}}}$$

$$(crf - r - 30)_{1\text{-basic}} = \frac{2260 - 1890}{2000}$$

$$(crf - r - 30)_{1\text{-basic}} = \frac{370}{2000}$$

$$(crf - r - 30)_{1\text{-basic}} = 0.185$$

By interpolation $(crf - r - 30) = 0.185$ gives $r = 18.4$ percent. Using alternative 1 as the defender, incremental analysis is carried out for alternative 2.

$$(crf - r - 30)_{2\text{-}1} = \frac{\text{Savings due to incremental investment}}{\text{incremental investment}}$$

$$(crf - r - 30)_{2\text{-}1} = \frac{1890 - 1710}{2000}$$

$$(crf - r - 30)_{2\text{-}1} = \frac{180}{2000}$$

$$(crf - r - 30)_{2\text{-}1} = 0.090$$

By interpolation $(crf - r - 30) = 0.90$ gives $r = 8.5$ percent.

Assuming a 7 percent minimum attractive rate of return, alternative 2 is economically advisable. Alternative 3 is next evaluated against the new defender, alternative 2.

$$(crf - r - 30)_{3\text{-}2} = \frac{1710 - 1610}{1000}$$

$$(crf - r - 30)_{3\text{-}2} = 0.10$$

By interpolation $(crf - r - 30) = 0.10$ gives $r = 9.8$ percent, making alternative 3 economically attractive.

Finally, alternative 4 is evaluated against the current defender, alternative 3.

$$(crf - r - 30)_{4\text{-}3} = \frac{1610 - 1390}{6000}$$

$$(crf - r - 30)_{4\text{-}3} = \frac{420}{6000}$$

$$(crf - r - 30)_{4\text{-}3} = 0.07$$

By interpolation, $(crf - r - 30) = 0.07$ gives $r = 5.7$ percent. Since the incremental rate of return is less than 7 percent (the designated minimum attractive rate of return), alternative 4 is not economically advisable, and alternative 3 is the selected alternative.

## 10-5. STRUCTURE OF THE AASHTO PROCEDURE FOR COMPUTING ECONOMIC CONSEQUENCES OF HIGHWAY AND BUS TRANSIT IMPROVEMENTS [7]

The American Association of State Highway and Transportation Officials (AASHTO) sets out precise guidelines for the evaluation of highway and bus transit improvement schemes using a purely economic viewpoint. Figure 10-1

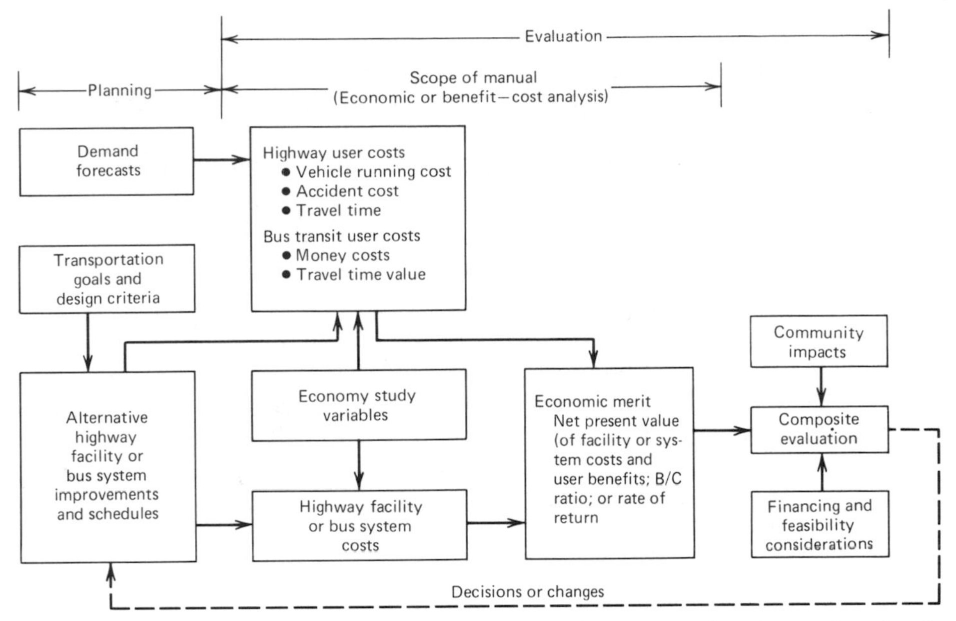


*Fig 10-1.* Relation of manual to total highway and bus system planning and evaluation process. (Source: AASHTO.)

shows the relationship of this process to the entire evaluation procedure. It is not seen to be a social benefit-cost analysis but merely an evaluation of the direct and quantifiable transportation benefits and costs. In summary the methodology allows the analyst to compute the net present worth or the benefit-cost ratio by:

1. Selecting basic study parameters: discount rate, unit value of time, project analysis period.
2. Computing highway user costs (with and without improvement) which are the sum of:

   - Basic running costs which are a function of user time and vehicle running costs. The latter are computed from vehicle type, highway type, design speed, volume/capacity ratio, grades and curvatures, and vehicle mix.
   - Accident costs.
   - Cost of transitioning movements between uniform highway sections.
   - Intersection delay costs.

3. Computing transit user costs which are a function of:

   - Walking time to the transit vehicle.

- Waiting time.
- In-vehicle time.
- Fares (including transfer costs).

4. Computing total user costs from the sum of highway user costs and transit user costs.
5. Computing project benefits from changes in total user costs.
6. Computing project highway costs from construction costs.
7. Computing capital and operating costs for transit improvements: Capital costs come from costs of roadway improvements; new roads for transit and busways; costs of terminals, shops, vehicle storage facilities, administrative facilities and other structures; shelters, signs, and other routeside equipment.

   Operating costs entailing drivers' wages and benefits costs of vehicle operation, bus maintenance costs; insurance, management and administrative labor costs; costs of vehicle rental; transit systems contribution to roadway maintenance and operating costs.
8. Computing total project costs from capital costs and changes in operating and maintenance costs.
9. Estimating salvage values of facilities and equipment at end of project life.
10. Determining net present worth of project costs and benefits discounted over the life of the project.

## 10-6. PROBLEMS OF BENEFIT-COST ANALYSIS [8]

In a great many countries benefit-cost analysis (or some other equivalent form of economic analysis covered in Section 10-3) is the preferred, and in some cases the required, method of project or plan evaluation. Occasionally the more extended and more complex form of social benefit-cost analysis is used in which monetary estimates are made of the costs and benefits of social consequences which cannot be priced directly. In all these analyses the discounted stream of money costs are compared with the discounted streams of benefits converted into monetary terms. Where benefit-cost ratios in excess of unity are obtained, the project or plan is considered worthwhile and where a number of options are to be evaluated, that with the highest marginal benefit-cost ratio is considered preferable. Extended social benefit-cost analysis in the past has led to calculations of great complexity, but of course the value of the calculations is limited by the accuracy with which the numerous factors being considered can be quantified. Because so very few of these factors currently are described in any detail, it is scarcely surprising that the results of social benefit-cost analysis often have appeared to be regarded with suspicion by some nontechnical decision maker and the public at large. A case in point was the political selection of the Maplin site for the third London airport in the face of strong benefit-cost arguments for all the other evaluated sites. The work of the Royal Commission on the third London airport must rank as one of the most extensive and costly benefit analyses ever carried out [9]. The findings of the commission were based on benefit-cost analysis, but they were overturned by the British government because it was apparent to an "intelligent layman" that the findings were obviously wrong. This was largely due to the fact that the terms of reference of the committee were interpreted too narrowly and the technique of economic evaluation was applied in an incomplete manner. The exclusion of unquantifiable amenity considerations may render a whole social benefit-cost exercise unreliable; furthermore, there are a number of basic and fundamental questions on which the analyst must be satisfied prior to entering into any evaluation exercise.

First it must be recognized that the treatment of externalities and some other quantifiable impacts in terms of additional monetary costs and benefits to be added into the direct cash flow stream is likely to be a process of great uncertainty which may lead to disagreement with other planners and confusion for decision makers. For example, there seems to be little community-wide agreement of the true costs of noise disturbance. This in the past has been related both to the amount an individual is willing to pay to avoid noise and to the amount that he is willing to receive as compensation for enduring it. [10] Similarly there is sharp disagreement on the true value of travel time savings and much debate about the question of whether very small savings have any value at all. [11] Often called notional values, indirect costs and benefits of this nature account for a large proportion of the total costs and benefits that are involved in the most sophisticated extended social benefit cost analyses. Even more debatable is the question of the stability of these notional values over the very long periods of time involved in strategic plans. A number of writers indicate that even basic social values may change as developed nations enter the postindustrial phase. Certainly some changes are discernable between the early 1960s and the 1980s.

The question of the discount rate itself poses severe philosophical questions when examined in the context of long-term planning. In many countries (e.g., the United States and Western Europe), it has been required for years that public investment should bring rates of return that reflect the opportunity cost of capital invested in the private sector. Under current conditions, the use of high discount rates might therefore appear rational in long-term planning. However, high discount rates tend to militate against the long term in favor of the short term and results in effectively nullifying the effect of long-term impacts due to the combined effects of the high rates and the very long-term planning periods. Some analysts have therefore argued that this type of analysis, which heavily discounts future states and gives preference to the short term, is most unsuitable for long-term planning. They claim that it is irrational to act on decisions made largely on the type of economic justification where the future is heavily discounted by the application of high current discount rates to notional values determined by current thinking.

A further fundamental difficulty arises from the recognized fact that society is pluralistic. While a single-benefit-cost ratio can indicate that as a whole society will gain a net benefit from a certain course of action, the single-benefit-cost ratio cannot indicate who is gaining the benefits and who is paying the cost. It is entirely possible that in the case of the location of an airport or the choice of the alignment of a freeway the costs incurred to the affected parts of the community may be much larger than the concomitant benefits to them.

These three basic problems all point out that while benefit-cost analysis may be very suitable for comparing two similar schemes (e.g., alternative freeway alignments), it is less suitable for providing a comprehensive evaluation of the multiplicity of factors that are implied by a long-term transportation plan. Planners too frequently make the mistake of striving for mathematical rather than philosophical rigor.

## 10-7. MULTIMODAL URBAN TRANSPORTATION ANALYSIS

When making economic analyses of transportation systems, computations must take consideration of both direct and indirect costs and benefits. While the AASHTO method is useful for a comparative evaluation of similar schemes, the omission of indirect impacts can lead to questionable conclusions. In the past there have been many dubious analyses which have included only direct user benefits and direct project costs. Because the construction of major transportation facilities has massive direct and indirect effects on both users and nonusers, this type of analysis can be relatively meaningless and has frequently led to misleading conclusions. In the area of urban transportation studies it is recommended that evaluation procedures should consider a range of likely impacts such as [12]:

1. Transit user effects.
2. Highway user effects, including savings in (a) travel time, (b) operating costs, (c) ownership costs, (d) accident costs, and (e) parking costs.
3. Unemployment effects.
4. Educational opportunity effects.
5. Business productivity effects.
6. Government productivity effects.
7. Real estate effects.
8. Life-style effects.
9. Environmental pollution effects.
10. Tax effects.
11. Disruptive effects, including those that are (a) temporary during construction, and (b) permanent in neighborhood division.

12. Construction labor effects.
13. Highway construction effects.
14. Aesthetic effects.
15. Property losses and relocation effects.
16. Regional and neighborhood growth effects.
17. Crime effects.
18. Civil defense effects.
19. Achievement of desired urban form.
20. Detailed modal studies and projections.
21. Implementation evaluation.
22. Financing effects.
23. Tourism effects.

Few studies in fact attempt such sophisticated analysis. More typical is the case of the preliminary engineering analysis of the Puget Sound study which included the following [13]:

*Costs*

1. Motor vehicle operating costs for fuel, tires, oil, maintenance and repairs, and depreciation. The additional operating and time costs for stopping, idling, and resuming speed at intersections (or as a result of traffic congestion delays) over uniform speed operation should be included. Motor vehicle operating costs in engineering economy analyses should exclude fuel taxes, since they are transfer payments used for highway construction.
2. Maintenance costs of all highway facilities. In the case of the Cross-Sound Bridge this included maintenance of the bridge, toll-booth operation, and insurance premiums.
3. Motor vehicle accident costs, including fatalities, injuries, and property damage on highways with various levels of access control.
4. Operating costs of bus transit facilities, including items for payrolls, maintenance, insurance, and overhead.
5. Operating costs of ferries, including items for payrolls, maintenance, insurance, and overhead.
6. Operating costs of parking facilities, including maintenance, insurance, and overhead.
7. Operating costs of rapid transit facilities including payrolls, maintenance, insurance, and overhead.

*Benefits.* Benefits were taken as net reduction in:

Total annual maintenance on all systems.

Total operating costs of all systems.

Total accident costs on all systems.

Total travel time costs on all systems.

Engineering analysis based on these parameters is highly user oriented and is hard to relate to overall development goals of the community [14].

## 10-8. ECONOMIC EVALUATION OF TRANSPORTATION PLANNING AT THE NATIONAL AND THE REGIONAL LEVELS

At the regional or national level of transportation planning, economic analysis of transport facilities and networks centers around comparison of all accruing benefits and project investment costs [15, 16, 17].

Benefits are due to both reductions in the cost of transportation and to increases in regional output due to transport investment. Reductions in transport costs that need consideration should include:

1. Decreases in vehicular operating costs such as depreciation, repairs, fuel, maintenance, and operators' wages.
2. Decreases in facility costs due to depreciation, maintenance, and operation.

Benefits accruing from increased regional output are related directly to anticipated levels of production. It is essential that such benefits are not double counted in several analyses. Increased national output can be due to added investment in parallel facilities, such as water and power distribution systems. It is possible that several schemes could be erroneously justified independently, using the same benefits. In underdeveloping and developing countries the lack of reliable underlying statistical data makes the evaluation of benefits from decreased vehicle and facility costs difficult.

As the national level of planning, some authorities indicate that benefits should reflect true growth in national income [18]. This may be difficult to estimate without firm data support. More easily projected are specific benefits such as reduced user expenses, lower maintenance costs, fewer accidents, time savings, increased levels of comfort and convenience, and stimulation of economic development. On the cost side of the comparison, Adler points out that accounting at the national level should reflect "shadow" prices which differ from apparent costs by [19]:

1. The exclusion of taxes, fees, import duties, and so on.
2. The reduction of the level of wages which does not reflect the true cost of labor to the country.

3. The difference between the financial costs of money and the opportunity costs available in alternate investments.

Other errors in national accounting can be introduced by the consideration of inflationary trends which cause no real costs, and by failure to include the costs of all necessary support systems not directly financed in the project costs, for example, access roads, power distribution, and water supply.

## OTHER EVALUATION PROCEDURES

### 10-9. GOALS AND OBJECTIVES—ACHIEVEMENT MATRICES [20]

One of the techniques that has been used in attempting to retain a multidimensional approach to evaluation evolved from more traditional systems analysis techniques. A weighted matrix methodology, which is suitable for a variety of metropolitan and regional plan types, has been used [21]. The procedure is as follows:

1. Determine regional and metropolitan goals.
2. Determine objectives, within the scope of the transportation plan, that are related to the goal statements.
3. Determine by such means as attitude surveys the relative importance of the objectives so that some weighting can be put on each objective.
4. Select quantitative criteria which can be used to evaluate how well each objective is satisfied.
5. Evaluate each scheme by determining how well each criterion satisfied the overall objective. Total system effectiveness is determined in reference to all objectives by an effectiveness measure which is the sum of each criterion evaluation weighted by the importance of the objective to which it relates.
6. Select the system with maximum effectiveness.

*Example:* Assuming that regional and metropolitan goals have been determined, the following objectives have been stated:

a. System should provide economy.
b. Minimum disruption of individuals by relocation.
c. Transit should provide a high level of comfort and convenience.
d. The central area should be made highly accessible.
e. Transit should be available to low-income areas.

The following criteria are selected to provide a measure of each objective:

a. Benefit-cost ratio.
b. Number of persons relocated.
c. Load factor on transit vehicles in peak hour.
d. Accessibility index of core areas.
e. Transit accessibility index to low-income traffic zones.

From attitude surveys and rating panels objectives *a* through *e* have the following ratios of relative importance: 40 percent, 20 percent, 20 percent, 10 percent, 10 percent.

| *Evaluation Matrix* | *Possible Effectiveness Score* | *Score for Plan A* | *Score for Plan B* | *Score for Plan C* |
|---|---|---|---|---|
| 1. Benefit-cost ratio | 40 | 35 | 25 | 30 |
| 2. Persons relocated | 20 | 10 | 20 | 5 |
| 3. Transit load factor | 20 | 10 | 15 | 3 |
| 4. Core accessibility | 10 | 3 | 5 | 10 |
| 5. Low-income transit availability | 10 | 2 | 10 | 8 |
| Total effectiveness score | 100 | 60 | 75 | 56 |

Plan B is selected as being most responsive to the regional goal statements that the transportation plan is meant to strengthen. The method hinges on the ability of the planner to specify a definitive number of goals to be achieved by the plan and to be able to determine and apply weights to individual criteria with goal areas. There is an apparent ability to include a variety of heterogeneous considerations while reducing the evaluation output to a single quantative score [22].

One of the greatest drawbacks to this form of goals achievement matrix, which has been used in a number of studies [23, 24], is the failure to take note of the social justice of the solution that is selected. The maximization of the implied objective function takes place without examining the impact of low levels of individual goal achievement [8].

## 10-10. RATIONAL MODELS

The problems of the goals achievement methods are overcome to some degree by the use of an adaptation of the rational model. Threshold analysis first defines and selects the criteria by which a plan or technology is to be evaluated. In each of the criteria areas minimum acceptable levels of achievement are defined. Plans that fail to meet these minima are rejected or are redesigned in such a way that all prescribed threshold levels are attained. Selection of the best option is carried out by some unidimensional analysis technique such as benefit-cost analysis or by the optimization of some objective function [25]. This is a special case of the rational model of conventional systems analysis which seeks to:

1. Define system objectives and construct from these objectives a system utility function.
2. Enumerate all possible optional courses of action.
3. Identify the consequences of each option.
4. Compute system consequences in terms of effect on the utility function.
5. Select the system that produces the best utility function.

However, it has been found that sociopolitical decisions such as the provision of transportation are not amenable to such an oversimplifying approach for various reasons. System objectives are neither clear nor are they unidimensional. It is not possible to define all feasible optional courses of action; nor is it possible to estimate system performance and consequences to the degree of precision that this sort of analysis requires. Finally, it must be recognized that should it be possible to construct some form of system utility function, it could not be unidimensional. Therefore, in the absence of known or even achievable social benefit-cost trade-offs between various factors, system selection on this type of basis remains dubious.

## 10-11. PLANNING BALANCE SHEETS

A modification of the benefit-cost evaluation technique which seriously attempts to deal with the multidimensional impacts of a planned change is the planning balance sheet [26]. Its application has been largely in the evaluation of urban and regional plans, but it also has been applied to long-term transportation planning. Essentially, the technique recognizes that whereas it is impossible to declare a proposal as optimal from the viewpoint of all affected individuals, it should be feasible to divide the community into affected groups. Having defined the various groups, or actors, it is then possible to compile and compose the costs and benefits that will accrue to them as a result of the proposed plan options. Nonmonetary considerations (for example, disruption and time impacts) are quantified and taken into account. It has been stated that "the balance sheet cannot, and does not, aim to provide a conclusion in terms of rate of return or net profit measured by money values as is the case in some typical cost benefit studies. Its value lies in exposing the implications of each set of proposals to the whole community and in indicating how the alternatives might be improved to produce a better result." [26] The strength of the method lies in the identification of the principal actor groups affected by plans and the inclusion of nonquantifiable impacts. Figure 10-2 shows a planning balance sheet developed for an airport development.

## 10-12. FACTOR PROFILES

One method of assessment that attempts to retain the multidimensional nature of the evaluation procedure is the factor profile. The basic method is shown in Fig. 10-3, which indicates how the individual evaluating factors are listed and the level of effect marked for each individual factor on a visual scale. The principal advantage of the method is the simultaneous visual display of all factors and their impacts. Dominant and redundant solutions are self-evident. Unfortunately the method tends to convey the impression that all factors are equally important. The example shows that system B outperforms system A over the whole range of evaluation factors and therefore is always preferable. The relative value of system C is obviously less apparent since it outperforms system B for some factors, but is a considerably worse choice when other factors are considered. Without some knowledge of either the absolute relationships between the factors (i.e., weightings) or the relative relationships (i.e., ranking), the method is really only a visual statement of the evaluation problem, and not a solution procedure.

In California, a refinement of this procedure has been used to evaluate a number of similar transportation options. There, the problem was to select a "best" freeway alignment from a number of available locations [27]. Beneficial changes in the factor profile were compared using incremental benefit-cost analysis on a systematic basis. This approach enabled the planner to determine whether or not improvements to the factor profile were cost effective. The method works less well where the systems under comparison are radically different in nature (e.g., when comparing a freeway system to serve a multinuclear city form with a radial rapid transit system serving a strong CBD core concept).

## 10-13. PLAN RANKING

Implicit in both the planning balance sheet and factor profile methods of evaluation is the concept that, while the weighting of plan objectives may not be possible because this further implies inelastic social trade-offs, planners are able to rank objectives and hence plans from experience and knowledge. This is an entirely reasonable conclusion. If the analyst were unable either to weight or rank individual or multiple objectives, then he would be indifferent to all solutions, which is clearly not the case. Based on this premise a number of methods of evaluation have been developed that use rankings rather than

| Actors | Impact Type | Level of Impact | | | |
|---|---|---|---|---|---|
| | | *Option I* | *Option II* | *Option III* | *Option IV* |
| A. Airport Authority | 1. Required Capital | $120m | $140m | $200m | $350m |
| | 2. Benefit/cost ratio | 2.4 | 3.2 | 2.3 | 3.3 |
| | 3. Skilled labor requirement | 2200 | 3200 | 3000 | 3800 |
| | 4. Ease of operation | Low | Medium | Low | High |
| B. Airlines | 1. Required Capital | $0.5m | $1.2m | $5m | $7m |
| | 2. Annual aircraft operating cost | $2m | $3m | $5m | $4m |
| | 3. Staffing requirements | 15200 | 16400 | 20000 | 24000 |
| | 4. Adverse noise abatement procedures costs | $2m | $4.2m | $3m | $2m |
| | 5. Ranking of Safety of noise abatement procedures | 1 | 3 | 4 | 2 |
| C. Air Traveller | 1. Ranking of Ease of Using Terminal | 3 | 2 | 1 | 4 |
| | 2. Ranking of Ease of Using Parking | 4 | 1 | 2 | 3 |
| | 3. Estimated time required in terminal for departing passengers | 85 min | 80 min | 65 min | 60 min |
| | 4. Estimated time required in terminal for arriving passengers | 15 min | 20 min | 24 min | 15 min |
| | 5. Aesthetic level of terminal design | Low | Medium | High | Medium |
| | 6. Estimated % of Transfer flights missed by failing to make connection | 2% | 1.5% | 1% | 1.5% |
| | 7. Average walking distance for embarking passenger | 1500 ft. | 1000 ft. | 700 ft. | 180 ft. |
| D. Non-traveller in Neighborhood of Airport | 1. Average increase in $L_{dn}$ level | 25 | 15 | 12 | 10 |
| | 2. Effect on average housing values | −$10,000 | −$8000 | NIL | +$2000 |
| | 3. % increase in road traffic on arterial routes and freeways | 5% | 4% | 1% | 2% |
| | 4. Loss of amenity | High | High | Medium | Low |
| | 5. Effect on local job market | Low | Medium | High | High |
| | 6. Effect on local taxes/capita/annum | −$20 | −$15 | +5 | NIL |
| | 7. No. of properties taken by eminent domain | 200 | 300 | 200 | 450 |
| | 8. Loss of public land (acres) | 25 | 200 | 300 | 750 |
| | 9. Total land take (acres) | 300 | 600 | 750 | 3500 |

*Fig 10-2.* Example of a planning balance sheet for an airport development.

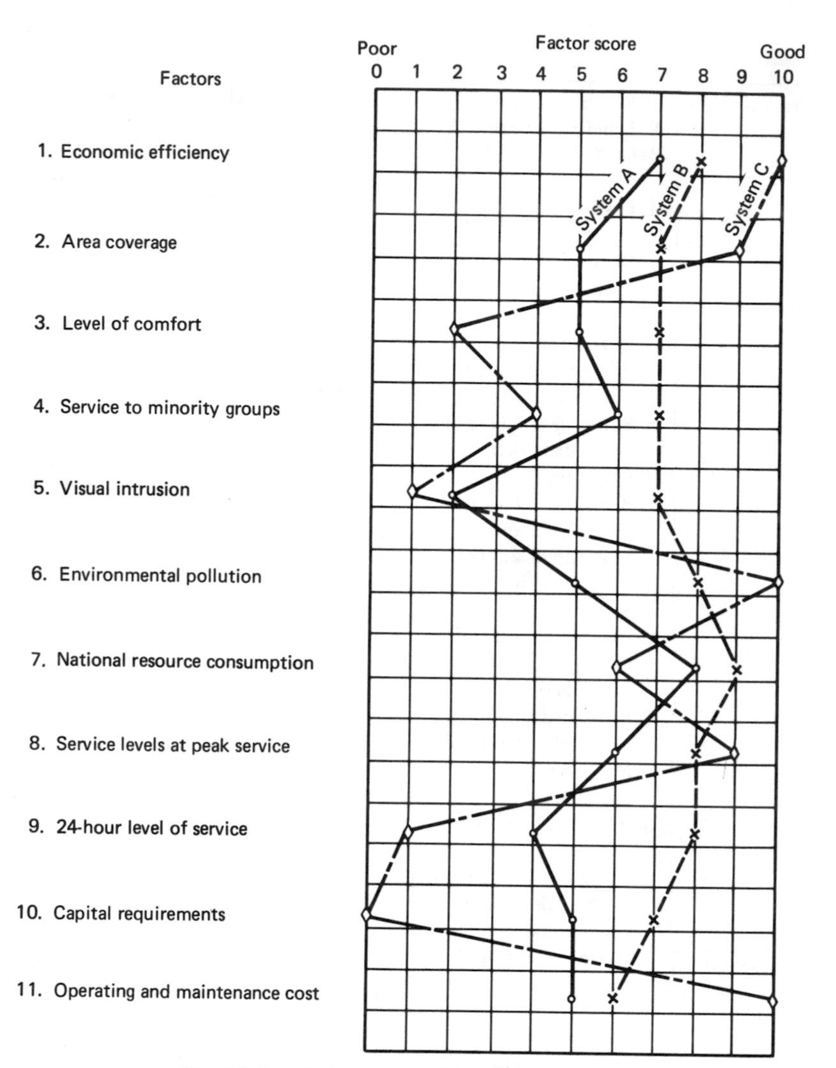


*Fig 10-3.* A factor profile chart.

weightings. Figure 10-4 shows a procedure used in Wisconsin [28]. The method is carried out in the following manner.

1. All objectives, $n$ in number, are ranked in order of importance and are assigned values of $n$, $n$-1, $n$-2, and so on, in descending rank order.
2. The plan options, $m$ in number, are ranked under each of the specified objectives in the order that they meet that objective, with assigned values $m$, $m$-1, $m$-2, and so on, in descending rank order.
3. A probability $p$ of successful implementation is assigned to each of the plans being ranked.

| Plan | Probability of Implementation | Provide a Balance Transportation System<br>Rank Order of Objective<br>$n = 2$ | Provide for an Approprate Spatial Distribution of Land Use<br>Rank Order of Objective<br>$n = 3$ | Economic Efficiency<br>Rank Order of Objective<br>$n = 1$ | Plan Value<br>$v = p\Sigma\,(n_1 m_1 + n_2 m_2 + n_3 m_3)$ |
|---|---|---|---|---|---|
| | | Rank Order Value of Plan ($m$) | Rank Order Value of Plan ($m$) | Rank Order Value of Plan ($m$) | |
| Continue existing trends with controls | $p = 0.6$ | 3 | 1 | 3 | 0.6 [(2 X 3) + (3 X 1) + (1 X 3)] = 7.2 |
| Corridor development | $p = 0.5$ | 2 | 2 | 1 | 0.5 [(2 X 2) + (3 X 2) + (1 X 1)] = 5.5 |
| Satellite city concept | $p = 0.9$ | 1 | 3 | 2 | 0.9 [(2 X 1) + (3 X 3) + (1 X 2)] = 11.7 |

*Fig 10-4.* **Rank based expected value method.**

4. The value of each plan option $V$ is computed by summing the product $n \times m \times p$ for each of the specified objectives:

$$V = p \, \Sigma \, (n_1 m_1 + n_2 m_2 + \ldots + n_n m_n)$$

A combination of ranking and goals achievement matrix evaluation was used with success in the San Diego-Los Angeles Corridor Study [29].

## 10-14. MATHEMATICAL PROGRAMMING APPROACHES

A number of approaches have been made to evaluating plans through mathematical programming techniques. Typical are linear programming methods which seek to minimize or maximize some cost or benefit function. The general formulation is [30]:

$$\text{Maximize or minimize } \sum_i c_i x_i$$

subject to

$$\sum_i a_{ij} x_i \geq b_j \qquad j = 1, 2, \ldots, m$$

$$x_i \geq 0 \qquad \text{all } i$$

where

$x_i$ = variable to be evaluated
$c_i$ = objective function coefficient for $x_i$
$a_{ij}$, $b_j$ = parameters of the constraint functions

Approaches of this nature generally are not useful or widely applicable in that they imply the ability to calculate accurately factor weights and assume that all factors can be dealt with in terms of a one-dimensional objective function which is usually expressed in monetary terms. A more useful procedure is the goal programming approach, which uses linear programming to minimize the total deviations of performance in subgoal areas, based on the relative importance placed on these subgoals [31]. This form of mathematical programming retains the multidimensional character of the problem.

## 10-15. APPROACHES WITH INTERACTIVE COMMUNITY PARTICIPATION

Some planners have attempted to introduce plan evaluation procedures that are much more interactive with the community than the methods described previously. While requiring more time and resources than the more passive procedures, such methods generally have produced plans that are more in accord with general community goals as expressed through the organs of community participation. An approach used by the San Diego Regional Comprehensive Planning Organization is shown in Fig. 10-5 [20]. The method used had a number of significant variations from more conventional methods:

1. Two-tier evaluation: one at the *regional,* or *macrolevel,* one at the *community,* or *microlevel.*
2. Presentation of simple impact summaries to decision makers and citizen participants.
3. The systematic subjective assessment of plan impacts by both lay groups (decision makers and citizen participants).
4. The iterative cycling of lay assessments among plan evaluators and plan designers.

## 10-16. DESIGN-SYNTHESIS APPROACH [32]

The conventional alternative-directed approach to planning shown in Fig. 10-6 focuses on evaluating how well a number of generated solution options meet the stated goals and objectives of the community. In general objectives

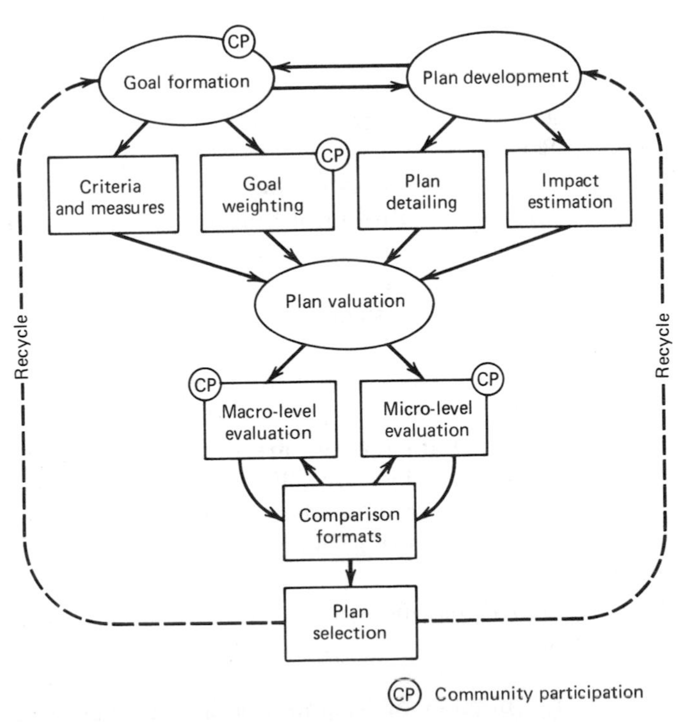


*Fig 10-5.* Interactive regional planning process.

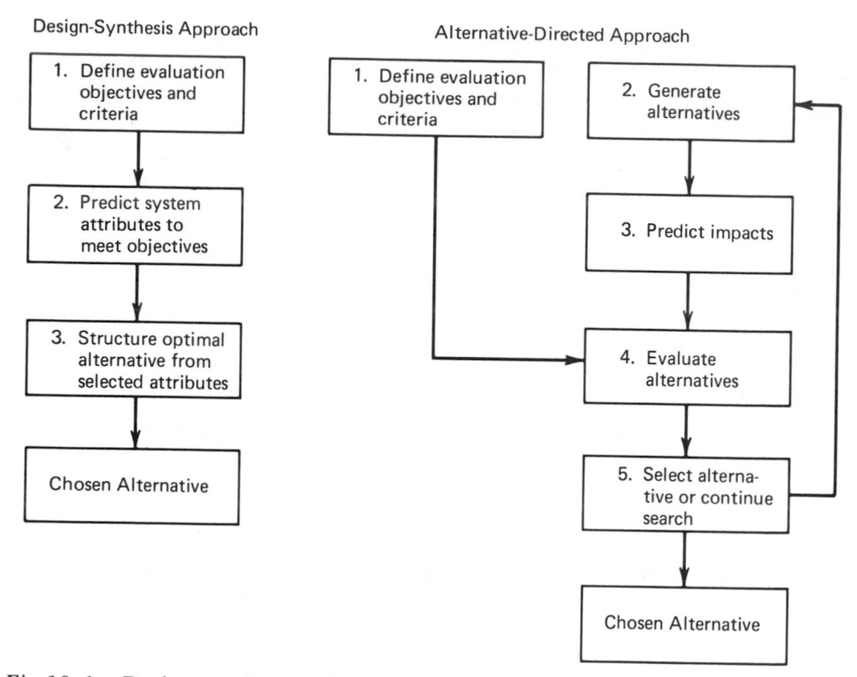


*Fig 10-6.* Design synthesis and alternative-directed approaches to transportation planning.

are used only to evaluate alternatives. Some planners have indicated that this is an inefficient way of achieving optimal plans and have proposed instead a *design-synthesis* approach, which has been used with success in transit planning, for example in both San Diego and Denver [33, 34]. Figure 10-6 compares design synthesis to the basic procedures of the alternative-directed approach. The newer approach allows for infusion of objectives directly into the planning process. Consequently, a transportation system is generated that incorporates characteristics that will optimize the attainment of specific objectives. The procedure has three principal stages:

1. A set of regional and local transportation objectives are determined. These typically fall into two categories: service and cost.
2. A number of system characteristics are determined which are necessary for meeting system objectives. For complicated systems, mathematical programming techniques can be used to examine a wider range of system characteristics simultaneously and to select the best combination.
3. A system is designed that incorporates the optimal characteristics obtained in step 2.

Where there are a large number of feasible options, the mathematical programming approach can be time consuming and costly. An alternative manner of moving toward the optimal solution is to use heuristic techniques which examine the effects of varying individual components of the transportation system separately. This is generally more cost effective, allowing the intuitive experience of the planner to shortcut the mathematical programming approach. With the heuristic approach the planner plays a more creative and active role in balancing the different system components.

## 10-17. IDEALIZED EVALUATION PROCEDURE

From an examination of the shortcomings of the methods already discussed, it is possible to hypothesize the structure of an idealized plan evaluation procedure which would allow for interaction between the technical staff making up the evaluation analysis team and the lay decision makers while providing a methodology capable of multidimensional analysis. Such a procedure has been suggested [8] and is shown in Fig. 10-7; it would have the following salient steps:

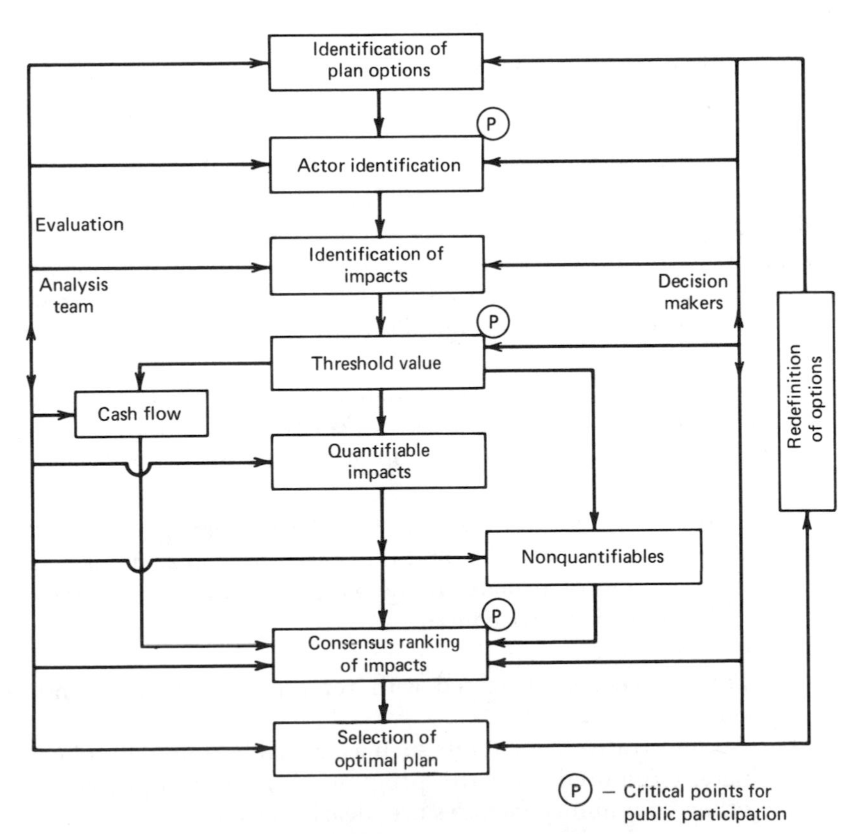


*Fig 10-7.* An idealized evaluation procedure.

- Identification of a range of plans for evaluation. This range would include all feasible options.
- Identification of the major actors affected by the adoption of the individual plans. All major groups and organizations should be included.
- Identification of the scale of impacts on the major actors.
- Establishment of threshold values (maximum and minimum acceptable values of impact).
- Estimation of cash flows, quantifiable impacts, and nonquantifiable effects of each option in a comparable manner.
- Establishment of a consensus ranking based on the impact assessment.
- Selection of the plan for implementation or modification of the option range for a recycling through the process.

## IMPLEMENTATION

### 10-18. CAPITAL PROGRAMMING AND IMPLEMENTATION [35]

The work carried out in the evaluation phase of the comprehensive study indicates the particular system that best meets the goals and objectives of the study area. This is sometimes rather loosely referred to as the optimal system. Further work, however, is necessary to break this system down into individual projects and to schedule those projects and their inherent costs into the capital expenditures estimates of the area. Only when year-by-year estimates of capital expenditures are calculated does the ease or difficulty of implementation really become apparent. It is suggested that the planner and engineer should carry out capital programming in the following manner.

1. *Project breakdown.* The selected system should be broken down into individual contract-size projects. Project size and boundaries should reflect the scale of construction projects normally carried out in the area.
2. *Project evaluation and priority scheduling.* Each project is evaluated and based on its judged priority, and is programmed in sequence subject to the constraints of available funds. Programming can be carried out using the following procedures.

    a. Each project is estimated with reference to its time and cost requirements.
    b. Administrative constraints such as available funding and coordination required with the programming of transportation and other community facilities are clearly set out.

c. Service considerations of existing facilities are evaluated to determine their maximum continued useful life, for example, pavement conditions for roads and airports, useful life of rail roadbed, and structural condition of bridges.
d. Evaluation of preliminary priority rankings based on benefit-cost ratios and community benefits is made.
e. Establishment of the time period over which the capital programming is feasible. If a curtailed period is used, the design system is available at an earlier date at relatively high annual capital budgeting costs. If the catch-up period for construction is over the full period to the design year, annual costs are minimized but the interim level of service of the system is also minimized.
f. Final programming of individual projects based on preliminary priority rankings and other considerations which include:

   Coordination with other public improvements.
   Equable geographic distribution of improvements.
   Continuity of minimum levels of service.
   Engineering considerations.

   Minimum levels of service are verified by assigning interim year traffic demands on interim facilities. Where severe problems become apparent in the human interim years, it is necessary to shuffle projects to maintain adequate service.

g. Calculation of final annual costs based on the accepted capital program.

## UPDATING THE TRANSPORTATION PLAN

### 10-19. UPDATING PROCEDURES

Modern transportation plans are conceived by their authors as dynamic entities. Faced with evolving technology and constantly changing social and economic conditions, a static plan has no chance of implementation. Urban planners, therefore, speak of their work in the context of a continuing process in which the details constantly are modified subject to the overall directions laid down by regional and community objectives. The design of any transportation study should incorporate procedures for continuous plan updating.

Updating procedures call for a day-to-day maintenance of data that affects the elements of the plan. These procedures require the establishment of an areal information system which can accommodate a continuous data flow. The information system should be either partially or totally computerized, and should be keyed to the lowest possible data collection unit, which is usually the individual land parcel. A general surveillance system would record changes in:

Land use
Transportation facilities
Social economic data
Population

Through the United States, the large comprehensive plans were finished by the mid-1970s or earlier. Long-term planning now concentrates on studies that serve to update the original study. Many planners now believe that the publication of a single long-range land use transportation plan is anachronistic. Rather than attempting major formal updates of the original strategic plans, updating studies are carried out on a periodic basis. The transportation plan generated in this manner is a situation-responsive assembly of documents, plans and reports rather than a single document.

## PROBLEMS

1. Use the benefit-cost ratio method to select the most economic scheme from the following three proposals, with a minimum attractive rate of return of 10 percent and an analysis period of 30 years.

| *Scheme* | *Initial Capital Cost* × *$1000* | *Maintenance Costs* × *$1000* | *User Costs* × *1000* | *Salvage Value at 30 years* × *1000* |
|---|---|---|---|---|
| Do nothing | — | 250 | 8000 | — |
| A | 6,500 | 170 | 6500 | 500 |
| B | 10,200 | 100 | 5600 | 1200 |
| C | 17,500 | 90 | 4500 | 2500 |

2. Using the figures of Problem 1 conduct an internal rate of return analysis for the selection of the most economic scheme.
3. Compute the net present value of the following three projects, analyzed over a 40-year period, with a 12 percent minimum attractive rate of return.

| *Project* | *Initial Investment (× $ million)* | *Salvage Value (× $ million)* | *Service Life (years)* | *Annual Decrease in Travel Costs (× $ million)* |
|---|---|---|---|---|
| A | 10 | 0.5 | 10 | 0.8 |
| B | 22 | 1.5 | 20 | 0.75 |
| C | 42 | 3.6 | 40 | 0.9 |

**4.** Compute the equivalent uniform net annual return of the following rapid transit project. Use a discount rate of 8 percent.

***Costs***

| *Item* | *Initial Cost (× $ million)* | *Salvage Value (× $ million)* | *Service Life* |
|---|---|---|---|
| Right-of-way | 200 | — | infinite |
| Main structures | 250 | 25 | 50 years |
| Drainage | 70 | 10 | 25 years |
| Earthwork | 50 | nil | 50 years |
| Signalization | 100 | 25 | 25 years |
| Roadbed | 20 | nil | 25 years |
| Track | 40 | 8 | 10 years |

***Annual Benefits (assume over an infinite period)***

| | |
|---|---|
| Decreased direct user costs | $40 million |
| Indirect community benefits | $55 million |

What is the benefit-cost ratio of this scheme?

**5.** In attempting to evaluate a new airport at a "green field" site, what factors should enter into the evaluation? How would you quantify those that are quantifiable? How would you take account of qualitative factors?

**6.** What are the main differences in factors that must be accounted for in the evaluation process for the construction of an urban rapid transit line and radical improvement to an existing surface bus system.

## REFERENCES

1. Puget Sound Regional Transportation Study, *Summary Report,* Seattle: 1967.
2. Atlanta Metropolitan Regional Transportation Study, Unpublished Staff Manual on System Evaluation, Atlanta, 1967.

3. Thomas, E.N., and J.L. Schofer, *Strategies for the Evaluation of Alternative Transportation Plans, NCHRP Report No. 96,* Highway Research Board, Washington, D.C., 1970.

4. Beimborn, E.A. "Structured Approach to the Evaluation and Comparison of Alternative Transportation Plans," *Transportation Research Record No. 619,* Transportation Research Board, Washington, D.C., 1976.

5. Grant, E.L., *Principles of Engineering Economy,* Seventh Edition, John Wiley, New York, 1985.

6. Winfrey, Robley, *Economic Analysis for Highways,* International Textbook Company, Scranton, PA, 1969.

7. *A Manual on User Benefit Analysis of Highway and Bus-Transit Improvements 1977,* American Association of State Highway and Transportation Officials, Washington, D.C., 1978.

8. Ashford, N.J., and J.M. Clark, "An Overview of Transport Technology Assessment," *Transportation Planning and Technology,* Vol. 3, No. 1, 1975.

9. Committee on the Third London Airport, Chairman Hon. Mr. Justice Roskill, *Report,* Her Majesty's Stationery Office, London, 1971.

10. Ollerhead, J.B., and R.M. Edwards, *A Further Study of Some Effects of Aircraft Noise in Residential Communities near London Heathrow Airport, Transport Technology Report No. 7705,* Dept. of Transport Technology, Loughborough University, Loughborough, June 1977.

11. Heggie, I.G. (ed), *Modal Choice and the Value of Travel Time,* Clarendon Press, Oxford, 1976.

12. Haney, D.G., "Problems, Misconceptions and Errors in Benefit Cost Analyses of Transit Systems," *Record No. 314,* Highway Research Board, Washington, D.C., 1970.

13. Niebur, H.H., "Preliminary Engineering Economy Analysis of Puget Sound Regional Transportation Systems," *Record No. 180,* Highway Research Board, Washington, D.C., 1967.

14. Nijkamp, P., *Multidimensional Spatial Data and Decision Analyses,* John Wiley, Chichester, 1979.

15. *Introduction to Transport Planning,* United Nations, New York, 1967.

16. Barker, P., and K. Button, *Case Studies in Cost Benefit Analysis,* Heinemann, London, 1975.

17. Thompson, J.M., *Modern Transport Economics,* Penguin Education, London, 1974.

18. Adler, H.A., "Economic Evaluation of Transport Projects," *Transport Investment and Economic Development,* Gary Fromm (ed.), The Brookings Institution, Washington, D.C., 1965.

19. Adler, H.A., *Economic Appraisal of Transport Projects,* Indiana University Press, Bloomington, 1971.

20. Schofer, J.L., and D.G. Stuart, "Evaluating Regional Plans and Community Impacts," *Journal of the Urban Planning and Development Division,* American Society of Civil Engineers, March 1974.

21. Jessiman, W., et. al., "A Rational Decision-Making Technique for Transportation

Planning," *Highway Research Record No. 180,* Highway Research Board, Washington, D.C., 1967.

**22.** Hill, M., "A Goals-Achievement Matrix for Evaluating Alternative Plans," *Journal of the American Institute of Planners,* Vol. 34, January 1968.

**23.** "Guidelines for Long Range Transportation Planning in the Twin Cities Region," Barton-Aschman Associates Inc., Twin Cities Metropolitan Council, Minneapolis-St. Paul, December 1971.

**24.** "Land Use/Transportation Study: Recommended Regional Land Use and Transportation Plans," Planning Report No. 7, Vol. III, Southeastern Wisconsin Planning Commission, Waukesha, WI, November 1966.

**25.** Irwin, N.A. "Criteria for Evaluating Alternative Transportation Systems," *Highway Research Record No. 148,* Highway Research Board, Washington, D.C., 1966.

**26.** Lichfield, N., "Cost Benefit Analysis in City Planning," *Journal of the American Institute of Planners,* Vol. 26, 1968, and 'Cost Benefit Analysis in Urban Expansion; A Case Study, Peterborough," *Regional Studies, Vol. 3,* 1969.

**27.** Oglesby, C.H., B. Bishop, and G.E. Willeke, "A Method for Decisions Among Freeway Location Alternatives Based on User and Community Consequences," *Highway Research Record No. 305,* Highway Research Board, Washington, D.C., 1970.

**28.** Schlager, K., "The Rank-Based Expected Value Method of Plan Evaluation," *Highway Research Record No. 238,* Highway Research Board, Washington, D.C., 1968.

**29.** Stuart, D.G., and W.D. Weber, "Accommodating Multiple Alternatives in Transportation Planning," *Transportation Research Record 639,* Transportation Research Board, Washington, D.C., 1977.

**30.** Eash, R.W., and E.K. Morlock, "Development and Application of a Model to Evaluate Transportation Improvements in Urban Corridors," *Transportation Research Record 639,* Transportation Research Board, Washington, D.C., 1977.

**31.** Yu, J.C., and R.C. Hawthorne, "Goal-Programming Approach to Assessing Urban Transit Systems," *Transportation Research Record 574,* Transportation Research Board, Washington, D.C., 1976.

**32.** Scheibe, M.H., and G.W. Schultz, "Design-Synthesis Approach to Transit Planning," Transportation Research Board, 1977.

**33.** *Systems Analysis Approach to Transit Route and Service Design,* R.H. Pratt Assoc. Inc., San Diego Comprehensive Planning Organization, August 1974.

**34.** *Transit Concept Comparison, Long Range Transit Development Analysis,* Denver Regional Transportation District, April 1975.

**35.** *Developing Project Priorities for Transportation Improvements,* Procedure Manual 10A, National Committee on Urban Transportation, Public Adminisration Service, Chicago, 1959.

# *Chapter 11*

# *Traffic Impact Analysis*

## 11-1. INTRODUCTION

It has been recognized for a number of years that land use changes affect levels of traffic demand, in terms of trip generation, and that increased demand requires the provision of greater transportation capacity, which can influence further changes in land use. So much has already been discussed in Section 7-2. In recent years, transportation planners have paid increasing interest to one side of the land use transportation cycle shown in Fig. 7-1. This recent interest centers upon the recognition that land use changes not only change demand but also affect individual facilities, particularly at the local level. The implications of the land use–traffic demand interaction are important, therefore, not only in the strategic context discussed in Section 7-2, but also in the short and medium term. The interaction between land use changes, (either in terms of new or replacement uses) is often termed *traffic impact analysis*. A more generic term favored by the FHWA is *site impact traffic evaluation (SITE)*; the terminology *site access study* is also sometimes used. The techniques covered by these various terminologies enable the analyst to assess how well an existing or future road network will serve a proposed change in land use.

Because traffic impact analyses are site specific, they are detailed and vary considerably in content. Some, in FHWA terminology are *basic,* that is, they apply to the development or redevelopment of small sites which have a limited propensity to generate traffic and therefore a limited potential impact on the road network. Others are termed *complex,* in that they concern large traffic generators or extremely complex traffic conditions. An example of a basic condition might well be a small retail store located on a one acre site with the development to be completed over a time span of one year. A complex condition on the other hand could well involve a development of several million square feet of mixed land use on a 100-acre site, with development in stages over a 10-year period [1].

The need for traffic impact analysis stems from the fact that the demand for urban development is increasing rapidly and that the redevelopment of derelict and decaying inner city sites has become more attractive. At the same time, the availability of public funds for construction of highway transportation facilities has continued to decrease at a time when it is recognized that significant development or redevelopment of land use can cause serious deterioration of the transportation network. Another factor which has arisen in recent years is the greater interest paid by developers to the matter of site access. Developers for the 1980s and 1990s are well informed of the importance of ensuring adequate site access; also there is a more general recognition that development costs may have to bear some share of providing facilities that ensure adequate site access.

Traffic impact analysis is therefore carried out to permit the following:

1. Determination of the travel demand and traffic generated by the proposed development.
2. Identification of deficiencies in the existing and proposed transportation systems.
3. Identification of improvements necessary to maintain acceptable levels of service [2].

## 11-2. THE FHWA SITE IMPACT TRAFFIC EVALUATION PROCESS

Because the procedure varies between sites and because the manner of proceeding may be governed by the recommendations and standards of the particular community involved, there are no detailed guidelines on how traffic impact analysis is to be handled. However, the generalized procedure shown in Fig. 11-1 has been put forward by the Federal Highway Administration. Seven separate generalized phases are involved:

1. Determine the scope of the study, the background traffic situation, and the existing peak hour level of service of the traffic network within the study area.
2. Project the future background traffic without development of the site and compute levels of service.
3. Compute the site specific generated traffic for the form of development which is planned.
4. Determine the peak hour situation with the site fully developed and occupied. The total traffic estimated at this stage is obtained by combining the traffic estimations made in steps 2 and 3.

   Compute levels of service in the "site-developed" condition and compare these with the "without-site-development" levels calculated in step 2.
5. Identify and analyze the various options for site-access-related improvements in terms of peak hour effectiveness. These changes may simply be minor alterations such as a roadway or traffic operations

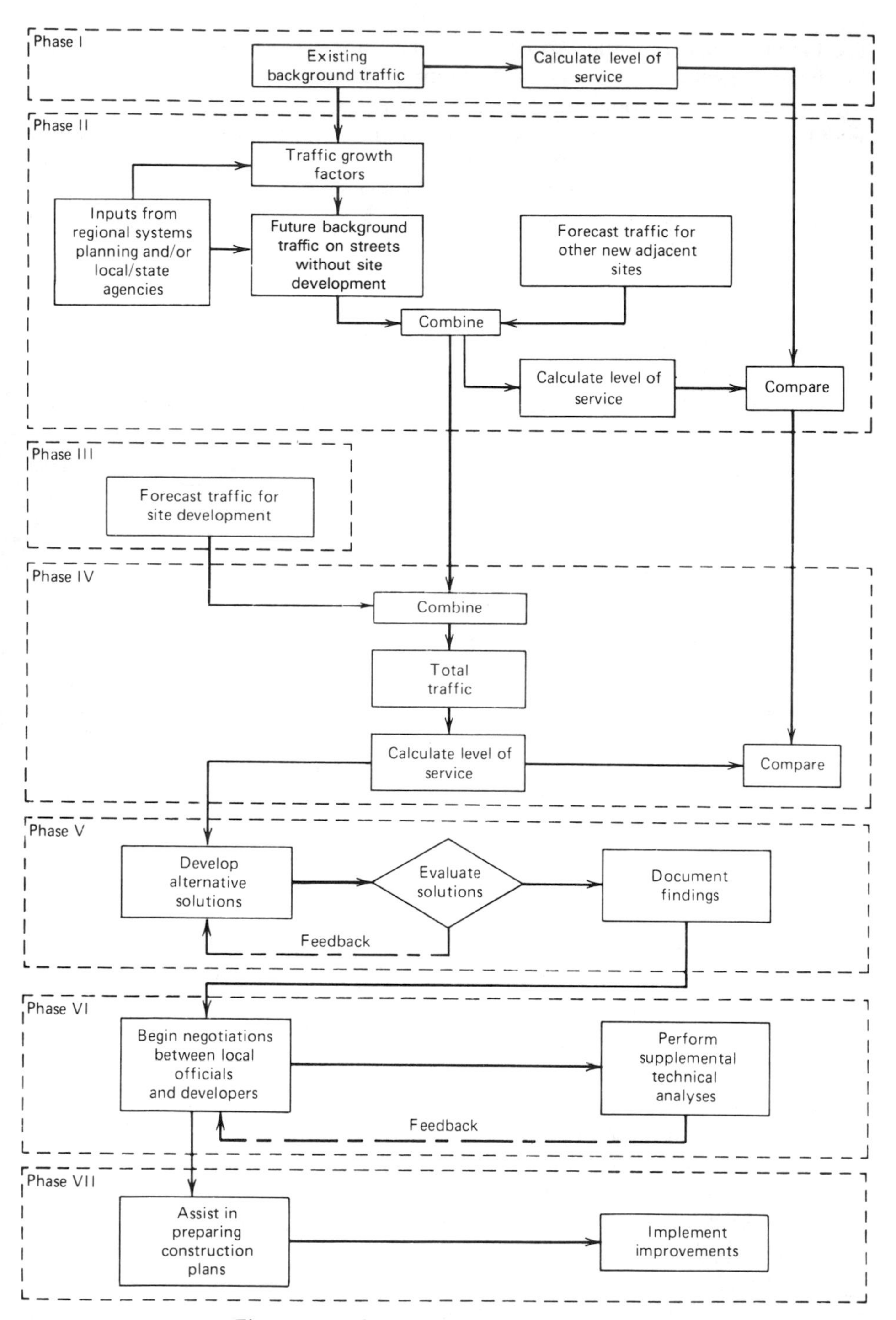


*Fig 11-1.* The site access study process.

improvements. They may, however, include major roadway improvements and the adoption of transportation systems management (TSM) measures to limit or reduce peak hour traffic. At this stage the findings are documented for review by local officials.

6. Negotiations are carried out between local officials and the developers. As discussions continue, further issues with traffic impact implications may be raised, requiring further technical analysis, for example, the introduction of the possibility of developer ride share programs, shuttle, or bus/van service, or the adoption of TSM measures which may not have been considered in the original analysis.
7. The final step is implementation, which occurs after agreement is reached between local officials and the developers. Implementation usually follows upon authorization in the form of a permit to construct the traffic-related improvements and to provide funds for all or part of those improvements. TSM actions such as ride sharing or the shuttle bus service may also be provided by the developer.

   Final site plans, road construction plans, and items such as installation plans for signalization are then prepared by the appropriate engineers and architects.

## 11-3. SCOPE OF DEVELOPMENT INVOLVED IN SITE STUDIES

Virtually any major traffic generator must be regarded as potentially suitable for traffic impact analysis. Generally small developments of low-density residential units can be excluded but the following types of development are prime candidates for this type of assessment [3]:

- High-density residential
- Office
- Retail/commercial
- Hotel
- Business park
- Hospital
- School
- Industrial
- Stadium

In recent years, local governments and regional planning agencies have put considerable thought into determining guidelines which define the size and types of projects which require traffic impact analysis. Although these guidelines vary from area to area they are in general remarkably similar. The

Atlanta Regional Commission has indicated that it will review all projects which exceed the following thresholds [3]:

- Office development: 500,000 net ft$^2$ or more
- Retail/commercial: 700,000 net ft$^2$ or more
- Hospital: 600 beds or more
- Residential: 1000 dwelling units or more
- Hotel: 1000 rooms or more
- Industrial: 500 acres or more
- Mixed use: 500,000 net ft$^2$ or more.

## 11-4. FORMAT OF SITE IMPACT STUDIES

There is no standard format of site impact studies. Each study tends to be site specific; furthermore each local governmental authority has its own particular requirements. However, extensive experience with impact studies has meant that the general format is also remarkably similar from jurisdiction to jurisdiction. The following guidelines were put out by the city of Boca Raton for the structuring impact reports for developments which will generate more than 1000 trips per day [3].

1. *Letter to accompany the final report.* This letter should:

   a. Identify the persons to whom the report is addressed.
   b. State principal findings and recommendations.

   The report itself should contain the following information:

2. *Proposed development and access routes*

   a. Zoning, type, shape, size, and area of development.
   b. Map of site in relation to street network.
   c. Functional classification of street network with average daily traffic (ADT) volumes.

3. *Existing traffic conditions*

   a. Number of lanes, amount of existing right-of-way of access routes.
   b. A.M. and P.M. peak hour lane counts at critical intersections.
   c. Queue lengths at controlled intersections which may affect project.

4. *Projected site traffic*

   a. Generation rates

      Total ADT and Peak hour traffic[1]

      The guidelines state in the form of footnotes that acceptable sources of trip generation data in order of preference are:
      i. Palm Beach County's Reports of 1972, 1975, and 1981.
      ii. Studies of similar sites within neighboring Broward and Palm Beach Counties—not more than three years old.
      iii. Trip Generation, An Informational Report, Institute of Transportation Engineers (ITE), Washington D.C., 1982 [4].

   b. Traffic distribution.
   c. Illustration of ADT and peak hour turning movements at each driveway.

5. *Traffic impact/capacity analysis*

   a. Capacity analysis of critical roadway links and intersections under present conditions. Level of service to be identified.
   b. Capacity analysis of critical road way links and intersections including driveways with added developmental traffic and approved future development.
   c. Identification of required improvements and analysis of effects.
   d. Summary of recommended improvements.

6. *City code compliance*

   a. Review of the development's driveway and off-street parking design for compliance with applicable city codes.
   b. Review of the on-site circulation.
   c. Review of the ingress/egress sight distances, capacity, and safety.

## 11-5. TECHNIQUES USED IN TRAFFIC ESTIMATION

The scale of traffic impact studies does not permit, or even necessitate, the use of the full-scale techniques of trip generation, trip distribution, modal split,

[1] Peak hour data will be calculated from the following formula:

$$\text{Palm Beach peak hour traffic} = \text{Palm Beach total ADT} \times \frac{\text{peak hour unit (ITE)}}{\text{total ADT unit (ITE)}}$$

and traffic assignment which were earlier described in the context of strategic planning in Chapter 9. Shortcut methods have therefore been devised which are considered to be sufficiently accurate for the intended purpose but which are known to be reasonably reliable. These methods are based on results drawn from the vast background of transportation planning experience gained from the many comprehensive studies carried out over the last 30 years. The following techniques vary from the crude to the sophisticated. The complexity of the method selected will depend on the nature and importance of the development being analyzed. Basic sites usually warrant only simple techniques; complex sites involve more elaborate reports and warrant the use of more elaborate procedures.

*Trip generation.* Rather than carry out extensive data-based trip generation analysis, the number of trips generated by a development can be estimated in a variety of ways. These include:

1. Determining trip generation rates for similar developments in the area for the same time of day and multiplying the rate per unit area on a *pro rata* basis.
2. Using trip generation rates from a similar area.
3. Referring to the trip generation rates published by the Institute of Transportation Engineers. These rates are derived from data collected from a large number of sites around the United States [4]. Similar data are available from Transportation Research Board documents [5]. Table 11-1 gives an example of the trip generation rates published by the ITE.
4. Using the techniques available on microcomputer software through the quick response traffic estimation program which is documented in Reference 6. The QRS system utilizes generation rates derived from the ITE data previously mentioned. It also has the facility to derive trip attractions from the following relationship:

$$A_{ik} = b_k E_i^r + c_k E_i^n + d_k H_i$$

where

$b_k, c_k, d_k$ = trip attraction parameters for purpose $k$. These are stored in the software and do not require calibration.
$E_i^r$ = retail employees in zone $i$
$E_i^n$ = nonretail employees in zone $i$
$H_i$ = dwelling units in zone $i$
$A_{ik}$ = attractions for purpose $k$ at zone $i$

Trip productions can also be estimated from the following equation:

$$P_{ik} = P_{it} W_{ik}$$

where

$P_{it}$ = total trips produced in zone $i$

$W_{ik}$ = percent of trips produced at zone $i$ for purpose $k$. This factor is dependant on income levels and auto ownership.

*Trip distribution* can also be achieved by methods of varying complexity. These include:

1. Determining an "area of influence" for the proposed development based on previous experience with similar developments and distributing traffic by some impedence based method.
2. Surveying zip codes for employees or users of similar facilities in the area or in other areas to determine likely major origins/destinations, or rates of decay of travel with distance.
3. Using available zonal accessibility indices as a measure of impedence for applying a gravity model. Such indices may be available from previous comprehensive studies in the same area or may be synthesized from the results of other areas.
4. Using the gravity model formulation available with the QRS microcomputer program software. Calibration of the model is avoided by estimating friction factors based on variables of city size, arterial network parameters, and trip purpose.

*Modal split* is best achieved by a close analysis of existing modal splits for similar developments in similarly located sites with similar transit service levels. In the absence of such available information, mathematical models can be used to forecast the probability a tripmaker will use transit. These models are usually defined in terms of a *disutility function* that accounts for the generalized costs that are associated with modal choice. Its magnitude depends on variables such as travel time, time spent waiting, transfers, and costs of fares and fuel. The QRS procedure can estimate modal split using the following relationship [6]:

$$p_{ijk}{}^{t} = S_i S_j \left( \frac{1 - Y_{ik}}{1 + \exp\left(-MD_{ijk}{}^{d}\right)} \right) + Y_{ik}$$

where

$P_{ijk}{}^{t}$ = probability that a person trip from $i$ to $j$ takes transit mode $k$

$S_i$ = sum of area splits for zone $i$

$S_j$ = sum of the area splits for zone $j$

$Y_{ik}$ = captivity for zone $i$ and for purpose $k$ (expressed as a fraction)

$M$ = mode split multiplier

$D_{ijk}{}^{d}$ = disutility difference

$$D_{ijk}{}^{d} = \frac{t_{ij} + t_{ji} - D_{ij} - D_{ji}}{2} - D_k^{c} - D_i^{p} - D_j^{a}$$

*Table 11-1*
**Example of ITE Trip Generation Rates**

SUMMARY OF TRIP GENERATION RATES

Land Use/Building Type General Office Building ITE Land Use Code 710
Independent Variable—Trips per 1,000 Gross Square Feet

| | | | *Average Trip Rate* | *Maximum Rate* | *Minimum Rate* | *Correlation Coefficient* | *Number of Studies* | *Average Size of Independent Variable/Study* |
|---|---|---|---|---|---|---|---|---|
| Average Weekday Vehicle Trip Ends | | | 12.30 | 43.5 | 3.6 | | 35 | 233 |
| Peak Hour of Adjacent Street Traffic | A.M. Between 7 and 9 | Enter | 1.86 | 3.3 | 1.3 | | 8 | 210 |
| | | Exit | 0.35 | 1.0 | 0.1 | | 8 | 210 |
| | | Total | 2.32 | 5.9 | 1.2 | | 25 | 150 |
| | P.M. Between 4 and 6 | Enter | 0.27 | 0.7 | 0.1 | | 4 | 125 |
| | | Exit | 1.36 | 2.6 | 0.7 | | 4 | 125 |
| | | Total | 2.20 | 6.4 | 0.8 | | 24 | 122 |
| Peak Hour of Generator | A.M. | Enter | 1.86 | 3.3 | 1.3 | | 8 | 210 |
| | | Exit | 0.35 | 1.0 | 0.1 | | 8 | 210 |
| | | Total | 2.32 | 5.9 | 1.2 | | 25 | 150 |

| | | | | | | | | |
|---|---|---|---|---|---|---|---|---|
| | P.M. | Enter | 0.27 | 0.7 | 0.1 | | 4 | 125 |
| | | Exit | 1.36 | 2.6 | 0.7 | | 4 | 125 |
| | | Total | 2.20 | 6.4 | 0.8 | | 24 | 122 |
| Saturday Vehicle Trip Ends | | | 3.34 | 8.9 | 1.0 | | 13 | 81 |
| Peak Hour of Generator | | Enter | 0.19 | 0.2 | 0.1 | | 2 | 88 |
| | | Exit | 0.16 | 0.2 | 0.1 | | 2 | 88 |
| | | Total | 0.50 | 2.2 | 0.2 | | 13 | 81 |
| | | | | | | | | |
| Sunday Vehicle Trip Ends | | | | | | | | |
| Peak Hour of Generator | | Enter | | | | | | |
| | | Exit | | | | | | |
| | | Total | | | | | | |
| | | | | | | | | |
| Source Numbers 2, 5, 19, 20, 21, 51, 53, 54, 72, 88, 89, 92, 95, 97 | | | | | | | | |
| ITE Technical Committee 6A-6—Trip Generation Rates<br>Date: 1975, Rev. 1979 | | | | | | | | |

SOURCE: Reference 4.

where

$t_{ij}$, $t_{ji}$ = auto travel time from $i$ to $j$ and $j$ to $i$, respectively
$D_{ij}$ = weighted average disutility from zone $i$ to zone $j$ by transit
$D_{ji}$ = weighted average disutility from zone $j$ to zone $i$ by transit
$D_k^c$ = transit specific constant for purpose $k$
$D_i^p$ = production extra disutility for zone $i$
$D_j^a$ = attraction extra disutility for zone $j$

The values to the various parameters are available in the software.

A simpler modal split formulation is available through the UTPS program UMODEL. This is of the form:

$$T_{ijm} = P_i \frac{A_j e^{-\theta I_{ijm}}}{\sum_j \sum_m A_j e^{-\theta I_{ijm}}}$$

where

$T_{ijm}$ = trips from $i$ to $j$ by mode $m$
$\theta$ = calibrated constant varying by trip purpose
$I_{ijm}$ = measure of impedence by mode $m$
= (1.0 × in-vehicle time) + (2.5 × out-of-vehicle journey time) + (trip cost) / (0.33 × income/min)

*Auto occupancy* can also be modeled from relationships developed in the QRTE handbook. Average automobile occupancy rates can be estimated with the following inputs [5]:

1. Time of day.
2. Trip length.
3. Income level.
4. Parking cost.
5. Size of urban area.

*Peak to daily trip ratio* can also be estimated directly from the same reference, based on [5]:

1. Urbanized population.
2. Trip purpose.
3. Balance between internal drivers and total traffic in a corridor.
4. Orientation of the route (radial or peripheral to the center of the urban area).

*Traffic assignment* can be carried out in a number of ways. The particular method chosen must reflect the complexity of the problem to be solved.

1. An all-or-nothing assignment can be used to add the trip distribution of the new generator to the volumes of an existing assignment.
2. Gruen's method can be used to estimate through corridors or sectors, traffic volumes caused by special land use generators. The procedure starts with either an assignment, a traffic map, or an estimate of the suburban traffic situation. Primary estimates are adjusted for:

   Density and project size
   Level of service
   Auto ownership
   Transit usage
   Projected nonresidential/residential mix
   Freeway diversion

   The orientation of trips is determined and the rate of decay with distance from the generator is estimated.

**Level of service (LOS) sensitivity analysis**

Traffic level of service is usually determined using the Transportation Research Board's classification of six categories A through F, see Section 4-7. There are no universally recognized criteria for design. The FHWA indicates that minimum level of service D is used for design in urban areas and level C for rural areas [1]. The planning analysis method uses three descriptive categories: below capacity (levels of service A, B, or C), near capacity (levels of service D and E), and above capacity (level of service F). Obviously local and state ordinances, where they prescribe design levels, must be used to relate the level of service to design and to improvement needs.

In carrying out any analysis of the effect of development on levels of service, the analyst must bear in mind that the procedures involved are imprecise. It is therefore essential that some form of sensitivity analysis be conducted to evaluate just how sensitive level of service changes are to changes in assumptions in the various stages of the analytical procedure. Assumptions must be made at each stage of the analysis. Typically, the sensitivity analysis could take the form of examining the effect of changes in assumptions of 10 and 20 percent at each stage. In cases where the projected level of service is found to be particularly sensitive to the estimated traffic impact, there is a need to attempt to refine the procedures of estimation. It may also be necessary to change where possible the recommendations for operational and design modifications to be less sensitive to variations in estimation.

Another widely used method of testing the sensitivity of the level of service to assumptions is to vary the underlying assumptions of trip generation, trip distribution, and background growth rates, and to examine the effect on levels of traffic service. Typically this can be done in the following way:

| | |
|---|---|
| Generation: | Use ITE rates/NCHRP 187 rates/rates from similar developments within the region. |
| Distribution: | Use QRS gravity model formulation/zonal accessibility indices. |
| Background growth factor: | Use local/regional/national estimates. |

Variation of the technique used for modal split does not usually bring about major changes in the level of service. The general principles involved in traffic impact analysis are illustrated in the following example.

## EXAMPLE 11-1

It was proposed to develop a previously unused site as a new office building of 100,000 gross ft$^2$. Access to the site was from a two-lane minor collector which itself led to a four-lane minor arterial and a four-lane major collector. The major collector itself connected with a six-lane major arterial. The general layout of the proposal is shown in Fig. 11-2. Development was to be complete in two years. The site traffic impact analysis was carried out in the following way:

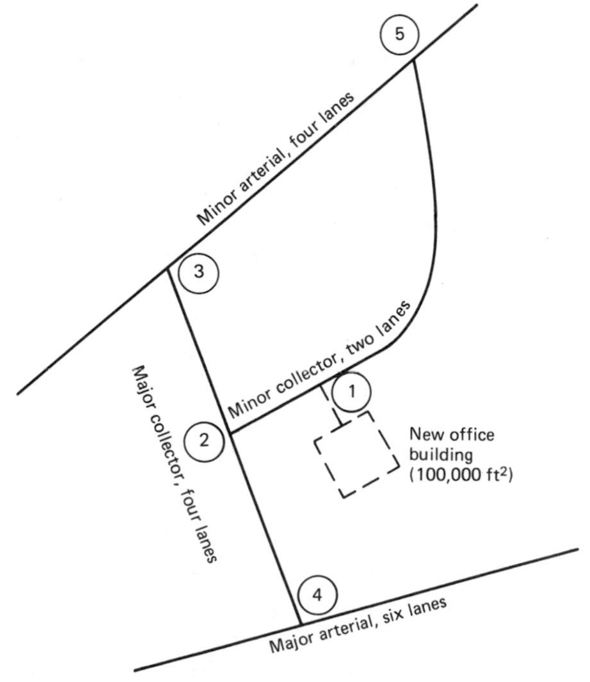


*Fig 11-2.* General layout of the access to the proposed new office building.

***Phase I:***
***Establish study design and verify existing peak hour traffic situation***

1. Obtain development plans.
2. Confirm scope and program of site development.
3. Examine existing traffic situation. The background traffic situation was examined and it was found that the critical peak hour for existing traffic occurred in the P.M. peak period. As the P.M. peak would also be more critical for the office-site-generated traffic, analysis could be restricted to peak hour conditions. Critical lane volumes for the current background traffic situation are shown in Fig. 11-3.
4. Analysis was limited to intersections 1 to 5 in Fig. 11-2.

***Phase II:***
***Project future peak hour traffic without site development***

This phase involved establishing a growth rate or rates for the analysis. Discussions with the FHWA and the State DOT officials indicated that a statewide growth rate of traffic had averaged 6 percent in similar urban areas over the last three years. However, local data on nearby arterials indicated a higher growth rate, 10 percent, for a similar period. It was decided therefore to carry out the analysis for two growth rates: 6 percent and 10 percent. Figures 11-4*a* and 11-4*b* show the projected peak hour

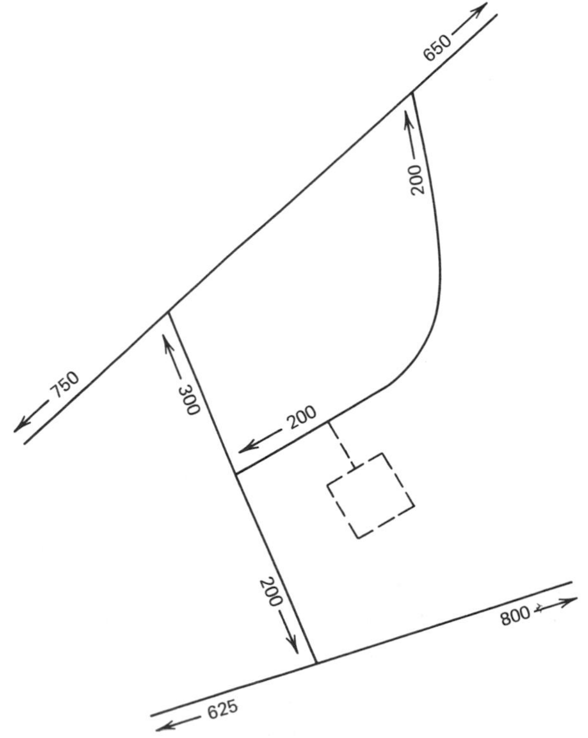


*Fig 11-3.* Background traffic—current year.

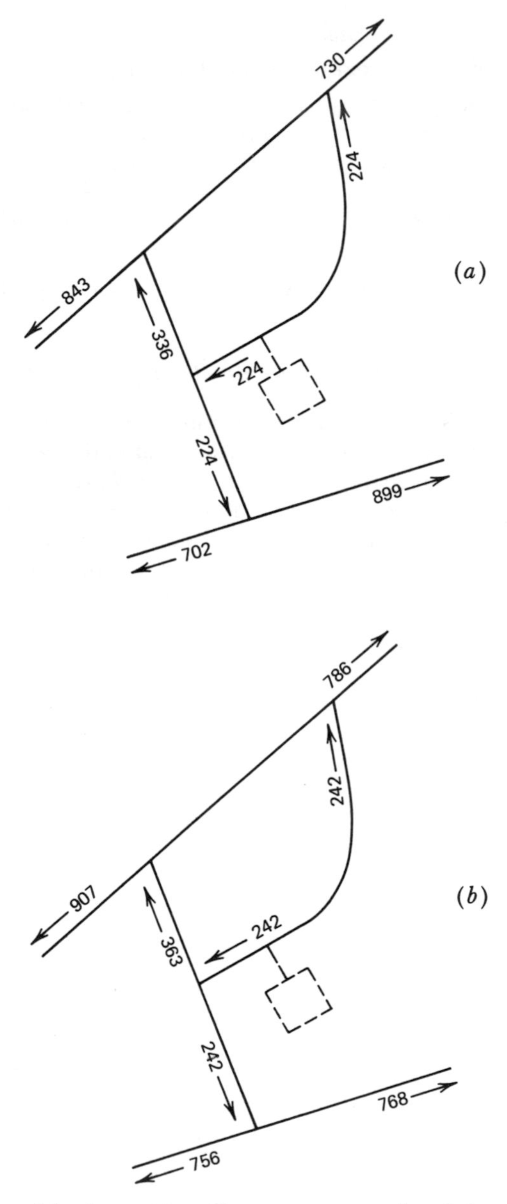


*Fig 11-4.* (*a*) Projected backgound traffic—two years ahead 6 percent growth per annum. (*b*) Projected background traffic—two years ahead 10 percent growth per annum.

background traffic in critical lanes at the end of the site development period (i.e. two years ahead), without the site development traffic added in.

### *Phase III:*
### *Project and distribute peak hour site development traffic*

Site development traffic can be estimated in a number of ways. In this case there were no available local data which were sufficiently current or similar in nature to be directly usable. Estimates of traffic generation were therefore made from nationally published sources: the Institute of Transportation Engineers [4] and the National Cooperative Highway Research Program [7]. The appropriate peak hour trip generation rates for gross office space are shown in Table 11-2.

Two different trip distribution patterns were used. Case A utilized the QRS microcomputer software; case B was a "hand simulation" using zonal accessibility indices available from a recent transportation study. These two distributions were then assigned on an all-or-nothing basis to the network giving the percentages of generated traffic indicated in Figs. 11-5*a* and 11-5*b*.

### *Phase IV:*
### *Project future peak hour traffic with site developed*

The future peak hour traffic was projected based on a number of assumptions. The various assumptions were used in a number of combinations in order to give the analyst some feel for the sensitivity of the solution to the particular assumption. The volumes for critical lanes were then compared with level of service (LOS) volume ranges to determine the effect of development on service conditions. Level of service ranges are shown in Table 11-3 and the results of the analysis for the five intersections are tabulated in Table 11-4.

The various conditions shown in this table are as follows:

| *Condition* | *Assumptions* |
|---|---|
| 1. | Background traffic, current status. |
| 2. | Background traffic, no development, 6 percent growth rate, two years ahead. |
| 3. | Background traffic, no development, 10 percent growth rate, two years ahead. |
| 4. | Site developed, ITE generation rates, distribution pattern A, 6 percent growth rate. |
| 5. | Site developed, ITE generation rates, distribution pattern A, 10 percent growth rate. |
| 6. | Site developed, ITE generation rates, distribution pattern B, 6 percent growth rate. |
| 7. | Site developed, ITE generation rates, distribution pattern B, 10 percent growth rate. |

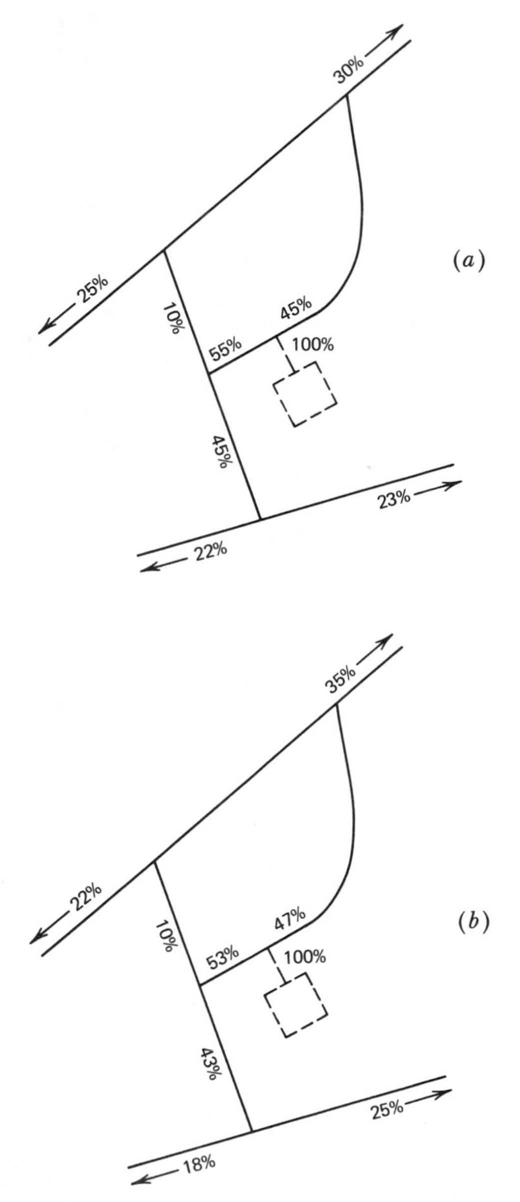


*Fig 11-5.* (*a*) Trip distribution using QRS method. (*b*) Trip distribution using hand simulation with accessibility indices.

*Table 11-2*
**Trip Generation Rates for Traffic Projection**

| *Hourly Trip Generation Rates per 1000 Gross ft² of Office Space* | | |
|---|---|---|
| | ITE[a] | NCHRP[b] |
| A.M. peak | 2.50 | 1.98 |
| P.M. peak | 2.82 | 1.93 |

[a]Reference 4.
[b]Reference 7.

*Table 11-3*
**Level of Service *vs* Volume Ranges, Criteria for Traffic volumes on Critical Lane**

| *Level of Service* | *Volume* |
|---|---|
| A | 0– 855 |
| B | 855–1000 |
| C | 1000–1140 |
| D | 1140–1275 |
| E | 1275–1425 |
| F | Over 1425 |

*Table 11-4*
**Comparison of Critical Lane Volumes and Levels of Service (LOS) Under Current and Future Conditions**

| | *Condition* | | | | | | | | |
|---|---|---|---|---|---|---|---|---|---|
| *Intersection* | *1* | *2* | *3* | *4* | *5* | *6* | *7* | *8* | *9* |
| 1 | 200<br>LOS A | 224<br>LOS A | 242<br>LOS A | 1024<br>LOS C | 1042<br>LOS C | 934<br>LOS B | 954<br>LOS B | 724<br>LOS A | 724<br>LOS A |
| 2 | 200<br>LOS A | 224<br>LOS A | 242<br>LOS A | 541<br>LOS A | 569<br>LOS A | 431<br>LOS A | 450<br>LOS A | 440<br>LOS A | 451<br>LOS A |
| 3 | 750<br>LOS A | 843<br>LOS A | 907<br>LOS B | 918<br>LOS B | 1082<br>LOS C | 998<br>LOS B | 1062<br>LOS C | 964<br>LOS B | 949<br>LOS B |
| 4 | 800<br>LOS A | 900<br>LOS B | 968<br>LOS B | 1008<br>LOS C | 1076<br>LOS C | 1020<br>LOS C | 1088<br>LOS C | 975<br>LOS B | 1001<br>LOS C |
| 5 | 650<br>LOS A | 730<br>LOS A | 786<br>LOS A | 942<br>LOS B | 998<br>LOS B | 979<br>LOS B | 1035<br>LOS C | 875<br>LOS B | 899<br>LOS B |

8. Site developed, NCHRP generation rates, distribution pattern A, 6 percent growth rate.
9. Site developed, NCHRP generation rates, distribution pattern B, 10 percent growth rate.

It will be seen that all intersections opened at level of service A in the current year and only two intersections would fall to level of service B under 10 percent background traffic growth conditions. Under conditions 5 and 7 three intersections would drop to level of service C.

### *Phase V*

Based on calculated volumes, and with a view to recognizing the sensitivity of the solution to assumptions the transport planner would discuss medium-term traffic engineering and TSM measures to ensure the best possible service level with the anticipated development.

## REFERENCES

1. *Site Impact Traffic Evaluation (S.I.T.E) Handbook,* Federal Highway Administration, U.S. Department of Transportation, Washington, D.C., January 1985.
2. "Guidelines for Transportation Impact Assessment of Proposed New Development," *Site Impact Traffic Evaluation,* PD-014, Institute of Transportation Engineers, Washington, D.C., 1986.
3. Voorhees, Kenneth O., "Guidelines on Producing a Traffic Impact Study for Urban Developments," in *Site Impact Traffic Evaluation,* Educational Foundation on Seminars, Institute of Transportation Engineers, Washington, D.C., 1986.
4. *Trip Generation, An Informational Report,* Institute of Transportation Engineers, Washington, D.C., 1982.
5. "Quick Response Urban Travel Estimation Techniques," *National Co-operative Highway Research Program Report 187,* Transportation Research Board, Washington, D.C., 1978.
6. Horowitz, Alan J., *Quick Response System II Reference Manual,* Federal Highway Administration, U.S. Department of Transportation, Washington, D.C., June 15, 1987.
7. *Development and Application of Trip Generation Rates HHP-22,* Department of Transportation, Federal Highway Administration, Washington, D.C., January 1985.

# *PART 4*

# *DESIGN OF LAND TRANSPORTATION FACILITIES*

*The remainder of the book deals with one of the most important aspects of transportation: the design of the physical elements of a transportation system. It describes design criteria, engineering standards, and creative procedures necessary to translate conceptual transportation plans into detailed engineering drawings and specifications. The five chapters that comprise Part Four describe the design of land transportation facilities (roadways, railways, transit guideway systems, pipelines, and terminals).*

# *Chapter 12*

# *Design of Roadways, Railways, and Guideway Systems: Location and Route Layout*

This chapter focuses on the location and route layout techniques for streets and highways, railways, and transit guideway systems.

## 12-1. OVERVIEW OF THE PLANNING AND DESIGN PROCESS

The planning and design process usually consists of several sequential steps or phases. Consider, for example, the planning and design of a roadway or a fixed transit link. If we follow the planning process described previously, the need for a new route is established and the basic features of the link are determined (e.g. termini, general location and size or class, level of service, etc).

The next step is to establish the location of the facility. The location normally is selected from among several alternatives situated within a corridor connecting the termini. After a preliminary location has been chosen, the facility may be designed. This involves specifying the precise horizontal and vertical alignment, dimensions, slopes, construction standards, and the quantities and types of materials.

## 12-2. LOCATION OF TRANSPORTATION LINKS

The location of highways, railways, and other land transport links is an extremely important part of the design process. Decisions made during the location selection process not only determine the cost and operational efficiency of the facility but also influence the disbenefits to or negative impact on nearby communities and the environment. In recognition of the importance of this task, many transportation agencies rely on teams of professionals to perform location studies. These professionals usually include engineers, planners, economists, and social, environmental, and ecological specialists.

The location procedure is an iterative one in which an approximate location selected on the basis of available maps, charts, photographs, and data gradually is narrowed and finally fixed as additional information and data become available and more detailed study is done.

Information for the location study comes in large part from two or more surveys which provide information on the topography, soil conditions, water courses, property lines, utilities, land uses, and so on. Such information is gathered by conventional ground survey methods or by remote sensing techniques, most commonly aerial photogrammetry. While the older techniques still are used on projects of medium and small size, the use of photogrammetric surveys on large works is almost universal and is becoming more frequent on smaller projects. The adaptability of photogrammetry to computer operations has enabled substantial savings in time and money over conventional methods.

By means of aerial or ground surveys, topographic maps are prepared which serve as a basis for the selection of a preliminary and final location. For the preliminary location selection process, a base map is used that typically has a scale of 1 in.=200 ft and a 2- or 5-ft contour interval. By a study of the map and a stereoscopic examination of aerial photographs, the location team can choose and evaluate possible routes on the basis of traffic service, directness, suitability of terrain, adequacy of crossings of streams and other transport routes, and extent of adverse social, environmental, and ecological effects.

Possible alternative alignments usually are plotted on the base map and, from these alignments with preliminary gradelines, the alternatives are compared. Examples of criteria used in selecting a location are shown in Table 12-1.

For the final design and location, additional mapping is usually necessary. The selected preliminary alignment is used as a guide for the strip area to be mapped. It is common practice, especially for large projects, to utilize low-level aerial photogrammetry to produce base maps along the preliminary alignment, typically at a scale of 1 in. = 100 ft and with a 1- or 2-ft contour interval.

A coordinate system is used to indicate the location of key points along an alignment. Usually, horizontal and vertical controls are established precisely along a baseline. Distances between baseline control points often are determined by electronic measuring devices such as the Electrotape or Geodimeter, which have excellent accuracy over long distances, ranging from a few hundred feet to 20 or 30 mi. Such devices measure distances by precisely measuring the time required for electromagnetic waves to travel the distance being measured.

The process of the design of the final alignment requires great skill and judgment. The alignment is fitted by hand to the topography and land use

*Table 12-1*
**Examples of Criteria To Be Used in Facility Location Decisions**

| *Criteria* | *Influencing Factors* |
|---|---|
| Construction costs | Functional classification/design type |
| | Topography and soil conditions |
| | Current land use |
| User costs | Traffic volume |
| | Facility design features (e.g., gradients, intersections) |
| | Operating conditions (e.g., speeds, traffic control systems) |
| Environmental impacts | Proximity to sensitive areas |
| | Design features to mitigate impacts |
| Social impacts | Isolation or division of neighborhoods |
| | Aesthetics of design |
| | Fostering of desired development patterns |
| Acceptance by various interest groups | Government agencies |
| | Private associations and firms |
| | Neighborhood groups and the general public |

until the designer is satisfied that no better fitting can be achieved. Most designers use specially scaled templates and a flexible plastic guide called a spline. Once established, the hand-fitted alignment is converted into a defined line by precisely computing tangent and curve lengths, transition spirals, and the location of control points for the alignment.

## 12-3. A ROAD LOCATION EXAMPLE

A location study for a 5.7 mi four-lane controlled access parkway near Atlanta illustrates the wide variety of factors that must be considered in the location selection process. The purposes of the proposed highway were to improve the accessibility to and from eastern Douglas County and south central Fulton County and to link Interstate Route I-20 west of Atlanta with Route I-85 near the Hartsfield/Atlanta International Airport.

Originally, two alternate locations were considered, designated *A* and *B* on Fig. 12-1. After a public hearing, alternate *D* was developed and evaluated along with the other two alternates and *C*, the no-build alternate. Nine federal, state, and local agencies were asked to comment on the project.

Alternate *B* was chosen as the preferred location. Although this alternate was the most costly, it caused no families to be relocated and caused no

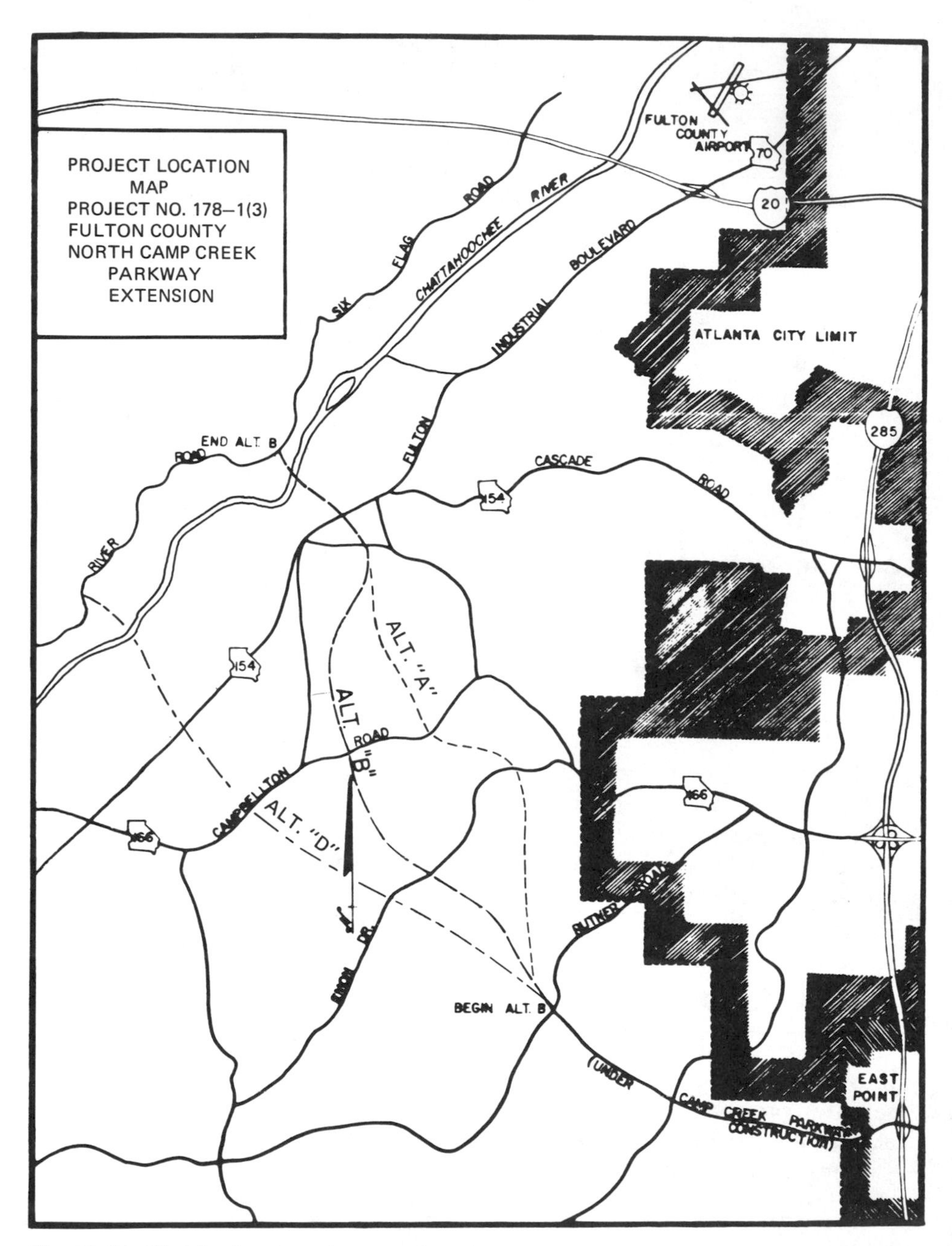


*Fig 12-1.* Sketch showing alternate locations in a road location example. (Courtesy Georgia Department of Transportation.)

significant impact due to noise. Like alternate *A*, it crossed the Chattahoochee river just southwest of Buzzard Roost Island, at one of the narrowest widths of the river's flood plain. Table 12-2 summarizes the impacts for each of the alternate locations. It will be noted that many of the impacts were common for the three alternate locations.

## 12-4. GEOMETRIC DESIGN

The essential design features of a roadway or guideway are its location and its cross section. In the horizontal plane, the location of points are referenced to a coordinate system in which the positive $y$-axis is north and the positive $x$-axis is east. Positions along the $y$-axis are called latitudes and those along the $x$-axis are called departures. Customarily, points along the route are identified by *stations,* the distance in feet from some reference point,[1] commonly the beginning point for the project. The location of points in the vertical plane (or along the $z$-axis) is given as the elevation above mean sea level.

The cross section of a roadway or guideway is described by its dimensions at a right angle to the direction of the alignment, including widths, clearances, slopes, and so on. Typical dimensions of roadway and guideway cross sections are given in Chapter 13. The features of design are directly affected by traffic volume and speeds, as well as by vehicle weights and dimensions. Many of the geometric design standards and procedures for railroads were developed several decades ago to meet the needs of an expanding railroad system. Most of these criteria, which were based on fundamental concepts of force and motion, have been refined and improved by empirical studies and experience. Studies in the dynamics of highway movements by planning surveys, along with other factual data pertaining to traffic, form the basis of the newer geometric standards in highway design. Research and experience are also reflected in the highway design standards currently in use.

## 12-5. DESIGN CONTROLS AND CRITERIA FOR STREETS AND HIGHWAYS

The elements of highway design are influenced by a wide variety of design controls and criteria. Such factors include:

1. Functional classification of the roadway being designed.
2. Traffic volume and composition.
3. Design speed.
4. Topography.

[1]For example, station 42 + 00.00 refers to a point 4200.00 ft from the reference point.

*Table 12-2*

**A Summary of Impacts for Example Location Study**

| *Type of Impact/Condition* | *Alternate* A | *Alternate* B | *Alternate* D |
|---|---|---|---|
| Length | 5.8 mi | 5.7 mi | 5.3 mi |
| Average daily traffic, 2001 | ← | 26,500-33,800 | → |
| Economic impacts | | | |
| Expenditure of public funds | $7,498,000 | $8,303,000 | $7,327,000 |
| Annual loss in tax revenues | $4,778 | $4,778 | $4,778 |
| Relocations | 5 residences | 0 | 1 abandoned home |
| Social impacts | Bisects residential area | Minor disruption in interaction | Minor disruption in interaction |
| Historical and archeological | 4 sites | 7 sites | 11 sites |
| Air quality | | | |
| Carbon monoxide | 2.7–4.1 ppm | 2.7–4.1 ppm | 2.7–4.1 ppm |
| Hydrocarbons | Overall decrease in production of HC over no-build, all alternates | | |
| Noise | 5 residences (>70 dBA) | 0 | 2 residences (>70 dBA) |
| Geology and soils | None | None | None |
| Hydrology | ← Increase in impervious ground cover → | | |
| Life systems | Loss of naturally occurring vegetation; restriction of movement of land animals; no rare, threatened, or endangered species | | |
| Construction impacts | Increase in daytime noise levels; increase in dust; slight erosion; rearrangement in utilities | | |
| Transportation impacts | Relief of traffic loads on smaller roads; provision of more direct route from airport and I-285 to Fulton Industrial Blvd. complex | | |

SOURCE: *Draft Environmental Statement, Administrative Action for The Camp Creek Parkway Extension,* Georgia Department of Transportation.

5. Cost and available funds.
6. Human sensory capacities of drivers, bikers, and pedestrians.
7. Size and performance characteristics of the vehicles that will use the facility.
8. Safety considerations.
9. Social and environmental concerns.

These considerations are not, of course, independent. The functional class of a proposed highway is largely determined by the volume and composition of the traffic to be served. It may also depend on availability of funds and social and environmental considerations. For a given class of highway, the choice of design speed is governed primarily by topography, which in turn is a determinant of facility cost. Once a design speed is chosen, many of the elements of design may be established on the basis of fundamental sensory capabilities of drivers and other users and on the performance characteristics of vehicles.

The design features of a highway influence its capacity and efficiency, its safety performance, and its social acceptability to highway users, owners of abutting property, and the general public.

The principal design criteria for highways are traffic volume, design speed, the physical characteristics of the vehicles, and the proportions of vehicles of various sizes which use the highways. These criteria are discussed in more detail in the following paragraphs.

## 12-6. THE RELATIONSHIP OF TRAFFIC TO HIGHWAY DESIGN [1]

The major traffic elements that influence highway design are average daily traffic (ADT), design hour volume (DHV), directional distribution ($D$), percentage of trucks ($T$), and design speed ($V$).

As Section 8-14 describes, traffic volume studies normally involve the measurement or estimation of the average daily traffic. Knowledge of ADT is important for many purposes, but it is not appropriate to use in the geometric design of highways because it does not account for the variation in traffic that occurs in the various seasons of the year, days of the week, and hours of the day. Highway engineers commonly use hourly traffic volumes for purposes of design. Considerable thought has been given to determining which hourly traffic volume would be most appropriate for design. If the peak hour traffic volume of the year were used, the highway would be grossly overdesigned most of the time, and public funds would be wasted. On the other hand, if the average hourly traffic volume were selected for design, the facility would be inadequate, and intolerable congestion would occur many hours of the year.

In selecting an appropriate design hourly traffic volume, it is helpful to consider a curve such as Fig. 12-2 in which the hourly volumes during the year

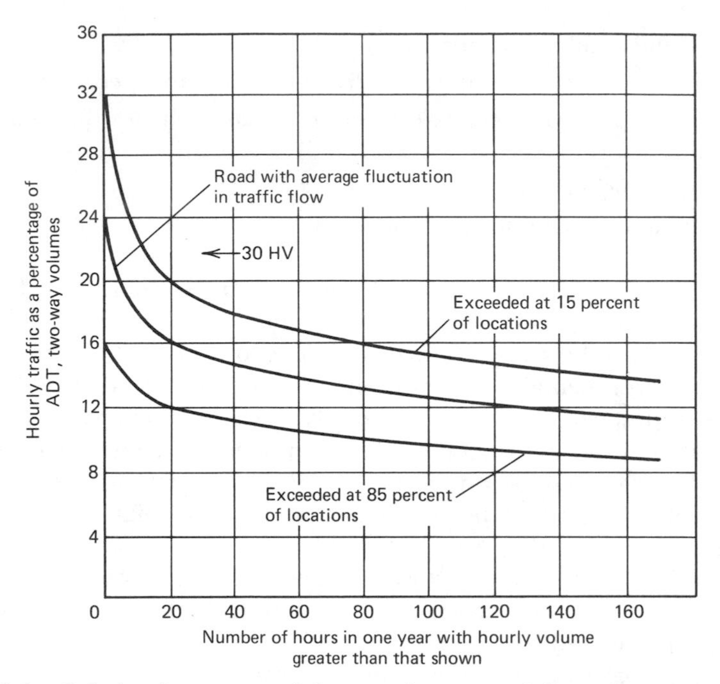


*Fig 12-2.* Relation between peak-hour and average daily traffic volumes on rural arterials. (Source: *A Policy on Geometric Design of Highways and Streets,* American Association of State Highway and Transportation Officials, 1984.)

are arranged in descending order of magnitude. In this illustration, which is typical of rural highway traffic, it will be observed that the maximum hourly volume is 25 percent of the ADT. If the traffic were uniformly distributed, the maximum hourly volume would be only one-twenty-fourth or about 4 percent of the ADT. Note that the "knee" of the curve occurs at about the thirtieth highest hourly volume. To the left of this point, the slope of the curve increases sharply; to the left, the curve flattens gradually. The volume corresponding to this point is commonly used as the design hour volume. It represents a compromise which allows for some degree of congestion during a few hours during the year. Thus, if the thirtieth highest hour is used as the design hourly volume, the designer is willing to tolerate unfavorable operating conditions 29 hours during the year which is only about 0.3 percent of the total hours during the year.

Studies have shown that the thirtieth highest hour of traffic expressed as a percentage of ADT does not vary much from year to year, in spite of significant changes in ADT. Thus, it is possible to measure and forecast volumes in terms of ADT and to multiply the ADT by a representative percentage to arrive at a design hourly volume. This percentage $K$ is typically 8 to 12 percent for urban highways, 10 to 15 percent for suburban highways, and 12 to 18 percent for rural facilities.

Directional distribution of traffic is also necessary for geometric design. The directional distribution, *D,* is the one-way volume in the predominant direction of travel, expressed as a percentage of the two-way design hourly volume. This may average about 67 percent in rural areas and ranges from about 55 near city centers to as much as 70 percent in suburban areas.

Composition of traffic (*T*) is usually expressed as the percentage of trucks (exclusive of light delivery trucks) during the design hour. That percentage typically varies from about 5 to 10 percent. In urban areas, the percentage of trucks during peak hours tends to be considerably less than percentages on the daily basis.

It will be recalled from Chapter 8 that the traffic volumes that can be served at each level of service are termed "service volumes." Once a level of service has been chosen for design, the corresponding service volume logically becomes the design service volume. This implies that if the traffic volume using the facility exceeds that value, the operating conditions will be inferior to the level of service for which the roadway was designed. The level of service appropriate for the design of various types of highways is shown by Table 12-3.

Table 8-2 generally describes the operating characteristics of various types of highways operating at each level of service. Such information is useful in determining the number of lanes required to serve the estimated design year traffic. A more extensive capacity analysis, based on data and procedures given in Reference 2, may be required to identify potential bottlenecks and to facilitate the design of intersections, ramp terminals, weaving sections, and the like.

*Assumed Design Speed.* The assumed design speed for a highway, according to the American Association of State Highway and Transportation Officials, may be considered as "the maximum safe speed that can be maintained over a specified section of highway when conditions are so favorable that the design features of the highway govern."

The main factor that affects the choice of a design speed is the character of the terrain through which the highway or street is to pass. Thus, a 40-mph design speed for a state two-lane arterial might be tolerated in mountainous terrain, while a 70-mph design speed might be specified in flat terrain. Choice of design speed should also be logical with respect to the type of highway being designed. In contrast to high-speed arterials, local residential streets should be designed so as to discourage excessive speeds. A design speed as low as 30 mph, therefore, is recommended for local streets in ordinary terrain, while 20 mph is specified for hilly terrain. Other considerations determining the selection of an assumed design speed are economic factors based on traffic volume, traffic characteristics, costs of rights-of-way, and other factors that may be of an aesthetic nature. Consideration must also be given to the speed capacity of the motor vehicle. Approved design speeds, as adopted by the American Association of State Highway and Transportation Officials, are 20, 30, 40, 50, 60, and 70 mph.

*Table 12-3*
**Guide for Selection of Design Levels of Service**

| | Type of Area and Appropriate Level of Service[a] | | | |
|---|---|---|---|---|
| *Highway Type* | *Rural Level* | *Rural Rolling* | *Rural Mountainous* | *Urban and Suburban* |
| Freeway | B | B | B | C |
| Arterial | B | B | C | C |
| Collector | C | C | D | D |
| Local | D | D | D | D |

[a]A, free flow, with low volumes and high speeds; B, stable flow, but speeds beginning to be restricted by traffic conditions; C, in stable flow zone, but most drivers restricted in freedom to select their own speed; D, approaching unstable flow, drivers have little freedom to maneuver; E, unstable flow, may be short stoppages.

SOURCE: *A Policy on Geometric Design of Highways and Streets.* American Association of State Highway and Transportation Officials, Washington, D.C., 1984.

The design speed chosen is not necessarily the speed attained when the facility is constructed. Speed of operation will be dependent upon the group characteristics of the drivers under prevailing traffic conditions. Running speeds on a given highway will vary during the day depending upon the traffic volume. The relationship between the design speed and the running speed will also vary.

*Design Designation.* The design designation indicates the major controls for which a highway is designed. As already discussed, these include in a broad sense traffic volume, character or composition of traffic, and design speed.

Where access is fully or partially controlled, this should be shown in the design designation; otherwise, no control of access is assumed.

An example of such designations is given below; the tabulation on the left is for a two-lane highway; the tabulation on the right is for a multilane highway.

| Two-lane | | Multilane | |
|---|---|---|---|
| | | Control of access = | full |
| ADT (1988) = | 2,500 | ADT (1988) = | 10,200 |
| ADT (2008) = | 5,200 | ADT (2008) = | 22,000 |
| DHV = | 720 | DHV = | 2,950 |
| $D$ = | 65 percent | $D$ = | 60 percent |
| $T$ = | 12 percent | $T$ = | 8 percent |
| $V$ = | 60 mph | $V$ = | 70 mph |

*Vehicle Design.* The dimensions of the motor vehicle also influence design practice. The width of the vehicle naturally affects the width of the traffic lane; length has a bearing on roadway capacity and affects the turning radius; the height of the vehicle affects the clearance of the various structures. Weight affects the structural design of the roadway. Vehicle weights, dimensions, and operating characteristics are given in Chapter 4.

## 12-7. THE NATURE OF RAILROAD AND TRANSIT GUIDEWAY TRACK DESIGN

In the United States, the major system of railways has been completed, for all practical purposes. With the exception of new passenger rail lines such as those planned or being constructed in Atlanta and Miami, the construction of extensive mileage of rail trackage is no longer needed.

In other parts of the world, railroad development continues, including high-speed passenger lines providing speeds up to 270 km/hr (168 mph). A notable example of this type of development is the Train à Grande Vitesse (TGV) or high-speed train which has been constructed to improve railway service from Paris to southeastern France (see Fig. 12-3.) Similar systems have been built in Japan, and studies have been conducted in the United States to examine the feasibility of high-speed rail passenger lines in several high-density corridors.

With the freight-hauling railroads, the emphasis is on such activities as the construction of spur tracks to new industrial sites and newly developed mines and power plants. As railroad companies merge, short connecting lines are being constructed to improve movements over the new system. Considerable emphasis is also being placed on improving geometric design features and the riding quality of existing tracks.

In recent years, many financially troubled railroad companies have found it necessary to defer needed track maintenance. These companies, with the encouragement and support of the federal government, are searching for ways to upgrade the tracks in order to make them suitable for heavier loads and higher speeds. The rapid increase in axle loads, especially in the United States where 125-ton cars are being introduced into service, has led to excessive rail wear and an increase in rail failures. Derailments have increased sharply, many caused by track-related deficiencies such as defective rails and errors in track geometry. During the next decade, the reconstruction and renovation of tracks along vital railway links will be a major engineering challenge.

As with highways, the design of new rail lines is influenced strongly by such factors as topography and manmade developments. The nature of the traffic, whether it be freight or passenger, is of great importance. The factors of train operating speeds, freight tonnages, and type of rolling stock also significantly control the elements of geometric design for new railroad lines.

*Fig 12-3.* The TGV (Train à Grande Vitesse) provides high-speed passenger service in central France. The line extends from Paris to Lyon, a distance of 425 km, serving trains that reach speeds of 270 km/hr (168 mph). Each trainset consists of two electrically driven power units and eight articulated trailers with a first class seating capacity of 111 and a second class capacity of 275. The train can be coupled in multiple trainset units. (Courtesy Societe National des Chemins de fer Francais (SNCF), Centre Audio Visuel (CAV), Photo by Jean-Marc Fabbro.)

## 12-8. ALIGNMENT

An ideal and most interesting roadway is one that generally follows the existing natural topography of the country. This is the most economical to construct, but there are certain aspects of design that must be adhered to which may prevent the designer from following this undulating surface without making certain adjustments in a vertical and horizontal direction.

The designer must produce an alignment in which conditions are consistent. Sudden changes in alignment should be avoided as much as possible. For example, long tangents should be connected with long sweeping curves, and short sharp curves should not be interspersed with long curves of small curvature. The ideal location is one with consistent alignment where both grade and curvature receive consideration and satisfy limiting criteria. The final alignment will be that in which the best balance between grade and curvature is achieved.

Terrain has considerable influence on the final choice of alignment.

Generally, the topography of an area is fitted into one of the following three classifications: level, rolling, or mountainous.

In level country, the alignment, in general, is limited by considerations other than grade, that is, cost of right-of-way, land use, waterways requiring expensive bridging, existing roads, railroads, canals and power lines, and subgrade conditions or the availability of suitable borrow.

In rolling country, grade and curvature must be considered carefully. Depths of cut and heights of fill, drainage structures, and number of bridges will depend on whether the route follows the ridges, the valleys, or a cross-drainage alignment.

In mountainous country, grades provide the greatest problem, and in general the horizontal alignment (curvature) is conditioned by maximum grade criteria.

## 12-9. CIRCULAR CURVES

Circular curves may be described by giving either the radius or the degree of curve. In highway design, the degree of curve is defined as the central angle subtended by an arc of 100 ft. This is known as the arc definition. Books on railroad location define the degree of curve as the central angle subtended by a chord of 100 ft, and some highway departments follow this procedure. According to the arc definition, it can be shown that the curve radius, $R$, and the degree of curve $D$, are related by the following equations:

$$\frac{2\pi R}{360^\circ} = \frac{100 \text{ ft}}{D}$$

and

$$D = \frac{5729.58}{R} \tag{12-1}$$

The chord definition of degree of curve leads to a different equation:

$$\sin \tfrac{1}{2}D = \frac{50}{R} \tag{12-2}$$

Arc or chord measurements can be considered alike for all curves less than 4 degrees without appreciable error. An examination of tables will show that the following chords may be assumed to be equal to the arcs without appreciable error.

100-ft chords up to 4 degrees
50-ft chords up to 10 degrees
25-ft chords up to 25 degrees
10-ft chords up to 100 degrees

Figure 12-4 shows a simple circular curve and its components, with the necessary formulas for finding the values of the various elements. A combination of simple curves can be arranged to produce compound or reverse curves.

## 12-10. HORIZONTAL ALIGNMENT DESIGN CRITERIA FOR STREETS AND HIGHWAYS

In general, the maximum desirable degree of curvature for highways is from 5 to 7 degrees in open country and 10 degrees in mountainous areas. Many states limit curvature to 3 degrees on principal highways. In contrast, a minimum centerline radius of 350 ft is permitted for collector streets in ordinary terrain, while local residential streets in hilly terrain may have a centerline radius as short as 100 ft.

Two curves in the same direction connected with a short tangent known as "broken back" curves, should be combined into one continuous curve. This

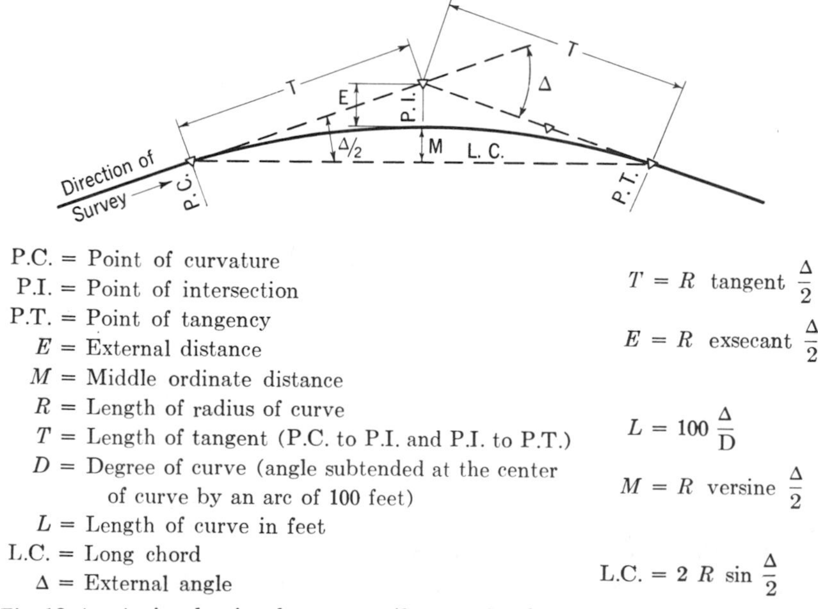


*Fig 12-4.* A simple circular curve. (Source: Paul H. Wright and Radnor J. Paquette, *Highway Engineering,* Fifth Edition, John Wiley, 1987.)

may be accomplished by compounding the curve, provided that the difference between the two branches of the curve does not exceed 5 degrees.

Where reversed curves are necessary, they should be separated by a tangent distance of at least 200 ft in order to allow proper easement from one curve to another.

For small deflections, a curve should be at least 500 ft long for a 5-degree central angle, with an increase of 100 ft for each decrease of one degree in the central angle.

## 12-11. HORIZONTAL ALIGNMENT DESIGN CRITERIA FOR RAILWAYS AND GUIDEWAYS

Sudden changes in horizontal alignment cannot be tolerated in railroad design. Horizontal curvature limits the speed of trains and increases the probability of derailments and overturning accidents. Tracks on curves also require greater horizontal clearances and may create difficulties in coupling. These undesirable effects of curvature stem from the necessity to design for a vehicle unit (the passenger or freight car) that is typically 85 ft long and measures 60 ft between truck centers.

The operation of trains around curves causes wear on the rails and increases train resistance. Curve resistance increases in direct proportion to the degree of curve and amounts to about 0.8 pounds per ton per degree of curve. It has been estimated that a 12-degree curve approximately doubles the train resistance likely to be experienced on a straight level track [3].

When horizontal curvature is imposed on a section that has steep gradient, it may be necessary to compensate for the curvature by lessening the grade. Since a 1-percent grade results in a resistance of 20 lb/ton, a decrease in grade of 0.8/20 or 0.04 percent will compensate for the resistance of a 1-degree horizontal curve.

Generally speaking, 1-degree to 3-degree railroad curves are considered relatively flat curves while 8-degree to 10-degree curves are considered relatively sharp. Curves greater than about 10 degrees are seldom used for main railroad lines, although curves of this sharpness have sometimes been utilized in mountainous areas. Even in mountainous topography, seldom will a horizontal curve greater than about 24 degrees be used. Curves as sharp as 40 degrees, however, have been used in railroad yards.

France's TGV high-speed railroad is designed with a normal minimum radius of 4000 m (13,120 ft or less than 0.5 degrees); however, the design allows for a minimum radius of 3200 m (10,496 ft) for exceptional conditions [4].

Minimum curve radii for urban passenger systems are exemplified by the data in Table 12-4. The differences in the criteria are explained in part by variations in the size of car. For example, the 75-ft-long car used for Atlanta's

*Table 12-4*
**Minimum Recommended Curve Radii for Urban Passenger Systems**

| System/Criteria Source | Minimum Radius of Curvature, ft[a] | |
|---|---|---|
| | Main Lines | Yards and Secondary Tracks |
| Metropolitan Atlanta Rapid Transit Authority (MARTA) | 750[b] | 350 |
| Montreal Bureau de Transport Metropolitain | 459 | 170 |
| Italian Transport Organization (UNIFER) | 492 | 246 |

[a]1 ft = 0.3048 m.

[b]Minimum desirable = 1000 ft.

SOURCES: *MARTA System Design Criteria, Vol. 1,* prepared by Parsons, Brinckerhoff, Quade and Douglas, Inc./Tudor Engineering Company, rev. March 2, 1977; and correspondence with M. Pierre-Paul Arbic, Superintendent of Projects, Montréal Bureau de Transport Metropolitan, September, 20, 1979.

system requires a longer minimum radius of curvature than does Montreal's system, which has a 52.5-ft car.

## 12-2. SUPERELEVATION OF HIGHWAY CURVES

On rural highways, most drivers adopt a more or less uniform speed when traffic conditions will permit them to do so. When making a transition from a tangent section to a curved section, if the sections are not designed properly, the vehicle must be driven at reduced speed for safety as well as for the comfort of the occupants. This is due to the fact that a force is acting on the vehicle that tends to cause an outward skidding away from the center of the curve. Most highways have a slight crowned surface to take care of drainage. It can be readily seen that when these crowns are carried along the curve, the tendency to slip is retarded on the inside of the curve because of the banking effect of the crown. The hazard of slipping is increased on the outside of the curve, however, due to the outward sloping of the crown. In order to overcome this tendency to slip and to maintain average speeds, it is necessary to superelevate the roadway sections; that is, raise the outside edge or bank the curve.

Analysis of the forces acting on the vehicle as it moves around a curve of constant radius indicates that the theoretical superelevation is equal to:

$$e = \frac{V^2}{15R} - f \qquad (12\text{-}3)$$

where

$e$ = rate of superelevation in ft/ft
$f$ = side-friction factor
$V$ = velocity in mph
$R$ = radius of curve, ft

Research and experience have established limiting values for $e$ and $f$. Use of the maximum $e$ with the safe $f$ value in the formula permits determination of minimum curve radii for various design speeds. Safe side-friction factors to be used in design range from 0.17 at 20 mph to 0.10 at 70 mph. Present practice suggests a maximum superelevation rate of 0.10 ft/ft. Where snow and ice conditions prevail, the maximum superelevation should not exceed 0.08 ft/ft. On low-volume gravel surfaced roads a superelevation rate of 0.12 may be used.

The maximum safe degree of curvature for a given design speed can be determined from the maximum rate of superelevation and the side-friction factor. The minimum safe radius $R$ can be calculated from the formula given above:

$$R_{min} = \frac{V^2}{15\,(e_{max} + f)} \tag{12-4}$$

Using $D$ as the degree of curve (arc definition), it follows from Eq. 12-1:

$$D_{max} = \frac{85{,}900\,(e + f)}{V^2} \tag{12-5}$$

The relationship between superelevation and degree of curve is illustrated in Table 12-5.

## 12-13. SUPERELEVATION OF RAILWAY AND TRANSIT GUIDEWAY CURVES

As shown by Fig. 12-5, a train car traversing a curve is subjected to forces of its weight, the resisting forces exerted by the rails, and the centrifugal force:

$$F = \frac{Wv^2}{gR} \tag{12-6}$$

where

$W$ = weight of car, lb
$v$ = velocity in ft/sec
$R$ = radius of curve, ft
$g$ = acceleration due to gravity, ft/sec$^2$

*Table 12-5*

**Maximum Degree of Curve and Minimum Radius Determined for Limiting Values of *e* and *f***

| *Design Speed (mph)*[a] | *Maximum e* | *Maximum f* | *Total (e + f)* | *Maximum Degree of Curve* | *Rounded Maximum Degree of Curve* | *Minimum Radius (ft)*[2] |
|---|---|---|---|---|---|---|
| 20 | 0.04 | 0.17 | 0.21 | 44.97 | 45.0 | 127 |
| 30 | 0.04 | 0.16 | 0.20 | 19.04 | 19.0 | 302 |
| 40 | 0.04 | 0.15 | 0.19 | 10.17 | 10.0 | 573 |
| 50 | 0.04 | 0.14 | 0.18 | 6.17 | 6.0 | 955 |
| 60 | 0.04 | 0.12 | 0.16 | 3.81 | 3.75 | 1528 |
| 20 | 0.06 | 0.17 | 0.23 | 49.25 | 49.25 | 116 |
| 30 | 0.06 | 0.16 | 0.22 | 20.94 | 21.0 | 273 |
| 40 | 0.06 | 0.15 | 0.21 | 11.24 | 11.25 | 509 |
| 50 | 0.06 | 0.14 | 0.20 | 6.85 | 6.75 | 849 |
| 60 | 0.06 | 0.12 | 0.18 | 4.28 | 4.25 | 1348 |
| 65 | 0.06 | 0.11 | 0.17 | 3.45 | 3.5 | 1637 |
| 70 | 0.06 | 0.10 | 0.16 | 2.80 | 2.75 | 2083 |

| | | | | | | |
|---|---|---|---|---|---|---|
| 20 | 0.08 | 0.17 | 0.25 | 53.54 | 53.5 | 107 |
| 30 | 0.08 | 0.16 | 0.24 | 22.84 | 22.75 | 252 |
| 40 | 0.08 | 0.15 | 0.23 | 12.31 | 12.25 | 468 |
| 50 | 0.08 | 0.14 | 0.22 | 7.54 | 7.5 | 764 |
| 60 | 0.08 | 0.12 | 0.20 | 4.76 | 4.75 | 1206 |
| 65 | 0.08 | 0.11 | 0.19 | 3.85 | 3.75 | 1528 |
| 70 | 0.08 | 0.10 | 0.18 | 3.15 | 3.0 | 1910 |
| 20 | 0.10 | 0.17 | 0.27 | 57.82 | 58.0 | 99 |
| 30 | 0.10 | 0.16 | 0.26 | 24.75 | 24.75 | 231 |
| 40 | 0.10 | 0.15 | 0.25 | 13.38 | 13.25 | 432 |
| 50 | 0.10 | 0.14 | 0.24 | 8.22 | 8.25 | 694 |
| 60 | 0.10 | 0.12 | 0.22 | 5.23 | 5.25 | 1091 |
| 65 | 0.10 | 0.11 | 0.21 | 4.26 | 4.25 | 1348 |
| 70 | 0.10 | 0.10 | 0.20 | 3.50 | 3.5 | 1637 |

NOTE: In recognition of safety considerations, use of $e_{max} = 0.04$ should be limited to urban conditions.

[a] 1 mi = 1.6093 km, 1 ft = 0.3048 m.

SOURCE: *A Policy on Geometric Design of Highways and Streets*, AASHTO, 1984.

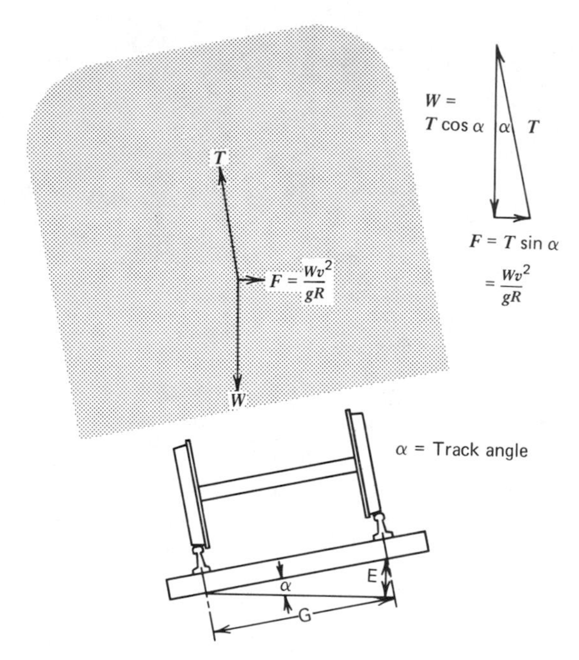


*Fig 12-5.* Forces on a car body traversing a curve at equilibrium speed. (Source: *Proceedings,* American Railway Engineering Association, Vol. 56, 1955.)

A state of equilibrium is said to exist when both wheels bear equally on the rails. Under these conditions $E$, the equilibrium elevation, is just sufficient to cause the resultant force, $T$, to be perpendicular to the plane of the top of the rails. Using an effective gage (center to center of rails), $G$, of 60 in., it can be shown by similar triangles that

$$\frac{E}{60} = \frac{F}{T}$$

For small track angles where the sine and tangent are approximately equal, the following relationship is essentially correct:

$$\frac{E}{60} = \frac{F}{T} = \frac{v^2}{gR}$$

Recalling that $R = \dfrac{5730}{D}$, it can be shown that:

$$E = 0.0007V^2D \qquad (12\text{-}7)$$

where

$E$ = equilibrium elevation of the outer rail, in.
$V$ = speed, mph
$D$ = degree of curve

Values of equilibrium elevations for various speeds and degrees of curvature are given in Table 12-6.

It has been found that a rail car will ride comfortably and safely around a curve at a speed that requires an elevation about three inches[2] higher for equilibrium. It will be noted from Table 12-6, for example, that for a 60-mph speed and a 2-degree curve, the outer rail should be elevated 5.04 in. Experience indicates that a train will ride safely and comfortably around the curve at about 76 mph, the value corresponding to an elevation of 8.04 in.

Not all trains, of course, travel at the same speed. Even high-speed trains must occasionally slow down or stop on horizontal curves because of traffic interferences or other reasons. Because passengers experience discomfort

[2]Modern cars equipped with specially designed suspension systems tend to have less car body roll and can negotiate curves comfortably at 4- or 4½-in. unbalanced elevation [7, 8].

*Table 12-6*
**Equilibrium Elevation for Various Speeds on Curves**

| | *E = Equilibrium Elevation for Various Speeds on Curves* *V = Speed in mph*[a] | | | | | |
|---|---|---|---|---|---|---|
| *D = Degree of Curve* | *30* | *40* | *50* | *60* | *70* | *80* |
| 0°30′ | 0.32 | 0.56 | 0.88 | 1.26 | 1.72 | 2.24 |
| 1°00′ | 0.63 | 1.12 | 1.75 | 2.52 | 3.43 | 4.48 |
| 1°30′ | 0.95 | 1.68 | 2.63 | 3.78 | 5.15 | 6.72 |
| 2°00′ | 1.26 | 2.24 | 3.50 | 5.04 | 6.86 | 8.96 |
| 2°30′ | 1.58 | 2.80 | 4.38 | 6.30 | 8.58 | 11.20 |
| 3°00′ | 1.89 | 3.36 | 5.25 | 7.56 | 10.29 | |
| 3°30′ | 2.21 | 3.92 | 6.13 | 8.82 | | |
| 4°00′ | 2.52 | 4.48 | 7.00 | 10.08 | | |
| 5°00′ | 3.15 | 5.60 | 8.75 | | | |
| 6°00′ | 3.78 | 6.72 | 10.50 | | | |
| 7°00′ | 4.41 | 7.84 | | | | |
| 8°00′ | 5.04 | 8.96 | | | | |
| 9°00′ | 5.67 | 10.08 | | | | |
| 10°00′ | 6.30 | 11.20 | | | | |
| 11°00′ | 6.93 | | | | | |
| 12°00′ | 7.56 | | | | | |

$E$ in in. $= 0.0007V^2D$

[a]1 mi = 1.6093 km.

SOURCE: American Railway Engineering Association, *Manual of Recommended Practice*, 1986.

when the train stops or moves slowly around superelevated curves, a maximum elevation of the outer rail of 8 in. is recommended and a maximum elevation value of 6 or 7 in. is desired. For urban passenger rail lines and guideways, a maximum track elevation of about 6 in. usually is specified [5, 6].

Trains traveling at speeds less than the equilibrium speed tend to cause excessive wear on the inside rail. Similarly, trains that exceed the equilibrium speed cause excessive wear on the outer rail, and in extreme cases, can cause overturning. The velocity that would cause overturning can be computed by Eq. 12-8 which is based on the assumption that the center of gravity of the car is 7.0 ft above the top of the outer rail.

$$V = \frac{170}{\sqrt{D}} \tag{12-8}$$

## 12-14. SPIRALS OR TRANSITION CURVES FOR HIGHWAYS

Transition curves serve the purpose of providing a gradual change from the tangent section to the circular curve, and vice versa. A vehicle that enters a circular curve with transitions travels smoothly and naturally along a curve that gradually changes from zero to curvature of some finite value, $D$, which is maintained throughout the length of the circular curve. As the vehicle emerges from the circular curve, the curvature is gradually diminished from $D$ to zero. The most commonly used transition curve is the spiral, the radius of which at any point is inversely proportional to its length.

Being unrestrained laterally, automobiles and trucks are free to shift across the traffic lane allowing the driver to effect artificially a transition from a tangent section (of infinite radius) to a curve section (of finite radius). For this reason, transition curves are not used universally for highway design. Although there is no consistent practice on the use of transition curves by the various highway agencies, spiral transition curves normally are used only on high-volume highways where the degree of curvature exceeds about 3 degrees.

When used in combination with superelevated sections, the superelevation should be attained within the limits of the transition.

The minimum length of the transition curve is given as:

$$L_s = 1.6\frac{V^3}{R} \tag{12-9}$$

where

$L_s$ = length of the transition, ft
$V$ = speed in mph
$R$ = radius, ft

Barnett [9] has published tables from which required highway transition curves can be chosen and located without extensive calculations. These tables are still referenced by many road designers.

## 12-15. SPIRALS OR TRANSITION CURVES FOR RAILWAYS AND TRANSIT GUIDEWAYS

A railroad car is restrained strictly between the two rails and, therefore, is not free to shift laterally and to artificially effect a transition in horizontal curvature. For this reason, spiral transition curves are used extensively in railroad design. The use of spiral curves is recommended by the American Railway Engineering Association [7] on all mainline tracks between tangent and curve and between different degrees of curvature where compound curves are used.

According to the AREA [7]:

> The desirable length of the spiral for main tracks where the alignment is being entirely reconstructed or where the cost of the realignment of the existing track will not be excessive should be such that when passenger cars of average roll tendency are to be operated the rate of change of the unbalanced lateral acceleration acting on a passenger will not exceed 0.03 g per sec. Also, the desirable length in this case needed to limit the possible racking and torsional forces produced should be such that the longitudinal slope of the outer rail with respect to the inner rail will not exceed 1/744, which is based on an 85-ft-long car.

Two formulas are given by AREA to achieve these results:

$$L = 1.63E_uV \tag{12-10}$$

where

$L$ = desirable minimum length of spiral, ft
$E_u$ = unbalanced elevation, in.
$V$ = maximum train speed, mph

and

$$L = 62E_a \tag{12-11}$$

where

$E_a$ = actual elevation, in.

The maximum length of spiral computed by the two formulas should be used.

Somewhat shorter spiral lengths are specified for conventional rail urban transit systems. For example, the larger of the lengths computed by the following equations is recommended for Atlanta's rail transit system:[3]

$$L = 1.4\ E_a V \tag{12-12}$$

$$L = 60\ E_a \tag{12-13}$$

The AREA *Manual for Railway Engineering* [7] lists the various formulas for the calculation of spiral curve data.

## 12-16. ATTAINMENT OF SUPERELEVATION IN HIGHWAY DESIGN

The transition from the tangent section to a curved superelevated section must be accomplished without any appreciable reduction in speed and in such a manner as to ensure safety and comfort of the vehicle and occupants.

In order to effect this change, it will be readily seen that the normal road cross section will have to be tilted to the superelevated cross section. This tilting usually is accomplished by rotating the section about the centerline axis. The effect of this rotation is to lower the inside edge of the pavement and to raise the outside edge without changing the centerline grade. Another method is to rotate about the inner edge of the pavement as an axis so that the inner edge retains its normal grade but the centerline grade is varied or rotation may be about the outside edge. Rotation about the centerline is used by a majority of the states, but for flat grades too much sag is created in the ditch grades by this method. On grades below 2 percent, rotation about the inside edge is preferred. Regardless of which method is used, care should be exercised to provide for drainage in the ditch sections of the superelevated areas.

The roadway on full superelevated sections should be a straight inclined section. When a crowned surface is rotated to the desired superelevation, the change from a crowned section to a straight inclined section should be accomplished gradually. This may be done by first changing the section from the centerline to the outside edge to a level section. Second, the outside edge should be raised to an amount equal to one-half the desired superelevation, and at the same time the inner section changed to a straight section and the whole section put in one inclined plane. Third, rotation about the centerline should be continued until full superelevation is reached.

The distance required for accomplishing the transition from a normal to a superelevated section, sometimes called the transition runoff, is a function of the design speed, degree of curvature, and the rate of superelevation.

[3]An even shorter length is permitted where physical restrictions or higher speed requirements make the use of the lengths required by the equations prohibitive or impracticable [5].

Recommended minimum lengths of superelevation runoff for two-lane pavements are shown in Table 12-7. On multilane undivided roads or roads with wide medians, increased runoff lengths are necessary, typically 1.2 to 2.0 times the lengths shown in Table 12-7.

Superelevation usually is started on the tangent at some distance before the curve starts, and full superelevation generally is reached beyond the P.C. of the curve. In curves with transitions the superelevation can be attained within the limits of the spiral. In curves of small degree where no transition is

*Table 12-7*

**Length Required for Superelevation Runoff—Two-Lane Pavements**

| *Superelevation Rate, e* | *L—Length of Runoff (ft)[a] for Design Speed (mph)[b] of* | | | | | | |
|---|---|---|---|---|---|---|---|
| | *20* | *30* | *40* | *50* | *60* | *65* | *70* |
| | | *12-ft Lanes* | | | | | |
| 0.02 | 30 | 35 | 40 | 50 | 55 | 60 | 60 |
| 0.04 | 60 | 70 | 85 | 95 | 110 | 115 | 120 |
| 0.06 | 95 | 110 | 125 | 145 | 160 | 170 | 180 |
| 0.08 | 125 | 145 | 170 | 190 | 215 | 230 | 240 |
| 0.10 | 160 | 180 | 210 | 240 | 270 | 290 | 300 |
| | | *10-ft Lanes* | | | | | |
| 0.02 | 25 | 30 | 35 | 40 | 45 | 50 | 50 |
| 0.04 | 50 | 60 | 70 | 80 | 90 | 95 | 100 |
| 0.06 | 80 | 90 | 105 | 120 | 135 | 145 | 150 |
| 0.08 | 105 | 120 | 140 | 160 | 180 | 190 | 200 |
| 0.10 | 130 | 150 | 175 | 200 | 225 | 240 | 250 |
| Design minimum length regardless of superelevation | 50 | 100 | 125 | 150 | 175 | 190 | 200 |

[a] 1 ft = 0.3048 m.

[b] 1 mph = 1.6093 km/hr.

SOURCE: *A Policy on Geometric Design of Highways and Streets,* American Association of State Highway and Transportation Officials, Washington, D.C., 1984.

used, between 60 and 80 percent of the superelevation runoff is put into the tangent.

In order to obtain smooth profiles for the pavement edges, it is recommended that the breaks at cross sections be replaced by smooth curves (see Fig. 12-6).

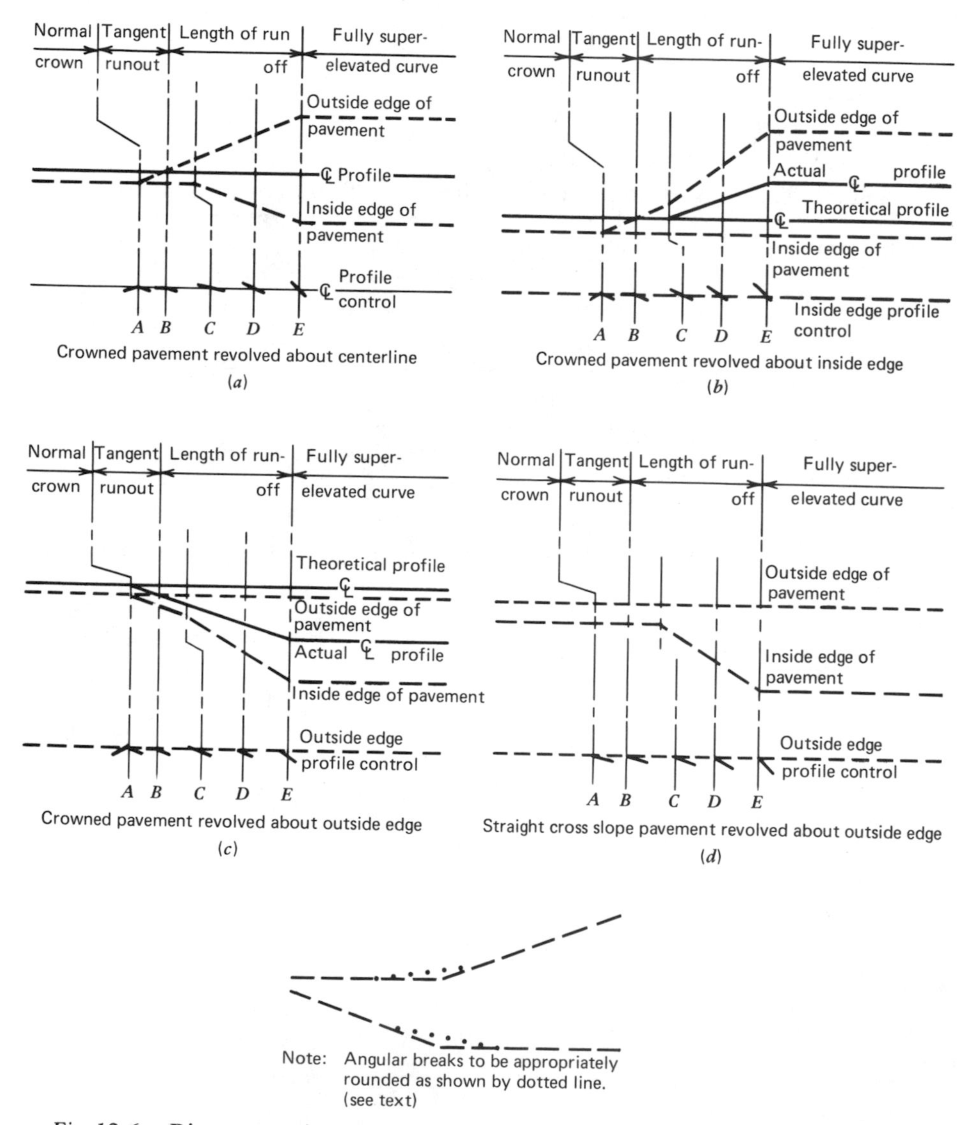


*Fig 12-6.* Diagrammatic profiles showing methods of attaining superelevation for a highway curve to the right. (Courtesy American Association of State Highway and Transportation Officials.)

## 12-17. ATTAINMENT OF SUPERELEVATION IN RAILWAY AND TRANSIT GUIDEWAY DESIGN

As in the case of highway curves with transitions, the superelevation of railway curves is attained or run out uniformly over the length of the spiral. This is accomplished by elevating the outer rail. The inner rail is normally maintained at grade.

## 12-18. GRADES AND GRADE CONTROL

The vertical alignment of the roadway and its effect on the safe and economical operation of the vehicle constitutes one of the most important features of highway and railway design. The vertical alignment, which consists of a series of straight lines connected by vertical parabolic or circular curves, is known as the grade line. When the grade line is increasing from the horizontal, it is known as a plus grade, and when it is decreasing from the horizontal it is known as a minus grade. In analyzing grade and grade controls, the designer usually studies the effect of change in grade on the centerline profile.

In the establishment of a grade, an ideal situation is one in which the cut is balanced against the fill without a great deal of borrow or an excess of cut to be wasted. All hauls should be downhill if possible, and not too long. Ideal grades have long distances between points of intersection, with long vertical curves between grade tangents to provide smooth riding qualities and good visibility. The grade should follow the general terrain and rise and fall in the direction of the existing drainage. In rock cuts and in flat, swampy areas it is necessary to maintain higher grades. Further possible construction and the presence of grade separations and bridge structures also control grades.

Change of grade from plus to minus should be placed in cuts, and changes from a minus grade to a plus grade should be placed in fills. This generally will give a good design, and many times it will avoid the appearance of building hills and producing depressions contrary to the general existing contours of the land. Other considerations for determining the grade line may be of more importance than the balancing of cuts and fills.

Urban projects usually require a more detailed study of the controls and a fine adjustment of elevations than do rural projects. It is often best to adjust the grade to meet existing conditions because of additional expense of doing otherwise.

## 12-19. VERTICAL CURVES

The parabolic curve is used almost exclusively in connecting grade tangents because of the convenient manner in which the vertical offsets can be computed. This is true for both highway and railway design. Typical vertical curves are shown in Fig. 12-7. Offsets for vertical curves may be computed by means of Eqs. 12-14 and 12-15.

The distance from the P.I. to the middle of the parabolic curve:

$$E = \frac{gL}{8} \tag{12-14}$$

where

$E$ = the external distance, ft

$g$ = the algebraic difference in grade in percent of the intersecting grades

$L$ = the length of the curve in stations

The vertical offset at any point on the curve

$$y = \left(\frac{x}{l}\right)^2 E \tag{12-15}$$

where

$y$ = offset, ft

$x$ = any distance from the P.C., ft

$l = \frac{L}{2} \times 100$ ft or half the length of the curve

$E$ = external distance, ft

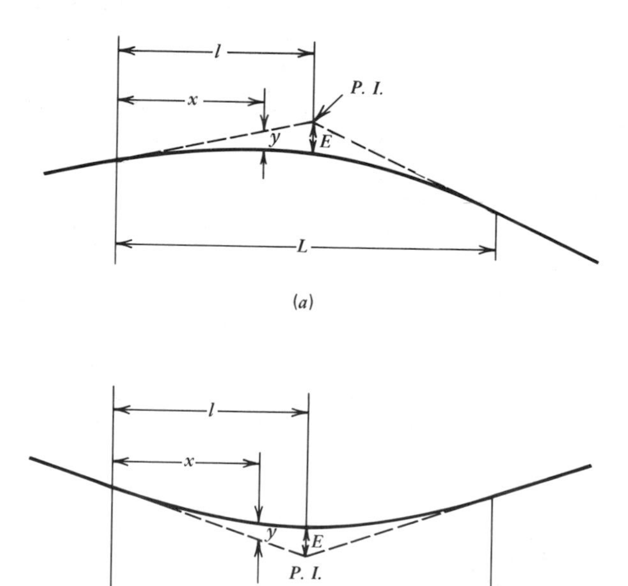


*Fig 12-7.* Typical vertical curves: (*a*) crest, (*b*) sag.

It is usually necessary to make calculations at 50-ft stations, while some paving operations require elevations at 25-ft intervals or less. In order to ensure proper drainage or clearances, it is often necessary to compute other critical points on the vertical curve, such as its sags and crests. The low point or high point of a parabolic curve is not usually vertically above or below the vertex of the intersecting grade tangents, but is to the left or right of this point. The distance from the P.C. of the curve to the low or high point is given as:

$$x = \frac{Lg_1}{g_1 - g_2} \tag{12-16}$$

and the difference in elevation is given as:

$$Y = \frac{Lg_1^2}{2(g_1 - g_2)} \tag{12-17}$$

where

$x$ = distance from P.C. to the turning point in stations
$L$ = length of the curve in stations
$g_1$ = grade tangent from the P.C. in percent
$g_2$ = grade tangent from the P.T. in percent
$Y$ = the difference in elevation between the P.C. and the turning point, ft

The above formulas apply only for the symmetrical curve, that is, one in which the tangents are of equal length. The unequal tangent or unsymmetrical vertical curve is a compound parabolic curve. Its use generally is warranted only where a symmetrical curve cannot meet imposed alignment conditions.

## 12-20. VERTICAL ALIGNMENT DESIGN CRITERIA FOR STREETS AND HIGHWAYS

In the analysis of grade and grade control, one of the most important considerations is the effect of grades upon the operating costs of the motor vehicle. An increase in the gasoline consumption and a reduction of speed are apparent when grades are increased. An economical approach would be to balance the added annual cost of grade reduction against the added annual cost of vehicle operation without grade reduction. An accurate solution to the problem depends on the knowledge of traffic volume and type, which can be obtained only by means of a traffic survey.

While maximum grades vary a great deal in various states, AASHTO recommendations make maximum grades dependent on design speed and topography. Present practice limits grades in flat country to 6 percent at

30 mph design speed and 3 percent for 80 mph. In mountainous terrain 9 percent at 30 mph and 5 percent at 70 mph should be used.

Residential streets, which have the primary function of providing access to land areas and only the secondary function of traffic service, tend to conform to and blend with undulations in the existing terrain. Maximum grades as steep as 15 percent may be permitted for local streets in hilly terrain.

Whenever long sustained grades are used, the designer should not substantially exceed the critical length of grade without the provision of climbing lanes for slow-moving vehicles. Critical grade lengths vary from 1700 ft for a 3 percent grade to 500 ft for an 8 percent grade.

Long-sustained grades should be less than the maximum grade used on any particular section of a highway. It is often preferred to break the long-sustained uniform grade by placing steeper grades at the bottom and lightening the grades near the top of the ascent. Dips in the profile grade in which vehicles may be hidden from view should also be avoided.

Minimum grades are governed by drainage conditions. Level grades may be used in fill sections in rural areas when crowned pavements and sloping shoulders can take care of the pavement surface drainage. It is preferred, however, to have a minimum grade of at least 0.5 percent under most conditions in order to secure adequate drainage.

*Sight Distance.* Safe highways must be designed to give drivers a sufficient distance of clear vision ahead so that they can avoid hitting unexpected obstacles and can pass slower vehicles without danger.

Sight distance is the length of highway visible ahead to the driver of a vehicle. When this distance is not long enough to permit passing an overtaken vehicle, it is termed stopping (or nonpassing) sight distance. The stopping distance is the minimum distance required for stopping a vehicle traveling at or near the design speed before reaching an object in its path. This stationary object may be a vehicle or some other object on the roadway. When the sight distance is long enough to enable a vehicle to overtake and pass another vehicle on a two-lane highway without interference from an oncoming vehicle, it is termed passing sight distance.

*Minimum Stopping Sight Distance.* Sight distance at every point should be as long as possible, but never less than the minimum stopping sight distance. The minimum stopping sight distance is based upon the sum of two distances: one, the distance traveled from the time the object is sighted to the instant that the brakes are applied, and two, the distance required for stopping the vehicle after the brakes are applied. The first of these two distances is dependent upon the speed of the vehicle and the perception time and brake reaction time of the operator. The second distance depends upon the speed of the vehicle; condition of brakes, tires, and roadway surface; and the alignment and grade of the highway.

*Braking Distance.* There is a wide variation among vehicle operators as to the time that it takes to react and to apply the brakes after an obstruction is sighted. Investigations seem to indicate that the minimum value of perception time can be assumed to vary from 2 sec at 30 mph to 1 sec at 70 mph. Brake reaction time is less than perception time, and tests indicate that the average brake reaction time is about ½ sec. To provide a factor of safety for operators whose brake reaction time is above average, a full second is assumed as the total brake reaction time. Some investigators feel that perception and brake reaction time should be combined and have assigned 1½ sec for this value.

The approximate braking distance of a vehicle on a level highway is determined by:

$$d = \frac{v^2}{2fg} \tag{12-18}$$

where

$d$ = braking distance, ft
$v$ = velocity of the vehicle, ft/sec, when the brakes are applied
$g$ = acceleration due to gravity
$f$ = coefficient of friction between tires and roadway

Changing $v$ in ft/sec to $V$ in mph, and substituting 32.2 for $g$, we have:

$$d = \frac{V^2}{30f} \tag{12-19}$$

It is assumed that the friction force is uniform throughout the braking period. This is not strictly true; it varies as some power of the velocity. Other physical factors affecting the coefficient of friction are the condition and pressure of tires, type and condition of the surface, and climatic conditions such as rain, snow, and ice. Friction factors for skidding are assumed to vary from 0.62 at 30 mph to 0.55 at 70 mph for dry pavements. For wet pavements these are lower, as shown in Table 12-8. Recommended minimum stopping sight distances are shown in Table 12-8. In this table, perception and brake reaction time are combined.

*Effect of Grade on Stopping Distance.* When a highway is on a grade, the formula for braking distance is modified as follows:

$$d = \frac{V^2}{30(f \pm g)} \tag{12-20}$$

in which $g$ is the percentage of grade divided by 100. The safe stopping distances on upgrades are shorter and on downgrades longer than horizontal stopping distances. Where an unusual combination of steep grades and high speed occurs the minimum stopping sight distance should be adjusted to provide for this factor.

*Measuring Minimum Stopping Sight Distance.* It is assumed that the height of eye of the average driver, except drivers of buses and trucks, is about 3.50

*Table 12-8*

**Stopping Sight Distance on Wet Pavements**

| Design Speed (mph) | Assumed Speed for Condition (mph)[b] | Brake Reaction | | Coefficient of Friiction f | Braking Distance on Level[a] (ft)[c] | Stopping Sight Distance | |
|---|---|---|---|---|---|---|---|
| | | Time (sec) | Distance (ft)[c] | | | Computed[a] (ft)[c] | Rounded for Design (ft)[c] |
| 20 | 20–20 | 2.5 | 73.3–73.3 | 0.40 | 33.3–33.3 | 106.7–106.7 | 125–125 |
| 25 | 24–25 | 2.5 | 88.0–91.7 | 0.38 | 50.5–54.8 | 138.5–146.5 | 150–150 |
| 30 | 28–30 | 2.5 | 102.7–110.0 | 0.35 | 74.7–85.7 | 177.3–195.7 | 200–200 |
| 35 | 32–35 | 2.5 | 117.3–128.3 | 0.34 | 100.4–120.1 | 217.7–248.4 | 225–250 |
| 40 | 36–40 | 2.5 | 132.0–146.7 | 0.32 | 135.0–166.7 | 267.0–313.3 | 275–325 |
| 45 | 40–45 | 2.5 | 146.7–165.0 | 0.31 | 172.0–217.7 | 318.7–382.7 | 325–400 |
| 50 | 44–50 | 2.5 | 161.3–183.3 | 0.30 | 215.1–277.8 | 376.4–461.1 | 400–475 |
| 55 | 48–55 | 2.5 | 176.0–201.7 | 0.30 | 256.0–336.1 | 432.0–537.8 | 450–550 |
| 60 | 52–60 | 2.5 | 190.7–220.0 | 0.29 | 310.8–413.8 | 501.5–633.8 | 525–650 |
| 65 | 55–65 | 2.5 | 201.7–238.3 | 0.29 | 347.7–485.6 | 549.4–724.0 | 550–725 |
| 70 | 58–70 | 2.5 | 212.7–256.7 | 0.28 | 400.5–583.3 | 613.1–840.0 | 625–850 |

[a]Different values for the same speed result from using unequal coefficients of friction.

[b]1 mi = 1.6093 km.

[c]1 ft = 0.3048 m.

SOURCE: *A Policy on Geometric Design of Highways and Streets,* American Association of State Highway and Transportation Officials, Washington, D.C., 1984.

ft above the pavement. The height of the stationary object may vary, but a height of 6 in. is assumed in determining stopping sight distance.

Engineers have combined the minimum stopping sight distance required with the sight distance provided by the roadway geometry and produced design graphs that show the minimum vertical curve lengths for any combination of design speed and algebraic difference in grades. Such graphs, shown as Figures 12-8*a* and 12-8*b*, give design controls for crest vertical curves and sag vertical curves, respectively. To use these graphs, enter the ordinate scale at the algebraic difference in grades, project a horizontal line to its intersection with the design speed line, then project a vertical line downward to the abscissa scale and read the minimum length of vertical curve. The value $K$ shown on the graphs indicates the rate of vertical curvature expressed in length of curve, feet per percent of algebraic difference in grades.

*Minimum Passing Sight Distance.* The majority of our highways carry two lanes of traffic moving in opposite directions. In order to pass slower moving vehicles, it is necessary to use the lane of opposing traffic. If passing is to be accomplished safely, the vehicle driver must be able to see enough of the highway ahead in the opposing traffic lane to permit him to have sufficient time to pass and then return to the right traffic lane without cutting off the passed vehicle and before meeting the oncoming traffic. The total distance required for completing this maneuver is the passing sight distance.

When computing minimum passing sight distances, various assumptions must be made relative to traffic behavior. On two-lane highways it may be assumed that the vehicle being passed travels at a uniform speed, and that the passing vehicle is required to travel at this same speed when the sight distance is unsafe for passing. When a safe passing section is reached a certain period of time elapses in which the driver decides whether or not it is safe to pass. When he or she decides to pass, it is assumed that the speed is increased during the entire passing operation. It is also assumed that the opposing traffic appears the instant the passing maneuver starts and arrives alongside the passing vehicle when the maneuver is completed.

Accepting these assumptions, it can be shown that the passing minimum sight distance for a two-lane highway is the sum of four distances:

$d_1$ = distance traveled during perception and reaction time and during the initial acceleration to the point where the vehicle will turn into the opposite lane.

$d_2$ = distance traveled while the passing vehicle occupies the left lane.

$d_3$ = distance between the passing vehicle at the end of its maneuver and the opposing vehicle.

$d_4$ = distance traveled by the oncoming vehicle for two-thirds of the time the passing vehicle occupies the left lane.

The preliminary delay distance $d_1$ is computed from the following formula:

$$d_1 = 1.47 t_1 \left( V - m + \frac{a t_1}{2} \right) \tag{12-21}$$

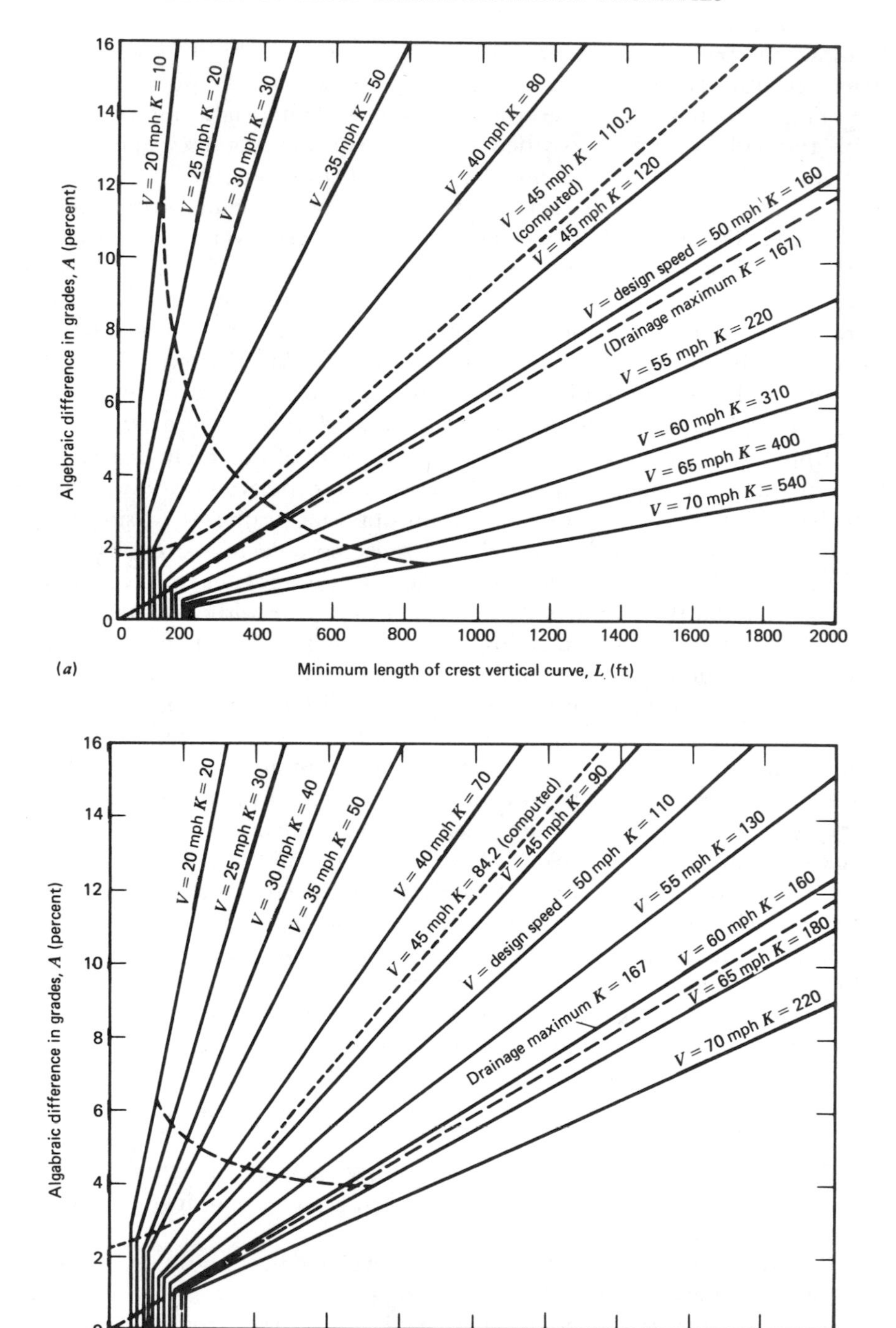


*Fig 12-8.* Design controls for vertical curves. (*a*) Crest vertical curves. (*b*) Sag vertical curves. (Courtesy American Association of State Highway and Transportation Officials.)

where

$t_1$ = time of preliminary delay, sec
$a$ = average acceleration rate, miles per hour per second
$V$ = average speed of passing vehicle, mph
$m$ = difference in speed of passed vehicle and passing vehicle, mph

The distance

$$d_2 = 1.47Vt_2 \tag{12-22}$$

where

$t_2$ = time passing vehicle occupies the left lane, sec
$V$ = average speed of passing vehicle, mph

The clearance distance $d_3$ varies from 110 to 300 ft. These distances are adjusted as shown in Fig. 12-9. The distance $d_4 = 2d_2/3$. The relationship of $d_1$, $d_2$, $d_3$, and $d_4$ is illustrated in Fig. 12-9.

Minimum passing sight distances for the design of two-lane highways are shown in Table 12-9. Here the passing speed is related to the design speed of the highway.

*Sight Distances for Four-Lane Highways.* A four-lane highway should be designed so that the sight distance at all points is greater than the stopping minimum. The crossing of vehicles into the opposing traffic lane on a four-lane highway should be prevented by a median barrier and prohibited by enforcement agencies.

*Extra Lanes for Passing.* Inability to pass indicates that the traffic volume is greater than the capacity of the road at the assumed design speed and that a wider road is indicated. The capacity of the road may be increased by adding an extra lane at safe passing sections. The added lane should be at least as long as the minimum length of safe passing section to which it is added. Under such conditions it must be remembered that the primary purpose of the added lane is to provide a passing lane at the safe passing section. When additional lanes for passing are used on steep grades, the length of the additional lane should have as great a sight distance as possible.

*Measuring Minimum Passing Sight Distance.* As previously stated, the height of eye of the average driver is assumed to be about 3.50 ft above the pavement. As vehicles are the objects that must be seen when passing, it is assumed that the height of object for passing sight distance is 4.25 ft.

*Horizontal Sight Distances.* Horizontal sight distances may be scaled directly from the plans with a fair degree of accuracy. However, where the distance from the highway centerline to the obstruction is known, it is possible to determine the sight distance for a known radius or degree of curve from the following relations:

$$R = \frac{m}{1 - \cos \dfrac{SD}{200}} \tag{12-23}$$

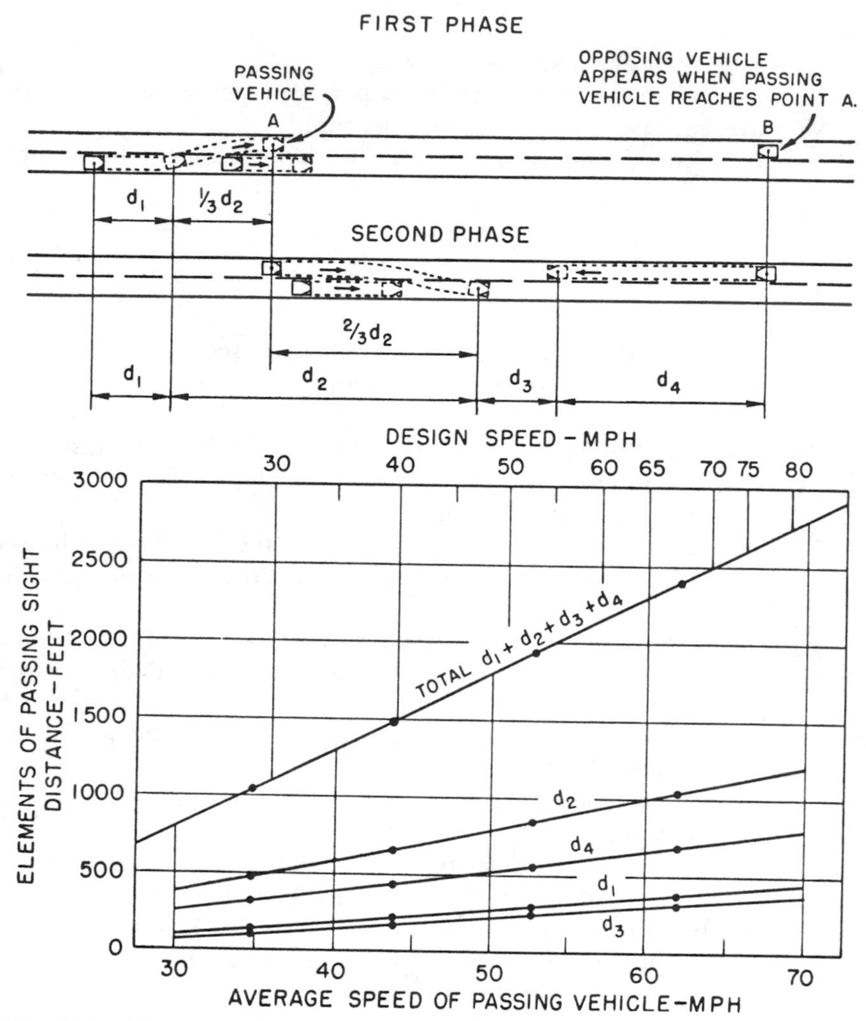


*Fig 12-9.* Elements of and total passing sight distance, two-lane highways. (Courtesy American Association of State Highway and Transportation Officials.)

$$S = \frac{200}{D} \cos^{-1} \frac{R - m}{R} \tag{12-24}$$

where

$m$ = distance from center of roadway to edge of obstruction, ft
$S$ = stopping distance along the center of the inside lane, ft
$R$ = radius of curvature, ft
$D$ = degree of curvature

*Table 12-9*

**Elements of Safe Passing Sight Distance—Two-Lane Highways**

| *Speed Group (mph)*[a] <br> *Average Passing Speed (mph)* | *30–40* <br> *34.9* | *40–50* <br> *43.8* | *50–60* <br> *52.6* | *60–70* <br> *62.0* |
|---|---|---|---|---|
| Initial maneuver: | | | | |
| $a$ = average acceleration (mphps)[b] | 1.40 | 1.43 | 1.47 | 1.50 |
| $t_1$ = time (sec)[b] | 3.6 | 4.0 | 4.3 | 4.5 |
| $d_1$ = distance traveled (ft)[c] | 145 | 215 | 290 | 370 |
| Occupation of left lane: | | | | |
| $t_2$ = time (sec)[b] | 9.3 | 10.0 | 10.7 | 11.3 |
| $d_2$ = distance traveled (ft) | 475 | 640 | 825 | 1030 |
| Clearance length: | | | | |
| $d_3$ = distance traveled (ft)[b] | 100 | 180 | 250 | 300 |
| Opposing vehicle: | | | | |
| $d_4$ = distance traveled (ft) | 315 | 425 | 550 | 680 |
| Total distance, $d_1 + d_2 + d_3 + d_4$ (ft) | 1035 | 1460 | 1915 | 2380 |

[a] 1 mile = 1.6093 km.

[b] For consistent speed relation, observed values adjusted slightly.

[c] 1 ft = 0.3048 m.

SOURCE: *A Policy on Geometric Design of Highways and Streets,* American Association of State Highway and Transportation Officials, 1984.

The formulas do not apply when the length of the circular curve is less than the sight distance, or when the radius of curvature is not constant throughout.

Figure 12-10 shows a horizontal sight distance chart developed from the above formula.

*Marking and Signing Nonpassing Zones.* The marking of pavements for nonpassing zones is intended to show the motorist that sight distance is restricted. This restriction may be caused by vertical or horizontal alignment or by a combination of both.

The *Manual on Uniform Traffic Control Devices* [10] shows standards for the marking of pavements at horizontal and vertical curves. The values for nonpassing shown there are substantially less than design distances and were based on different assumptions than those for highway design.

## 12-20. VERTICAL ALIGNMENT DESIGN FOR RAILROADS AND TRANSIT GUIDEWAYS

Grade design for railroads is similar in many respects to that for highways. As in highway design, there is a need to provide smooth and consistent vertical alignment and to consider controlling elevations of crossing and connecting

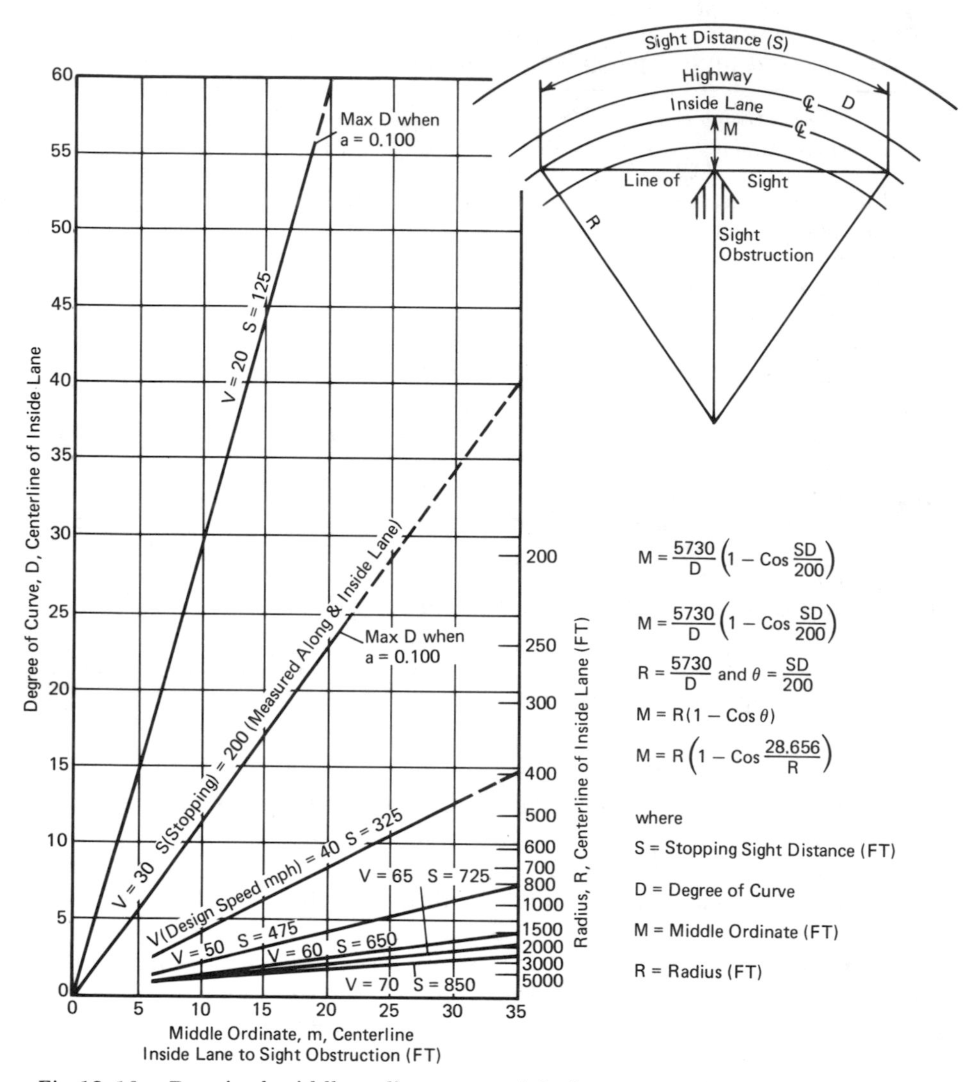


*Fig 12-10.* Required middle ordinates to satisfy the upper values of stopping sight distances (Table 12-8) for curves of various degrees under open conditions. (Source: *A Policy on Geometric Design of Highways and Streets,* AASHTO, 1984.)

railroads, highways, bridges and drainage structures. Vertical parabolic curves are used to connect intersecting railroad grade lines, and the calculation of curve elevations is accomplished as described in Section 12-19.

Railroad vertical alignment design differs significantly in several respects from the profile grade design of highways. These differences arise from inherent vehicle differences and result in more stringent design criteria for railroads. The need for stricter design criteria for railroads is principally attributed to two considerations.

1. The much longer and heavier railroad vehicle.
2. The relatively low coefficient of friction between the driver wheels and the rails.

Railroad design is characterized by much smaller maximum grades and much longer vertical curves than are highways. Generally, steep grades cannot be tolerated in railroad design. The maximum grade for most main lines is about 1.0 percent, although grades as high as 2.5 percent may be used in mountainous terrain. This is especially true for railroads that accommodate freight trains. Slightly greater grades can be tolerated for railroads designed exclusively for passenger service. For example, a maximum gradient of 3.0 percent has been specified for Atlanta's conventional rail transit system. France's TGV railroad is designed for a maximum grade of 3.5 percent. In Montreal, where rubber-tired vehicles are used, a maximum gradient of 6.5 percent was used for the guideway.

A minimum gradient of about 0.3 percent may be required in underground and on aerial line structures to accommodate the drainage.

Table 12-10 gives the specifications of the AREA [7] for the calculation of the lengths of vertical curves for mainline railroad tracks. Generally, much shorter curves are permitted along urban rail transit lines, as Fig. 12-11 and 12-12 demonstrate.

### EXAMPLE 12-1

A plus 0.8 percent intersects a minus 0.3 percent grade on a high-speed main track. What minimum length of vertical curve should be used?

This curve is on a crest. The total change of grade is 1.1 percent.

$$\text{Length of vertical curve} = \frac{1.1}{0.1} = 11 \text{ stations or } 1100 \text{ ft}$$

*Table 12-10*
**Vertical Curve Criteria for Railroads**

| | Maximum Rate of Change of Gradient, Percent per Station[a] | |
|---|---|---|
| | *In Sags* | *On Crests* |
| High-speed main tracks | 0.05 | 0.10 |
| Secondary main tracks | 0.10 | 0.20 |

[a]1 station = 100 ft.

SOURCE: *Manual for Railway Engineering,* Vol. 1, American Railway Engineering Association, 1986.

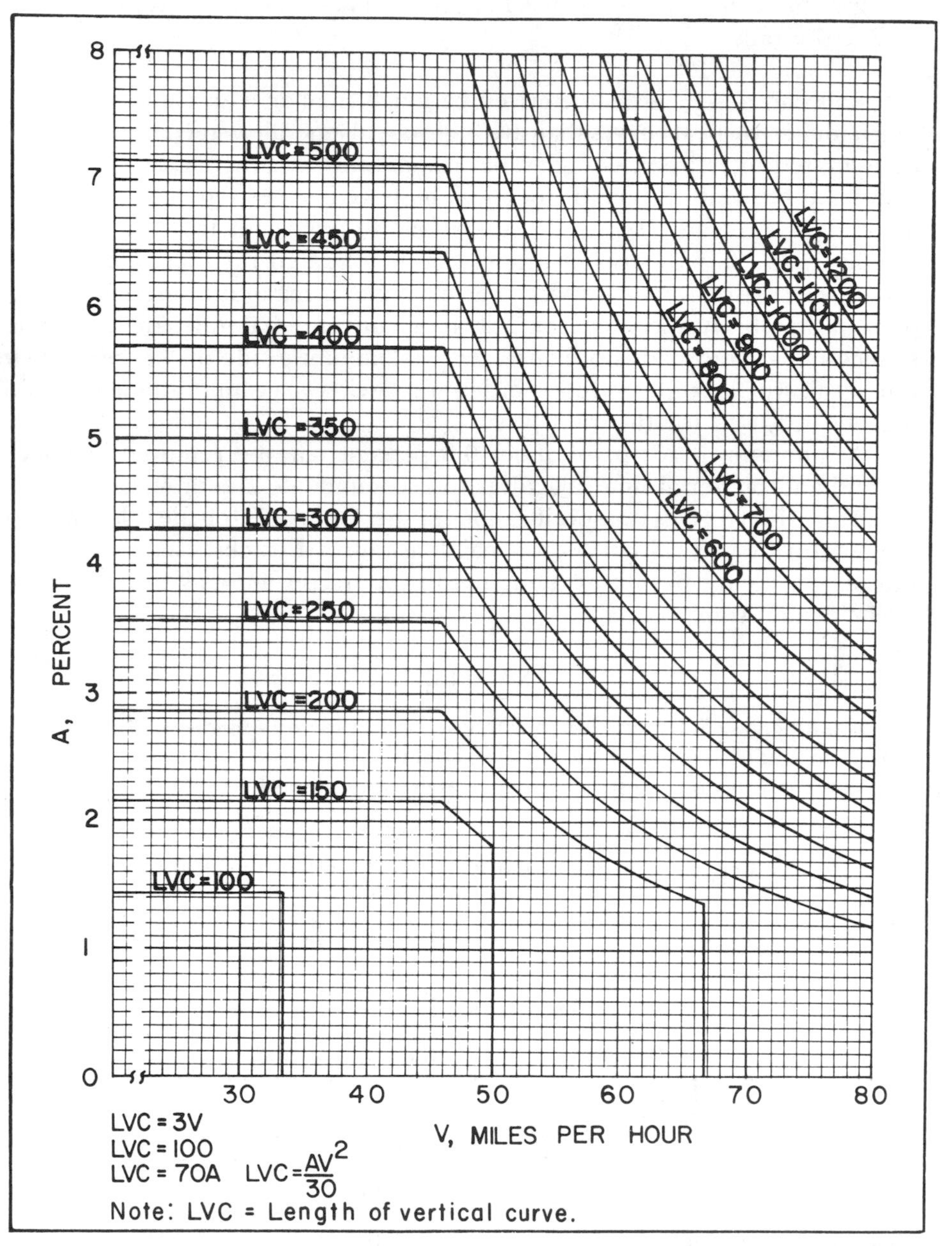


*Fig 12-11.* Design graph for crest vertical curves for a rail transit line. (Source: MARTA System Design Criteria, Vol. 1, prepared by Parsons, Brinckerhoff, Quade and Douglas, Inc./Tudor Engineering Co., rev. March 2, 1977.)

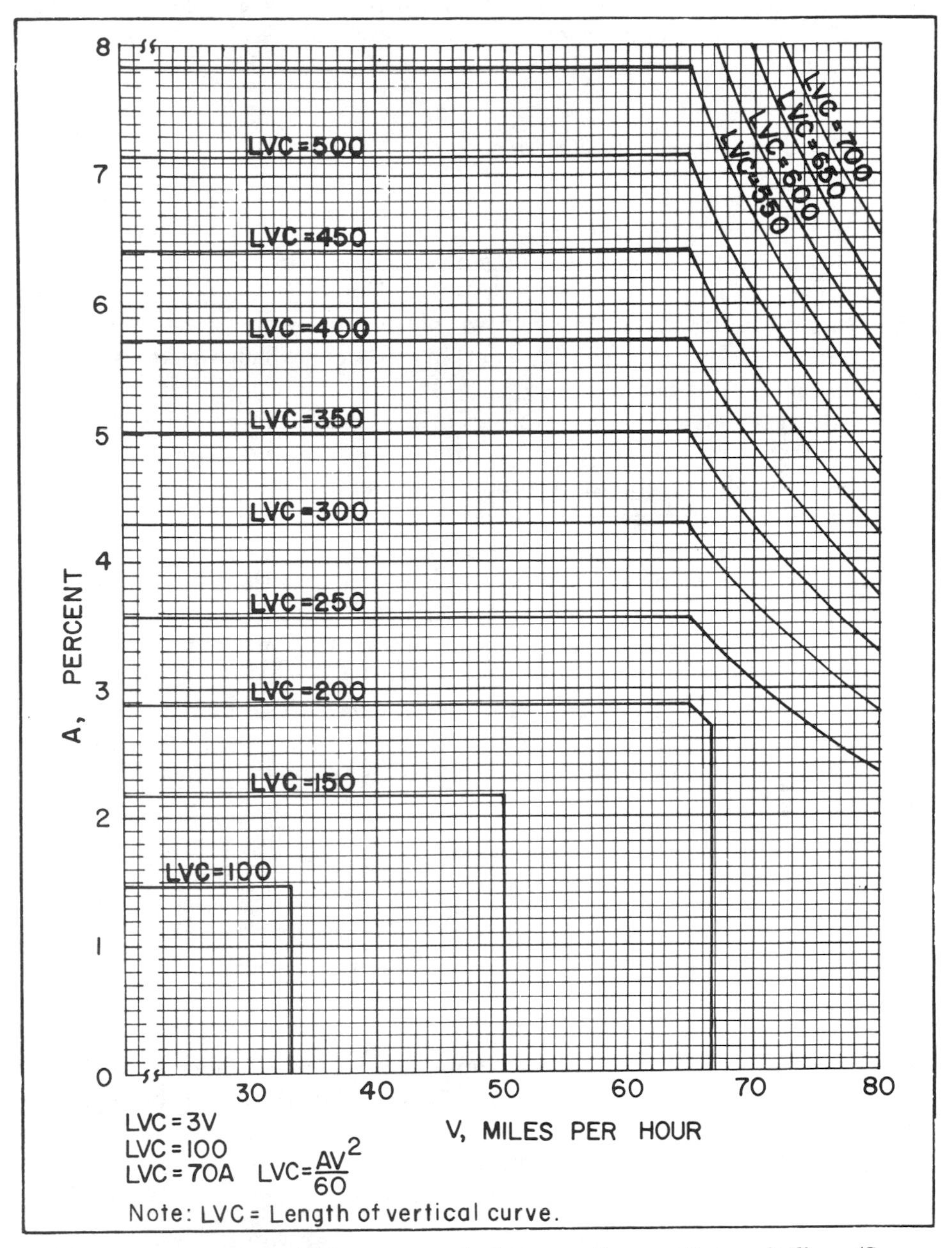


*Fig 12-12.* Design graph for sag vertical curves for a rail transit line. (Source: MARTA System Design Criteria, Vol. 1, prepared by Parsons, Brinckerhoff, Quade and Douglas, Inc./Tudor Engineering Co., rev. March 2, 1977.)

EXAMPLE 12-2

A minus 0.4 percent grade intersects a plus 1.2 percent grade on a high-speed main track. What minimum length of vertical curve is required?

This curve occurs in a sag. The total change of grade is 1.6 percent.

$$\text{Length of vertical curve} = \frac{1.6}{0.05} = 32 \text{ stations or } 3200 \text{ ft}$$

It will be noted that the criterion for length of vertical curve is more critical for sags than for crests. Longer vertical curves are required for sag curves because of the tendency of undesirable slack to develop in the couplings as the cars in the front are slowed down by the change in grade. The subsequent removal of the slack may cause jerking to occur which, in extreme cases, may cause a train to break in two.

Finally, the suitability of a railroad profile grade design depends on the ability of trains to operate over the line smoothly and economically and whether trains of a given size (tonnage) will stall on the maximum grade.

In studying performance characteristics of a locomotive and train of cars over a particular stretch of track, the construction of a *velocity profile* may be helpful. The velocity profile is based on the fact that the total energy of a moving train is the sum of the elevation head (potential energy) and the velocity head (kinetic energy). An example of a velocity profile is shown in Fig. 12-13. In this figure, the solid line represents the actual profile of the track, the elevation head. The dashed line represents the virtual or velocity profile, the elevation head plus the velocity head.

The velocity head, $h$, of a moving train is approximately equal to that of a freely falling body:

$$h = \frac{v^2}{2g} \tag{12-25}$$

where

$v$ = train velocity, ft/sec

According to the AREA, this value should be increased about 6 percent to allow for energy stored in the rotating wheels. This results in the equation:

$$h = 0.035V^2$$

where

$V$ = train speed, mph

Line $AB'$ represents the maximum grade the particular locomotive and train of cars can negotiate at a speed corresponding to point $B$. This grade is

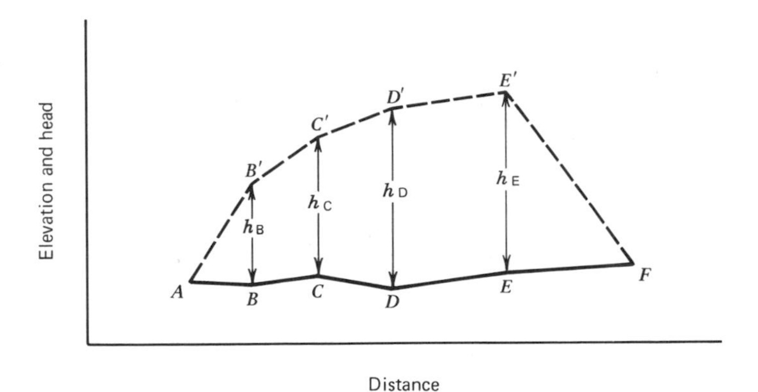


*Fig 12-13.* A velocity profile.

called the *acceleration grade.* Acceleration grade is computed by the following steps:

1. Determine the net tractive effort available for acceleration[4] by subtracting the train resistance (on tangent level track) from the drawbar pull of the locomotive.
2. Express the net tractive effort available for acceleration in lb/ton by dividing by the total weight of the train.
3. Divide the value obtained in step 2 by 20 lb/ton/percent grade, the resistance due to grade. The result is the acceleration grade.

It will be noted that acceleration grade decreases with increase in speed because of similar decreases in drawbar pull.

The general procedure for the construction of a velocity profile is to lay out on a graph of the actual profile the equivalent acceleration grades for speed increments of 0 to 5 mph, 5 to 10 mph, and so on. For example, consider first the 0 to 5 mph increment. From the first point on the survey profile (point *A*), a slope equal to the acceleration grade is laid out until the distance between the two grade lines is equal to the velocity head corresponding to 5 mph, $h_B = 0.035\ (5)^2 = 0.88$. From point *B,* a slope equal to the acceleration grade until the distance between the velocity grade line and the actual grade line is equal to the velocity head corresponding to 10 mph, $h_c = 0.035\ (10)^2 = 3.5$. This procedure is repeated until the maximum speed is reached. When this occurs, the velocity profile is parallel to the actual profile and continues to be parallel until a steeper grade is reached or the locomotive decelerates due to braking or a reduction of the throttle.

Given a freight train of a certain tonnage and the minimum desirable

[4]This effort may also be used for climbing grades.

climbing speed, the velocity profile can be used to determine the ruling grade. The ruling grade is defined as the maximum gradient over which a given locomotive pulling maximum tonnage can be hauled at a given constant speed. On a ruling grade, the velocity profile is parallel to the track profile as line $D'$ $E'$ in the figure.

On a grade greater than the ruling grade, termed a momentum grade, a part of the momentum of the locomotive and train is used in ascending the grade, resulting in a reduction in speed. The extent of this reduction can be determined by the velocity profile. On a momentum grade, the train decelerates, and the actual and velocity profiles converge.

An extensive discussion of the velocity profile and its application to railroad vertical alignment design has been given by Hay [3].

## PROBLEMS

1. Given an intersection angle $\Delta = 23° \ 42'$ right and a degree of curve $D = 1° \ 45'$, compute the tangent distance, curve length, and station of the P.T. of a circular horizontal curve by the chord definition. The station of the point of intersection, P.I., is 27 + 85.50.

2. Solve Problem 1 in SI (metric) units. The metric station[5] of the P.I., would become M8 + 49.030. Compute the curve data and metric station of the P.T.

3. A plus 1.7 percent grade intersects a minus 4.3 percent grade at station 150 + 60 at an elevation of 657.88. Calculate the centerline elevation for every 100-ft station for a 500-ft vertical curve.

4. A minus 2.3 percent grade intersects a minus 1.6 percent grade at metric station M68 + 50 and an elevation of 190.587 m. Calculate the centerline elevations in meters at metric stations M68 + 00 and M69 + 50 for a 300-m curve.

5. A minus 4.3 percent grade intersects a plus 3.7 percent grade at station 50 + 00 and elevation 532.20. To provide proper cover for a culvert, it is desired that the elevation of the vertical curve at station 50 + 00 be 539.00 or higher. What minimum length of vertical curve will provide the necessary cover? If a 700-ft curve is used, determine the station and the elevation of the low point of the curve.

6. A vertical parabolic curve is to be used under a railroad grade separation structure. The curve is 300 m long. The minus grade from left to right is 4.7 percent, and the plus grade is 5.6 percent. The intersection of the two grades is at metric station M135 + 25 and an elevation of 483.571 m. Calculate the station and elevation of the low point of the curve.

[5]As used here, 1 metric station = 100 m.

7. Determine the equilibrium elevation for a 1° 30′ railroad curve given a design speed of 70 mph. What would be the overturning speed for these conditions?
8. Develop an equation similar to Eq. 12-7 using SI (metric) units.
9. It is the policy of a certain urban railroad to use a 7-in. maximum elevation of the outer rail. Given a 75-mph design speed and a 2.5 degree curve, determine the superelevation that should be used and the minimum length of spiral.
10. Develop a formula similar to Eq. 12-10 using SI (metric) units.
11. A minus 1.75 percent grade intersects a minus 0.45 percent grade on a high-speed main railroad track. What minimum length of vertical curve should be used?
12. Express the braking distance formula in Eq. 12-18 in equivalent SI (metric) units.
13. Explain why spiral transition curves are used extensively in railroad design but rarely for highway design.

## REFERENCES


1. *A Policy on Geometric Design of Highways and Streets,* American Association of State Highway and Transportation Officials, Washington, D.C., 1984.
2. *Highway Capacity Manual,* Transportation Research Board Special Report No. 209, 1985.
3. Hay, William W., *Railroad Engineering,* Second Edition, Wiley Interscience, New York, 1982.
5. *MARTA System Design Criteria, Vol. 1,* prepared by Parsons, Brinckerhoff, Quade, and Douglas, Inc./Tudor Engineering Company, rev. March 2, 1977.
4. *La Linde à Grande Vitesse Paris-Sud-Est, The Paris-Southeast High-Speed Line,* Société Nationale des Chemins de fer Français (SNCF), Paris, 1985.
6. *Project UNIFER NO. 18.03, Metropolitan Railroads—Geometry of the Track Layout,* International Union of Public Transportation, Bruxelles, Belgique, 1977.
7. *Manual for Railway Engineering,* American Railway Engineering Association, 1986.
8. "Passenger Ride Comfort on Curved Track," *Proceedings,* American Railway Engineering Association, Vol. 56, 1955.
9. Barnett, Joseph, *Transition Curves for Highways,* U.S. Government Printing Office, Washington, D.C., 1940.
10. *Manual on Uniform Traffic Control Devices for Streets and Highways, including Revisions 1-4,* U.S. Department of Transportation, Federal Highway Administration, Washington, D.C., 1986.

# Chapter 13

# Design of Roadways, Railways, and Guideway Systems: Sections and Intersections

This chapter is concerned with two major aspects of the design of roadways, railways, and guideways systems: the establishment of the dimensional features of the cross section, and the treatment of the intersection of two or more transport links.

## THE ROADWAY CROSS SECTION

### 13-1. NUMBER OF LANES

The number of lanes should be capable of accommodating the anticipated type and volume of traffic. Roads currently in use include two-lane, three-lane, multilane undivided, and multilane divided.

Over 90 percent of rural roads in the United States have two lanes. These facilities vary from remote unpaved roads that follow the natural ground surface to heavily traveled, high-speed primary highways with paved surfaces and stabilized shoulders.

Many three-lane roads have been built in previous years and are still in use. The great advantage seemed to stem from the large improvement in capacity over the two-lane road with only a moderate increase in construction and right-of-way costs. Three-lane roads tend to have high accident rates, and the construction of these roads is no longer recommended. However, many agencies add a passing lane to two-lane roadways along long slopes in hilly or mountainous areas.

Although the four-lane highway is the basic multilane type, traffic volumes may warrant the use of highways having six or even eight lanes, particularly in urban areas.

The *Highway Capacity Manual* [1] recommends a simple planning analysis to determine the probable number of lanes required and whether a multilane highway is appropriate for the expected conditions. The planning procedure requires information on:

1. The general terrain through which the highway is to be built.
2. The annual average daily traffic (AADT) for the design year.
3. The percent of traffic in the peak direction of flow *(D)*.
4. The peak hour factor (PHF) for the design year.
5. The percent of trucks in the traffic stream *(T)*.
6. The type of multilane highway and its development environment.

As the first step, the annual average daily traffic is converted to a directional design hourly volume (DDHV) by using the following equation:

$$\text{DDHV} = \text{AADT} \times K \times D \tag{13-1}$$

The $K$ factor is the percent of AADT that occurs in the peak hour, typically ranging from about 10 percent for urban environments to as much as 20 percent for rural environments. If possible, the $K$ factor used in Eq. 13-1 should be determined by empirical studies performed in the vicinity of the proposed development.

The directional factor, $D$, depends on the type of route being planned. Typical values of $D$ are 0.50 for an urban circumferential, 0.55 for an urban radial, and 0.65 for a rural highway. Preferably, the value of $D$ used in Eq. 13-1 should be based on local data.

In the next step of the planning analysis, a per lane service flow rate, $\text{SFL}_i$, for a specified level of service $i$ is chosen from Table 13-1.

Finally, the number of lanes, $N$, is estimated by the equation:

$$N = \frac{\text{DDHV}}{\text{SFL}_i \times f_E \times \text{PHF}} \tag{13-2}$$

where

$f_E$ = an adjustment factor for the type of multilane highway and amount of development (see Table 13-2)

PHF = the peak hour factor, defined as the ratio of the total hourly volume to the maximum 15-min rate of flow within the hour

Stated as an equation,

$$\text{PHF} = \frac{V}{4 \times V_{15}} \tag{13-3}$$

*Table 13-1*

**Service Flow Rate per Lane for Planning Applications (Design Speed = 70 mph)**

| Level of Service | Percent Trucks | | | | | | | |
|---|---|---|---|---|---|---|---|---|
| | *0* | *2* | *4* | *6* | *8* | *10* | *15* | *20* |
| | | | | *Level Terrain*[a] | | | | |
| A | 700 | 700 | 700 | 650 | 650 | 650 | 650 | 600 |
| B | 1100 | 1100 | 1050 | 1050 | 1050 | 1000 | 1000 | 1000 |
| C | 1400 | 1400 | 1350 | 1350 | 1350 | 1300 | 1250 | 1250 |
| D | 1750 | 1750 | 1700 | 1650 | 1650 | 1650 | 1600 | 1550 |
| E | 2000 | 2000 | 1950 | 1900 | 1900 | 1850 | 1800 | 1750 |
| | | | | *Rolling Terrain*[a] | | | | |
| A | 700 | 650 | 600 | 600 | 550 | 550 | 500 | 500 |
| B | 1100 | 1050 | 1000 | 950 | 900 | 850 | 800 | 700 |
| C | 1400 | 1300 | 1250 | 1200 | 1150 | 1100 | 1000 | 900 |
| D | 1750 | 1650 | 1550 | 1500 | 1400 | 1350 | 1250 | 1100 |
| E | 2000 | 1900 | 1800 | 1700 | 1600 | 1550 | 1450 | 1250 |
| | | | | *Mountainous Terrain*[a] | | | | |
| A | 700 | 600 | 550 | 500 | 450 | 400 | 350 | 300 |
| B | 1100 | 950 | 850 | 700 | 700 | 650 | 550 | 450 |
| C | 1400 | 1250 | 1100 | 1000 | 900 | 850 | 700 | 600 |
| D | 1750 | 1550 | 1350 | 1250 | 1100 | 1050 | 850 | 750 |
| E | 2000 | 1750 | 1550 | 1400 | 1250 | 1200 | 1000 | 850 |

[a]All values rounded to the nearest 50 veh/hr/lane.

SOURCE: *Highway Capacity Manual,* Special Report 209, Transportation Research Board, Washington, D.C., 1985.

*Table 13-2*

**Adjustment Factor for Type of Multilane Highway and Development Environment, $f_E$**

| *Type* | *Divided* | *Undivided* |
|---|---|---|
| Rural | 1.00 | 0.95 |
| Suburban | 0.90 | 0.80 |

SOURCE: *Highway Capacity Manual,* Special Report 209, Transportation Research Board, Washington, D.C., 1985.

where

$V$ = hourly volume in veh/hr

$V_{15}$ = volume during the peak 15 min of the peak hour, in veh/15 min

EXAMPLE 13-1

ESTIMATION OF NUMBER OF HIGHWAY LANES. An undivided highway is to be built in a suburban environment with rolling terrain. The forecast AADT is 18,000 veh/day, with 10 percent trucks. The fraction of traffic in the peak direction of flow is 0.60. The peak hour factor is 0.95, and the desired level of service is B. Assume that the $K$ factor is 0.15. Determine the number of lanes that will be needed in each direction.

The directional design hourly volume is calculated by multiplying the AADT by the $K$ factor and the directional factor, using Eq. 13-1:

$$\text{DDHV} = 20{,}000 \times 0.15 \times 0.60 = 1800 \text{ veh/hr}$$

From Table 13-1, the service flow rate per lane for LOS B in rolling terrain and with 10 percent trucks is:

$$\text{SFL}_{\text{B}} = 850$$

From Table 13-2, the adjustment factor, $f_E$, for an undivided highway in a suburban environment is 0.80.

By Eq. 13-2, the number of lanes required in each direction is:

$$N = \frac{1800}{850 \times 0.80 \times 0.95} = 2.8 \text{ lanes}$$

To maintain a level of service B, three lanes will have to be built in each direction.

## 13-2. WIDTHS OF HIGHWAY PAVEMENTS

While most design standards permit the lane width to be less than 12 ft, there is general agreement that the 12-ft width is more desirable. On high-speed and high-volume roads 13-ft and 14-ft widths have been used. Widths in excess of 14 ft are not recommended because some drivers will use the roadway as a multilane facility. On some multilane divided highways combinations of 12-ft and 13-ft lanes are used especially where large truck combinations are likely to occur.

In urban areas, recommended pavement width varies primarily with the classification of street or highway and the volume of traffic served. The latter factor depends primarily on the development density. Local residential streets vary in width from about 27 to 36 ft, providing safe movement of one lane traffic in each direction, even where occasional curb parking can be expected to occur.

A minimum pavement width of 36 ft is recommended for residential collector streets except in high-density developments where at least 40 ft of pavement is specified. These widths provide two traffic lanes, one in each direction, plus space for curb parking on each side. By prohibiting curb parking in the vicinity of intersections, an additional lane may be provided to facilitate turning movements.

## 13-3. DIVIDED HIGHWAYS

To provide protection against the conflict of opposing traffic, highways are frequently divided by a median strip.

The width of these median strips varies from 4 to 60 ft, or more. A median strip less than 4 to 6 ft in width is considered to be little more than a centerline stripe and its use, except for special conditions, should be discouraged. The narrower the median, the longer must be the opening in the median to give protection to vehicles making left turns at points other than intersections. Where narrow medians must be used, many agencies install median barriers to separate physically opposing streams of traffic to minimize the number of head-on collisions.

A variety of median barriers have been employed successfully, including steel W-beam guard rail, box beam, and steel cable [2]. Concrete barriers with sides specially sloped to redirect errant vehicles are recommended for narrow medians.

While medians of 14 to 16 ft are sufficient to provide most of the separation advantage of opposing traffic, medians of 16 to 60 ft are now recommended. The median should also be of sufficient width to maintain vegetation and to support low-growing shrubs that reduce headlight glare of opposing traffic. Median strips at intersections should receive careful consideration and should be designed to permit necessary turning movements.

Divided highways need not be of constant cross section. The median strip may vary in width; the roads may be at different elevations; and the superelevation may be applied separately in each pavement. In rolling terrain, substantial savings may be affected in construction and maintenance costs by this variation in design. This type of design also tends to eliminate the monotony of a constant width and equal grade.

## 13-4. PARKING LANES AND SIDEWALKS

A parking lane is a lane separate and distinct from the traffic lane. Parking should be prohibited on rural highways but in some rural areas parking adjacent to the traffic lane cannot be avoided. Parallel parking in this case should be permitted and extra lanes provided for this purpose. Where it is desirable to provide parking facilities in parks, scenic outlooks, or other points of interest, off-the-road parking should be provided. In urban and suburban locations the parking area often includes the gutter section of the roadway and may vary from 6 to 8 ft.

The minimum width of parking lane or parallel parking is 8 ft, with 10 ft preferred. For angle parking, the width of the lane increases with the angle. When the angle of parking exceeds 45 degrees, it is necessary to use two moving traffic lanes for maneuvering the vehicle into position. Angle parking should be used only in low-speed urban areas where parking requirements take precedence over the smooth flow of traffic. Parking at the approaches to intersections should be prohibited.

The use of sidewalks in the highway cross section is accepted as an integral part of city streets. In rural areas little consideration has been given to their construction since pedestrian traffic is very light. Serious consideration should be given to the construction of sidewalks in all areas where the number of pedestrians using the highway warrants it. The Institute of Transportation Engineers recommends that sidewalks be provided along subdivision streets where the development density exceeds two dwellings per acre. Where the residential density warrants the use of sidewalks, a sidewalk width of 4 to 6 ft is recommended.

## 13-5. SPECIAL CROSS-SECTIONAL ELEMENTS

In the design of multilane highways and expressways, it is necessary to separate the through traffic from the adjacent service or function roads. The width and type of separator is controlled by the width of right-of-way, location and type of overpasses or underpasses, and many other factors. The border strip, which is that portion of the highway between the curbs of the through highway and the frontage or service road, is generally used for the location of utilities.

Special consideration for off-the-road parking facilities, mailbox turnouts, or additional lanes for truck traffic on long grades causes corresponding changes in the cross sections. Highways involving these elements of the cross section usually are treated as special cases and receive consideration at the time of their design.

## 13-6. RIGHT-OF-WAY

Right-of-way requirements are based on the final design of the cross-sectional elements of the facility. On two-lane secondary highways with an average daily traffic volume of 400 to 1000 vehicles, a minimum of 66 ft is required, with 80 ft desirable. On the Interstate System minimum widths will vary depending on conditions from 150 ft without frontage roads to 250 ft with frontage roads. An eight-lane divided highway without frontage roads will require a minimum of 200 ft, while the same highway with frontage roads will require a minimum of 300 ft. In rural areas of high-type two-lane highways, a minimum width of 100 ft with a desirable width of 120 ft is recommended. A minimum of 150 ft and a desirable width of 250 ft is recommended for divided highways.

A right-of-way width of 60 ft is generally recommended for local subdivision streets, while collector streets should have 70 ft.

Right-of-way should be purchased outright or placed under control by easement or other means. When this is done, sufficient right-of-way is available when needed. This eliminates the expense of purchasing developed property or the removal of other encroachments from the highway right-of-way.

## 13-7. PAVEMENT CROWN OR SLOPE

Another element of the highway cross section is the pavement crown or slope. This is necessary for the proper drainage of the surface to prevent ponding on the pavement. Pavement crowns have varied greatly through the years. On the early low-type roads, high crowns were ½ in. or more per foot. Present day high-type pavements with good control of drainage now have crowns as low as ⅛ in./ft. This has been made possible with the improvement of construction materials, techniques, and equipment which permit closer control. Low crowns are satisfactory when little or no settlement is expected and when the drainage system is of sufficient capacity to remove the water quickly from the traffic lane. When four or more lanes are used, it is desirable to provide a higher rate of crown on the inner lanes in order to expedite the flow of water.

## 13-8. SHOULDERS

Shoulders should be provided continuously along all highways. This is necessary to provide safe operation and to develop full traffic capacity. Well-maintained, smooth, firm shoulders increase the effective width of the traffic lane as much as 2 ft, as most vehicle operators drive closer to the edge of the pavement in the presence of adequate shoulder. Shoulders should be wide enough to permit and encourage vehicles to leave the pavement when stopping. The greater the traffic density the more likely will the shoulder be put to emergency use. A shoulder width of at least 10 ft and preferably 12 ft clear of all obstructions is desirable for all heavily traveled and high-speed highways. In mountainous areas where the extra cost of providing wide shoulders may be prohibitive, a minimum width of 4 ft may be used, but a width of 6 to 8 ft is preferable. Under these conditions, however, emergency parking strips should be provided at proper intervals. In terrain where guard rails or retaining walls are used, an additional 2 ft of shoulder should be provided.

The slope of the shoulder should be greater than that of the pavement. A shoulder with a high-type surfacing should have a slope at least ⅜ in./ft. Sodded shoulders may have a slope as high as 1 in./ft in order to carry water away from the pavement.

## 13-9. SLOPES AND ROADSIDES

The downward slope from the edge of the shoulder toward the ditch is called the side slope[1]. The slope from the edge of the ditch upward is called the back slope. Side slopes and back slopes may vary a great deal depending upon the type of material and the terrain through which the highway is being constructed.

Side slopes of 6 : 1 (6 horizontal to 1 vertical) or flatter can be negotiated safely by a vehicle and should be provided wherever practical. A 6 : 1 slope usually can be provided with little extra expense on fills up to 15 ft in height. On higher fills, some agencies specify a recovery area having a 6 : 1 slope for about 20 ft then a steeper slope. This design is known as a barn roof cross section. An alternative is to provide a 4 : 1 slope downward from the shoulder. Much steeper slopes, as steep as 1.5 : 1, are often used along high fill sections for economical reasons. In such instances, roadside barriers must be placed along the edge of the fill.

The back slope in cut areas may vary from 1 : 6 to vertical in rock or loose formations to 1.5 : 1 in normal soil. It is advisable to have back slopes as flat as 4 : 1 when side borrow is needed. Slope transitions from cut to fill should be gradual and should extend over a considerable length of roadway.

Since a large percentage of crashes involve vehicles that run off the roadway, a great deal of care should be given to the design of the roadsides and slopes. The roadside area (including the shoulder) should be designed so as to give the errant motorist as much chance as possible to regain control of the vehicle. Research has shown that properly designed roadside areas of about 30-ft width should permit recovery of about 85 percent of vehicles out of control. Where feasible, a 30-ft-wide, obstacle-free zone adjacent to the roadway should be provided. Signs and other potential hazards that must be placed close to the roadway should provide energy attenuation or be of breakaway design. Where neither removal nor safety modification of a fixed object hazard is feasible, the installation of a highway guard rail may be warranted.

## 13-10. HIGHWAY GUARDRAIL

Generally speaking, guardrail barriers are warranted in fills more than 8 ft in height with slopes steeper than 3 : 1. Guardrail may also be required along the edges of deep roadside ditches with steep banks and in areas of limited

[1]Some agencies term this slope the *downslope* in fill (embankment) sections and the *inslope* in cut (excavation) sections.

rights-of-way or steep terrain where steep side slopes must be used. Where guardrail is used, the width of the shoulders should be increased by approximately 2 ft to allow space for placing the posts.

Various types of guardrail systems are in use at the present time. Three types of guardrail that have performed satisfactorily are shown in Fig. 13-1. References 3, 4, and 5 provide more detailed information on the selection, location, and design of guardrail, median barriers, and energy attenuation devices.

## 13-11. CURBS, CURB AND GUTTER, AND DRAINAGE DITCHES

The use of curbs generally is confined to urban and suburban roadways. The design of curbs varies from a low, flat, lip-type to a nearly barrier-type curb. Curbs adjacent to traffic lanes, where sidewalks are not used, should be low and very flat. The face of the curb should be no steeper than 45 degrees so that vehicles may mount them without difficulty. Curbs at parking areas and adjacent to sidewalks should be 6 to 8 in. in height, with faces nearly vertical. Clearances should be sufficient to clear fenders and bumpers and to permit the opening of doors. When a barrier or nonmountable curb is used, it should be offset a minimum of 10 ft from the traffic lane.

Drainage ditches, in their relation to the highway cross section, should be located within or beyond the limits of the shoulder and under normal conditions be low enough to drain the water from under the pavement. The profile gradient of the ditch may vary greatly from that of the adjacent pavement. A rounded ditch section has been found to be safer than a V-type ditch, which also may be subject to severe washing action. Ditch maintenance is also less on the rounded ditch sections.

Typical cross-section dimensions for arterial streets and rural highways are shown, respectively, in Fig. 13-2 and Fig. 13-3.

## 13-12. LIMITED-ACCESS HIGHWAYS

A limited-access highway may be defined as a highway or street, designed especially for through traffic, to which motorists and abutting property owners have only a restricted right of access. Limited or controlled access highways may consist of (1) freeways that are open to all types of traffic, and (2) parkways from which all commercial traffic is excluded. Most of our present expressway systems have been developed as freeways.

Limited-access highways may be elevated, depressed, or at grade. Many examples of the various types may be found in the United States in both rural and urban areas.

The control of access is attained by limiting the number of connections to

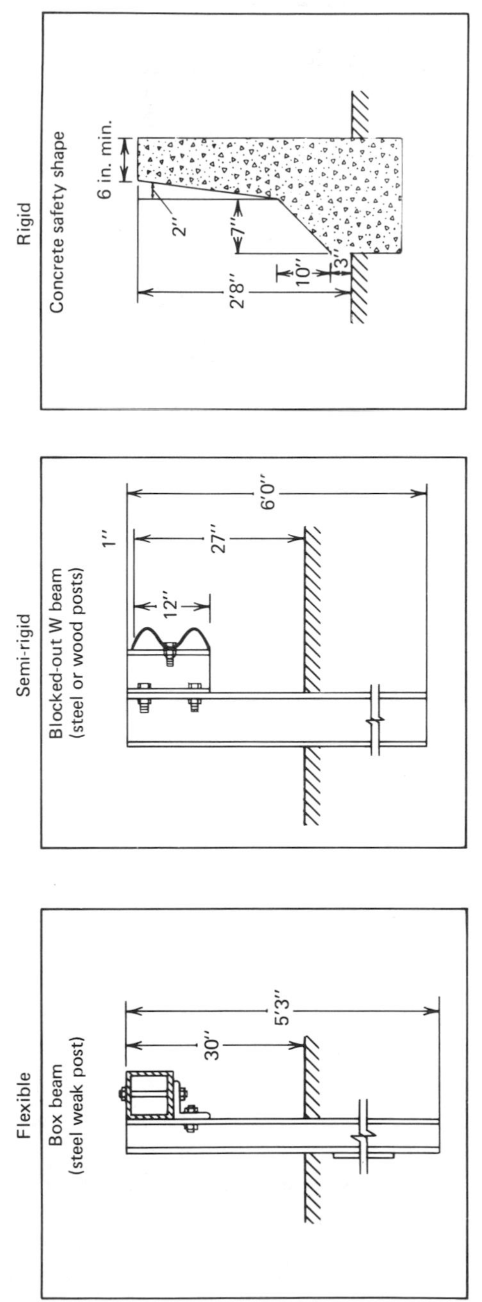


*Fig 13-1.* Examples of guardrails that have performed satisfactorily. (Source: *Handbook of Highway Safety Design and Operating Practices*, Federal Highway Administration, 1979.)

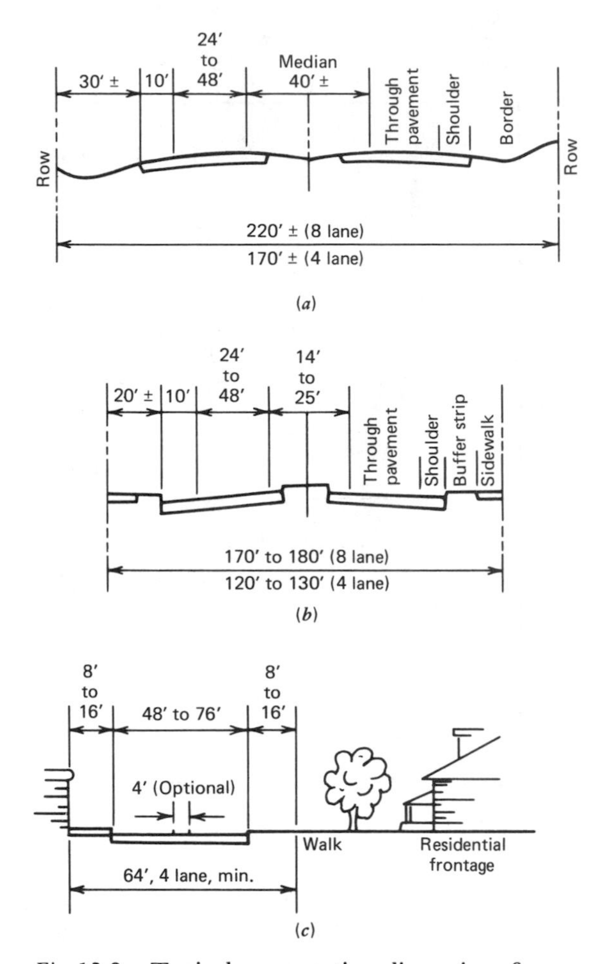


*Fig 13-2.* Typical cross-section dimensions for arterial streets. (Courtesy American Association of State Highway and Transportation Officials.) (*a*) Desirable. (*b*) Intermediate. (*c*) Restricted without frontage roads.

and from the highway, facilitating the flow of traffic by separating cross-traffic with overpasses or underpasses, and by eliminating or restricting direct access by abutting property owners.

The design of limited-access routes should provide for adequate width of right-of-way, adequate landscaping, prohibition of outdoor advertising in the controlled access proper, and provisions for controlling abutting service facilities such as gas stations, parking areas, and other roadside appurtenances.

In urban areas, the design of a limited-access facility is usually accompanied by the design of frontage roads, parallel to the facility, which may serve local traffic and provide access to adjacent property. Such roads may be

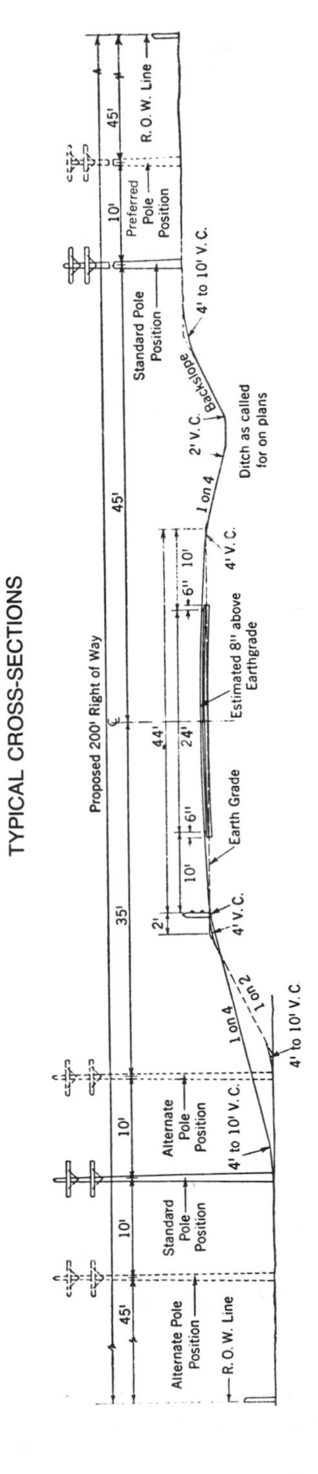


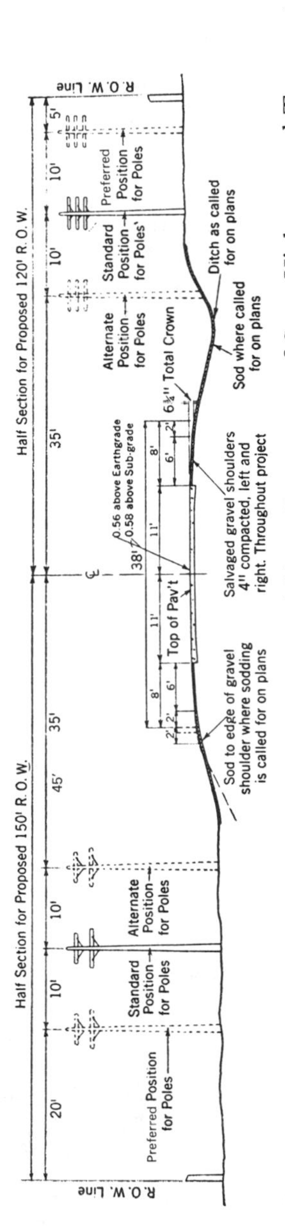


*Fig 13-3.* Typical highway cross sections. (Courtesy Michigan Department of State Highways and Transportation.)

designed for either one-way or two-way operation. Reasonably convenient connections should be provided between through-traffic lanes and frontage roads.

## 13-13. HIGHWAY DESIGN STANDARDS

Because of functional differences, a wide array of design standards must be used for the many types of facilities comprising a highway system. Design standards vary widely for different functional classes. For example, freeways are designed predominantly for traffic movement, and freeway standards are characterized by high-design speeds, wide lanes, and straight horizontal and vertical alignments. At the other end of the spectrum, standards for subdivision streets reflect an emphasis on land access function. Such standards, exemplified by Tables 13-3 and 13-4, are based on design speeds of 20 to 35 mph. In fact, local streets should be designed to discourage excessive speeds through the use of curvilinear alignment and discontinuities in the street system [4, 6].

Within a functional class, design standards may vary with the type of terrain, anticipated traffic to be served, and whether the highway is to be in an urban or rural area. It is not surprising, therefore, to find some variation in standards among the various highway agencies.

# THE RAILWAY CROSS SECTION

The railway cross section of today is the result of more than a century of evolutionary change. In the early days of railroading, both the longitudinal-tie-track and the crosstie-track were used. For the U.S. intercity railroad system, the crosstie-track has prevailed and, predominantly, wood crossties have been used.

The cross-sectional area of the rail has been increased continuously, and the tie spacing has been decreased to accommodate increasing wheel loads. However, the spacing of the ties cannot be reduced indefinitely, nor can the rail section be increased without limit. For these reasons, some thought now is being given to eliminating the tie spaces altogether by using instead a continuous reinforced concrete slab. In this design, the rails would be secured by fasteners that are anchored in the slab [7].

The paragraphs that follow describe the various components of a railroad cross section. Looking to the future it should be remembered that other design concepts and approaches to track design may be needed to serve heavier loads and increased traffic and to adapt to technological change.

Standards for the width of the subgrade are determined for mainline, secondary line, and light traffic branch lines and spurs. Many railroads have

*Table 13-3*
**Collector Street Design Guidelines**

| Terrain classification | Level | | | Rolling | | | Hilly | | |
|---|---|---|---|---|---|---|---|---|---|
| Development density | Low | Medium | High | Low | Medium | High | Low | Medium | High |
| Right-of-way width (ft)[a] | 70 | 70 | 70 | 70 | 70 | 70 | 70 | 70 | 70 |
| Pavement width (ft) | 36 | 36 | 40 | 36 | 36 | 40 | 36 | 36 | 40 |
| Type of curb (v = vertical face) | V | V | V | V | V | V | V | V | V |
| Sidewalk width (ft) | ← | 4–6 | — | — | 4–6 | — | — | 4–6 | → |
| Sidewalk distance from curb face (ft) | ← | 10 | — | — | 10 | — | — | 10 | → |
| Minimum sight distance (ft) | ← | 250 | → | ← | 200 | → | ← | 150 | → |
| Maximum grade | ← | 4 | → | ← | 8 | → | ← | 12 | → |
| Minimum spacing along major traffic route (ft) | ← | 1300 | — | — | 1300 | — | — | 1300 | → |
| Design speed (mph)[b] | ← | 35 | → | ← | 30 | → | ← | 25 | → |
| Minimum centerline radius (ft)[c] | ← | 350 | → | ← | 250 | → | ← | 175 | → |
| Minimum tangent between Reverse curves (ft) | ← | 100 | — | — | 100 | — | — | 100 | → |

[a]1 ft = 0.3048 m.

[b]1 mph = 1.6093 km/hr.

[c]Assumes superelevation.

SOURCE: *Recommended Guidelines for Subdivision Streets: A Recommended Practice,* Institute of Transportation Engineers, Washington D.C., 1984.

*Table 13-4*
**Local Street Design Guidelines**

| Terrain classification → | Level | | | Rolling | | | Hilly | | |
|---|---|---|---|---|---|---|---|---|---|
| Development density → | Low | Medium | High | Low | Medium | High | Low | Medium | High |
| Right-of-way width (ft)[a] | 50 | 60 | 60 | 50 | 60 | 60 | 50 | 60 | 60 |
| Pavement width (ft) | 22–27 | 28–34 | 36 | 22–27 | 28–34 | 36 | 28 | 28–34 | 36 |
| Type of curb (V = vertical face; R = roll-type; 0 = none) | 0/R | V | V | V | V | V | V | V | V |
| Sdewalks and bicycle paths (ft) | 0 | 4–6 | | 0 | | 4–6 | 0 | | 4–6 |
| Sidewalk distance from curb face (ft) | — | 6 | 6 | — | 6 | 6 | — | 6 | 6 |
| Minimum sight distance (ft) | ← | 200 | → | ← | 150 | → | ← | 110 | → |
| Minimum grade (percent) | ← | 4 | → | ← | 8 | → | ← | 15 | → |
| Maximum cul-de-sac length (ft) | 1000 | 700 | 700 | 1000 | 700 | 700 | 1000 | 700 | 700 |
| Minimum cul-de-sac radius (right-of-way) (ft) | ← | 50 | — | — | 50 | — | — | 50 | → |
| Design speed (mph)[b] | ← | 30 | → | ← | 25 | → | ← | 20 | → |
| Minimum centerline radius of curves (ft) | ← | 250 | → | ← | 175 | → | ← | 110 | → |
| Minimum tangent between reverse curves (ft) | ← | 50 | — | — | 50 | — | — | 50 | → |

[a] 1 ft = 0.3048 m.

[b] 1 mph = 1.6093 km/hr.

SOURCE: *Recommended Guidelines for Subdivision Streets: A Recommended Practice,* Institute of Transportation Engineers, Washington, D.C., 1984.

adopted 20 ft as the width of single-lane main lines. This width will also vary according to the height of fill. A width of 22 ft has been used for fills under 20 ft; 24 ft for fills 20 to 50 ft; and 26 ft for fills over 50 ft. Standard widths in cuts including side ditches is 30 ft. In some locations 40 ft is used to permit the use of off-track equipment for maintenance purposes. Common widths of rights-of-way in open country are 50, 60, 80, 100, and sometimes 200 and 400 ft.

There are seven main elements of a railroad cross section: (1) ballast, (2) crossties, (3) rails, (4) tie plates, (5) fastenings, (6) rail anchors, and (7) rail joints.

## 13-14. BALLAST

Track ballast is a key structural element of the railroad permanent way. Its prime function is to transmit and distribute the wheel loadings from the base of the crossties to the subgrade at pressures that will not cause subgrade failure. In addition, ballast serves to anchor the track, preventing longitudinal and transverse track movements under dynamic train loading, to provide immediate drainage of the permanent way under the ties, and to provide a road material that inhibits vegetation growth and minimizes dust.

Open graded materials that can perform satisfactorily the required functions of ballast are crushed stone, washed river or pit run gravel, and furnace slags. Typically, material varies in grain size from 1½ to 1¾ in. Where ballast material is expensive or in short supply, or where subgrade strength is sufficiently low that excessive depths of ballast would be required, a layer of sub-ballast frequently is used. Material for the sub-ballast layer of the permanent way can be a less openly graded material meeting less stringent quality requirements.

The depth of the ballast may vary from 6 to 30 in. or more depending on wheel loads, traffic density and speed, and the type and condition of the foundation. The thickness of the sub-ballast may also vary, but excellent results have been obtained with a thickness of about 12 in. [8].

The AREA has set quality standards on ballast with reference to the following criteria:

1. *Wear resistance.* Under the Los Angeles abrasion test, percentage of wear of any ballast material is limited normally to 40 percent.
2. *Cleanliness.* Deleterious substances are limited in prepared ballasts to the following amounts:

| | |
|---|---|
| Soft and friable pieces | 5 percent |
| Material finer than No. 200 Sieve | 1 percent |
| Clay lumps | 0.5 percent |

3. *Frost resistance.* Ballast must be capable of resisting freeze-thaw

cycles. AREA requires an average weight loss of not more than 7 percent after 5 cycles of the sodium sulfate soundness test.

4. *Unit weight.* Specifications require compacted weights of not less than 70 and 100 lb/ft$^3$ for blast furnace and open hearth slags, respectively.

## 13-15. CROSSTIES OR SLEEPERS

The crosstie serves several functions, including (1) spreading the horizontal and vertical loadings to the ballast; (2) maintaining the correct gage between the rails; (3) providing, in conjunction with the ballast, a means of anchoring the track against longitudinal and lateral movements; and (4) providing a convenient means for making needed adjustments of the vertical profile of the track [8].

Vertical loadings are applied to the ties by the train weight. Horizontal longitudinal loadings occur as trains accelerate and decelerate, while transverse loadings are applied as the vehicles transverse curved sections. Additional transverse loads are present due to the "barreling" effect of locomotives at high speed. To permit the horizontal transfer of forces from the tie, the ballast is tamped mechanically between the ties as shown in Fig. 13-4.

Typically, ties are made of wood which is treated with both preservative and coating materials for protection against weathering and splitting while in service. Tie sections vary from 6 in. thick × 6 in. wide to 7 in. thick × 9 in. wide. Tie lengths also vary. Standard sizes are 8, 8.5, and 9 ft long. Tie replacement, which averages about 3 percent of all ties yearly, accounts for a large proportion of total track maintenance.

Many railroads have had satisfactory experience with concrete crossties (or *sleepers* as they are sometimes called). Reference 9 reports that there are more than 3000 million sleepers in the world of which more than 400 million are concrete sleepers. In fact, concrete sleeper demand accounts for more than half of the sleeper demand in many parts of the world, especially in most of Europe, in the USSR, in Japan, and in some parts of Africa [9].

Two types of design systems have been used for concrete sleepers to accommodate the wide range of positive and negative bending movements. In the first system, there is one rigid concrete block under each rail connected by a central flexible (e.g., steel) piece. See Fig. 13-5. In the other system, a single rigid concrete beam is used. The monoblock sleepers are almost always prestressed to resist the dynamic bending moment distribution [9]. Concrete sleepers must also be renewed, typically at a rate of 2–5 percent per year.

Statistics of the Association of American Railroads indicate that there are approximately 850 million crossties in service in the United States, an average of about 3000 ties/mi of track [10]. This gives an average spacing of approximately 21 in. However, most railroads space ties more closely on main line tracks and more widely on branch lines and in yards.

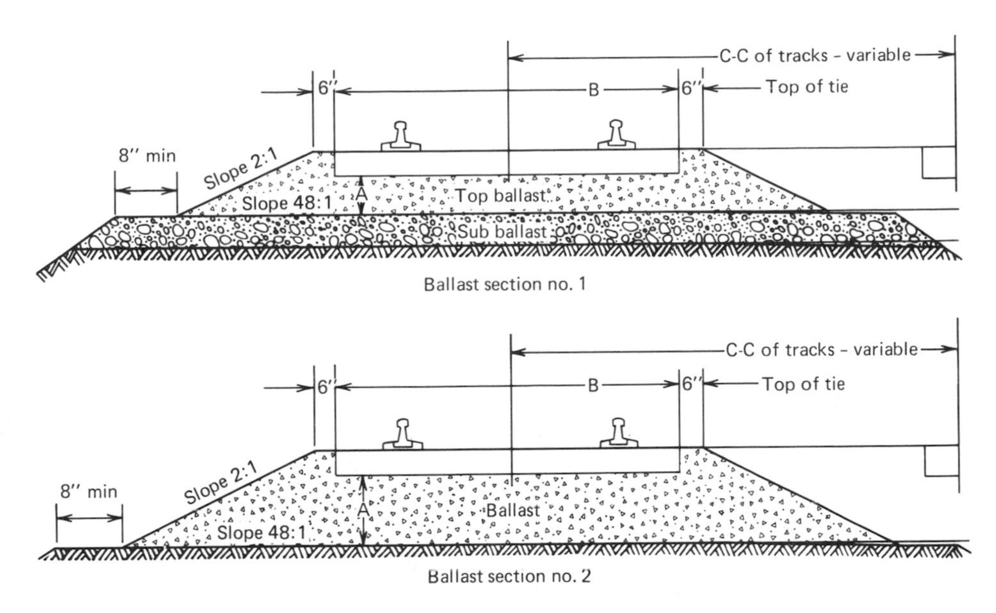


Notes:
Depth of ballast section to be used will depend on conditions peculiar to each railroad or location.
Sections apply to all types of ballast.
Sections for use with jointed or continuous rail.
Top of ballast determined by the use of various mechanized ballast distributing operations.

*Fig 13-4.* Typical roadbed section for single main track. (Courtesy Southern Railway System.)

## 13-16. RAILS

The steel rail has a characteristic inverted T-shape. It functions as a continuous steel beam, transmitting vertical loads and horizontal shears to the ties via the tieplates and fastenings. Rail sections come in standard lengths of 39 ft, a length chosen because it can be transported conveniently in a single 40-ft car. In recent years, continuous welded rail in lengths of 1440 ft has been used increasingly by a number of railroads. Advantages claimed for the use of continuous welded rail include decreased maintenance of way costs, higher permissible operating speeds, less likelihood of damage to lading, and a smoother ride resulting in less wear and tear on equipment [10].

Rails are designated in standard sections by weight in lb/yd (e.g., 115 lb/yd). The AREA [8] recommends seven rail sections ranging from 90 to 140 lb/yd. There is no established formula for selecting the size of rail. Railroad officials generally base the selection of a rail section on the annual gross tons of traffic to be transported and on the train speeds over the track. A typical rail section is shown by Fig. 13-6.

Rail gage, the distance between the inside rail heads, is a standard 4 ft

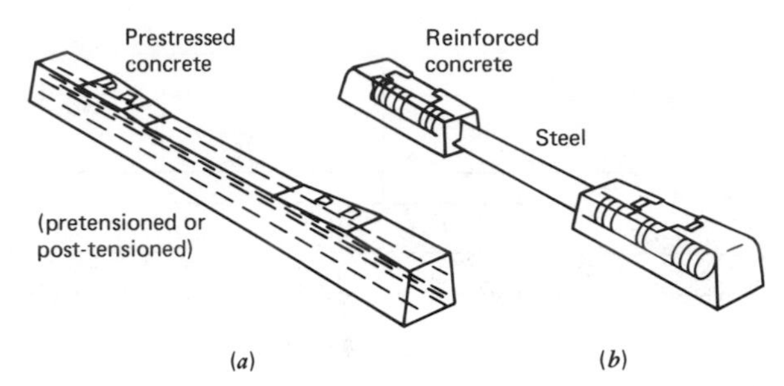


*Fig 13-5.* Basic types of concrete sleepers. (*a*) Monoblock. (*b*) Twin-block. (Source: *Concrete Railway Sleepers, State of the Art Report,* Thomas Telford, London, 1987.)

8½ in. in the United States. Gage standardization permits free interchange of rolling stock on a nationwide basis.

## 13-17. TIE PLATES

The rail is laid on tie plates that are secured to the crossties by spikes or other fastenings. A tie plate may be seen in Fig. 13-10.

Tie plates have three principal functions [10]:

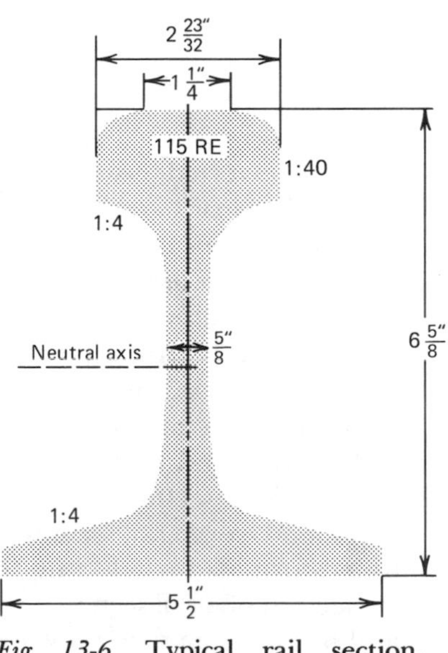


*Fig 13-6.* Typical rail section. (Courtesy Southern Railway System.)

1. To prevent damage to the wood crosstie by distributing the wheel loads over a larger area.
2. To help hold the rail to proper gage.
3. To partially offset the outward lateral thrust of the wheel loads by tilting the rails slightly inward.

Tie plates are typically 7 to 8 in. wide, 10 to 14 in. long, and 9/16 to 1 in. thick.

Special rubber, neoprene, or plastic pads sometimes are placed between the tie plates and crossties to provide softer-riding track and reduce tie wear.

## 13-18. FASTENINGS

Tie plates must be anchored firmly to the tie to prevent destructive abrasion from plate movement. Several types of fastenings may be used to secure the tie plates including common cut spikes (Fig. 13-7), spring spikes, screw spikes, compression clips (Fig. 13-8), and clamps.

## 13-19. RAIL ANCHORS

Rail anchors, exemplified by those shown in Fig. 13-9, are employed to reduce or stop the longitudinal movement of rails under traffic and to control temperature-induced expansion of rails. Their main purpose is to hold the rail in a fixed position with respect to the tie. Without anchorage, rails tend to expand unevenly, and local concentrations of expansion forces may cause the track to buckle or warp. This can result in twisted ties, tight gage, and broken welds at rail joints.

The rail anchor usually is attached to the base of the rail with one of its vertical surfaces bearing against the side of the tie or tie plate, or both. Creeping pressures of the rail are transmitted to the tie and ultimately to the ballast.

Rail anchors are used in great numbers. For mainline tracks carrying traffic essentially in one direction, the AREA recommends that 8 forward anchors and 2 backup anchors be used per 39 ft of rail length. For mainline tracks carrying traffic in both directions, 16 anchors per 39 ft of rail length are recommended. Where continuous welded rail is used, each rail usually is anchored at alternate ties [10].

## 13-20. RAIL JOINTS

Rail joints are used to provide smooth continuity of alignment and surface where two rail ends meet and to transfer the wheel load from one rail end to

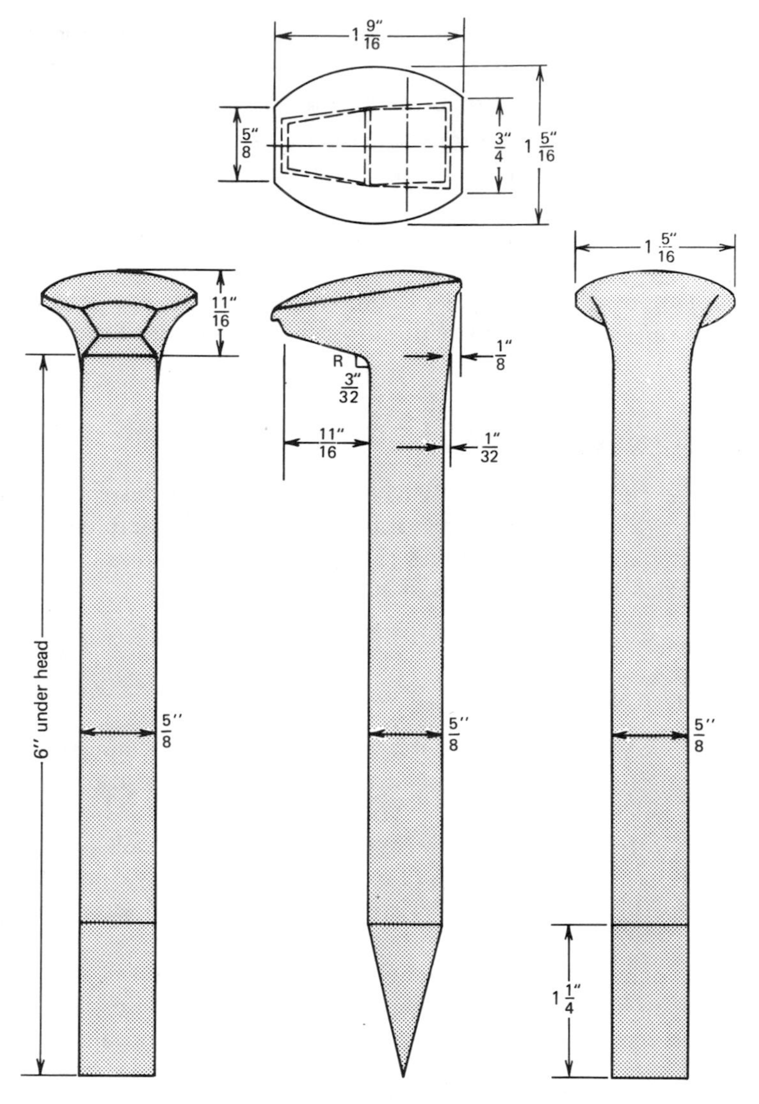


*Fig 13-7.* Design of a ⅝-in. reinforced throat track spike. (Source: *AREA Manual of Railway Engineering,* pp. 5-2-6.)

the other. Rail joints, illustrated by Fig. 13-10, consist of two steel members that fit on each side of the rail and span the gap between the two rails. These bars are typically 24 or 36 in. in length and usually are held in place by bolting through holes in the rail flange. The 24-in. bars have four bolt holes, and the 36-in. bars have six.

Where it is necessary to electrically isolate the track circuit carried by the

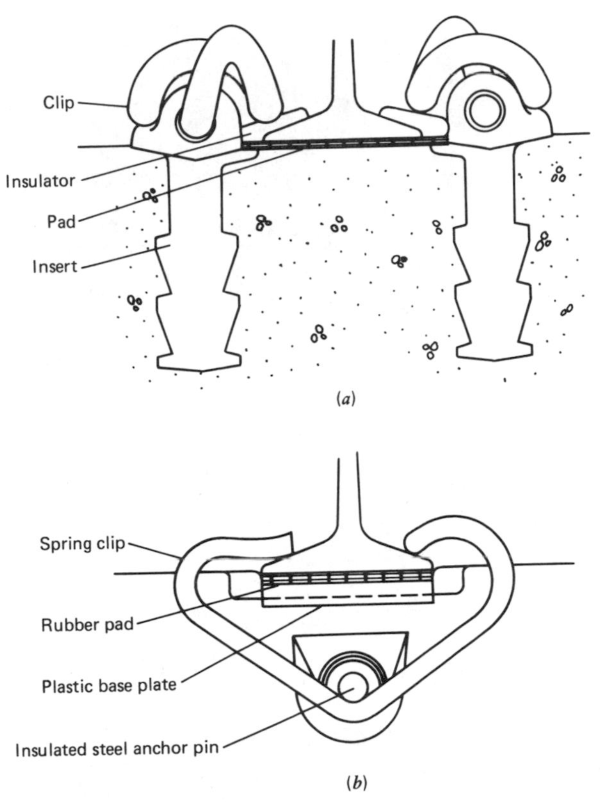


*Fig 13-8.* Examples of clip fastening systems. (*a*) Pandrol. (*b*) Fist. (Source: *Concrete Railway Sleepers, State of the Art Report.* Thomas Telford, London, 1987.)

rails, insulated joints are used. (See Chapter 5 for a discussion of automatic block signaling.) This type of joint has an insulating material placed between the bars and the rail. Some railroads now use a glued or bonded insulated joint, in which the joint bars are fastened rigidly to the rail by means of structural adhesives.

## URBAN RAIL TRANSIT CROSS SECTION

The cross sections of urban rail transit systems are essentially similar to conventional railroad systems where duorail systems are used. However, with lighter cars and less demanding service requirements, urban systems may employ prestressed concrete ties with specially designed tie plates, fastenings, anchors, and joints. Atlanta's rail system (MARTA), for example, has 115-lb

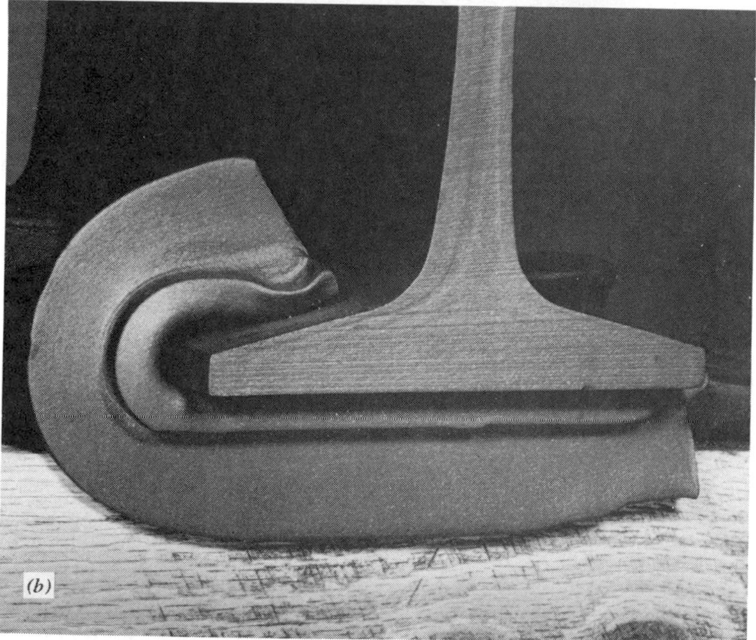


*Fig 13-9.* Typical rail anchors. (Photos courtesy of (*a*) Woodings-Verona, (*b*) Portec, Inc.)

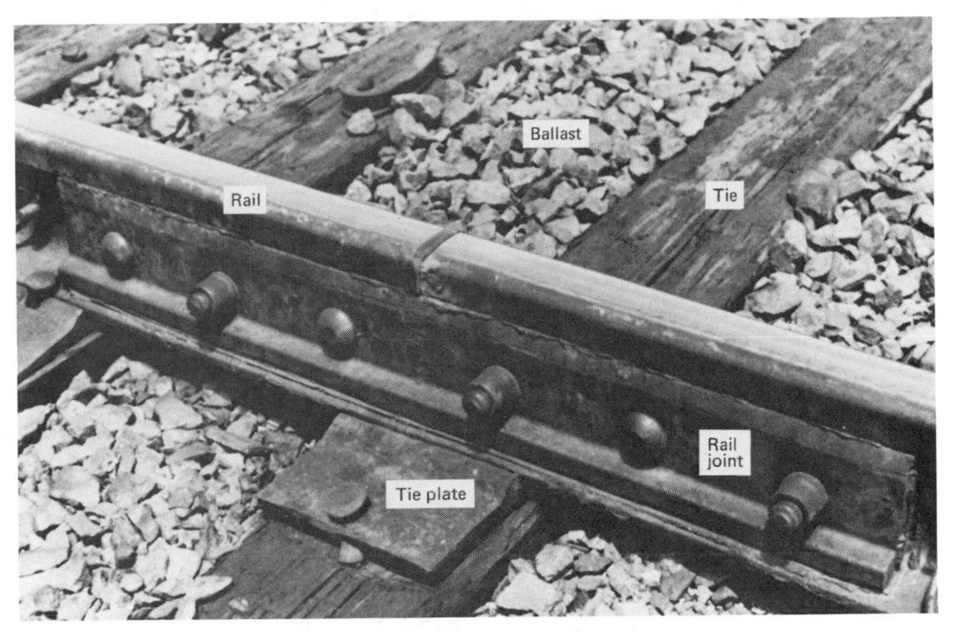


*Fig 13-10.* A rail joint. (Courtesy, Portec, Inc.)

rail attached to prestressed concrete ties by Pandrol rail clips (illustrated in Fig. 13-8). In the stations, each rail is attached to 2-ft 2-in.-wide concrete slabs by means of Hixon direct fixation fasteners. The concrete slabs are typically 10 ft in length, and the fasteners are spaced in 30-in. centers. A typical section of the MARTA primary track at grade is shown as Fig. 13-11, while typical station sections are shown as Fig. 13-12.

Special cross sections are necessary for nonconventional designs such as suspended and rubber-tired systems. For example, Montreal's transit system utilizes rubber-tired cars that travel along a 10-in. wide slab while being guided by small horizontally mounted rubber-tired wheels. A cross section of track for the Montreal system is shown in Fig. 13-13.

In designing the alignment of rail transit and limited-access freeways, the engineer faces the choice of elevated, at-grade, depressed, and subway sections.

At-grade alignments being the cheapest to construct have large economic advantages. This type of facility causes the most community disruption from the viewpoint of severing existing street patterns. Noise and fume levels can also be high.

Depressed systems tend to cut noise and fume levels and, depending on the number of bridge structures provided, can decrease the disruption of existing traffic patterns. Depressed facilities are more expensive to construct than at-grade alignments.

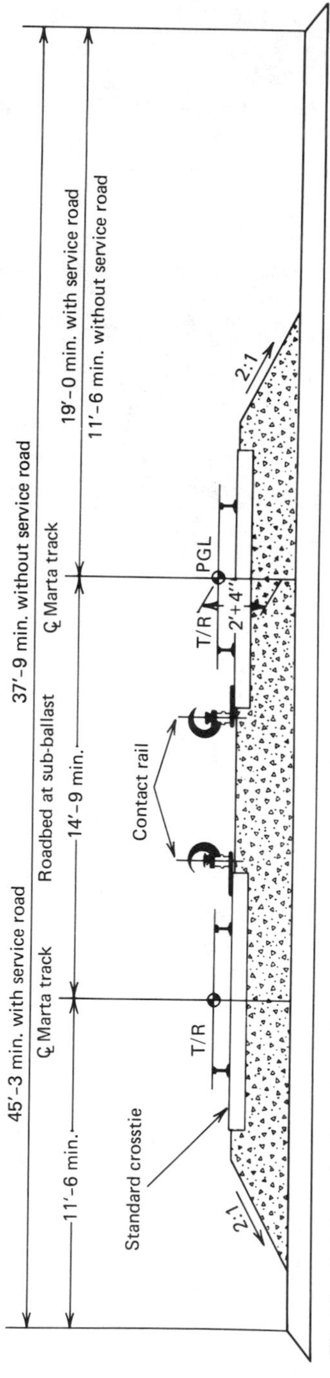


*Fig 13-11.* A typical section of an urban rail transit track, at grade. (Courtesy Metropolitan Atlanta Rapid Transit Authority.)

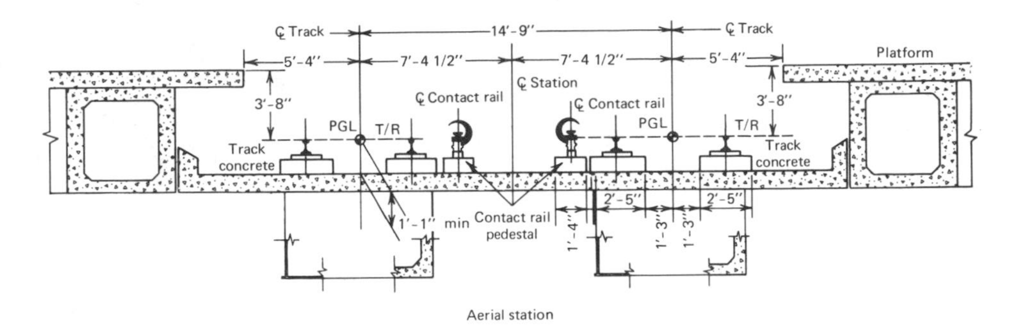


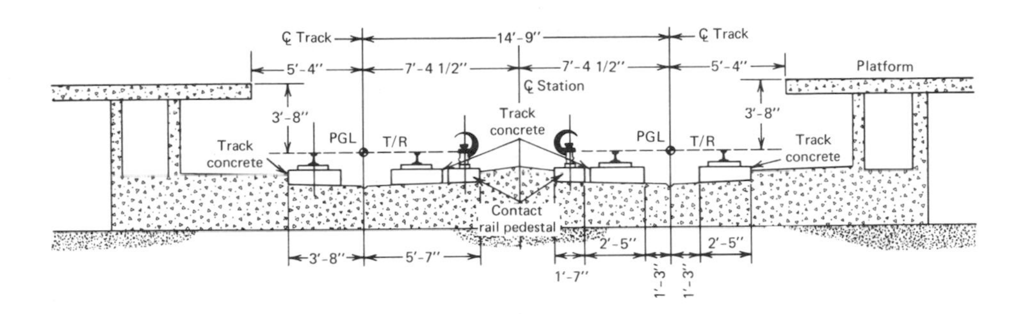


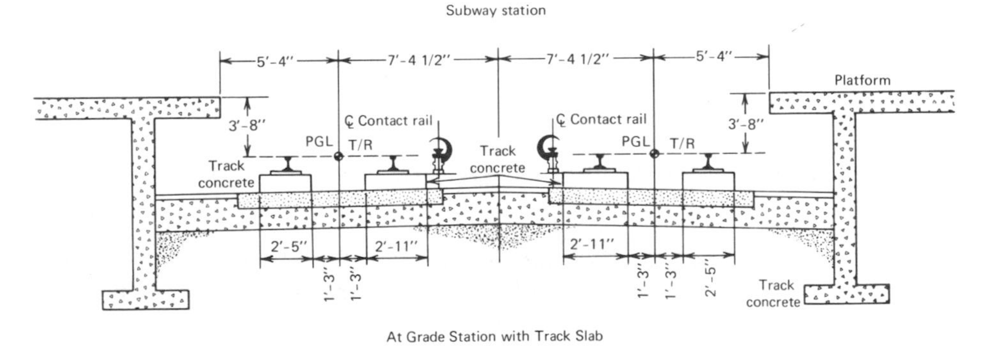


*Fig 13-12.* Typical sections of urban rail transit tracks in stations. (Courtesy Metropolitan Atlanta Rapid Transit Authority.)

Elevated facilities can eliminate almost completely surface traffic disruption. Noise, fume, and aesthetic pollution levels tend to be high in spite of the designer's best efforts to make the structures acceptable. In high-density areas these facilities become uneconomic due to the difficulty of providing station space for rail systems and ramp spacing for freeways.

Subway alignments cause minimum disruption of existing circulation. Their chief disadvantages lie in the very large cost of construction, and the increased noise levels within the vehicles.

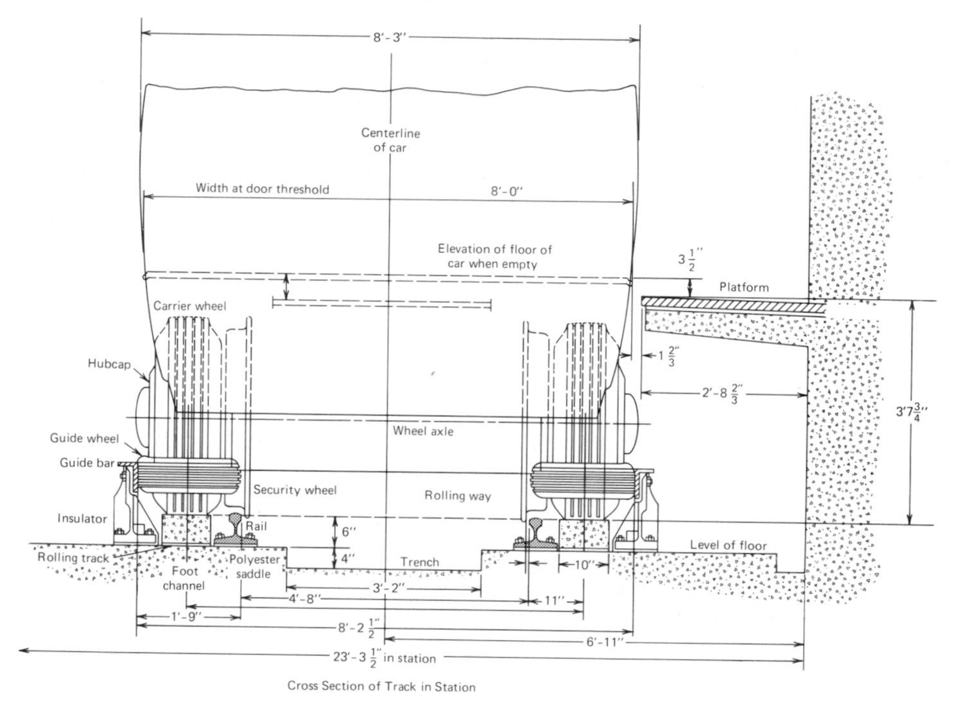


*Fig 13-13.* Cross section of track in station, for Montreal's rubber-tired transit system.

## HIGHWAY INTERSECTIONS AND INTERCHANGE DESIGN

An important part of a highway is the intersection. This is the place where two or more highways meet and provides an area for the cross movement of vehicle traffic. The efficiency, safety, speed, cost of operation, and capacity are dependent upon its design.

There are three general types of intersections: (1) intersection at grade, (2) grade separations without ramps, and (3) interchanges.

### 13-21. HIGHWAY INTERSECTIONS AT GRADE

Most highways intersect at grade, and the intersection area should be designed to provide adequately for turning and crossing movements, with appropriate consideration given to alignment, grades, sight distance, and traffic control.

Simple intersections at grade consist of three, four, or more road approaches. A junction of three approaches forms a "branch," T, or Y. A branch is a minor roadway that intersects a main highway at a small deflection angle. A T-intersection is one in which two roads intersect to form a continuous highway and the third road intersects at or nearly at right angles. A Y-intersection is one in which three roads intersect at nearly equal angles. In addition to these types, a flared intersection may be used, which has additional traffic lanes at the intersection area. Figure 13-14 shows examples of the general types of intersections at grade.

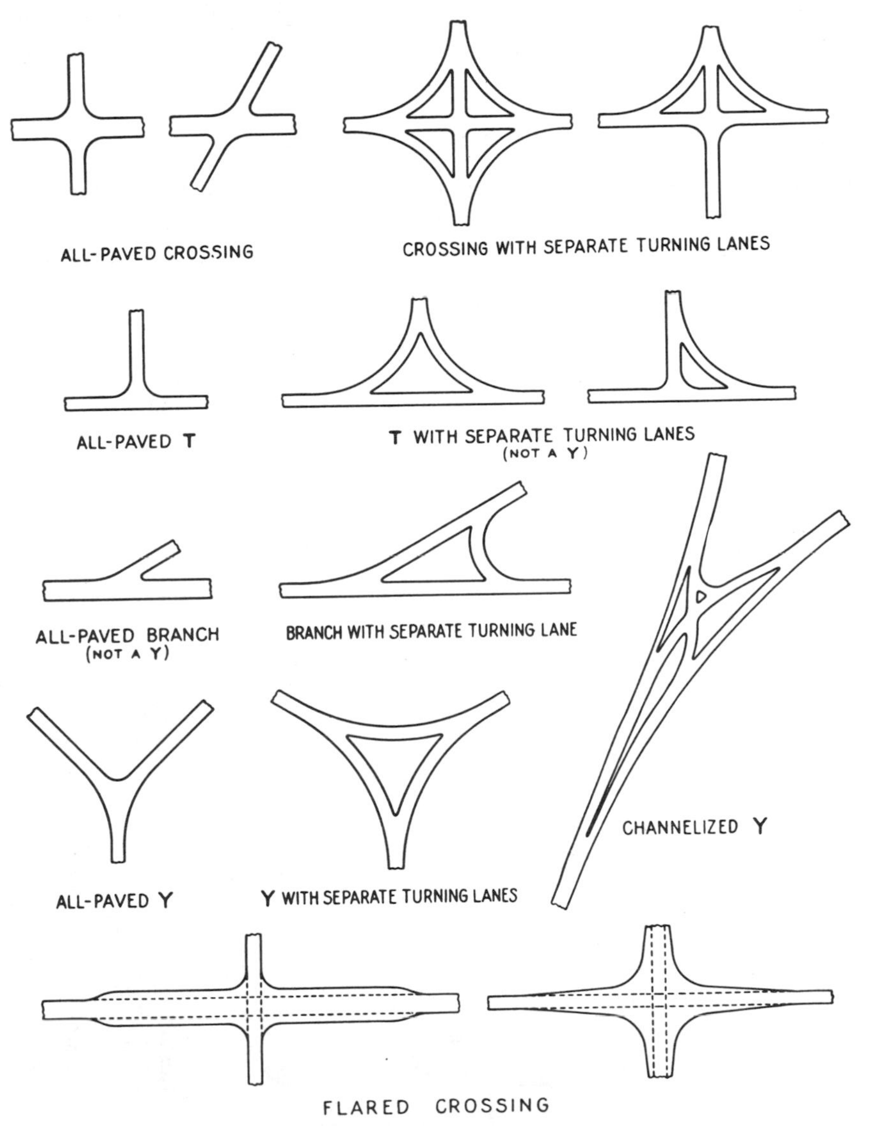


*Fig 13-14.* General types of at-grade intersections. (Source: *A Policy on Geometric Design of Rural Highways,* 1965, AASHO, p. 388.)

The design of the edge of pavement for a simple intersection should provide sufficient clearance between the vehicle and the other traffic lanes. It is frequently assumed that all turning movements at intersections are accomplished at speeds of less than 20 mph, and the design is based on the physical characteristics of the assumed vehicle. For 90-degree turns, a minimum curb radius of 30 ft for passenger car traffic is recommended, and a radius of at least 50 ft is required for single-unit trucks. Larger vehicles are best served by three-centered compound curves. Tables 13-5 and 13-6 give recommended minimum edge of pavement designs for turns at intersections using simple curves and three-centered compound curves, respectively.

## 13-22. CHANNELIZED INTERSECTIONS

Channelization is defined as the separation of conflicting traffic movements into definite paths of travel by means of markings, raised islands, or other suitable means to facilitate the safe and orderly movements of both vehicles and pedestrians.

*Table 13-5*
**Minimum Edge of Pavement Designs for Turns at Intersections—Simple Curves and Tapers**

| Angle of Turn (degrees) | Design Vehicle | Simple Curve Radius (ft) | Simple Curve Radius with Taper | | |
|---|---|---|---|---|---|
| | | | Radius (ft)[a] | Offset (ft) | Taper (ft:ft) |
| 60 | P | 40 | — | — | — |
| | SU | 60 | — | — | — |
| | WB-40 | 90 | — | — | — |
| | WB-50 | — | 95 | 3.0 | 15:1 |
| | WB-55 | — | 110 | 3.0 | 15:1 |
| 90 | P | 30 | 20 | 2.5 | 10:1 |
| | SU | 50 | 40 | 2.0 | 10:1 |
| | WB-40 | — | 45 | 4.0 | 10:1 |
| | WB-50 | — | 60 | 4.0 | 15:1 |
| | WB-55 | — | 75 | 4.0 | 15:1 |
| 120 | P | — | 20 | 2.0 | 10:1 |
| | SU | — | 30 | 3.0 | 10:1 |
| | WB-40 | — | 35 | 5.0 | 8:1 |
| | WB-50 | — | 45 | 4.0 | 15:1 |
| | WB-55 | — | 50 | 6.0 | 15:1 |

[a]1 ft = 0.3048 m.

SOURCE: *A Policy on Geometric Design of Highways and Streets,* American Association of State Highway and Transportation Officials, Washington, D.C., 1984.

*Table 13-6*
**Minimum Edge of Pavement Designs for Turns at Intersections—Three-Centered Compound Curves**

| | | *Three-Centered Compound* | | *Three-Centered Compound* | |
|---|---|---|---|---|---|
| *Angle of Turn (degrees)* | *Design Vehicle* | *Curve Radii (ft)*[a] | *Symmetric Offset (ft)* | *Curve Radii (ft)* | *Asymmetric Offset (ft)* |
| 60 | P | — | — | — | — |
| | SU | — | — | — | — |
| | WB-40 | — | — | — | — |
| | WB-50 | 200- 75-200 | 5.5 | 200- 75-275 | 2.0- 6.0 |
| | WB-55 | 200- 85-200 | 5.0 | 120- 80-240 | 1.0- 7.5 |
| 90 | P | 100- 20-100 | 2.5 | — | — |
| | SU | 120- 40-120 | 2.0 | — | — |
| | WB-40 | 120- 40-120 | 5.0 | 120- 40-200 | 2.0- 6.0 |
| | WB-50 | 180- 40-180 | 6.0 | 120- 40-200 | 2.0-10.0 |
| | WB-55 | 200- 65-200 | 7.0 | 100- 55-260 | 2.0-10.0 |
| 120 | P | 100- 20-100 | 2.0 | — | — |
| | SU | 100- 30-100 | 3.0 | — | — |
| | WB-40 | 120- 30-120 | 6.0 | 100- 30-180 | 2.0- 9.0 |
| | WB-50 | 180- 40-180 | 8.5 | 150- 35-220 | 2.0-12.0 |
| | WB-55 | 400- 40-400 | 10.0 | 100- 35-500 | 6.0-15.0 |

[a]1 ft = 0.3048 m.

SOURCE: *A Policy on Geometric Design of Highways and Streets,* American Association of State Highway and Transportation Officials, Washington, D.C., 1984.

Islands in an intersection can separate conflicting movements or control the angle at which conflict may occur. They can regulate traffic, indicate the proper use of the intersection, and often reduce the amount of pavement required. They provide for the protection of motorists, protection and storage of turning and crossing vehicles, and space for traffic control devices.

Islands are generally grouped into three major classes: directional, divisional, and refuge. General types and shapes of islands are shown in Fig. 13-15.

Directional islands are designed primarily to guide the motorists through the intersection by indicating the intended route. The placing of directional islands should be such that the proper course of travel is immediately evident and easy to follow. A complicated system of islands where the desired course of travel is not immediately evident may result in confusion and may be more hindrance than help in maintaining a steady traffic flow. Islands should be so

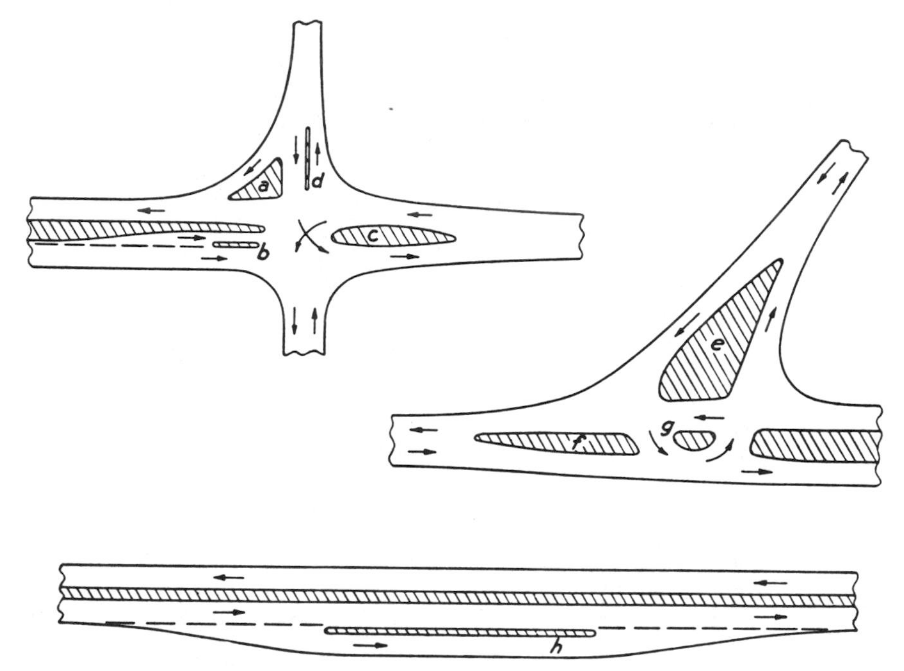


*Fig 13-15.* General types and shapes of islands. (Source: *A Policy on Geometric Design of Rural Highways,* AASHO, 1965.)

placed that crossing streams of traffic will pass at approximately right angles and merging streams of traffic will converge at flat angles. By the use of such angles there will be less hindrance to traffic in the thoroughfare and possibility of accidents in the intersection will be decreased.

Divisional islands are used most frequently in individual highways approaching intersections. They serve to alert the driver to the intersection and regulate the flow of traffic into and out of the intersection. Their use is particularly advantageous for controlling left-turning traffic at skewed intersections.

A refuge island is located at or near crosswalks to aid and protect the pedestrian. These islands are used most generally on wide streets in urban areas for loading and unloading transit riders. The design of refuge islands is the same as that of other types of islands, except that a higher barrier curb is necessary.

A good approach in the design of channelization is to make a comprehensive study of field conditions. Then, with the use of pavement markings, observation of the traffic patterns is done. This is followed by the placing of sand bags and another observation of the traffic pattern, which is then followed by the placing of the permanent channelization.

## 13-23. ROTARY INTERSECTIONS

One approach to the channelization of traffic is by the use of a rotary intersection. A rotary intersection is one in which all traffic merges into and emerges from a one-way road around a central island. Rotary intersections provide continuous traffic movement but at relatively low speeds. However, experience has shown that a rotary can handle no more traffic than a well-designed channelized intersection. Furthermore, rotaries require relatively large expanses of flat land and are not suitable where large amounts of pedestrian traffic exist. Because of these and other disadvantages, rotary intersections are seldom used in the United States for new construction.

## 13-24. GRADE SEPARATIONS AND INTERCHANGES

Intersections at grade can be eliminated by the use of grade-separation structures which permit the cross-flow of traffic at different levels without interruption. The advantage of such separation is the freedom from cross-interference with resultant savings of time and increase in safety for traffic movements.

Grade separations and interchanges may be warranted (1) as a part of an express highway system designed to carry heavy volumes of traffic, (2) to eliminate bottlenecks, (3) to prevent accidents, (4) where the topography is such that other types of design are not feasible, and (5) where the volumes to be catered to would require the design of an intersection, at grade, of unreasonable size.

An interchange is a grade separation in which vehicles moving in one direction of flow may transfer direction by the use of connecting roadways. These connecting roadways at interchanges are called ramps.

Many types and forms of interchanges and ramp layouts are used in the United States. These general forms may be classified into four main types.

1. T- and Y-interchanges.
2. Diamond interchanges.
3. Partial and full cloverleafs.
4. Directional interchanges.

*T- and Y-Interchanges.* Figure 13-16 shows typical layouts of interchanges at various junctions. The geometry of the interchange can be altered in favor of certain movements by the provision of large turning radii, and to suit the topography of the site. The trumpet interchange has been found suitable for orthogonal or skewed intersections. Figure 13-16*a* favors the left turn on the freeway by the provision of a semidirect connecting ramp. Figure 13-16*c* indicates an intersection where all turning movements are facilitated in this way.

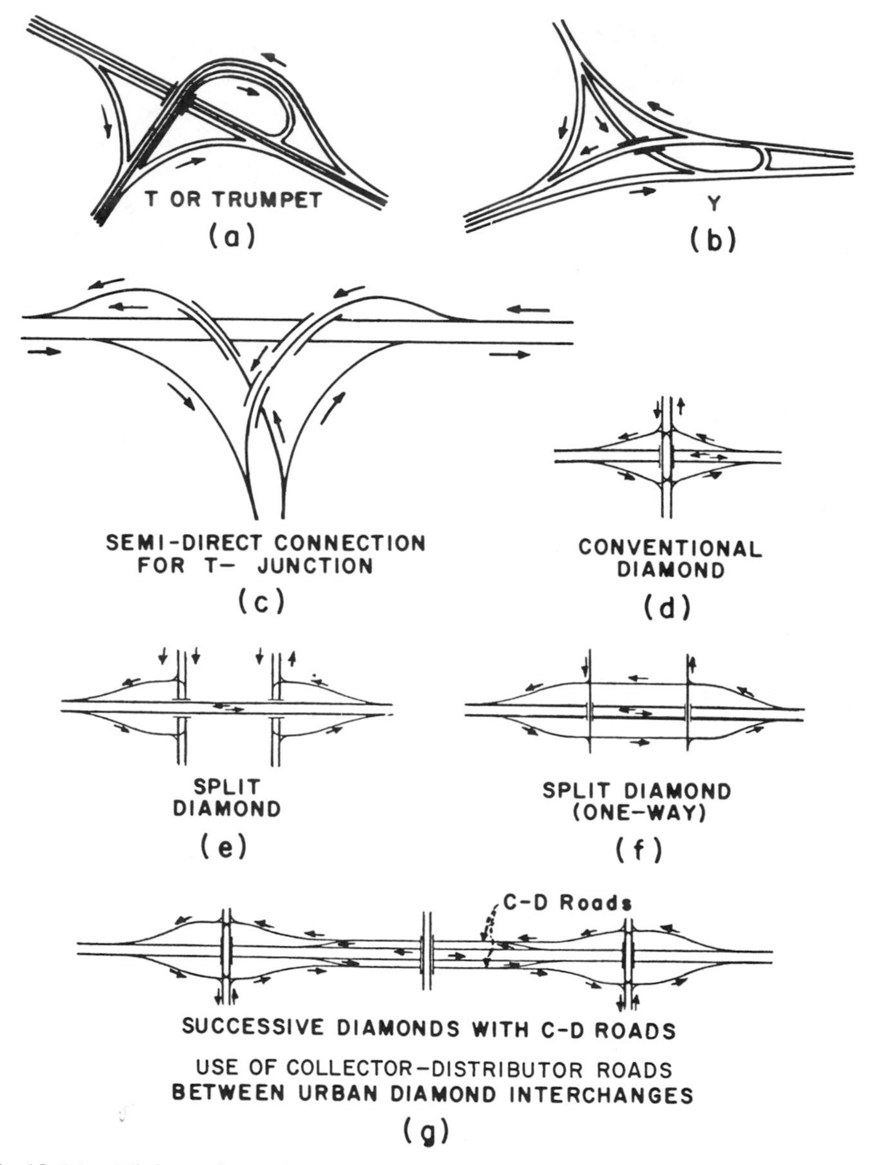


*Fig 13-16.* Highway interchanges. (Partially adapted from "Adaptability of Interchanges to Interstate Highways," *Transactions,* American Society of Civil Engineers, Volume 124, page 558.)

*Diamond Interchanges.* The diamond interchange is adaptable to both urban and rural use. The major flow is grade separated, with turning movements to and from the minor flow achieved by diverging the merging movements with through traffic on the minor flow. Only the minor flow directions have intersection at grade. In rural areas this is generally acceptable, owing to the light traffic on the minor flow. In urban areas, the at-grade intersections generally will require signalized control to prevent serious interference of ramp traffic and the crossing arterial street. The design of the intersection should be such that the signalization required does not impair the capacity of the arterial street. To achieve this, widening of the arterial may be necessary in the area of the interchange. Care must also be taken in the design of the ramps, so that traffic waiting to leave the ramp will not back up into through lanes of the major flow.

One disadvantage of the diamond interchange is the possibility of illegal wrong-way turns, which can cause severe accidents. Where the geometry of the intersection may lead to these turns, the designer can use channelization devices and additional signing and pavement marking. Wrong-way movements are, in general, precluded by the use of cloverleaf designs.

Figure 13-16*d* shows the conventional diamond interchange. Increased capacity of the minor flows can be attained by means of the arrangement shown in Fig. 13-16*e* or Fig. 13-16*f*. The arrangement shown in Fig. 13-16*g* is suitable where two diamond ramps are in proximity. Weaving movements which, in this case would inhibit the flows of the major route, are transferred to the parallel collector–distributor roads. Figure 13-17 shows a typical depressed expressway in an urban area.

*Partial and Full Cloverleafs.* The partial cloverleaf shown in Fig. 13-18 is sometimes adopted in place of the diamond interchange. Traffic can leave the major flow either before or after the grade-separation structure, depending on the quadrant layout. The intersections at grade for the minor road are present as for the diamond interchange, but the probability of illegal turning movements can be reduced. By the provision of two on-ramps for each direction of the major route as in Fig. 13-18*c*, left-turn traffic on the minor route can be eliminated. The more conventional arrangement of the full cloverleaf, which can be adapted to nonorthogonal layouts, eliminates at-grade crossings of all traffic streams for both major and minor roads. The ramps may be one-way, two-way separated, or two-way unseparated roads. Although all crossing movements are eliminated, the cloverleaf design has some disadvantages: (1) the layout requires large land areas, and (2) decelerating traffic wishing to leave the through lanes must weave with accelerating traffic entering the through lanes. Figure 13-18*e* is a layout using collector–distributor roads to overcome this second disadvantage.

*Directional Interchanges.* Directional interchanges are used whenever one freeway joins or intersects another freeway. The outstanding design characteristic of this type of interchange is the use of a high-design speed throughout, with curved ramps and roadways of large radius. The land requirements for a directional interchange, therefore, are very large. In cases where volumes for certain turning movements are small, design speeds for these movements are reduced and the turnoff is effected within a loop. In the highest type of design weaving sections are eliminated. Figure 13-19 shows a multilevel interchange.

*Fig 13-17.* A typical depressed expressway in an urban area. (Courtesy, Federal Highway Administration.)

## RAILROAD INTERSECTIONS

Railroad tracks intersect at *turnouts, crossovers,* and *crossings.* Turnouts are curved sections of track that permit the diversion of rolling stock from one track to another. Where the turnout provides an intersection with another continuous parallel or nonparallel track it is called a crossover.

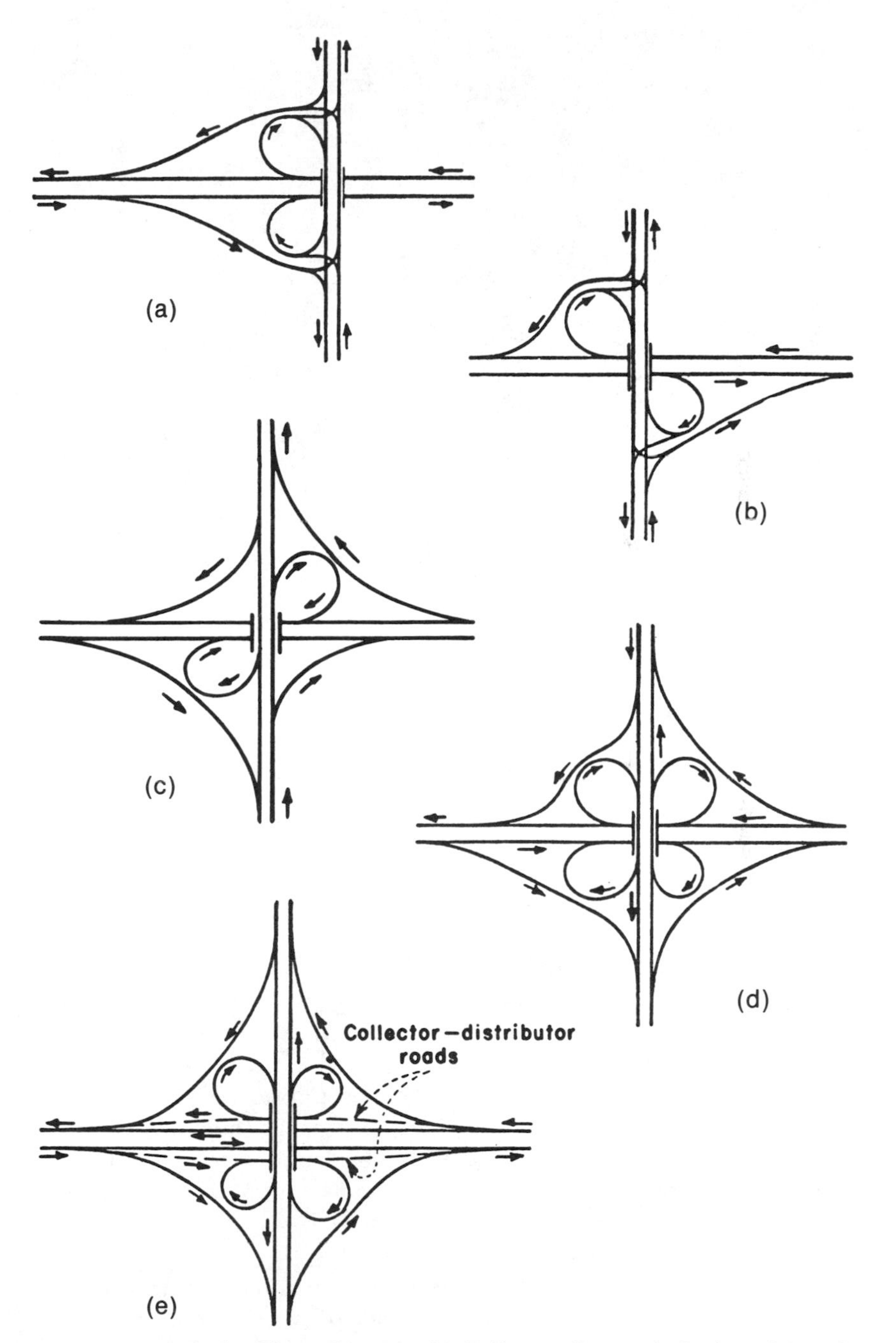


*Fig 13-18.* Cloverleaf interchanges. (*A Policy on Geometric Design of Rural Highways,* AASHO, 1965.)

*Fig 13-19.* The Tom Moreland interchange, Atlanta, Georgia. (Photo by Nick Arroyo.)

At crossings, tracks intersect permitting movement of the rolling stock on one track across the alignment of the other. Figure 13-20 shows schematic arrangements for the various types of railroad intersections discussed in the following sections.

## 13-25. SWITCHES

The device that determines the diversion of rolling stock movement through a turnout is known as a switch. A switch is designated as a left-hand or right-hand switch depending on the direction of diversion of the rolling stock into the turnout. Switches are relatively simple devices principally composed of switch rails, rods that hold the points in their proper position and relationship, gage and switchplates that support the switch rails at their proper elevation, and heel blocks that effect a rigid joint at the head of the switch.

Various types of switches are available, only one of which is of standard railroad use in the United States.

1. Stub switches are used to some extent in industrial tramways. Both switch rails are mainline rails. This type of switch is not considered safe for high-speed train movements. Therefore, its use is extremely limited. It is found in a few switch yards and other relatively low-speed areas.

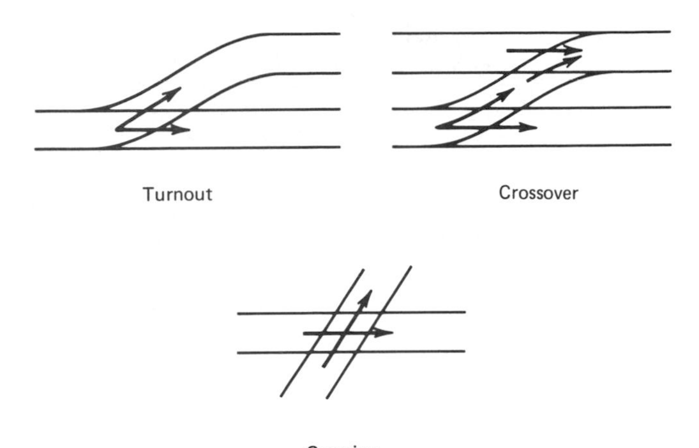


*Fig 13-20.* Schematics of simple turnout, crossover, and crossings.

2. Tongue switches are used in paved street locations. Designed for slow-moving traffic, they consist of a movable tongue on one side of the track with either another movable tongue or a fixed mate on the other side.
3. Spring switches permit point movement to allow the passage of trailing wheels through the switch in the reverse direction of movements prevented by the points for facing movements. After the passage of the trailing wheels, the points are moved back into position by spring devices.
4. Split switches are the standard switch in use on American railroads. They have proven safe for very high-speed movements. Switching is carried out by the use of one mainline rail and one turnout rail for switch rails as shown in the schematic diagram in Fig. 13-21.

The arrangement of a typical left-hand turnout is shown in Fig. 13-22.

## 13-26. FROGS

The turnout frog is a device that permits rolling stock wheels on one rail to cross the rail of a diverging track. It performs two functions, supporting the wheel over the intersection of the flangeways and providing continuous channels for the wheel flanges. Two principal frog types are in common use, the rigid frog and the spring frog. Figure 13-23 shows typical arrangements of rigid and spring frogs. In the rigid frog, both flangeways are always open and both wing rails are bolted to the body of the frog. In the design of the conventional spring frog, the main flangeway only is always open. The turnout

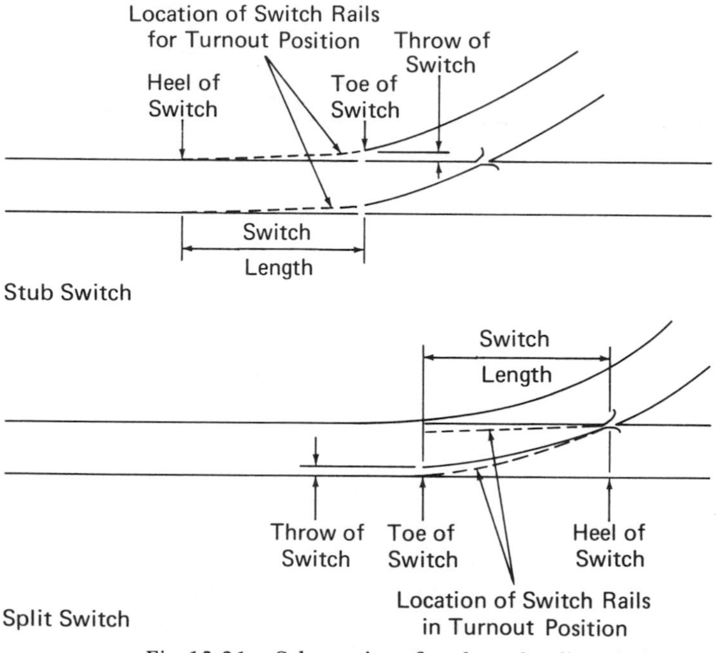


*Fig 13-21.* Schematics of stub and split switches.

direction is opened when one wing moves, establishing a flangeway for the wheels.

Frogs normally are designated by a frog number, which is defined as one-half the cotangent of one-half the frog angle or the ratio of the spread at any point to the length of a bisecting line between that point and any theoretical point of frog:

$$n = \frac{1}{2} \cot \frac{\phi}{2} \tag{13-4}$$

where

$n$ = frog number
$\phi$ = frog angle

Frog angles in common use in railwork are between 9° 32′ (No. 6 frog) and 2° 52′ (No. 20 frogs). No. 18 and No. 20 frogs are in standard use on large railroad systems with mainline tracks. Slower movements can be accommodated with No. 10 and No. 12 frogs while No. 6 and No. 8 frogs are in use in sidings and industrial tramways.

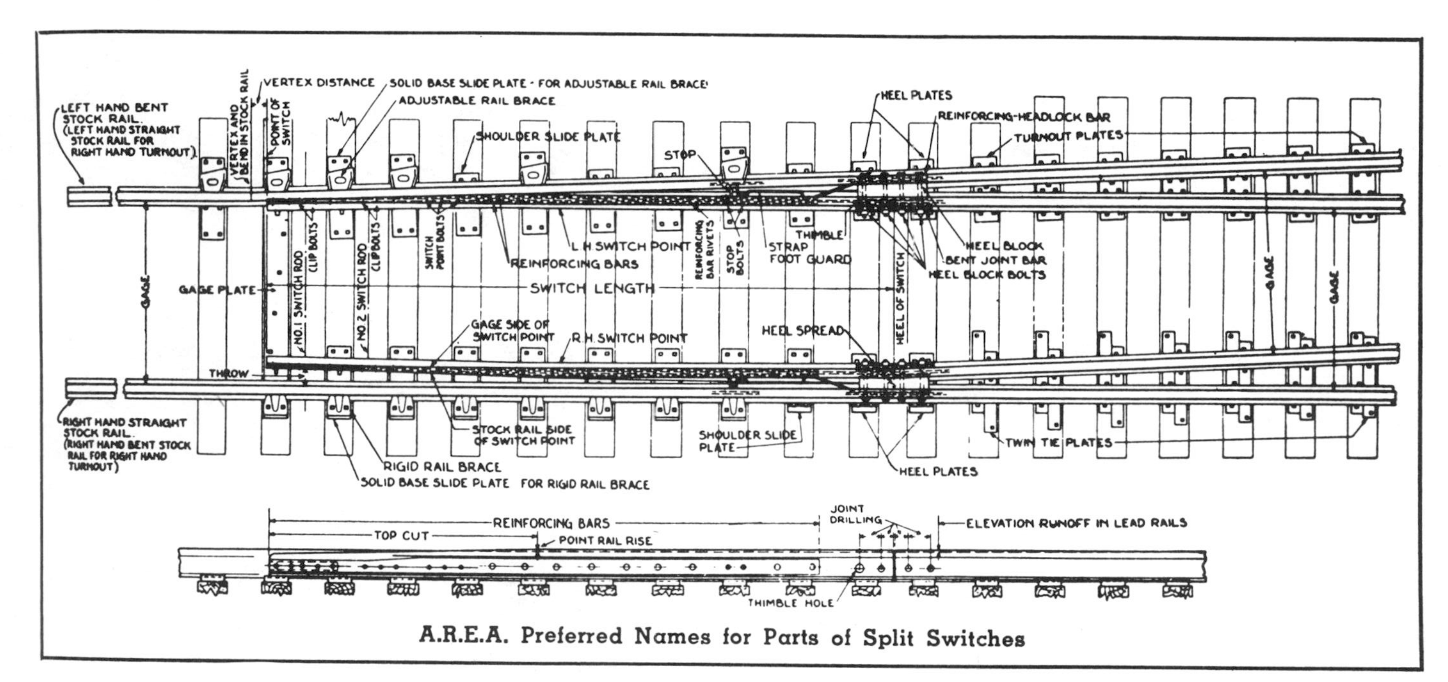


**A.R.E.A. Preferred Names for Parts of Split Switches**

*Fig 13-22.* A typical split switch left-hand turnout. (Courtesy American Railway Engineering Association.)

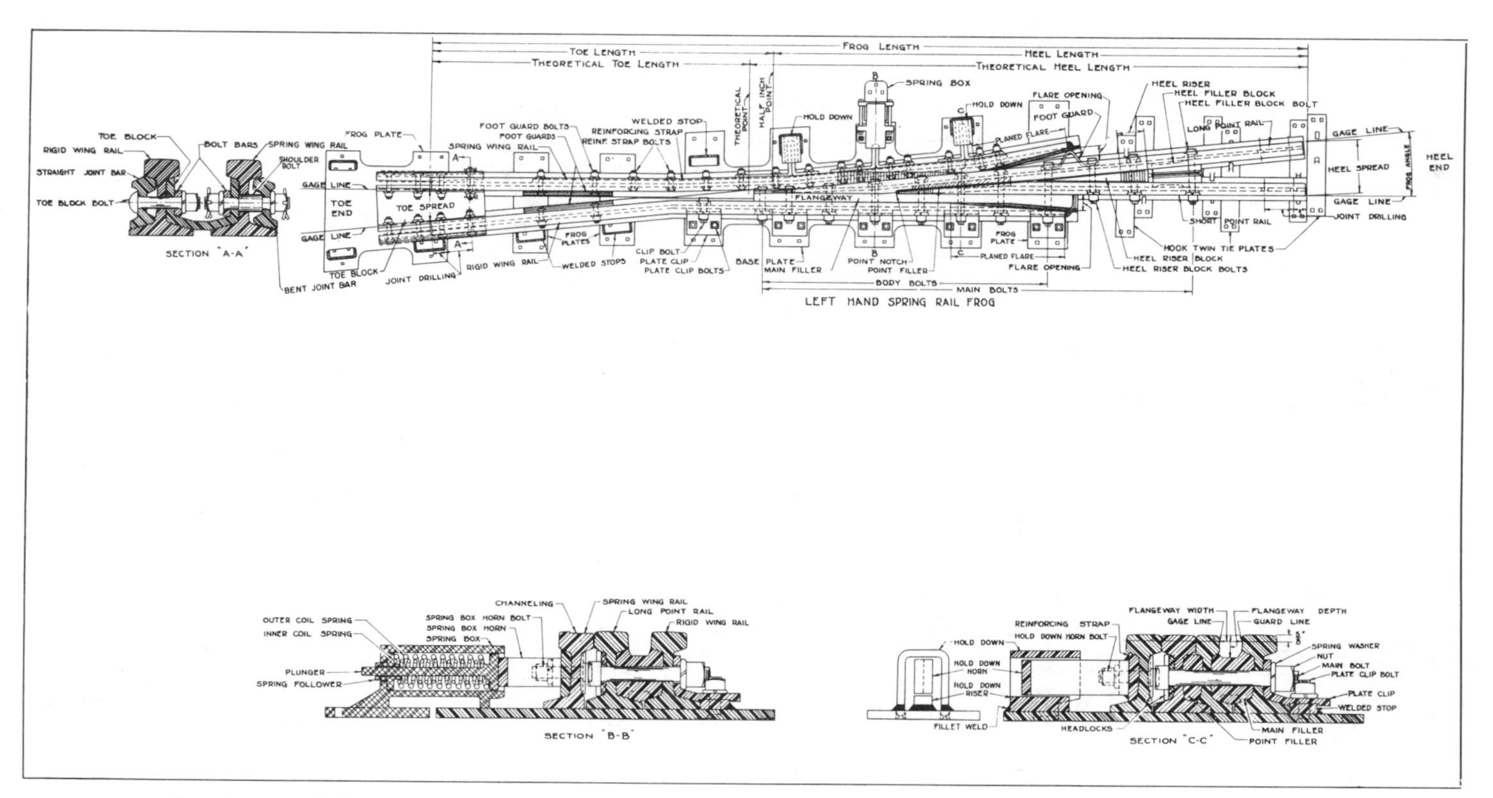


*Fig 13-23.* Rigid and spring frog arrangements. (Courtesy American Railway Engineering Association.)

## 13-27. CROSSINGS

Where two tracks intersect and cross, specially designed and fabricated crossings are necessary. Since the angle of intersection is usually nonstandard, specially designed frogs are necessary. Crossings can be designated into four general classifications.

1. Bolted rail crossings. All members are heat treated or open hearth rails bolted together.
2. Manganese steel insert crossings. Manganese steel cast inserts are fitted into rolled rails to form the wing and points of frogs.
3. Solid manganese steel crossings in which each frog is a single solid casting.
4. Double slip switches with movable point crossings.

The choice of crossing type is dependent on speeds of operation, the angle of intersection, and the degree of curvature of the track.

## 13-28. RAILROAD GRADE INTERSECTIONS

In the design of a highway that intersects a railroad at grade, consideration must be given to approach grades, sight distance, drainage, volume or vehicular traffic, and the frequency of regular train movements at the particular intersection. The particular type of surfacing and kind of construction at railroad crossings at grade will depend upon the class of railroad and kind of roadway improvement.

All railroad intersections at grade require proper advance warning signs. At crossings on heavily traveled highways where conditions justify, automatic devices should be installed. Recommended standards for railroad-highway-grade-crossing protection have been adopted by the Association of American Railroads.

The use of grade separations at railroad crossings is recommended at all mainline railroads that consist of two or more tracks and at all single-line tracks when regular train movements consist of six or more trains per day. Other considerations for separating railroad and highway traffic are the elements of delay and safety.

Railroad grade-separation structures may consist of an overpass on which the highway is carried over the railroad or an underpass which carries the highway under the railroad. The selection of the type of structure will depend in large part upon the topographical conditions and a consideration of initial cost. Drainage problems at underpasses can be serious. Pumping of surface and subsurface water may have to be carried on a large part of the time and the failure of power facilities sometimes causes flood conditions at the underpass with the resultant stoppage of traffic.

## PROBLEMS

1. Two intersecting highways cross at an angle of 80 degrees. Make a sketch of a cloverleaf grade separation for complete traffic flow.
2. What would be the result on the geometric design of a primary two-lane road if truck and tractor-trailer combinations are permitted to be larger than the present design vehicle?
3. Discuss the different philosophies that form the basis for the radically different design criteria for a local residential subdivision street and an expressway.
4. How does the United States railroad gage and other appurtenances compare with that of European countries? Why are concrete ties used extensively in Europe but not in the United States?
5. Suppose a subdivision developer has a 3500-ft by 1200-ft rectangular piece of property that has streets abutting on two of the adjacent sides. Sketch a street layout and residential lot layout plan using the design criteria given here:

   Minimum lot size — 1/3 acre
   Maximum lot size — 1/2 acre
   Minimum lot frontage — 100 ft
   Minimum set-back of houses from right-of-way line — 50 ft

6. Suppose a subdivision developer has a 1000-m by 400-m rectangular piece of property which has streets abutting on two of the adjacent sides. Sketch a street layout and residential lot layout plan using the design criteria given here:

   Minimum lot size — 0.2 hectare
   Maximum lot size — 0.3 hectare
   Minimum lot frontage — 35 m
   Minimum set-back of houses from right-of-way line — 18 m

## REFERENCES

1. *Highway Capacity Manual,* Special Report No. 209, Transportation Research Board, Washington, D.C., 1985.
2. *Handbook of Highway Safety Design and Operating Practices,* Federal Highway Administration, Washington, D.C., 1978.
3. *A Policy on Geometric Design of Highways and Streets,* American Association of State Highway and Transportation Officials, Washington, D.C., 1984.
4. Wright, Paul H., and Radnor J. Paquette, *Highway Engineering,* Fifth Edition, John Wiley, New York, 1987.

**5.** *Guide for Selecting, Locating, and Designing Traffic Barriers,* American Association of State Highway and Transportation Officials, Washington, D.C., 1977.

**6.** *Recommended Guidelines for Subdivision Streets,* Institute of Transportation Engineers, Washington, D.C., 1984.

**7.** Kerr, Arnold D. (ed.), *Railroad Track Mechanics and Technology,* Pergamon Press, New York, 1978.

**8.** *Manual for Railway Engineering,* American Railway Engineering Association, Chicago, 1986.

**9.** *The Track Cyclopedia,* compiled and edited in cooperation with the Association of American Railroads, H.C. Archdeacon, Editor-in-Chief, Simmons-Boardman Publishing Corp., Omaha, NE, 1978.

# Chapter 14

# *Design of Streets, Highways, and Railways: Drainage, Earthwork and Pavements*

Three important aspects of railway and roadway design will be discussed in this chapter: the provision of adequate drainage, earthwork operations, and pavements.

## DESIGN OF DRAINAGE STRUCTURES AND FACILITIES

In this chapter, the discussion of street, highway, and railway drainage will be limited to a discussion of surface drainage, which is the process of controlling and removing excess water from the traveled way. The specialized problem of dealing with subsurface drainage will not be covered but is discussed in Reference 1.

Three major topics will be discussed in this section.

1. Estimation of runoff.
2. Hydraulic design of culverts and drainage ditches.
3. A comparison of alternative drainage systems.

The first of these topics deals with techniques for estimating the quantity of runoff water on the bases of certain rainfall data and land use characteristics. Culvert design essentially encompasses a discussion of nonuniform open channel flow, while the design of drainage ditches involves uniform open channel flow. Approaches used in the design of underground storm drainage systems such as those used for city streets are given in Section 19-7.

In the sections that deal with hydraulic design, emphasis will be placed on the use of empirical data rather than on fluid flow theory. For the reader who is more interested in theoretical approaches to drainage design, reference should be made to one or more of the numerous textbooks in fluid mechanics.

## 14-1. SURFACE DRAINAGE

Before proceeding with a discussion of the techniques for estimating runoff quantities, it should first be noted that consistent with other design objectives, every effort should be made to remove precipitation from the traveled way as expeditiously as possible. Water that is not removed quickly from a highway or railway nearly always is harmful to the load-carrying capability of the pavement system. Furthermore, flood waters serve as a deterrent to free traffic movements and create unnecessary perils for the users of the facility. Uncontrolled water movements may weaken, damage, or even destroy transportation structures and the pavement system. For these reasons, highway designers provide pavement crown and shoulder slopes to expedite the removal of surface water.[1] Similarly, railroad design engineers specify an open-graded ballast material and a sloped subgrade to insure adequate and quick drainage. Well-designed culvert and bridge structures must be provided for railways and highways alike to prevent destructive back waters and roadway overtopping from occurring.

To ensure adequate drainage, unpaved side ditches should have a slope of at least 0.5 percent. Experience has shown that paved ditches with an average slope of about 0.3 percent will drain satisfactorily.

## 14-2. STREAMFLOW RECORDS

It may be possible, especially in the case of design of large culverts and bridges, to base estimates of peak flow on historical empirical data. Flood-frequency studies based on statistical analysis of streamflow records have been published by a number of states as well as the U.S. Geological Survey.

> Where suitable streamflow records are not available, useful information may be gained from observations of existing structures and the natural stream. Drainage installations above and below the proposed location may be studied and a design based upon those that have given satisfactory service on other portions of the same stream. Lacking this information, an examination of the natural channel may be made, including the evidences left by flood crests which have occurred in the past, and an estimate made of the quantity of water which has been carried by the stream during flood periods. Measurements may be made and values assigned to the slope, area, wetted perimeter, and roughness coefficient of the flood channel and the quantity may be estimated by the Chezy formula. The Chezy formula is of the form,

$$Q = Ca\sqrt{rS} \tag{14-1}$$

[1]Cross section design criteria to ensure good drainage are given in Chapter 13.

where

$Q$ = quantity of flow, $ft^3/sec$

$C$ = roughness coefficient, varying from 30 to 80, depending on the condition and nature of the channel

$a$ = area of the flow cross section, $ft^2$

$S$ = slope of the channel, ft/ft

$r$ = hydraulic radius, feet, $= a/p$, where $p$ = wetted perimeter, the length of the boundary of the cross section of the flow channel in contact with the water [1].

Of course, streamflow records are not often available, and a detailed physical examination of the existing channel may not be feasible. These conditions commonly prevail, in fact, for small channels that drain areas of only a few hundred acres. A number of formulas and analytical procedures have been developed for estimating runoff from small drainage areas. Before we describe some of these analytical tools, a brief discussion of the factors that influence the magnitude of surface runoff will be given.

## 14-3. THE NATURE OF THE PROBLEM

It should be noted at the outset that the estimation of surface runoff is not an exact science. When two or more methods are used to estimate runoff, discrepancies in the estimates of 50 percent or more are not uncommon. Difficulties in the estimation of runoff stem from the fact that there are a large number of factors that affect the answer. Certain of these factors are difficult to quantify and may be subject to change with time. In the final analysis, no matter what technique of estimation is used, the application of sound engineering judgment is required to produce a reliable answer.

## 14-4. COEFFICIENT OF RUNOFF

Runoff results from precipitation that falls on the various surfaces of the watershed. A part of the precipitation evaporates and some of it may be intercepted by vegetation. A portion of the precipitation may infiltrate the ground or fill depressions in the ground surface. The storm runoff for which roadside ditches and drainage structures must be designed, then, is the precipitation minus the various losses that occur.

These losses and, thus, the runoff are strongly dependent upon the slope, vegetation, soil condition, and land use of the watershed. The designer should remember that certain of these factors, notably vegetation and land use, will nor remain constant with time. For example, a drainage structure that is designed to accommodate a watershed used for agricultural purposes may prove totally inadequate if the land is later developed into a residential subdivision.

Most analytical procedures for estimating runoff involve the use of a coefficient of runoff to take into consideration the hydrologic nature of the drainage area. Values of the coefficient of runoff for one method of estimating the runoff quantity are given by Table 14-1. If the drainage area under consideration consists of several land use types, a coefficient should be chosen for each such subarea. The runoff coefficient for the entire area then should be taken as the weighted average of the coefficients for the individual areas.

## 14-5. RAINFALL INTENSITY, DURATION, AND FREQUENCY

Rainfall intensity is the rate at which the rain falls, typically expressed in inches per hour. Because of the capriciousness of weather, it is necessary to discuss rainfall intensity in the context of rainfall frequency and duration.

Rainfall intensity-duration data have been collected and published by the National Weather Service [2] for various sections of the United States. Typical rainfall intensity-duration graphs are shown in Fig. 14-1. To utilize such data, the designer is confronted at once with the decision of which curve should be used. In making this decision, the designer must weigh the physical (as well as social) damages that might occur from a flood of a given frequency against the additional costs of designing the structure so as to lessen the risk of these damages. Referring to the hypothetical data in Fig. 14-1, for example, the

*Table 14-1*
**Coefficients of Runoff To Be Used in the Rational Formula**

| *Description of Area* | *Runoff Coefficients* |
|---|---|
| Business: | |
| Downtown areas | 0.70 to 0.95 |
| Neighborhood areas | 0.50 to 0.70 |
| Residential: | |
| Single-family areas | 0.30 to 0.50 |
| Multiunits, detached | 0.40 to 0.60 |
| Multiunits, attached | 0.60 to 0.75 |
| Residential (suburban) | 0.25 to 0.40 |
| Apartment dwelling areas | 0.50 to 0.70 |
| Industrial: | |
| Light areas | 0.50 to 0.80 |
| Heavy areas | 0.60 to 0.90 |
| Parks, cemeteries | 0.10 to 0.25 |
| Playgrounds | 0.20 to 0.35 |
| Railroad yard areas | 0.20 to 0.40 |
| Unimproved areas | 0.10 to 0.30 |

SOURCE: *Concrete Pipe Handbook,* American Concrete Pipe Association, 1981.

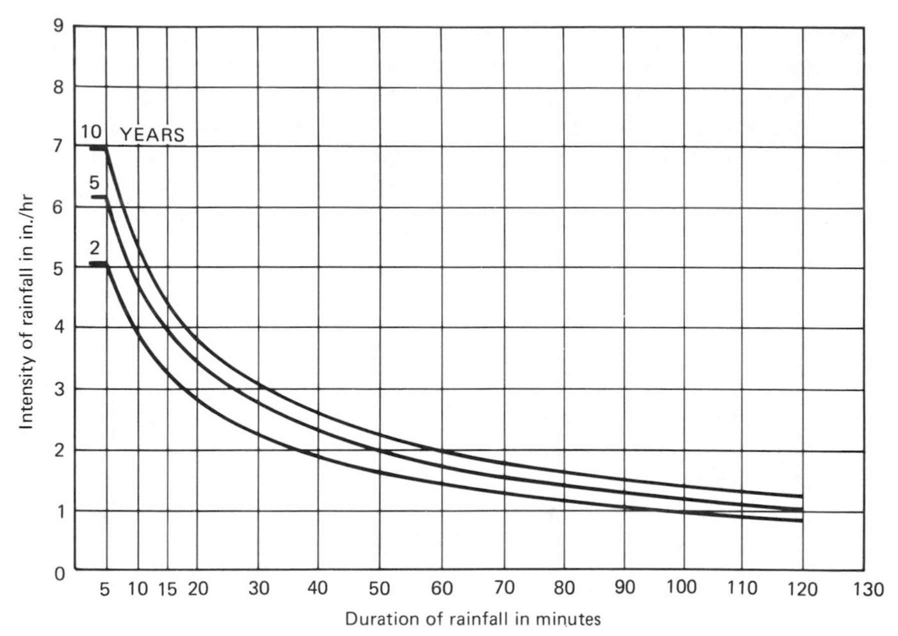


*Fig 14-1.* Typical rainfall intensity-duration curves. (Courtesy Federal Aviation Administration.)

choice of the 10-year curve, instead of the 5-year curve, would mean designing for a more severe storm but at a higher cost. Conversely, the choice of a 5-year frequency would result in a less costly system but at the risk of more frequent runoffs which exceed the capacity of the drainage facility. The selection of rainfall frequency is to a large extent a matter of experience and judgment, although agency or departmental policy may dictate this choice.

As can be seen from Fig. 14-1, rainfall intensity varies greatly with the duration of rainfall. The average rainfall intensity for short periods of time is much greater than for long periods. In the design of most railroad and highway drainage facilities, a duration equal to the *time of concentration* is chosen.

## 14-6. TIME OF CONCENTRATION

The time of concentration is defined as the time required for a particle of water to flow from the most remote point in the drainage area to the outlet end of the drainage structure. In other words, it consists of the time of overland flow plus the time of flow in the drainage system. For most railroad and highway culverts, the time of flow in the drainage system is negligible in comparison to the time of overland flow. Thus, for the purpose of this chapter, the terms time of overland flow and time of concentration may be used synonymously.

Time of concentration varies with the land slope, type of surface, rainfall intensity, the size and shape of the drainage area, and many other factors. A number of empirical studies have been made relating time of concentration to the slope and dimensions of the drainage area. An estimate of time of concentration can be obtained from Fig. 14-2 which is based on a study of six watersheds in Tennessee which varied in size from 1.25 to 112 acres.

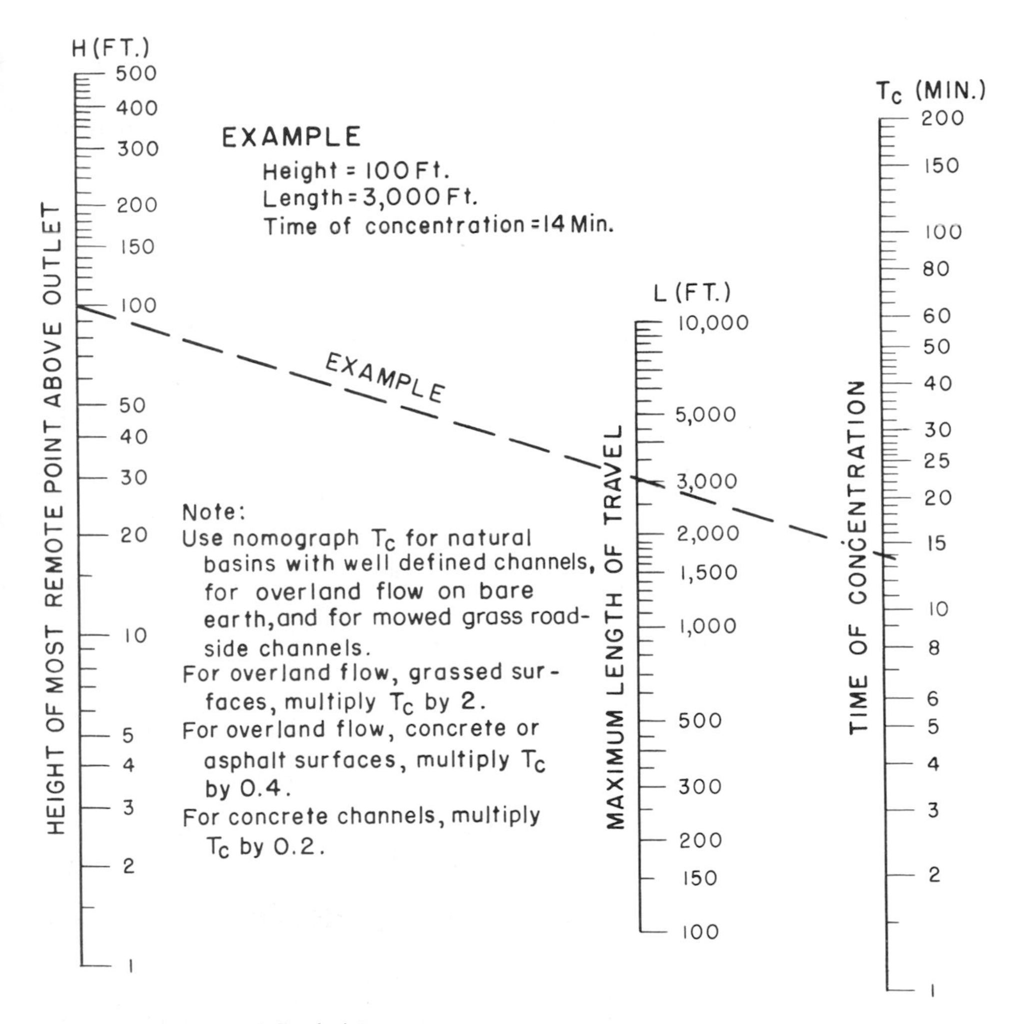


*Fig 14-2.* Time of concentration of small drainage basins. (Source: *Design of Roadside Drainage Channels,* 1965.)

## 14-7. THE RATIONAL METHOD

By far, the most popular method for estimating runoff from small drainage areas is the rational method. In the rational formula, the quantity of water falling at a uniform rate is related by simple proportion to the total quantity that appears as runoff.

$$Q = CIA \tag{14-2}$$

where

$Q$ = runoff, in $ft^3/sec$
$C$ = a coefficient representing the ratio of runoff to rainfall. Typical values of $C$ are given in Table 14-1.
$I$ = intensity of rainfall, in inches per hour for the estimated time of concentration.
$A$ = drainage area in acres. The area may be determined from field surveys, topographical maps, or aerial photographs.

The designer should contact the nearest office of the National Weather Service for rainfall intensity data. In the event that suitable local rainfall intensity data are not available, approximate data may be obtained from Fig. 14-3, a chart published by the Federal Highway Administration. This map shows rainfall intensity values in inches per hour for various areas of the contiguous United States for a two-year, 30-minute rainfall. The two-year rainfall intensity for other durations may be obtained by multiplying by the following factors.

| *Rainfall Duration (mins)* | *Factor* |
|---|---|
| 5 | 2.22 |
| 10 | 1.71 |
| 15 | 1.44 |
| 20 | 1.25 |
| 40 | 0.80 |
| 60 | 0.60 |
| 90 | 0.50 |
| 120 | 0.40 |

To obtain rainfall intensities corresponding to their intervals of recurrence, the values of Fig. 14-3 should be multiplied by the following factors:

| *Recurrence Internal (Years)* | *Factor* |
|---|---|
| 1 | 0.75 |
| 2 | 1.00 |
| 5 | 1.30 |
| 10 | 1.60 |
| 25 | 1.90 |
| 50 | 2.20 |

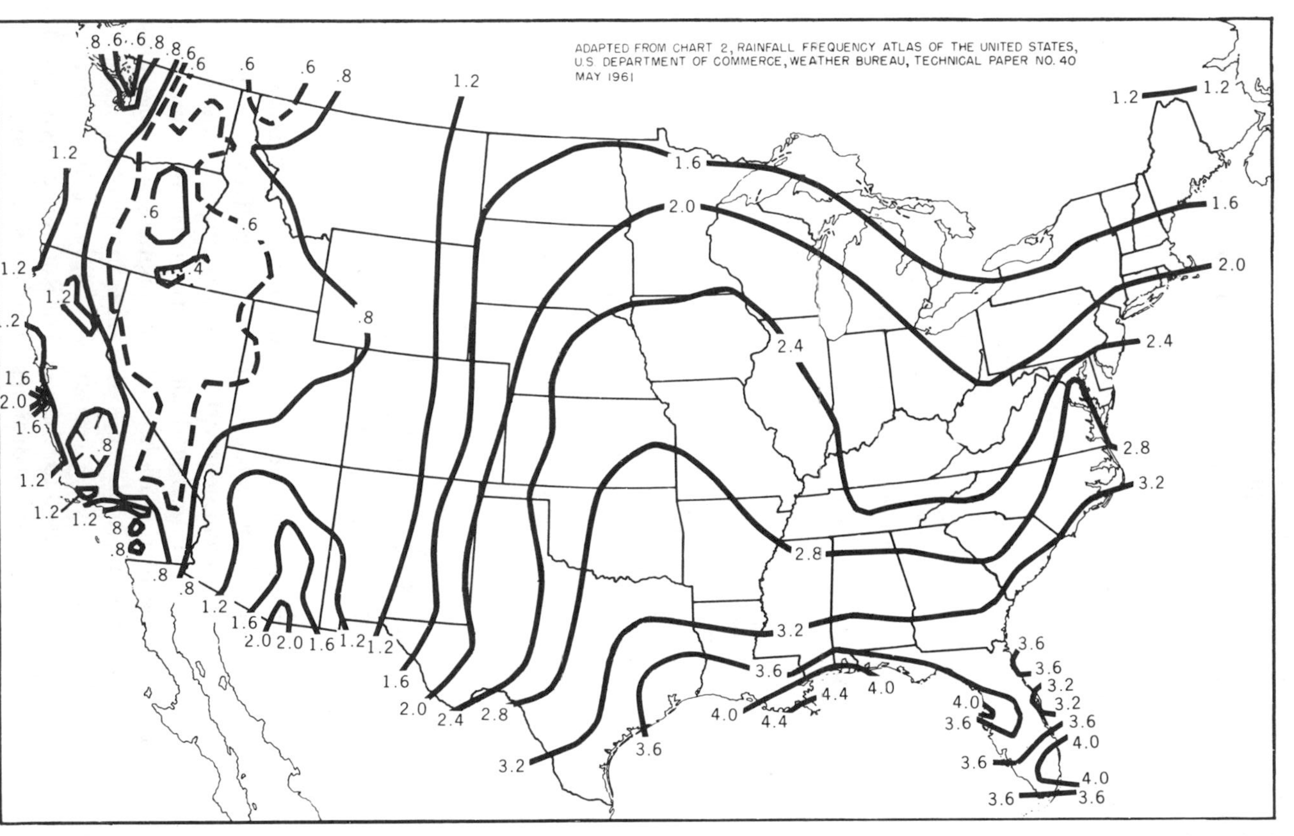


*Fig 14-3.* Map of the contiguous United States showing two-year, 30-min rainfall intensity. (Source: *Design of Roadside Drainage Channels*, 1965.)

It may be noted that the rational formula is not dimensionally correct. It happens, however, that a 1-in. depth of rainfall applied at a uniform rate to an area of one acre during one hour will produce 1.008 $ft^3$/sec of runoff if there are no losses. Thus, the runoff coefficient represents the approximate fraction of the total water that falls and reaches the lowest point in the drainage basin.

## 14-8. THE BURKLI-ZIEGLER FORMULA

Another runoff formula that has gained wide acceptance is the Burkli-Ziegler formula:

$$Q = AIC\sqrt[4]{\frac{S}{A}} \qquad (14\text{-}3)$$

where

$Q$ = quantity of water reaching drainage structure in $ft^3$/sec
$A$ = drainage area, in acres
$I$ = average rate of rainfall in in./hr during the heaviest rainfall (2.75 in./hr is commonly used in the Middle West)
$S$ = average slope of ground, in ft/1000 ft
$C$ = a coefficient depending upon the character of the surface drained and determined approximately as follows:
$C$ = 0.75 for paved streets and built up business blocks
$C$ = 0.625 for ordinary city streets
$C$ = 0.30 for villages with lawns and macadam streets
$C$ = 0.25 for farming country

# HYDRAULIC DESIGN OF CULVERTS AND DRAINAGE CHANNELS

In the preceding paragraphs we have discussed concepts and design procedures relating to the estimation of the quantity of runoff from a drainage basin. The following sections will deal with principles and techniques for the hydraulic design of culverts and other drainage structures for streets, highways, and railways. Some of the applicable fundamental principles govern fluid flow in conduits flowing under pressure and in open channels. This material admittedly is sketchy. The reader who wishes a more extensive treatment of this subject is referred to Chapter 3 of Reference 4 where a brief but well written discussion on the hydraulics of culverts is given.

Following a brief review of hydraulic principles, various types of culvert flow will be discussed after which typical design charts for culverts and open channels will be introduced and described.

## 14-9. FUNDAMENTAL PRINCIPLES—CONDUITS FLOWING FULL

Two fundamental principles form the basis for the theory of conduits accommodating fluid flow under pressure: the conservation of mass (expressed as the continuity equation), and the conservation of energy (expressed as the Bernoulli equation).[2]

The continuity equation merely states that the quantity of flow throughout a given flow system is constant.

$$Q = va \tag{14-4}$$

where

$Q$ = rate of flow, $ft^3/sec$
$v$ = average velocity of flow, ft/sec
$a$ = cross-sectional area, $ft^2$

The Bernoulli equation states that the total energy (pressure, kinetic, and potential) at a selected section of a flow system is equal to the energy at some previous section provided allowance is made for any energy added to or taken from the system.

$$z_1 + \frac{P_1}{w_1} + \frac{v_1^2}{2g} - \Sigma H_L = z_2 + \frac{P_2}{w_2} + \frac{v_2^2}{2g} \tag{14-5}$$

where

$\frac{P}{w}$ = pressure energy
$\frac{v^2}{2g}$ = kinetic energy
$z$ = potential energy
$\Sigma H_L$ = summation of energy losses (head losses)

The head losses, $\Sigma H_L$, usually are expressed in terms of velocity of flow. These losses are due primarily to friction losses within the conduit and entrance losses.

*Friction Losses.* Friction losses in a conduit flowing full are most commonly obtained by the Darcy-Weisbach equation:

$$H_f = f \frac{L}{D} \frac{v^2}{2g} \tag{14-6}$$

[2]These principles are treated in more depth in Chapter 15, which deals with pipeline transportation.

where $L$ and $D$ refer, respectively, to the length and diameter of the pipe; $v^2/2g$ is the velocity head; and $f$ is the friction factor. The friction factor is a dimensionless measure of pipe resistance which depends on the characteristics of the pipe and the flow. For culverts the friction factor is most commonly given in terms of the coefficient of the Manning equation, $n$. The relationship between $n$ and $f$ is given by the following equation:

$$f = 185 \frac{n^2}{D^{1/3}} \tag{14-7}$$

*Entrance Losses.* Another major source of head loss occurs when flow is constricted in a culvert entrance. Entrance losses are due mainly to the expansion of flow following the entrance constriction. These losses may be computed by multiplying the velocity head of the full pipe times a constant value called the entrance loss coefficient, $K_e$.

$$H_L = K_e \frac{v^2}{2g} \tag{14-8}$$

Entrance loss coefficients vary widely with different types of entrance geometry. Typical values of $K_e$ for various types of entrances are given by Table 14-2.

*Table 14-2*

**Entrance Loss Coefficient, $K_e$; Outlet Control; Full or Partly Full**

| *Type of Structure and Design of Entrance* | *Coefficient $K_e$* |
|---|---|
| Pipe—concrete | |
| Projecting from fill, socket end (groove-end) | 0.2 |
| Projecting from fill, square cut end | 0.5 |
| Headwall, square edge | 0.5 |
| Beveled edges, 33.7 or 45 degree bevels | 0.2 |
| Pipe or pipe arch—corrugated metal | |
| Projecting from fill (no headwall) | 0.9 |
| Headwall or headwall and wingwalls, square edge | 0.5 |
| Beveled edges, 33.7 or 45 degree bevels | 0.2 |
| Box—reinforced concrete | |
| Headwall parallel to embankment | |
| Square-edged on three edges | 0.5 |
| Wingwalls at 30 to 75 degrees to barrel | |
| Square-edged at crown | 0.4 |
| Side or slope tapered inlet—all culvert types | 0.2 |

SOURCE: *Hydraulic Design of Improved Inlets for Culverts,* Federal Highway Administration, Washington, D.C., 1972.

For circular culverts flowing full, the difference in elevation between the upstream and downstream water surface $H$ is equal to the velocity head, plus the energy lost at the entrance and in the culvert:

$$H = \left(1 + K_e + \frac{185\, n^2 L}{D^{4/3}}\right) \frac{v^2}{2g} \qquad (14\text{-}9)$$

where

$H$ = difference in elevation between the headwater and tailwater surfaces, illustrated in Fig. 14-6*a* or between the headwater surface and the crown of the culvert at the outlet, as shown Fig. 14-6*b*.
$K_e$ = the entrance loss coefficient. See Table 14-2.
$n$ = Manning's roughness coefficient. See Table 14-3.
$L$ = length of culvert, ft
$D$ = culvert diameter, ft
$v^2/2g$ = velocity head, ft

## 14-10. FUNDAMENTAL PRINCIPLES—OPEN CHANNEL FLOW

In open channels, the water surface, which is exposed to atmospheric pressure, serves as a flow boundary. Flow in open channels must, therefore, adjust itself so that the pressure at the water surface is equal to the pressure of the atmosphere. Open channel flow exists in conduits flowing part full as well as in open drainage ditches.

The basic laws of continuity of flow and conservation of energy also underlie the analysis of flow in open channels. In this case, however, the term for pressure is eliminated from the Bernoulli equation since the flow occurs under atmospheric pressure, which is assumed to be constant. The equation becomes:

*Table 14-3*
**Values of Manning's Roughness Coefficient (Open Channels)**

| *Type of Lining* | *Values of n* |
|---|---|
| Smooth concrete | 0.013 |
| Rough concrete | 0.022 |
| Riprap | 0.030 |
| Asphalt, smooth texture | 0.013 |
| Good stand, any grass—depth of flow more than 6 in. | 0.09–0.30 |
| Good stand, any grass—depth of flow not less than 6 in. | 0.07–0.20 |
| Earth, uniform section, clean | 0.016 |
| Earth, fairly uniform section, no vegetation | 0.022 |
| Channels not maintained, dense weeds | 0.08 |

$$z_1 + d_1 + \frac{v_1^2}{2g} - \Sigma H_L = z_2 + d_2 + \frac{v_2^2}{2g} \tag{14-10}$$

where

$z$ = elevation of the bottom of the channel above a horizontal datum
$d$ = depth of water in the channel
$v$ = average velocity of flow

The energy relationships for open channel flow are shown by Fig. 14-4. Uniform flow occurs when the total energy line shown in the figure is parallel to the channel slope. Uniform flow is not often attained in highway and railway culverts.

## 14-11. TYPES OF CULVERT FLOW

The type of flow occurring in a culvert depends upon the total energy available between the inlet and outlet. Naturally occurring flow is one that will completely expend all of the available energy. Energy is thus expended at entrances, in friction, in velocity head, and in depth.

The flow characteristics and capacity of a culvert are determined by the location of the control section [5]. A control section in a culvert is similar to a control valve in a pipeline. The control section may be envisioned as the section of the culvert that operates at maximum flow; the other parts of the system have a greater capacity than actually is used.

Laboratory tests and field studies have shown that highway and railway culverts operate with two major types of control: inlet control and outlet control. Examples of flow with inlet control and outlet control are shown, respectively, by Figs. 14-5 and 14-6.

Under inlet control, the discharge capacity of a culvert depends primarily

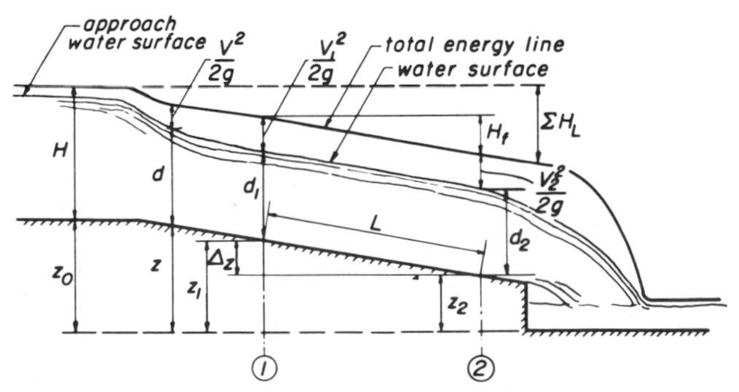


*Fig 14-4.* Definition sketch for open channel flow. (Source: *Handbook of Concrete Culvert Pipe Hydraulics,* 1964.)

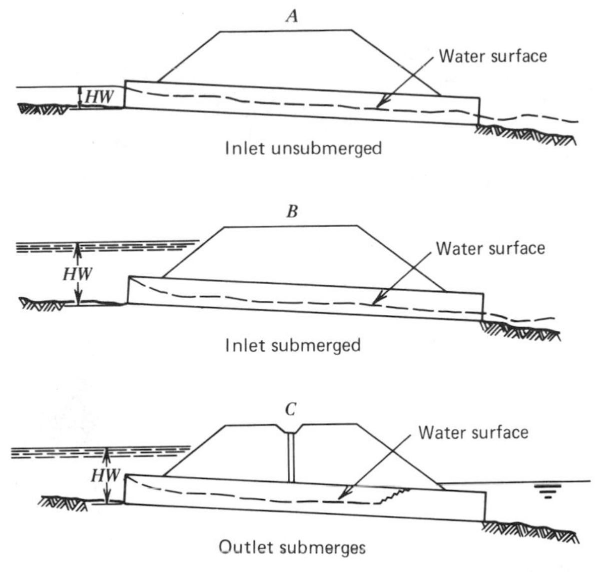


*Fig 14-5.* Inlet control for culverts. (Source: *Hydraulic Design for Improved Inlets for Culverts,* August 1972.)

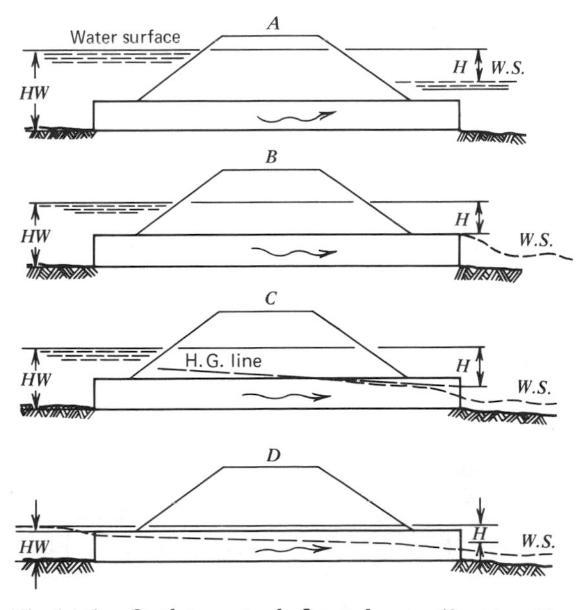


*Fig 14-6.* Outlet controls for culverts. (Source: *Hydraulic Design for Improved Inlets for Culverts,* August 1972.)

on the depth of headwater at the entrance and the entrance geometry (barrel shape, cross-sectional area, and type of inlet edge). Inlet control commonly occurs when the slope of the culvert is steep and the outlet is not submerged.

Maximum flow in a culvert operating with outlet control depends on the depth of headwater and entrance geometry and the additional considerations of the elevation of the tailwater in the outlet, the slope, roughness, and length of the culvert. This type of flow most frequently occurs on flat slopes, especially where downstream conditions cause the tailwater depth to be greater than the critical depth.

## 14-12. CULVERT DESIGN CHARTS

It is possible by involved hydraulic computations to determine the probable type of flow under which a given culvert will operate and to estimate its capacity. These computations may often be avoided by using design charts and nomographs published by the Federal Highway Administration. From these charts and graphs, the headwater depths for both inlet control and outlet control may be determined for practically all combinations of culvert size, material, entrance geometry, and discharge.

*Capacity Charts.* An example of capacity charts published in the FHWA publication *Hydraulic Engineering Circular Number 10* [6] is given as Fig. 14-7. The solid line curves, which were plotted from model test data, represent inlet control. The dashed-line curves were computed for culverts of various lengths on relatively flat slopes and operating under outlet control. The dotted line, stepped across the curves, indicates the upper headwater limit for the unrestricted use of the charts. For greater headwaters, more reliable results are obtained by the nomographs described here.

*Nomographs.* Hydraulic charts in the form of nomographs have been published by the Federal Highway Administration as *Hydraulic Engineering Circular Number 5.* Examples of these nomographs are shown as Fig. 14-8 and Fig. 14-9. The inlet control charts are based on laboratory research conducted by the National Bureau of Standards and the U.S. Coast and Geodetic Survey. The nomographs for outlet control were prepared from computations based on fundamental energy relationships.

## 14-13. IMPROVED CULVERT INLET DESIGN

With inlet control, flow in the culvert barrel is very shallow and the potential capacity of the barrel generally is wasted. Since the barrel is usually the most expensive component of the structure, flow under inlet control tends to be uneconomic. Surveys of culvert design practices by highway agencies indicate that millions of dollars could be saved each year by the use of improved inlet design concepts. An article by Normann [7] in *Civil Engineering,* abstracted

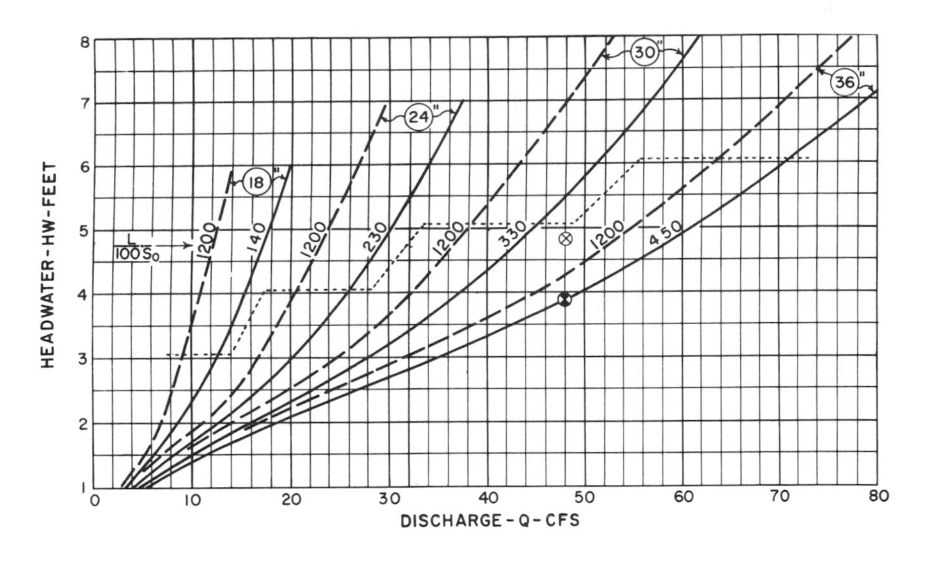


*Fig 14-7.* Culvert capacity chart for circular concrete pipes with square-edged entrance. (Source: *Capacity Charts for the Hydraulic Design of Highway Culverts*, March 1965.)

here, describes fundamental concepts of improved inlet design. For more detailed information on this important subject, the reader should refer to the FHWA Publication *Hydraulic Design of Improved Inlets for Culverts* [8].

Three basic improved inlet designs have been proposed by the FHWA: (1) bevel-edged inlets (Fig. 14-10*a*), (2) side-tapered inlets (Fig. 14-10*b*), and (3) slope-tapered inlets (Fig. 14-10*c*). These inlets improve hydraulic performance in two ways: (1) by reducing the flow contraction at the culvert inlet and more nearly filling the barrel and (2) by lowering the inlet control section and thus increasing the effective head exerted at the control section for a given headwater pool elevation.

*Bevel-Edged Inlet.* The bevel-edged inlet is the least sophisticated inlet improvement and the least expensive. The degree of improvement is recommended for use on all culverts, both in inlet and outlet control. For concrete pipe culverts, the groove end, facing upstream, will serve essentially the same purpose as the bevel-edged inlet.

*Side-Tapered Inlet.* The side-tapered inlet increases hydraulic efficiency by further reducing the contraction at the inlet control section, normally located near the throat section. The face section is designed to be large enough so as

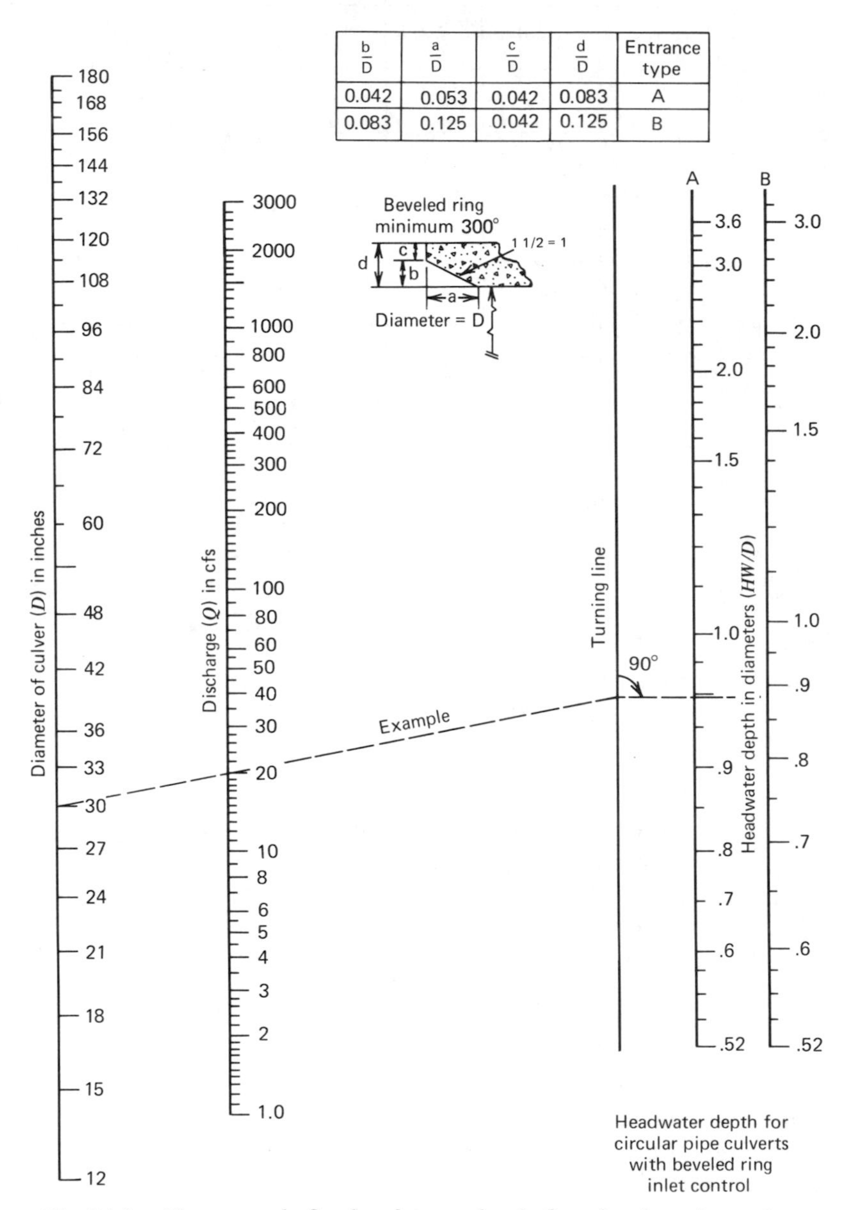


*Fig 14-8.* Nomograph for headwater depth for circular pipe culverts with beveled ring inlet control. (Source: *Hydraulic Design of Improved Inlets for Culverts.* August 1972.)

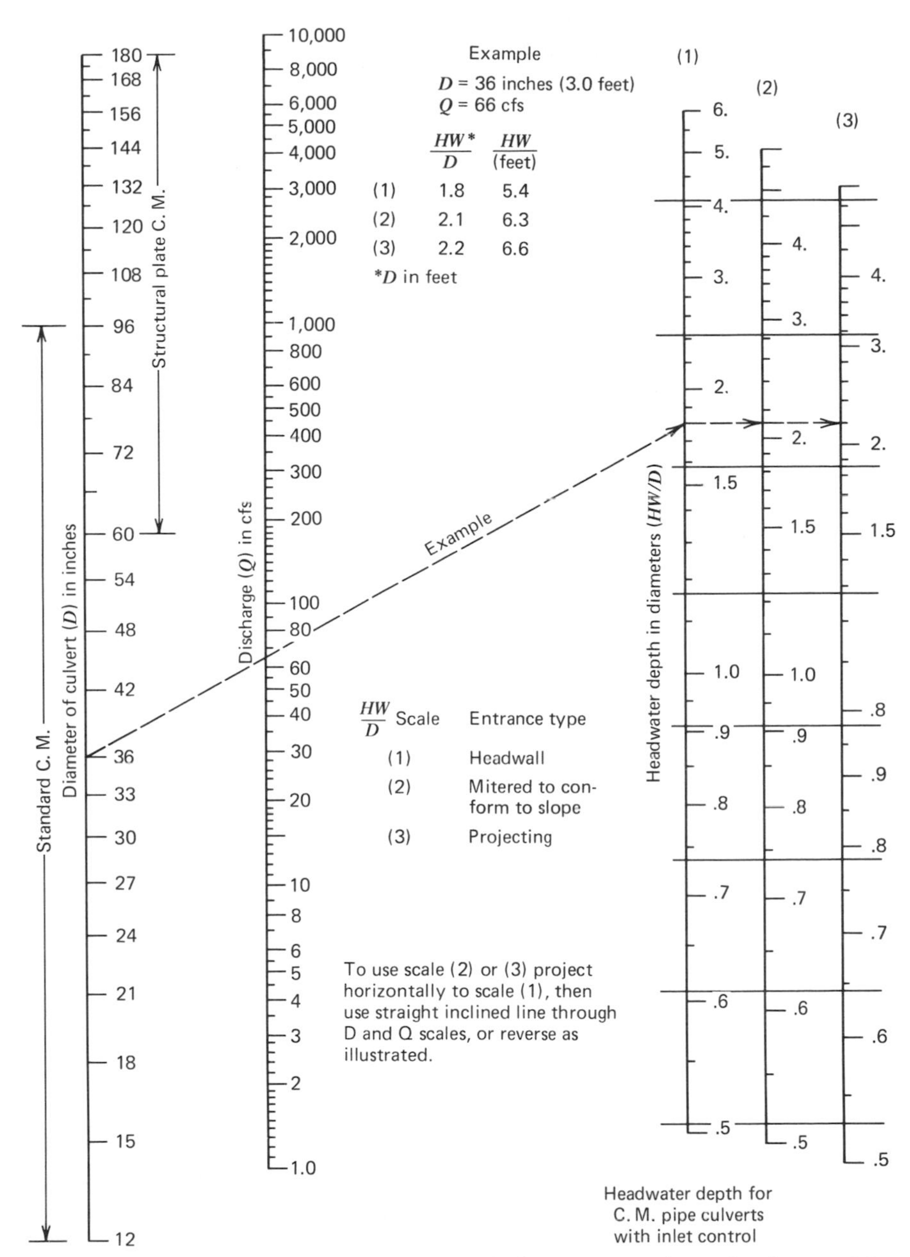


*Fig 14-9.* Nomograph for headwater depth for corrugated metal culverts with inlet control. (Source: *Hydraulic Design of Improved Inlets for Culverts.* August 1972.)

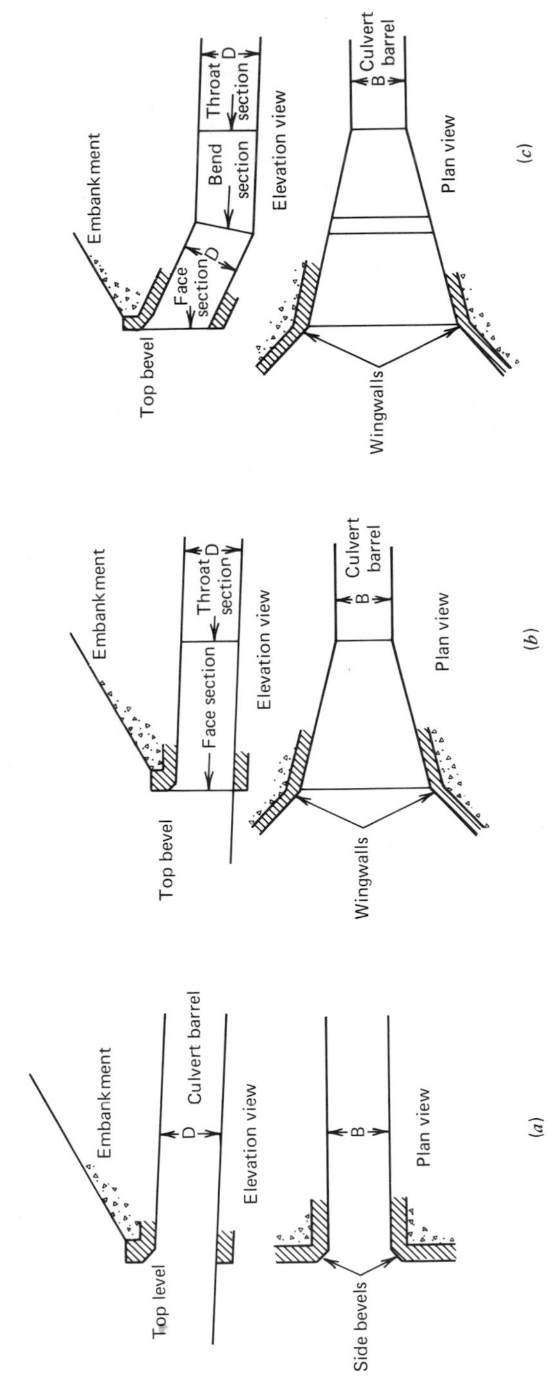


*Fig 14-10.* Improved culvert inlet designs. (*a*) Bevel-edge inlet. (*b*) Side-tapered inlet. (*c*) Slope-tapered inlet.

to not restrict the flow. The roof and floor of the inlet are straight-line extensions of the culvert roof and floor, and the tapered side walls meet the barrel walls at a smaller angle than the beveled edges (9.5 to 14 degrees *vs* 33 to 35 degrees). Because of the slope of the structure, the throat section is somewhat lower than the face, thus concentrating more head on the control section for a given headwater elevation.

*Slope-Tapered Inlet.* The slope-tapered inlet incorporates both methods of increasing hydraulic performance: reducing the entrance contraction and lowering the control section. This design provides an efficient control section at the throat, similar to that provided by the side-tapered inlet, and further increases the concentration of head on the throat. The face section remains near the streambed elevation, and the throat is lowered by incorporating a fall within the inlet structure. This fall reduces the slope of the barrel and increases the required excavation.

In addition to their use on new installations improved inlets, especially beveled-edge and side-tapered inlets, may be added to existing barrels to increase hydraulic performance if the existing inlet is operating in inlet control. Many times this will preclude the construction of a new barrel when the existing culvert is undersized.

A FHWA survey of 66 drainage installations indicated that benefits from improved inlet structures ranged from $500 to $482,000, and savings of greater than $50,000 were quite common [8].

## 14-14. DESIGN OF DRAINAGE CHANNELS

The simplest type of open channel flow occurs in long channels. In this case, equilibrium is established such that the energy losses due to friction are counter-balanced by the gain in energy due to slope. Discharge of this type, which is known as uniform flow, can be computed by Manning's equation.

$$Q = \frac{1.486}{n} aR^{2/3}S^{1/2} \tag{14-11}$$

where

$n$ = channel friction factor (see Table 12-3)
$a$ = cross-sectional area of flow, ft$^2$
$R$ = hydraulic radius = $\frac{a}{\text{wetted perimeter}}$, ft and S = channel slope, ft/ft

The Manning equation can be solved for discharge in a given channel if the depth of flow is known. The more common problem of solving for the depth of flow corresponding to a known discharge requires repeated trials. Charts have been published, however, providing a direct solution to the Manning equation for various sizes of rectangular, triangular, circular, and trapezoidal cross sections [9].

The Manning equation most commonly applies to the underground storm drainage systems (discussed in Section 19-7) and to the design of open drainage channels.

An example of one of the channel charts available in the literature is shown as Fig. 14-11. This chart is applicable to a trapezoidal channel with 2:1 side slopes and a constant 2-ft bottom width.

> Depths and velocities shown in the chart apply accurately only to channels in which uniform flow at normal depth has been established by sufficient length of uniform channel on a constant slope when the flow is not affected by backwater.
>
> Depth of uniform flow for a given discharge in a given size of channel on a given slope and with $n = 0.030$ may be determined directly from the chart by entering on the *Q*-scale and reading normal depth at the appropriate slope line (or an interpolated slope). Normal velocity may be read on the *V*-scale opposite this same point. This procedure may be reversed to determine discharge at a given depth of flow.
>
> For channel roughness other than $n = 0.030$, compute the quantity *Q* times *n* and use the $Q \cdot n$- and $V \cdot n$-scales for all readings, except those which involve values of critical depth or critical velocity. Critical depth for a given value of *Q* is read by interpolation from the depth lines at the point where the *Q*-ordinate and the critical curve intersect, regardless of channel roughness. Critical velocity is the reading on the *V*-scale at this same point. Where $n = 0.030$, the critical slope is read at the critical depth point. Critical slope varies with *n*; therefore, in order to determine the critical slope for values of *n* other than 0.030, it is first necessary to determine the critical depth. Critical slope is then read by interpolation from the slope lines at the intersection of this depth with the $Q \cdot n$-ordinate [1].

Suppose, for example, one wishes to determine the depth and velocity of flow in a trapezoidal channel ($n = 0.030$) with 2:1 side slopes and a 2-ft bottom width given a flow of 100 ft$^3$/sec and a slope of 4.0 percent. The chart is entered at $Q = 100$ ft$^3$/sec and a line is projected vertically until it intersects the slope line $S_O = 0.04$. At the point of intersection, the normal depth $d_n =$ 1.6 ft and the corresponding normal velocity $V_n = 10$ ft/sec is read.

To find the critical depth, velocity, and slope for these conditions, the line $Q = 100$ ft$^3$/sec is projected upward to its intersection with the critical curve. At the point of intersection, the following values are read

Critical depth $d_c = 2.3$ ft
Critical velocity $V_c = 6.6$ ft/sec
Critical slope $S_c = 0.014$

The normal depth is less than the critical depth indicating that the flow is rapid and not affected by backwater conditions.

Suppose $n = 0.012$ and other conditions are the same as given above. In this case, the product $Q \cdot n$ is determined and the $Q \cdot n$ scale is used.

$$Q \cdot n = 100(0.012) = 1.2$$

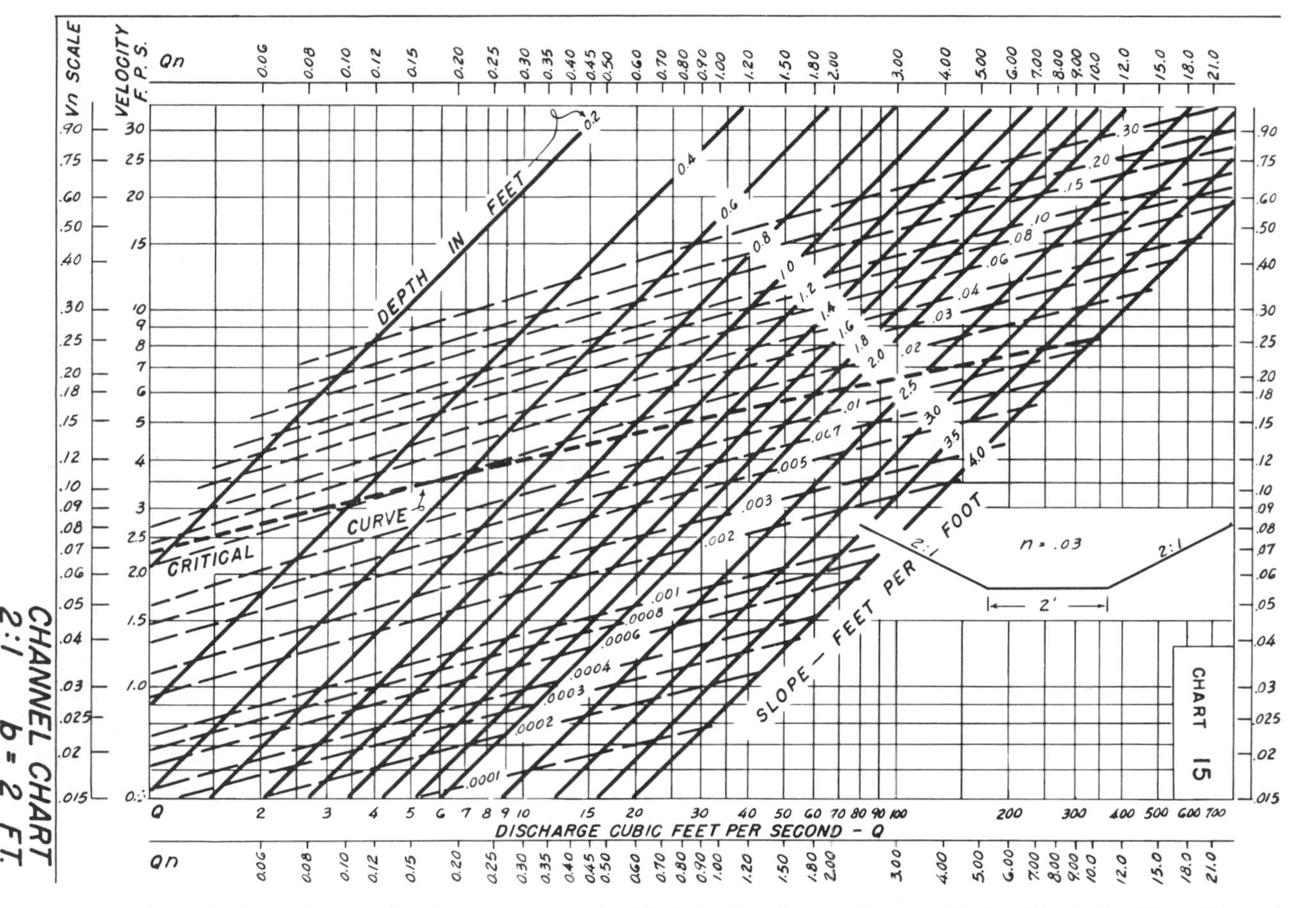


*Fig 14-11.* Channel chart for two-foot bottom channel with 2 : 1 side slopes. (Source: *Design Charts for Open Channel Flow*, August 1961.)

Entering the $Q \cdot n$ scale at a value of 1.2 and proceeding as before yields the following values:

$$d_n = 1.2 \text{ ft}$$

$$V_n = \frac{0.24}{0.012} = 20 \text{ ft/sec}$$

$$d_c = 1.5 \text{ ft}$$

$$V_c = \frac{0.17}{0.012} = 14 \text{ ft/sec}$$

$$S_c = 0.016$$

*Erosion Control.* The designer should remember that erosion of open channels is not only aesthetically offensive, but it may also increase roadway maintenance costs and, in its most drastic form, may create roadway hazards. There are several steps that can be taken to minimize the probability and lessen the effects of erosion. In the first place, erosion channel velocities should be avoided whenever possible. The magnitude of stream velocity that causes erosion varies, of course, with the type of channel lining. Unlined ditches in certain fine-grained soils may be eroded by streams flowing as slow as 1.5 to 2.0 ft/sec, while other untreated soils may resist erosive forces caused by streams flowing up to 6.0 ft/sec.

To avoid erosion, it may be necessary to provide protective linings in the bottom and along the sides of drainage channels. A variety of protective linings are used including portland cement concrete, soil-cement, rock riprap, and vegetation. To prevent erosion during the construction process, various temporary linings are utilized to protect seeded channels until vegetation has been established [10].

*Culvert Selection.* While a variety of materials has been utilized, the large majority of culverts in common use fall into two classes;

1. reinforced concrete culverts;
2. corrugated metal culverts.

Reinforced concrete culverts fall into two general classes: pipes and box culverts. Concrete pipes come in a variety of diameters ranging from 12 to 108 in. Box culverts usually are formed and poured in place with rectangular or square cross section.

Concrete culverts are durable and able to withstand large stresses imposed by heavy wheel loads. In many localities, concrete culverts cost less than comparable sizes of corrugated metal culverts. Possessing a smaller roughness coefficient, concrete culverts are more efficient hydraulically than corrugated metal culverts. On the other hand, concrete culverts are heavy and not installed easily, particularly in cases involving steep slopes and large sizes.

Corrugated metal culverts are made of galvanized steel of varying thicknesses. These culverts are manufactured in diameters from 8 to 96 in. Corrugations, typically 2⅔ in. from crest to crest and ½ in. deep, are formed in the sheet metal for added strength.

Corrugated metal culverts are easy to handle and install, even in large sizes and steep slope. Where very large pipe sizes are required, corrugated metal culverts may be formed from heavy plates which are bolted together at the field site.

Corrugated metal culverts may be purchased as pipe arches, the vertical dimension (rise) being about 0.6 the horizontal dimension (span). Arch culverts are used commonly and advantageously in low-fill areas where the headroom is limited. Although corrugated metal culverts are subject to deterioration in certain severe exposure conditions, this may be largely overcome by the use of a pipe in which the invert has been covered with a heavy bituminous mixture.

## EARTHWORK OPERATIONS

### 14-15. INTRODUCTION

Practically all highway and railroad construction jobs involve a considerable amount of earthwork. Earthwork operations in general are those construction processes that involve the soil or earth in its natural form and that precede the building of the pavement structure itself. These processes may include everything that pertains to the grading and drainage structures, which will also include clearing and grubbing, roadway and borrow excavation, the formation of embankments and the finishing operations for the preparation of the highway or runway pavement or railroad ballast. Any or all of these construction processes may be performed on a given project and they may overlap to some extent.

Clearing is the removal of trees, shrubs, brush, and so on, from within designated areas, while grubbing refers to the removal of roots, stumps, and similar obstacles to a nominal depth below the existing ground surface. Frequently, clearing and grubbing comprise a single contract item and may include the removal of topsoil to a shallow depth. Excavation refers to the removal of earth from its natural resting place to a different place for the highway or railroad foundation. Embankments required in construction usually are formed in relatively thin layers or lifts of soil and compacted to a high degree of density. Such embankments are called rolled earth embankments or rolled earth fills. Hydraulic fills also may be required in construction.

Finishing operations include such items as trimming and finishing of slopes and the fine grading operations required to bring the grade to the desired final elevation.

Broadly speaking, earthwork operations may include all the operations involved in bringing the foundation to the point where the surface or ballast is to be applied.

## 14-16. EARTHWORK EQUIPMENT

In modern practice earthwork operations are accomplished largely by the use of highly efficient and versatile machines. Machines have been developed that are capable of performing every form of earthwork operation efficiently and economically. These include tractors, bulldozers, scraper units, shovel crane units for the excavation and moving of earth, and the sheepsfoot, pneumatic-tired, and steel-wheel rollers for compacting the earth as well as vibratory compactors and pneumatic tampers. Motor graders, trucks, and other specialized equipment may be used for earthwork operations.

A more detailed discussion of earth-moving equipment may be found in Reference 1.

## 14-17. EXCAVATION

Excavation may be classified into two types: (1) rock excavation and (2) common excavation. Rock excavation will include boulders that may be over 0.5 $yd^3$ or more in volume and all hard rock that has to be removed by blasting. In some cases the requirement relative to the removal by blasting is not specified and a phrase such as solid well-defined ledges of rock is substituted in the definition of rock excavation. Even though a contractor may choose to rip some materials, this operation may still be called rock excavation by some agencies.

Common excavation consists of all excavation not included in the rock definition.

Borderline cases frequently arise in which there may be some question as to the proper classification of a portion of the work. To eliminate this condition most agencies use the term unclassified excavation to describe the excavation of all materials, regardless of their nature. Many feel that the unclassified term tends to put the risk of uncertainty on the bidder, which results in higher prices than when rock and common excavation units are used.

If there is not sufficient material within the cross section of construction for completion of the grade, additional material may be obtained from borrow pits. This is called borrow excavation. Sometimes additional excavation may be obtained by changing the cross section by widening. This is called side borrow. All excavation is paid for on a cubic yard basis, measured in place before excavation occurs.

## 14-18. CONSTRUCTION OF EMBANKMENTS

Embankments are used in construction when it is required to maintain a grade for the roadway, runway, or railway. Usually the grade is built up in a fill or embankment section from material in a cut or excavation section and is termed a rolled-earth embankment. The excavated material may be obtained from within the construction limits or from borrow pits.

Rolled-earth embankments are constructed in relatively thin layers of loose soil. Each layer is rolled to a satisfactory degree of density before the next layer is placed and the fill or embankment thus is built up to the desired height by the formation of successful layers or lifts. Most agencies at the present time require layers to be from 6 to 12 in. thick before compaction begins, when normal soils are encountered. Specifications may permit an increase in layer thickness where large rocks are used in the lower portion of a fill, up to a maximum thickness of 24 in.

The layers are required to be formed by spreading the material to uniform thickness before compaction is permitted. End dumping from trucks without spreading definitely is not permitted. The only exception to this rule may be when the embankment foundation is such that it cannot support the weight of the spreading and the compacting equipment. In such cases end dumping may be permitted until sufficient thickness can support the equipment.

A close relationship between the compaction procedure and the type of soils used for embankment purposes has to be evaluated carefully. The treatment of the various types of soils is not considered in this text.

## 14-19. CONTROL OF COMPACTION

Practically all soils exhibit a similar relationship between moisture content and density (dry unit weight) when subjected to dynamic compaction. Practically every soil has an optimum moisture content at which the soil attains maximum density under a given compactive effort. In the laboratory this relationship usually is performed under the Standard Proctor or the Standard AASHTO Method (T99). Briefly stated, this procedure uses the soil that passes the No. 4 sieve which is placed in a 4-in. diameter mold having a volume of $1/30$ ft$^3$. The soil is placed in three layers of about equal thickness and each layer is subjected to 25 blows from a hammer weighing 5.5 lbs, having a striking face 2 in. in diameter and falling through a distance of 12 in. (12,375 ft-lb/ft$^3$).

Due to the use of heavier compaction equipment in recent years and in order to correlate more effectively laboratory procedures with field conditions, the procedure was modified and is now known as the Modified Proctor of Modified AASHTO (T180) compaction. Under the modified procedure the same mold is employed using 25 blows from a 10-lb hammer dropping a distance of 18 in. on 5 equal layers (56,250 ft-lb/ft$^3$).

Regardless of which method is used, the optimum moisture and maximum density are usually found in the laboratory by a series of determinations and the results are plotted. Figure 14-12 shows the moisture-density relationship for a typical soil under dynamic compaction. The zero air voids curve shown in Fig. 14-12 represents the theoretical density that this soil would attain at each moisture content if all the void spaces were filled with water, namely if the soil were saturated completely.

After the laboratory density has been determined, recommendations are

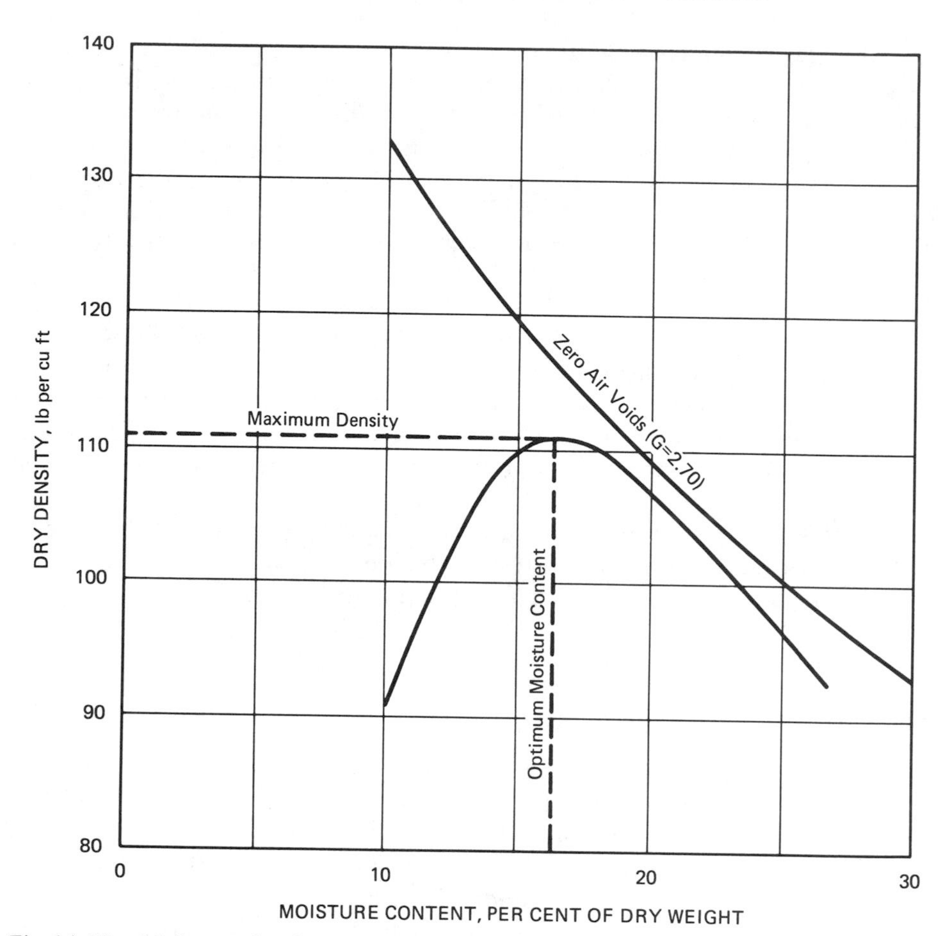


*Fig 14-12.* Moisture-density relationships for a typical soil under dynamic compaction. (Source: Paul H. Wright and Radnor J. Paquette, *Highway Engineering*, Fifth Edition, John Wiley, 1987.)

made for field conditions. A large majority of agencies require compaction to a certain percentage of the maximum density, as determined in the laboratory. This may range from 90 to 100 percent.

In order to determine if compaction meets laboratory requirements, field testing is required. Three methods have been used widely, which are designated as the sand, the balloon, and the heavy oil methods. Nuclear devices to measure in-place density and moisture content are being used increasingly. The details of these tests will not be discussed here.

## 14-20. SPECIAL EMBANKMENT FOUNDATIONS

In swampy areas, particularly where peat and other highly organic soils are encountered, special treatment may be required in the construction of embankments to prevent failure of the embankment. These methods in general use may be classified as (1) gravity subsidence, (2) partial or total excavation, (3) blasting, (4) jettying, (5) vertical sand drains, and (6) reinforcement with engineering fabrics.

In gravity subsidence a fill simply may be placed on the surface of an unsatisfactory foundation soil and allowed to settle at will, with no special treatment of the underlying soil. A temporary surface may be placed and, as settlement occurs, the temporary surface may be replaced.

Under the partial or total excavation, the undesirable soil may be removed partially or completely and backfilled with suitable material. Total excavation of the undesirable material is expensive, but has its advantages in that the pavement surface may be placed immediately. Under partial excavation, gravity subsidence takes place and the temporary surface has to be replaced.

When blasting is used in swampy areas, a fill is usually placed first by the end-filling method. When the fill is complete, dynamic charges are used and so timed that the swampy material is thrust sideways and the fill material settles in place.

In jettying, the fill is made in a similar way as for blasting. The process of jettying involves the pumping of water into the underlying soil in order to liquefy it and thus aid in its displacement by the weight of the fill or embankment.

Vertical sand drains have been used increasingly in recent years. Vertical sand drains generally consist of circular holes or shafts from 18 to 24 in. in diameter, which are spaced from 6 to 20 ft apart on centers beneath the embankment section and are carried beneath the layer of compressible soil. The holes are backfilled with suitable granular material. A sand blanket from 3 to 5 ft thick is placed on top of the drains of the embankment. The embankment is then constructed by normal methods on top of the sand blanket. The weight of the embankment forces the water out of the sand drains and consolidation takes place.

Recently, engineers have found that engineering fabrics can be used to advantage in the construction of low fills over swampy or marshy areas. These products are permeable textile cloths or mats made from a variety of artificial fibers. Typically, the engineering fabric is placed on the weak foundation and overlain with the embankment fill. This increases the bearing capacity of the foundation and allows a higher fill to be constructed on it [11].

## 14-21. COMPUTING EARTHWORK QUANTITIES

The amount of earthwork on a project is usually one of the most important features in its design. Earthwork includes the excavation of material and any hauling and compaction required for completing the embankment. Payment for earthwork is based on excavated quantities only and generally includes the costs of hauling and compaction. However, an additional item of payment called *overhaul* may be used to provide for hauling of the excavated material beyond a specified *freehaul distance.*

In order to determine earth excavation and embankment requirements before construction, the grade line for the proposed highway or railway will have to be determined and cross sections will have to be made of the original ground. Earthwork quantities then are determined by placing templates of the proposed grade over the original ground. The areas in cut and the areas in fill are determined and the volumes between the sections are computed. Figure 14-13 shows template sections and original ground in cut and fill. The terms cut and fill are used for areas of the section and the terms excavation and embankment generally refer to volumes.

Cross sections are plotted on standard cross section paper to any convenient scale. A scale of 1 in. equals 5 ft vertically and horizontally is common practice. Each cross section should show the location or station of the original ground section and template section, the elevation of the proposed grade at that station, and the areas of cut and fill for each section. The computed volumes of excavation and embankment may be placed on the sheet between two successive cross sections to facilitate the tabulation of earthwork quantities.

The areas of cut and fill may be measured by the use of a planimeter, a computation method using coordinates, or some other suitable method.

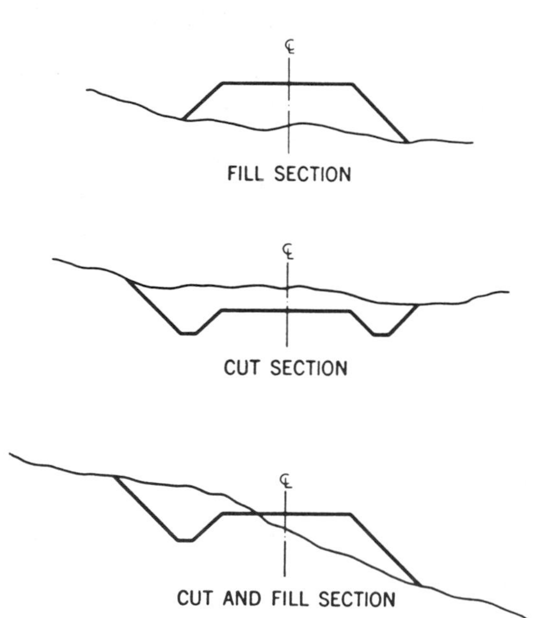


*Fig 14-13.* Original ground line and template sections. (Source: Paul H. Wright and Radnor J. Paquette, *Highway Engineering,* Fifth Edition, John Wiley and Sons, Inc. 1987.)

Volumes may be computed by the average end area method or by the prismoidal formula. The average end area method is based on right prisms and the computed volumes are slightly in excess of those computed by the prismoidal formula. This error is small when the sections do not change rapidly; however, when sharp curves are used, prismoidal correction should be applied. The average end area method is generally used by a majority of agencies. The formula for the average end area method is as follows:

$$V = \frac{\frac{1}{2}L(A_1 + A_2)}{27} \qquad (14\text{-}12)$$

where

$V$ = volume in $yd^3$
$A_1 + A_2$ = area of end sections in $ft^2$
$L$ = distance between end sections in ft

When a section changes from a cut section to a fill section, a point is reached where zero cut and zero fill occur. At this point, it will be necessary to take additional cross sections so that the proper volumes may be computed. Manual methods described above are being supplanted rapidly by methods based on electronic computers. In typical cases and for earthwork quantities at the design state, the computer input includes the centerline data, the cross section data, template data, and other pertinent information. The computer produces a tabulation of the cut and fill volumes at each station and the difference between them adjusted for shrinkage and swell, the cumulative volumes of cut and fill, mass diagram coordinates, and slope stake coordinates. Figure 14-14 shows such a tabulation.

## 14-22. SHRINKAGE AND SWELL

When the freshly excavated material is hauled to an embankment, the material increases in volume; however, during the construction process of compacting the embankment, the volume decreases below that of its original volume. This is known as shrinkage. In estimating earthwork quantities, this factor must be taken into consideration. The amount of shrinkage varies with the type of soil, the depth of the fill, and the amount of compactive effort. An allowance of 10 to 15 percent frequently is made for high fills with from 20 to 25 percent for shallow fills. The shrinkage may be as high as 40 or 50 percent for some soils. This generally also allows for shrinkage due to loss of material during the hauling process and loss of material at the toe of the slope.

When rock is excavated and placed in the embankment, the material will occupy a large volume. This increase is called swell and may amount to 30 percent or more. The amount of swell is not important when small amounts of loose rock or boulders are placed in the embankment.

ROADWAY DESIGN SYSTEM
EARTHWORK QUANTITIES CALCULATION PROCESS
IDENTIFICATION DALLASB

EARTHWORK QUANTITIES LIST FOR ROADWAYS A B

| BASELINE STATION NUMBER | PRISM SHRINK/ SWELL FACTOR | STATION CUT (SQ-FT) | STATION CUT (CU-YD) | ADJUSTED STATION CUT (CU-YD) | STATION FILL (SQ-FT) | STATION FILL (CU-YD) | ADJUSTED STATION FILL (CU-YD) | MASS ORDINATE (CU-YD) |
|---|---|---|---|---|---|---|---|---|
| 220+00.00 | 0.8500 | 0.00 | 0 | 0 | 5429.53 | 13630 | 13630 | -29423 |
| 221+00.00 | 0.8500 | 0.00 | 0 | 0 | 4767.00 | 18882 | 18882 | -48306 |
| 222+00.00 | 0.8500 | 0.00 | 0 | 0 | 4520.67 | 17199 | 17199 | -65505 |
| 223+00.00 | 0.8500 | 0.00 | 0 | 0 | 3466.71 | 14791 | 14791 | -80297 |
| 224+00.00 | 0.8500 | 0.00 | 0 | 0 | 3253.31 | 12444 | 12444 | -92741 |
| 225+00.00 | 0.8500 | 0.00 | 0 | 0 | 3831.79 | 13121 | 13121 | -105862 |
| 226+00.00 | 0.8500 | 0.00 | 0 | 0 | 4146.29 | 14774 | 14774 | -120636 |
| 227+00.00 | 0.8500 | 0.00 | 0 | 0 | 4837.34 | 16636 | 16636 | -137272 |
| 228+00.00 | 0.8500 | 0.00 | 0 | 0 | 4933.10 | 18093 | 18093. | -155366 |
| 229+00.00 | 0.8500 | 0.00 | 0 | 0 | 3915.22 | 16386 | 16386 | -171751 |
| 230+00.00 | 0.8500 | 71.95 | 133 | 113 | 2092.53 | 11125 | 11125 | -182764 |
| 231+00.00 | 0.8500 | 0.81 | 135 | 115 | 1076.44 | 5868 | 5868 | -188517 |
| 232+00.00 | 0.8500 | 401.91 | 746 | 634 | 213.84 | 2389 | 2389 | -190273 |
| 233+00.00 | 0.8500 | 791.34 | 2210 | 1878 | 75.88 | 537 | 537 | -188931 |
| 234+00.00 | 0.8500 | 840.49 | 3022 | 2569 | 0.00 | 141 | 141 | -186503 |
| 235+00.00 | 0.8500 | 1108.37 | 3609 | 3068 | 0.00 | 0 | 0 | -183436 |
| 236+00.00 | 0.8500 | 1558.94 | 4939 | 4199 | 0.00 | 0 | 0 | -179237 |
| 237+00.00 | 0.8500 | 1334.00 | 5357 | 4554 | 0.00 | 0 | 0 | -174683 |
| 238+00.00 | 0.8500 | 1634.14 | 5497 | 4672 | 0.00 | 0 | 0 | -170011 |
| 239+00.00 | 0.8500 | 1368.45 | 5560 | 4726 | 0.00 | 0 | 0 | -165285 |
| 240+00.00 | 0.8500 | 535.73 | 3526 | 2997 | 0.00 | 0 | 0 | -162288 |
| 241+00.00 | 0.8500 | 776.52 | 2430 | 2065 | 0.00 | 0 | 0 | -160222 |

*Fig 14-14.* Mass diagram data.

## 14-23. THE MASS DIAGRAM

A mass diagram is a graphical representation of the amount of earth excavation and embankment involved on a project and the manner in which earth is to be moved. It shows the location of balance points, the direction of haul, and the amount of earth taken from or hauled to any location. It is a valuable aid in the supervision of grading operations and is helpful in determining the amount of overhaul and the most economical distribution of material.

Figure 14-15 is a partially completed mass diagram generated from the data tabulated in Figure 14-14. Columns 5 and 8 in the table show the cut volumes and full volumes, respectively, that have been adjusted for shrinkage. The cut and full volumes are summed algebraically to the cumulative shown for the previous station. The resulting mass ordinates are plotted for each station as shown in Fig. 14-15.

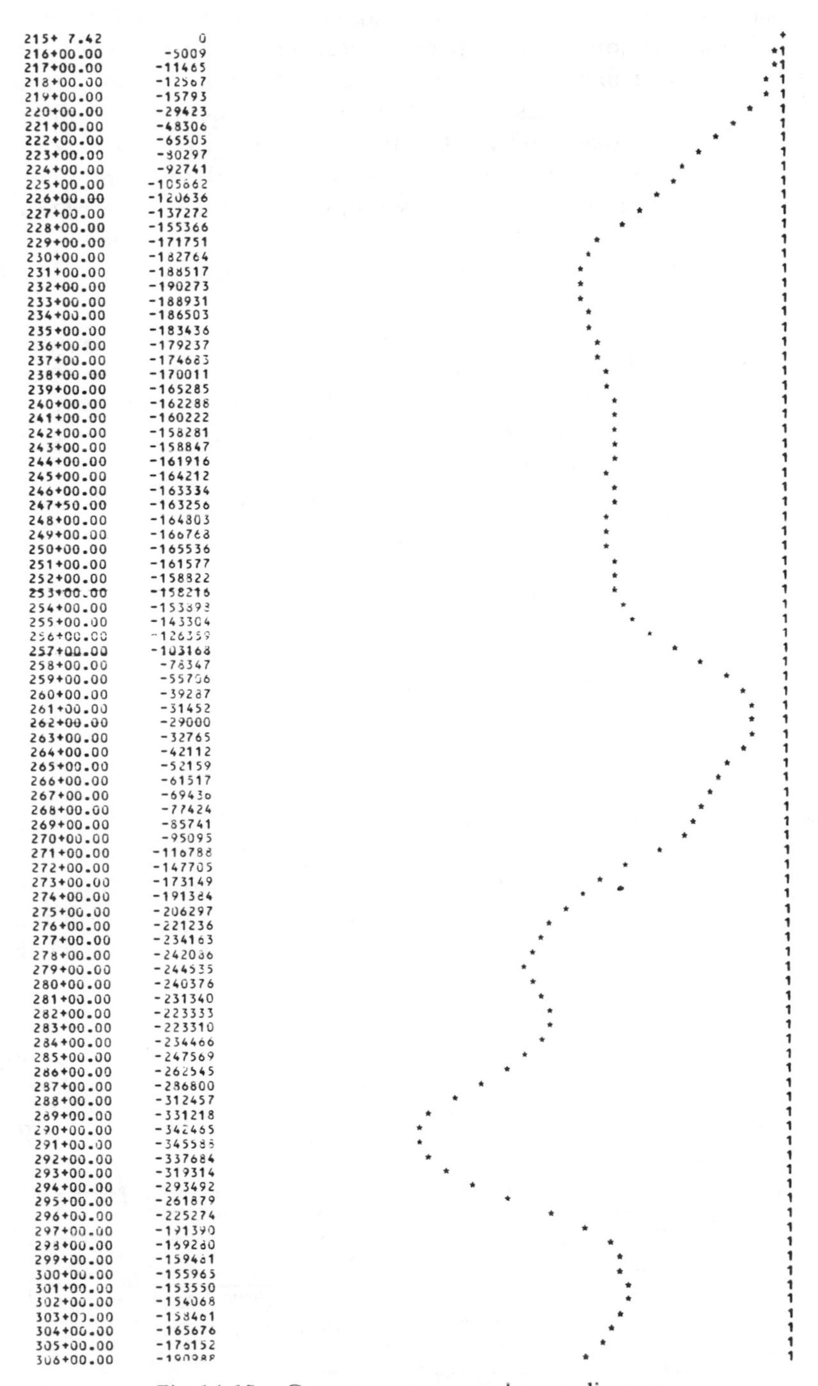


*Fig 14-15.* Computer-generated mass diagram.

*Overhaul.* The overhaul distance may be defined as the length of haul beyond a certain distance known as freehaul. This freehaul distance may be as low as 500 ft and as long as 3000 ft or more. Some agencies do not consider any overhaul which means that the cost of excavation and hauling of the material is included in the cost of earth excavation. The overhaul distance is found from the mass diagram by determining the distance from the center of mass of the excavated material to the center of mass of the embankment. This distance may be measured in stations or in miles. Thus, a cubic yard station is the hauling of 1 $yd^3$ of excavation one station beyond the freehaul distance and 1 cubic yard mile is the hauling of 1 $yd^3$ of material a distance of 1 mi beyond the freehaul limits.

Several methods for determining overhaul are in use. The graphical method, the method of movements, and the planimeter method are a few. Various computer programs have been developed to perform these calculations, which result in a large saving of time. The graphical method will be illustrated here to give some idea as to the approach to the problem and theory involved. For a more detailed explanation of the other methods, reference is made to Chapter 13 of *Highway Engineering* by Wright and Paquette [1].

*Graphical Method of Determining Overhaul.* Consider the example mass diagram shown as Figure 14-16.

The balance points and direction of haul are as indicated. To determine the overhaul for each balance, a line equal to the freehaul distance, 1000 ft in this case, is drawn parallel to the base line *AJ*. The shaded area indicates freehaul and is eliminated from further consideration. The next step is to drop perpendiculars from *C* and *D* to the base line. These perpendiculars are bisected and extended to intersect the mass diagram curve at *E* and *F*. The distance *EF* is the average haul distance. The vertical distance *CG* or *DH* indicates the number of cubic yards. To compute the overhaul for this balance, it will be the average distance *EF* in stations less the freehaul distance in stations multiplied by the volume in cubic yards. This is done for each balance.

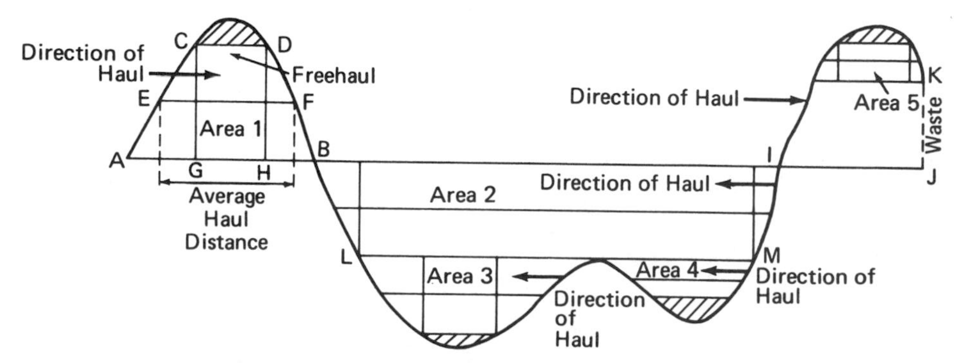


*Fig 14-16.* A mass diagram.

Where the curve changes within a balance the area has to be divided as indicated by the line *LM* and the process is repeated. The summation of each area will give the total overhaul for the project. *KJ* indicates that so many cubic yards of material has to be wasted. If this were below the base line *AJ*, it would indicate that borrow would have to be made.

This method gives the average haul and it has its limitations. If the slopes of the curve are more or less uniform, a minimum of error occurs. If the slopes are irregular, the method of movements or some other method should be used.

*Length of Economical Haul.* Where it is necessary to haul material long distances it is sometimes more economical to waste material excavated from the roadway or railway section and borrow material from a borrow pit from within the freehaul distance. The length of economical haul can be determined by equating the cost of the excavation plus the cost of overhaul to the cost of excavation in the road or railway plus the cost of excavation from the borrow pit.

If $h$ equals the length of haul in stations beyond the freehaul distance, $e$ equals the cost of excavation, and $o$ equals the cost of overhaul, then to move 1 $yd^3$ of material from cut to fill, the cost will be $e + ho$ and the cost to excavate from cut, waste the material, borrow, and place 1 $yd^3$ in the fill will equal $2e$. Assuming that the cost of the roadway excavation is equal to the cost of borrow excavation, then:

$$e + ho = 2e$$

and

$$h = \frac{e}{o} \text{ stations}$$

## DESIGN OF PAVEMENTS

Generally speaking, pavements (and bases) may be divided into two broad classifications or types: flexible and rigid. A flexible pavement structure maintains intimate contact with and distributes loads to the subgrade and depends on aggregate interlock, particle friction, and cohesion for stability [12]. Thus, the classical flexible pavement includes primarily those pavements that are composed of a series of granular layers topped by a relatively thin high-quality bituminous wearing surface. As commonly used in the United States, the term "rigid pavement" is applied to wearing surfaces constructed of portland-cement concrete. A pavement constructed of concrete is assumed to possess considerable flexural strength that will permit it to act as a beam and allow it to bridge over minor irregularities, which may occur in the base or subgrade on which it rests.

## 14-24. ELEMENTS OF A FLEXIBLE PAVEMENT

The principal elements of a flexible pavement include a wearing surface, base, subbase (not always used), and subgrade. (See Fig. 14-17.) The wearing surface may range in thickness from less than 1 in., in the case of a bituminous surface treatment used for low-cost, light-traffic roads, to 6 in. or more of asphalt concrete used for heavily traveled routes. The wearing surface must be capable of withstanding the wear and abrasive effects of moving vehicles and must possess sufficient stability to prevent it from shoving and rutting under traffic loads. In addition, it serves a useful purpose in preventing the entrance of excessive water into the base and subgrade.

The base is a layer (or layers) of natural or processed material, such as graded crushed rock, gravel, asphalt concrete, or cement-treated materials. Its principal purpose is to distribute or spread the stresses created by wheel loads acting on the wearing surface so that the stresses transmitted to the subgrade will not cause excessive deformation or displacement of that foundation layer. The base must also be resistant to damage by capillary water and/or frost action. Locally available materials are extensively used for base construction, and materials preferred for this type of construction vary widely in different sections of the country.

A subbase of granular material or stabilized material may be used in areas where frost action is severe or in locations where the subgrade is extremely weak. It may also be used in the interests of economy, in locations where suitable subbase materials are cheaper than base materials of higher quality.

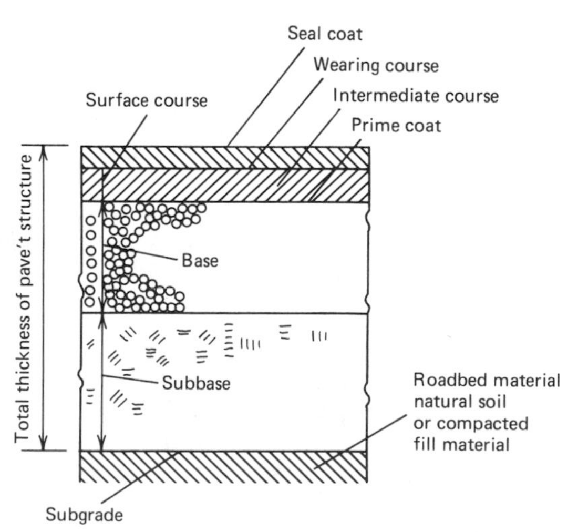


*Fig 14-17.* Typical flexible pavement cross section. (Courtesy Transportation Research Board.)

The subgrade is the foundation layer, the structure that must eventually support all of the loads that come onto the pavement. Sometimes, this layer will simply be the natural earth surface. In other instances it will be the compacted soil existing in a cut section or the upper layer of an embankment section.

## 14-25. FACTORS THAT AFFECT FLEXIBLE PAVEMENT DESIGN

The principal factors that affect the thickness design of a flexible pavement are

1. Traffic loading.
2. Material characteristics.
3. Climate or environment.

The primary loading factors that are important in flexible pavement design are

1. Magnitude of axle (and wheel) loads.
2. Volume and composition of axle loads.
3. Tire pressure and contact area.

The magnitude of maximum loading is normally controlled by legal load limits. Traffic surveys and loadometer studies are often used to establish the relative magnitude and occurrence of the various loadings to which a pavement is subjected. The prediction or estimation of the total traffic that will use a pavement during its design life is a very difficult but important task. Most design procedures provide for an increase in traffic volume on the basis of experience by using some estimated growth rate.

Proper design of flexible pavement systems requires a thorough understanding of the important characteristics of the materials of which the pavement is to be composed and on which it is to be supported. The relevant material characteristics may include gradation, strength or stability, and resistance to the effects of repeated loadings. Various standard test methods are available for determining the desired properties and are documented elsewhere [13, 14].

The climate or environment in which a flexible pavement is to be established has an important influence on the behavior and performance of the various materials in the pavement and subgrade. Probably the two climatic factors of major significance are temperature and moisture.

The magnitude of temperature and its fluctuations affect the properties of certain materials. For example, high temperatures cause asphaltic concrete to lose stability whereas at low temperatures asphaltic concrete becomes very hard and stiff. Low temperature and temperature fluctuations are also associated with frost heave and freeze-thaw damage.

Moisture also has an important influence on the behavior and performance of many materials. Moisture is an important ingredient in frost-related damage. Subgrade soils and other pavement materials weaken appreciably when saturated, and certain clayey soils exhibit substantial moisture-induced volume change.

## 14-26. THE AASHO ROAD TEST

The single most costly element of the nation's highway system is the pavement structure [15]. Seeking to control this cost, state and federal highway agencies have been involved in a continuous program of pavement research since the late 1950s. The most significant pavement research initiative was the AASHO[3] road test, a large-scale project undertaken cooperatively by the various states, the federal government, and industry groups in 1958. In this project, special test sections of variable thicknesses were constructed in Ottawa, Illinois and subjected to repeated loadings from traffic that included both single- and tandem-axle vehicles. Each test section was subjected to thousands of load repetitions. The research project dealt with both flexible and rigid pavements.

The AASHO road test and subsequent research led to the publication of a series of guides for the design of pavements, first in 1961, then in 1972, 1981, and 1986. The following section summarizes the design approach for flexible pavements recommended by the AASHO Design Committed in the 1986 edition of the guide [15].

## 14-27. THE AASHTO DESIGN METHOD FOR FLEXIBLE PAVEMENTS

One of the products of the AASHO road test was the "pavement serviceability concept." In essence, this involves the measurement, in numerical terms, of the behavior of the pavement under traffic, its ability to serve traffic at some instant during its life [16].

Such an evaluation can be made on the basis of a systematic, but subjective rating of the riding surface by individuals who travel over it. Or pavement serviceability can be evaluated by means of certain measurements made on the surface, as was done on the test road. The AASHTO design method employs a

[3]At the time of the test, AASHTO was called the American Association of State Highway Officials (AASHO).

*present serviceability index* (PSI) that is measured on a scale of 0 to 5 with 0 corresponding to an impossible road and 5 representing a perfect road.

The AASHTO method considers the change in the PSI rating over it performance period:

$$\Delta \text{PSI} = P_o - P_t$$

where

$P_o$ = original or initial serviceability index
$P_t$ = the terminal or lowest allowable serviceability index

The original or initial PSI rating for flexible pavements generally falls between 4.0 and 4.5. A value of 4.2 was reported for the flexible pavements tested at the AASHO test road project.

The value of $P_t$ is the lowest allowable PSI rating that will be tolerated before rehabilitation, resurfacing, or reconstruction is required. AASHTO recommends $p$ values of 2.5 or higher for the design of major highways and 2.0 for highways with low traffic volumes.

*Time Constraints.* With the AASHTO method, the designer must select two time-related variables or constraints: the performance period and the analysis period. The performance period is the period of time that an initial pavement structure will last before it needs rehabilitation. In other words, it is the time elapsed as a new, reconstructed, or rehabilitated structure deteriorates from its initial serviceability to its terminal serviceability.

The analysis period is the period of time for which the analysis is conducted. It is similar to the term "design life" used by designers in the past. It is suggested that consideration be given to extending the analysis period to include one rehabilitation, or longer in the case of high volume urban freeways.

The approach allows the designer to consider strategies ranging from the initial structure lasting the entire analysis period to stage construction with overlays at planned intervals.

*Traffic.* The AASHTO design procedure is based on the cumulative expected load applications during the analysis period. The design committee handled the problem of mixed traffic by first adopting an 18,000-lb single-axle load as a standard and then developing a series of "equivalence factors" for each axle weight group. If the estimated traffic to be used in design can be broken down into axle load groupings, the number of load applications in each group can be multiplied by the equivalence factor to determine the number of 18,000-lb axle loads that would have an equivalent effect on the pavement structure. The committee tabulated equivalence factors for a range of axle loads, structural numbers, and $p_t$ values [15].

For purposes of pavement design, an estimate must be made of the cumulative equivalent single-axle load applications *in the design lane.* It may be necessary, therefore, to multiply the estimated cumulative two-directional

18-kip single axle load applications by a directional factor, and if the facility has two or more lanes in one direction, by a lane distribution factor.

*Roadbed Soil.* In the AASHTO design procedure, the roadbed soil strength is characterized by its modulus of elasticity or *resilient modulus.* In the standard method of test for the resilient modulus of subgrade soils, a specially prepared and conditioned specimen is subjected to repeated applications of axial deviator stress of fixed magnitude, duration, and frequency. During the test, the specimen is subjected to a static all-around stress in a triaxial pressure chamber. The test is intended to simulate the conditions that exist in pavements subjected to moving wheel loads. Detailed procedures for the test are given by AASHTO Designation T274-82 [14].

Suitable equations have also been published to make it possible to estimate the resilient modulus of a soil from other measures of subgrade strength such as the CBR and the R-value.

*Environmental Effects.* Reference 15 provides a method for accounting for certain detrimental environmental effects on pavement performance. It recommends that if roadbed swelling or frost heave can lead to a significant loss of serviceability during the analysis period, the serviceability loss from these effects should be added to that resulting from the cumulative axle loadings.

*Reliability.* The structural design of a pavement is fraught with uncertainty. The design process is extremely complex, and its performance period will depend on the imposed traffic loadings, roadbed soil factors, climate, its structural design (e.g., layer types and thicknesses), and the intended serviceability.

The AASHTO design committee introduced the concept of reliability into the design process to incorporate some degree of certainty that the various design alternatives will last the analysis period.

The reliability concept allows the designer to choose a predetermined level of assurance that the pavement will serve the period for which it was designed. Generally speaking, as the volume of traffic, difficulty of rerouting traffic, and the public expectation of using the facility increases, the risk of failing to perform to expectations must be minimized. This may be accomplished by selecting higher levels of reliability. Recommended levels of reliability vary from 50 to 80 percent for local roads to 80 to 99.9 percent for interstate facilities.

In addition to a level of risk, the reliability concept takes into account the chance variation in traffic prediction and the normal variation in pavement performance prediction. This is accomplished by the selection of an overall standard deviation. The results of the AASHO road test and subsequent experience indicate that the overall standard deviation for flexible pavements is about 0.45 years.

*Design Chart.* The AASHTO design chart for flexible pavements is shown as Figure 14-18. This nomograph provides a *design structural number,* SN, required for specific conditions of reliability, overall standard deviation, estimated future traffic, the effective resilient modulus of the roadbed material, and the design serviceability loss.

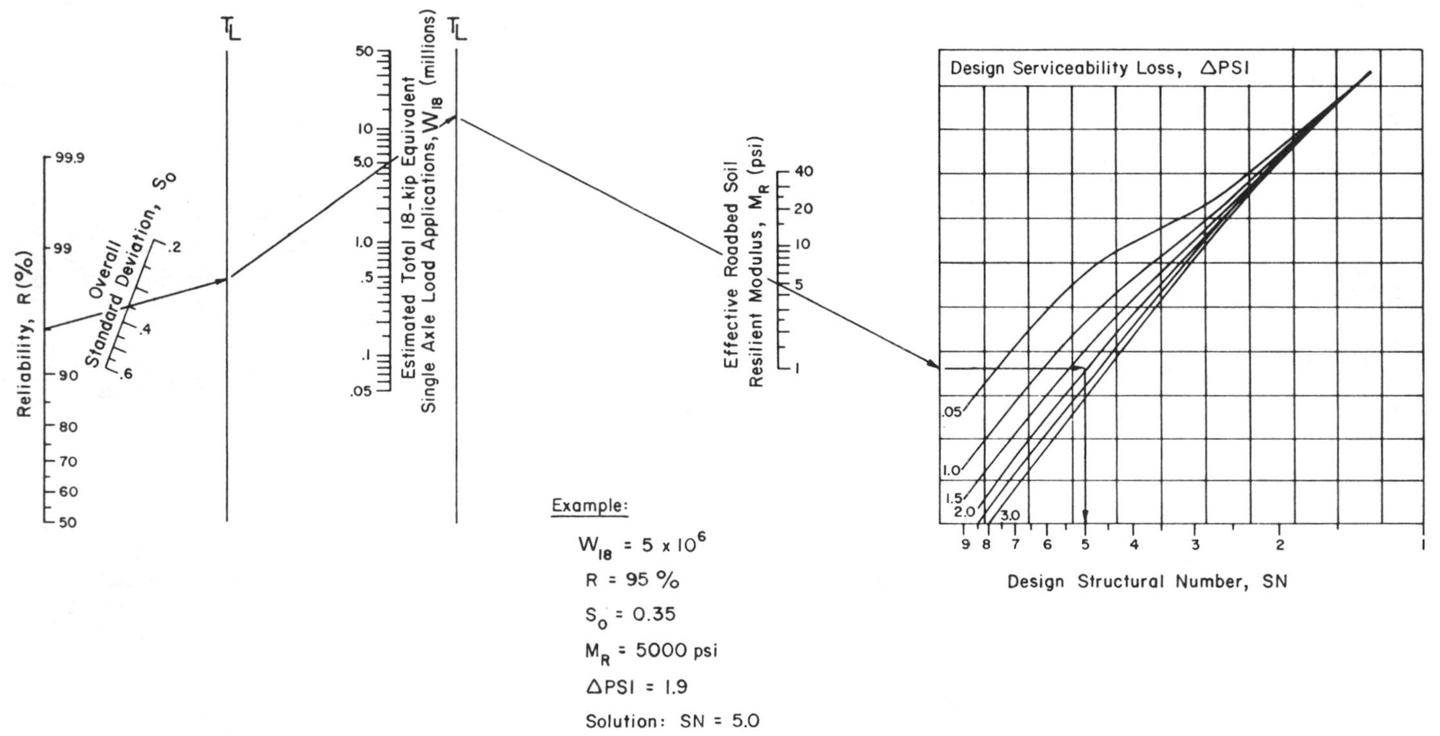


*Fig 14-18.* A design chart for flexible pavements that is based on using mean values for each input. (Courtesy American Association of State Highway and Transportation Officials.)

The design structural number is an index number that may then be converted to thickness of flexible pavement layers through the use of suitable layer coefficients related to the types of material being used. This is expressed mathematically as

$$SN = a_1 D_1 + a_2 D_2 + a_3 D_3 \tag{14-13}$$

where

$a_1, a_2, a_3$ = layer coefficients representative of the surface, base, and subbase courses, respectively,

$D_1, D_2, D_3$ = actual thicknesses (in inches) of the surface, base, and subbase courses, respectively.

In the AASHO road test, the layer coefficients were 0.44, 0.14, and 0.11, respectively, for asphaltic concrete, crushed limestone, and sand-gravel. Reference 15 gives specific guidance in the choice of layer coefficients based on laboratory test of layer materials.

The AASHTO guide [15] provides a method for accounting for the effects of base and subbase drainage on pavement performance. The method involves incorporating modifying coefficients, $m_2$ and $m_3$ into the structural number equation to allow for the drainage effects:

$$SN = a_1D_1 + a_2D_2m_2 + a_3D_3m_3 \tag{14-14}$$

Recommended values for $m_2$ and $m_3$ are given in the guide [15].

The structural number equation does not have a single solution. Many combinations of layer thicknesses and types will give satisfactory designs, and it is the designer's responsibility to determine the most appropriate design based on availability of materials, economy, and other factors.

## 14-28. ELEMENTS OF A RIGID PAVEMENT

Rigid pavements consist of a portland cement concrete slab placed on a uniform subbase or subgrade. Such pavements commonly are one of four types: (1) plain pavements, (2) plain-doweled pavements, (3) reinforced pavements, or (4) continuously reinforced pavements.

As portland cement concrete cures, there is a tendency for the slab to crack. The designer must recognize this tendency and provide for load transfer between adjacent slabs. Four approaches have been used to provide for load transfer. In plain pavements, closely spaced joints (with typical spacing of 15 ft) are provided to control cracking, and load transfer is obtained by aggregate interlock between the cracked faces of the joint. In plain-doweled pavements, relatively close joint spacings (commonly not more than 20 ft) are similarly used to control cracking, but smooth steel dowel bars are installed at each joint to ensure proper load transfer. Reinforced concrete pavements contain reinforcing steel mats as well as dowel bars and are built with joints spaced up to 40 ft. Some agencies have constructed continuously reinforced concrete pavements without contraction joints. These pavements tend to develop minute cracks at close intervals and provide load transfer by aggregate interlock at the crack faces held together by steel reinforcement.

Portland cement concrete pavements may be placed directly on a carefully prepared subgrade, but more commonly are constructed on a relatively thin subbase of sand, soil-cement, or some other subbase.

Typical sections of concrete pavements are shown in Fig. 14-19.

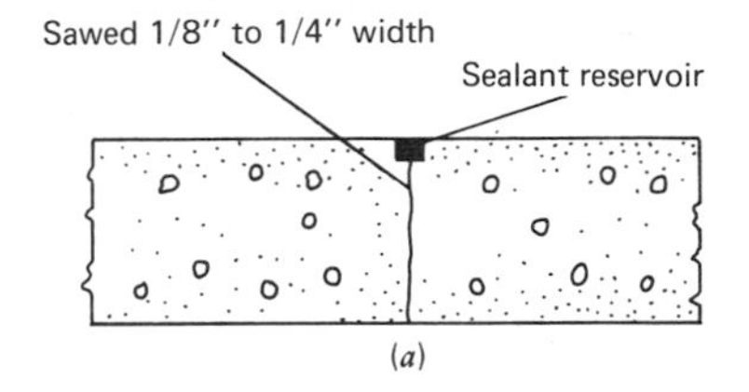


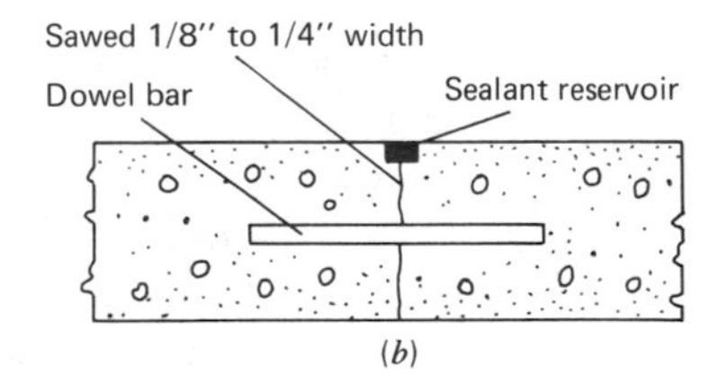


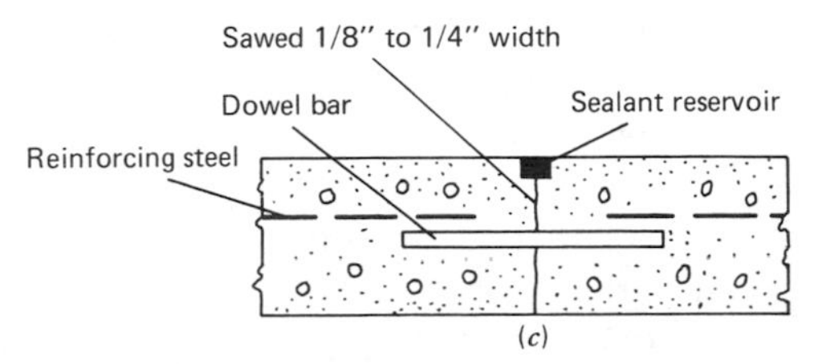


*Fig 14-19.* Types of concrete pavements. (*a*) Plain pavement. (*b*) Plain-doweled pavement. (c) Reinforced pavement.

## 14-29. FACTORS THAT AFFECT RIGID PAVEMENT DESIGN

Like flexible pavements, the thickness of rigid pavements that is required for satisfactory performance depends on traffic loading, material characteristics, and environmental or climatic factors. The numbers and weights of heavy axle loads that are expected during the pavement's service life are major factors in thickness design. Reliable estimates are needed not only of the volume of traffic expected during the design life of the pavement but also of the distribution of the traffic by various axle-load categories. These estimates are normally based on data routinely collected at the highway agency's loadometer weight stations or on special traffic studies on routes with a distribution of traffic similar to the project being designed. When specific axle-load data are not available, composite axle-load distributions that represent different categories of road and street types can be used.

The strength of the concrete is the most important material characteristic that affects the design of the pavement. Since direct compressive stresses in concrete pavements due to wheel loads are very small in relation to the compressive strength of the material, they can be ignored. Flexural strength is a key factor in thickness design, since the flexural stress produced by a heavy wheel load often is more than one half the flexural strength.

Flexural strength is measured by modulus of rupture tests on beams subjected to third-point loadings. Typically, 28-day to 90-day strengths are used for roads and streets, since very few stress repetitions occur during the first 90 days of pavement life, compared with the millions of repetitions that occur after that time. Also, the strength of concrete increases with age.

Another major factor in thickness design is subgrade or subbase support, which may be measured by the Westergaard modulus of subgrade reaction $k$. To determine $k$, a plate bearing test is performed, generally using a 30-in. plate. The $k$-value is equal to the load in pounds per square inch divided by the deflection in inches caused by that load. Its units are therefore psi/inch or pounds per cubic inch. Because plate bearing tests are expensive and time-consuming, the $k$-value is sometimes estimated by its correlation to simpler tests, such as the California Bearing Ratio or soil classification.

The $k$-value can be increased by using a treated or untreated subbase, although the use of a subbase for this purpose alone seldom would be economical.

The most troublesome environmental factor that can affect concrete pavement performance is the existence of a surplus of moisture in the subgrade soil. This condition under a pavement that is subjected to frequent heavy wheel loadings can cause a problem called "pumping." This phenomenon may be defined as follows. The movement of slab ends under traffic loads causes the extrusion or "pumping" of a portion of the subgrade material at joints, in cracks, and along the edges of the pavement. The amount of soil removed by pumping may be sufficient to cause a sizable reduction in subgrade support for the slab and may result in eventual failure of the pavement.

Where pavements are to be constructed on a fine-grained subgrade soil, a granular subbase 3 to 6 in. in thickness should be used. In areas where pumping may be a problem, it is desirable to minimize or eliminate expansion joints and ensure that contraction joints are adequately sealed to prevent the intrusion of surface water.

## 14-30. THE PCA THICKNESS DESIGN METHOD

Here we present the essentials of one method of design of rigid pavements, in the belief that this method is typical of modern approaches to this subject. The method presented is that contained in the publication, "Thickness Design for Concrete Pavements," published by the Portland Cement Association (PCA) in 1984 [17].

According to PCA engineers, design considerations that are vital to the satisfactory performance and long life of a concrete pavement are reasonably uniform support for the pavement, elimination of pumping by the use of a thin treated or untreated base course, adequate joint design, and a thickness that will keep load stresses within safe limits. The overall objective of the design procedure is to determine the minimum thickness that will give the least annual cost.

The PCA recommends two design criteria:

1. Fatigue, to keep pavement stresses from repeated loads within acceptable limits to prevent fatigue cracking.
2. Erosion, to limit the effects of pavement deflections at joints, corners, and the edges of slabs in order to control the erosion of foundation materials.

Concrete is subject to fatigue, as are other construction materials. In the PCA design concept, a fatigue failure occurs when a material ruptures under continued repetitions of loads that cause stress ratios of less than 1. The stress ratio is the ratio of flexural stress to the modulus of rupture.

Flexural fatigue research on concrete has shown that the number of stress repetitions to failure increases as the stress ratio decreases; studies show that, if the stress ratio is less than 0.55, concrete will withstand virtually unlimited stress repetitions without loss in load-bearing capacity. To be conservative, designers reduced this ratio to 0.50.

In the past, thickness design methods for concrete pavements included an allowance for impact of moving loads. The PCA method discards this concept, substituting "load safety factors" in the design. The following load safety factors are recommended: Interstate and other multiple-lane projects where there will be an uninterrupted flow of heavy truck traffic, 1.2; highways and streets with moderate volumes of truck traffic, 1.1; and streets and highways that carry small volumes of truck traffic, 1.0.

Comprehensive analyses of concrete stresses and deflections at pavement joints, corners, and edges have established that the most critical axle-load positions are

1. For stresses, when the truck wheels re placed at or near the pavement edge and midway between the joints.
2. For pavement deflections, at the slab corner when an axle load is placed at the joint with the wheels at the corner.

It is known that only a small percentage of trucks travel with their outside wheels placed at the edge. Most truck drivers drive with their outside wheels placed about 2 ft from the pavement edge. The Portland Cement Association design procedure is based on the assumption that 6 percent of the trucks travel with the outside tire at or beyond the pavement edge. The maximum stresses and deflections result from those loadings. At increasing distances inward from the pavement edge, the magnitudes of stresses and deflections decrease, but the frequency of load applications increases.

PCA engineers analyzed the distributions of truck placements and accounted for the variability of positioning of loadings in their design figures and tables. They computed the fatigue incrementally for various load placements inward from the edge of the slab. They expressed the results in terms of an equivalent load factor which, when multiplied by the edge-load stress, gives the same degree of fatigue consumption that would result from a given load distribution. A similar approach was used to account for the effects of load placements on deflections.

Procedures recommended by the PCA are based on a design period of 20 years.

The PCA design procedure takes into account the effects of the erosion of foundation and shoulder materials due to pavement deflections at slab edges, joints, and corners. This analysis recognizes modes of pavement distress such as pumping, faulting, and shoulder problems that are unrelated to fatigue.

PCA engineers worked out a number of design examples, one of which is summarized below.

EXAMPLE 14-1

Determine the thickness of a concrete pavement for a rural secondary road that will serve a design average daily traffic of 720. The design period is 40 years. Designers estimate that the truck traffic volume *(ADTT)* is 2.5 percent of the total. Thus, the *ADTT* is $720(0.025) = 18$. Truck traffic each way is $18/2 = 9$. For the design period of 40 years, there will be $9 \times 365 \times 40 = 131{,}400$ trucks using each lane. Based on a composite axle-load distribution for rural secondary roads (not shown here but given in Reference 17), the expected repetitions of single axles and tandem axles were computed and are shown on the design worksheet, Fig. 14-20.

## Calculation of Pavement Thickness

Project Design 2A, two-lane secondary road

Trial thickness 6.0 in.  Doweled joints: yes ___ no ✓

Subbase-subgrade k 100 pci  Concrete shoulder: yes ___ no ✓

Modulus of rupture, MR 650 psi  Design period 40 years

Load safety factor, LSF 1.0

no subbase

| Axle load, kips | Multiplied by LSF 1.0 | Expected repetitions | Fatigue analysis: Allowable repetitions | Fatigue analysis: Fatigue, percent | Erosion analysis: Allowable repetitions | Erosion analysis: Damage, percent |
|---|---|---|---|---|---|---|
| 1 | 2 | 3 | 4 | 5 | 6 | 7 |

8. Equivalent stress 411  10. Erosion factor 3.40

9. Stress ratio factor 0.632

**Single Axles**

| | | | | | | |
|---|---|---|---|---|---|---|
| 22 | 22 | 130 | 340 | 38.2 | 120,000 | 0.1 |
| 20 | 20 | 550 | 2,000 | 27.5 | 210,000 | 0.3 |
| 18 | 18 | 2,080 | 13,000 | 16.0 | 380,000 | 0.5 |
| 16 | 16 | 5,000 | 80,000 | 6.2 | 740,000 | 0.7 |
| 14 | 14 | 7,370 | 800,000 | 0.9 | 1,600,000 | 0.5 |
| 12 | 12 | 16,290 | Unlimited | 0 | 4,200,000 | 0.4 |
| 10 | 10 | 26,930 | " | 0 | 15,000,000 | 0.2 |
| 8 | 8 | 63,500 | " | 0 | Unlimited | 0 |
| 6 | 6 | 96,180 | | | " | 0 |
| | | | | | | |

11. Equivalent stress 348  13. Erosion factor 3.53

12. Stress ratio factor 0.535

**Tandem Axles**

| | | | | | | |
|---|---|---|---|---|---|---|
| 36 | 36 | 550 | 190,000 | 0.3 | 160,000 | 0.3 |
| 32 | 32 | 9,140 | 2,500,000 | 0.3 | 310,000 | 2.9 |
| 28 | 28 | 9,000 | Unlimited | 0 | 660,000 | 1.4 |
| 24 | 24 | 5,150 | " | 0 | 1,700,000 | 0.3 |
| 20 | 20 | 7,500 | " | 0 | 5,400,000 | 0.1 |
| 16 | 16 | 9,860 | | | 26,000,000 | 0 |
| 12 | 12 | 18,300 | | | Unlimited | 0 |
| 8 | 8 | 11,250 | | | " | |
| | | | | | | |
| | | | | | | |
| | | | Total | 89.4 | Total | 7.7 |

*Fig 14-20.* Design worksheet. (Courtesy Portland Cement Association.)

PCA engineers evaluated two designs (2A and 2B). For design 2A, the $k$-value for the subgrade was taken to be 100 lb/in$^3$. The load safety factor is 1.0, with a modulus of rupture of the concrete of 650 lb/in$^2$.

The engineers assumed a trial depth of 6.0 in. Key calculations are shown on the design worksheet.

For the fatigue analysis, equivalent stresses are determined from Table 14-4 for single- and tandem-axles and are entered on the design worksheets as items 8 and 11, respectively. The equivalent stresses are then divided by the concrete modulus of rupture and entered as items 9 and 12 (stress ratio factors). Next, the allowable repetitions are determined for each loading category from Figure 14-21. The percent-

*Table 14-4*

**Equivalent Stress—No Concrete Shoulder (Single Axle/Tandem Axle)**

| *Slab Thickness in.* | *k of Subgrade-Subbase, pci* | | | | | | |
|---|---|---|---|---|---|---|---|
| | *50* | *100* | *150* | *200* | *300* | *500* | *700* |
| 4 | 825/679 | 726/585 | 671/542 | 634/516 | 584/486 | 523/457 | 484/443 |
| 4.5 | 699/586 | 616/500 | 571/460 | 540/435 | 498/406 | 448/378 | 417/363 |
| 5 | 602/516 | 531/436 | 493/399 | 467/376 | 432/349 | 390/321 | 363/307 |
| 5.5 | 526/461 | 464/387 | 431/353 | 409/331 | 379/305 | 343/278 | 320/264 |
| 6 | 465/416 | 411/348 | 382/316 | 362/296 | 336/371 | 304/246 | 285/232 |
| 6.5 | 417/380 | 367/317 | 341/286 | 324/267 | 300/244 | 273/220 | 256/207 |
| 7 | 375/349 | 331/290 | 307/262 | 292/244 | 271/222 | 246/199 | 231/186 |
| 7.5 | 340/323 | 300/268 | 279/241 | 265/224 | 246/203 | 224/181 | 210/169 |
| 8 | 311/300 | 274/249 | 255/223 | 242/208 | 225/188 | 205/167 | 192/155 |
| 8.5 | 285/281 | 252/232 | 234/208 | 222/193 | 206/174 | 188/154 | 177/143 |
| 9 | 264/264 | 232/218 | 216/195 | 205/181 | 190/163 | 174/144 | 163/133 |
| 9.5 | 245/248 | 215/205 | 200/183 | 190/170 | 176/153 | 161/134 | 151/124 |
| 10 | 228/235 | 200/193 | 186/173 | 177/160 | 164/144 | 150/126 | 141/117 |
| 10.5 | 213/222 | 187/183 | 174/164 | 165/151 | 153/136 | 140/119 | 132/110 |
| 11 | 200/211 | 175/174 | 163/155 | 154/143 | 144/129 | 131/113 | 123/104 |
| 11.5 | 188/201 | 165/165 | 153/148 | 145/136 | 135/122 | 123/107 | 116/98 |
| 12 | 177/192 | 155/158 | 144/141 | 137/130 | 127/116 | 116/102 | 109/93 |
| 12.5 | 168/183 | 147/151 | 136/135 | 129/124 | 120/111 | 109/97 | 103/89 |
| 13 | 159/176 | 139/144 | 129/129 | 122/119 | 113/106 | 103/93 | 97/85 |
| 13.5 | 152/168 | 132/138 | 122/123 | 116/114 | 107/102 | 98/89 | 92/81 |
| 14 | 144/162 | 125/133 | 116/118 | 110/109 | 102/98 | 93/85 | 88/78 |

SOURCE: *Thickness Design for Concrete Highway and Street Pavements,* Portland Cement Association, Skokie, Ill., 1984.

age of fatigue is then calculated by dividing the expected repetitions (column 3) by the allowable repetitions (column 4). The sum of the percentages of fatigue used is then totaled for all loading categories, and for this example is 89.4 percent.

In a similar way, the erosion analysis involves determining the allowable repetitions to ensure that harmful erosion of the foundation and shoulder materials does not occur from pavement deflections at slab edges, joints, and corners.

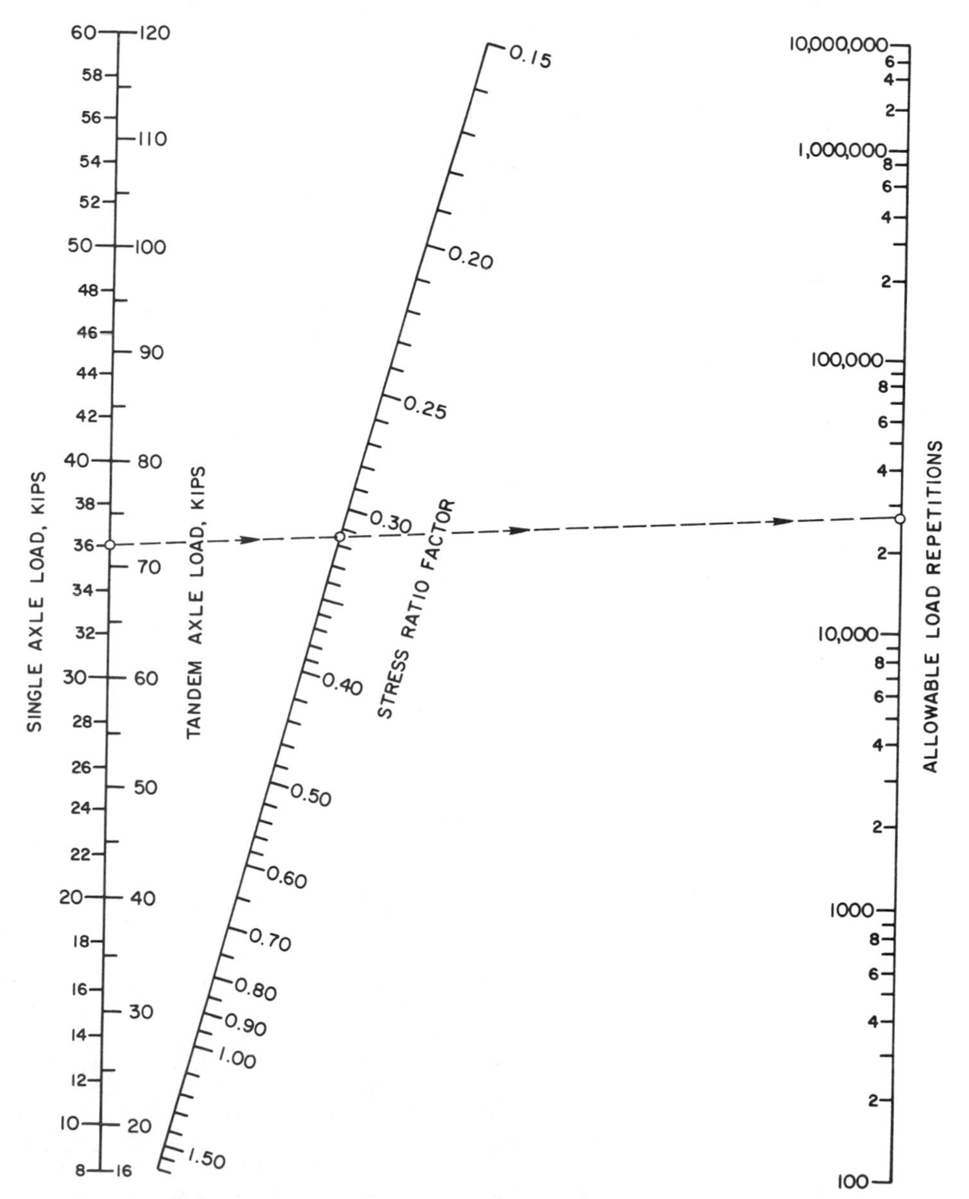


*Fig 14-21.* A nomograph for fatigue analysis—allowable load repetitions based on stress ratio factor (with and without concrete shoulder). (Courtesy Portland Cement Association.)

To perform this analysis, erosion factors are determined from Table 14-5 and recorded as items 10 and 13 for single- and tandem-axle loading, respectively. Allowable repetitions are then determined from Figure 14-22 and are recorded in column 6 for each loading category.

The damage due to erosion is then compiled for each loading condition by dividing the expected repetitions (column 3) by the allowable repetitions (column 6). The total erosion damage (percent) is determined by summing the incremental damages (column 7). For this example, the erosion damage is 7.7 percent.

The totals of fatigue use and erosion damage use of 89.4 and 7.7 percent, respectively, show that the 6.0-in. thickness is satisfactory for the given conditions.

*Table 14-5*

**Erosion Factors for Aggregate-Interlock Joints (Single Axle/Tandem Axle)**

| *Slab Thickness in.* | *k of Subgrade-Subbase, pci* | | | | | |
|---|---|---|---|---|---|---|
| | *50* | *100* | *200* | *300* | *500* | *700* |
| 4 | 3.94/4.03 | 3.91/3.95 | 3.88/3.89 | 3.86/3.86 | 3.82/3.83 | 3.77/3.80 |
| 4.5 | 3.79/3.91 | 3.76/3.82 | 3.73/3.75 | 3.71/3.72 | 3.68/3.68 | 3.64/3.65 |
| 5 | 3.66/3.81 | 3.63/3.72 | 3.60/3.64 | 3.58/3.60 | 3.55/3.55 | 3.52/3.52 |
| 5.5 | 3.54/3.72 | 3.51/3.62 | 3.48/3.53 | 3.46/3.49 | 3.43/3.44 | 3.41/3.40 |
| 6 | 3.44/3.64 | 3.40/3.53 | 3.37/3.44 | 3.35/3.40 | 3.32/3.34 | 3.30/3.30 |
| 6.5 | 3.34/3.56 | 3.30/3.46 | 3.26/3.36 | 3.25/3.31 | 3.22/3.25 | 3.20/3.21 |
| 7 | 3.26/3.49 | 3.21/3.39 | 3.17/3.29 | 3.15/3.24 | 3.13/3.17 | 3.11/3.13 |
| 7.5 | 3.18/3.43 | 3.13/3.32 | 3.09/3.22 | 3.07/3.17 | 3.04/3.10 | 3.02/3.06 |
| 8 | 3.11/3.37 | 3.05/3.26 | 3.01/3.16 | 2.99/3.10 | 2.96/3.03 | 2.94/2.99 |
| 8.5 | 3.04/3.32 | 2.98/3.21 | 2.93/3.10 | 2.91/3.04 | 2.88/2.97 | 2.87/2.93 |
| 9 | 2.98/3.27 | 2.91/3.16 | 2.86/3.05 | 2.84/2.99 | 2.81/2.92 | 2.79/2.87 |
| 9.5 | 2.92/3.22 | 2.85/3.11 | 2.80/3.00 | 2.77/2.94 | 2.75/2.86 | 2.73/2.81 |
| 10 | 2.86/3.18 | 2.79/3.06 | 2.74/2.95 | 2.71/2.89 | 2.68/2.81 | 2.66/2.76 |
| 10.5 | 2.81/3.14 | 2.74/3.02 | 2.68/2.91 | 2.65/2.84 | 2.62/2.76 | 2.60/2.72 |
| 11 | 2.77/3.10 | 2.69/2.98 | 2.63/2.86 | 2.60/2.80 | 2.57/2.72 | 2.54/2.67 |
| 11.5 | 2.72/3.06 | 2.64/2.94 | 2.58/2.82 | 2.55/2.76 | 2.51/2.68 | 2.49/2.63 |
| 12 | 2.68/3.03 | 2.60/2.90 | 2.53/2.78 | 2.50/2.72 | 2.46/2.64 | 2.44/2.59 |
| 12.5 | 2.64/2.99 | 2.55/2.87 | 2.48/2.75 | 2.45/2.68 | 2.41/2.60 | 2.39/2.55 |
| 13 | 2.60/2.96 | 2.51/2.83 | 2.44/2.71 | 2.40/2.65 | 2.36/2.56 | 2.34/2.51 |
| 13.5 | 2.56/2.93 | 2.47/2.80 | 2.40/2.68 | 2.36/2.61 | 2.32/2.53 | 2.30/2.48 |
| 14 | 2.53/2.90 | 2.44/2.77 | 2.36/2.65 | 2.32/2.58 | 2.28/2.50 | 2.25/2.44 |

SOURCE: *Thickness Design for Concrete Highway and Street Pavements,* Portland Cement Association, Skokie, Ill., 1984.

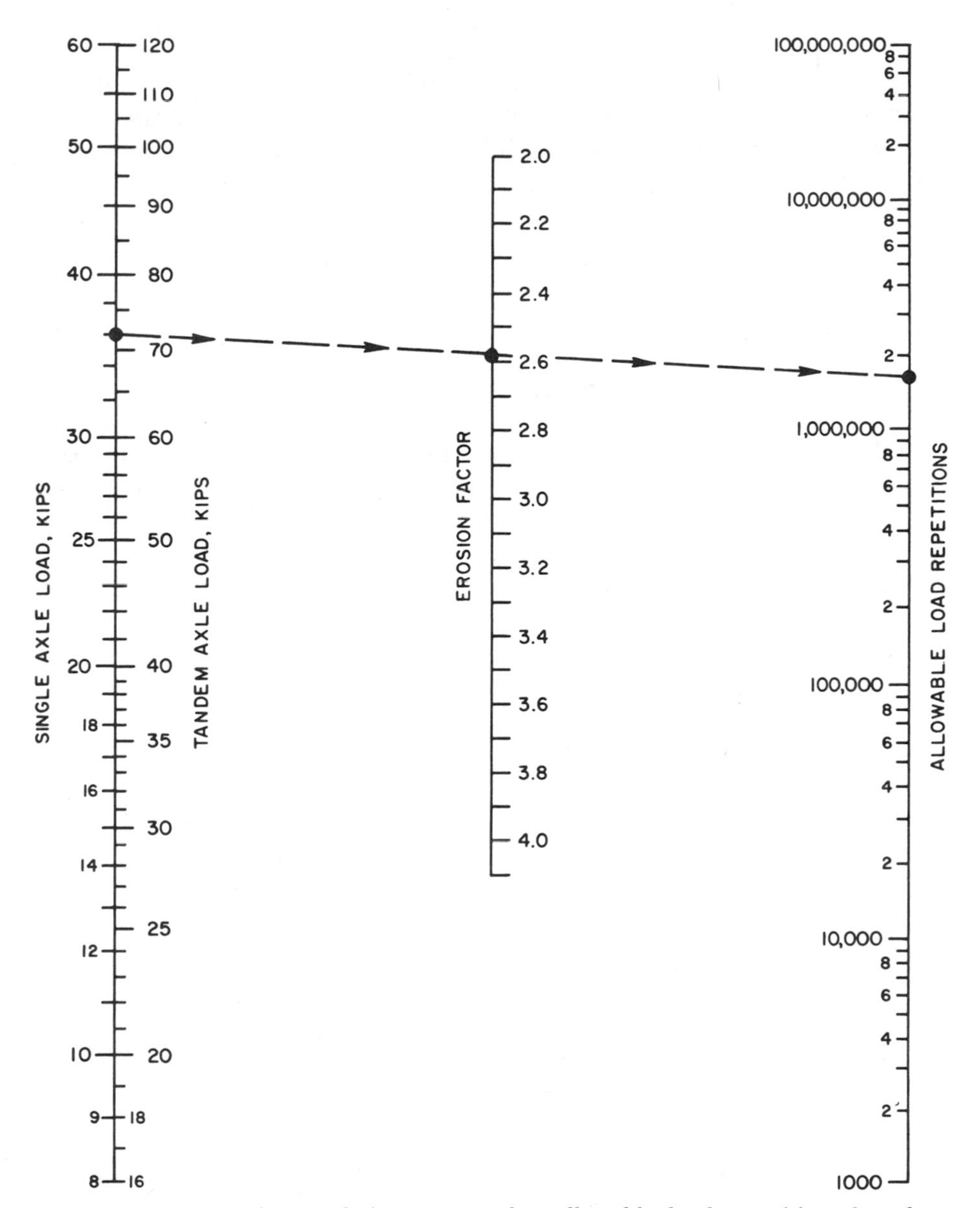


*Fig 14-22.* An erosion analysis nomograph—allowable load repetitions based on erosion factor without concrete shoulder. (Courtesy Portland Cement Association.)

In the preceding example, we have illustrated the Portland Cement Association design procedure for one set of conditions: plain joints, no subbase, and no concrete shoulder. Reference 17 provides similar illustrations and examples of other combinations of subbase support, shoulder treatment, and joint design.

## PROBLEMS

1. A highway culvert is to be designed for a 105-acre drainage area near Chicago, Illinois. The height of the most remote point above the outlet is 20 ft, and the maximum length of water travel is 2400 ft. The drainage area consists of rolling farmland with extremely impervious soil. Using a 10-year recurrence interval, determine the runoff in $ft^3/sec$.
2. A railroad culvert is being designed for a 64-hectare drainage area near St. Louis, Missouri. The time of concentration is 10 min. It is estimated that 70 percent of the rainfall will infiltrate the ground, evaporate, or otherwise not show up as runoff. Estimate the runoff in $m^3/sec$, assuming a 25-year recurrence interval.
3. A 24-in. circular corrugated metal culvert ($n = 0.024$) is to be placed under a roadbed for a rail transit line. The estimated flow is 22 $ft^3/sec$. The culvert is 68 ft long and has a square-edged headwall. Determine the difference in elevation between the upstream and downstream water surfaces.
4. A smooth concrete 2-ft flat bottom channel with a 2 : 1 side slopes has a slope of 0.4 percent, and the depth is 1.5. Determine the quantity of flow in $m^3/sec$.
5. Given an allowable headwater depth of 8 ft and a runoff of 52 $ft^3/sec$, determine the required size of circular concrete pipe and the actual headwater depth. The slope for the location is 0.3 percent and the length of culvert is 106 ft. The pipe has a square-edged entrance.
6. Develop an equivalent expression for Eq. 14-1 using SI (metric) units.
7. Determine the depth and velocity of flow in a trapezoidal channel ($n = 0.03$) with 2 : 1 side slopes and a 2-ft bottom width, given a flow of 200 $ft^3/sec$ and a slope of 2.0 percent. Determine the depth and velocity of flow if $n = 0.012$.
8. Given an allowable headwater depth of 10 ft and a runoff of 150 $ft^3/sec$, determine the required size of circular corrugated metal pipe and the actual headwater depth. Assume that the pipe has a projecting entrance and will operate with inlet control.

**9.** Assuming a freehaul distance of 2000 ft, the cost of excavation \$1.15/ $yd^3$, and the price of overhaul \$0.18/cubic station yard, what is the limit of economical haul?

**10.** Given below are the ordinates of a portion of a mass diagram for a hypothetical highway project. No haulage is permitted back of station 20 or beyond station 40, the end of the project. The freehaul distance is 800 ft. Draw the mass diagram, and determine: (a) the volume of excavation and embankment, (b) the quantities of overhaul (station yards), waste, and borrow; and (c) the direction of haul.

**11.** On the basis of the mass diagram from Problem 10, make a sketch of the profile of the project showing the beginning points and relative heights of cuts and fills.

| *Station* | *Ordinate* | *Station* | *Ordinate* | *Station* | *Ordinate* |
|---|---|---|---|---|---|
| 20 | 0 | 27 | −28000 | 34 | +4000 |
| 21 | −4000 | 28 | −32000 | 35 | +10000 |
| 22 | −8000 | 29 | −26000 | 36 | +8000 |
| 23 | −12000 | 30 | −20000 | 37 | +6000 |
| 24 | −16000 | 31 | −14000 | 38 | +4000 |
| 25 | −20000 | 32 | −8000 | 39 | +2000 |
| 26 | −24000 | 33 | −2000 | 40 | 0 |

**12.** The design structural number of a certain pavement, SN = 4.20. A 4-in. asphaltic concrete surface has been chosen for the pavement. Crushed limestone is to be used for the base, and sand-gravel will serve as the subbase. The policy of the highway agency is to use a minimum base thickness of 6 in. What thickness should be used for the subbase?

**13.** Given the traffic roadway conditions described in Example 14-1. Determine the thickness of concrete pavement required if the subgrade has a modulus of subgrade reaction, $k = 50$ pci.

## REFERENCES

1. Wright, Paul H., and Radnor J. Paquette, *Highway Engineering,* Fifth Edition, John Wiley, 1987.
2. *Rainfall Frequency Atlas of the United States,* Technical Paper No. 40, U.S. Weather Bureau (now National Weather Service), Department of Commerce, May 1961.
3. *Design of Roadside Drainage Channels,* Hydraulic Design Series No. 4, Bureau of Public Roads, 1965.
4. *Handbook of Concrete Culvert Pipe Hydraulics,* Portland Cement, Association, 1964.

5. *Hydraulic Charts for the Selection of Highway Culverts,* Hydraulic Engineering Circular No. 5, Bureau of Public roads, December 1965.
6. *Capacity Charts for the Hydraulic Design of Highway Culverts,* Hydraulic Engineering Circular No. 10. Federal Highway Administration, Washington, D.C., 1972.
7. Normann, Jerome M., "Improved Design of Highway Culverts," *Civil Engineering,* March 1975.
8. Harrison, L.J., J.L. Morris, J.M. Normann, and F.L. Johnson, *Hydraulic Design of Improved Inlets for Culverts,* Hydraulic Engineering Circular No. 13, Federal Highway Administration, Washington, D.C., 1975.
9. *Design Charts for Open Channel Flow,* Hydraulic Design Series, No. 3, Federal Highway Administration, Washington, D.C., 1973.
10. *Design of Stable Channels with Flexible Linings,* Hydraulic Engineering Circular No. 15, Federal Highway Administration, Washington, D.C., 1975.
11. Haliburton, T. Allan, Jack D. Lawmaster, and Verne C. McGuffey, *Use of Engineering Fabrics in Transportation-Related Applications,* prepared for Federal Highway Administration, Washington, D.C., October 1981.
12. *Standard Nomenclature and Definitions for Pavement Components and Deficiencies.* Special Report No. 113, Highway Research Board, Washington, D.C. (1970).
13. *Annual Book of Standards: Bituminous Materials, Soils, Skid Resistance,* Part 11. American Society for Testing and Materials, Philadelphia, Pa.
14. *Standard Specifications for Transportation Materials and Methods of Sampling and Testing.* American Association of State Highway and Transportation Officials, Washington, D.C.
15. *AASHTO Guide for Design of Pavement Structures.* American Association of State Highway and Transportation Officials, Washington, D.C., 1986.
16. *The AASHO Road Test: Pavement Research.* Special Report No. 61E, Highway Research Board, Washington, D.C., 1962.
17. *Thickness Design for Concrete Highway and Street Pavements.* Portland Cement Association, Skokie, Illinois, 1984.

# Chapter 15

# Pipeline Transportation

## 15-1. INTRODUCTION

In 1865, Samuel Van Syckel built a two-inch pipeline to transport oil from a northwestern Pennsylvania oil field to a railroad terminal six miles away. This seemingly inauspicious event, probably more than any other, ushered in the era of modern-day pipeline transportation. Prior to this time, oil was transported by barrels in horsedrawn wagons at a cost of about $0.30 per barrel-mile. In the decade that followed, longer and larger pipelines were constructed and, in 1878, Byron D. Benson completed the first long-distance pipeline, a six-inch line extending 108 miles from Coryville to Williamsport, Pennsylvania. Today, pipelines transport about 24 percent of the total intercity freight (ton-miles) hauled in the United States in a more than 228,000-mile system of pipelines that interlace the country. In this modern system, a gallon of petroleum can be transported 275 miles at a cost of about a penny.

## 15-2. THE NATURE OF PIPELINE TRANSPORTATION

Because pipelines are virtually noiseless and unseen, it is easy to overlook their contribution to the progress and well-being of the nation. Cities depend on pipelines for safe and adequate water supply and for removal of human and industrial wastes. By pipelines, shorelines are reshaped into new land on which rise factories, homes, and centers of shopping and recreation. Serious research is currently underway that could advance solids pipeline flow technology to make it possible to transport packages or capsules of industrial raw materials, grains, and even people through pipes.

The pipeline system of today is utilized predominately for the transportation of petroleum products and natural gas. The vital importance of pipelines is understood when it is realized that approximately 75 percent of the energy needs of the nation are supplied by petroleum and about 60 percent of the total ton-miles of petroleum shipments are transported by pipelines.

In the United States, the large majority of commercial pipelines are owned by petroleum refining companies. Notable exceptions to this statement are the Pipeline Division of the Southern Pacific Railroad, the Little Inch Division of the Texas Eastern Transmission Company, and the Mid American Pipeline Company.

Interstate oil and natural gas pipelines that serve as common carriers are subject to regulation by the Federal Energy Regulatory Commission (FERC). Many pipelines also operate on an intrastate level and, therefore, are subject to state regulation in addition to that of the FERC.

## TECHNOLOGY OF FLUID FLOW IN PIPELINES

The theory of fluid flow in pipelines is based on two fundamental principles: *the conservation of mass* and *the conservation of energy.*

### 15-3. THE PRINCIPLE OF CONSERVATION OF MASS

The principle of the conservation of mass states that matter can be neither created nor destroyed. This law leads to the conclusion that the mass of fluid passing one section of pipe per unit of time must simultaneously equal the mass per unit of time passing every other section. This principle is expressed by the *equation of continuity:*

$$G = A_1 v_1 w_1 = A_2 v_2 w_2 \tag{15-1}$$

where

$G$ = the weight rate of flow typically expressed in lb/sec
$A$ = the cross-sectional area of flow, $ft^2$
$v$ = average velocity of flow, ft/sec
$w$ = the specific weight of the fluid, $lb/ft^3$

The subscripts refer to two arbitrarily selected sections along the pipeline.

In the flow of liquids, incompressibility can be assumed ($w_1 = w_2$) and the equation of continuity may be expressed in terms of $Q$, the rate of flow with typical units of cubic feet per second.

$$Q = A_1 v_1 = A_2 v_2 \tag{15-2}$$

### 15-4. THE PRINCIPLE OF CONSERVATION OF ENERGY

Consider a system that undergoes a process during which energy is added to or removed from it in the form of work or heat. According to the principle of conservation of energy, none of the energy added is destroyed within the system and none of the energy removed is created within the system. In the case of fluid flow in pipelines, one may express the total energy possessed by the fluid at any given flow section as a sum of the energy due to molecular agitation (internal energy), velocity, pressure, and height. By taking into account any energy added to or taken from the system, the conservation of energy principle makes it possible to draw conclusions about the energy possessed by the fluid at other sections along the pipeline.

If one considers one pound of fluid that flows between two sections in a pipeline system, the conservation of energy principle may be expressed by the following general equation:

$$I_1 + \frac{p_1}{w_1} + \frac{v_1^2}{2g} + z_1 + 778E_H + E_M = I_2 + \frac{p_2}{w_2} + \frac{v_2^2}{2g} + z_2 \tag{15-3}$$

where

$I$ = internal energy

$\frac{p}{w}$ = pressure energy, the ratio of pressure to the specific weight of the fluid

$\frac{v^2}{2g}$ = kinetic energy, obtained from the general kinetic equation, one-half the product of the mass and velocity squared

$z$ = potential energy, the height above a known datum

$E_H$ = heat energy added to the fluid, in Btu

$E_M$ = the mechanical energy added to the fluid

$g$ = 32.2 ft/sec$^2$, acceleration due to gravity

Each term in Eq. 15-3 has the units of foot-pounds per pound or simply feet of *head* and the subscripts refer to two arbitrarily selected sections along the pipeline.

In most pipeline applications, it can be assumed that no heat is added to or taken from the system ($E_H = 0$) and that there is no significant change in internal energy ($I_1 = I_2$). If, in addition, no mechanical energy is added to the fluid ($E_M = 0$) and there is no change in the specific weight of the fluid ($w_1 = w_2$), Eq. 15-3 reduces to the familiar Bernoulli equation:

$$\frac{p_1}{w} + \frac{v_1^2}{2g} + z_1 = \frac{p_2}{w} + \frac{v_2^2}{2g} + z_2 \tag{15-4}$$

Equations 15-3 and 15-4 apply to an ideal system. In a real fluid system, fluid friction will result in a dissipation of energy which must be accounted for by the addition of another term, $h_L$, to allow for these *head losses.*

$$\frac{p_1}{w_1} + \frac{v_1^2}{2g} + z_1 = \frac{p_2}{w_2} + \frac{v_2^2}{2g} + z_2 + h_L \tag{15-5}$$

Equation 15-5 may be represented graphically as shown in Fig. 15-1. It should be noted that for the usually assumed conditions of uniform flow, the loss in head is reflected in a loss in pressure and drop in the total energy line and the *hydraulic grade line.*

For uniform flow in long pipes of constant cross section ($v_1 = v_2$), it can be seen from Eq. 15-5 that:

$$h_L = \frac{p_1}{w_1} + z_1 - \frac{p_2}{w_2} - z_2 \tag{15-6}$$

Under these conditions that head loss is the decrease in *piezometric head* between Sections 1 and 2.

## 15-5. THE DARCY-WEISBACH EQUATION

The head loss in a pipeline was shown by Henri P. G. Darcy and Julius Weisbach to depend on the pipe length ($L$) and diameter ($D$), and the velocity of flow ($v$). In independent nineteenth century research, these scientists developed the following relationship[1]:

[1]This relationship is sometimes expressed as $f\frac{4L}{D}\frac{v^2}{2g}$, the Fanning equation. Values of $f$ used in the Fanning equation are one-fourth the Darcy-Weisbach friction coefficients.

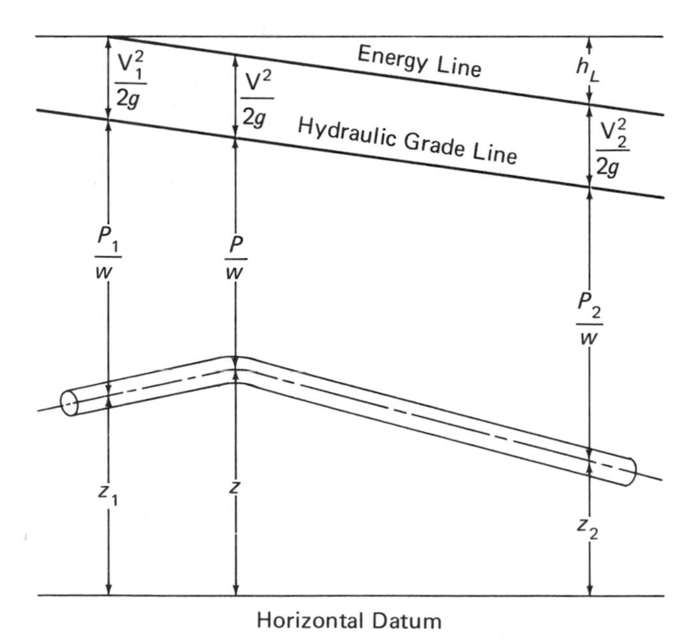


*Fig 15-1.* Graphical representation of energy relationships in a pipeline system.

$$h_L = f\frac{L}{D}\frac{v^2}{2g} \tag{15-7}$$

where

$f$ = the friction factor, a dimensionless measure of pipe resistance that depends on the characteristics of the pipe and the flow

Researchers have shown that the friction factor is a function of the relative pipe roughness and the Reynolds number, $R$. The Reynolds number is a dimensionless ratio involving the averge fluid velocity ($v$), pipe diameter ($D$), and the viscosity ($\mu$) and density ($\rho$) of the fluid:

$$R = \frac{vD\rho}{\mu} = \frac{vDw}{\mu g} \tag{15-8}$$

The variation of the friction coefficient with Reynolds number and pipe roughness is shown by an engineering chart developed by Moody [1]. See Fig. 15-2.

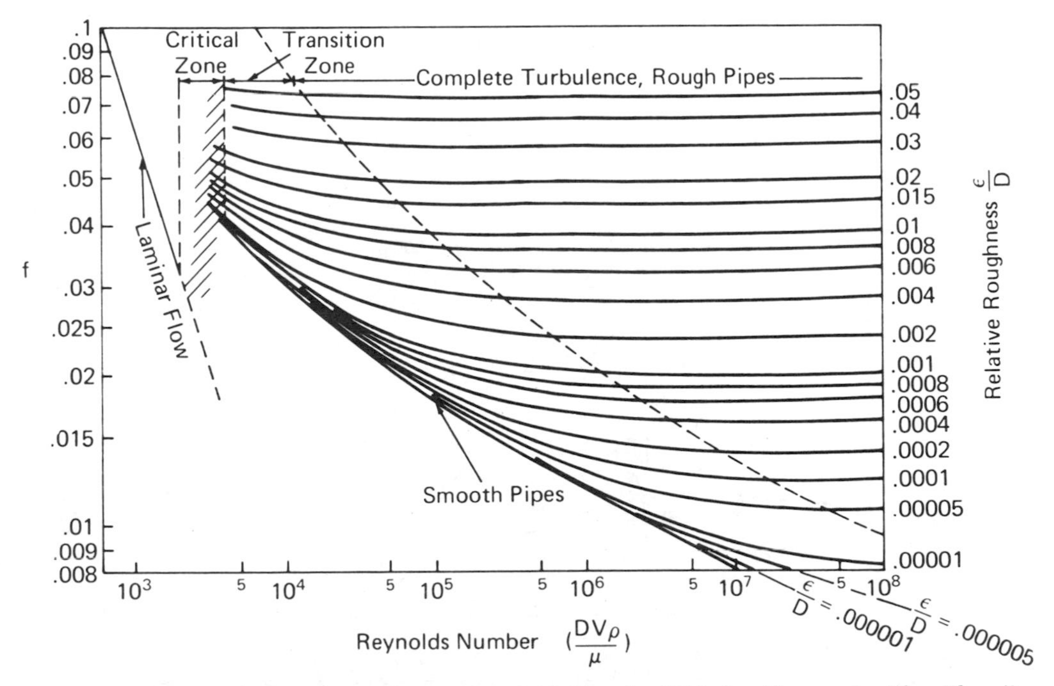


*Fig 15-2.* Friction factor chart. (Source: Lewis F. Moody, "Friction Factors for Pipe Flow," *Transactions,* ASME, November 1944.)

### EXAMPLE 15-1

**LIQUID PIPELINES PROBLEM.** A 22-in. pipeline is being designed to transport 200,000 bbl/day of petroleum (specific weight = 43 lb/ft³) between points X and Y (see Fig. 15-3). The total length of the pipeline is to be 360 mi. The pipe friction factor is 0.02, and the allowable working pressure is 800 psi. Determine the number of pumping stations required and the maximum distances between adjacent stations.

$$Q = \frac{200{,}000 \text{ bbl/day} \times 42 \text{ gal/bbl}}{(24 \times 3600) \text{ sec/day} \times 7.48 \text{ gal/ft}^3} = 13.0 \text{ ft}^3\text{/sec}$$

$$Q = Av = \frac{\pi}{4}\left(\frac{22}{12}\right)^2 v = 13.0 \text{ ft}^3\text{/sec}$$

$$v = 4.92 \text{ ft/sec}$$

$$h_L = f\frac{L}{D}\frac{v^2}{2g} = 0.02\,\frac{5280}{22/12}\,\frac{(4.92)^2}{2g} = 21.65 \text{ ft/mi}$$

This head loss is equivalent to:

$$21.65 \times \frac{43 \text{ lb/ft}^3}{144 \text{ in.}^2\text{/ft}^2} = 6.47 \text{ psi/mi}$$

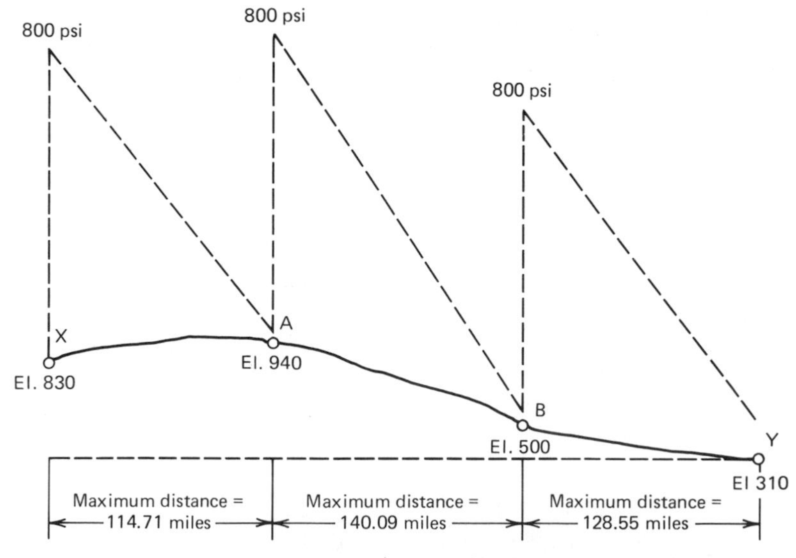


*Fig 15-3.* Hydraulic gradient profile for Example 15-1.

The total pressure drop = 6.47 × 360 mi = 2329 psi. The difference in elevation between X and Y is 520 ft which is equivalent to:

$$520 \times \frac{43 \text{ lb/ft}^3}{144 \text{ in.}^2/\text{ft}^2} = 155 \text{ psi}$$

The pumps need to supply only 2329 − 155 = 2174 psi.

$$\text{Number of stations required} = \frac{2174}{800} = 2.72 \text{ (say, 3 stations)}$$

As a matter of practice, the pressure is not allowed to drop below about 25 psi. The hydraulic grade line, therefore, in this example will drop 775 psi between stations.

$$\text{Maximum distance between } X \text{ and } A = \frac{775 - (110 \times 43/144)}{6.47} = 114.71 \text{ mi}$$

Similarly, maximum distances $AB$ = 140.09 mi and $BY$ = 128.55 mi. Since these are maximum distances, the sum exceeds the total length of the pipeline, 360 mi. The precise location of pumping stations, of course, will be governed by practical considerations such as suitable topography, accessibility, cost of land, and so forth.

## 15-6. FLOW IN GAS PIPELINES

Although the flow of gas in pipelines obeys the same basic laws outlined earlier in this chapter, design computations basically differ from those for liquid pipeline flow. These differences result from the need to consider the thermodynamic effects inherent in compressible fluid flow.

In the following paragraphs, general equations of gas flow will be introduced along with the results of recent research evaluating frictional resistance at the large Reynolds numbers typically experienced for flow in commercial gas lines.

In its most general form the equation for flow or *throughput* for a natural gas pipeline is:

$$Q_b = 38.77 \left(\frac{T_b}{P_b}\right) \sqrt{\frac{1}{f}} \left[\frac{p_1^2 - p_2^2 - 0.0375G\,\Delta h P_{\text{avg}}^2/TZ}{SLTZ}\right]^{0.500} D^{2.5} \qquad (15\text{-}9)$$

This equation is based on the assumptions that the flow is steady state and isothermal and that the kinetic energy change in the length of pipe is negligible. Nomenclature and typical units for this equation are given in Table 15-1.

*Table 15-1*
**Nomenclature and Symbols for Gas Flow Equations**

| *Symbol* | *Value* | *Units* |
|---|---|---|
| $T_b$ | Absolute temperature at reference or base condition | °R |
| $P_b$ | Absolute pressure at reference or base condition | psia |
| $f$ | Friction factor | None |
| $p_1$ | Pressure at inlet of line section | psia |
| $p_2$ | Pressure at outlet line section | psia |
| $S$ | Specific gravity of gas, dry air - 1.000 | None |
| $L$ | Length of section | mi |
| $T$ | Temperature of flowing gas | °R |
| $Z$ | Compressibility factor at base conditions | None |
| $D$ | Inside diameter of pipelines | in. |
| $\Delta h$ | Elevation change | ft |
| $R$ | Reynolds number | None |
| $k$ | Average height of roughness peak | micro-inches |
| $V$ | Average velocity | ft/sec |
| $\rho$ | Density of gas | lb-sec$^2$/ft$^4$ |
| $\mu$ | Viscosity of gas | lb-sec/ft$^2$ |
| $Q_b$ | Volume of gas flow | ft$^3$/day |

If the terminal points of the pipeline are at the same elevation, the general equation reduces to:

$$Q_b = 38.77 \left(\frac{T_b}{P_b}\right) \sqrt{\frac{1}{f}} \left[\frac{p_1{}^2 - p_2{}^2}{SLTZ}\right]^{0.500} D^{2.5} \qquad (15\text{-}10)$$

The most troublesome part of these equations is the evaluation of the term $\sqrt{1/f}$, which is called the *transmission factor.* A number of empirical equations have been proposed over the years for the computation of the transmission factor. These equations that were developed under various field conditions have yielded a confusing variety of flow equations[2] which are found in pipeline industry journals.

Research has shown that at high Reynolds numbers the transmission factor curves tend to level off indicating that under these conditions the transmission factor does not depend on the Reynolds number. At very high Reynolds numbers, the transmission factor is correlated better in terms of a ratio of the wall roughness to the pipe diameter, $k/D$.

Early research indicated that laminar flow is rare in gas pipelines and that

[2]A discussion of the most popular of these equations, including the Panhandle A, the New Panhandle, and the Weymouth is given by Reference 2.

two major types of turbulent flow may occur. The type of flow that develops at low to moderate Reynolds numbers was termed *smooth-pipe flow* because of the absence of wall effects. For this type of flow, the flow resistance is a function only of the Reynolds number.

The type of flow existing at very high Reynolds numbers usually is termed *fully turbulent flow* and, in this case, the flow resistance depends on the pipe diameter and the wall roughness and not on the rate of flow.

For turbulent flow near *smooth* boundaries, the transmission factor can be expressed by the following equation:

$$\sqrt{\frac{1}{f_{SP}}} = 4 \log\left(\frac{\mathrm{R}}{\sqrt{1/f}}\right) - 0.6 \qquad (15\text{-}11)$$

For turbulent flow near *rough* boundaries, the following equation yields an accurate value for the transmission factor:

$$\sqrt{\frac{1}{f_{RP}}} = 4 \log\left(\frac{3.7D}{k}\right) \qquad (15\text{-}12)$$

Research of the Institute of Gas Technology [3] has indicated that truly smooth pipe flow rarely is achieved in practice because of the influence of pipe fittings and bends. To allow for this finding, IGT classed gas pipeline flow into two categories: partially turbulent flow and fully turbulent flow.

Computation of the transmission factor for partially turbulent flow can be made by use of the following equation:

$$\sqrt{\frac{1}{f_{PT}}} = F_f 4 \log\left(\frac{\mathrm{R}}{1.4\sqrt{\frac{1}{f_{SP}}}}\right) \qquad (15\text{-}13)$$

Use of Eq. 15-13 requires the following steps:

1. Compute the *bend index* which is defined as the total degrees of bend in a pipe section divided by the length of the section in miles.
2. Determine the *drag factor*, $F_f$. (See Table 15-2).
3. Select the value of $\sqrt{1/f_{SP}}$ from Table 15-3 or compute it by Eq. 15-11.
4. Compute $\sqrt{1/f_{PT}}$ by Eq. 15-13.

The transmission factor for fully turbulent flow is given by:

$$\sqrt{\frac{1}{f_{FT}}} = 4 \log\left(\frac{3.7D}{k_e}\right) \qquad (15\text{-}14)$$

This equation is identical to Eq. 15-13 except that the roughness is expressed as the operating or effective roughness of the interior surface rather than the absolute roughness of the pipe wall alone.

*Table 15-2*
**Drag Factor, $F_f$, as a Function of Bend Index**

| Material | *Bend Index* 5° to 10° (Extremely Low) | 60° to 80° (Average) | 200° to 300° (Extremely High) |
|---|---|---|---|
| Bare steel | 0.975–0.973 | 0.960–0.956 | 0.930–0.900 |
| Plastic lined | 0.979–0.976 | 0.964–0.960 | 0.936–0.910 |
| Pig-burnished | 0.982–0.980 | 0.968–0.965 | 0.944–0.920 |
| Sand-blasted | 0.985–0.983 | 0.976–0.970 | 0.951–0.930 |

Note: Values are for a line with 40-foot joints and 10-mile valve spacing.

SOURCE: *Pipeline Design for Hydrocarbon Gases and Liquids,* American Society of Civil Engineers, 1975.

*Table 15-3*
**Examples Values of Transmission Factor, $F_t$ for Various Reynolds Numbers**

| R Millions | $F_t$ | R Millions | $F_t$ | R Millions | $F_t$ |
|---|---|---|---|---|---|
| 0.100 | 14.727 | 11.000 | 22.182 | 31.000 | 23.855 |
| 0.150 | 15.359 | 12.000 | 22.322 | 32.000 | 23.907 |
| 0.200 | 15.809 | 13.000 | 22.451 | 33.000 | 23.956 |
| 0.300 | 16.444 | 14.000 | 22.570 | 34.000 | 24.005 |
| 0.400 | 16.897 | 15.000 | 22.682 | 35.000 | 24.052 |
| 0.500 | 17.249 | 16.000 | 22.786 | 36.000 | 24.097 |
| 0.600 | 17.537 | 17.000 | 22.884 | 37.000 | 24.142 |
| 0.700 | 17.781 | 18.000 | 22.976 | 38.000 | 24.185 |
| 0.800 | 17.992 | 19.000 | 23.063 | 39.000 | 24.227 |
| 0.900 | 18.179 | 20.000 | 23.146 | 40.000 | 24.268 |
| 1.000 | 18.346 | 21.000 | 23.225 | 41.000 | 24.308 |
| 2.000 | 19.449 | 22.000 | 23.300 | 42.000 | 24.347 |
| 3.000 | 20.096 | 23.000 | 23.372 | 43.000 | 24.385 |
| 4.000 | 20.556 | 24.000 | 23.441 | 44.000 | 24.423 |
| 5.000 | 20.914 | 25.000 | 23.507 | 45.000 | 24.459 |
| 6.000 | 21.207 | 26.000 | 23.570 | 46.000 | 24.495 |
| 7.000 | 21.454 | 27.000 | 23.631 | 47.000 | 24.530 |
| 8.000 | 21.669 | 28.000 | 23.690 | 48.000 | 24.564 |
| 9.000 | 21.858 | 29.000 | 23.747 | 49.000 | 24.597 |
| 10.000 | 22.028 | 30.000 | 23.802 | 50.000 | 24.630 |

SOURCE: Arthur E. Uhl, "Steady Flow in Gas Pipe Lines," *Pipe Line Industry.*

It is not always clear which flow regime will exist for a given set of conditions. The Reynolds number for the point of transition from partially turbulent flow to fully turbulent flow is given by:

$$R = 5.65\left(\frac{3.7D}{k_e}\right)^{1/F_f} \log\frac{3.7D}{k_e} \tag{15-15}$$

If the Reynolds number for the given conditions is less than that computed by Eq. 15-15, partially turbulent flow will exist; otherwise, the flow will be fully turbulent.

The equations that have been introduced make it possible to solve all but the most involved pipeline flow problems. However, use of these equations is cumbersome at best and may involve trial-and-error solutions in certain instances. If repetitive flow calculations are required, a great saving in time may be realized by referring to the American Gas Association's publication TR-10 [4]. In this manual, various terms in the general flow equation are tabulated in factorial form providing a rapid means for the solution of simple flow problems associated with planning and preliminary design work. For more complicated design problems, electronic computer programs are used by the various pipeline companies.

EXAMPLE 15-2

GAS PIPELINES PROBLEM. Natural gas (specific gravity = 0.62) is to be transported from point A to point B, a distance of 150 mi, at a rate of 100 million $ft^3$/day. The pipeline, which is to be constructed of bare steel, is to have a bend index of 60 degrees/mi and an effective roughness of 0.3 mil. The inlet pressure is 700 psia and the outlet pressure is 100 psia. Determine the size of pipe required assuming a base pressure of 14.73 psia and a base temperature and flowing temperature of 520° absolute. The specific weight of air at 60° F is 0.0763 lb/$ft^3$. Assume $\mu_{60°F} = 2.5 \times 10^{-7}$ and use a compressibility factor of 0.891. Since the Reynolds number depends on the rate of flow and, thus, on the pipe size, the solution must be done by trial.

A first estimate of pipe size can be made by use of the Weymouth equation, which is applicable to fully turbulent flow.

$$Q_b = 433.5\frac{T_b}{P_b}D^{8/3}\left(\frac{p_1^2 - p_2^2}{STL}\right)^{1/2} \tag{15-16}$$

$$100{,}000{,}000 = 433.5\frac{520}{14.73}D^{8/3}\left[\frac{(700)^2 - (100)^2}{(0.62)(520)(150)}\right]^{1/2}$$

$$D = 17.6 \text{ in. (say 18 in.)}$$

The weight rate of flow,

$$G = \frac{100{,}000{,}000 \times 0.0763 \times 0.62}{3600 \times 24}$$

$$G = 54.63 \text{ lb/sec}$$

Combining Eqs. 15-1 and 15-8,

$$R = \frac{GD}{\mu Ag} = \frac{54.63 \times 1.5}{2.5 \times 10^{-7} \times (\pi/4)(1.5)^2 \times 32.2} = 5.76 \times 10^6$$

From Table 15-2, the drag factor $F_f = 0.960$.

The Reynolds number for the point of transition from partially turbulent flow to fully turbulent flow is given by Eq. 15-15:

$$R = 5.65 \left(\frac{3.7 \times 18}{0.0003}\right)^{1/0.96} \log \frac{3.7 \times 18}{0.0003}$$
$$R = 11.32 \times 10^6$$

The flow is partially turbulent.

From Table 15-3,

$$\sqrt{\frac{1}{f_{SP}}} = 21.141$$

By Eq. 15-13.

$$\sqrt{\frac{1}{f_{PT}}} = (0.96)4 \log \frac{5.76 \times 10^6}{1.4 \times 21.141}$$

$$\sqrt{\frac{1}{f_{PT}}} = 20.3$$

Using the general flow equation,

$$100{,}000{,}000 = 38.77 \frac{520}{14.73}(20.3)\left[\frac{(700)^2 - (100)^2}{0.62 \times 520 \times 150 \times 0.891}\right]^{1/2} D^{2.5}$$
$$D = 16 \text{ in.}$$

Since this value differs from the assumed, the computations should now be repeated using an assumed value of $D = 16$ in. When this is done for this example, however, the answer is not changed significantly.

## 15-7. TWO-PHASE FLOW

All of the discussion thus far has been confined to single-phase flow, that is, flow of either gas or a liquid. There are situations in which it is economically advantageous to transport simultaneously both a gas and a liquid in a single pipeline. This is called two-phase flow. An example of two-phase flow involves certain offshore oil operations where it is extremely expensive to separate the liquid and gas phases in deep water. As the technology advances, two-phase flow is likely to become commonplace in situations such as this.

Two-phase flow may occur over a wide range of liquid to gas ratios and manifests a variety of flow patterns. (See Fig. 15-4.)

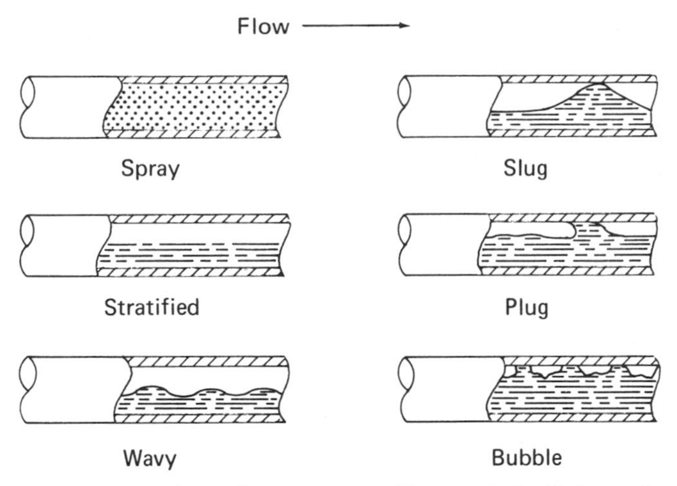


*Fig 15-4.* Two-phase flow patterns. (Source: J. L. Huitt and N. Marusov, "Where We Stand in Flow Technology," *Pipe Line Industry*, May 1964.)

Two-phase flow technology is much more complex than single-phase flow. However, research accomplished to date suggests that the two-phase design problem is not insurmountable. Indeed, Flanigan [5] has published a design technique that should yield satisfactory results under a wide range of flow conditions. Publication of the details of this method is beyond the scope of this chapter and the interested student is referred to papers by Flanigan [5] and Huitt and Marusov [6].

Experimental studies have revealed that two-phase flow results in large drops in pressure far beyond that indicated by single-phase technology. Flanigan [5] suggests that there are two components of pressure drop in two-phase flow:

1. The pressure drop due to friction which increases with increase in gas flow rate. This is the only component in horizontal lines.
2. The pressure drop due to the head of liquid in inclined lines. Practically all of this pressure drop occurs in the uphill section of the pipeline.

Flanigan describes a technique for computing these components which, when added, will yield a satisfactory estimate of the total pressure drop in the transportation of natural gas and condensate.

## 15-8. TRANSPORTATION OF SOLIDS IN PIPELINES

Another form of two-phase pipeline flow involves the transportation of solids in a liquid or gaseous medium. Although solids pipelines have been used for more than a century, most of the activity in this field has occurred since World War II.

There are three general classes of solids pipeline systems: (1) slurry pipelines, (2) pneumatic pipelines, and (3) capsule pipeline systems.

*Slurry Pipelines.* In this system, the freight is in bulk form and is mixed with water or another conveying liquid. The most common class of solids pipelines, slurry pipeline systems have been used primarily in manufacturing, mining, and construction activities. Such pipelines have most often been used for loading and unloading operations, in-plant movement of materials, and transportation of construction materials for short distances. While the vast majority of slurry pipelines are less than 15 mi long, a few exceed 150 mi in length.

In order for solids to be transported successfully and economically by slurry pipelines, the following conditions must generally be satisfied [7]:

1. The solid material should not react in any undesirable way with the carrying fluid or become otherwise contaminated within the pipeline system.
2. Attrition during transport should either be beneficial to or its effect be negligible on subsequent operations.
3. The top particle size should be such that it can be handled in commercially available pumps, pipes, and preparation equipment.
4. The solid material should mix easily with, and separate easily from, the carrying fluid at the feeder and discharge terminals, respectively.
5. The solid material should not be corrosive or become so in the carrying fluid.

Researchers have demonstrated that slurry pipeline flow may occur in at least four flow regimes:

1. *Flow as a homogeneous suspension.* For velocities experienced in most pipeline applications, this regime involves suspension of particles of diameter of less than 30 microns. Provided the flow is turbulent, the suspension flows like a homogeneous fluid. In this case, head losses may be computed as previously described using density and viscosity values corresponding to the suspension.
2. *Flow as a heterogeneous suspension.* In a heterogeneous suspension, the concentration of particles in a vertical plane is not uniform. This type of flow involves transportation of particles slightly larger than those that flow in a homogeneous suspension. In heterogeneous transport, the solid particles travel with a velocity slightly less than that of the liquid.
3. *Flow by saltation.* In solids pipeline flow involving relatively large particles and low velocities, particles collect at the bottom of the pipe and form a stationary bed. There is virtually no movement of solids in this

case. At higher velocities, movement of particles at the interface occurs. This phenomenon is known as saltation.

4. *Flow with a moving bed.* At high-liquid velocities, particles of large diameter may slide forward along the bottom of the pipe in a single mass or as a moving bed. This movement occurs at a much lower velocity than that of the liquid.

Which of these flow regimes occurs in a given instance will depend on the mean velocity, particle size, pipe diameter, and specific gravity of the material. The effect of velocity and particle size on the flow regime is illustrated by Fig. 15-5. The location of boundaries separating the various regimes for a given case, of course, will depend on the nature of the solids-liquid mixture.

Basically, the problem is designing a solids pipeline is to find a flow velocity that will prevent settling, yet minimize the friction loss. At the same time, the percent of solids and diameter of pipe must be sufficiently large to transport the required dry tons per day.

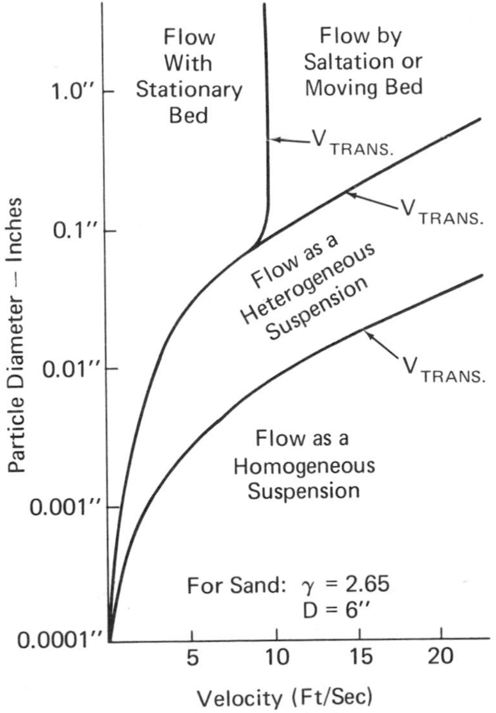


*Fig 15-5.* Classifications of flow regimes. (Source: Colorado School of Mines Research Foundation, *The Transportation of Solids in Steel Pipelines,* 1963.)

*Pneumatic Pipelines.* In pneumatic pipelines, the conveying fluid normally is air. These pipelines are used most often to transport fine powders, light and granular materials, or solids wastes over short distances, typically, a few miles or less [8]. Typical examples of the use of such systems include unloading bulk commodities from ships, removing dust, sawdust, and other waste material from manufacturing plants, and collecting solid waste from recreational parks, fairs, and apartment projects.

Pneumatic pipelines can operate either as a pressure system or a suction system. The pressure system is employed when products are to be transported from one origin to several destinations. The suction system is used to transport products from many origins to one destination.

Although pneumatic pipelines are technologically feasible for transporting many products, the technology for the design of such systems is not well established. The design usually must be empirically based with reliance on limited experience.

*Capsule Pipeline Systems.* During the 1970s, research demonstrated that solids encased in large capsules can be carried in a pipeline by a flowing fluid. This research suggests that it may be feasible to transport a wide variety of commodities in capsule pipeline systems including mail, grain, chemicals, and other objects that, if not capsulated, could be damaged by the transporting fluid.

The capsules may be in the form of cylinders or spheres with diameters nearly as large as the diameter of the pipeline. The transporting fluid can be either a gas or a liquid, the resulting systems being called pneumatic and hydraulic capsule pipelines, respectively.

Although the technological feasibility of capsule pipelines has been shown, the commercial feasibility of these systems remains to be demonstrated.

Certain characteristics of various types of solids pipeline are given in Table 15-4. Additional information on these systems, including equations for the calculation of energy losses due to the flow, is given in Reference 8.

## OPERATION OF PETROLEUM PIPELINE SYSTEMS

In the following paragraphs, the journey of a petroleum product is traced from the oil well to its ultimate destination. This information, which was abstracted from a publication of the Petroleum Extension Service, University of Texas [9], should provide the reader with a more complete understanding of the problems and practices involved in the operation of pipeline systems. For a more extensive examination of how U.S. oil pipelines operate and public policy issues concerning their ownership, Reference 10 is recommended.

## 15-9. TRANSPORTATION OF CRUDE OIL FROM WELL HEAD TO TANK BATTERY

The average oil well produces only about 20 bbl/day while a typical 20-in. pipeline is capable of transporting about 200,000 bbl/day. This means that a single pipeline is capable of carrying the output of up to 10,000 wells. Thus, before great quantities of crude oil can be transported in high-capacity lines to the refinery, it must first be concentrated in sufficient quantities.

In the interests of simplicity and efficiency, shipments of oil are made in "batches" consisting of several thousand barrels. The oil is usually moved in *gathering lines* from the well head to temporary storage tanks and accumulated in pipeline company tanks until the entire shipment is ready for delivery to the main pipeline. Prior to this time, some preliminary treatment is required to separate gas from the oil and to remove water and other contaminants.

While temporarily stored in the pipeline tank battery, the oil is measured (gauged) and its specific gravity and water content are determined. In modern installations, these measurements may be made automatically as the oil flows through special pipeline sections instrumented with electronic measuring and recording devices.

## 15-10. TRANSPORTATION OF CRUDE OIL TO REFINERY

Efficient operation of a pipeline system requires careful planning and detailed scheduling of oil movements. Every effort is made to operate the pipeline at capacity and at the same time minimize the use of storage tanks.

The typical oil pipeline has a small number of shippers, each of which is continually receiving deliveries and adding oil to the line. The oil remains the property of the shippers and, since the line must remain full, the shipper always has a *balance* of oil in the line. A scheduler canvasses the shippers during the last week in the month to determine the customers' shipping needs and desires. From this information, the monthly barrel-per-day line pumping rates are established.

Once the schedule is prepared, it is important that it is executed according to plan. This is the duty of the dispatcher. The dispatcher's post is manned 24 hours per day. By means of modern and varied communications equipment, the dispatcher is able to monitor line pressures and rates of flow, receipts and deliveries, and to detect any leaks and breaks in the line that may occur.

Changes in the custody of oil shipments are recorded and controlled by means of a run ticket, which is signed by the receiver and the deliverer. This document describes the oil in quality and quantity based on readings of meters, gauges, and other such devices, and legally accounts for receipts and deliveries.

*Table 15-4*
**Solids Pipeline Characteristics**

| *Characteristics* | *Type of Freight Pipeline* | | | |
|---|---|---|---|---|
| | *Slurry Pipeline* | *Pneumatic Pipeline* | *Pneumocapsule Pipeline* | *Hydrocapsule Pipeline* |
| State of technology | Commercial utilization is practiced | Commercial utilization is practiced | Commercial utilization is practiced | It is now in pilot plant stage |
| Size of aggregates | Small, usually less than 1 in. | Medium, usually less than several inches | All sizes, but less than the size of capsule | All sizes, but less than the size of capsule |
| Speed of conveyance | 3–6 mph | 20–60 mph | 25 mph | 3–6 mph |
| Nature of solid | (1) It must be inert relative to the water or other conveying fluid<br>(2) Its property and value must not be damaged by attribution. | (1) Its property and value must not be damaged by attribution.<br>(2) Must not produce explosive conditions | It can be any solid | It can be any solid |

| | | | | |
|---|---|---|---|---|
| Specific environmental consideration | (1) Water or other conveying liquid is required (2) Water treatment facility is required at the end point | Dust must be removed for exhaust | None | (1) Water or other conveying liquid is required (2) Water can be recirculated for double line and water treatment is required at the destination for single line |
| Typical shipments | Bulk materials such as coal, mineral, ores, gravel, etc. | Bulk materials such as solid waste, wheat, powder chemicals, etc. | Almost everything including mail, industrial samples, manufactured products | Almost everything |
| Shipping distance | Long, on the order of several hundred miles | Short, usually less than several miles | Long or short, up to tens of miles | No commercial experience exists |

SOURCE: *Transport of Solid Commodities Via Freight Pipeline,* Vol. II, U.S. Department of Transportation, 1976.

## 15-11. TRANSPORTATION FROM REFINERY TO MARKET

From the refinery, petroleum products must be transported to areas of high market demand. This final movement is made in product pipelines. In these lines, different petroleum products from different shippers flow in adjacent batches.

Products handled in these pipelines include various grades of gasolines, turbine fuel, burning or heating kerosene, and diesel fuels. Figure 15-6 shows a schematic representation of the typical cycle of batches. In certain cases, contiguous batches are segregated by means of rubber spheres called batch separators. Even when batch separators are not used, very little commingling occurs.

Products are shipped to delivery terminals and stored in break-out tanks and later relayed to points on the stub lines and to shippers' tanks through smaller lines.

At the delivery terminals, the "interface" between adjacent batches is allowed to pass, and stocks are drawn from the center of the batch. If the adjacent batches are similar, the commingled material will be used to upgrade the lower quality product. When two dissimilar products mix, the material is placed in special tanks where it is reblended under strict laboratory procedures.

## PROBLEMS

1. A 24-in. pipeline is being designed to transport 160,000 bbl/day of crude oil (specific weight = 55 lb/ft$^3$) from point $A$ to point $B$, a distance of 290 mi. The pipe friction factor is 0.016, and the allowable working pressure is 650 psi. Point $B$ is 770 ft higher than $A$. Determine the number of

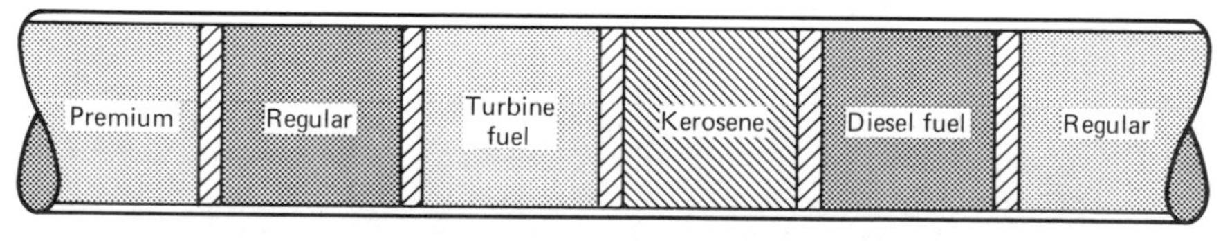


1. Premium—regular commingling can be split both ways with no loss of degradation.
2. Turbine fuel commingling usually must be cut out and blended to regular.
3. Turbine fuel—kerosene commingling can be cut directly to turbine fuel within certain limits.
4. Kerosene—diesel fuel commingling can be cut directly to diesel fuel within certain limits.
5. The kerosene buffer batch between diesel fuel and turbine fuel is cut both ways.

*Fig 15-6.* Typical cycle of batches in products pipelines. (Source: Petroleum Extension Service, The University of Texas, *Introduction to the Oil Pipeline Industry,* May 1966.)

pumping stations between points $A$ and $B$ and the minimum distance between the first two stations. It may be assumed that the ground slope between points $A$ and $B$ is constant.

2. A 45-cm pipeline is being designed to transport 15,000 $m^3$/day of crude oil (specific weight = 840 kg/$m^3$) from point $X$ to point $Y$, a distance of 460 km. The pipe friction factor is 0.018, and the allowable working pressure is $4.8 \times 10^6$ N/$m^2$. Point $Y$ is 250 m higher than $X$. Determine the number of pumping stations required.

3. Solve Problem 1 assuming that the oil is to be pumped from point $B$ to point $A$.

4. Solve Problem 2 assuming that the oil is to be pumped from point $Y$ to point $X$.

5. Natural gas (specific gravity = 0.62) is to be transported from point $C$ to point $D$, a distance of 205 mi, at a rate of 125 million $ft^3$/day. The pipeline is to be constructed of bare steel and will have an effective roughness of 0.3 mils. It will have a bend index of 70 degrees/mi. The inlet pressure is 750 psia and the outlet pressure is 100 psia. Determine the size of pipe required assuming a base pressure of 14.73 psia and a base temperature and flowing temperature of 520° absolute. Assume $\mu_{60°F} = 2.5 \times 10^{-7}$ and use a compressibility factor of 0.890.

## REFERENCES

1. Moody, Lewis F., "Friction Factors for Pipe Flow," *Transactions,* ASME, November 1944.
2. *Pipeline Design for Hydrocarbon Gases and Liquids,* American Society of Civil Engineers, New York, 1975.
3. Uhl, Arthur E., "Steady Flow in Gas Pipe Lines," five parts, *Pipe Line Industry,* August 1966, September 1966, January 1967, March 1967, and April 1967.
4. Institute of Gas Technology Publication TR-10, Chicago, 1966.
5. Flanigan, Orin, "Two-Phase Gathering Systems," *The Oil and Gas Journal,* March 10, 1958.
6. Huitt, J. L., and N. Marusov, "Where We Stand in Flow Technology," *Pipe Line Industry,* May 1964.
7. Colorado School of Mines Research Foundation, *The Transportation of Solids in Steel Pipelines,* 1963.
8. *Transportation of Solid Commodities Via Freight Pipeline,* Volumes I through V, prepared the University of Pennsylvania, U. S. Department of Transportation, Office of University Research, Washington, D. C., July 1976.
9. Petroleum Extension Service, The University of Texas, *Introduction of the Pipeline Industry,* May 1966.
10. Wolbert, Jr., George S., *U. S. Oil Pipe Lines,* American Petroleum Institute, Washington, D. C., 1979.

# Chapter 16

# Design of Land Transportation Terminals

Terminals often constitute the most important component of a transport system. Terminal costs comprise a significant if not dominant portion of the total costs of transportation. An inadequately designed terminal facility may cause inordinate delays to the movement of passengers or freight and ultimately may contribute to the failure of the system to remain competitive with other companies or modes.

An average railway freight car travels only 57 mi/day, a journey requiring only about one hour [1]. During the other 23 hours of the day, the car sits idle at a yard or other terminal facility. Similarly, it is not unusual for ships, motor carriers, and aircraft to spend more than half of the time unproductively at a terminal facility.

The physical features of land transportation terminals vary a great deal depending on the transport mode, the type of commodity, and the amount of traffic it serves. Clearly, major design differences are created by the differing sizes and operational characteristics of the transporting vehicles. That trucks, buses, railways, and pipelines have inherently different terminal needs is so obvious that it hardly bears mentioning. Even essentially similar passenger modes such as commuter rail and intercity rail facilities vary significantly because of the different needs of the two types of travelers.

## 16-1. FUNCTIONS OF TERMINALS

Land transportation terminals may be required to perform eight basic functions:

1. *Traffic concentration.* Passengers arriving in continuous flows are grouped into batch movements; small shipments of freight are grouped into larger units for more efficient handling.
2. *Processing.* This function includes ticketing, checking in, and baggage handling for passengers and preparation of waybills and other procedures for freight.

3. *Classification and sorting.* Passengers and freight units must be classified and sorted into groups according to destination and type of commodity.
4. *Loading and unloading.* Passengers and freight must be moved from waiting rooms, loading platforms, temporary storage areas, and the like to the transportation vehicle at the origin, and the process must be reversed at the destination.
5. *Storage.* Facilities for short-term storage such as waiting rooms for passengers and transit sheds for freight commodities are required to permit loads to be assembled by concentration and classification.
6. *Traffic interchange.* Passengers and freight arriving at a terminal often are destined for another location and must transfer to a similar or different mode of travel to complete the journey.
7. *Service availability.* Terminals serve as an interface between the transport user and the carrier, making the transportation system and its services available to the shipper and the traveling public.
8. *Maintenance and servicing.* Terminals often must include facilities for fueling, cleaning, inspection, and repair of vehicles.

## 16-2. THE NATURE OF THE TERMINAL PLANNING PROCESS

Although the major focus of this chapter is on terminal design, it seems appropriate to consider some of the approaches recommended for the planning of terminal facilities. The objective of the terminal planner is to define the elusive "optimum" design—one that is sufficient in size and complexity to provide a suitable level of service but not so elaborate that it could involve wasteful outlays for construction and operation of facilities that would be idle much of the time.

The planner must first forecast the future level of activity at the terminal: the number of passengers to be accommodated by passenger terminals, as well as their patterns and modes of arrival and departure and their needs while at the terminal; the volume of freight, classified by commodity type, and again, the patterns and modes of shipment to and from the terminal. The procedures used for forecasting terminal demand vary a great deal depending upon terminal type and size. In certain instances, forecasts can be based on historical data, empirical studies, and extrapolation of trends. In forecasting passenger terminal requirements, planners may need to perform surveys of parkers and travelers to determine current travel deficiencies and desires. Models such as those described in Chapter 9 also may be helpful in estimating levels of passenger terminal activity. Planners of freight terminal facilities sometimes base forecasts on known or assumed relationships between the tonnage of freight and the volume of wholesale or retail sales, gross regional product, or some other measure of economic growth. Freight terminal planners may find it necessary to perform special studies of vehicle arrival rates and times, loading and unloading rates, processing procedures, and work habits and rules.

Usually, a terminal facility is designed to provide for 5 to 10 years in the future. However, forecasts in demand must appropriately account for fluctuations over time including seasonal, daily, and hourly variations. In most instances, it is not advisable to design for the absolute peak day or peak hour demand expected over the forecast period. Adjustments should be made for a continuous peak period but not necessarily the highest day's or highest hour's activity. Rather, a typical peak hour demand is usually chosen for passenger movements similar to the thirtieth highest hourly traffic volume recommended for highway design. A typical peak daily traffic is recommended for estimating the level of freight terminal activity.

If the number of vehicles or passengers arriving at a terminal were precisely predictable, if their times of arrival could be scheduled, and if their service times could be anticipated, the problem of planning the most economical facilities could possibly be determined by an elementary arithmetical analysis. However, arrivals at a terminal are not regular but tend to be characterized by chance variability about the mean arrival rate. Furthermore, the time required for processing or servicing vehicles or passengers at a terminal is not constant but also tends to have a random component.

Because of the probabilistic nature of terminal operations, an elementary arithmetical analysis of such problems is not possible. For the same reason, certain aspects of terminal operations are subject to analysis by queuing or waiting-line theory.

## 16-3. QUEUING THEORY

An extensive treatment of queuing theory is beyond the scope of this book. This chapter will sketch some of the most elementary applications of these methods to terminal planning. Generally speaking, queuing theory is most useful for analyses of the behavior of simple waiting lines or for studies of some component of more complex operations.

Several characteristics of a queue must be known or assumed if queuing problems are to be solved analytically.

1. The mean rate at which units (vehicles or people) arrive for service and the probability distribution of the arrivals.
2. The mean service rate and the probability distribution of the services.
3. The number of channels or servers (e.g., truck loading spots, toll booths, etc.) and whether the channels are arranged parallel (as a toll booth) or in a series (as in a vehicle repair facility).
4. The queue discipline, the order in which arriving units will be served.

Both arrivals and service times are random variables. Arrivals are discrete random variables, and service times are continuous random variables. It is often appropriate to describe units arriving at a terminal by the Poisson probability distribution:

$$P(n) = \frac{(\lambda t)^n e^{-\lambda t}}{n!} \tag{16-1}$$

where

$P(n)$ = probability of $n$ arrivals in a period $t$
$\lambda$ = mean arrival rate or volume
$e$ = Napierian logarithmic base

It may be advantageous to focus on the time intervals or headways between successive arrivals rather than on the number of arrivals occurring during a stated interval of time. For a Poisson process, it can be shown that the probability density function[1] of interrarrival times is:

$$f(t) = \lambda e^{-\lambda t} \tag{16-2}$$

This equation, known as the negative exponential distribution, commonly is expressed as a cumulative distribution function, expressing the probability of a headway, $h$, being greater than or equal to $t$:

$$P(h \geq t) = \int_t^{\infty} f(t) = e^{-\lambda t} \tag{16-3}$$

In many queuing situations, the distribution of service times is also best described by a negative exponential distribution:

$$P(s \geq t) = e^{-\mu t} \tag{16-4}$$

where

$P(s \geq t)$ = the probability that a randomly chosen service time, $s$, will be equal to or greater than $t$
$\mu$ = the mean service time

Other probability distributions have been used to describe arrivals and service times for transportation queues, but limitations in space preclude a description of these models being given.

To characterize a waiting line, we refer to it as being in a certain "state." The queuing system is said to be in state $n$ if there are $n$ units (vehicles, people)

[1]A probability density function is a non-negative function $f(t)$ whose integral over the entire $t$-axis is unity.

in the system, including those being served. State probabilities, indicating the fraction of time the system should operate with a specified number in the system, are useful in evaluating the effectiveness of various choices of terminal design features. Various other performance measures are used including the average number of units in the system, average length of queue, and the average time spent in the system. Tables 16-1 and 16-2 list various queuing relationships for the following assumed conditions: Poisson arrivals; negative exponential service times; first-come, first-served queue discipline; and no limitation on the length of queue. The equations given are only applicable for steady-state conditions, giving results that would be expected after the system had operated for such a long time that the performance measures would not change with further passage of time. It should be noted further that the equations do not apply for the conditions where the arrival rate exceeds the service rate. This condition, if continued, would cause the waiting line to grow without limit and prevent the existence of a steady state. Use of the state probability equations from the tables are illustrated by Example 16-1.

*Table 16-1*

**Relationships for Single-Channel Queues with Poisson Arrivals and Negative Exponential Service Times**

1. Probability of having exactly $n$ units[a] in system,

$$P(n) = \left(\frac{\lambda}{\mu}\right)^n \left(1 - \frac{\lambda}{\mu}\right)$$

2. Average number of units in system,

$$\bar{n} = \frac{\lambda}{\mu - \lambda}$$

3. Average length of queue,

$$\bar{m} = \frac{\lambda^2}{\mu(\mu - \lambda)}$$

4. Average waiting time of arrival,

$$\bar{w} = \frac{\lambda}{\mu(\mu - \lambda)}$$

where

$\lambda$ = average number of arrivals per unit of time

$\mu$ = average service rate, number of services per unit of time.

[a]Vehicles or people.

*Table 16-2*
**Relationships for Multi-Channel Queues with Poisson Arrivals and Negative Exponential Service Times**

1. Probability of having exactly zero units[a] in system,

$$P(0) = \frac{1}{\left[\sum_{n=0}^{k-1} \frac{1}{n!}\left(\frac{\lambda}{\mu}\right)^n\right] + \frac{1}{k!}\left(\frac{\lambda}{\mu}\right)^k \frac{k\mu}{k\mu - \lambda}}$$

2. Probability of having exactly $n$ units in system,
   (a) For $n < k$:

$$P(n) = \frac{1}{n!}\left(\frac{\lambda}{\mu}\right)^n P(0)$$

   (b) For $n \geq k$:

$$P(n) = \frac{1}{k!k^{n-k}}\left(\frac{\lambda}{\mu}\right)^n P(0)$$

3. Average number of units in system,

$$\bar{n} = \frac{\lambda\mu(\lambda/\mu)^k}{(k-1)!(k\mu - \lambda)^2} P(0) + \frac{\lambda}{\mu}$$

4. Average length of queue,

$$\bar{m} = \frac{\lambda\mu(\lambda/\mu)^k}{(k-1)!(k\mu - \lambda)^2} P(0)$$

5. Average waiting time of an arrival,

$$\bar{w} = \frac{\mu(\lambda/\mu)^k}{(k-1)!(k\mu - \lambda)^2} P(0)$$

where
$\lambda$ = average number of arrivals per unit of time
$\mu$ = average service rate for each channel, number of units per unit of time
$k$ = number of channels or service stations

[a] Vehicles or people.

EXAMPLE 16-1

QUEUING AT A PARKING GARAGE. Vehicles arriving at a parking garage at an average of 45 veh/hr are parked on a first-come, first-served basis by one attendant at an average rate of one per minute. The arrivals may be described by a Poisson distribution and the service times by a negative exponential distribution. What is the probability the attendant will be idle?

Solution. By Table 16-1, Eq. 1,

$$P(0) = \left(\frac{45}{60}\right)^0 \left(1 - \frac{45}{60}\right) = 0.25 \quad \text{Answer}$$

If a second attendant is employed, what fraction of the time will one or both of the attendants be idle?

Solution. The fraction of the time at least one attendant will be idle is $P(0) + P(1)$. By Table 16-2, Eq. 1,

$$P(0) = \frac{1}{\left[\frac{1}{0!}\left(\frac{45}{60}\right)^0 + \frac{1}{1!}\left(\frac{45}{60}\right)^1\right] + \frac{1}{2!}\left(\frac{45}{60}\right)^2 \frac{2(60)}{2(60) - 45}} = 0.45$$

By Table 16-2, Eq. 2a,

$$P(1) = \frac{1}{2!}\left(\frac{45}{60}\right)^1 P(0) = 0.34$$

$$P(0) + P(1) = 0.45 + 0.34 = 0.79 \text{ Answer}$$

## 16-4. SIMULATION

Simulation literally means to assume the appearance of, without the reality. Digital computer simulation often serves as an attractive approach to the study of terminal systems too complex to be studied analytically.

Generally speaking, there are five steps involved in conducting a simulation study:

1. *Definition of the problem.* This involves setting forth a clear statement of the problem and establishing the objectives of the study.
2. *Formulation of the model and writing the computer program.* This step involves establishing rules of behavior of each component of the system. Here, decisions should be made regarding how arrivals and services are to be modeled, the nature of the queue discipline, and what measures of effectiveness are to be used. It may be appropriate to describe arrival and service distributions by analytical queuing models such as Eqs. 16-1, 16-2, and 16-3, but field studies may be needed to justify the use of these equations or else to indicate another approach. In certain instances, empirical data can be utilized directly by means of a Monte Carlo model. See Section 16-5 for a discussion of the Monte Carlo technique. To facilitate the writing of the computer program it is recommended that a block or functional diagram be prepared showing the sequencing and interrelationships of events.
3. *Validation of the model.* It is always advisable to observe the behavior of the model and compare typical results from the model (e.g., average queue lengths, waiting times, etc.) to empirical data. Only then can one be confident that the model reasonably describes the system being studied.

4. *Design of experiments.* Once the model is validated, a series of experiments should be designed to test the behavior of the terminal system under various levels of traffic demand and combinations of design layouts and operating conditions.
5. *Evaluation of results.* Simulation models provide a means of studying the consequences of design and management decisions on the operational efficiency of a terminal. Typically, the analyst wishes to evaluate the effect of such decisions on queue lengths, delays, and other measures of effectiveness and the impact of long queues, delays, and the like on the system, its users, and its environs.

## 16-5. EXAMPLE OF SIMULATION OF WAREHOUSE LOADING FACILITIES

Consider the example of a simulation study to determine the requirements for warehouse dock facilities for trucks [2]. A chain of department stores decided to consolidate its warehouse facilities at a single location. The problem was to determine the desired number of loading docks required to handle the volume of truck traffic formerly accommodated by the three separate locations plus that occasioned by future growth.

Empirical studies of truck arrivals and servicing revealed that truck arrival rates were not significantly different among half-hour intervals during the morning, and that variations within half-hour periods followed a pattern closely corresponding to the Poisson law.

Since the truck arrival rates did not directly apply to the consolidated warehouse, it was necessary to break them down into elements that could be identified as being present or absent in the new operational scheme and to recombine the elements appropriately. Allowance for future growth of truck traffic was based on the assumption that it would be proportional to the expected future dollar volume of the retail stores involved.

Two types of trucks were identified: vendor trucks that would unload only, and delivery trucks that would load only. The vendor trucks were divided into two classes each having service times described by an exponential distribution but with different average servicing times. Specifically, 75 percent of the vendor trucks required an average servicing time for 10 min, and the remaining 25 percent of the trucks required 45 min.

It was found that the service times of the delivery trucks could be modeled by a shifted exponential distribution that accounted for an average of 18 min each driver spent with paperwork and preloading cargo assembly.

A Monte Carlo simulation procedure was employed. The procedure for each truck involved the assignment of a time of arrival and a servicing time. These quantities were selected randomly from stocks of numbers (frequently distributions) that reflected the properties established by the empirical studies. For each specified number of service docks, the movement of trucks through

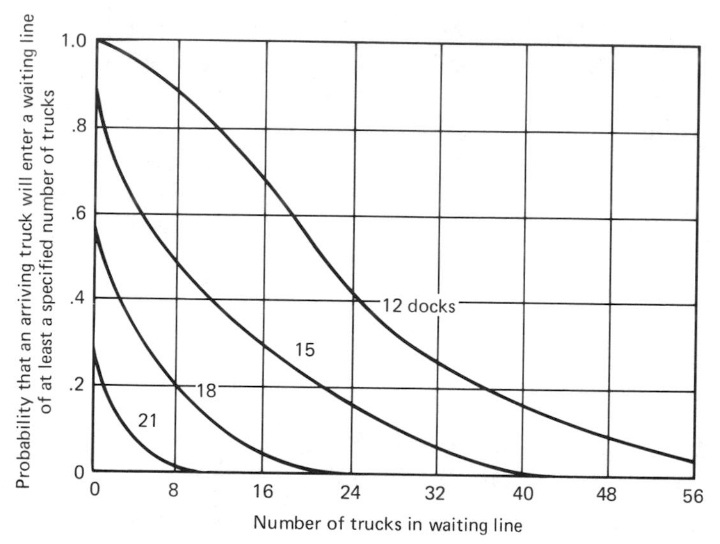


*Fig 16-1.* Results from example simulation study.

the docking system was simulated, and the length of waiting line and the delay until service were determined. Repeated simulations of a day's operations were conducted, and the results were averaged for each set of conditions. Figure 16-1 illustrates one of the results from this example simulation study.

## TERMINAL DESIGN CONTROLS AND CRITERIA

In the remainder of this chapter design controls and criteria for the principal land transportation terminals are described in the order outlined here.

- Passenger terminals
  - Automobile parking facilities
  - Bus terminals
  - Rail passenger stations
  - Urban rail transit stations
- Freight terminals
  - Truck terminals
  - Rail classification yards
  - Pipeline tank farms

### Automobile parking facilities

Widespread parking problems exist in central business districts, on college campuses, at shopping centers, and in industrial parks and other highly developed areas. Shortages of parking spaces in close proximity to major traffic generators have worsened as automobile registrations have increased and more and more people have chosen to travel by automobile. Parking problems have been aggravated in many cities as curb parking has been eliminated to improve flow.

The procedures for conducting parking surveys are described in Chapter 8. In the following paragraphs, the principles for the location, layout, and design of curb and off-street parking are discussed. [3]

## 16-6. CURB PARKING

Curb parking tends to seriously impede traffic flow and contribute to conflicts and crashes. Many highway and traffic engineers therefore recommend that curb parking be prohibited along major streets. It should also be prohibited within bus stops and pedestrian crosswalks, adjacent to fire plugs, and in the vicinity of intersections, alleys, and driveways. Where permitted, curb parking should be regulated to minimize its effects on accidents and congestion and to ensure that available parking spaces are used appropriately and efficiently.

From a safety and operational viewpoint, parallel parking is preferred over angle parking, and curb parking functions best when stalls are marked properly. Properly marked parking stalls tend to make it easier for drivers to park, lessening the likelihood of congestion and crashes. One study [4] found a 43 percent reduction in the average time required to park after suitably marked parking stalls were painted. Stalls are especially recommended for locations where there is great demand for curb parking spaces and where parking meters are used.

Three basic types of parallel stalls are used [4]: (1) end stalls, (2) interior stalls, and (3) paired parking stalls. End stalls are situated adjacent to intersections, alleys, driveways, and other restricted areas. End stalls are typically 20 ft in length. Interior stalls are usually 22 ft long, providing about 4.5 ft between adjacent cars for maneuvering. As Fig. 16-2 illustrates, paired parking consists of pairs of contiguous 18-ft stalls separated by an 8-ft open space that may be used for maneuvering. In this arrangement, the open spaces must be marked

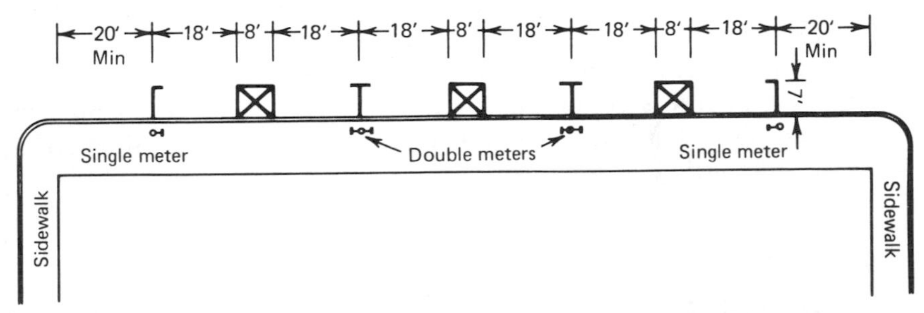


*Fig 16-2.* Paired curb parking layout. (Courtesy Transportation Research Board.)

clearly to prevent drivers from parking there. Paired parking is used frequently in conjunction with double parking meters, that is, two meters installed on a single post.

Curb parking stalls are designated by white lines extending out from the curb a distance, typically, of 7 ft. The end row of curb parking spaces normally is marked with an inverted L-shaped line, and interior stalls are designated by a T-shaped line.

Curb parking stalls should not be placed closer than 20 ft to the nearest sidewalk edge at nonsignalized intersections. At signalized intersections, a clearance to the sidewalk edge of 50 ft and preferably 100 ft should be observed. Parking stalls should be placed not closer than 15 ft away from fire hydrants and driveways. [4]

## 16-7. LOCATION OF OFF-STREET PARKING FACILITIES

One of the most important aspects of parking facility planning and design is the choice of a site. Experience has shown that an improperly located lot or garage is likely to fail or have limited use, even if located only a few blocks away from the center of parking demand.

Most parkers are reluctant to walk even short distances from the parking location to their ultimate destination. The maximum walking distance a parker will tolerate depends on trip purpose, city size, and the cost of parking. Office workers may be willing to walk a maximum distance of about 2000 ft (5 or 6 blocks) if the rates are attractively low [4]. Short-term parkers such as shoppers will usually resist walking more than 1 or 2 blocks. Generally speaking, the smaller the city and the higher the parking cost, the less distance a parker is willing to walk.

Preferably, parking lots and garages should be located on or near major arterials; garages should have access to two or more streets. It is also desirable that both lots and garages be accessed by right-hand turning movements.

Safe and convenient pedestrian access to parking facilities should be provided. To alleviate parkers' fears for their personal security, pedestrian walkways should be well lighted, suitably marked, and free of blind corners. Most pedestrians prefer overhead bridges to tunnels. To the extent feasible, pedestrian conflicts with vehicular traffic should be avoided.

In summary, parkers prefer parking facilities near their destination, that are easily accessible, that can be used without fear for one's personal safety, and that will cost little or no money.

## 16-8. LAYOUT OF PARKING LOTS AND GARAGES

Ideally, a parking lot or garage should be rectangular with cars parked on both sides of access aisles. Parking sites should be wide enough to provide two or more bays approximately 60 ft wide. Relatively long aisles, 250 ft or longer, function well. Aisles running with the long dimension of the lot or garage provide a better search pattern and yield more spaces per unit area.

Ninety-degree parking is generally used for two-way traffic. Right-angled parking tends to require slightly less area per parked car than do other configurations.

Parking stalls oriented at angles of 45 to 75 degrees to the aisles are often used with one-way circulation. Where such an arrangement is used to serve a building along the end of the parking lot, drivers must circulate at least once next to the building during entry or exit, causing more vehicle-pedestrian conflict than the two-way, 90-degree layout.

Where space permits, it is desirable to provide pedestrian sidewalks between adjacent rows of parked cars. When sidewalks are provided, it is usually necessary to install wheel stops to prevent vehicle encroachment. For 90-degree pull-in parking, the wheel-stop setback from the edge of the curb should be about 2.5 ft. For back-in parking, a setback of at least 4.0 ft is recommended [4].

Stall widths of 8.5 to 9.0 ft commonly are used for parking facilities in the United States. Widths of 9.5 or even 10.0 ft are sometimes used for supermarket lots and other areas where packages are being placed in cars [4]. Stall depths of 18.0 to 18.5 ft are used most often. For future layouts designers will need to consider the effects of vehicle downscaling and use smaller stalls.

The most common and preferred layout pattern is the bumper-to-bumper interlocked pattern illustrated by Fig. 16-3. The herringbone, or nested, interlock is sometimes used with 45-degree parking. In that configuration, the bumper of one car faces the fender of another car, necessitating the installation of wheel stops and increasing the probability of vehicular damage. Parking layout dimensions for 8.5-ft by 18-ft stalls arranged at various angles are given by the table that accompanies Fig. 16-3. The best parking layout for a given site, usually determined by a process of trial, will depend primarily on:

1. The size and shape of the available area.
2. The type of facility (self-park, attendant).
3. The type of parker (short-term, long-term).
4. The type of operation (pull-in, back-in, one-way, two-way, etc.).

## 16-9. PARKING GARAGE DESIGN CRITERIA

Special design criteria for parking garages, abstracted from Ref. 5, are given in the following paragraphs.

Single entrances and exits, with multiple lanes, are preferable to several openings. Entrances and exits should be located away from street intersections to prevent traffic congestion. Lane widths of 12 to 14 ft typically are used for entrances and exits. Lanes should be tapered to a width of 9 to 10 ft for approaches to ticket dispensers and cashiers' booths.

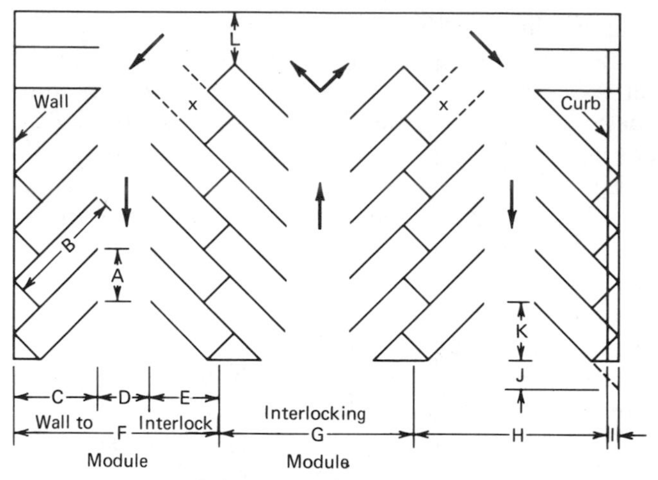


x = Stall not accessible in certain layouts

Parking Layout Dimensions (in feet) for 9-ft Stalls at Various Angles

| | On Diagram | 45° | 60° | 75° | 90° |
|---|---|---|---|---|---|
| Stall width, parallel to aisle | A | 12.7 | 10.4 | 9.3 | 9.0 |
| Stall length of line | B | 25.0 | 22.0 | 20.0 | 18.5 |
| Stall depth to wall | C | 17.5 | 19.0 | 19.5 | 18.5 |
| Aisle width between stall lines | D | 12.0 | 16.0 | 23.0 | 26.0 |
| Stall depth, interlock | E | 15.3 | 17.5 | 18.8 | 18.5 |
| Module, wall to interlock | F | 44.8 | 52.5 | 61.3 | 63.0 |
| Module, interlocking | G | 42.6 | 51.0 | 61.0 | 63.0 |
| Module, interlock to curb face | H | 42.8 | 50.2 | 58.8 | 60.5 |
| Bumper overhang (typical) | I | 2.0 | 2.3 | 2.5 | 2.5 |
| Offset | J | 6.3 | 2.7 | 0.5 | 0.0 |
| Setback | K | 11.0 | 8.3 | 5.0 | 0.0 |
| Cross aisle, one-way | L | 14.0 | 14.0 | 14.0 | 14.0 |
| Cross aisle, two-way | – | 24.0 | 24.0 | 24.0 | 24.0 |

*Fig 16-3.* Parking stall layout elements. (Courtesy Transportation Research Board.)

When self-parking is used there is little need for reservoir space at the entrance. Studies indicate that the capacity of an entrance lane with an automatic ticket dispenser is approximately 400 veh/hr. One entrance lane for every 300 to 500 spaces usually is provided.

If attendant parking is used, a storage reservoir will be needed. The reservoir space required will depend on the passenger unloading time and the time required for parking each vehicle, as well as the arrival rate.

If possible, clear span construction should be provided with column spacing equal to the unit parking depth (module width). A clear ceiling height of 7 ft (preferably 7.5 ft) should be provided.

Vehicular access between floors can be provided by sloped floors or by ramps. For self-park facilities, the floor slopes should not exceed 3 to 4 percent. Floor slopes up to 10 percent can be used for attendant park facilities;

ramp slopes should preferably not exceed 10 percent. Straight ramps should be at least 9 ft wide; curved ramps are usually at least 12 to 13 ft in width. A minimum radius of ramp curvature of 30 ft is recommended, measured at the face of the outer curb of the inside lane.

Counterclockwise circulation is preferred for parking garages. When helical ramps are used, the down ramp should be placed inside and the up ramp should be outside. [3]

**Bus terminals**

The layout and design of intercity bus terminals vary a great deal in size and complexity depending on the number of passengers to be accommodated and whether the facilities are to serve one, two, or more carriers. Figure 16-4 shows a typical layout for a transportation center designed to serve two carriers. The layout features common waiting area, restaurant, and toilets, but has separate ticketing, passenger loading, and baggage handling areas.

Angle parking is commonly provided for the buses, typically arranged as shown by Fig. 16-5. The spatial requirements depend primarily on the size, physical features, and operating characteristics of the buses and the angle of parking. Table 16-3 gives recommended loading platform dimensions required for a 40-ft by 8.5-ft bus using a 45-degree parking angle. To facilitate efficient baggage handling and passenger loading, a minimum clearance of 6 ft between buses is desirable. A protective canopy overhang should be provided and should extend at least beyond the loading door of parked buses.

Suitable provisions must be made for package express handling, temporary storage, pickup, and dropoff.

All new or substantially renovated terminals must conform to the minimum design standards in the "American National Standard Specifications for Making Buildings and Facilities Accessible to and Usable by the Physically Handicapped." These specifications require the provision of reserved parking spaces for the handicapped, entrances usable by persons in wheelchairs, ramps sloped not steeper than 8.33 percent with handrails at least on one side, and a number of other requirements [6].

The amount of space required for waiting area, ticket counters, toilets, and other public areas is based on empirical relationships between spatial needs and typical peak-hour passenger volumes. For further guidance on this aspect of the problem, the reader should refer to Chapter 18.

**Rail passenger stations**

Rail passenger travel in the United States declined sharply following World War II, reaching a low of 254 million passengers transported in 1973. That figure, which included urban commuters, represented 31 percent of all passengers traveling by public carrier between U.S. cities; however, excluding urban commutation, rail travel accounted for only 0.7 percent of the total intercity travel expressed in passenger miles. Since 1973, there have been increases in both the number of passengers traveling by rail and the passenger

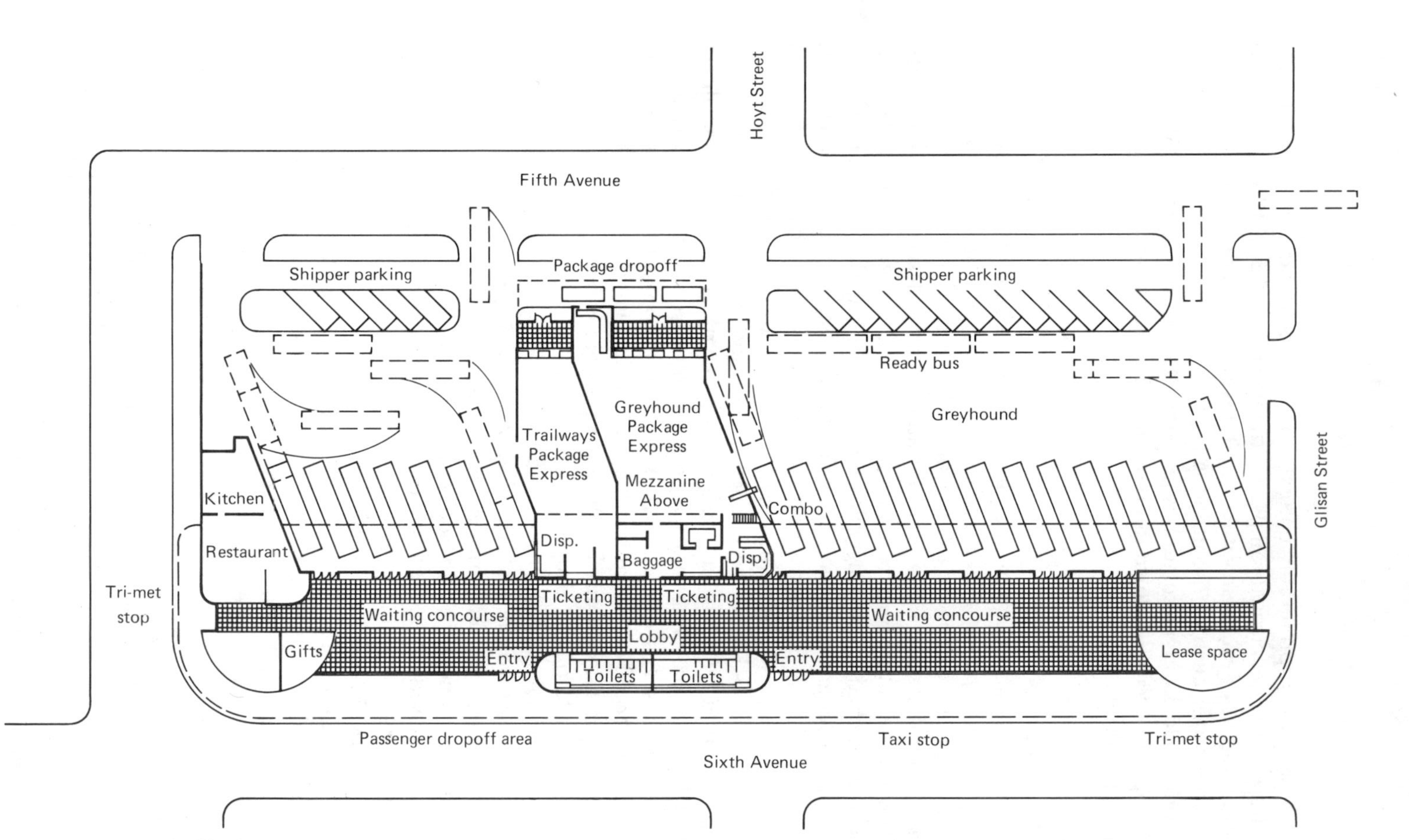


*Fig 16-4.* Typical layout of bus transportation center. (Courtesy Trailways Bus System.)

Terminal Loading Clearances – 45° Angle Parking – Average conditions
Figures given do not apply to special conditions or angles other than 45°.
Loading with buses arranged in reverse direction not recommended.

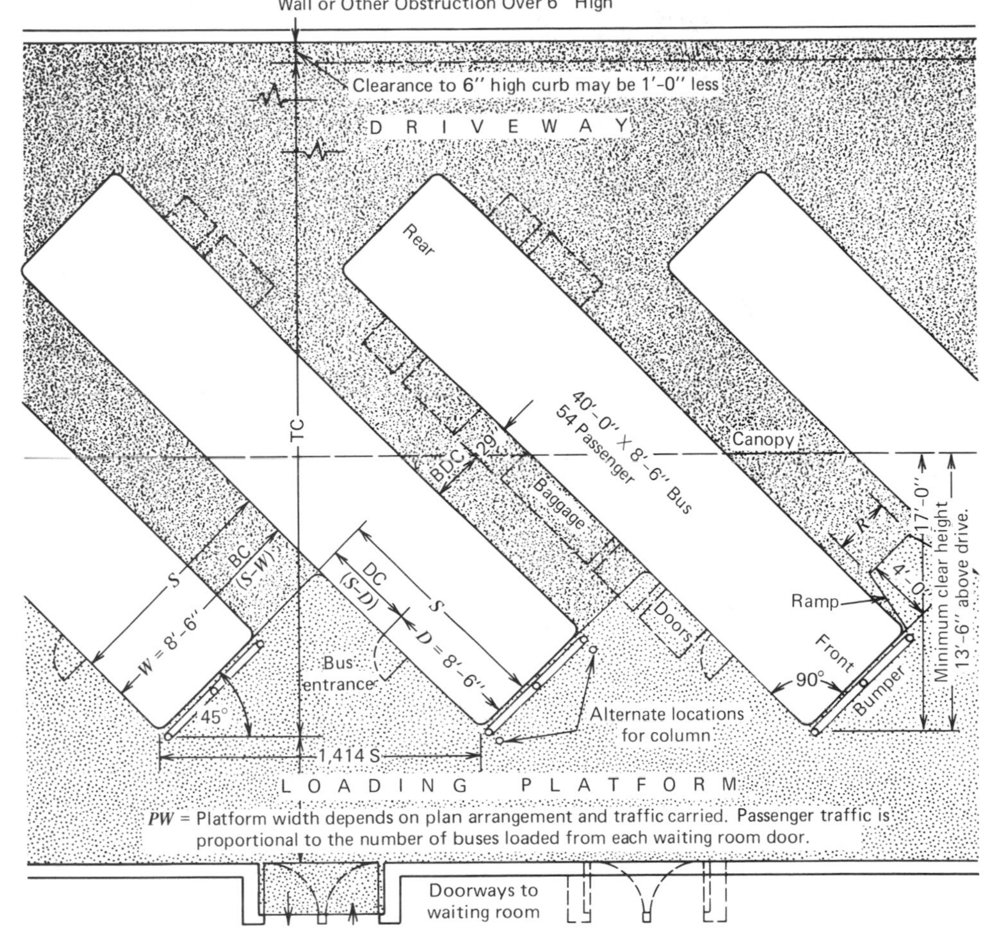


Tabulation:

| Clearance | | S = Spacing center to center of buses | | | | | |
|---|---|---|---|---|---|---|---|
| | | For reserve parking | | For passenger loading | | | |
| | | 11′ | 12′ | 14′ | 15′ | 16′ | 17′ |
| *TC* = Recommended minimum turning clearance for typical positions* | to curb | 63′-6″ | 60′-6″ | 55′-3″ | 52′-9″ | 52′-0″ | 51′-6″ |
| | to wall | 64′-6″ | 61′-6″ | 56′-3″ | 53′-9″ | 53′-0″ | 52′-6″ |
| *BC* = Clearance between buses<br>*DC* = Door clearance | | 2′-6″ | 3′-6″ | 5′-6″ | 6′-6″ | 7′-6″ | 8′-6″ |
| *BDC* = Baggage door clearance | | 1″ | 1′-1″ | 3′-1″ | 4′-1″ | 5′-1″ | 6′-1″ |
| *R* = Ramp width | | 1′- | 2′-6″ | 3′-10″ | 4′-6″ | 5′-0″ | 5′-6″ |

*For clearance at intermediate spacings see graph. Special conditions require special clearances. When buses enter with right turn from street, first few loading positions will require additional turning clearance.

*Fig 16-5.* Parking layout for buses at terminal. (Courtesy Greyhound Lines.)

*Table 16-3*
**Recommended Bus Loading Platform Dimensions for a 45-Degree Parking Angle[a]**

| *Spacing, Center to Center of Buses, S (ft[b])* | *Minimum Turning Clearance (TC)* | | *Clearance Between Buses (BC) and Door Clearance (DC)* | *Baggage Door Clearance (BDC)* |
|---|---|---|---|---|
| | *To Curb* | *To Wall* | | |
| 11[c] | 63 ft 6 in. | 64 ft 6 in. | 2 ft 6 in. | 1 in. |
| 12[c] | 60 ft 6 in. | 61 ft 6 in. | 3 ft 6 in. | 1 ft 1 in. |
| 13 | 57 ft 9 in. | 58 ft 9 in. | 4 ft 6 in. | 2 ft 1 in. |
| 14 | 55 ft 3 in. | 56 ft 3 in. | 5 ft 6 in. | 3 ft 1 in. |
| 15 | 52 ft 9 in. | 53 ft 9 in. | 6 ft 6 in. | 4 ft 1 in. |
| 16 | 52 ft 0 in. | 53 ft 0 in. | 7 ft 6 in. | 5 ft 1 in. |
| 17 | 51 ft 6 in. | 52 ft 6 in. | 8 ft 6 in. | 6 ft 1 in. |

[a]Data apply to a 40-ft by 8.5-ft bus.

[b]1 ft = 0.3048 m.

[c]Recommended for reserve parking only.

SOURCE: *Planning Standards for Terminals and Garages,* Architectural Engineering and Properties Department, Greyhound Lines, Inc.

mileage, reflecting public concerns about the availability and cost of automobile fuel. Yet Amtrak, the National Railroad Passenger corporation established to provide improved rail passenger service, continues to experience annual operating deficits of about $700 million. [7].

Although few new railroad passenger stations are being constructed, Amtrak has developed a standard stations program that includes five classes of facilities. The salient features of these designs are given in Table 16-4.

*Table 16-4*
**Features of AMTRAK's Standard Station Designs**

| *Design* | *Design Volumes, Typical Peak-Hour Passengers* | *Number of Ticket Counters* | *Overall Station Area ($ft^2$)[a]* | *Auto Parking, Number of Cars* | *Land Area (Acres)[b]* |
|---|---|---|---|---|---|
| 300A | 300–1,000 | 4 | 18,000 | 225 | 15.4 |
| 150B | 150–350 | 3 | 8,220 | 110 | 9.3 |
| 50C | 50–175 | 2 | 2,000 | 30 | 5.5 |
| 25D | 25–75 | 1 | 1,150 | 20 | 2.5 |
| E | <25 | 0 | 240 | 15 | 1.0 |

[a]1 $ft^2$ = 0.0929 $m^2$.

[b]1 acre = 0.404 hectares.

SOURCE: *Standard Stations Program, Executive Summary,* AMTRAK, Washington, D.C., 1978.

## 16-10. RAIL STATION DESIGN

A rail station includes the station tracks, the platforms, connecting thoroughfares (passageways, ramps, stairways, etc.), and the terminal building.

With respect to the track layout, three types of passenger stations may be used:

1. *Through station.* With this common arrangement, the trains stop on a mainline track while passengers board and disembark. This type of station is satisfactory where traffic is light and only through service is provided.
2. *Loop station.* Where traffic is heavy, station tracks may be provided alongside the mainline tracks. Trains enter these tracks through "throat tracks," with 2.5 to 3.0 station tracks being provided for each throat track. For turning trains, loop tracks generally provide faster and more economical service.
3. *Stub station.* With the stub station, trains reverse direction of travel when leaving the station. This type normally would be used for the end of a line.

Passenger platforms and corridors should be at least 12 ft in width. Combined passenger and trucking platforms at heavily trafficked stations should be at least 20 ft in width [8].

Spatial requirements for passengers in the main building are based on empirical studies, and the areas required for various functions in the station usually are based on the number of typical peak-hour passengers. These requirements are not unlike those described in Chapter 18 for air passengers, but the facilities are generally much smaller in scale. Typical requirements for various facilities in rail stations are given in Table 16-5.

The number of ticket counter positions will depend on the mode of operation. At positions dedicated to ticket sales only, approximately 35 passengers per hour can be accommodated. Agents accepting baggage only can process about 55 passengers per hour. In most AMTRAK stations it can be assumed that about 25 percent of the peak-hour passengers have been ticketed [9].

Rail passenger stations and the facilities in the stations must be accessible to and usable by the physically handicapped.

### Urban rail transit stations

Urban rail transit stations differ from conventional rail passenger stations in several respects. For example, in urban rail transit stations:

1. Little or no provision must be made for the handling of baggage.
2. There are higher densities of passengers and heavier flows requiring more rapid entry and egress. The vehicles therefore are designed with

*Table 16-5*

**Spatial Requirements for Various Facilities in Rail Stations**

| | | |
|---|---|---|
| *Passenger Areas* | | |
| Waiting area | | |
| For peak-hour passengers | <150 | 14 ft²/passenger[a] |
| For peak-hour passengers | 150–500 | 12 ft²/passenger |
| For peak-hour passengers | >500 | 10 ft²/passenger |
| Baggage area | | 7–9 ft²/passenger |
| Coffee shop/eating areas | | |
| For peak-hour passengers | <300 | Vending machines |
| For peak-hour passengers | ≥300 | 3–4 ft²/passenger in dining room/lunch room; 1 seat for every 7–10 passengers |
| Kitchen | | |
| For peak-hour passengers | ≥300 | 500–800 ft² |
| Ticket counter | | |
| With conveyor, depth | | 8 ft[b] |
| Nonconveyor, depth | | 5 ft |
| Length, per agent position | | 7.5 ft |
| Length of queuing line | | 15 ft |
| *Employee Areas* | | |
| Employee lounge | | 100 ft² plus 10 ft² per employee on duty |
| Cash accounting office | | 60 ft² plus 40 ft² for each employee above 2 |
| Station services office | | 120 ft² |
| Supervisor's office | | 80 ft² |
| Station manager's office | | 120 ft² |
| Secretarial area | | 80 ft² |
| Red cap ready room | | 100 ft² plus 10 ft² for each employee above 5 |
| *District Offices* | | |
| District superintendent's office | | 150 ft² |
| Manager's office | | 120 ft² |
| Supervisor's office | | 80 ft² |
| Secretary's office | | 80 ft² |
| Clerk's office | | 60 ft² |
| Conference/hearing room | | 150 ft² |

[a] 1 ft² = 0.0929 m².

[b] 1 ft = 0.3048 m.

SOURCES: *Standard Stations Program, Executive Summary,* AMTRAK, 1978; *Manual for Railway Engineering,* AREA, 1978.

wide, automatic doors, and platforms are built at the level of the vehicle floors.

3. The trains operate at short headways, typically 10 min during off-peak periods and as little as 90 sec during peak periods. There is relatively little need for separate waiting areas with seats or benches for waiting patrons.
4. Automated ticketing and fare collection systems are used and must be provided for in the station design.
5. Since many urban rail stations are built as subways or elevated facilities, special attention must be paid to vertical circulation and the planning and design of street-level entrances and exits.
6. Urban rail transit stations are often built in high crime areas, and special security measures such as monitored closed-circuit television must be employed to ensure public safety and to allay patrons' fears of crime.
7. At suburban rail transit stations, facilities may need to be provided for feeder bus loading and unloading, passenger car loading and unloading (kiss-and-ride), and for short-term and long-term parking.

Figure 16-6 illustrates a modern urban rail transit station.

*Fig 16-6.* A modern urban rail transit station. (Courtesy American Consulting Engineers Council.)

## 16-11. TYPICAL RAIL TRANSIT STATION DESIGN PROCEDURES AND CRITERIA

In the following paragraphs, design procedures and criteria for the Metropolitan Atlanta Rapid Transit Authority (MARTA) system are briefly described [10]. The approach used for that system is presented as typical of that employed for urban rail transit stations.

Estimates of station patronage were made for the A.M. peak-hour. The P.M. peak-hour patronage was assumed to be equal to but in opposite direction to that of the A.M. peak-hour.

Station elements were sized on the basis of the *ultimate station capacity,* which was 150 percent of the station patronage. However, parking lot capacity was based on the requirements projected in the patronage analysis.

Certain elements of design were based on the peak 5-min patronage, taken to be 12.5 percent of the ultimate station capacity. Others were based on the peak minute patronage, assumed to be 2.5 percent of the ultimate station capacity.

MARTA stations have been designed on the basis of levels of service as recommended by John Fruin and as described in Section 6-8. During normal operating conditions, the levels of service were assumed to be between C and D. For emergency conditions, MARTA designers assumed that the patron flow is all exiting the station, and the level of service would be between D and E.

The sizing of station platforms was based on the *station accumulation,* which was assumed to be equal to the maximum number of patrons waiting on the platform if a train is up to 90 sec late during the ultimate peak 5-min period.

Two basic types of platforms were used: a center platform between tracks, and a side platform at one side of a single track or on each side of an adjacent pair of tracks.

MARTA platforms were designed for lengths of at least 600 ft, the length of the eight-car trains which are expected to operate ultimately at peak-hours. At above-ground stations, wind and weather screens were provided along approximately 300 ft of each side platform to shelter waiting patrons from the elements.

The platforms were sized to allow 8 ft$^2$ of unobstructed walking surface per patron of the station accumulation on the platform. An additional width of 1 ft from walls and 1.5 ft from the platform edge were provided.

The designers adhered to several fundamental principles of station circulation, namely:

1. Right-hand flows were preferred.
2. Whenever possible, cross-flows were avoided.
3. Passenger flows moving in opposite directions were separated.
4. Dead-end conditions were avoided.

5. Design features were chosen and arranged so as to eliminate or minimize delays.

The part of the rail transit station that facilitates the movement of large numbers of people in and out of the stations is known as the concourse. It functions as a transition area between the points of entry into the station and the access ways to the train platforms. It provides space for a variety of functions including fare collection, directional and informational signing, and a location for a comfortable waiting area, especially during off-peak hours. The concourse may be located at street level, platform level, or at a mezzanine area between these levels.

The design of the concourse and access ways is highly dependent on the requirements for fire and panic safety. The MARTA design standards allow for exiting requirements to accommodate the maximum plausible emergency conditions, specifically, a station load equal to the ultimate passenger load of a train during the peak 15 min plus the ultimate peak minute entraining patron load awaiting that train on the platform. The standards specify that at least two platform exits must be provided a minimum of 100 ft apart. The standards further provide [10]:

1. There shall be a minimum of two exits from each concourse, located as far apart as possible.
2. In subway stations, no point on the platform(s) or concourse shall be located more than 200 ft from an exit.
3. In above-ground stations, no point on the platform(s) or concourse shall be located more than 300 ft from an exit.

Sufficient exit lanes were required to evacuate the exiting occupant load in 4 min or less. Emergency exiting capacities were evaluated on the basis of 24-in lane widths, a lane capacity of 40 persons per minute, and a travel speed of 200 ft/min. In calculating the number of exit lanes available, 1 ft was deducted from the width of exit corridors and ramps to allow for side walls, and 2 ft were deducted to allow for a buffer zone at the platform edge.

MARTA transit stations were designed to ensure accessibility and usability for the handicapped. To this end, MARTA stations are designed with:

1. At least one primary entrance accessible to individuals in wheel chairs.
2. Walks at least 48 in. wide with gradients not more than 5 percent.
3. Ramp slopes not more than 8.33 percent with hand rails on at least one side 32 in. in height.
4. Wall fixtures appropriately mounted to make them reachable by the handicapped.
5. Designated automobile parking spaces for use by individuals with physical disabilities.

**Truck terminals**

The design of a truck terminal depends on the area to be served, the volume of freight to be handled, and the level of service to be provided. An adequate number of doors must be provided to assure that the terminal meets its schedules and freight handling requirements. The doors must have suitable dimensions and allow for proper clearance and maneuvering space for equipment. Adequate yard space must also be provided for parking and access to the loading platform and operating areas. [11]

A minimum width of 11 ft, and preferably 12 ft, is recommended for each spot in order to allow proper clearances on each side of 8.5-ft-wide vehicles. The dock approach length is the length of the largest tractor-trailer combination using the dock plus the apron length necessary to maneuver the vehicle into and out of the parking spot. Generally speaking, the dock approach should be at least twice the length of the longest tractor-trailer combination. Table 16-6 shows the minimum recommended apron and dock approach lengths assuming 12-ft widths for dock spots. If 11-ft widths are used, the apron lengths shown should be increased by about 2 ft.

The apron dock approach area should be nearly level. Although most trucks are designed to negotiate a 15 percent grade, the startup grade for pulling away from the dock should not exceed 3 percent [11].

The dock height should be approximately equal to the bed height of the vehicles using the dock. Typically, trailer bed heights vary from about 48 to 52 in., while pickup and delivery vehicle bed heights range from about 44 to 48 in. In order to ensure that vehicle doors can be opened and closed, it is best to select a dock height that is too low rather than too high.

Vertical clearances of at least 15 ft are recommended in order to accommodate 13.5-ft-high trailers.

*Table 16-6*

**Minimum Recommended Apron Lengths and Dock Approach Lengths for Truck Terminals**

| *Overall Length of Tractor-Trailer (ft)*[a] | *Apron Length (ft)* | *Dock Approach Length (ft)* |
|---|---|---|
| 40 | 43 | 83 |
| 45 | 49 | 94 |
| 50 | 57 | 107 |
| 55 | 62 | 117 |
| 60 | 69 | 129 |
| 65 | 75 | 140 |

[a] 1 ft = 0.3048 m.

SOURCE: *Shipper-Motor Carrier Dock Planning Manual.* The Operations Council, The American Trucking Associations, Washington, D.C., 1970.

Traffic circulation plans should encourage vehicles to circulate in a counterclockwise direction. For two-way traffic, service roads should be a minimum of 23 ft and preferably 24 ft in width. For one-way roads, the minimum widths on straight sections should be 12 ft. On curves, the pavement must be widened to conform to the wheel paths of the turning vehicles.

The width of parking spaces for trucks should be about 12 ft, and the length should be the vehicle length plus 20 percent.

**Railroad freight terminals and yards**

Railroad yards provide shops and facilities for the maintenance of the rolling stock and storage of idle cars. The most important need, however, is for classification—the making up of trains for the distribution of freight to various parts of the country.

A railroad freight terminal facility usually contains at least three components.

1. A receiving yard, where incoming trains are directed from the main line and are stored temporarily before being sorted and classified.
2. The classification yard, where cars are sorted and classified into blocks of common destination.
3. The departure yard, where the sorted blocks of cars are made into trains and stored while arrangements are made for mainline movement.

A large terminal facility may also contain a repair yard and a local yard, the latter being for classification of cars scheduled for local deliveries. A typical layout of a large railroad freight terminal facility is shown in Fig. 16-7. Small yards may be considerably less elaborate, containing only one general yard with certain tracks being assigned for receiving and departure activities.

The layout of the various yards depends primarily on the dimensions of the cars and locomotives, the length of the trains, the volume of traffic, and

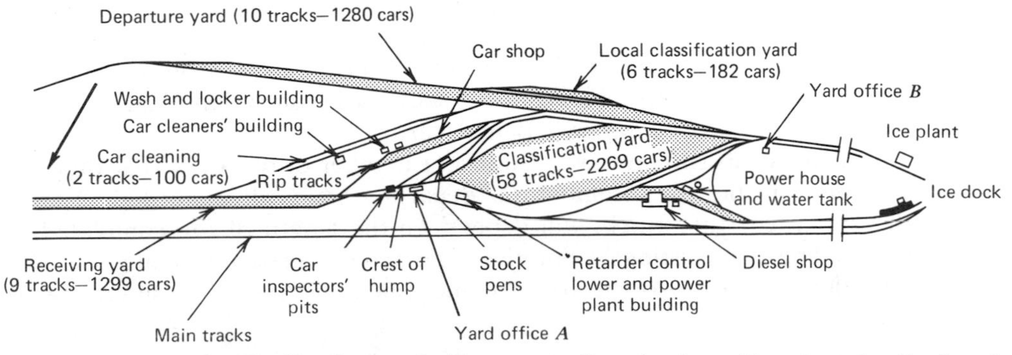


*Fig 16-7.* Plan of a classification yard. (Source: *An Introduction to Transportation Engineering,* Second Edition, by Wm. W. Hay, John Wiley, 1977.)

the rate at which cars are processed. Generally speaking, a minimum of 50 ft per car should be allowed for all freight car tracks except for repair tracks (which require 55 ft per uncoupled car) and for special equipment tracks [8]. Body tracks[2] should be spaced not less than 14 ft center to center. When parallel to a main track, the first body track should be spaced not less than 20 ft from that track center to center. Parallel ladder tracks[2] are usually spaced not closer than 18 ft center to center.

A sufficient number of receiving tracks should be provided so that there will be one available whenever an arriving train needs to enter the yard. The type of classification yard and the number of tracks depends on the volume and scheduling of traffic to be handled. Flat yards are sometimes used where the number of switching cuts per train is small. In such facilities, the sorting of cars is performed by switching engines at a rate of 30 to 60 cars/hr [1].

At large and busy terminals, the sorting of cars is accomplished by gravity or hump switching. The cars are pushed by an engine to the top of a hump located between the receiving and classification yards. The gradients leading to the classification yard are designed carefully so that each car will roll to and stop at the far end of the yard or will roll to a coupling at a speed of not more than 4 mph. The speed of a car on a grade can be calculated at any point by use of fundamental energy relationships. The change in velocity that occurs depends on the magnitude and length of slope. The calculations, described in more detail by Reference 8, are similar to those for calculating the velocity of a freely falling body. Suitable allowance must be made for energy losses due to rolling resistance and that due to the rotation of the wheels.

In modern classification yards, retarders are placed along the tracks leading to the classification yards to control the speed of the entering cars. These devices bear against the flanges of the wheels to slow or stop the car as it moves through the retarder. The amount of braking pressure that will be applied by the retarder may be controlled by an operator in a nearby tower by means of a push-button system. In newer yards the devices may be controlled automatically by computerized systems that measure the weight and speed of each car and calculate the needed retardation after appropriately accounting for variations due to wind resistance, rolling resistance, and the coupling or stopping distance.

Classification rates may vary from 100 to 300 cars/hr depending on the degree of automation used. Figure 16-8 shows a classification yard at the Union Pacific Railroad Company's Bailey Yard.

Departure yards should have a sufficient number of tracks so that there will be one available for assembling a departing train when needed. The length of the departure tracks will depend on the length of the completed trains including the assisting locomotives. To avoid excessive starting resistance, gradients adverse to the forward movement of the trains should be avoided.

[2] Body tracks are those tracks in the yard where cars are parked, while ladder tracks provide access to the yard.

*Fig 16-8.* Union Pacific Railroad Company, Bailey Yard, North Platte, Nebraska.

**Tank farms**

In pipeline transportation, the terminals take the form of groupings of strategically arranged storage tanks called tank farms. Depending on the nature of the product being transported, a wide variety of storage tanks are used. For example, certain gases such as propane must be stored in high-pressure tanks. Here, the discussion is limited to terminal facilities for liquid petroleum products that can be stored in atmospheric tanks, that is, at atmospheric pressure.

Three general types of tanks are used: fixed roof tanks, floating roof tanks, and tanks that combine features of both fixed and floating roof tanks. Most liquids can be stored in floating roof tanks. Liquids that have a high vapor pressure are usually stored in floating roof tanks to prevent excessive evaporation. Some pipeline companies prefer hybrid tanks that have a fixed roof combined with an internal lightweight floating pan to control evaporation.

Generally, storage tanks are built above ground, although special circumstances may dictate their placement underground. For example, they are often built underground at major airports to facilitate safe aircraft movements.

The minimum size of tank used depends on the minimum batch size transported by the pipeline. The available storage capacity of a tank farm is determined by the number and size of all the tanks, while the required storage capacity is a function of the turnover.

Pipelines companies that transport flammable and combustible liquids conform to strict standards for the location of storage tanks with respect to public areas. The companies generally follow the recommendations of Reference 12, which gives minimum distances from storage tanks to adjacent property lines and public transport ways.

Adjoining property and waterways are protected further by means of impounding by diking around the tanks. Reference 12 recommends that the ground surface around a tank be sloped at least 1 percent away from the tank for at least 50 ft or to the base of the dike, whichever is less. The volumetric capacity of the diked area should not be less than the greatest amount of liquid that can be released from the largest tank within the diked area, assuming a full tank [11].

The dike walls may be made of earth, steel, concrete, or solid masonry but must be liquid-tight and able to withstand a full hydrostatic head. Earthen walls 3 ft or more in height should have a flat section at the top at least 2 ft wide.

Storage tanks usually are built directly on the ground or on foundations made of concrete, masonry, piling, or steel. The foundations should be designed so as to minimize the likelihood of uneven settlement of the tank. The tank may need to be specially treated to resist corrosion.

A typical tank farm is shown in Fig. 16-9.

*Fig 16-9.* Aerial view of a tank farm. (Courtesy Chicago Bridge and Iron Co., Inc.)

## PROBLEMS

**1.** How many cars could be placed in a commercial parking lot 140 ft wide (street frontage) and 240 ft deep? Make a sketch showing the recommended arrangement. Assume two driveways, one-way circulation, and self-parking.

**2.** How many cars can be placed in a commercial parking lot on land that is 40 m wide (street frontage) and 100 m deep? Make a sketch showing the design using SI (metric) units. Assume two driveways, one-way circulation, and self-parking.

**3.** Vehicles arrive at a toll booth at an entrance to a state park at a rate of 250 per hour. One attendant is on duty and serves the vehicles at a rate of one every 12 sec.
   a. Estimate the percent of time that the attendant will be idle.
   b. What is the average number of vehicles in the queue (including the one being served)?
   c. Estimate the average waiting time of an arrival.

**4.** The park described in Problem 3 has a reservoir space for only 3 vehicles plus the vehicle being served. If more than 3 vehicles join the queue, the traffic backs up and blocks traffic on a busy state highway.
   a. What is the probability that traffic will be blocked?
   b. What is the probability that traffic will be blocked if two attendants are used? (Note: In this state, five vehicles can be in the system without traffic being blocked.)
   c. What percent of the time will one or both of the attendants be idle if two attendants are used?

**5.** Gasoline (specific weight = 53.4 $lb/ft^3$) arrives at a terminal at a rate of 50,000 bbl/hr. Two alternatives are being considered for shipping the product to its ultimate destination. Alternative A would utilize inland waterway barges with 3000-ton capacity in tows of 10 barges each and 6 shipments per day. Alternative B would rely on unit trains with 7500-ton capacity leaving once each hour. Contrast the size of storage facilities required for these two alternatives. If the gasoline is stored at the terminal for 24 hours, how many 125,000-barrel tanks would be required? Note: There are 42 gal/bbl and 7.48 $gal/ft^3$.

**6.** The following are the predicted volumes of entering and departing automobiles at a proposed parking garage during a typical weekday. Draw a cumulative flow diagram of vehicles in the garage and indicate the minimum number of spaces required.

| *Period* | *Arriving Volume, (veh/hr)* | *Departing Volume, (veh/hr)* |
|---|---|---|
| 6–7 A.M. | 14 | 6 |
| 7–8 | 16 | 7 |
| 8–9 | 120 | 10 |
| 9–10 | 138 | 15 |
| 10–11 | 43 | 23 |
| 11–12 | 20 | 29 |
| Noon–1 P.M. | 15 | 25 |
| 1–2 | 14 | 13 |
| 2–3 | 14 | 15 |
| 3–4 | 10 | 120 |
| 4–5 | 6 | 100 |
| 5–6 | 4 | 26 |

## REFERENCES

1. Hay, William W., *An Introduction to Transportation Engineering,* Second Edition, John Wiley, New York, 1977.
2. Schiller, Donald H., and Marvin M. Lavin, "The Determination of Requirements for Warehouse Dock Facilities," *Operations Research,* April 1956.
3. Wright, Paul H., and Radnor J. Paquette, *Highway Engineering,* Fifth Edition, John Wiley, New York, 1987.
4. *Parking Principles,* Special Report No. 125, Highway Research Board, Washington, D.C., 1971.
5. *Traffic Design of Parking Garages,* The Eno Foundation for Highway Traffic Control, Saugatuck, CT, 1957.
6. *Trailways Design Standards Manual, Volume 2,* 1980.
7. *Railroad Facts, 1987 Edition,* Association of American Railroads, Washington, D.C., 1988.
8. *Manual for Railway Engineering,* American Railway Engineering Association, Chicago, 1978.
9. *Standard Stations Program, Executive Summary,* AMTRAK, Office of the Chief Engineer, Washington, D.C., 1978.
10. *Metropolitan Atlanta Rapid Transit Authority Architectural Standards,* Atlanta, GA, 1975.
11. *Shipper-Motor Carrier Dock Planning Manual,* The Operations Council, The American Trucking Associations, Washington, D.C., 1970.
12. *Flammable and Combustible Liquids Code, NFPA 30,* National Fire Protection Association, Boston, MA, 1977.

# *PART 5*

# *DESIGN OF AIR TRANSPORTATION FACILITIES*

*Here we focus more closely on the air transportation mode and describe airport master planning and airport layout and design. The three chapters in this part of the book discuss the location of airports and the determination of runway orientation and length; the function and layout of airport aprons and terminal buildings; the geometric design of runways and taxiways; the design of airport drainage systems; and airport marking and lighting systems.*

# Chapter 17

# Airport Planning and Layout

## 17-1. INTRODUCTION

It happened at midmorning on a windy beach at Kitty Hawk, North Carolina, on December 17, 1903. A fragile looking two-winged craft with a man at the controls was propelled on a little trolley along a wooden rail. It rose from the rail, surged forward into the wind, and settled back into the sand. Man's first flight in a heavier-than-air craft had lasted 12 seconds and covered a distance of 120 feet. On that morning, Wilbur and Orville Wright were to make three additional flights, the longest lasting for almost a minute. The air age had begun, and the world would never be the same.

Man's ancient aspirations to fly had been, in Wilbur Wright's words, "handed down to us by our ancestors who, in their gruelling travels across trackless lands in prehistoric times, looked enviously on the birds soaring freely through space, at full speed, above all obstacles, on the infinite highway of the air [1]."

The Wright's success had been preceded by centuries in which inaccurate and often ludicrous theories of flight were proposed, and abortive and sometimes disastrous attempts were made to leap and glide and soar.

A groundswell of interest in air transportation developed in 1783 when Joseph and Etienne Montgolfier demonstrated in Annonay, France, that man could travel by balloons filled with hot air. Later that same year, a French physicist, J. A. C. Charles, made a successful flight in a balloon filled with hydrogen.

Still later, when it became possible to steer these *montogolfières* and *charlières*, they came to be known as dirigibles. Strong interest in transportation by the slow and cumbersome dirigibles continued until the 1930s, when several spectacular disasters occurred to the lighter-than-air craft. These dramatic tragedies, along with progress in heavier-than-airflight technology, resulted in the assignment of the dirigible to very limited and specialized uses.

The Wright brothers' flight came at a time when there was arduous and seemingly frenetic activity by other aerial pioneers. The conviction that humanity was on the brink of successful heavier-than-air flight encouraged widespread study and experiment.

The Englishman, George Cayley, has been called the father of aerial

*Fig 17-1.* The first heavier-than-air flight, December 17, 1903. (Courtesy The Library of Congress.)

navigation. His experiments during the first half of the nineteenth century, with small- and full-scale gliders, demonstrated the feasibility of flight in heavier-than-air craft. In 1866, another Englishman F. H. Wenham, in a report to the first meeting of the Aeronautical Society of Great Britain, did much to advance the state of knowledge of aerodynamics and the design of wings.

A German, Otto Lilienthal, made over 2000 successful glides during the last decade of the nineteenth century, some of which covered several hundred feet. In 1894, Octave Chanute, a successful civil engineer in the United States, who was spurred by Lilienthal's successful experiments, published a historical account of man's attempts to fly. He later designed gliders on his own, using his knowledge of bridge building to improve their structural design. Chanute developed a close friendship with Orville and Wilbur Wright and was a source of help and encouragement during their years of experiments at Kitty Hawk.

The airplane made a spectacular, if not decisive, contribution to the outcome of World War I, and after the war more serious attention was given to airplanes as an effective means of transporting people and goods.

Acceptance of air transportation was increased by Charles Lindbergh's dramatic solo flight from New York to Paris in 1927. Air transportation was nearing the threshold of maturity at the advent of World War II, and few can dispute its vital contribution to the war effort.

*Fig 17-2.* The Concorde supersonic transport aircraft. (Courtesy British Airways.)

## 17-2. GROWTH OF AVIATION ACTIVITY

The growth of aviation activity since the World War II has been unprecedented and extraordinary. United States air travel grew from 4.3 billion passenger miles in 1945 to 307.6 billion in 1986 [2]. From 1975 to 1985 intercity travel (in terms of passenger miles) by air carrier doubled, and travel by public air carriers now represents 17.0 percent of all intercity travel [2]. During that period, intercity travel by private aircraft increased from 11.4 to 12.7 billion passenger miles. Air freight has grown even faster than passenger travel, increasing an average of more than 9 percent annually since 1947 [2].

Trends in the factors that are accepted to be indicators of air travel activity (population, wealth, education, etc.) indicate that air travel will continue to grow in the years ahead. However, as the air transportation industry matures, it is likely that the rapid growth rate will stabilize to a lower steadier rate that is close to the rate of population growth, modified to some degree by socioeconomic changes [3].

## 17-3. NATURE OF THE PROBLEM

In the pages that follow, some of the problems and techniques of planning and designing an airport will be described. The remainder of this chapter will be devoted to a discussion of airport planning, site selection, runway orientation, obstruction clearance standards, and typical runway configurations. Chapter 18 will be concerned with the planning and design of the terminal area, while Chapter 19 will relate more specific design criteria and procedures for the runway and taxiway system.

Although these topics will be discussed separately, it should be recognized that there is an interaction between the various components of the problem, and that goals and requirements of one component of the problem will often conflict with the needs of another. The overall task is an iterative one which seeks an optimum airport plan and design that best satisfies the various needs, constraints, and controls of the system.

## 17-4. AIRPORT DEMAND

Before engineering plans and designs for a new airport or improvements to an existing facility are made, a study should be made to determine the extent of future needs for airport facilities. Airport demand studies include forecasts of annual, peak-day, and peak-hour volumes of passengers and aircraft, types of forecast usage (i.e., business, commercial, passenger and freight, pleasure, etc.), as well as factors concerning size of community to be served and economic and population growth trends.

While the detailed techniques for making these forecasts are beyond the scope of this text, the following general observations and relationships are given as a matter of interest.

Experience has shown that a community's aviation activity is primarily dependent on: (1) the population and population density of the city being served, (2) the economic character of the city, and (3) the proximity of the airport to other airports.

Annual passenger enplanements and aircraft operations are directly related to city size; however, two cities approximately equal in population may have significantly different airport needs due to differences in the social and economic character of the city. Generally speaking, industrial cities tend to generate much less aviation activity than centers of government, education, and finance.

Airport demand studies must take into account the existence of nearby airports which may compete for passengers and freight. Passengers may elect to drive great distances to a competing airport in order to take advantage of cheaper fares, nonstop connections, or a more attractive schedule of flights. For example, a survey of Columbus, Georgia travel agencies revealed that about 45 percent of airline passengers from that city elected to drive 100 mi to Atlanta's Hartsfield Airport in lieu of originating their air travel at the Columbus Metropolitan Airport.

A variety of approaches are used to forecast aviation demand. Such forecasts may be based on:

1. The judgment of the forecaster.
2. Predictions of knowledgeable persons from airline companies or the aircraft manufacturing industry.
3. National forecasts such as Reference 4.
4. Analytical models which relate air travel to some combination of variables such as air fares, travel time, city population, and social and economic characteristics of a city or pair of cities [5]. Such models are conceptually similar to the planning models for surface transportation described in Chapter 9.

Reference can also be made to the *National Airport System Plan,* a publication prepared and maintained by the Federal Aviation Administration which lists recommendations for future airport needs to promote the development of an adequate national system of airports. Airport development projects must first be included in the *National Airport System Plan* before being considered for federal-aid airport funds.

Once forecasts of annual air passengers are made, these must be converted to a peak hourly flow by using empirically based relationships, for example [3]:

Average monthly passengers = 0.08417 X annual passenger flow
Average daily passengers = 0.03226 X average monthly flow
Peak-day flow = 1.26 X average daily flow
Peak-hour flow = 0.0917 X peak daily flow

Peak-hour passenger flows can be used to estimate peak hourly aircraft movements using estimates of average passenger load per aircraft.

## 17-5. SELECTION OF AIRPORT SITE

Perhaps the single most important aspect of the planning and design of airports is the selection of an airport site. Mistakes made in this phase of the airport development program can result in the failure or early obsolescence of the facility.

Several contemporary trends have complicated the problem of selection of airport site:

1. Urban sprawl has occurred around most U.S. cities and has been accompanied by increasing scarcity of land and rising land costs.
2. Faster and larger aircraft have appeared requiring longer runways and more aircraft service space along with increased automobile parking and circulation space.
3. The requirements and desires of the public regarding air passenger services have become more elaborate and sophisticated.

Guidelines regarding procedures for making a study of alternate airport sites, along with a discussion of the major factors influencing site selection have been promulgated by the FAA [6]. While the critical investigation and location of airport sites is a responsibility of the airport sponsor,[1] FAA endorsement of the site is required if federal aid is contemplated. An airspace review by the FAA is required in any event.

After airport needs been established, the FAA recommends that the following procedure be followed in selecting an airport site.

*Desk Study of Area.* Before a field investigation is made, a great deal can be learned from a desk study which includes:

1. A review of existing comprehensive land use plans and other community and area plans.
2. An analysis of available wind data to determine the desired runway orientation. (This is discussed in Section 17-6).
3. A study of National Geodetic Survey quadrangle sheets, road maps, and aeronautical charts to select feasible sites for further evaluation.
4. A study of general land costs in the areas of interest.

In this study, special attention should be given to the location of other airports and land transportation facilities, obstructions, topographic features, and atmospheric peculiarities.

*Physical Inspections.* Actually, two physical field inspections should be made of the potential sites: preliminary and final.

1. The preliminary inspection should be made jointly by sponsor representatives of all existing airports and potential sites. After these inspections, sketches of the various sites should be made, as well as an overall sketch or small-scale map showing all of the sites under consideration. If possible, an aerial inspection of the various sites should be made and aerial photographs should be taken.
2. On the final inspection, those sites selected during the preliminary inspection are visited. If federal-aid funds are involved, an FAA airports representative should accompany the inspection group.

*Evaluation and Recommendations.* The FAA recommends that a final report be prepared including a rough cost estimate for each site. The report should list the sites in order of preference and indicate the advantages and disadvantages of each site. In cases involving federal aid, this report is submitted to the FAA for endorsement, and a site mutually agreeable to the sponsor

[1]The airport sponsor is usually a state, city, or other local body, although it may be a private organization or individual.

and the FAA is chosen. Written endorsement by the FAA does not imply a commitment of federal funds, but it is a necessary step in obtaining funds under the federal-aid airports program.

If no federal funds are requested, the FAA should be contacted for further information regarding the initiation of a request for airspace review.

There are at least 10 factors that should be considered when analyzing potential airport site:

1. Convenience to users.
2. Availability of land and land costs.
3. Design and layout of the airport.
4. Airspace obstructions.
5. Engineering factors.
6. Social and environmental factors.
7. Availability of utilities.
8. Atmospheric conditions.
9. Hazards duc to birds.
10. Coordination with other airports.

*Convenience to Users.* If it is to be successful, an airport must be conveniently located to those who use it. From this viewpoint, the airport ideally would be located near the center of most cities. The obvious problems of air obstructions and land costs rule out this possibility, and most cities have found it necessary to locate the airport several miles from the city center. In major U.S. cities, the average central-city-to-airport distance is about 10 mi.

Urban sprawl and increasing scarcity and costs of land have resulted in airports being located farther and farther from the city center.[2] At the same time, air speeds have increased with the result that an increasing percentage of air passengers spend more time in the ground transportation portion of the trip than in the air.[3] Yet is it known that the amount of airport use is very sensitive to the ratio of the ground travel time to the total journey time, and as this ratio increases, the air traffic can be expected to decrease precipitously.

In view of the airport user, travel time is a more important measure of convenience than distance. Thus, a relatively remote potential airport site

[2]An Arthur D. Little study of 11 cities indicated that in the case of the old airport, the average central city to airport distance was 7.0 mi. For these cities, the average distance from the central city to a proposed or new airport was 15.8 mi. The most remote new airport was Dulles International Airport, located 27.0 mi from downtown Washington, D.C. [7].

[3]The ratio of ground travel time to total trip time decreased with increases in the total length of trip. A study of the fifty most heavily traveled city-to-city routes indicated that trips in which the airport-to-airport mileage was 250 mi or less, 51 to 65 percent of the total trip time was spent in ground travel. For trips in which the airport-to-airport mileage was 1000 or more, this percentage ranged from 22 to 32 percent [7].

should not be ruled out if it is conveniently located to a major highway or other surface transportation facility.

*Availability of Land and Land Costs.* Vast acreages are required for major airports, and it is not uncommon for new airports in large cities to require 10,000 acres or more. However, the smallest general aviation airports may require less than 100 acres. Table 17-1 gives the minimum recommended land requirements for four classes of utility airports.[4] The amount of land required will depend on the length and number of runways, lateral clearances, and areas required for buildings, aprons, automobile parking and circulation, and so forth.

Since desirable airport site land is also in demand for other purposes, land costs are high, and real estate can be expected to appreciate with the planning and development of a new airport facility. Land costs for large air carrier airports often amount to hundreds of millions of dollars.

It is important that sufficient land be acquired for future expansion. Failure to do so could mean that a convenient and otherwise desirable airport site would have to be abandoned due to limitations in aircraft operations from an unexpandable runway or inadequate space for aircraft or passenger handling.

The requirements for large acreages, the need for convenient location, and the high land costs may lead to novel approaches to the problem of site selection. For example, consideration was given in the late 1970s to building an airport situated approximately 6 mi offshore on Lake Erie that would serve Cleveland [8]. That proposal was later found to be infeasible.

*Design and Layout of the Airport.* In considering alternate potential airport sites, the basic layout and design essentially should be constant. One should avoid making major departures from the desired layout and design to fit a particular site. In this connection, one consideration is especially important. Runways should be oriented so as to take advantage of prevailing winds, and variations in runway alignment from optimum orientation more than ±10 degrees normally should not be made. (See Section 17-6 for FAA standards on runway orientation.)

*Airspace and Obstruction.* To meet essential needs for in-flight safety, two requirements must be met:

1. Adjacent airports must be located so that traffic using one in no way interferes with traffic using the other. An airspace analysis should be made to ensure that this requirement is met. It is desirable that the assistance of the FAA be sought in conducting this analysis, especially when an airport is to be located in a highly developed terminal complex.
2. Physical objects such as towers, poles, buildings, mountain ranges, and so on must not penetrate navigable airspace. Criteria on "Objects

[4]The FAA defines a utility airport as one which serves general aviaiton aircraft of 12,500 lb or less.

*Table 17-1*
**Minimum Land Requirements for Utility Airports**

| *Runway Type* | *Runway Length in ft (m)* | *Landing Area (acres)* | *Approach Area (acres)* | *Building Area (acres[a])* | *Total Area (acres)* |
|---|---|---|---|---|---|
| Basic utility stage I | 2000 (600) | 23 | 21 | 8 | 52 |
| | 3000 (900) | 32 | 21 | 8 | 61 |
| | 4000 (1200) | 41 | 21 | 8 | 70 |
| | 5000 (1500) | 50 | 21 | 8 | 79 |
| Basic utility stage II | 2500 (750) | 27 | 21 | 12 | 60 |
| | 3500 (1050) | 36 | 21 | 12 | 69 |
| | 4500 (1350) | 45 | 21 | 12 | 78 |
| | 5500 (1650) | 54 | 21 | 12 | 87 |
| General utility stage I | 3000 (900) | 39 | 30 | 24 | 93 |
| | 4000 (1200) | 51 | 30 | 24 | 105 |
| | 5000 (1500) | 63 | 30 | 24 | 117 |
| | 6000 (1800) | 75 | 30 | 24 | 129 |
| General utility stage II | 3500 (1050) | 90 | 125 | 24 | 239 |
| | 4500 (1350) | 113 | 125 | 24 | 262 |
| | 5500 (1650) | 136 | 125 | 24 | 285 |
| | 6500 (1950) | 159 | 125 | 24 | 308 |

[a]These figures vary due to assumed higher degree of activity at the higher type of airport.

SOURCE: *Utility Airports*, FAA Advisory Circular AC 150/5300-4B, including Change 8, July 3, 1985.

Affecting Navigable Airspace," given in Federal Aviation Regulation, Part 77, should be consulted prior to beginning the site selection process. (See Section 17-7.)

*Engineering Factors.* An airport site should have fairly level topography and be free of mountains, hills, and so on. Further, the terrain should have sufficient slope that adequate drainage can be provided. Areas that require extensive rock excavation should be avoided, as should sites containing peat, muck, and otherwise undesirable foundation materials.

An adequate supply of aggregates and other construction materials should be located within a reasonable distance of the site.

A desirable airport site will be relatively free of timber, although a border of timber along the airport periphery may suppress undesirable noise.

*Social and Environmental Factors.* One of the most difficult social problems associated with airport location is that of noise. With the advent of the jet aircraft engine airport noise has worsened and, despite efforts by industry and government groups, the development of a quiet aircraft engine does not seem likely within the near future.

Airports are not good neighbors, and some control in the development of land surrounding an airport should be exercised. In selecting an airport site, proximity to residential areas, schools, and churches should be avoided, and the runways should be oriented so that these land uses do not fall in the immediate approach-departure paths.

As a result of several federal environmental laws passed during the period from 1966 to 1970, applicants for federal aid for airport construction must now prepare environmental impact statements for all airport developments that would significantly affect the quality of the environment. References 9 and 10 provide guidance for the preparation of environmental impact statements for U.S. airports. Similar guidelines have been published by the International Civil Aviation Organization for airports throughout the world [11].

*Availability of Utilities.* With rare exceptions (e.g., the Dulles International Airport, Washington, D.C.), airports must depend upon existing utilities. The site should be accessible to water, electrical service, telephones, gas lines, and so on, and these utilities should be of proper type and size.

*Atmospheric Conditions.* Peculiar atmospheric conditions, such as fog, smoke, snow, or glare may rule out the use of some potential airport sites.

*Hazards Due to Birds.* Aircraft impact with birds and bird ingestion into turbine engines have caused numerous air disasters. Airports should not be situated near bird habitats or natural preserves and feeding grounds. At certain potential sites, special work such as filling of ponds and closing of dumps may be required to ensure that birds will not present a hazard to aircraft flights.

*Coordination with Other Airports.* Studies of aviation activity in heavily populated metropolitan areas indicate that more than one major airport will be required in order to meet future air travel needs. Clearly, where two or more large cities are closely spaced (e.g., Seattle and Tacoma, Baltimore and Washington, D.C., etc), individual airport requirements must be determined in relation to the needs of the entire metropolitan area and each airport must be considered as a part of a total system.

## 17-6. RUNWAY ORIENTATION

Because of the obvious advantages of landing and taking off into the wind, runways are oriented in the direction of prevailing winds. Aircraft may not maneuver safely on a runway when the wind contains a large component at right angle to the direction of travel. The point at which this component (called the crosswind) becomes excessive will depend upon the size and operating characteristics of the aircraft. Recommended limiting crosswinds for aircraft operations are 12 mph for propeller-driven aircraft 12,500 lb or less, and 15 miles per hour for all other aircraft [12].

According to FAA standards, runways should be oriented so that aircraft may be landed at least 95 percent of the time without exceeding the allowable crosswinds. Where a single runway or set of parallel runways cannot be oriented to provide 95 percent wind coverage, one or more crosswind runways should also be provided.

*Wind Rose Method.* A graphical procedure utilizing a wind rose typically is used to determine the "best" runway orientation insofar as prevailing winds are concerned. (See Fig. 17-3.)

For U.S. airports, wind data are usually available from the National Oceanic and Atmospheric Administration, National Climatic Center, Asheville, North Carolina. A record of 10 consecutive years of wind observations should be utilized if available. If suitable weather records are not available accurate wind data for the area should be collected. (Another alternative would be to form a composite wind record from nearby wind-recording stations.) The wind data are arranged according to velocity, direction, and frequency of occurrence as shown by Table 17-2. This table indicates the percentage of time wind velocities within a certain range and from a given direction can be expected. For example, the table indicates that for the hypothetical site, northerly winds in the 4- to 15-mph range can be expected 4.8 percent of the time.

These data are plotted on wind rose by placing the percentages in the appropriate segment of the graph. On the wind rose, the circles represent wind velocity in miles per hour, and the radial lines indicate wind direction. The data from Table 17-2 have been plotted properly on Fig. 17-3.

The wind rose procedure makes use of a transparent template on which three parallel lines have been plotted. The middle line represents the runway center line and the distance between it and each of the outside lines is equal to the allowable crosswind component (e.g., 15 mph).

The following steps are necessary to determine the "best" runway orien-

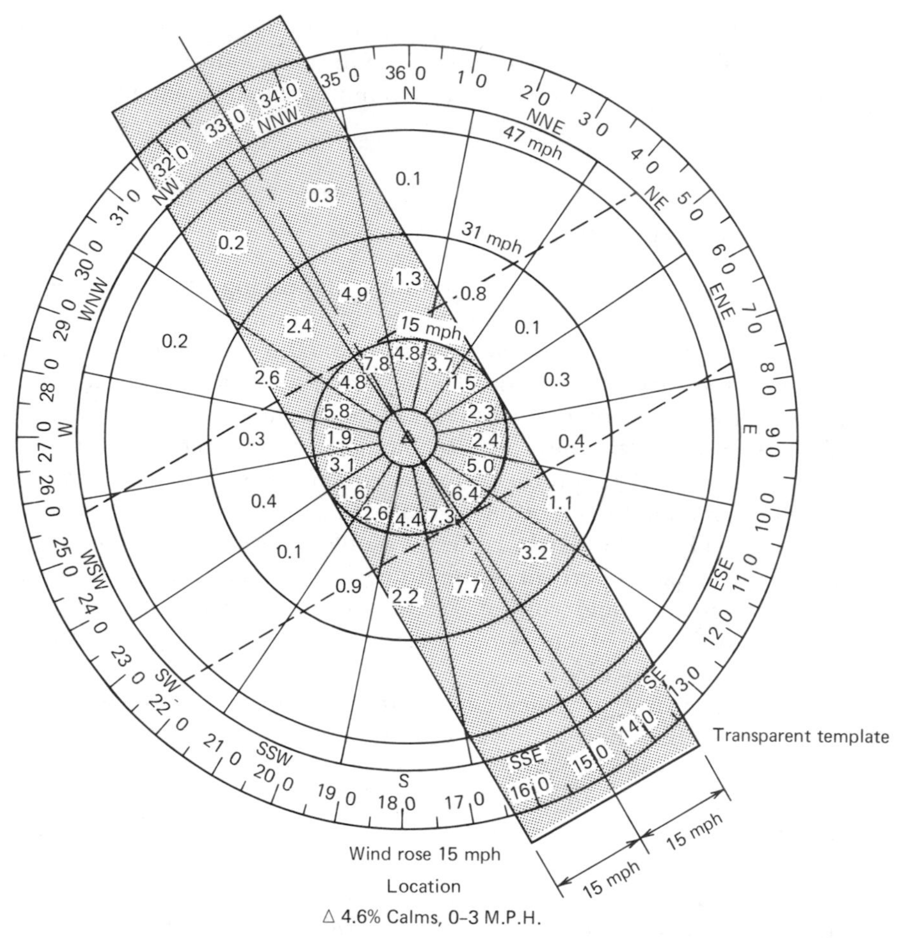


*Fig 17-3.* A typical wind rose.

tation and to determine the percentage of time that orientation conforms to the crosswind standards.

1. Place the template on the wind rose so that the middle line passes through the center of the wind rose.
2. Using the center of the wind rose as a pivot, rotate the template until the sum of the percentages between the outside lines is a maximum. When the template strip covers only a fraction of a segment, a corresponding fractional part of the percentage shown should be used.
3. Read the true bearing for the runway on the outer scale of the wind rose beneath the center line of the template. In the example the best orientation is 150°–330° or S 30° E, true.

*Table 17-2*
**Typical Wind Data**

| | Percentage of Winds | | | |
|---|---|---|---|---|
| *Wind Direction* | *4–15 mph* | *15–31 mph* | *31–47 mph* | *Total* |
| N | 4.8 | 1.3 | 0.1 | 6.2 |
| NNE | 3.7 | 0.8 | — | 4.5 |
| NE | 1.5 | 0.1 | — | 1.6 |
| ENE | 2.3 | 0.3 | — | 2.6 |
| E | 2.4 | 0.4 | — | 2.8 |
| ESE | 5.0 | 1.1 | — | 6.1 |
| SE | 6.4 | 3.2 | 0.1 | 9.7 |
| SSE | 7.3 | 7.7 | 0.3 | 15.3 |
| S | 4.4 | 2.2 | 0.1 | 6.7 |
| SSW | 2.6 | 0.9 | — | 3.5 |
| SW | 1.6 | 0.1 | — | 1.7 |
| WSW | 3.1 | 0.4 | — | 3.5 |
| W | 1.9 | 0.3 | — | 2.2 |
| WNW | 5.8 | 2.6 | 0.2 | 8.6 |
| NW | 4.8 | 2.4 | 0.2 | 7.4 |
| NNW | 7.8 | 4.9 | 0.3 | 13.0 |
| Calms | | 0–4 mph | | 4.6 |
| Total | | | | 100.0 |

4. The sum of percentages between the outside lines indicates the percentage of time that a runway with the proposed orientation will conform with cross wind standards.

It is noted that wind data are gathered and reported with true north as a reference, while runway orientation and numbering are based on the magnetic azimuth. (See Section 19-12.) The true azimuth obtained from the wind rose analysis should be changed to a magnetic azimuth by taking into account the magnetic variation[5] for the airport location. An easterly variation is subtracted from the true azimuth, and a westerly variation is added to the true azimuth.

[5]The magnetic variation can be obtained from aeronautical charts.

## 17-7. OBJECTS AFFECTING NAVIGABLE AIRSPACE

Part 77 of the of the Federal Aviation Regulations [13] establishes standards for determining obstructions in navigable airspace, sets forth the notice requirements of certain proposed construction or alteration, provides for aeronautical safety studies of obstructions, and provides for public hearings on the hazardous effect of proposed construction or alteration. FAA's regulations describe imaginary surfaces that define airspace that should be free of objects hazardous to air navigation. If an obstacle (for example, a building, radio tower, or mountain range) penetrates any of these surfaces, it is classified as an obstruction, requiring further study in collaboration with the FAA to determine if it would constitute a hazard to air navigation.

*Notification Requirements.* FAR Part 77 [13] sets forth a procedure by which the FAA must be notified of proposed construction or alteration activities which might affect navigable airspace. Depending on the effect that the obstacle might have on air navigation, the FAA is authorized to convene formal hearings and, on the basis of evidence presented at such hearings, to determine appropriate measures to be applied for continued safety of air navigation. Such measures include requiring that an obstacle be removed or be marked and lighted properly [14], or that the plans for construction or alteration be modified or canceled.

Generally speaking, an airport sponsor must notify the FAA of any construction or alteration of any airport that is available for public use and is in the Airport Directory of the current Airman's Information Manual [15]. Similarly, the FAA must be notified of new civilian airports that are under construction and are expected to become available for public use and also must be notified of changes to airports operated by one of the military services.

In addition, an airport sponsor must notify the FAA of any construction or alteration:

- of more than 200 ft in height above the ground level at the site;
- that penetrates the surfaces shown in Fig. 17-4;
- within an instrument approach area where available information indicates that an obstacle might exceed a standard of Subpart C of FAR Part 77, which gives standards for evaluating obstructions to air navigation.

The height of highways, railroads, and waterways must be adjusted upward when determining if the notification requirements are met. The adjustments are:

| | |
|---|---|
| For Interstate highways | 17 ft |
| For other public roadways | 15 ft |
| For railroads | 23 ft |

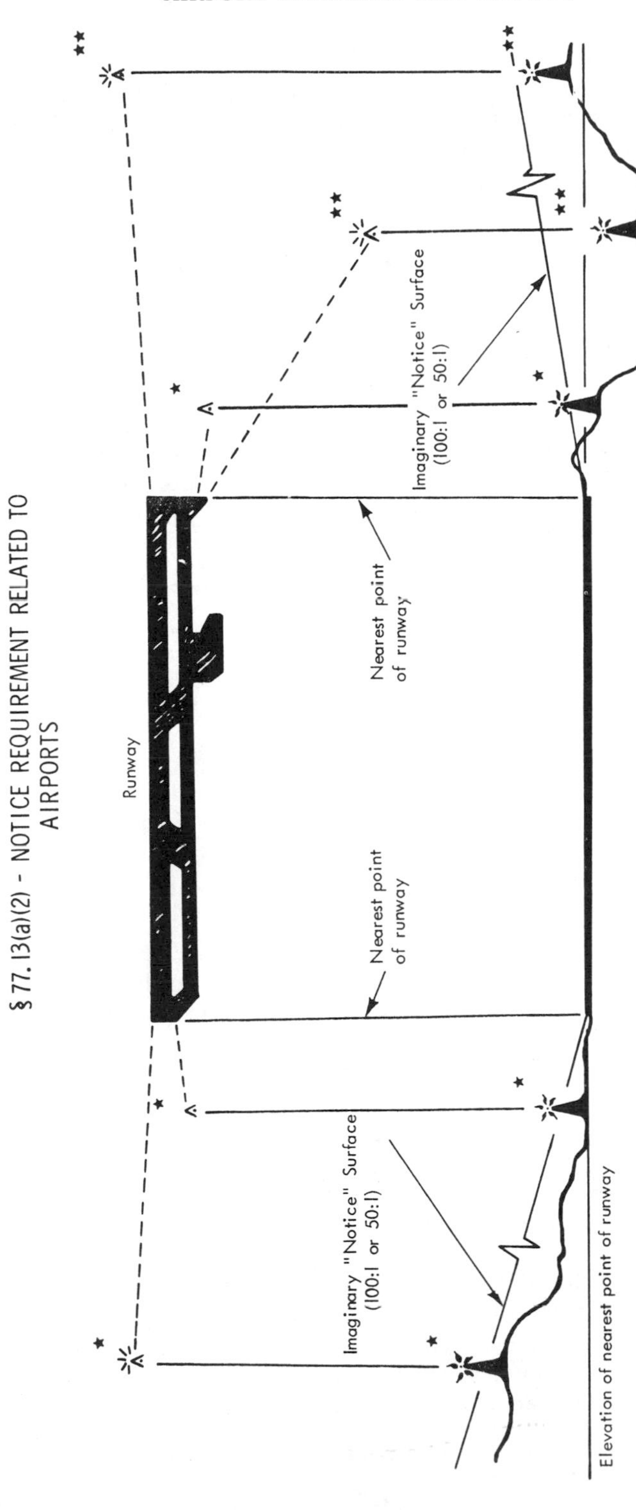


*Fig 17-4.* FAA notice requirements for airport construction or alteration. (Source: *Federal Aviation Regulations, Part 77,* 1975.)

For private roads and waterways, the height usually must be adjusted by the height of the highest vehicle that will use the traverse way.

*Obstruction Standards.* According to Subpart C of FAR Part 77, an object is an obstruction to air navigation if it is of greater height than any of the following heights of surfaces:

1. A height of 500 ft above ground level at the site of the object.
2. A height that is 200 ft above ground level or above the established airport elevation, whichever is higher, within 3 nautical miles of the established reference point of an airport excluding heliports, with its longest runway more than 3200 ft in actual length, and that height increases in the proportion of 100 ft for each additional nautical mile of distance from the airport up to a maximum of 500 ft. (See Fig. 17-5.)
3. A height within a terminal obstacle clearance area, including an initial approach segment, a departure area, and a circling approach area, that would result in the vertical distance between any point on the object and an established minimum instrument flight altitude within that area or segment to be less than the required obstacle clearance.
4. A height within an en route obstacle clearance area, including turn and termination areas, of a federal airway or approved off-airway route, that would increase the minimum obstacle clearance altitude.
5. The surface of a takeoff and landing area of an airport or any imaginary surface described in the following paragraphs. (See Fig. 17-6.)

*Civil Airport Imaginary Surfaces.* The FAA has established imaginary surfaces as a means of checking the effect of objects in the vicinity of airports and approaches. These surfaces are illustrated in Fig. 17-6. Their dimensions are described in the following paragraphs, quoted from Reference 13, with but minor editorial changes.

> The size of each such imaginary surface is based on the category of each runway according to the type of approach available or planned for that runway. The slope and dimensions of the approach surface applied to each end of a runway are determined by the most precise approach existing or planned for that runway end.
>
> (a) Horizontal surface—a horizontal plane 150 ft above the established airport elevation,[6] the perimeter of which is constructed by swinging arcs of specified radii from the center of each end of the primary surface of each runway of each airport and connecting the adjacent arcs by lines tangent to those arcs. The radius of each arc is:
>
> (1) 5000 ft for all runways designated as utility or visual;
> (2) 10,000 ft for all other runways.
>
> The radius of the arc specified for each end of a runway will have the same arithmetical value. That value will be the highest determined for either end of the runway. When a 5000-ft arc is encompassed by tangents connecting two adjacent

[6]The established airport elevation is usually taken to be the highest elevation on the runway system.

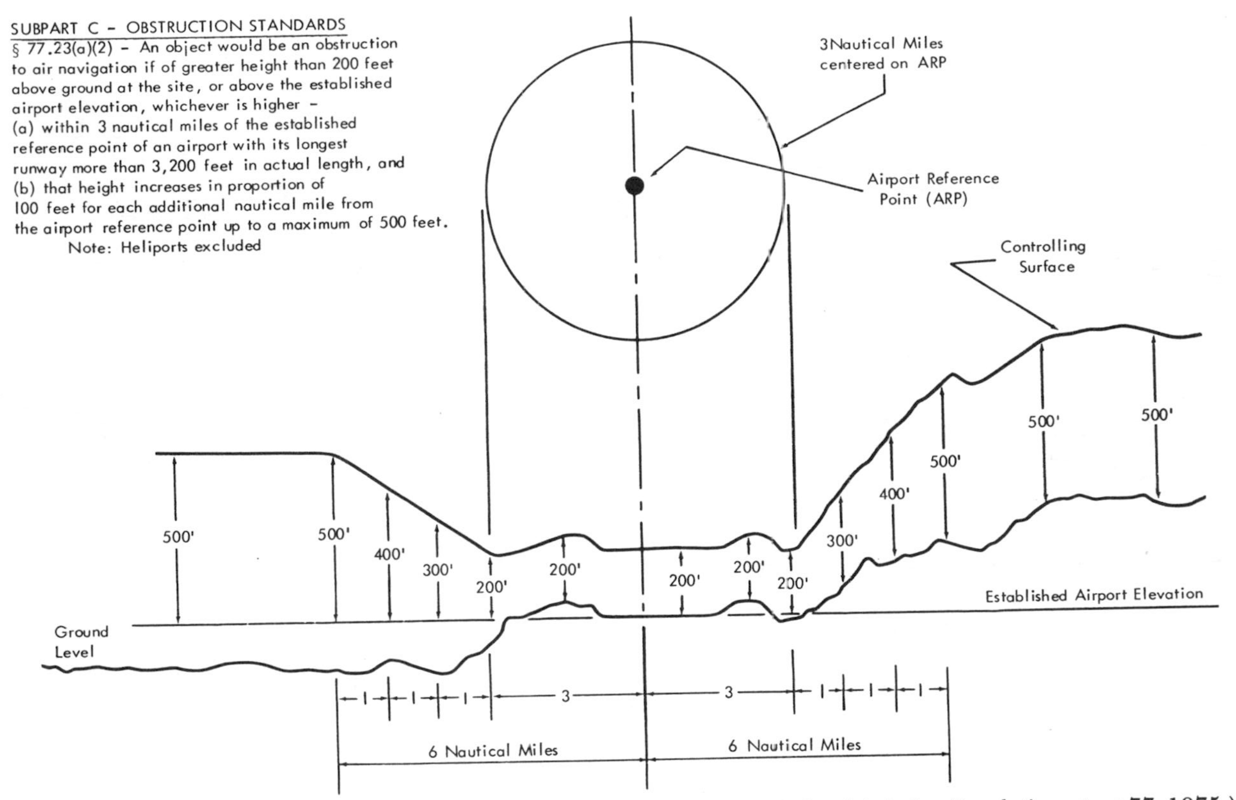


*Fig 17-5.* Obstruction standards in the vicinity of airports. (Source: *Federal Aviation Regulations, part 77,* 1975.)

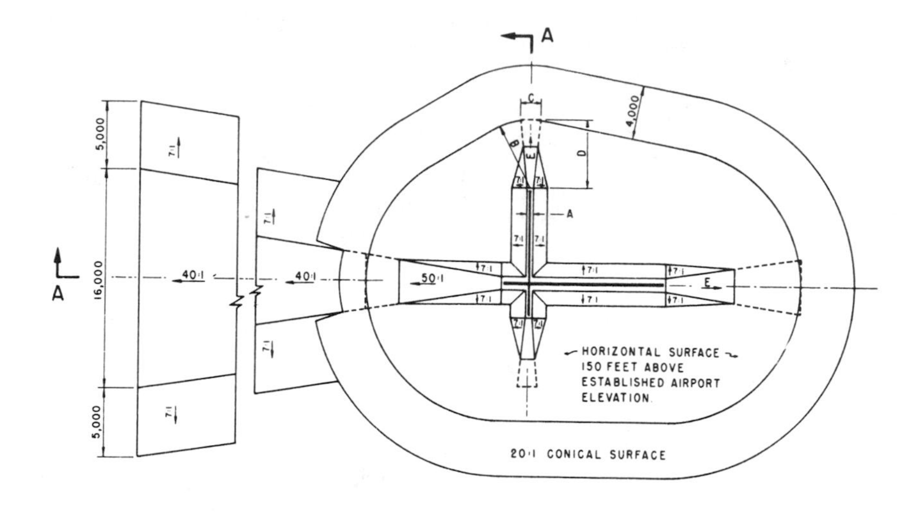


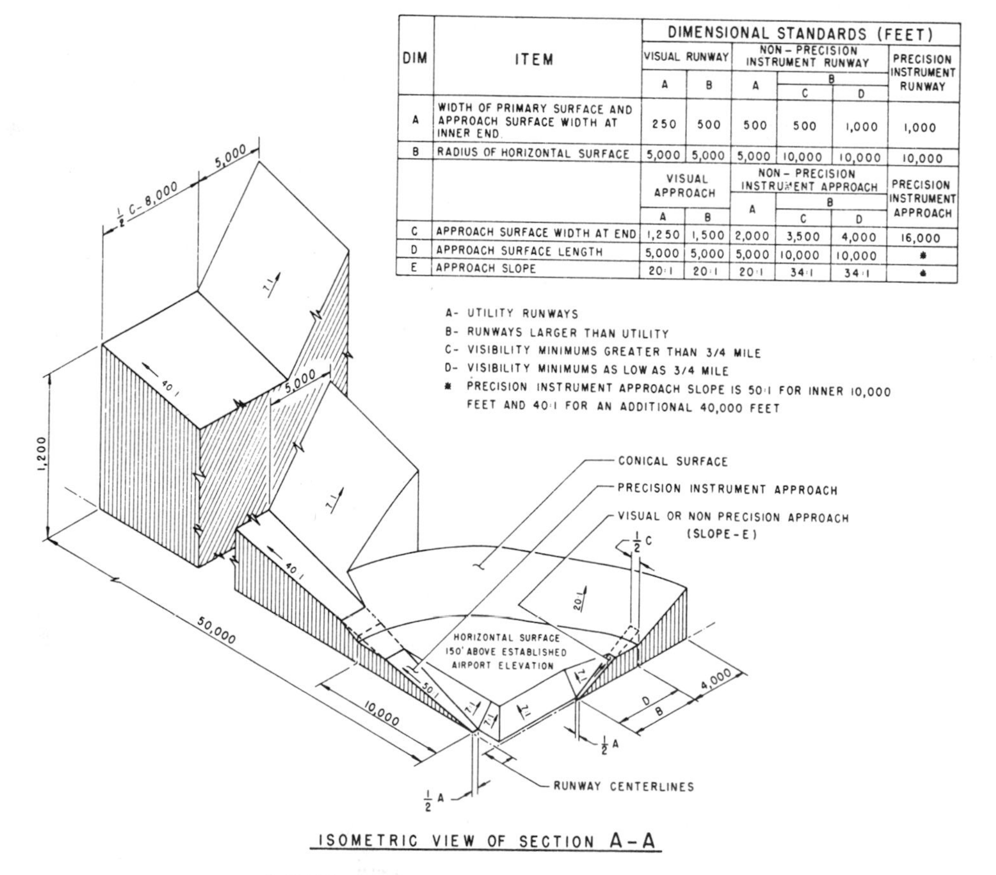

| DIM | ITEM | DIMENSIONAL STANDARDS (FEET) | | | | | |
|---|---|---|---|---|---|---|---|
| | | VISUAL RUNWAY | | NON-PRECISION INSTRUMENT RUNWAY | | | PRECISION INSTRUMENT RUNWAY |
| | | A | B | A | B | | |
| | | | | | C | D | |
| A | WIDTH OF PRIMARY SURFACE AND APPROACH SURFACE WIDTH AT INNER END | 250 | 500 | 500 | 500 | 1,000 | 1,000 |
| B | RADIUS OF HORIZONTAL SURFACE | 5,000 | 5,000 | 5,000 | 10,000 | 10,000 | 10,000 |
| | | VISUAL APPROACH | | NON-PRECISION INSTRUMENT APPROACH | | | PRECISION INSTRUMENT APPROACH |
| | | A | B | A | B | | |
| | | | | | C | D | |
| C | APPROACH SURFACE WIDTH AT END | 1,250 | 1,500 | 2,000 | 3,500 | 4,000 | 16,000 |
| D | APPROACH SURFACE LENGTH | 5,000 | 5,000 | 5,000 | 10,000 | 10,000 | * |
| E | APPROACH SLOPE | 20:1 | 20:1 | 20:1 | 34:1 | 34:1 | * |

A- UTILITY RUNWAYS
B- RUNWAYS LARGER THAN UTILITY
C- VISIBILITY MINIMUMS GREATER THAN 3/4 MILE
D- VISIBILITY MINIMUMS AS LOW AS 3/4 MILE
* PRECISION INSTRUMENT APPROACH SLOPE IS 50:1 FOR INNER 10,000 FEET AND 40:1 FOR AN ADDITIONAL 40,000 FEET



§ 77.25 CIVIL AIRPORT IMAGINARY SURFACES

*Fig 17-6.* Civil airport imaginary surfaces.

10,000-ft arcs, the 5000-ft arc shall be disregarded on the construction of the perimeter of the horizontal surface.

(b) Conical surface—a surface extending outward and upward from the periphery of the horizontal surface at a slope of 20 to 1 for a horizontal distance of 4000 ft.

(c) Primary surface—a surface longitudinally centered on a runway. When the runway has a specially prepared hard surface, the primary surface extends 200 ft beyond each end of that runway: but when the runway has no specially prepared hard surface, or planned hard surface, the primary surface ends at each end of that runway. The elevation of any point on the primary surface is the same as the elevation of the nearest point on the runway centerline. The width of a primary surface is:

(1) 250 ft for utility runways having only visual approaches.

(2) 500 ft for utility runways having nonprecision instrument approaches.

(3) For other than utility runways the width is:

(i) 500 ft for visual runways having only visual approaches.

(ii) 500 ft for nonprecision instrument runways having visibility minimums greater than three-fourths statute mile.

(iii) 1000 ft for a nonprecision instrument runway approach having a nonprecision instrument with a visibility minimums as low as three-fourths of a statute mile, and for precision instrument runways.

The width of the primary surface of a runway will be that width prescribed in this section for the most precise approach existing or planned for either end of that runway.

(d) Approach surface—a surface longitudinally centered on the extended runway center line and extending outward and upward from each end of the primary surface. An approach surface is applied to each end of each runway based upon the type of approach available or planned for that runway end.

(1) The inner edge of the approach surface is the same width as the primary surface and it expands uniformly to a width of:

(i) 1250 ft for that end of a utility runway with only visual approaches;

(ii) 1500 ft for that end of a runway other than a utility runway with only visual approaches;

(iii) 2000 ft for that end of a utility runway with a nonprecision instrument approach;

(iv) 3500 ft for that end of a nonprecision instrument runway other than utility, having visibility minimums greater than three-fourths of a statute mile;

(v) 4000 ft for that end of a nonprecision instrument runway, other than utility, having a nonprecision instrument approach with visibility minimums as low as three-fourths statute mile; and

(vi) 16,000 ft for precision instrument runways.

(2) The approach surface extends for a horizontal distance of:

(i) 5000 ft at a slope of 20 to 1 for all utility and visual runways;

(ii) 10,000 ft at a slope of 34 to 1 for all nonprecision instrument runways other than utility; and,

(iii) 10,000 ft at a slope of 50 to 1 with an additional 40,000 ft at a slope of 40 to 1 for all precision instrument runways.

(3) The outer width of an approach surface to an end of a runway will be that

width prescribed in this subsection for the most precise approach existing or planned for that runway end.

(e) Transitional surface—these surfaces extend outward and upward at right angles to the runway centerline and the runway centerline extended at a slope of 7 to 1 from the sides of the primary surface and from the sides of the approach surfaces. Transitional surfaces for those portions of the precision approach surface that project through and beyond the limits of the conical surface, extend a distance of 5000 ft measured horizontally from the edge of the approach surface and at right angles to the runway centerline.

## 17-8. RUNWAY CAPACITY

Runway capacity, which is normally the determining element of airport capacity, refers to the ability of a runway system to accommodate aircraft operations (i.e., landings and takeoffs). It is expressed in operations per unit time, typically operations per hour or operations per year. Airport capacity research [16] has provided procedures for the determination of the *ultimate* or *saturation* airfield capacities. Saturation capacity is based on the assumption of a continuous backlog of aircraft waiting to take off and land. It does not represent a desirable mode of operation but can be useful in comparing alternative layouts and designs.

Previously, the FAA has employed a concept of practical capacity, an empirical-based measure that corresponds to a specified "reasonable" or "tolerable" delay. The preferred measure and that employed in this chapter, is the ultimate or saturation capacity, the maximum number of aircraft operations that can be handled during a given period under conditions of continuous demand.

*Factors That Affect Capacity.* Fundamentally, the capacity of a runway system is limited by air traffic control measures, especially the need to dictate a minimum safe allowable separation between aircraft during landing and takeoff.

At a given airport the capacity of a runway system depends to a considerable extent on the types and relative percentages of various groups of aircraft (termed aircraft mix). Generally, the greater the percentage of large aircraft, the smaller the runway capacity.

Runway capacity also varies widely with weather conditions, and it is necessary to distinguish between operations in VFR and IFR conditions. Landings and takeoffs at an airport may be made under visual flight rules (VFR) or instrument flight rules (IFR). It will be recalled from Section 5-25 that VFR operations are made in good weather conditions and the aircraft is operated by visual reference to the ground. IFR operations are made in periods of inclement weather and poor visibility, and under these conditions, positive traffic control is maintained by radar and other electronic devices. Runway capacity under IFR conditions is normally less than under VFR conditions.

Finally, a major determinant of airport capacity is the overall layout and design of the system. Foremost in this group of factors is runway configuration, discussed in Section 17-9.

*Procedures for Estimating Runway Capacities.* A discussion of detailed procedures for estimating capacities is beyond the scope of this book. Such matters are treated extensively in other publications [17, 18].

The FAA has published approximate hourly and annual service volumes (ASV) for various runway configurations. Some relatively simple runway configurations are shown by Table 17-3 along with hourly capacities and ASV.

In the table, the aircraft mix is expressed in terms of four aircraft "classes":

Class A: small single-engine aircraft, 12,500 lb or less.
Class B: small multiengine aircraft, 12,500 lb or less and Learjets.
Class C: large aircraft, 12,500 lb and up to 300,000 lb.
Class D: heavy aircraft, more than 300,000 lb.

The "mix index" is determined by the percentages of aircraft in classes C and D.

$$\text{Mix index} = (\text{percent aircraft in class C}) + 3\,(\text{percent aircraft in class D})$$

It should be noted that the data given in Table 17-3 are based on average traffic and airport conditions. Procedures for making adjustments to the capacity values for local variations in traffic, airport layout, and weather are given by Reference 17. The capacities shown in Table 17-3 are recommended only for long-range planning purposes.

## 17-9. RUNWAY CONFIGURATION

Inherent in the layout of an airport is the need to arrange a given runway efficiently in relation to other runways and service facilities such as the terminal building, aprons, hangars, and other airport buildings. As was indicated earlier, runway configuration is a principal factor affecting capacity. Practically speaking, it is the only factor an airport planner can change and bring about an increase in capacity to serve future demand.

Typical runway configurations are shown schematically by Fig. 17-7 and in the left margin of Table 17-3.

The simplest runway configuration is a single-runway system, which has an hourly capacity of 51 to 98 operations in VFR conditions and 50 to 59 operations under IFR conditions.

Frequently, a second runway is added to take advantage of a wider range of wind direction. This system has a higher capacity than a single-runway

*Table 17-3*

**Airport Capacities for Long-Range Planning Purpose**

| No. | Runway-Use Configuration | Mix Index (Percent) (C + 3D) | Hourly Capacity (Operations) VRF | IFR | Annual Service Volume (Operations/yr) |
|---|---|---|---|---|---|
| 1. |  | 0 to 20 | 98 | 59 | 230,000 |
| | | 21 to 50 | 74 | 57 | 195,000 |
| | | 51 to 80 | 63 | 56 | 205,000 |
| | | 81 to 120 | 55 | 53 | 210,000 |
| | | 121 to 180 | 51 | 50 | 240,000 |
| 2. | 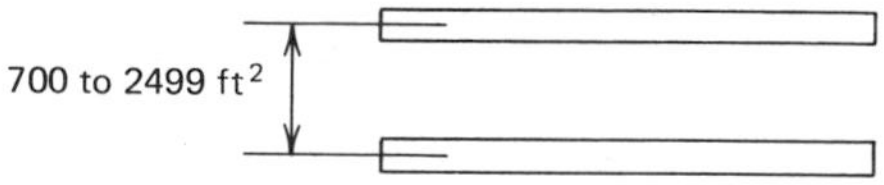 700 to 2499 ft² | 0 to 20 | 197 | 59 | 355,000 |
| | | 21 to 50 | 145 | 57 | 275,000 |
| | | 51 to 80 | 121 | 56 | 260,000 |
| | | 81 to 120 | 105 | 59 | 285,000 |
| | | 121 to 180 | 94 | 60 | 340,000 |
| 3. | 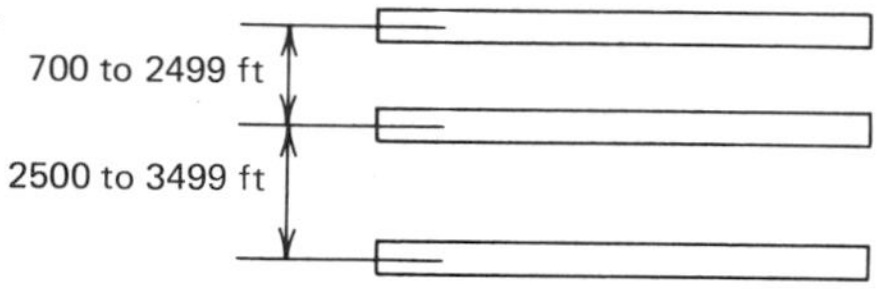 700 to 2499 ft; 2500 to 3499 ft | 0 to 20 | 295 | 62 | 385,000 |
| | | 21 to 50 | 219 | 63 | 310,000 |
| | | 51 to 80 | 184 | 65 | 290,000 |
| | | 81 to 120 | 161 | 70 | 315,000 |
| | | 121 to 180 | 146 | 75 | 385,000 |
| 4. | 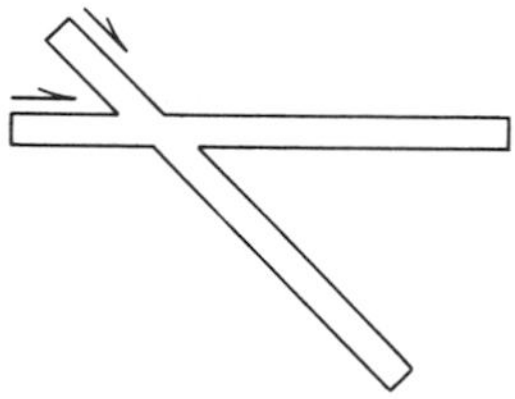 | 0 to 20 | 98 | 59 | 230,000 |
| | | 21 to 50 | 77 | 57 | 200,000 |
| | | 51 to 80 | 77 | 56 | 215,000 |
| | | 81 to 120 | 76 | 59 | 225,000 |
| | | 121 to 180 | 72 | 60 | 265,000 |
| 5. | 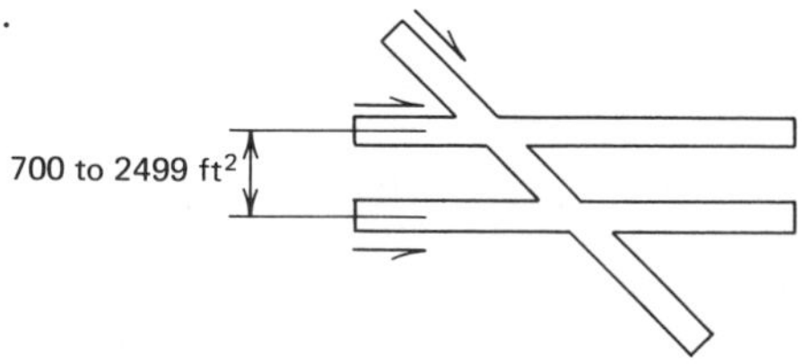 700 to 2499 ft² | 0 to 20 | 197 | 59 | 355,000 |
| | | 21 to 50 | 145 | 57 | 275,000 |
| | | 51 to 80 | 121 | 56 | 260,000 |
| | | 81 to 120 | 105 | 59 | 285,000 |
| | | 121 to 180 | 94 | 60 | 340,000 |

*continued*

*Table 17-3*

**Airport Capacities for Long-Range Planning Purpose —continued**

| No. | Runway-Use Configuration | Mix Index (Percent) (C + 3D) | Hourly Capacity (Operations) VRF | Hourly Capacity (Operations) IFR | Annual Service Volume (Operations/yr) |
|---|---|---|---|---|---|
| 6. | 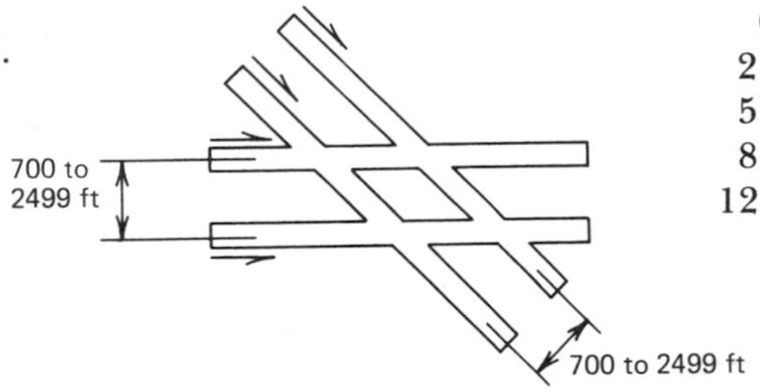 700 to 2499 ft; 700 to 2499 ft | 0 to 20 | 197 | 59 | 355,000 |
| | | 21 to 50 | 147 | 57 | 275,000 |
| | | 51 to 80 | 145 | 56 | 270,000 |
| | | 81 to 120 | 138 | 59 | 295,000 |
| | | 121 to 180 | 125 | 60 | 350,000 |
| 7. | 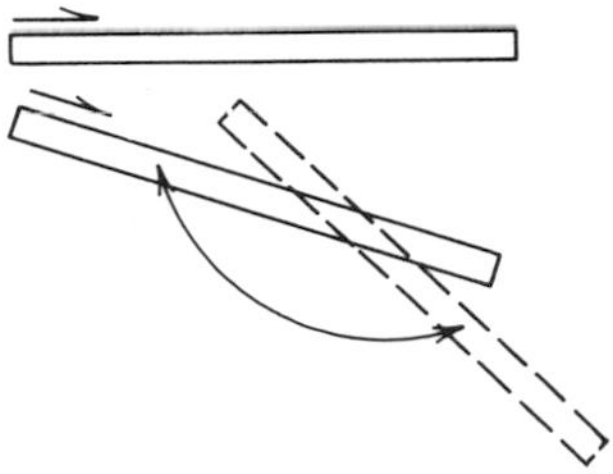 | 0 to 20 | 150 | 59 | 270,000 |
| | | 21 to 50 | 108 | 57 | 225,000 |
| | | 51 to 80 | 85 | 56 | 220,000 |
| | | 81 to 120 | 77 | 59 | 225,000 |
| | | 121 to 180 | 73 | 60 | 265,000 |
| 8. | 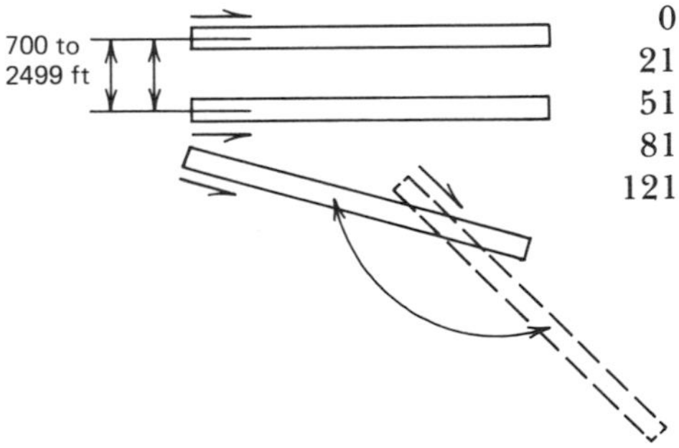 700 to 2499 ft | 0 to 20 | 295 | 59 | 385,000 |
| | | 21 to 50 | 210 | 57 | 305,000 |
| | | 51 to 80 | 164 | 56 | 275,000 |
| | | 81 to 120 | 146 | 59 | 300,000 |
| | | 121 to 180 | 129 | 60 | 355,000 |

[a]Staggered threshold adjustments may apply

SOURCE: *Airport Capacity and Delay,* AC 150/5060.5, Federal Aviation Administration.

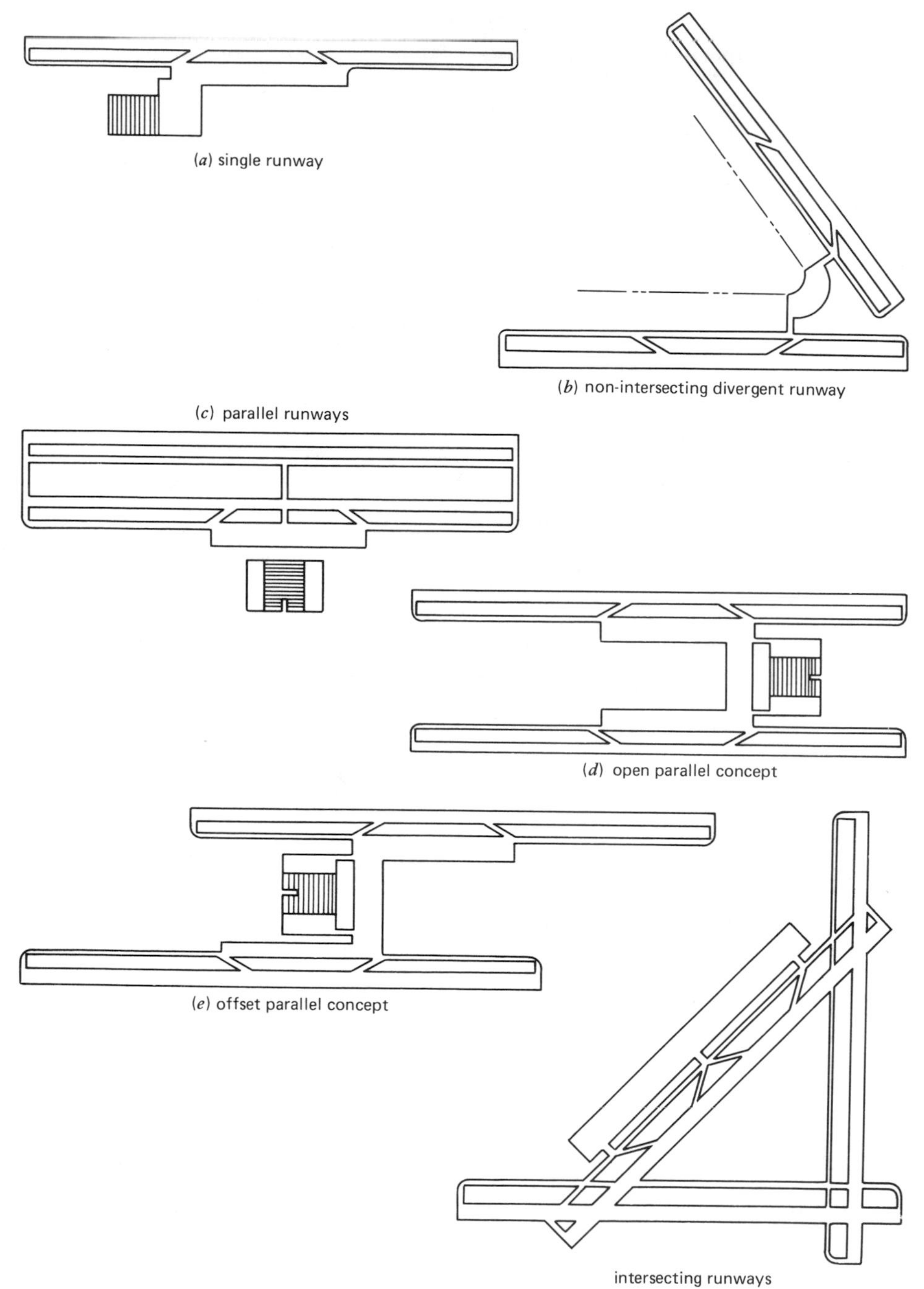


*Fig 17-7.* Typical airport configurations. (*a*) Single runway. (*b*) Nonintersecting divergent runways. (*c*) Parallel runways. (*d*) Open parallel concept. (*e*) Offset parallel concept. (*f*) Intersecting runways.

system, provided winds are not strong. In conditions of high winds and poor visibility, this system operates as a single runway.

An airport's runway capacity can be increased by adding a second parallel runway. With a separation of at least 700 ft, simultaneous landings in the same direction[7] may be made on the parallel runways under VFR conditions with a resulting capacity of 94 to 197 operations per hour. With a runway separation of at least 4300 feet, simultaneous operation may be made under IFR conditions, and the IFR capacity of this system is 99 to 119 operations per hour.

The parallel runway system shown in Fig. 17-7*c* has the disadvantage of requiring aircraft using the outboard runway to taxi across the runway adjacent to the terminal area. This disadvantage may be overcome by placing the terminal facilities between the two runways as shown by Fig. 17-7*d*.

In a location where there are prevailing winds from one direction a large percentage of the time, the parallel runways may be staggered or placed in tandem as shown by Fig. 17-7*e*. This makes it possible to reduce taxiway distances by using one runway exclusively for takeoff operations and the other runway for landings. This configuration, however, requires a great deal of land.

Large airports may require three or more runways. The best configuration for a multiple runway system will depend on the minimum spacing required for safety, prevailing wind directions, topographic features of the airport site, shape and amount of available space, and the space requirements for aprons, the terminal and other buildings.

## PROBLEMS

**1.** Using the wind data given here, construct a wind rose and indicate what would be the best orientation for a runway based on these prevailing winds.

| | *Percentage of Winds* | |
|---|---|---|
| *Wind Direction* | *15–31 mph* | *31–47 mph* |
| N | 4.3 | 0.1 |
| NNE | 1.3 | — |
| NE | 1.3 | 0.3 |
| ENE | 0.7 | 0.4 |
| E | 3.1 | 0.1 |
| ESE | 5.4 | 0.2 |
| SE | 4.2 | 0.5 |

*continued*

[7]Simultaneous VFR operations in opposite directions on parallel runways require 1400-ft distance between runway center lines during daylight hours and 2800-ft distance during periods of darkness [12].

| | *Percentage of Winds* | |
|---|---|---|
| *Wind Direction* | *15–31 mph* | *31–47 mph* |
| SSE | 1.3 | — |
| S | 0.4 | 0.1 |
| SSW | 3.1 | 0.5 |
| SW | 1.8 | 0.7 |
| WSW | 0.6 | 0.4 |
| W | 2.5 | 0.8 |
| WNW | 2.0 | — |
| NW | 0.2 | 0.2 |
| NNW | 1.0 | 0.5 |
| Percent of winds at 4–15 mph = 58.2 | | |
| Percent of winds at 0–4 mph = 3.8 | | |

2. Obtain for a nearby airport wind data similar to these given in Problem 1. Construct a wind rose using SI (metric) units and determine optimum runway orientation based on prevailing winds. How does this orientation compare with the existing airport's runway(s)?

3. A utility airport is being planned to serve a city of 35,000 people. The visual runway is to be 3600 ft long. Indicate whether or not the following objects will be considered obstructions to air navigation by the FAA.

   a. A 220-ft radio tower that is not in the landing approach, located 3.0 mi from the airport reference point. The ground elevation at the tower is 25 ft higher than the established airport elevation.

   b. A planed 80-ft high office building within the landing approach $\frac{1}{2}$ mi from the end of the runway.

4. A new air carrier airport is being planned to serve a city of 300,000 people. The airport will have dual 8000-ft precision instrument runways. Indicate whether or not the following objects constitute an obstruction to air navigation. The established airport elevation is 1220 ft.

   a. A railroad within the approach path located 1 mi from the end of the runway with an elevation of 1310 ft. Assume the elevation of the end of the runway is 1220 ft.

   b. A 298-ft radio tower not in the approach path located 3.5 nm from the airport reference point. The ground elevation at the tower is 1275 ft.

5. Estimate the capacity of dual close parallel runways that are to serve the traffic mix shown here under VFR and IFR conditions.

| | |
|---|---|
| 18 percent | 4-engine jet aircraft |
| 38 percent | 2-engine jet and 4-engine piston aircraft |
| 22 percent | executive jet aircraft |
| 22 percent | single-engine piston aircraft |

## REFERENCES


1. Josephy, Alvin M., Jr. (ed.), *The American Heritage History of Flight,* American Heritage, New York, 1962.
2. *Transportation in America,* Fifth Edition, Transportation Policy Associates, Washington, D.C. 1987.
3. Ashford, Norman, and Paul H. Wright, *Airport Engineering,* Second Edition, John Wiley, New York, 1984.
4. *Terminal Area Forecasts, Fiscal Years 1981–1992,* FAA-APO-80-10, Federal Aviation Administration, Washington, D.C., February, 1981.
5. Brown, S.L., and W.S. Watkins, *Highway Research Record 213,* Transportation Research Board, Washington, D.C., 1968.
6. *Airport Master Plans,* FAA Advisory Circular AC 150/5070-6A, June 1985.
7. Deem, Warren H., and John S. Reed, *Airport Land Needs,* Arthur D. Little, Inc., Communication Service Corporation, Washington, D.C.
8. Crawford, Herbert R., "Lake Erie Airport Study," *Transportation Journal of ASCE,* Vol. 103, No. TE2, March 1977.
9. *Policies and Procedures for Considering Environmental Impacts,* FAA Order 1050.1D, Federal Aviation Administration, Washington, D.C., December 21, 1983.
10. *Airport Environmental Handbook,* FAA Order 5050.4A, Federal Aviation Administration, Washington, D.C., October 8, 1985.
11. *Airport Planning Manual,* Part 2: *Land Use and Environmental Control,* First Ed, International Civil Aviation Organization, Montreal, 1977.
12. *Airport Design Standards—Transport Airports,* FAA Advisory Circular 150/5300-12, February 28, 1983.
13. *Objects Affecting Navigable Airspace,* Federal Aviation Regulations, Part 77, U.S. Department of Transportation, January 1975.
14. *Obstruction Marking and Lighting,* FAA Advisory Circular 70/7460-1G, October 22, 1985.
15. *Airman's Information Manual: Basic Flight Information and ATC Procedures,* Federal Aviation Administration, Washington, D.C., September 26, 1985.
16. Harris, Richard M., *Models for Runway Capacity Analysis,* Report No. MTR-4102, Rev. 2, The Mitre Corporation, Washington, D.C. May 1974.
17. *Airport Capacity and Delay,* FAA Advisory Circular 150/5060.5, September 23, 1983.
18. *Techniques for Determining Airport Airside Capacity and Delay,* prepared for the FAA by Douglas Aircraft Company et al., Report No. FAA-RD-74-124, June 1976.

# Chapter 18

# The Airport Passenger Terminal Area

## 18-1. INTRODUCTION

This chapter will present planning procedures and design criteria for the airport passenger terminal area. For purposes of this chapter, the airport passenger terminal area generally consists of that portion of the airport other than the landing and takeoff areas. It includes the automobile parking lots, aircraft parking aprons, the passenger terminal building, and facilities for interterminal and intraterminal transportation.

The importance of a well-conceived airport terminal area design can be seen by considering the numerous and varied component movements that a typical airline passenger makes (see Fig. 18-1). A passenger leaves his or her origin and travels to the airport by automobile or one of a variety of public travel modes. From the automobile parking lot or vehicle-unloading platform, the passenger and the baggage move to the ticket and passenger service counter by walking, by moving sideways, or by other means. From the gate-loading position, passengers usually walk the short distance to the plane. Additional travel time is involved as the aircraft taxis to the runway holding apron where it waits for control tower clearance to take off. This procedure is essentially reversed at the destination end of the trip. For airline passengers who make enroute stops or transfers, the movements are more numerous and complex.

Each component movement in a typical trip involves possibilities of congestion and delay. It follows that each service facility within the terminal area must be planned carefully and designed to accommodate peak-hour traffic volumes if unacceptable delays are to be avoided. It is also apparent that an integrated layout and design of the airport terminal area is required to provide a smooth uninterrupted flow of people, baggage, and freight. This design must be sufficiently flexible to allow for orderly expansion of service areas without prohibitive costs.

Terminal facilities vary widely in size, design, and layout depending primarily upon the airport type and size and the volume and nature of air traffic.

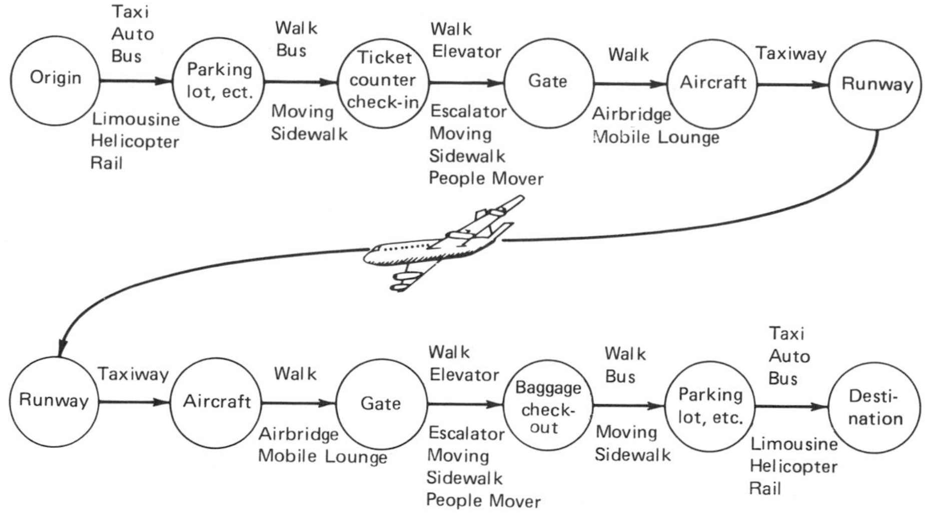


*Fig 18-1.* A typical air trip.

These facilities at a utility airport, for example, may consist only of two small buildings: (1) a fixed-based operators' (FBO) building which provides space for commercial activities such as aircraft maintenance and repair, air charter, and so on, and (2) an administration building to accommodate pilots, passengers, and visitors and to house the airport manager's office. At the other end of the spectrum there are cities making plans for airport facilities to accommodate more than 60 million annual passengers by providing elaborate terminal complexes with as many as 100 wide bodied parking positions for commercial jet aircraft.

The primary focus of this chapter will be on hub airports (see Table 1-4). Additional information for planning and development of utility airports and nonhub airports is given by References 1 and 2, respectively.

## 18-2. THE AIR TERMINAL PLANNING AND DESIGN PROCESS

The design process for a passenger terminal at a commercial carrier airport is a complex one involving at least four organizations:

*The Airport Owner.* Commercial carrier airports typically are owned by a municipality or airport authority. The owner is particularly concerned with the financing of the terminal and its operation when completed.

*The Federal Government.* Prior to 1961, the federal government took an active interest in the planning and design of terminal buildings and furnished federal aid for terminal construction. Between 1961 and 1976, when significant amendments were made to the Airport and Airways Development Act of

1970, terminal buildings were no longer eligible for federal aid. In 1976, Public Law 94-353 permitted the use of federal funds for nonrevenue producing areas of the passenger terminal. The federal government also planned for its operational role within the terminal area where immigration, customs, and health inspections may be necessary and where air traffic control facilities are integrated into the terminal design.

*The Airlines.* As prime tenants, and indeed sometimes the owners of terminal facilities, airlines exert an important influence of the design features of the terminal building and its surroundings. Each airline provides the following estimates of its own terminal needs, usually in stages of 5, 10, 15, and 20 years,

1. Estimates of aircraft movements.
2. Estimates of emplaning, deplaning, and interlining passenger traffic.
3. Space requirements for ticket counters, baggage handling, services, operations, maintenance, and supply areas.
4. Number, type, and size of aircraft parking positions.
5. Special requirements such as telecommunications, flight information systems, and passenger transport systems.
6. Building and space requirements for air freight.
7. Maintenance facilities.
8. Fixed ramp facilities such as air conditioning and electric power for aircraft and air bridges.
9. Mobile ramp service requirements such as sanitary services, catering, startup, towing vehicles, deicing, water and aircraft steps, and apron transfer vehicles.
10. Fuel requirements.
11. Automobile parking requirements for employees.

*Concessionaires.* The airport terminal building is a commercial venture of considerable magnitude and consumer services and rentals often produce over half of the terminal revenues. Although most of the concessions are leased after the building is designed, design information on the size and location of restaurants, shops, car rental and other tenants is required.

Consulting architectural and engineering firms usually are employed by the airport owner to develop a terminal design to satisfy the varied requirements and wishes of the airport management, the federal government, the airlines, and other tenants of the airport.

## 18-3. AIR TERMINAL LAYOUT CONCEPTS

In the planning of the air terminal the designer weighs a number of objectives: (1) To provide a high level of service to passengers at an acceptable cost in the processing, waiting, and circulation areas by providing adequate space and minimizing the difficulties of transfer movements; (2) to provide a design with a high level of flexibility that can cope with changes in air vehicle technology and changes of passenger traffic levels; (3) to reduce both internal and access walking distances for passengers and taxiing requirements for aircraft; (4) to provide a terminal layout capable of generating potential terminal revenues; and (5) to provide an acceptable working environment for airport and airline staff.

Increasing space needs for both automobiles and aircraft have taxed the designer's ingenuity in recent years. This has been especially true, of course, at large airports. The predominant preference of passengers, visitors, and well-wishers to travel to the airport by automobile largely explains the growing demand for automobile parking spaces. The use of larger aircraft has complicated the designer's task in at least three ways. In the first place, the parking and service area required for these aircraft has increased making it more difficult to park large numbers of these aircraft within a reasonable walking distance from the processing areas of the terminal building. Secondly, the sudden deposition of batches of several hundred passengers results in surges of terminal passenger traffic, and space and logistic needs for these peak flows must be provided. Finally, the larger aircraft with higher tails and greater floor heights require higher vertical clearances and more flexibility in the heights of loading platforms and bridges.

## 18-4. AIR TERMINAL LAYOUT SCHEMES

To satisfy the various design objectives, various physical layout schemes have evolved over the years. Six basic terminal layouts are shown in Fig. 18-2. The oldest and simplest concept is the *frontal system,* shown in Fig. 18-2*a*, in which the aircraft park parallel to the terminal building. This system is adequate for small airports where the number of aircraft served is small and where there are few or no flights by commercial air carriers.

A second layout scheme, shown by Fig. 18-2*b*, involves the use of concourses or *fingers* which extend from the terminal building to the aircraft parking areas. These fingers may simply be fenced walkways but commonly are enclosed structures that are temperature controlled throughout. Enclosed concourses protect passengers from the elements, aircraft noise, and propeller and jet blast.

In the mid-1950s, airlines began to process the passengers at the gate by accepting their tickets at a holding area within the terminal concourse structure. Later, when it became the practice to increase these areas in size and to provide comfortable furnishings, they came to be known as departure lounges.

Commenting on the departure lounge loading concept, it has been stated that: "Aside from the increase in the number of aircraft gates required for the

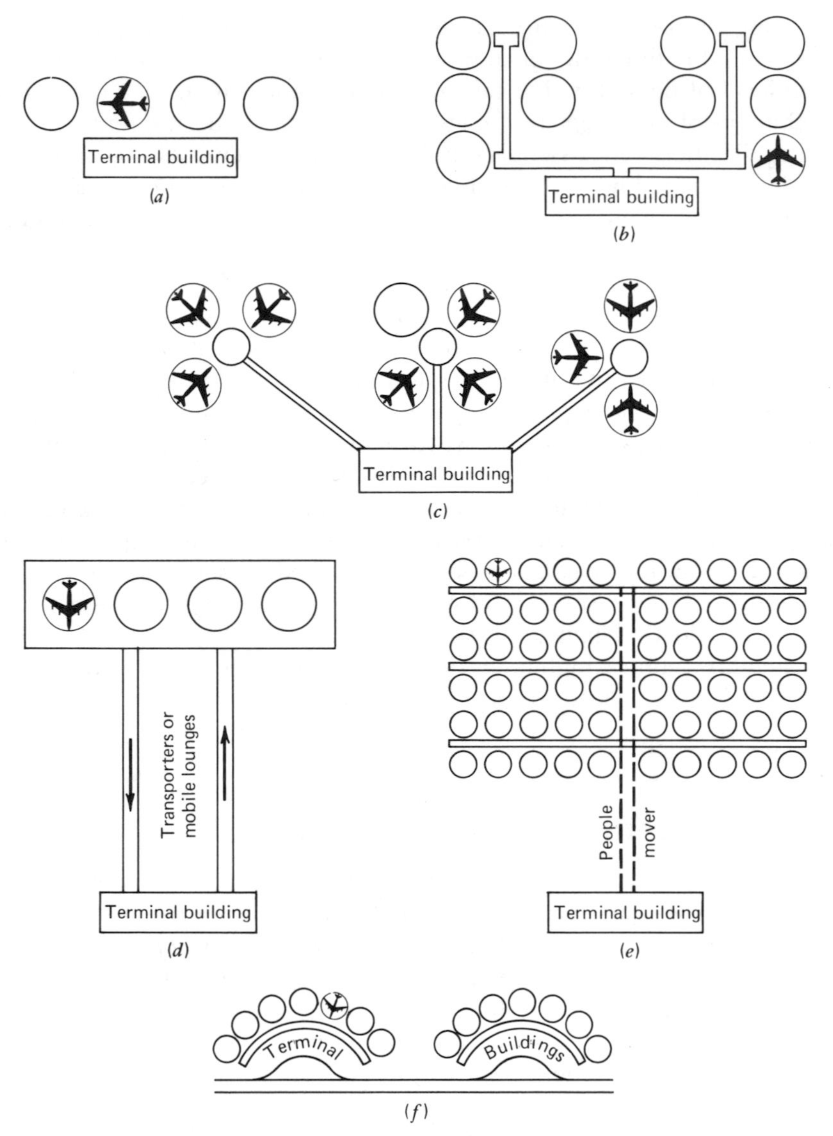


*Fig 18-2.* Terminal layout schemes.

increased schedules, no other single development had such an impact on terminal design (and, incidentally, cost) as this did." [3]

A logical extension of the departure lounge concept has been the recent development of *satellite enplaning structures*, sometimes called flight stations. (See Fig. 18-2*c*.) These satellite structures are self-contained areas with departure lounges, toilets, concessions, and limited food and beverage areas. These structures typically serve 5 to 10 loading gates and are connected to the terminal buildings in a number of ways. In some cases the satellite is connected by a concourse with departure lounges along its length (e.g., Paris-Orly), by a connecting tunnel beneath the apron (e.g., Los Angeles), or by a connector above the apron (e.g., Tampa).

As airports grow larger the size of terminals with pier figures and satellites become inconveniently large from the passengers' viewpoint, and the designs also require long taxiing distances for aircraft. The layouts shown in Fig. 18-2*d* and 18-2*e* are attempts to overcome this difficulty. The transporter airport separates the servicing apron of the aircraft from the passenger terminal. Passengers are brought to and from the aircraft by transporters which can range from inexpensive buses to elaborate mobile lounges that can move up or down to match aircraft or terminal floor heights. This form of design was first used at Dulles Airport, Washington, D.C., and subsequently has been adapted elsewhere (e.g., Mirabel, Montreal). As airports grow even larger this concept is modified to the remote stand arrangement exemplified by the Atlanta midfield terminal, where a high-capacity people-mover system transports passengers from the main terminal building to a series of linear gate buildings to which the aircraft taxi for service and turnaround.

The automobile is the predominant mode of access for many large airports throughout the world, especially in the developed countries of Western Europe and North America. Where passenger access is predominantly by the automobile, designers have developed the *gate arrival* concept, which minimizes the walking distance between the point at which the air travelers park their cars, or from where they are dropped off from either a car or taxi at the aircraft gate. This concept, shown in Fig. 18-2*f* employs a very long terminal building that is a concourse or finger with aircraft loading positions on one side and the terminal roadway, curb, and public entrances on the other. Because of its length, special ground transportation facilities must be provided for passengers moving from one point in the terminal to another, or between separate terminals in the airport terminal complex. In Kansas City shuttle buses are used, but in the larger Dallas-Fort Worth terminal a complicated automatic light guideway transit system is in operation.

With the gate arrival system, convenient automobile parking is provided, and abundant curb lengths are available for discharging and picking up passengers. There are, however, a number of disadvantages. There is some loss of cross-utilization of space and facilities. More rest rooms, concessions, and so on are required than for centralized terminal designs. Where there are a large number of interlines at the airport, the passenger's difficulty in transferring makes this design style unacceptably difficult. Figure 18-3 shows the ultimate design concept for Dallas-Fort Worth International Airport, a gate arrival design.

Many large airports have adopted the *unit terminal* approach, in which each airline has its own terminal building. An early example of this is the John F. Kennedy Airport in New York City, where each terminal is uniquely designed. A later example of the unit terminal design where each unit con-

*Fig 18-3.* Courtesy Dallas-Fort Worth International Airport. (Gate Arrival.)

forms to a similar design is Houston International Airport, (See Fig. 18-4.) A somewhat similar form of unit terminal repetition was originally planned at Charles de Gaulle Airport, Paris, and Toronto International Airport. In each case, the original circular unit concept was abandoned for a linear second terminal which was found to give more flexibility.

Richard de Neufville has correctly pointed out that airport designers frequently strive for overall designs that are aesthetically pleasing and symmetrical when viewed from the air [4]. In fact, designs should be functional and the most successful operations may be achieved by assymetrical and seemingly piecemeal designs that are tailored to the needs of the individual air traffic demands of different airline operations [5, 6]. Seattle is an example of an airport with a terminal that combines the open apron, satellite, and pier finger concepts.

## 18-5. INTRATERMINAL AND INTERTERMINAL TRANSPORTATION

Despite efforts to minimize walking distance by thoughtful air terminal layout and design, distance remains a problem at existing airports. A study [7] of nine

*Fig 18-4.* Ultimate design concept for Houston International Airport (Unit Terminal with Satellites).

major airports[1] revealed that for originating passengers the average maximum walking distance to the nearest gate was 565 ft, while the minimum walking distance to the farthest gate was 1342 ft. Passengers who transferred from one airline to another were confronted with an average walking distance of 4091 ft.

To overcome the problem of excessive walking distances, two basic approaches have been used: moving sidewalks and vehicle systems.

Moving sidewalks, which are usually limited to distances of about 1000 ft due to their slow speed and have been used at such major airports as San Francisco, London Heathrow, and Paris Charles-de-Gaulle. They are used principally to connect parking lots to the terminal and to connect central processing areas to the departure lounges. There are two types of moving sidewalks: belt conveyors and horizontal escalators using linked pallets. To prevent mishap, handrails must be provided which must move at the same

[1]Chicago (O'Hare), New York (John F. Kennedy), Los Angeles International, Atlanta, San Francisco International, Dallas, Miami International, Philadelphia International, and Detroit Metropolitan Airports.

speed as the sidewalk. At Birmingham airport in Britain, a Maglev train is used to connect the airport terminal to the mainline railroad station.

A wide variety of vehicular systems have been developed or are proposed for transportation within the terminal area. Conventional shuttle buses are used extensively on both the landside and airside of many terminals. Specially designed buses are used at a number of European airports and in Tokyo. Automatic transit vehicles are used to convey passengers to satellite terminals at Seattle, Miami, Atlanta, and Tampa, for example. Movement between terminals is provided by an automatic computer controlled light guideway system at Dallas-Fort Worth (see Fig. 18-5). Automatic systems normally are chosen to eliminate labor costs and to increase reliability. They are possible only when grade separated designs are possible.

Perhaps the most dramatic example of specialized vehicles for transportation in the terminal area is the mobile lounge that was originally developed for Dulles International Airport, Washington, D.C., and subsequently has been used in such airports as Baltimore, Kennedy, Mirabel, and Madrid. Mobile lounges are essentially holding areas on wheels that transport passengers from terminal buildings to the aircraft. These vehicles are about 50 ft long, 15 ft wide, with a seating capacity of 60 persons. They can be driven from either end. Most versions of mobile lounges have bodies that can be raised or

*Fig 18-5.* Automated people-mover system at the Dallas-Fort Worth Airport. (Courtesy Dallas-Fort Worth Airport.)

lowered to match aircraft floors or passenger loading ramps. Because of their complexity these vehicles are extremely expensive. However, studies have shown that where facilities are used sporadically in peak conditions it may be more economic to use mobile lounges rather than permanent buildings [4].

## 18-6. AUTOMOTIVE PARKING AND CIRCULATION NEEDS

Increasing airline activity and the growing popularity of the automobile have combined to create unprecedented demand for parking and circulation facilities at most airports. In small- and medium-sized cities, more than half of the air passengers travel to and from the airport by passenger car. Although this percentage may be only 20 to 25 percent for major airports, the total volume of passenger car traffic to these airports taxes the ingenuity of the airport designer in his efforts to provide adequate and convenient parking and circulation needs.

Ideally, the airline passenger, who is usually carrying baggage, should be provided a parking space within 300 to 400 ft of the terminal building. A maximum walking distance from parking lot to the terminal building of 1000 ft is recommended by IATA [5].

The classic parking plan has been a ground-level parking lot adjacent to the airport passenger terminal. At large airports, designers have found it impossible to provide sufficient parking on a single level, and the trend is toward the provision of multilevel parking structures. The design of these structures is similar to the design of a downtown parking garage (described in Chapter 16), the principal differences being the need to provide shorter walking distances and larger stalls for passengers carrying baggage. Parking garages at airports also tend to be larger than downtown garages, and several large airports are now designing ultimate garage capacities of 5000 to 12,000 spaces.

Most new U.S. airport parking facilities provide 300-500 parking spaces per million annual passengers. Typical parking stalls are 8.5 to 9.0 ft in width and 18 ft long. A popular design features angle parking (60 degrees) on each side of a central one-way aisle 22 ft in width. Thus, about 275 $ft^2$ of net parking area, including the aisle, is required. When space needs of baggage dropoff and pickup areas, sidewalks, elevators, and stairs are included, approximately 340 $ft^2$ per stall may be required.

Airport parking demand may be divided into several categories:

1. Passengers.
2. Visitors bringing passengers and well-wishers.
3. Employees.
4. Business callers.
5. Rental cars, taxis, limousines, and so on.

Since the parking characteristics of parkers in these various categories differ in time of occurrence and duration, separate parking analyses should be made for each category.

Preferably, parking spaces for short-time parkers should be located nearest the terminal, and airports charge higher fees for close-in parking spaces. It is the practice for most large airports to separate parking spaces for employees some distance from the terminal building. Shuttle buses may be required to transport employees to their destinations.

If possible, the airport should have direct connections to a controlled access highway system. Within the airport, vehicular circulation is generally counterclockwise and one-way. This permits passengers to be loaded and discharged safely from the right side of the vehicle. At grade intersections should be avoided in the circulation system, and traffic should be separated by destination at the earliest possible point. The use of overhead or tunnel crossings should be considered to prevent mixing of pedestrian and vehicular traffic.

Airports served by road modes also have a need for pickup and drop-down curb space. At U.S. airports it is usual to provide approximately 350 linear ft of curb space per million noninterlining passengers [6].

## 18-7. TERMINAL APRON SPACE REQUIREMENTS

The term apron or ramp refers to an area for the parking or holding of aircraft. In terms of operational efficiency of the airport, the terminal apron, which is situated adjacent to the terminal building, is most important.

There are three primary factors that determine the space requirements for a passenger terminal apron:

1. Size of gate positions
2. Number of gate positions
3. Aircraft parking configuration

*Size of Gate Positions.* The size of gate positions is principally determined by the size and maneuverability of aircraft, but it is also influenced by desirable wing-tip clearances and the manner in which the aircraft is moved into the gate position and serviced. Table 18-1 indicates the recommended apron dimensions required for maneuvering aircraft of various sizes, based on aircraft dimensions of the late 1980s [7].

The amount of space required for maneuvering and servicing the aircraft will vary depending upon airline operational procedures. The airport engineer should, therefore, consult with the various airlines on this matter during the early phase of the design process.

*Number of Gate Positions.* The number of gate positions required depends

*Table 18-1*
**Diagram and Summary of Push-Out and Taxi-Out Dimensions**

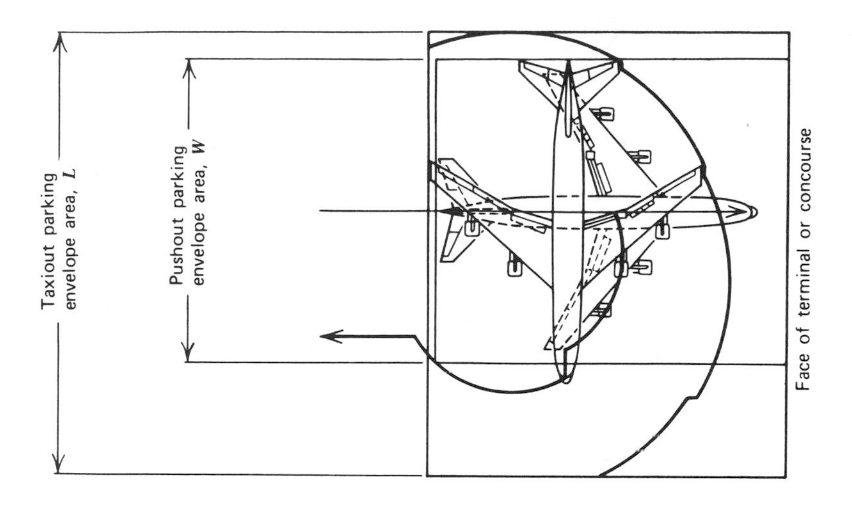


| Group A/C | Push-out[a] L | Push-out[a] W | Area (yd²) | Taxi out[b] L | Taxi out[b] W | Area (yd²) |
|---|---|---|---|---|---|---|
| A. FH-227 | 103 ft 1 in. | 115 ft 2 in. | 1319 | 148 ft 10 in. | 140 ft 2 in. | 2318 |
| YS-11B | 106 ft 3 in. | 124 ft 11 in. | 1474 | 171 ft 0 in. | 149 ft 11 in. | 2850 |
| BAC-111 | 123 ft 6 in. | 113 ft 6 in. | 1557 | 130 ft 0 in. | 138 ft 6 in. | 2001 |
| DC9-10 | 134 ft 5 in. | 109 ft 5 in. | 1634 | 149 ft 2 in. | 134 ft 5 in. | 2228 |
| B. DC9-21, 30 | 149 ft 4 in. | 113 ft 4 in. | 1880 | 149 ft 0 in. | 138 ft 4 in. | 2290 |
| 727 (all) | 173 ft 2 in. | 128 ft 0 in. | 2463 | 194 ft 0 in. | 153 ft 0 in. | 3298 |
| 737 (all) | 120 ft 0 in. | 113 ft 0 in. | 1507 | 145 ft 4 in. | 138 ft 0 in. | 2228 |
| C. B-707 (all) | 172 ft 11 in. | 165 ft 9 in. | 3188 | 258 ft 0 in. | 190 ft 9 in. | 5468 |
| B-720 | 156 ft 9 in. | 150 ft 10 in. | 2627 | 228 ft 0 in. | 175 ft 10 in. | 4454 |
| D. DC-8-43, 51 | 170 ft 9 in. | 162 ft 5 in. | 3081 | 211 ft 10 in. | 187 ft 5 in. | 4411 |
| DC 8-61, 63 | 207 ft 5 in. | 168 ft 5 in. | 3882 | 252 ft 4 in. | 193 ft 5 in. | 5423 |
| E. L-1011 | 188 ft 8 in. | 175 ft 4 in. | 3676 | 263 ft 6 in. | 200 ft 4 in. | 5865 |
| DC-10 | 192 ft 3 in. | 185 ft 4 in. | 3959 | 291 ft 0 in. | 210 ft 4 in. | 6801 |
| F. B-747 | 241 ft 10 in. | 215 ft 8 in. | 5795 | 328 ft 0 in. | 240 ft 8 in. | 8771 |

[a]Including clearances of 20-ft wing-tip, nose to bldg; 30-ft group A and B, 20-ft group C and D, 10-ft group E and F.

[b]Including clearance of 20-ft to other A/C and GSE: 45 ft.

Note: Length and width are based on the largest dimension in the group of aircraft.

on (1) the peak volume of aircraft to be served, and (2) how long each aircraft occupies a gate position.

Gate occupancy time will depend on:

1. Type of aircraft.
2. Number of deplaning and emplaning passengers.
3. Amount of baggage.
4. Magnitude and nature of other services required.
5. Efficiency of apron personnel.

From first principles, it can be deduced that the peak number of gate positions required at an airport is a function of the design peak hour aircraft movements, the length of time that the individual aircraft spend at the gates, and some utilization factor that takes into account the impossibility of having all gates filled for 100 percent of the peak period and the lack of suitabilitiy of all aircraft to all gates.

One formula that has been proposed for computing the required number of gates is [8]:

$$n = \frac{vt}{u}$$

where

$v$ = design hour volume for departures *or* arrivals, aircraft/hr
$t$ = weighted mean stand occupancy, hr
$u$ = utilization factor, suggested to be 0.6 to 0.8 where gates are shared

Another deterministic formula, which has been calibrated on European traffic is [12]:

$$n = mqt$$

where

$m$ = design hour volume for arrivals *and* departures, aircraft/hr
$q$ = proportion of arrivals to total movements
$t$ = mean stand occupancy, hr

Computation of gate requirements for an individual airport is now carried out frequently by computer simulation models which take into account the specific design of the apron area, the mix of traffic, and the handling times likely to be achieved in practice. Recently work has been carried out to generalize the work of simulation models in this area [9].

A study [10] of nine airports indicated that there is wide variation in the time and magnitude of peak gate occupancy, but each airport had a definite pattern that remained relatively constant over the years. It was further reported that there was a wide range of productivity per gate for various airports. It was proposed that the following formula for forecasting the number of future gate positions required for a given airport:

$$\text{Future gates} = \left[(\text{Present gates} - 2) \times \frac{\text{future passengers}}{\text{present passengers}}\right] + 2$$

This equation is applicable to any group of aircraft that has mutual use of gates, and separate calculations should be made for each such group. For example, if four groups of five gates are considered separated for traffic, and the future/present traffic ratio is three, for each group:

$$\text{Future gates} = (5 - 2) \times 3 + 2 = 11$$

Thus, the total gates required for four groups will be 44. It is recommended that in actual practice approximately 15 percent be added to the number of gates computed by the above formula to allow for contingencies of operations such as early arrivals or delayed departures.

This approach can be used only when there is expected to be no major change in airline operating procedures in the future.

*Aircraft Parking Configuration.* Parking configuration refers to the orientation of aircraft in relation to the adjacent building when the aircraft is parked. There are a variety of parking configurations that may be used. The aircraft may be nosed-in, nosed-out, parked parallel, or at some angle to the building or concourse. The parallel parking system, in which the longitudinal axis of the aircraft is parallel to the adjacent building has been used extensively and it provides the best configuration for passenger flow. At major airports, the trend is toward the use of a nose-in parking configuration with the aircraft being pushed away from the loading bridge or gangplank by a tractor after it is loaded. This parking configuration is especially suitable for use with circular satellite enplaning structures. In this case, the aircraft parks in the wedge. This permits flexibility in the provision of space for various sizes of aircraft since the amount of available space can be varied by moving aircraft out from the center of the satellite. This will imply of course air bridge reconfiguration.

## 18-8. THE TERMINAL BUILDING—TRANSPORT AIRPORTS

A well-designed terminal building is a vital element in the successful operation of an air carrier airport. It must provide for the smooth and efficient transfer of passengers and their baggage between surface transportation vehicles and aircraft.

The terminal building must provide ordered space and facilities for a variety of functions relating to air passenger service, air carrier operations, and operation and maintenance of the airport.

*Required Air Passenger Service Functions.* The terminal building usually houses a variety of air passenger service functions including ticket sales, restroom services, waiting and resting, and baggage checking and claiming. In addition, terminal facilities will normally be required for security protection, provision of flight information, passenger boarding and deplaning, and the handling and processing of mail and light cargo.

*Additional Facilities for Convenience of Passengers.* Most moderate-size to large airports provide numerous and varied facilities for the convenience of passengers and to generate net airport income. Such facilities usually include a newsstand, telephones, restaurants and coffee shops, gift shops, insurance sales, and car rental agencies. Large airports typically also have a bank, a barber shop, medical services, and hotel and motel accommodations.

*Air Carrier Operations.* Consideration must also be given to space requirements in the terminal building for air carrier operations including: (1) communications center, (2) ground crew and air crew ready rooms, and (3) operations room for crews.

*Airport Operations and Maintenance.* Finally, space may be required in the terminal building for the functions relating to the operation and maintenance of the airport. Certain of these functions may be housed in separate buildings.

1. Air traffic control.
2. Ground traffic control.
3. Airport administration.
4. FAA and other governmental administrative functions.
5. Airport maintenance.
6. Fire protection.
7. Employee cafeterias.
8. Utilities.

In view of the rapidly expanding and changing nature of air travel activity, it is especially important that terminal buildings be planned and designed to allow easy expansion and change. Generally, a rectangular configuration rather than an odd-shaped building is preferred. The design should be such that it will not be necessary to relocate kitchen facilities, toilets, and other such costly installations should expansion be required. Nonbearing partitions should be used whenever possible to allow for reallocation of space to meet changing requirements. In short, the terminal building design should be expansible and flexible, and long-range plans should be made to change it and add to it as traffic and economic conditions dictate. Because of diseconomies of scale, terminal buildings should be built in a modular manner with modules designed to handle annual throughputs of between 5 and 10 million passengers. Single terminal units begin to become very large with long walking distances when designed to handle in excess of 20 million annual passengers [11].

The terminal building design should provide for separation of service areas to prevent passenger and baggage congestion. Specifically, the lobby and waiting room activities should be separated from baggage handling activities.

The design should provide for ease of circulation of emplaning and deplaning passengers. Emplaning passengers should be able to move directly and smoothly to the ticket counter, thence through the waiting room area to the aircraft loading gate. Deplaning passengers should be able to follow a direct route from the aircraft to the baggage claim area and thence to the passenger loading platform. These movements are illustrated schematically by Fig. 18-6 for an airport with decentralized security and gate control and check-in.

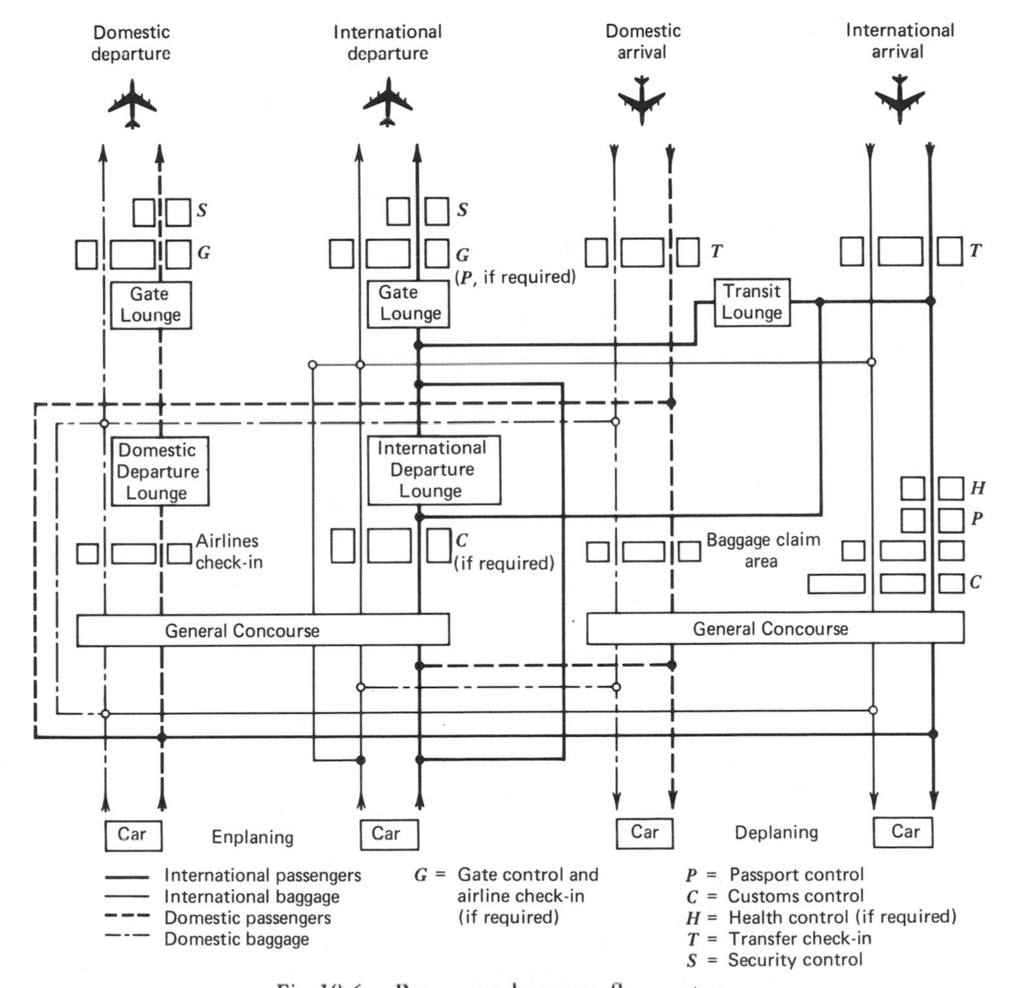


*Fig 18-6.* Passenger baggage flow system.

In the design of high-capacity terminal buildings, consideration should be given to the use of two- or three-level circulation systems. By providing two or more levels in the terminal building, a vertical separation of passenger and baggage flow can be realized. A multiple-level building also makes it easier to separate arriving and departing passengers, and to provide dropoff and pickup curb space on the land side of the terminal.

*Space Requirements for the Terminal Building.* Before one can obtain a reliable estimate of space requirements for the various functions of the terminal building, it is necessary to estimate the typical peak-hour passengers.[2] In the case of general aviation airports, typical peak-hour passengers have been shown to depend on the number of hourly aircraft operations [1]. (See Fig. 18-7.) At the larger air carrier airports, the number of peak hour-passengers can be obtained from Fig. 18-8, which relates typical peak-hour passengers to total annual passengers.

More precise estimates of the peak-hour population of airports can be obtained by making actual counts of passengers, employees, visitors and customers making use of existing terminal facilities and estimating future populations using FAA step-down passenger forecasts of future passenger demand for individual airport facilities.

The FAA has published design recommendations to aid those planning terminal buildings [7]. Using this publication, space requirements for the various activities in the terminal buildings may be estimated from graphs

[2]Typical peak-hour passengers are defined as the total of the highest number of passengers enplaning and deplaning during the busiest hour of a busy day of a typical week.

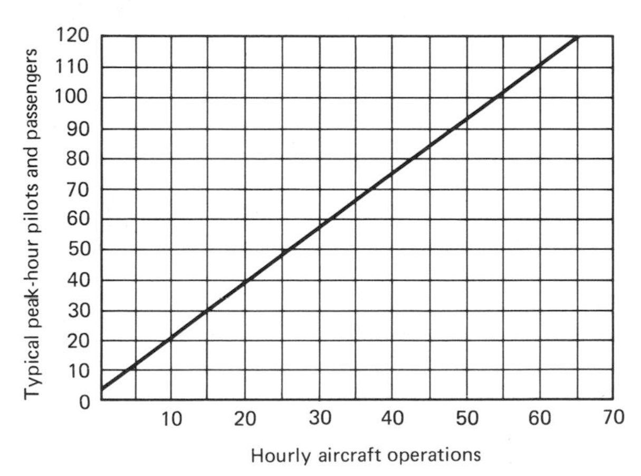


*Fig 18-7.* The relationship between typical peak-hour pilots and passengers and hourly aircraft operations at general aviation airports. (Source: *Utility Airports,* Federal Aviation Administration, including Change 8. July 3, 1985.)

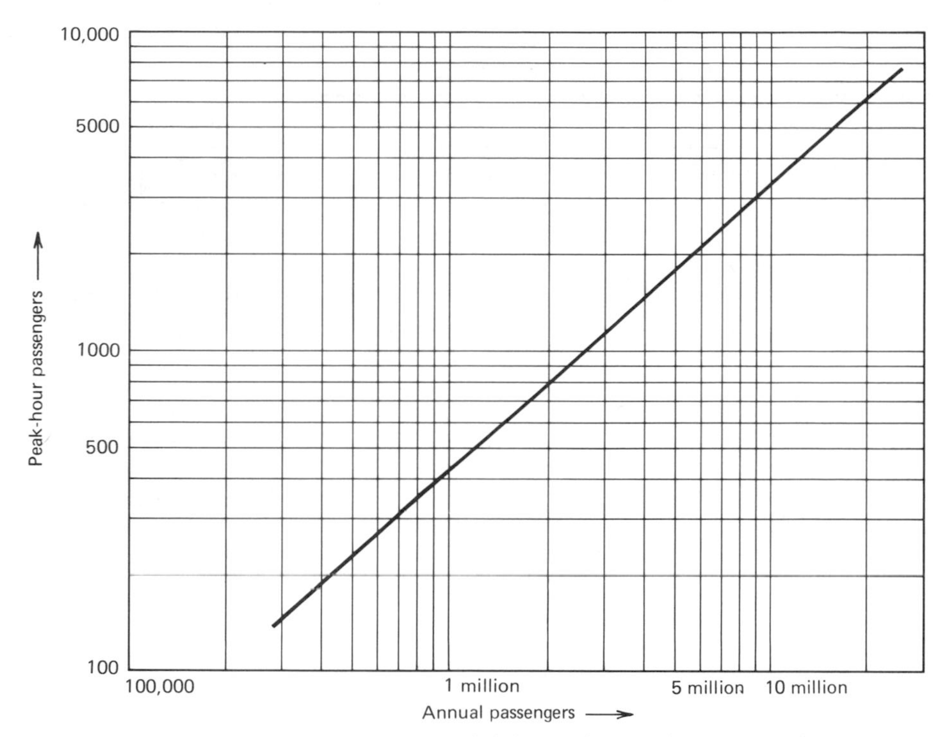


*Fig 18-8.* Relation between peak-hour and annual passenger flows.

relating space needs to typical peak-hour passenger throughput. While the graphs are useful for planning purposes, more detailed study may be required for design purposes to allow for local variations from average conditions [12].

Table 18-2 gives passenger space requirements according to the IATA-based design standards which have been adopted and in some cases slightly modified by a number of airport authorities [13].

## 18-9. THE GENERAL AVIATION TERMINAL AREA [1]

The requirements for general aviation terminal areas are relatively simple. Initially where there is little aviation activity a small maintenance hangar with an attached office will cater for the needs of the facility. A separate administrative terminal building should be built only when there are:

a. A minimum traffic volume of 10 operations, excluding touch and go, during the peak hours of a typically busy day.
b. One or more fixed base operators on the airport.
c. Airplane fuel available on the airport.

*Table 18-2*

**Passenger Terminal Space Standards[a] [13]**

| *Facility* | *Space Standard* | *Time Standard* |
|---|---|---|
| Check-in baggage drop | 0.8 $m^2$/passenger with baggage<br>0.6 $m^2$ for visitors | 95 percent of passengers <3 min; at peak times 80 percent <5 min |
| Departure concourse | None | None |
| Departure passport control | 0.6 $m^2$/passenger without hold baggage<br>0.8 $m^2$/passenger with hold baggage | 95 percent of passengers <1 min |
| Central security | | 95 percent of passengers <3 min; for high security flights 80 percent <8 min |
| Departure lounge | 1.0 – 1.5 $m^2$/seated passenger<br>1.2 $m^2$/standing pass. with trolley<br>1.0 $m^2$/standing passenger<br>Seating for 50 percent of throughput. | |
| Gate lounge | 0.6 $m^2$ for queuing passenger without hold baggage<br>0.8 $m^2$ for queueing passenger with hold baggage<br>1.0 $m^2$/passenger within gate lounge | 80 percent should queue less than 5 min for gate check-in |
| Immigration | 0.6 $m^2$/passenger | 95 percent of all passengers <12 min. 80 percent of nationals <5 min |
| Baggage reclaim | 0.8 $m^2$/domestic and short-haul international passenger<br>1.6 $m^2$ for long-haul passenger | Max of 25 min from first passenger in hall to last baggage from unit<br>90 percent of passengers wait <20 min for baggage |
| Customs | 2.0 $m^2$/passenger interviewed | None |

*Table 18-2*

**Passenger Terminal Space Standards[a] [13]—Continued**

| *Facility* | *Space Standard* | *Time Standard* |
|---|---|---|
| Arrivals concourse | 0.6 $m^2$/standing meeter; 1.0 $m^2$/seated meeter<br>0.8 $m^2$/short-haul passenger, 1.6 $m^2$/ long-haul passenger | None |

[a]Additional standards

| | |
|---|---|
| Forecourts: | 95 percent chance of finding space. |
| Piers: | Walking distances: <250 m unaided.<br><650 m with walkway (of which 200 m unaided).<br>Rapid transit for point-to-point journeys over 500 m. |
| Pier service: | Loading bridges for at least 75 percent of passengers. |

d. A hangar with repair facilities in operation.
e. A full time manager on duty during the normal day.
f. A need for a public waiting area and restrooms along with a telephone.

Before engaging in the design of a general aviation terminal a survey should be carried out to determine peak demand. For such terminals, this is the greatest number of pilots and passengers enplaning and deplaning during the busiest hour of the busy day of a typical week, *not* the absolute peak occurring on an abnormally busy day.

The terminal and administration building should be functional and designed to permit easy expansion. Figure 18-9 shows the simple juxtaposition of the runway, taxiway, apron, and terminal building at a general aviation building and Fig. 18-10 indicates how building and access routes can be positioned for a simple functional layout which minimizes walking distance.

Minimum terminal size is likely to be in the region of 2500 $ft^2$ with the airport manager's office taking up about 180 $ft^2$ and having a good view of the airfield operating area.

The waiting room, which should have a minimum size of 1000 $ft^2$ should also overlook the airfield and should provide comfortable seating arrangements. All buildings should have some concessions, which at small facilities may be coin-operated. There should also be a bulletin board for weather reports and notices to airmen. There should also be a space for mounting aeronautical charts. At larger terminals, eating facilities in the form of a dining room should be provided, in addition to public restrooms and a telephone. On the landside of the terminal, there should be adequate parking and road access provision. Figure 18-11 shows the inside of the waiting area of a medium-sized general aviation terminal.

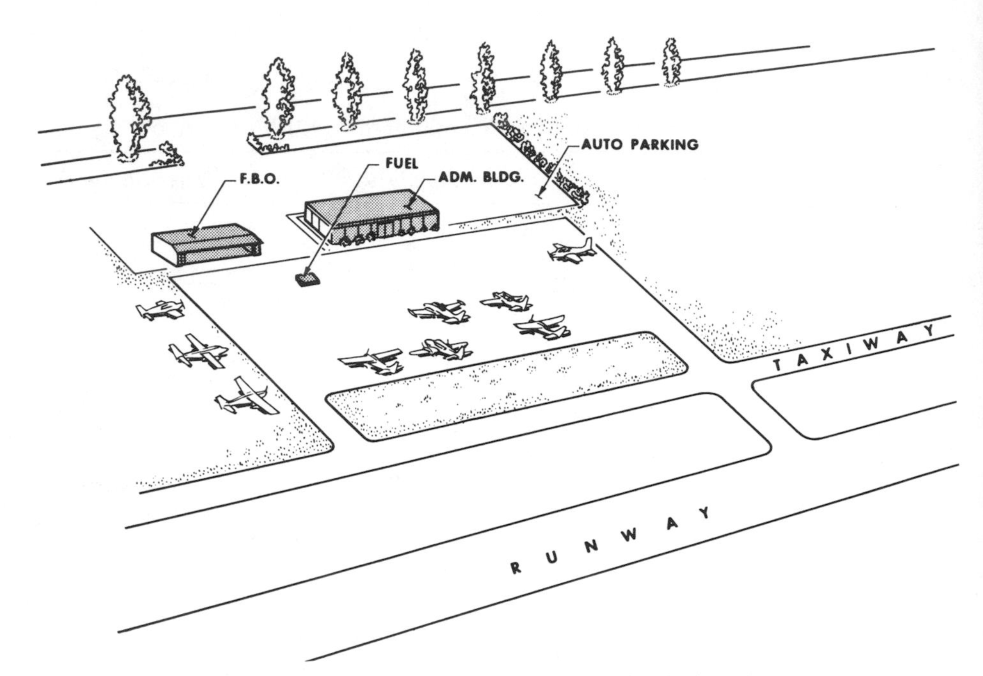


*Fig 18-9.* Apron terminal area at a general aviation airport.

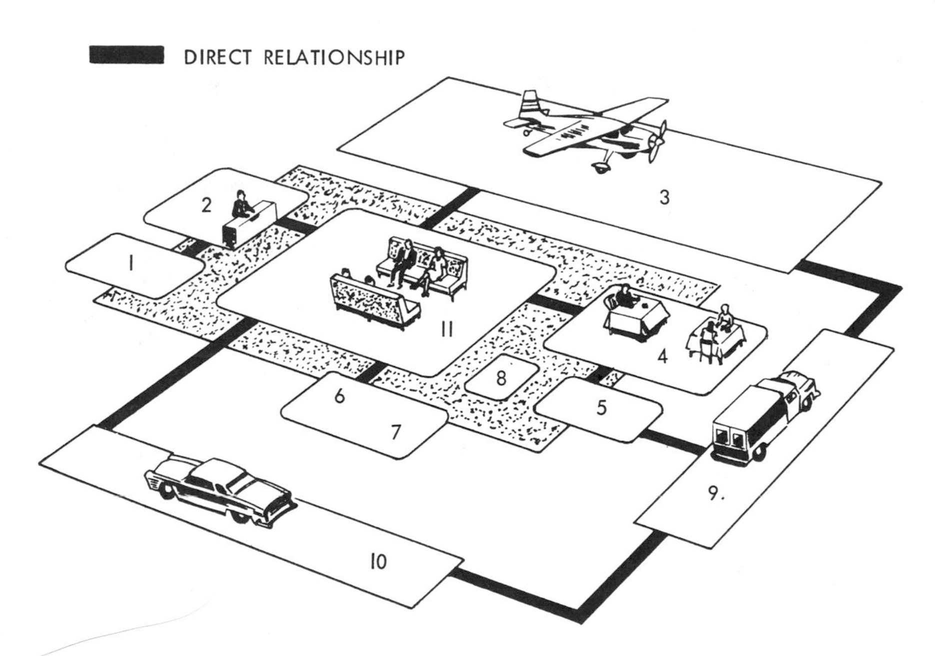


1. STORAGE
2. OPERATION MANAGEMENT
3. AIRPLANE LOADING APRON
4. DINING AREA
5. KITCHEN
6. REST ROOMS
7. JANITOR CLOSET
8. UTILITIES
9. SERVICE AND APRON ACCESS DRIVE
10. ADMINISTRATION BUILDING DRIVE
11. WAITING AREA

*Fig 18-10.* A general aviation terminal/administration building.

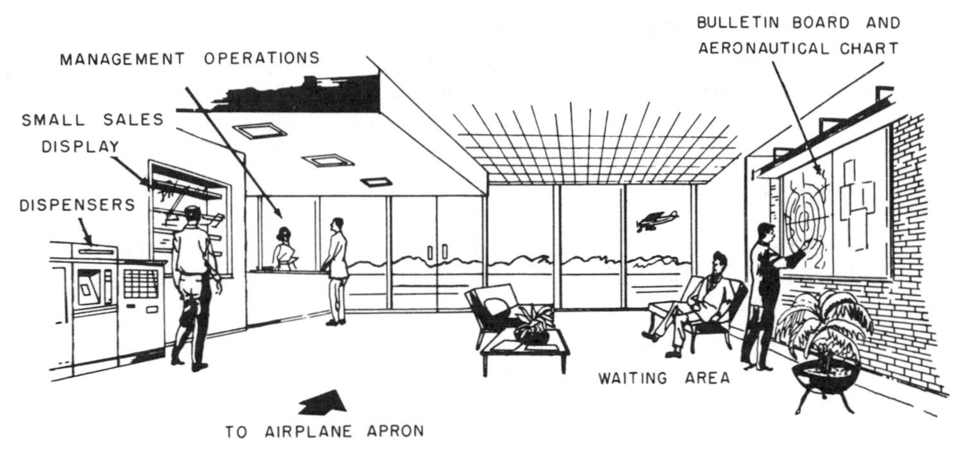


*Fig 18-11.* Waiting area in general aviation terminal.

## 18-10. SUMMARY

An airport terminal area is a complex and delicately balanced microcosm. Inevitable changes in passenger loads, actions by the airlines and airport management, and technological developments tend to upset this balance, sometimes resulting in undesirable and near-intolerable passenger and aircraft congestion and delays. Airport passenger loads typically fluctuate in the short run and increase in the long run. Airline companies react to heavier traffic by instituting innovations in passenger and baggage handling procedures and facilities. The introduction of larger and faster aircraft creates a need for more and larger waiting rooms and baggage pickup areas and may overload auto parking and circulation facilities. Airport planners and designers accept an awesome challenge in attempting to provide terminal facilities that resist obsolescence in the face of constant change.

## PROBLEMS

1. Compute the required number of aircraft gates from the following information:

| | |
|---|---|
| No. of arrivals in the peak hr | 30 |
| Average stand occupancy time | 45 min |
| Utilization factor | 0.8 |
| Proportion of arrivals to total movements | 0.6 |

2. The gate arrival system of airport terminal design has been promoted extensively as a good solution for the airports of the United States. Discuss how well this solution would work for airports with a large proportion of international and transfer passengers.

3. Sketch in schematic form the layout of a combined passenger terminal and administration building for a general aviation airport with 50 hourly airplane operations. Consult Reference 1 when preparing this answer.

## REFERENCES


1. *Utility Airports—Air Access to National Transportation,* Department of Transportation, Federal Aviation Administration, AC 150/5300-4B, including Change 9, September 18, 1987.
2. *Planning and Design of Airport Terminal Facilities at Nonhub Locations,* Federal Aviation Administration Advisory Circular AC 150/5360-9, April 4, 1980.
3. Thompson, Arnold W., "Evolution and Future of Airport Passenger Terminals," *Journal of the Aerospace Transport Division, Proceedings of the American Society of Civil Engineers,* October 1964.
4. de Neufville, R., *Airport Systems Planning,* McMillan, London, 1976.
5. *Airport Terminals Reference Manual,* Sixth Edition, International Air Transport Association, Montreal, 1976.
6. Ashford, Norman, and Paul H. Wright, *Airport Engineering,* 2nd Edition, Wiley Interscience, New York, 1984.
7. Planning and Design Considerations for Airport Terminal Building Development, FAA Advisory Circular AC 150/5360-7, September 5, 1976.
8. Horonjeff, R., and Francis X. McKelvey, *Planning and Design of Airports,* Third Edition, McGraw-Hill, New York, 1983.
9. Hamzawi, S.G., "Management and Planning of Airport Gate Capacity: A Microcomputer-Based Gate Assignment," *Transportation Planning and Technology,* Vol. 11, No. 3, 1986.
10. Stafford, Paul H., and D. Larry Stafford, "Space Criteria for Aircraft Aprons," *Transportation Engineering Journal, Proceedings of the American Society of Civil Engineers,* May 1969.
11. Ashford, Norman, H.P.M. Stanton, and C.E. Moore, *Airport Operations,* Wiley Interscience, New York, 1984.
12. Hart, Walter, *The Airport Passenger Terminal,* Wiley Interscience, New York, 1986.
13. Ashford, Norman, "Level of Service Design Concept for Airport Passengers," *Transportation Planning and Technology,* Vol. 12, No. 1, 1987.

# *Chapter 19*

# *Airport Design Standards and Procedures*

## 19-1. INTRODUCTION

In many respects, this chapter is the most important of the three chapters that deal with air transportation. It presents specific design standards and procedures that are required for the preparation of plans and specifications for an airport. Topics covered in this chapter will include runway lengths, geometric design of the runway system, earthwork, drainage, paving, and lighting and marking.

It should be remembered that the design standards given in this chapter are recommended standards rather than absolute requirements. Developed by the FAA for all parts of the nation, the standards are based on broad considerations. Local conditions and requirements must justify deviation from a particular standard in order to secure an advantage relating to another design feature. In such a case, designers should be prepared to justify their decisions to deviate from an accepted engineering design standard. In any event, designers would be well advised to check with the nearest office of the FAA.

## 19-2. RUNWAY LENGTH

One of the most important design features for an airport is runway length. Its importance stems from its dominant influence on air safety, and size and cost of the airport.

Design runway length is influenced most by the performance requirements of the aircraft using the airport, especially when operated with its maximum landing and takeoff loads. Variations in required runway length are caused by:

1. Elevation of the airport.
2. Average maximum air temperature at the airport.
3. Runway gradient.

The FAA publication *Utility Airports* [1] provides a family of curves, reproduced as Fig. 19-1, which give recommended runway lengths for three utility airport groups.

1. Basic utility—Stage I.
2. Basic utility—Stage II.
3. General utility—Stage I.

To use the curves in Fig. 19-1, one should enter the appropriate family of curves on the abscissa axis at the normal maximum temperatures.[1] From this point, a line is extended vertically until it intersects the slanted line corresponding to the airport elevation, interpolating if necessary. The point of intersection is extended horizontally to the right ordinate where the required runway length can be read.

EXAMPLE 19-1

USE OF UTILITY AIRPORT DESIGN CURVES. What length of runway is required for a general utility airport that is 5000 ft above sea level and has a normal maximum temperature of 80°F?

Entering the general utility airport family of curves at an abscissa value of 80°F, and projecting a line vertically to intersect the 5000-ft curve, a required runway length of 6100 ft is found at the right ordinate axis. This value is the answer. No correction is required for runway gradient or other such factors.

A much more precise runway length requirement can be determined if the design is made for a particular aircraft. The FAA publication *Runway Length Requirements for Airport Design* [2] provides design curves for landing and takeoff requirements for airplanes in common use in civil aviation. These curves are based on actual flight test and operational data.

Examples of FAA runway length curves are given as Figs. 19-2 and 19-3, which indicate the aircraft performance characteristics for a Boeing 707-300 series.

These and similar performance curves for takeoff are based on an effective runway gradient of zero percent. Effective runway gradient is defined as the maximum difference in runway centerline elevations divided by the runway length. The FAA specifies that the runway lengths for takeoff should be increased by the following rates for each 1 percent of effective runway gradient:

1. For piston and turboprop powered airplanes—20 percent.
2. For turbojet powered airplanes—10 percent.

[1]The normal maximum temperature is defined as the arithmetical average of the daily highest temperature during the hottest month. This information may be obtained from National Oceanic and Atmospheric Administration Technical Paper No. 81.

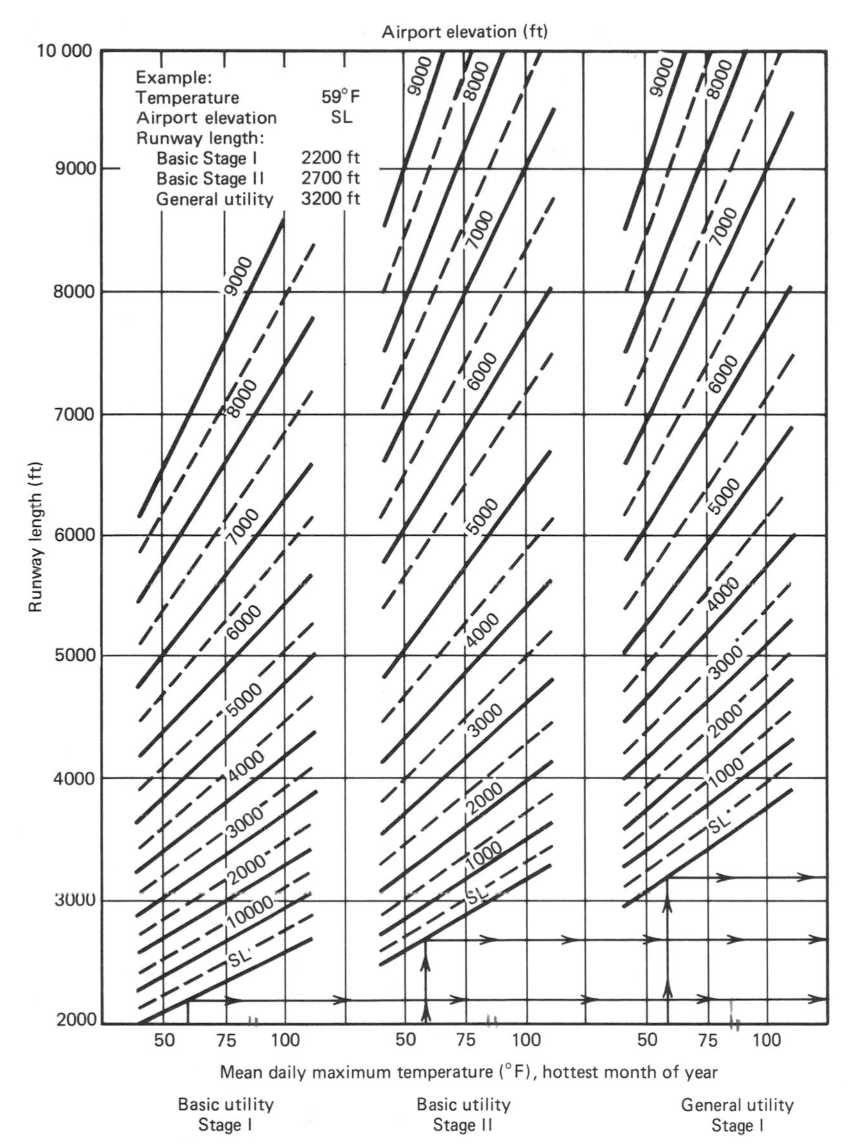


*Fig 19-1.* Runway length curves for utility airports. (Source: Utility Airports, Federal Aviation Administration, July 3, 1985.)

## EXAMPLE 19-2

RUNWAY LENGTH REQUIREMENT FOR A BOEING 707-300 SERIES AIRCRAFT. What length of runway is required for a Boeing 707-300 Series aircraft, given the following design conditions?

1. Normal maximum temperature—90°F.
2. Airport elevation—3000 ft.

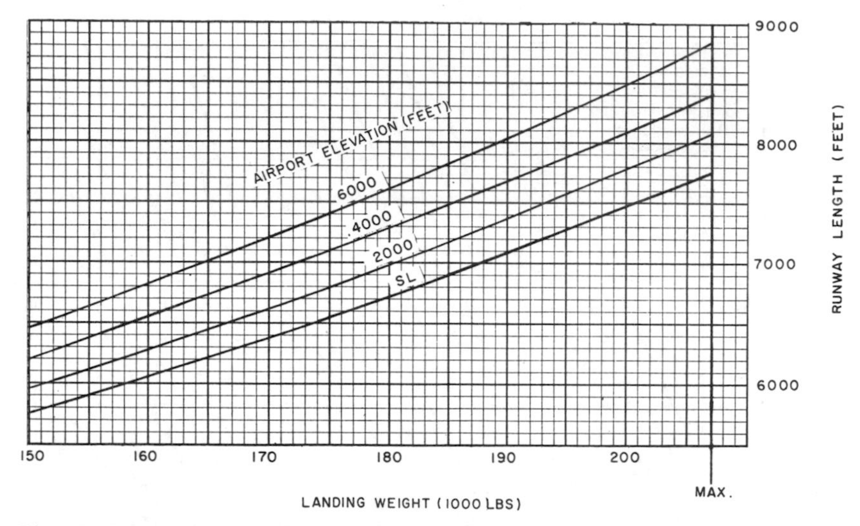


*Fig 19-2.* Landing performance curve for Boeing 707-300 Series aircraft. (Source: *Runway Length Design Requirements for Airport Design,* Federal Aviation Administration, September 27, 1978.)

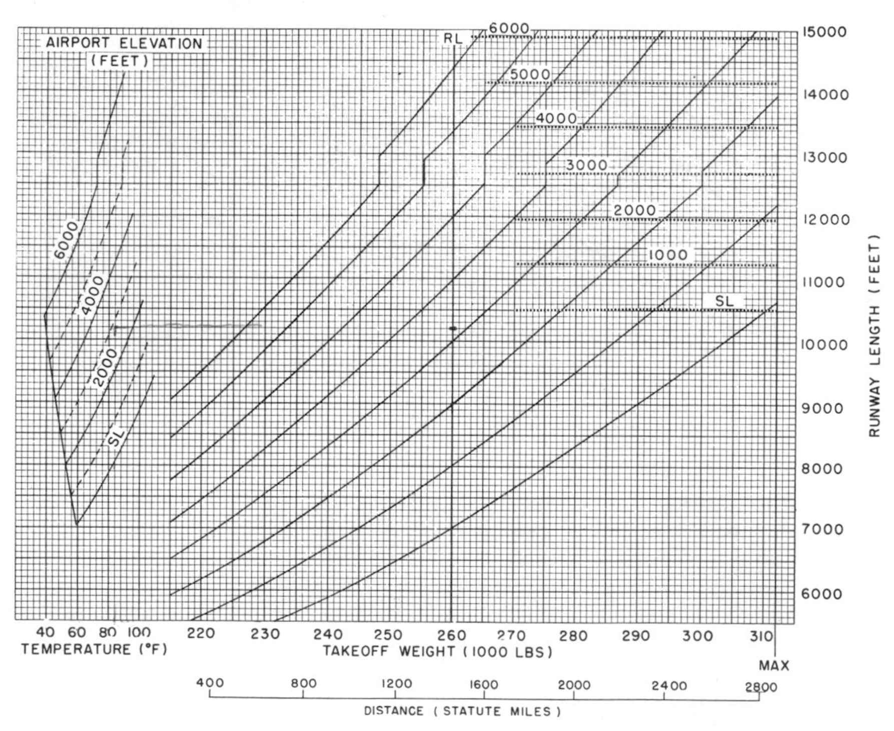


*Fig 19-3.* Takeoff performance curve for Boeing 707-300 Series aircraft. (Source: *Runway Length Design Requirements for Airport Design,* Federal Aviation Administration, September 27, 1978.)

3. Flight distance—1600 mi.
4. Maximum landing' weight—207,000 lb.
5. Effective runway gradient—0.4 percent.

***Runway length required for landing***

Figure 19-2 is entered on the abscissa axis at the maximum landing weight (207,000 lb) and this point is projected vertically to intersect with the 3000-ft airport elevation line (by interpolation). This point of intersection is extended horizontally to the right ordinate scale where a runway length required for landing of 8250 ft is read.

***Runway length required for takeoff***

The following steps are required to determine the runway length required for takeoff from Fig. 19-3:

1. Enter the temperature scale on the abscissa axis at the given temperature (85°F).
2. Project this point vertically to the intersection with the slanted line corresponding to the airport elevation (3000 ft).
3. Extend this point of intersection horizontally to the right until it coincides with the reference line (RL).
4. Then, proceed up and to the right or down and to the left parallel to the slanted lines to the intersection of the elevation limit line (in this case 3000 ft), or until reaching a point directly above the aircraft's takeoff weight or distance (e.g., 1600 mi), whichever occurs first.
5. Project this point horizontally to the right and read the required runway length for takeoff at the right ordinate scale. In this example, a length of 11,000 ft is required for takeoff.
6. Increase this runway length for effective gradient (0.4 percent).

$$\text{Runway length takeoff} = 11{,}000 + (11{,}000 \times 0.10 \times 0.4) = 11{,}440 \text{ ft}$$

$$\text{Design runway length} = 11{,}440 \text{ ft (largest of two values)}$$

Recently, the FAA has begun to provide aircraft performance data in tabular form, exemplified by Tables 19-1 and 19-2. The following simplified example illustrates the use of such tables.

EXAMPLE 19-3

RUNWAY LENGTH REQUIREMENT FOR A BOEING 727-200 AIRCRAFT. What length of runway is required for a Boeing 727-200 given the following design conditions?

1. Airport elevation—3800 ft.
2. Normal maximum temperature of hottest month—70°F.
3. Effective runway gradient—0.5 percent.
4. Length of haul—550 statute mi.
5. Average fuel consumption—19 lb/mi.

*Table 19-1*
**A Sample of**
**Aircraft Performance Data for a Boeing 727-200 Aircraft,**
**Landing, JT8D-7 Engine, 30° Flaps**

*a. Maximum Allowable Landing Weight (1000 lb)*

| *Temperature (°F)* | *Airport Elevation (ft)* | | |
|---|---|---|---|
| | *3000* | *4000* | *5000* |
| 65 | 148.0 | 148.0 | 148.0 |
| 70 | 148.0 | 148.0 | 148.0 |
| 75 | 148.0 | 148.0 | 148.0 |

*b. Runway Length for Landing (1000 ft)*

| *Weight (1000 lb)* | *Airport Elevation (ft)* | | |
|---|---|---|---|
| | *3000* | *4000* | *5000* |
| 140 | 5.75 | 5.88 | 6.02 |
| 145 | 5.93 | 6.07 | 6.21 |
| 150 | 6.11 | 6.25 | 6.40 |

SOURCE: *Runway Length Requirements for Airport Design,* FAA Advisory Circular AC 150/5325-4, including Change 14, September 27, 1978.

6. Passenger load—134 passengers at 200 lb/passenger = 26,800 lb.
7. Weight of empty aircraft plus reserve fuel—114,800 lb.

***Runway length required for landing***

By Table 19-1*a*, the maximum allowable landing weight is 148,000 lb. Entering Table 19-1*b* with this weight and interpolating for an elevation of 3800 ft, the runway length for landing is 6150 ft.

***Runway length required for takeoff***

The desired takeoff weight is the sum of the weight of empty aircraft (plus reserve fuel), the passenger load, and the weight of the fuel to be consumed:

$$\text{Desired takeoff weight} = 114{,}800 + 26{,}800 + (19 \times 550) = 152{,}050 \text{ lb}$$

This weight is checked to see if it exceeds the maximum allowable takeoff weight shown in Table 19-2*a* and, in this case, it does not exceed the allowable value.

Using the given temperature and elevation Table 19-2*b* is entered and, by interpolation, a reference factor $R$ is determined. For this example, $R = 53.98$. Using this reference factor and the desired takeoff weight, the runway length is found by interpolation from Table 19-2*c* to be 8908 ft. This length then must be corrected for gradient.

$$\text{Corrected runway length} = 8908 + 8908 \times 0.10 \times 0.5 = 9353 \text{ ft}$$
$$\text{Design runway length} = 9353 \text{ ft, say } 9350 \text{ ft}$$

*Table 19-2*
**A Sample of Aircraft Performance Data for a Boeing 727-200 Aircraft, Takeoff, JT8D-7 Engine, 15° Flaps**

*a. Maximum Allowable Takeoff Weight (1000 lb)*

| *Temperature (°F)* | *Airport Elevation (ft)* | | |
|---|---|---|---|
| | *3000* | *4000* | *5000* |
| 65 | 164.7 | 158.9 | 153.4 |
| 70 | 164.7 | 158.9 | 153.4 |
| 75 | 164.7 | 158.9 | 153.4 |

*b. Reference Factor* R

| *Temperature (°F)* | *Airport Elevation (ft)* | | |
|---|---|---|---|
| | *3000* | *4000* | *5000* |
| 65 | 50.3 | 54.3 | 58.8 |
| 70 | 50.7 | 54.8 | 59.3 |
| 75 | 51.2 | 55.2 | 59.9 |

*c. Runway Length (1000 ft)*

| *Weight (1000 lb)* | *Reference Factor R* | | |
|---|---|---|---|
| | *50* | *55* | *60* |
| 145 | 7.38 | 8.17 | 8.97 |
| 150 | 7.94 | 8.81 | 9.71 |
| 155 | 8.54 | 9.50 | 10.50 |

SOURCE: *Runway Length Requirements for Airport Design,* FAA Advisory Circular AC 150/5325-4, including Change 14, September 27, 1978.

## 19-3. LONGITUDINAL GRADE DESIGN FOR RUNWAYS AND TAXIWAYS

In the interests of safe and efficient aircraft operations, runway grades should be flat, and grade changes should be avoided. A maximum longitudinal grade of 1.5 percent generally is specified for air carrier airports [3] and a 2.0 percent maximum grade is recommended for utility airports [1].

Where grade changes are necessary, the recommended maximum changes in grade and minimum lengths of vertical curves given by Table 19-3 should be used. Careful thought should be given to what the future runway gradient requirements may be, and the runway gradient design should provide for future needs as well as present needs.

To minimize any hazard associated with objects on the runway, sight distance should be provided preferably for the full length of the runway. At airports without 24-hr traffic control, the FAA specifies that any two points 5 ft above the runway centerline must be mutually visible for the entire length of the runway. Generally speaking adherence to the longitudinal gradient standards for air carrier airports given in Table 19-3 will provide adequate line of sight.

Vertical curves should be provided when there is a change in grade as great as 0.4 percent. The length of vertical curve should be at least 1000 ft for each 1 percent of grade change at air carrier airports and 300 ft for each 1 percent of grade change at other airports.

Because aircraft movements along taxiways are relatively slow, longitudinal grade design standards for taxiways are not as rigorous as for runways. From an operational viewpoint, level taxiways are preferred. However, there is a need for taxiway gradients to harmonize with associated parallel runway gradients. Recommended design standards for longitudinal grades for taxiways are given in Reference 3.

## 19-4. RUNWAY AND TAXIWAY CROSS SECTION

Although much larger in scale, a runway cross section resembles that of a highway. The runway, a paved, load-bearing roadway, is typically 60 ft wide at small airports and 150 ft wide at large airports.

*Table 19-3*
**Longitudinal Grade Design Criteria for Runways**

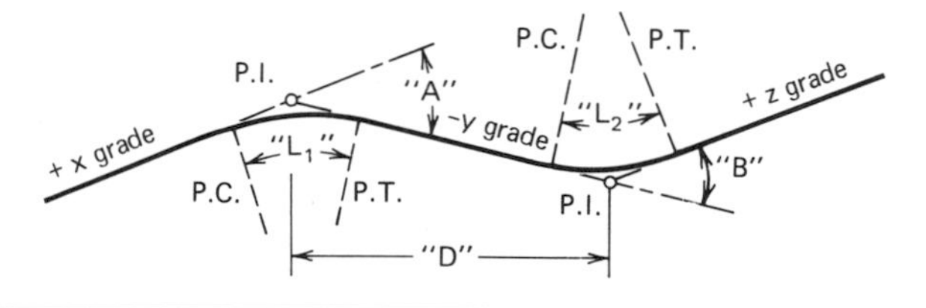


| | *Transport Airports* | *Utility Airports* |
|---|---|---|
| Maximum longitudinal grade, percent | 1.5 | 2.0 |
| Maximum grade change such as A or B, percent | 1.5 | 2.0 |
| Maximum grade, first and last quarter of runway, percent | 0.8 | — |
| Distance between points of intersection for vertical curves, such as D ft | $1000(A + B)^a$ | $250(A + B)^a$ |
| Lengths of vertical curve, such as $L_1$ or $L_2$, ft/1 percent grade change | 1000 | 300 |

[a]Use absolute values of *A* and *B*, expressed in percent.

Sources: *Utility Airports,* FAA Advisory Circular AC 150/5300-4B, April 11, 1978; *Airport Design Standards—Transport Airports* FAA Advisory Circular AC 150/5300-12, March 14, 1985.

Graded border areas are provided along each side of the runway as a safety measure should an aircraft lose control and veer from the runway. The border areas, which are typically stabilized earth with grass cover, vary in width from 25 ft at the smallest airports to 175 ft at the largest airports. The runway with adjacent borders is called the *runway safety area.* The width of the runway safety area varies from 120 ft to 500 ft, depending on airport class. It extends a minimum of 200 ft beyond each runway end.

The FAA uses the term shoulder to designate a relatively narrow paved or otherwise treated area adjacent to a runway or taxiway to resist jet erosion and/or to accommodate maintenance equipment.

The taxiway structural pavement is typically 25 to 50 ft wide at utility airports and 50 to 100 ft at air carrier airports. In the latter case, an additional 15 ft of pavement width should be provided on taxiway curves. In the interests of safety, taxiway centerlines are located 150 to 600 ft from runway centerlines.

A typical runway and taxiway cross-section is shown as Fig. 19-4. Table 19-4 gives recommended dimensional standards for utility airports, while similar standards for air carrier airports are given by Tables 19-5*a* and 19-5*b* [3]. The design standards are shown for nine *airplane design groups* which are determined primarily by the approach speed and wingspan of the critical aircraft. See. Fig. 19-5.

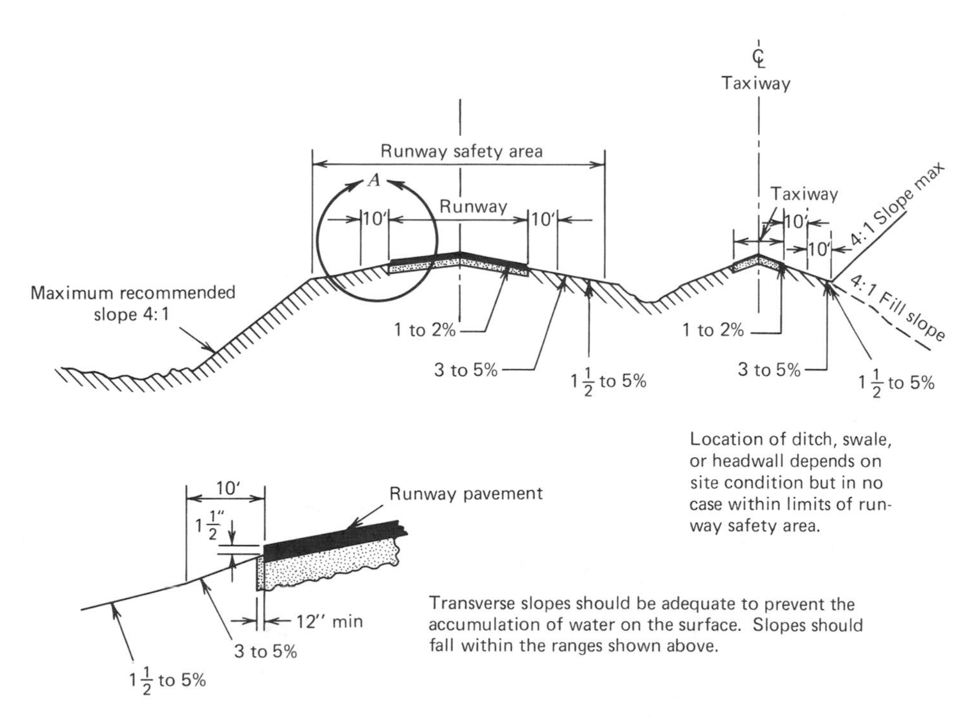


*Fig 19-4.* Typical runway and taxiway cross section. (Source: *Utility Airports,* Federal Aviation Administration, July 3, 1985.)

*Table 19-4*

**FAA Minimum Dimensional Standards for Utility Airports**

| Design Item | Nonprecision and Visual Runways Airplane Design Group | | | Precision Instrument Runways Airplane Design Group | | | |
|---|---|---|---|---|---|---|---|
| | *I*[b] | *I* | *II* | *I*[b] | *I* | *II* | *III* |
| Runway safety area width, ft[a] | 120 | 120 | 150 | 300 | 300 | 300 | 300 |
| Runway width, ft | 60 | 60 | 75 | 75 | 100 | 100 | 100 |
| Runway safety area length beyond runway end, ft[c] | 240 | 240 | 300 | 600 | 600 | 600 | 600 |
| Taxiway safety area width, ft | 49 | 49 | 79 | 49 | 49 | 79 | 118 |
| Taxiway width, ft | 25 | 25 | 35 | 25 | 25 | 35 | 50 |
| Runway centerline to: | | | | | | | |
| Taxiway centerline, ft | 150 | 225 | 240 | 200 | 250 | 300 | 350 |
| Building restriction line and aircraft tiedown area, ft | 125 | 200 | 250 | Refer to AC 150/5300-4B | | | |
| Taxiway centerline to: | | | | | | | |
| Taxiway centerline | 69 | 69 | 103 | 69 | 69 | 103 | 153 |

[a]1 ft = 0.3048 m.

[b]These dimensional standards are for facilities expected to serve only small airplanes.

[c]These distances may need to be increased to keep the stopway within the runway safety area.

SOURCE: *Utility Airports—Air Access to National Transportation,* FAA Advisory Circular AC 5300-4B, including Change 8, July 3, 1985.

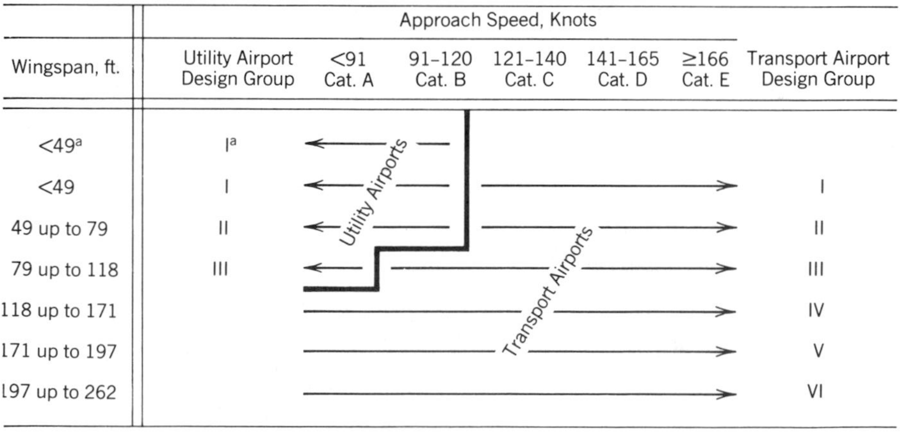

| Wingspan, ft. | Utility Airport Design Group | Approach Speed, Knots: <91 Cat. A | 91–120 Cat. B | 121–140 Cat. C | 141–165 Cat. D | ≥166 Cat. E | Transport Airport Design Group |
|---|---|---|---|---|---|---|---|
| <49[a] | I[a] | | | | | | |
| <49 | I | | | | | | I |
| 49 up to 79 | II | | | | | | II |
| 79 up to 118 | III | | | | | | III |
| 118 up to 171 | | | | | | | IV |
| 171 up to 197 | | | | | | | V |
| 197 up to 262 | | | | | | | VI |

Utility Airports

Transport Airports

[a]Applies to airports that are to serve only small airplanes.

*Fig 19-5.* FAA airplane design group concept. (Adapted from Reference 1.)

*Table 19-5a*
**FAA Runway Dimensional Standards for Transport Airports**

| Design Item | *Airplane Design Group* I | II | III | IV | V | VI |
|---|---|---|---|---|---|---|
| Runway safety area Width, ft[a,b] | 500 | 500 | 500 | 500 | 500 | 500 |
| Runway safety area Length, ft[c] | 1000 ft beyond each runway end | | | | | |
| Runway width, ft | 100 | 100 | 100[d] | 150 | 150 | 200 |
| Runway shoulder width, ft | 10 | 10 | 20 | 25 | 35 | 40 |
| Runway blast pad width, ft | 120 | 120 | 140[d] | 200 | 220 | 280 |
| Blast pad length, ft | 100 | 150 | 200 | 200 | 400 | 400 |
| Runway centerline to: | | | | | | |
| Taxiway centerline, ft | 400 | 400 | 400 | 400 | Varies[f] | 600 |
| Aircraft parking area, ft[e] | 500 | 500 | 500 | 500 | 500 | 500 |
| Property/building restriction line, ft[g] | 750 | 750 | 750 | 750 | 750 | 750 |

[a]1 ft = 0.3048 m.

[b]For airplanes in approach category C, the safety area width increases 20 ft for each additional 1000 ft of airport elevation greater than 8200 ft above sea level. For airplanes in approach category D, it increases 20 ft for each 1000 ft of airport elevation above sea level.

[c]For a runway with a stopway over 1000 ft in length, the runway safety area extends to the end of the stopway.

[d]For airports serving airplanes with maximum certificated weight greater than 150,000 lb, increase dimension by 50 ft.

[e]For airplane design groups I and II, with no precision instrument operations, this separation may be reduced to 400 ft when visibility minimums are 1 mi or greater.

[f]Dimension varies with airport elevation. See Reference 3.

[g]For airplane design groups I and II, with no precision instrument operations, this separation may be reduced to 500 ft when visibility minimums are 1 mi or greater.

SOURCE: *Airport Design Standards—Transport Airports,* FAA Advisory Circular AC 150/5300-12, March 3, 1985.

## 19-5. TAXIWAYS AND TURNAROUNDS

Taxiways are used to facilitate the movement of aircraft to and from the runways. Where air traffic warrants, the usual procedure is to provide a taxiway parallel to the runway centerline for the entire length of the runway. This makes it possible for landing aircraft to exit the runway more quickly and decreases delays to other aircraft waiting to use the runway.

At smaller airports, air traffic may not be sufficient to justify the construction of a parallel taxiway. In this case, taxiing is done on the runway itself and turnarounds should be constructed at the ends of the runways. A typical taxiway turnaround is shown as Fig. 19-6. When construction of a full parallel taxiway is not practicable, a partial parallel taxiway may be suitable.

*Table 19-5b*
**FAA Taxiway Dimensional Standards for Transport Airports**

| *Design Item* | *Airplane Design Group* | | | | | |
|---|---|---|---|---|---|---|
| | *I* | *II* | *III* | *IV* | *V* | *VI* |
| Taxiway safety area width, ft[a] | 49 | 79 | 118 | 171 | 197 | 262 |
| Taxiway width, ft | 25 | 35 | 50[b] | 75 | 75 | 100 |
| Taxiway edge safety margin, ft[c] | 5 | 7.5 | 10[d] | 15 | 15 | 20 |
| Taxiway shoulder width, ft | 10 | 10 | 20 | 25 | 35 | 40 |
| Taxiway centerline to: | | | | | | |
| Parallel taxiway centerline, ft | 69 | 103 | 153 | 225 | 251 | 340 |
| Fixed or movable object and to property line, ft | 44 | 65 | 94 | 139 | 153 | 205 |
| Fixed or movable object, ft | 39 | 54 | 80 | 118 | 131 | 172 |

[a]1 ft = 0.3048 m.

[b]For airplane design group III taxiways intended to be used by airplanes with a wheelbase equal to or greater than 60 ft, the standard taxiway width is 60 ft.

[c]The taxiway edge safety margin is the minimum acceptable distance between the outside of the airplane wheels and the pavement edge.

[d]For airplanes in design group III with a wheelbase equal to or greater than 60 ft, the taxiway edge safety margin is 15 ft.

SOURCE: *Airport Design Standards—Transport Airports*, FAA Advisory Circular AC 150/5300-12, March 3, 1985.

The design of the taxiway system will be determined by the volume of air traffic, the runway configuration, and the location of the terminal building and other ground facilities. The following general guidelines should be helpful in designing the taxiway system:

1. Taxiway routes should be direct and uncomplicated. Generally, taxiways should follow straight lines, and curves of long radius should be used when curves are required.
2. Whenever possible, taxiways should be designed so as not to cross active runways or other taxiways.
3. A sufficient number of taxiways should be provided in order to avoid congestion and complicated routes between runway exit points and the apron area.

At large and busy airports, the time an average aircraft occupies the runway frequently will determine the capacity of the runway system and the airport as a whole. This indicates that exit taxiways should be located conveniently so that landing aircraft can vacate the runway as soon as possible.

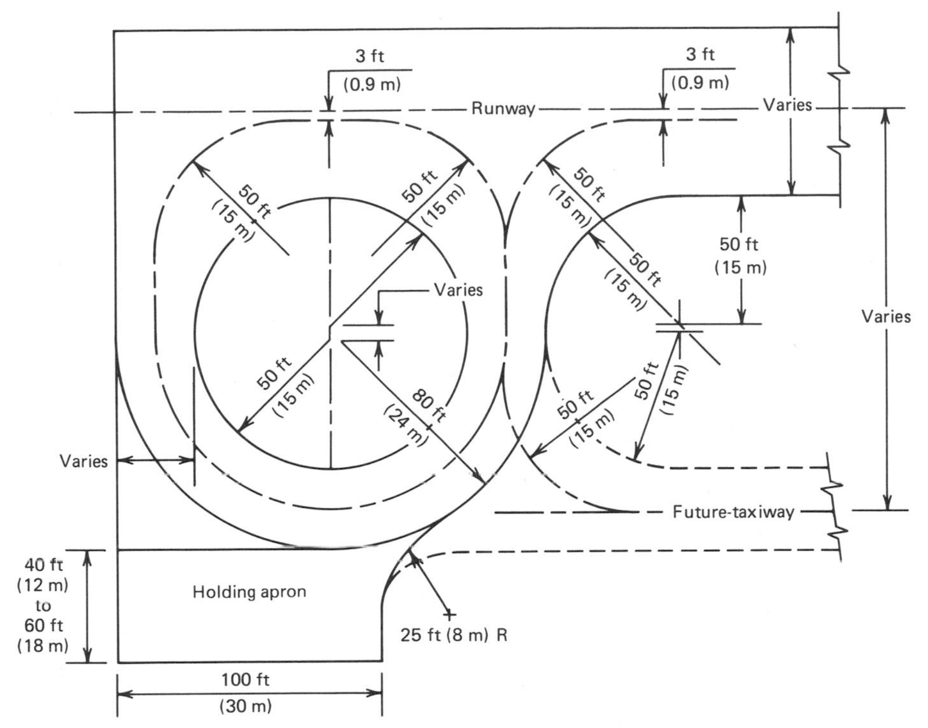


*Fig 19-6.* Circular taxiway turnaround for a utility airport. (Source: Utility Airports, Federal Aviation Administration, July 3, 1985.)

At utility airports, three exit taxiways will generally be sufficient: one at the center and one at each end of the runway.

Two common types of exit taxiways are illustrated by Fig. 19-7. Perpendicular exit taxiways may be used when the design peak-hour traffic is less than 30 operations per hour. To expedite the movement of landing aircraft from the runway, most modern air carrier airports provide exit taxiways that are oriented at an angle of about 30° to the runway centerline. This makes it possible for aircraft to leave the runway at speeds up to 60 mph, increasing the efficiency and capacity of the airport system.

The proper location of exit taxiways is dependent on the touchdown point and landing roll of the aircraft, as well as the configurations of the exits. The FAA [3] recommends that the points of curvature (P.C.s) of the angled type of exit be located at intervals beginning approximately 3000 ft from the threshold to approximately 2000 ft of the stop end of the runway. For the 90-degree type, the P.C.s of the taxiway exits should be located at intervals beginning about 3500 ft from the threshold to approximately 2000 ft from the stop end of the runway. Where the runway length exceeds 7000 ft, intermediate exits at intervals of approximately 1500 ft are recommended.

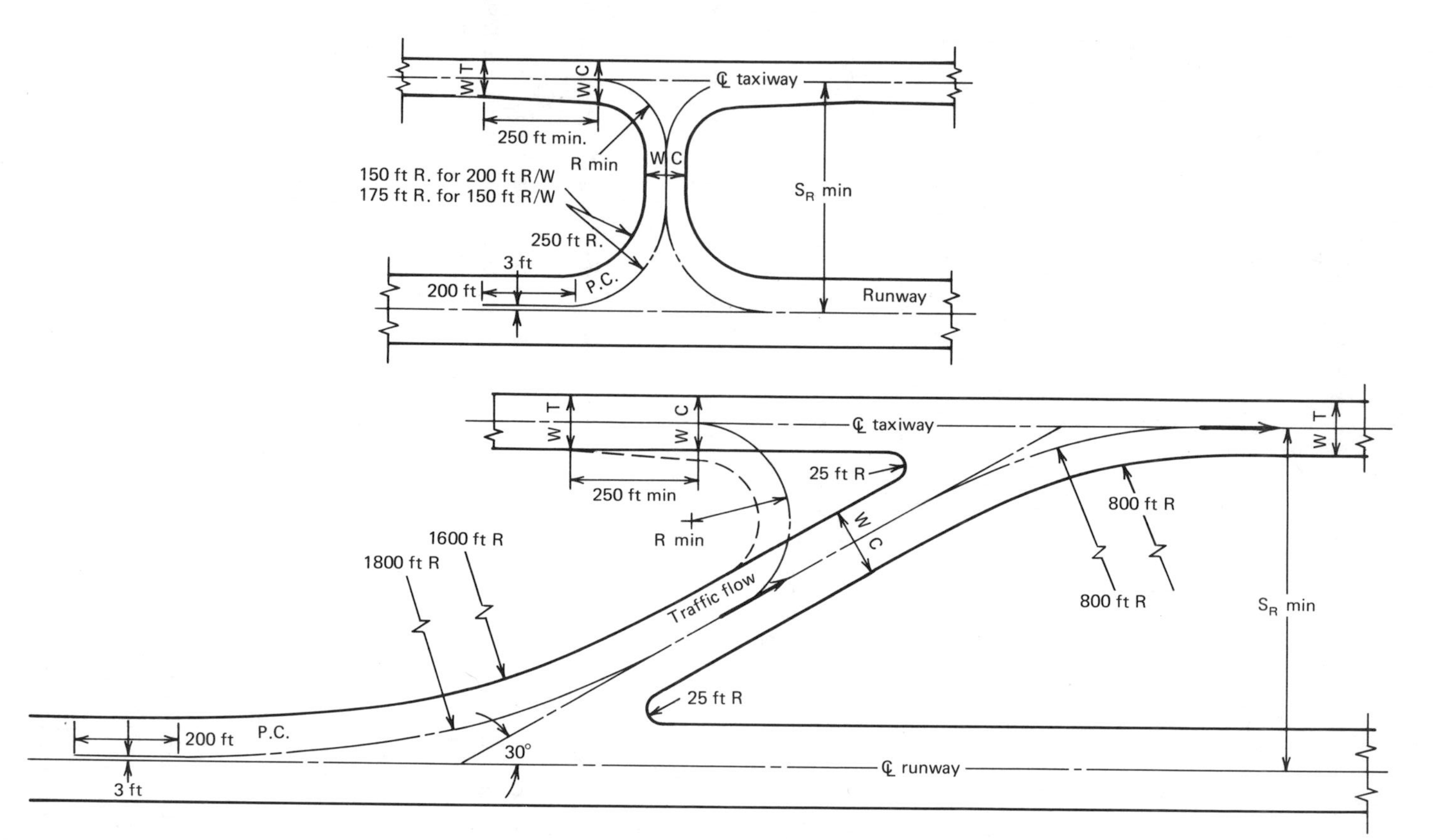


*Fig 19-7.* Recommended design for exit taxiways. (Source: Airport Design Standards-Transport Airports, Federal Aviation Administration, March 14, 1985.)

## 19-6. HOLDING APRONS

A holding apron is an area provided adjacent to the taxiway near the runway entrance for aircraft to park briefly while the cockpit checks and engine runups are made preparatory to takeoff. The use of holding aprons reduces interference between departing aircraft and minimizes delays at this portion of the runway system.

In the case of utility airports, the FAA [1] recommends that holding aprons be installed when air activity reaches 30 operations per normal peak-hour. Space to accommodate at least two but not more than four aircraft is recommended for small airports.

Approximate amounts of holding apron space can be determined by applying factors to the wingspans of aircraft that will be using the apron. To determine the diameter of the space required to maneuver and provide wing-tip clearance, the factors given in Table 19-6 should be multiplied by the aircraft wing span [4].

A sketch of a typical holding apron is shown in Fig. 19-6.

*Table 19-6*
**Factors for Determination of Holding Apron Space**

| *Aircraft Type* | *Factor* |
|---|---|
| Aircraft with single-wheel gear | 1.50 to 1.65 |
| Aircraft with dual-wheel undercarriages | 1.35 to 1.50 |
| Aircraft with dual-tandem gear | 1.60 to 1.75 |

SOURCE: *Airport Aprons,* FAA Advisory Circular AC 150/5335-2, January 27, 1965.

## 19-7. AIRPORT DRAINAGE

A well-designed drainage system is an essential requirement for the efficient and safe operation of an airport. Inadequate drainage facilities not only will result in costly damages due to flooding, but also may cause hazards to air operations and even result in the temporary closing of a runway or airport.

The design of a drainage system is based on the fundamental principles of open channel flow given in Chapter 14 and, in certain respects, the design procedures for airport drainage are identical to those for railway and highway drainage. The computation of runoff, for example, is accomplished by the rational formula, described in Chapter 14. On the other hand, an airport has certain peculiarities regarding its drainage requirements. Characterized by extensive areas and flat slopes and a critical need for the prompt removal of surface and subsurface water, airports usually are provided with an integrated drainage system. This system consists of surface ditches, inlets, and an underground storm drainage system. Typical drainage systems are shown in Reference 5.

The underground conduits are designed to operate with open channel flow, and because pipe sections in this system are long, uniform flow can be assumed. The hydraulic design of the channels and conduits, therefore, usually is accomplished by application of the Manning equation.

Storm drain inlets are placed as needed at low points and are typically spaced at 300-to 500-ft intervals. Manholes are provided to permit workers to inspect and maintain the underground system. Manholes are commonly placed at every abrupt change of direction and approximately every 300 to 400 ft on tangents. Typical inlets and manhole designs are given by Reference 5.

The design of a drainage system for an airport involves the following steps:

1. Using the proposed grading plan as a basis, a layout of the drainage system is made. The grading plan, which should show the proposed finished grade by 1-ft contour lines, will make it possible to select appropriate locations for drainage ditches and inlets and to determine the tentative layout of the underground pipe system.
2. Drainage structures and pipelines usually are identified by numbers or letters for easy reference in design computations.
3. For each drainage subarea, the runoff is computed by means of a rational formula. (See Chapter 14.) This involves the estimation of a runoff coefficient and a time of concentration (including flow time in the pipe system), and the selection of a design rainfall intensity from an intensity-duration curve similar to Fig. 14-1. In this connection, the FAA recommends a storm frequency of five years [5].
4. Beginning with the uppermost pipe section, the slope and pipe size is selected to carry the design flow. Design charts such as that shown as Fig. 19-8 are used for this purpose. As the design progresses along the line, each succeeding pipe section carries the water from its surface drainage area plus that contributed through its inlet structure.

Example problems for the actual design of a drainage system for a portion of an airport have been abstracted from the FAA publication *Airport Drainage* and are given here.

## EXAMPLE 19-4

DRAINAGE DESIGN WITHOUT PONDING. Suppose it is desired to design an underground drainage system to accommodate the surface flow from the apron and taxiways shown by Fig. 19-9. Inlets and line segments are first numbered and lengths are scaled from the map and recorded as shown by columns 1, 2, and 3 in Table 19-7.

Columns 4 through 10 record the data required for the calculation of runoff for various subareas in the system. These calculations are made by the rational formula which is described adequately in Chapter 14. It is noted that for a given inlet, time of concentration equals the inlet time (Column 4), or time required for water to flow

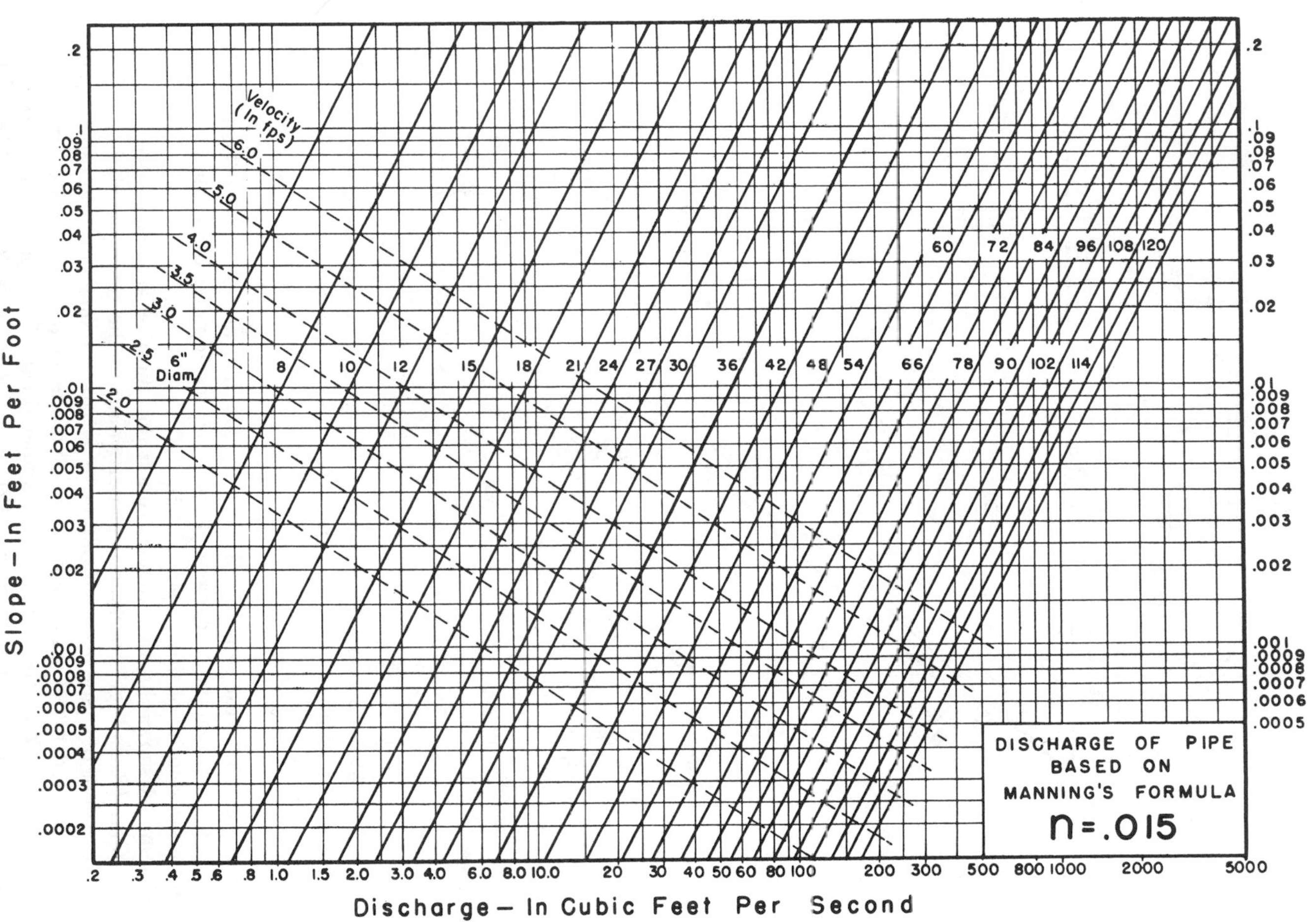


*Fig 19-8.* Design chart for uniform flow. (Source: *Airport Drainage,* Federal Aviation Administration, 1966.)

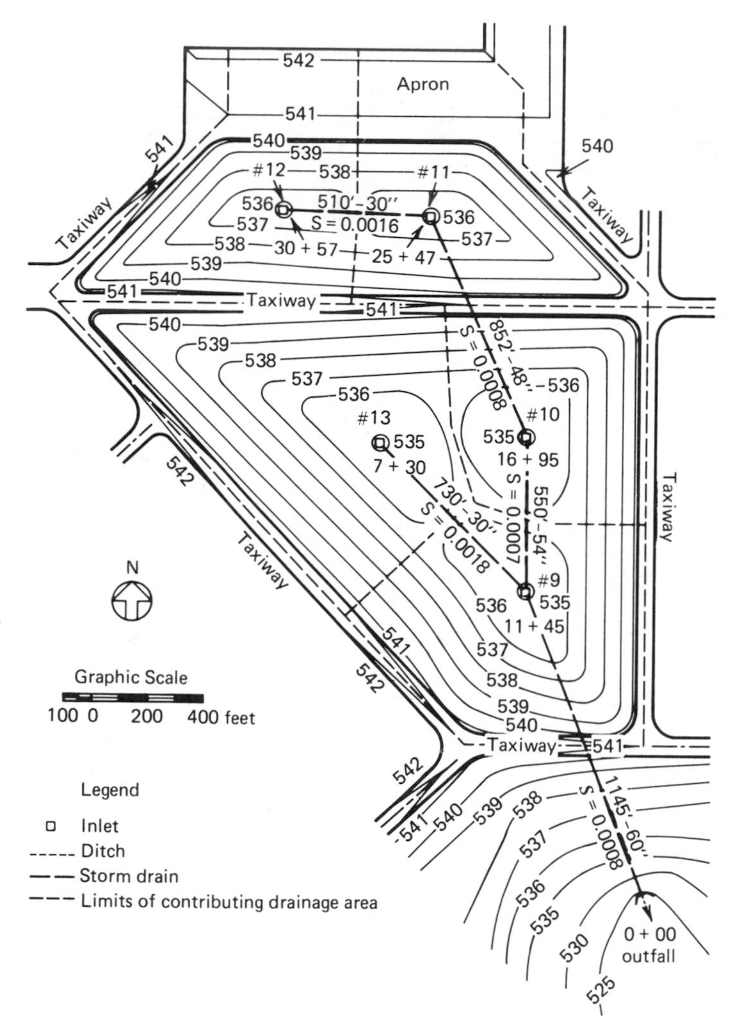


*Fig 19-9.* Portion of an airport showing drainage design. (Source: *Airport Drainage,* Federal Aviation Administration, July 1, 1970.)

overland from the most remote point in the subarea, plus flow time (Column 5) through the particular pipe segment. Flow time is computed by dividing the pipe length by the velocity of flow (Column 12).

Column 11 shows the accumulated runoff that must be accommodated.

Columns 12 through 16 show data pertaining to the hydraulic design of the system. The slope of a pipe section (Column 14) is based on such factors as topography, amount of cover, depth of excavation, elevation of the discharge basin or channel, and discharge velocity. With the slope and accumulated runoff, the size of pipe required (Column 13), velocity of flow (Column 12), and pipe capacity (Column 15) can be determined by means of a design chart for the Manning equation similar to Fig. 19-8. In this example, concrete pipe was used and a Manning roughness coefficient, $n$, of 0.015 was assumed.

*Table 19-7*
**Drainage System Design Data for Example 19-4**

| 1 | 2 | 3 | 4 | 5 | 6 | 7 | 8 | 9 | 10 | 11 | 12 | 13 | 14 | 15 | 16 | 17 |
|---|---|---|---|---|---|---|---|---|---|---|---|---|---|---|---|---|
| *Inlet* | *Line Segment* | *Length of Segment (ft)* | *Inlet Time (min)* | *Flow Time (min)* | *Time of Concentration (min)* | *Runoff Coefficient, C* | *Rainfall Intensity, I (in/hr)* | *Tributary Area, A (Acres)* | *Runoff, Q ($ft^3$/sec)* | *Accumulated Runoff ($ft^3$/sec)* | *Velocity of Drain (ft/sec)* | *Size of Pipe (in.)* | *Slope of Pipe (ft/ft)* | *Capacity of Pipe ($ft^3$/sec)* | *Invert Elevation* | *Remarks* |
| 12 | 12-11 | 510 | 52 | 3.4 | 52 | 0.49 | 1.98 | 14.69 | 14.25 | 14.25 | 2.8 | 30 | 0.0016 | 14.25 | 530.96 | ($n = 0.015$) |
| 11 | 11-10 | 852 | 53 | 5.0 | 55.4 | 0.53 | 1.90 | 14.72 | 14.82 | 29.07 | 2.8 | 48 | 0.0008 | 35.0 | 528.64 | See note below. |
| 10 | 10-9 | 550 | 39 | 3.3 | 60.4 | 0.35 | 1.78 | 11.97 | 7.46 | 36.53 | 2.8 | 54 | 0.0007 | 45.0 | 527.46 | See note below. |
| 13 | 13-9 | 730 | 62 | 3.9 | 62 | 0.35 | 1.76 | 21.50 | 13.24 | 13.24 | 3.1 | 30 | 0.0018 | 15.0 | 530.38 | |
| 9 | 9-out | 1145 | 42 | 4.2 | 65.9 | 0.35 | 1.70 | 16.05 | 9.55 | 59.32 | 3.3 | 60 | 0.0008 | 65.0 | 526.57 | |
| Out | | | | | | | | | | | | | | | 525.65 | |

Note: Time of concentration for inlet #11 is 55.4 min (52 + 3.4 = 55.4) which is the most time remote point for this inlet. Likewise time of concentration for inlet #10 is 60.4 min (52 + 3.4 + 5.0 = 60.4).

SOURCE: *Airport Drainage*, FAA Advisory Circular AC 150/5320-5B, July 1, 1970.

## EXAMPLE 19-5

**DRAINAGE DESIGN WITH PONDING.** Suppose we wish to drain the area shown in Fig. 19-10 with a single pipe to permit ponding of a short duration between the taxiways. It will be noted that this area is part of that shown in Fig. 19-9, except the contours have been changed to permit drainage by a single inlet. Suppose further that a 24-in. pipe is used to drain the area and that this pipe is to be installed on a 0.7 percent slope. Given the data shown below, what would be the maximum ponding expected within a 5-year period?

$$\text{Runoff coefficient} = 0.354$$
$$\text{Drainage area} = 49.52 \text{ acres}$$

From Fig. 19-8, it will be noted that the discharge for this pipe will be about 15.9 $ft^3$/sec. The runoff that can be accommodated by this pipe is a linear function of time and is plotted on Fig. 19-11.

Based on the rational formula, the amount of runoff ($ft^3$/sec) is:

$$Q = CIA$$
$$Q = 0.354 \times I \times 49.52$$

The rainfall intensity, $I$, which is dependent on duration, can be obtained for various durations from Fig. 14-1. In a 10-min period, for example, one would expect, for a 5-year flood frequency, a maximum runoff of

$$Q = 0.354 \times 4.68 \times 49.52 = 82.04 \text{ ft}^3/\text{sec}$$

Thus, the corresponding runoff value in $ft^3$ would be

$$\text{Runoff} = 82.04 \times 600 \text{ sec} = 49{,}224 \text{ ft}^3$$

Similarly, for a 30-min period,

$$Q = 0.354 \times 2.74 \times 49.52 = 48.03 \text{ ft}^3/\text{sec}$$

$$\text{Runoff} = 48.03 \times 1800 \text{ sec} = 86{,}485 \text{ ft}^3$$

Runoff values have been computed for other times and are plotted on Fig. 19-11. It will be noted from the graph that the maximum difference between the 24-in. pipe capacity line and the cumulative runoff curve occurs at a time of approximately 60 min, and that the maximum ponding value is:

$$P = 53{,}830 \text{ ft}^3$$

While this value is less than the storage capacity between the inlet and contour

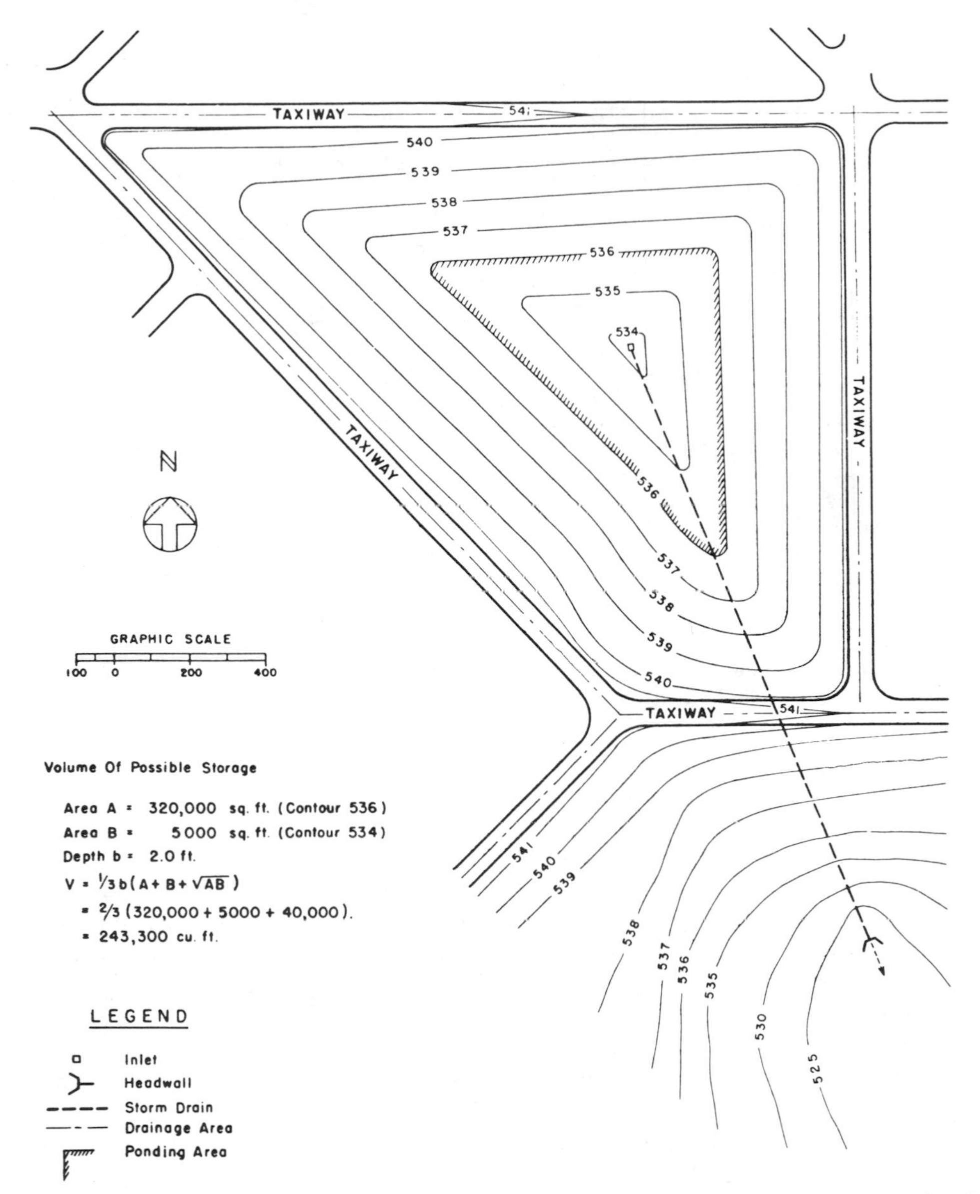


*Fig 19-10.* Example of providing for ponding area. (Source: *Airport Drainage,* Federal Aviation Administration, July 1, 1970.)

536, Fig. 19-11 indicates that it would require more than two hours for the 24-in pipe to empty the ponding area even when considering a 5-year flood. Since ponding over a long period of time is undesirable from the standpoint of safety and pavement performance, a larger culvert should be used. Fig. 19-11 shows that if a 30-in. pipe were used, ponding would occur for only slightly more than one hour.

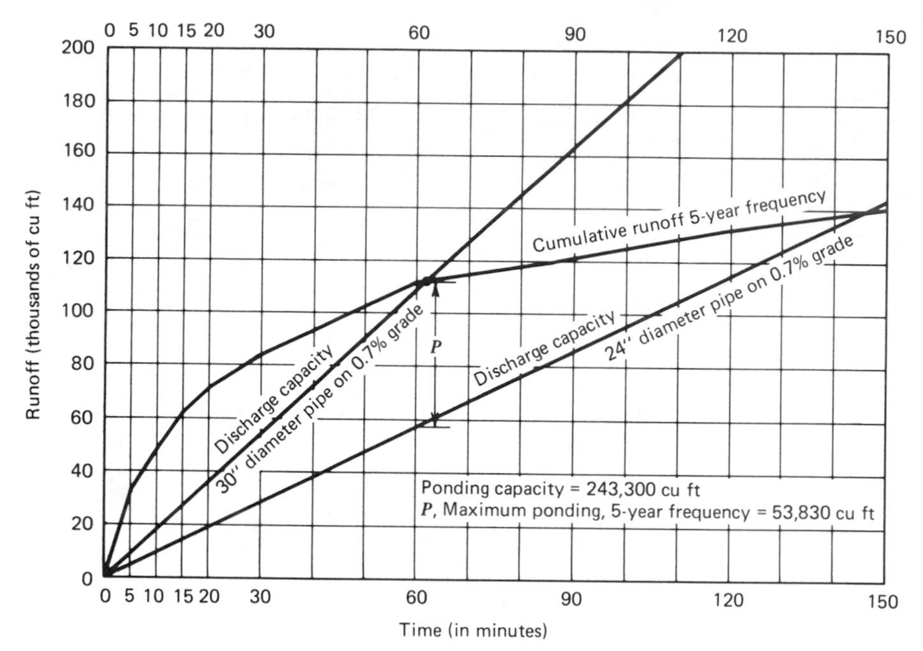


*Fig 19-11.* Cumulative runoff for ponding in Fig. 19-10. (Source: Adapted from *Airport Drainage,* Federal Aviation Administration, July 1, 1970.)

## 19-8. GRADING AND EARTHWORK

The proper grading of an airport is required to provide safe and efficient grades for aircraft operations to maintain good surface drainage, and to control erosion. Airport grading is characterized by wide, flat, and rounded slopes, with smooth transitions provided between graded and ungraded areas. Yet, because of the costs associated with earthwork operations, unnecessary grading should be avoided. Where future expansion of an airport is anticipated, the grading should be consistent with the ultimate proposed grades. Proposed grading operations usually are shown by means of a grading plan that shows original and proposed contour lines. (See Fig. 19-12.)

Grading quantities are usually computed by the average end area formula discussed in Chapter 14.

$$\text{Volume (yd}^3) = \frac{L}{27} \times \frac{(A_1 + A_2)}{2}$$

However, because of the relatively flat topography and large expanses of areas to be graded at airports, it may be advantageous to consider the areas enclosed by the original and final contour lines as end areas, $A_1$ and $A_2$. These areas can be measured with a planimeter and the contour interval becomes the length, $L$, used in the above formula. Where an embankment or excavation section ends between two contours, the vertical distance must be estimated.

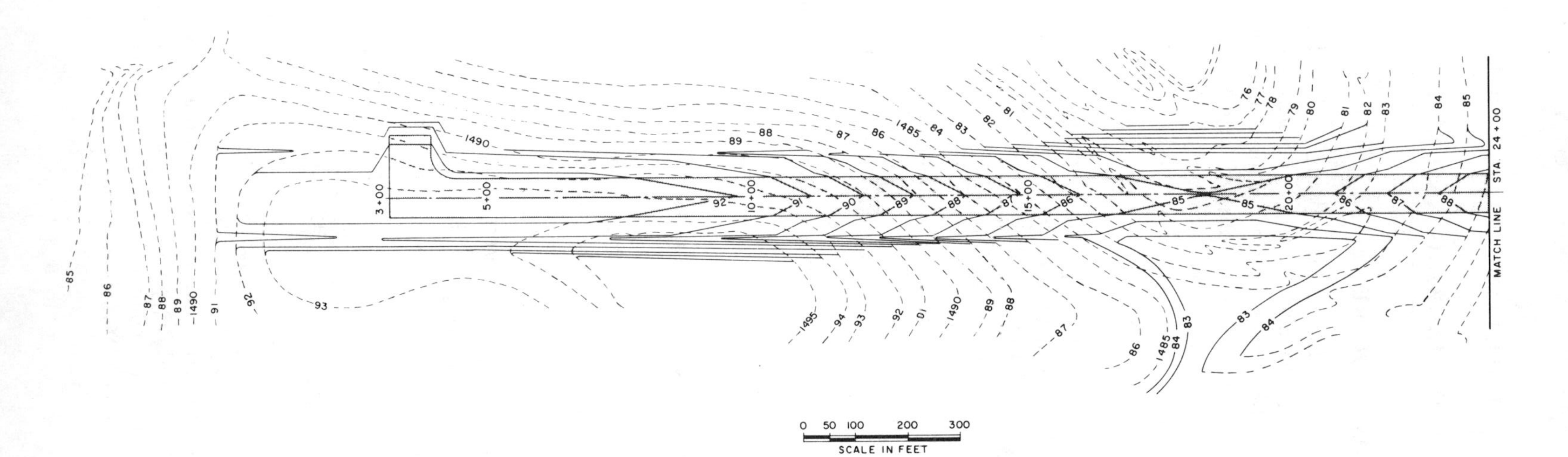


*Fig 19-12.* A typical grading plan. (Source: *Utility Airports,* Federal Aviation Administration, July 3, 1985.)

## AIRPORT MARKING AND LIGHTING

### 19-9. VISUAL AIDS REQUIREMENTS

During the major portion of flights while flying at altitude, pilots are assisted by magnetic compasses, gyros, and electronic devices. However, most instruments are not reliable when the aircraft is within about 200 ft of the ground. Thus, landings and takeoffs are accomplished largely "by eye," and during these critical operations various visual aids are placed at an airport to assist pilots. For operations in the daytime and in good weather, pilots are aided by airport markings. In inclement weather and at night they depend on airport lighting.

In addition to visual aids to help pilots locate and identify an airport or runway, they especially need assistance in properly approaching the runway and landing the aircraft. Walter and Roggenveen [6] have pointed out that, while landing, an airplane is a moving coordinate system that is approaching a stationary coordinate system, the runway. These coordinate systems are shown by Fig. 19-13. The aircraft not only may move about each of the three axes, but it also may rotate about them. Therefore, the pilot must rotate, orient, and translate the aircraft so that it coincides with the coordinate system of the runway.

To make a safe landing, a pilot must make correct judgments regarding:

1. Alignment—whether or not the plane is headed straight for the runway.
2. Roll—whether or not the aircraft is banked properly in relation to the ground surface.
3. The height of the aircraft above the runway.
4. Its distance from the end of the runway.

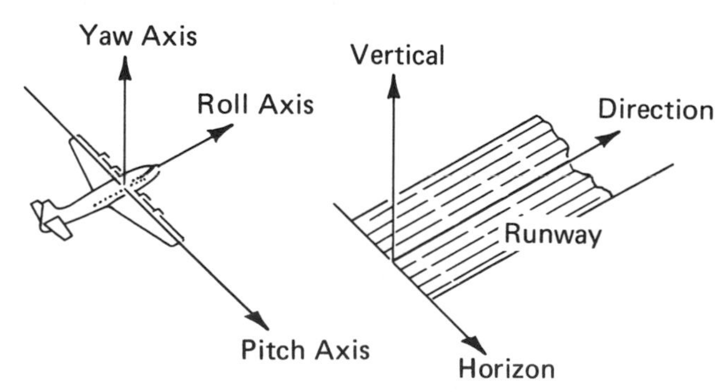


*Fig 19-13.* Runway and airplane coordinate systems. (Source: Walter and Roggeveen, "Airport Approach, Runway and Taxiway Lighting Systems," *Journal of the Air Transport Division, ASCE,* June 1958.)

In periods of good visibility these judgments can be made by reference to familiar objects on the ground such as trees and buildings. When the visibility is restricted due to inclement weather or darkness, the pilot requires visual aids in the form of airport markings and lights.

## 19-10. HANGAR AND STRIP MARKERS

At small airports that do not have paved runways, the landing strip may be identified by strip markings such as that shown by Fig. 19-14. For additional locational guidance, the name of the airport and an arrow showing true north may be provided as a hangar marker.

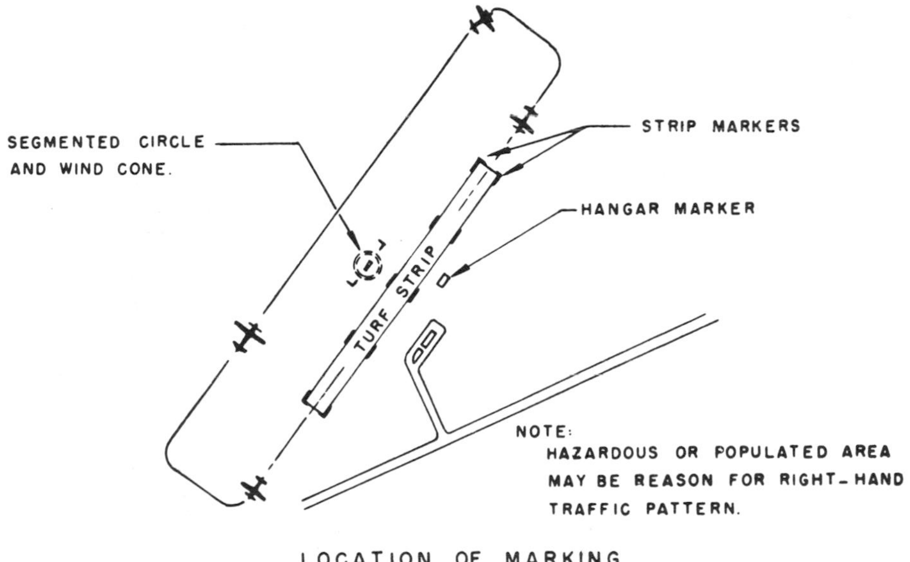

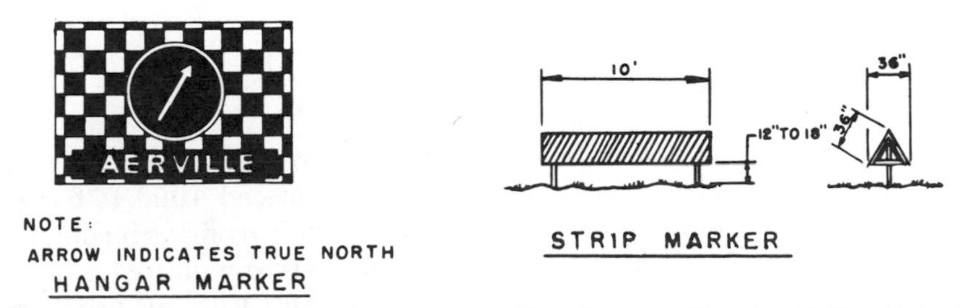


*Fig 19-14.* Hangar and strip markers for utility airports. (Source: *Utility Airports*, Federal Aviation Administration, July 3, 1985.)

## 19-11. THE SEGMENTED CIRCLE MARKER SYSTEM

Another visual aid commonly used at small airports is the segmented circle. It consists of a series of pointed markers arranged in the form of a circle of 50-ft radius. The segments of the circle are typically 3 ft in horizontal width and 6 to 12 ft in length. Typical details are shown by Fig. 19-15.

The segmented circle helps a pilot to identify an airport and provides a standard location for various signal devices. A wind cone or sock usually is placed at the center of the segmented circle. If the airport has more than one runway, a landing direction indicator may be provided at the center of the circle in the form of an arrow or tee. To indicate the landing pattern and orientation of landing strips, indicators may be placed at the periphery of the segmented circle.

## 19-12. RUNWAY AND TAXIWAY MARKING

Runway and taxiway marking consists of numbers and stripes that are painted on the pavement. Each end of a runway is marked with a number nearest one-tenth the magnetic azimuth of the runway centerline measured clockwise from the magnetic north. For example, a runway oriented N 10°E would be numbered 1 on the south end and 19 on the north end. Additional information is needed when two or more parallel runways are used, and the designations L, C, and R are used to identify left, center, and right runways, respectively. Where four or five parallel runways are used, two of the runways are assigned numbers of the next nearest one-tenth of the magnetic azimuth. The numbers and letters are about 60 ft tall and 20 ft wide.

For purposes of runway marking, the FAA groups runways into three classes: (1) visual runway, (2) nonprecision instrument runway, and (3) precision instrument runway. Recommended markings for these runway classes are shown by Fig. 19-16. In addition to runway numbers, the following runway markings may be used.

1. a dashed centered line stripe;
2. threshold markers;
3. side stripes;
4. markings indicating distance from the end of the runway.

The latter group of markings, which are used only on precision instrument runways, include: fixed distance markings placed 1000 ft from each runway end; touchdown zone markings placed 500 ft from each runway end; and additional markings provided at 500-ft intervals to indicate location with respect to the end of the runway. (See Fig. 19-16.) Runway markings are usually white but may be outlined in black to make them more conspicuous.

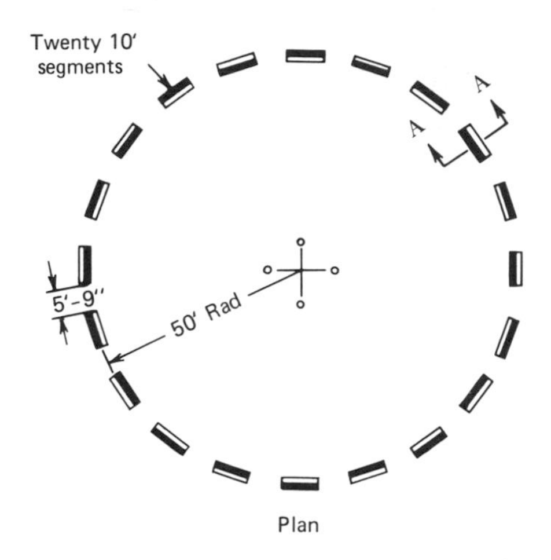


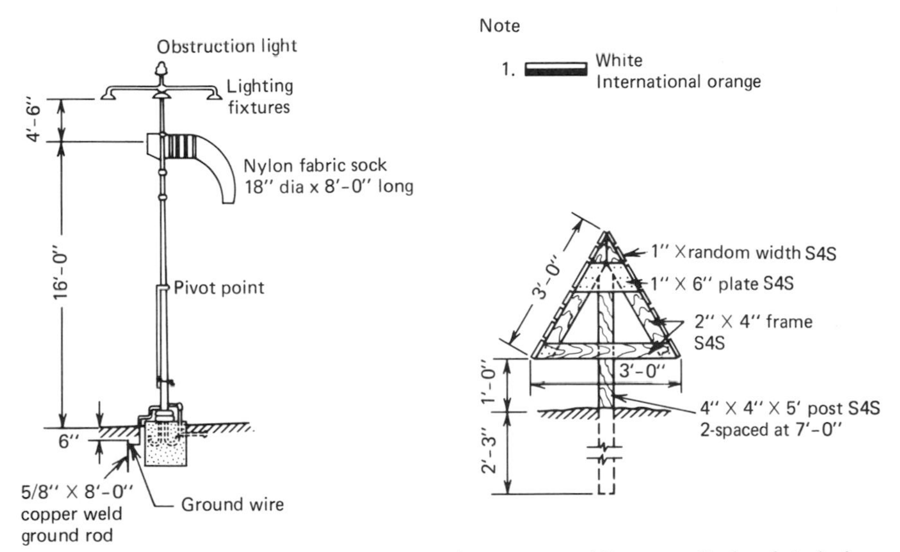


*Fig 19-15.* Segmented circle airport marker system. (Courtesy Federal Aviation Administration.)

Taxiway markings consist of a continuous centerline stripe and holding lines offset at least 100 ft from the runway edge. Taxiway markings are normally yellow.

## 19-13. AIRPORT LIGHTING

The lighting that should be provided for a given airport depends on the airport size, the nature and volume of air traffic at night and during periods of

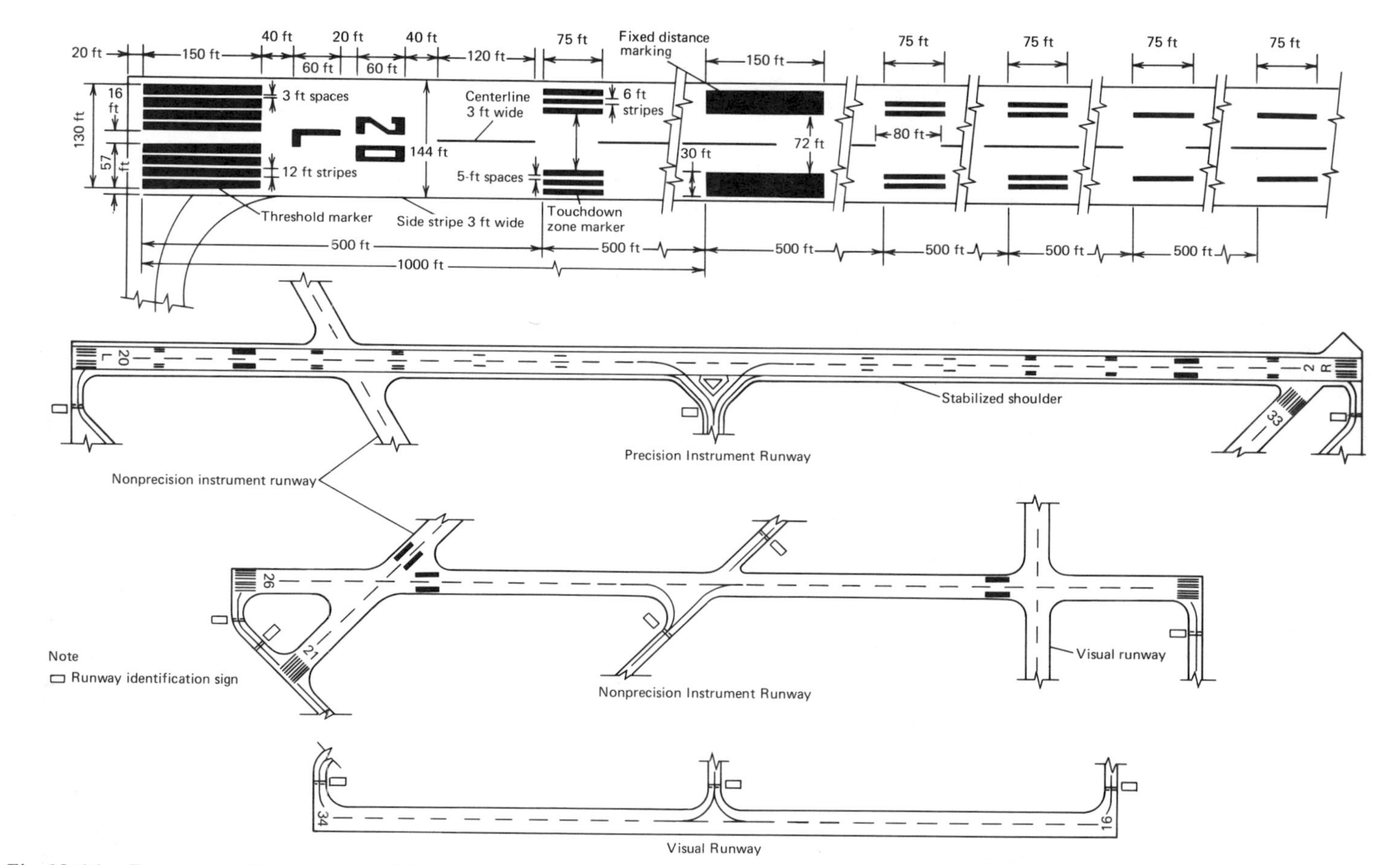


*Fig 19-16.* Runway and taxiway marking. (Source: *Marking of Paved Areas on Airports*, Federal Aviation Administration, November 14 1980.)

inclement weather, and local meteorological conditions. Five types of airport lighting are described in the following paragraphs:

1. Obstruction lighting.
2. Airport beacons.
3. Approach lighting.
4. Runway lighting.
5. Taxiway lighting.

This listing is not intended to be an exhaustive one, but it includes those major classes of airport lighting utilized to facilitate aircraft operations at night and in inclement weather. Illumination for wind cones and wind tees, ceiling light projectors, and lighting for aprons, hangars, and auto parking lots, though important, will not be described here due to limitations in space.

## 19-14. OBSTRUCTION LIGHTING

In the interests of air safety, obstruction lights must be placed on towers, bridges, smokestacks, and other structures that may constitute a hazard to air navigation. Single and double obstruction lights, flashing beacons, and rotating beacons are used to warn pilots during darkness or other periods of limited visibility of the presence of obstructions. A standard color, aviation red, is used for these lights.

The number, type, and placement of obstruction lights on a given structure will depend principally on its height. FAA standards for the lighting of obstructions are given in the publication *Obstruction Marking and Lighting* [7].

## 19-15. AIRPORT BEACONS

The location and presence of an airport at night is indicated by an airport beacon. While limited use has been made of a 10-in. "junior" beacon, a 36-in. beacon typically is used. The standard airport beacon rotates at a speed of 6 rpm and is equipped with an optical system that projects two beams of light 180 degrees apart. One light beam is green and the other is clear. A split beam is used to distinguish military airports.

## 19-16. APPROACH LIGHTING

Approach lights provide guidance to pilots during the few seconds it takes to travel from the approach area to the runway threshold. A pilot must be able to immediately identify and easily interpret approach lights. These lights provide the pilot with information regarding the alignment, roll, and height of the aircraft and its distance from the runway threshold.

The basic needs for approach lights have been determined objectively and thoughtfully on the basis of the glide angle, visual range, cockpit cutoff angle, and landing speed of the aircraft (taken to be 1.3 times the stalling speed). It is agreed, for example, that approach lights should extend at least 1400 ft from the runway threshold for nonprecision approaches and about 3000 ft for precision approaches. However, the FAA only recently has standardized approach lighting requirements, and a confusing variety of approach lighting systems remain in use. The FAA [8] recommends four standard approach lighting systems. These systems can be grouped into two categories.

1. Medium-intensity systems.
2. High-intensity systems.

All of the recommended FAA approach lighting configurations have a series of light bars installed perpendicular to the extended runway centerline at specified spacings. These light bars are composed of five lamps each. All of the recommended FAA configurations also feature a wide light bar installed at a distance of 1000 ft from the runway threshold. Sketches of FAA approach lighting systems are shown in Fig. 19-17. Except as noted, approach lights are white.

*Medium-Intensity Systems.* Medium-intensity approach lighting systems are economy systems that utilize 150-W lamps. These systems are recommended for utility airports. There are three recommended medium intensity approach configurations:

MALS—medium intensity approach lighting system.

MALSF—medium intensity approach lighting system with sequenced flashers. This is the same as MALS except it is equipped with three sequenced flasher lights. This system would be used where approach area identification problems exist.

MALSR—medium intensity lighting with runway alignment indicator lights. This system is similar to MALS except that five flashing lights are installed along the extended runway centerline at 200-ft spacings extending the total length of 2400 ft. Where the glide slope angle is less than 2.75 degrees, eight flashing lights are recommended providing a total length of 3000 ft.

*High-Intensity Systems.* These elaborate and expensive approach lighting systems are used most commonly at air carrier airports to facilitate ILS approaches. In addition to steady burning 300-W lights, these configurations feature a system of sequenced flashing lights. One such light is installed at each centerline bar starting 1000 ft from the threshold and extending outward to the end of the system. The sequenced flashing lights appear as a ball of light traveling at a high speed.

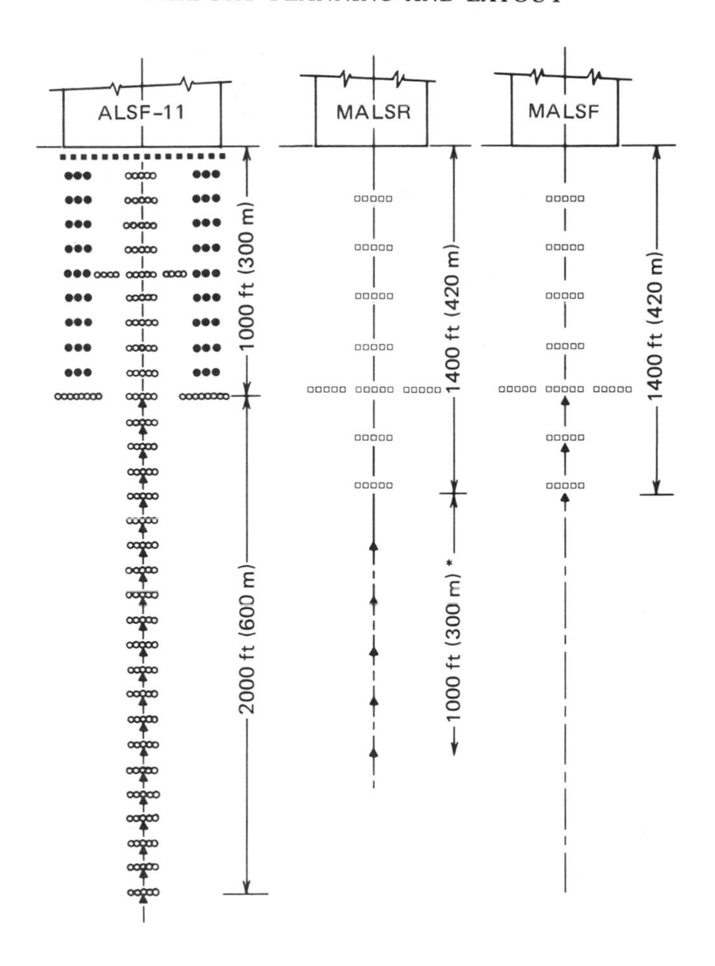


○ High-intensity steady burning white lights.
□ Medium-intensity steady burning white lights.
● Steady burning red lights.
▲ Sequenced flashing lights.
▪ ALS threshold light bar.

*Fig 19-17.* FAA approach lighting systems. (Source: *Airport Design Standards—Site Requirements for Terminal Navigational Facilities,* FAA Advisory Circular 150/5300-2D, including Change 1, 1980.)

Formerly, the FAA recommended two standard 3000-ft high intensity systems, ALSF-I and ALSF-II. The ALSF-I configuration was designed for Category I approach runways, that is, those that serve ILS landings under the least restrictive conditions of ceiling and visibility. The ALSF-I system is no longer recommended by the FAA.

The ALSF-II configuration was designed for operations in very poor visibility, Category II or lower. It is distinguished by two wide bands of barrettes or red lights located on either side of the centerline along the inner 1000 ft of the approach area. This configuration has a crossbar of white lights at 500 ft from the threshold in addition to the standard bar at 1000 ft. (See Fig. 19-17.)

## 19-17. OTHER VISUAL AIDS FOR AIRCRAFT APPROACHES

Additional visual aids systems used to facilitate easier and safer aircraft approaches are: (1) VASI, (2) PAPI, and (3) REIL.

*The Visual Approach Slope Indicator System (VASI).* This basically consists of light bars placed on each side of the runway, 600 ft from the runway end (downwind bars), and a second set of bars on each side of the runway, 1300 ft from the runway end (upwind bars). There are five basic VASI configurations, the 16-box, 12-box, 6-box, 4-box, and 2-box.

Each light box in the VASI system projects a split beam of light, the upper segment being white and the lower red. When a pilot makes an approach, he sees white lights if the approach is too high and red lights if the approach is too low. When a proper approach is made, the downwind bars appear white and the upwind bars appear red. This system is primarily intended for use in VFR weather conditions.

*Precision Approach Path Indicator (PAPI).* This two-color system consists of two or four identical light units in a row located on the left of the runway, perpendicular to the centerline. It is easily sited, set, and maintained, and capable of multipath interpretation. On approach the pilot will see the signal display shown in Fig. 19-18 indicating his position. PAPI systems will eventually replace VASI systems.

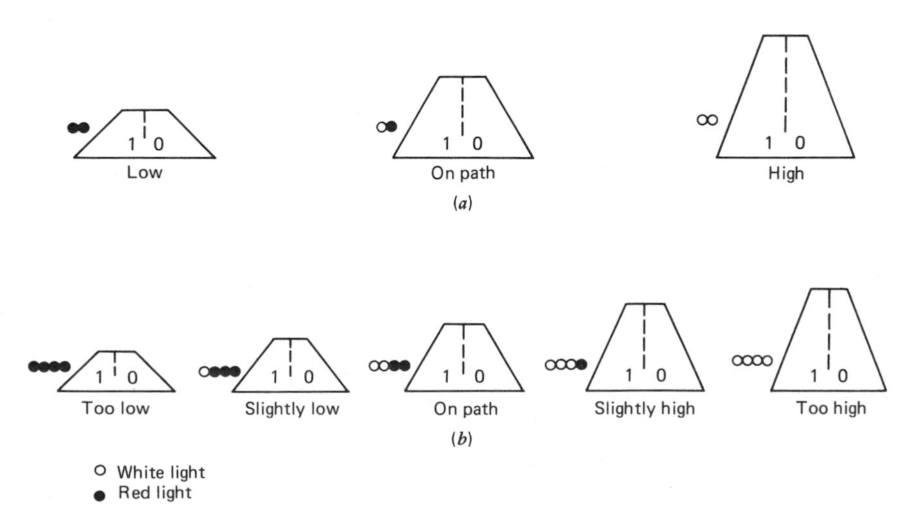


*Fig 19-18.* PAPI signal display. (*a*) Two-light system. (*b*) Four-light system. (Source: PAPI Systems, FAA Advisory Circular 150/5345-28D, May 23, 1985.)

*Runway End Identifier Lights (REIL).* These sometimes are placed at the ends of runways to provide rapid and positive identification of the approach end of a runway. The system consists of two synchronized flashing lights, one on each end of the runway threshold, the beams of which are aimed 10 to 15 degrees outside a line parallel to the centerline. The REIL system is used where there is a preponderance of confusing lights from off-airport sources such as motels, automobile lights, and so on. It normally would not be used if sequenced flashers are used in the approach lighting system.

## 19-18. THRESHOLD AND RUNWAY LIGHTING

Three classes of lights are installed in or near the runway to aid pilots during the final approach and to facilitate landings, rollouts, and takeoffs: (1) threshold lights; (2) runway edge lights; and (3) runway centerline and touchdown zone lights. Threshold lights, which consist of a line of green lights extending across the width of the runway, identify to the pilot the runway end and help him or her decide whether to complete the landing or to execute a missed approach.

Runway edge lights consist of lights installed not more than 10 ft from the pavement edge and spaced not more than 200 ft on centers. These lights are aviation white except that aviation yellow is used in the last 2000 ft of an instrument runway to indicate a caution zone.

Runway centerline and touchdown lighting systems are used in conjunction with electronic precision aids and high-intensity approach lighting systems under periods of limited visibility. These lights are built into the pavement surface. Runway centerline lights usually are installed at 50-ft centers along the runway centerline (offset a maximum of two feet to clear centerline marking). These lights are white except for the end 3000-ft sections of the runway.

In the zone from 3000 to 1000 ft to runway end, alternately red and white lights are displayed to a departing pilot. In the final 1000 ft, all red lights are displayed along the centerline. Runway centerline lights are bidirectional, so that red lights in the 3000-ft end zones appear as white lights to pilots approaching to land.

Touchdown zone lights consist of rows of transverse bars of white lights mounted flush in the pavement along 100-ft centers. Such lights are employed in conjunction with high-intensity approach lighting systems to provide pilots additional roll guidance along a 3000-ft length beyond the runway end.

## 19-19. TAXIWAY LIGHTING

Taxiway lighting provides guidance for pilots for maneuvering along the system of taxiways that connect the runways and the terminal and hangar aprons. The conventional taxiway lighting consists of omnidirectional blue lights located on each side of the taxiway pavement. These lights are offset not more than 10 ft from the pavement edge and spaced longitudinally not more than 200 ft apart. Much shorter spacings are required on short curves and at intersections.

Instead of taxiway edge lights, taxiway centerline lights may be installed in new construction, and may supplement taxiway edge lights where operations occur in low visibility or where taxiing confusion exists.

The taxiway centerline lighting system consists of single semiflush lights (with less than ⅜-in. protrusions above the pavement surface) inset in the taxiway pavement along the centerline. These lights are steady burning and have a standard color of aviation green.

Detailed information on the design, installation, testing, maintenance, and inspection of taxiway edge and centerline systems are given, respectively, by References 9 and 10.

## 19-20. AIRPORT PAVING

It is regretted that space limitations preclude the inclusion of a discussion of the subject of airport paving. The reader is referred to the FAA publication *Airport Paving* [11] for a complete discussion of this important topic.

## PROBLEMS

1. What length of runway is required for a general utility airport that is 6000 ft above sea level and has a normal maximum temperature of 75°F? The effective gradient is 1.2 percent.

2. What length of runway is required for a Boeing 707-300 series aircraft given the following design conditions?

| | |
|---|---|
| Normal maximum temperature | 75°F |
| Airport elevation | 4000 ft |
| Takeoff weight | 263,000 lb |
| Maximum landing weight | 180,000 lb |
| Effective runway gradient | 0.85 percent |

3. What length of runway is required for a Boeing 727-200 aircraft given the following design conditions?

| | |
|---|---|
| Airport elevation | 4000 ft |
| Normal maximum temperature of hottest month | 80°F |
| Effective runway gradient | 0.65 percent |
| Length of haul | 490 mi |

| | |
|---|---|
| Average fuel consumption | 19 lb/mi |
| Passenger load | 26,800 lb |
| Weight of empty aircraft plus reserve fuel | 114,800 lb |

**4.** Profile grade data for a proposed transport airport runway are given here. Does the proposed longitudinal grade design conform with the requirements of the FAA? The first vertical curve is 1600 ft long and the second is 1000 ft long.

| *Station* | *Grade (Percent)* | *Comment* |
|---|---|---|
| 0 + 00 | | Begin runway |
| | −0.85 | |
| 30 + 50 | | P.I. station No. 1 |
| | +0.65 | |
| 55 + 50 | | P.I. station No. 2 |
| | −0.40 | |
| 92 + 00 | | End runway |

**5.** Determine the following dimensions for an air carrier airport designed to accommodate a B-727-000 aircraft.

Runway centerline to taxiway centerline
Taxiway width on tangent
Taxiway width on curve
Radius of taxiway centerline on curves

**6.** Obtain rainfall data for your locality from the National Weather Service or from some other source. Prepare a rainfall intensity-duration curve similar to Fig. 14-1 (5-year curve). Using the other data given in Example 19-4, design a drainage system for the apron and taxiways shown in Fig. 19-9.

**7.** Design a drainage system for the apron and taxiways shown in Fig. 19-9 given the runoff values given here. Other required data are given by Table 19-7.

| *Inlet* | *Runoff, Q* |
|---|---|
| 12 | 29.4 |
| 11 | 23.1 |
| 10 | 10.7 |
| 13 | 17.4 |
| 9 | 14.8 |

## REFERENCES

1. *Utility Airports—Air Access to National Transportation,* FAA Advisory Circular 150/5300-4B, including Change 9, September 18, 1987.
2. *Runway Length Requirements for Airport Design,* FAA Advisory Circular 150/5325-4, including Change 14, September 27, 1978.
3. *Airport Design Standards-Transport Airports,* FAA Advisory Circular 150/5300-12, including Change 1, March 14, 1985.
4. *Airport Aprons,* FAA Advisory Circular 150/5335-2, January 27, 1965.
5. *Airport Drainage,* FAA Advisory Circular 150/5320-5B, July 1, 1970.
6. Walter, C. Edward, and Vincent J. Rogeveen, "Airport Approach, Runway and Taxiway Lighting Systems," *Journal of the Air Transport Division,* American Society of Civil Engineers, June 1978.
7. *Obstruction Marking and Lighting,* FAA Advisory Circular 70/7460-1, October 22, 1985.
8. *Airport Design Standards—Site Requirements for Terminal Navigation Facilities,* FAA Advisory Circular 150/5300-2D, including Change 2, January 30, 1987.
9. *Runway and Taxiway Edge Lighting System,* FAA Advisory Circular 150/5340-24, including Change 1, November 25, 1977.
10. *Taxiway Centerline Lighting System,* FAA Advisory Circular 150/5340-19, November 14, 1968.
11. *Airport Pavement Design and Evaluation,* FAA Advisory Circular 150/5320-6C, including Change 1, August 30, 1979.

# *PART 6*

# *DESIGN OF WATER TRANSPORTATION FACILITIES*

*The final part of this book is concerned with the layout and design of the infrastructure for water transportation—the harbors, the channels, and the ports. This part begins with a discussion of the nature of water movements and describes special problems in designing for the coastal environment. It lists desirable features for harbor sites and describes various structures used to construct artificial harbors and to protect coastal areas. The book then discusses various aspects of the planning, layout, and design of the aprons, buildings, and other components of modern ports.*

# Chapter 20

# Introduction to Water Transportation

## 20-1. INTRODUCTION

Since ancient times, water transportation has broadened humanity's horizons and has influenced profoundly the growth and development of civilization. Historians report that as early as 6000 B.C. the Egyptians had ships with masts and sails, and galleys were used on the Nile River as early as 3000 B.C. During the reign of King Solomon (circa 961–922 B.C.), Phoenician galleys sailed from Biblical Tyre and Sidon bringing copper from Cyprus, papyrus from Egypt, and ivory, gold and slaves from Africa. These large, low, and typically one-decked vessels were propelled by both sails and oars. Galleys, which often were manned by slaves, were used by Rome in her war with Carthage (140 A.D.). These vessels continued to be used throughout the Middle Ages, especially in the Mediterranean Sea.

In early America:

> River boats wre nothing more than huge rafts called *flatboats* or broad-horns. Generally, they were flat-bottomed and boxlike, covered from stem to stern. The flatboat was a one-way vessel, dependent entirely upon currents for propulsion with only occasional guidance from its handlers. At the end of its downstream run, it was usually broken up and the lumber sold.
>
> The *keelboat* began to appear on the rivers at about the turn of the nineteenth century. It was a long, narrow vessel with graceful lines, sturdily built to withstand many trips both downstream and upstream. The keelboat could carry as much as 80 tons of freight. It was floated downstream under careful guidance, and cordelled upstream. Cordelling took two forms: the crew walked along the bank and pulled the keelboat with ropes; or they literally pushed it upstream with iron-tipped poles which extended to the bottom of the river. One historian has estimated that there were as many as 500 keelboats operating on the Ohio River and its tributaries by 1819.
>
> The *steamboat* was invented in 1807; and in 1811 the river steamer *New Orleans* was launched in Pittsburgh and went into operation between there and New Orleans. By 1835 New Orleans was posting the arrival of over 1000 steamboats per year. Records indicate that by 1852 the public landing at Cincinnati was recording the arrival of steamboats at an annual rate of 8,000, about one per hour.

The tonnage handled by steamboats on the rivers of the United States at the height of the packet boat era, just before the War Between the States, is reported to have exceeded the tonnage handled by all the vessels of the British Empire [1].

Although the feasibility of steamboat travel in the open sea was demonstrated by Col. John Stevens in 1809, sailing vessels continued to dominate ocean transportation until shortly before the Civil War.

Scheduled ocean travel was initiated on January 5, 1818, when the *James Monroe,* a 424-ton ship, sailed from New York to Liverpool. This packet or liner service, which gave uncommon services to passengers and fast movement of freight, was highly successful. Larger, faster, and more expensive ships were built, additional lines were organized, and the packet service was extended.

The first half of the nineteenth century was a prosperous period for shipping companies. The quest to sail "at a fast clip" resulted in the design and building of over 400 clipper ships during the period from 1846 to 1855. These vessels traveled at speeds that rival some of the faster commercial ships in service today. Despite the speed and classic beauty of these vessels the clipper ships were soon replaced by steam ships, and following the Civil War, there was a sharp decrease[1] in shipping under the American flag. This decline in U.S. influence in ocean shipping is felt to the present day.

## 20-2. RECENT GROWTH IN WATER TRANSPORTATION

In recent years, substantial increases in water transportation activity have been noted. These increases are attributed to growing world population, the development of new products and new sources of raw materials, and general industrial growth, especially in the petroleum industry. Accompanying the increases in water transportation tonnage, larger ships have been built and innovative storage and loading facilities have been developed. These changes have created an ever-increasing need for the most modern and efficient port facilities.

Data furnished by the Maritime Administration, U.S. Department of Transportation, indicate that there has been substantial growth in the number and gross tonnage[2] of the merchant ships of the world in recent years:

[1]For all practical purposes the river fleet was destroyed during the Civil War and was not rebuilt. There was little progress in inland water transportation from the end of the Civil War until about 1920 [1].

[2]Gross tonnage is a cubic measurement: one gross ton equals 100 $ft^3$ of storage space. The data given are for ocean going ships of 1000 gross tons and over.

| *Year* | *Ships* | *Gross Tons* |
|---|---|---|
| 1958 | 16,966 | 112,314,000 |
| 1968 | 19,361 | 184,242,000 |
| 1978 | 24,512 | 378,909,000 |
| 1985 | 25,424 | 391,979,000 |

Gross tonnage has grown at a much greater rate than has the number of ships, reflecting the trend to larger ships.

Similar growth has been experienced in inland water transportation as indicated by the data furnished by the American Waterways Operators, Inc. [1, 2]:

| *Year* | *Towing Vessels* | *Barges* | *Capacity (net tons)* |
|---|---|---|---|
| 1958 | 4,169 | 15,221 | 13,771,757 |
| 1968 | 4,240 | 18,416 | 20,940,261 |
| 1978 est. | 4,500 | 30,750 | 42,500,000 |
| 1984 | 4,993 | 33,899 | 49,146,905 |

Today, nearly 99 percent of overseas freight tonnage is transported by ships and approximately 15 percent of the freight (ton-miles) within the continental United States moves on the inland waterway system.

## 20-3. THE NATURE OF WATER TRANSPORTATION

By its nature, water transportation is most suitable for bulky and heavy commodities that have to be moved long distances and for which time of transport is not a critical factor.

Table 20-1 lists the most common types of cargoes moved on U.S. oceanborne foreign trade routes by liner, nonliner, and tanker service. It is apparent from this table that liquid bulk cargo (especially petroleum and petroleum products) and dry bulk cargo (such as coal, cereal, and ores) are the dominant types of shipments.

Other common classes of waterborne cargo include machinery, fruits and vegetables, wood and timber, and motor vehicles and parts. In addition, a wide variety of manufactured commodities are shipped as packaged goods (general cargo) or in large sealed boxes called containers. (See Fig. 20-1.) The use of containers speeds dramatically the handling of certain types of freight while significantly decreasing shipping costs.

*Table 20-1*

**Top Ten Commodity Groups Shipped in 1984 by Liner, Nonliner, and Tanker Service**

| *Rank* | *Commodity Group* | *Thousands of Long Tons*[a] |
|---|---|---|
| *Liner Service* | | |
| 1 | Paper, paperboard, pulp, and waste paper | 6,370 |
| 2 | Iron and steel, manufactures of metal | 5,025 |
| 3 | Fruits and vegetables | 3,811 |
| 4 | Cereal and cereal preparations | 3,756 |
| 5 | Alcohol and nonalcoholic beverages | 2,296 |
| 6 | Textile fibers and their waste | 1,906 |
| 7 | Wood, lumber, and cork | 1,713 |
| 8 | Nonmetallic mineral manufactures | 1,700 |
| 9 | Synthetic resins, rubber, and plastic | 1,687 |
| 10 | Miscellaneous manufactured articles | 1,576 |
| *Nonliner Service* | | |
| 1 | Cereal and cereal preparations | 95,800 |
| 2 | Coal, coke, and briquets | 55,960 |
| 3 | Metalliferous ore and scraps | 40,380 |
| 4 | Crude fertilizer and minerals | 31,538 |
| 5 | Wood, lumber, and cork | 19,837 |
| 6 | Oilseeds, oilnuts, and kernals | 18,847 |
| 7 | Iron and steel | 18,276 |
| 8 | Petroleum and petroleum products | 12,222 |
| 9 | Manufactured fertilizers | 12,084 |
| 10 | Feed stuff for animals | 9,616 |
| *Tanker Service* | | |
| 1 | Petroleum and petroleum products | 241,192 |
| 2 | Organic chemicals | 7,824 |
| 3 | Inorganic chemicals | 6,402 |
| 4 | Gas—natural and manufactured | 3,462 |
| 5 | Vegetable oils and fats | 1,705 |
| 6 | Sugar, molasses, and honey | 1,485 |
| 7 | Chemical products and materials | 1,354 |
| 8 | Animal oils and fats | 1,256 |
| 9 | Manufactured fertilizers and materials | 671 |
| 10 | Cereal and cereal preparations | 557 |

[a] 1 long ton = 2240 lb.

SOURCE: *United States Oceanborne Foreign Trade Routes,* Maritime Administration, U.S. Department of Transportation, August, 1986.

*Fig 20-1.* A modern container ship. (Courtesy United States Lines.)

Some of the major commodities shipped on the inland waterways system are:

Petroleum and petroleum products
Coal and coke
Iron, steel, and aluminum ingots and plate
Scrap iron
Grains
Chemicals
Sugar, syrup, and molasses
Forest products and paper
Sand and gravel
Animal feeds and fertilizer

Freight movements on the Great Lakes consist predominantly of iron ore, steel, coal, and grain.

## 20-4. THE EFFECTS OF SHIP CHARACTERISTICS AND CARGO TYPE

The planning and design of port and harbor facilities is strongly dependent on the characteristics of the ships to be served and the types of cargo to be handled. The relevant characteristics of ships and the port characteristics they influence are [3]:

1. Main dimensions:

    a. Length, which governs the length and layout of single-berth terminals, the length of stretches of quay, and the location of transit sheds. The length also influences the widths and bends of channels and the size of port basins;
    b. Beam, which governs the reach of cargo-handling equipment and influences the width of channels and basins; and
    c. Draft, which governs the water depth along berths, in channels and in basins.

2. Cargo-carrying capacity, which governs the (minimum) storage requirement for full ship loads and influences the handling rate for loading/unloading installations;
3. Cargo-handling gear such as cranes and pumps, which govern cargo-handling rates, in particular for liquid bulk cargo, and which influence the need for port equipment, such as quay cranes and booster pumps. Cranes as well as derricks and winches, also restrict the ship motion, that may be tolerated without interruption of cargo-handling operations;
4. Types of cargo units, such as bulk, containers, pallets, and break bulk general cargo, which determine requirements for handling equipment (if ship's gear is not used) and storage;
5. Shape, hull strength and motion characteristics:

    a. Size and shape of hull and superstructure, which influence berth and fender system layout, as well as current and wind forces required for design of moorings and fenders;
    b. Ultimate strength of side shell and supporting members, which influences fender design;
    c. Ship motions induced by wind waves, swell, or long waves, which influence mooring and fender designs; and
    d. Superstructure configuration, which influences positioning and

design of port-handling equipment to avoid damage to superstructure and handling equipment.

6. Mooring equipment, such as ropes, wires, and constant tension winches, which influences the motion of ships and their mooring forces; and

7. Maneuverability at low speed, which influences channel, port entrance, and basin layout, as well as the need for harbor tugs.

Typical dimensions and operating characteristics of waterborne vessels are given in Chapter 4.

In the paragraphs that follow, coastal environmental conditions that complicate the engineering design and shorten the useful life of port and harbor facilities will be discussed.

Due to space limitations it will not be possible to discuss a number of topics that relate to water transportation. Specifically, the important topics of river hydraulics, and the design of locks, dams, and canals will not be covered. For information on these topics, the reader may refer to other textbooks or to the publications listed at the end of this chapter.

# DESIGN FOR THE COASTAL ENVIRONMENT

## 20-5. THE COASTAL ENVIRONMENT

The design of durable port and harbor facilities is one of the most challenging problems that faces the engineer. The environment of the seacoast is harsh and corrosive, and water transportation facilities must be designed to withstand the various destructive biological, physico-chemical, and mechanical actions that are inherent to the coastal environment. Enormous forces of winds, waves, and currents are imposed on port and harbor structures. Wood structures must withstand the forces of decay and the attack of termites and other biological life. Concrete structures must be designed and constructed to highest engineering standards to prevent rusting of the reinforcement and spalling of the concrete. Without protection, steel structures corrode and do not last long in the coastal environment.

## 20-6. WIND

Wind is the approximate horizontal movement of air masses across the earth's surface. Winds result from changes in the temperature of the atmosphere and the corresponding changes in air density. Wind exerts a pressure against objects in its path which depends on the wind velocity. The equation for the calculation of wind pressure on a structure is:

$$p = KV^2 \tag{20-1}$$

where $p$ is expressed in lb/ft$^2$, wind velocity, $V$, is expressed in mph, and $K$ is a factor that depends principally on the shape of the structure. The values of $K$ most frequently used range from 0.0025 to about 0.0040.

Considerable judgment is required in computing wind forces on coastal structures and port facilities. It should be remembered that loading equipment generally will not be used when winds exceed about 15 mph, and ships usually will not remain alongside a wharf during a severe storm.

## 20-7. WAVES

While the design of buildings and other structures must accommodate wind loads, our principal interest in wind in coastal design lies in its contribution to the formation of waves. Waves may be caused by earthquakes, tides, and manmade disturbances such as explosions and moving vessels; however, the waves of principal interest in coastal design are those formed by winds.

When wind moves across a still body of water, it exerts a tangential force on the water surface that results in the formation of small ripples. These irregularities tend to produce changes in the air stream near the water surface. Pressure differentials in the air stream are formed which cause the water surface to undulate. At the wind continues, this process is repeated and the waves grow.

The form and size of water waves have been the subject of considerable scientific observation and research. There is now general acceptance that the surface of deep-water waves is approximately trochoidal in form. According to the trochoidal theory, wave movement can be described by assuming that individual particles are not translated, but rotate in a vertical plane about a horizontal axis. This is consistent with the observed tendency of a floating object in deep water to rise and fall and oscillate but not to be translated by the waves.

Figure 20-2 shows the surface of an ideal deep-water wave as a trochoid formed by the rotating particles of water. The trochoid is described by a point on a circle that rotates and rolls in a larger concentric circle. The center of the circle moves along a line that lies above the still water level. The amount this line is elevated above the still water level depends on the wave steepness. The diameter of the smaller circle is equal to the wave height and the circumference of the larger circle equals the wave length. The line thus described represents the wave surface. Particles beneath the wave surface also tend to rotate similarly in circular paths, but the radii decrease rapidly and roughly exponentially with depth.

The speed of propagation of the wave form, or wave celerity, in ft/sec,

$$v = \frac{L}{T} \tag{20-2}$$

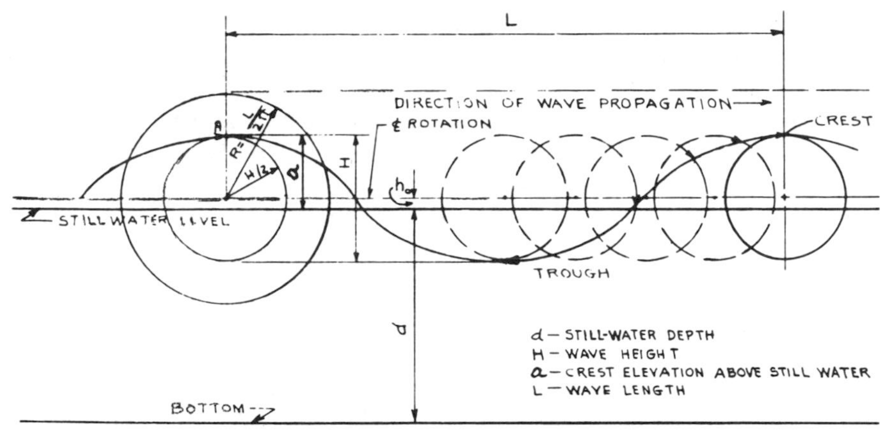


*Fig 20-2.* Deep-water wave characteristics. (Source: Alonzo Quinn, *Design and Construction of Ports and Marine Structures,* Second Edition, McGraw-Hill, 1972. By permission of the publisher.)

where

$L$ = wave length, or distance between consecutive crests, ft
$T$ = wave period, or time for wave to travel one wave length, sec

The speed of waves in deep water[3] is approximately equal to the velocity acquired by a body falling freely through a height equal to one-half the radius of a circle, the circumference of which equals the wave length. Thus:

$$v = \sqrt{2g\frac{1}{2}\frac{L}{2\pi}} = \sqrt{\frac{gL}{2\pi}} \tag{20-3}$$

In this equation, $g$ is acceleration due to gravity.

In shallow and transitional water, the paths of water particles are influenced by the frictional forces of the sea bed. This causes the orbit of the water particles to become approximately elliptical with the major axis horizontal.

Waves in shallow and transitional water are complex and many theories have been proposed to describe them. It is agreed that the velocity of such waves is a function of the water depth.

The following equation should give suitable estimates of wave velocities in depths less than 1/25 of the wave length:

$$v = \sqrt{gd} \tag{20-4}$$

[3]Deep-water waves are defined as those occurring where the depth is greater than one half the wave length.

where

$d$ = water depth, ft

A general equation for wave velocity in shallow and transitional water (applicable where $\frac{1}{25} < d/L < \frac{1}{2}$) has been proposed by G. B. Airy:

$$v = \sqrt{\frac{gL}{2\pi} \tanh \frac{2\pi d}{L}} \tag{20-5}$$

It is noted that Eq. 20-5 generalizes to Eq. 20-3 when $d$ becomes greater than $L$.

As a comparison of shallow- and deep-water equations indicate, the velocity of the wave decreases as it moves into shallow water. When waves approach the shore at an oblique angle, the portion of the wave nearest the shore slows down with the result that the wave swings around and tends to become parallel to the shore. At the same time, the wave lengths decrease as the wave period remains constant. This phenomenon is known as wave refraction. The U.S. Navy Hydrographic Office has published a graphical procedure for determining the direction of waves and the lengths of refracted waves [4].

When waves move into shallower depths, as along the coast, the orbits of the particles become distorted due to the friction exerted by the bottom. This causes the major axis of the elliptical path to tilt shoreward from the horizontal, and the wave gradually transforms from a purely oscillatory wave to a wave of translation. (See Fig. 20-3.) It is at this point that waves are capable of exerting great forces against bulkheads, breakwaters, and other coastal structures. Techniques for estimating these forces will be described in Chapter 21.

Wave heights have been found to vary with wind velocity and fetch, the straight-line stretch of open water available for wave growth. It is a region in which wind speeds and direction are reasonably constant. Several empirical equations have been proposed for the estimation of maximum wave height. Two such equations were published by Molitor [5] in 1934 for estimating maximum wave heights in inland lakes:

$$H_{max} = 0.17\sqrt{UF} \quad \text{for} \quad F > 20 \text{ mi} \tag{20-6}$$

$$H_{max} = 0.17\sqrt{UF} + 2.5 - \sqrt[4]{F} \quad \text{for} \quad F < 20 \text{ mi} \tag{20-7}$$

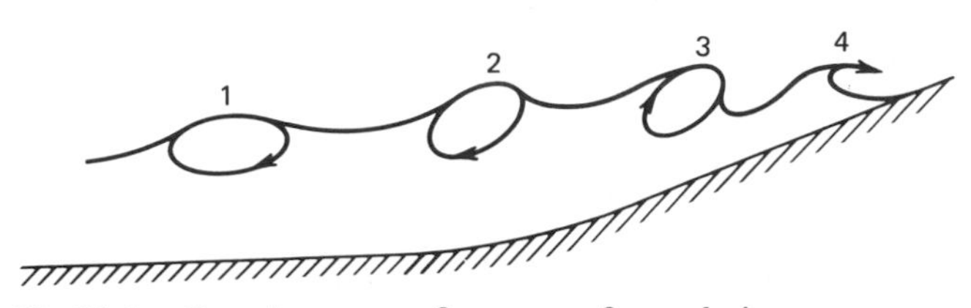


*Fig 20-3.* Development of a wave of translation.

where

$H_{max}$ = maximum wave height, ft

$F$ = fetch, statute mi

$U$ = wind velocity, statute mph

These equations were based on similar equations published by Thomas Stevenson in 1864.

Because of the difficulties inherent in measuring maximum wave heights, observers customarily refer instead to the significant wave height. Significant wave height is defined as the average height of the highest one-third of the waves for a stated interval. The maximum wave height, which should be used for design, is equal to approximately 1.87 times the significant height.

Several researchers have developed theoretically based, empirical equations relating significant wave height to fetch and wind velocity. For example, Wilson [6] fitted a curve to deep water height data from 14 sources and reported the following dimensionless relationship between the significant wave height and fetch and wind velocity:

$$\frac{gH}{U^2} = 0.26 \tanh\left[\frac{1}{10^2}\left(\frac{gF}{U^2}\right)^{1/2}\right] \tag{20-8}$$

In Eq. (20-8) $g$ is acceleration due to gravity.

The Corps of Engineers [7] has published the forecasting curves shown as Fig. 20-4 and 20-5 for predicting wave heights and wave periods, respectively. These idealized, dimensionless plots for wave growth include adjustments for shallow water to account for energy lost due to bottom friction and percolation. The curves are expressed in terms of the fetch, the water depth, and a function of the wind speed called the *wind stress factor*.

Below an elevation of about 1000 m, the frictional effects due to the presence of the ocean distort the wind field. In that zone, wind speed and direction depend on the elevation above the mean surface, air–sea temperature differences, and other factors. For example, it is usually assumed that the winds are measured at the 10-m elevation. If this is not the case, the wind speed must be adjusted accordingly.

The growth of wind-generated waves is most directly explained by the surface stress, which depends on the wind speed but may also be affected by a number of other factors. The wind effects on wave growth in the wave forecasting curves are therefore expressed in terms of $U_A$, the wind stress factor, which accounts for the nonlinear relationship between wind speed and wind stress. The Corps of Engineers [7] suggests that the following equation be used to convert the adjusted wind speed to the wind stress factor, $U_A$:

$$U_A = 0.589U^{1.23} \tag{20-9}$$

where

$U$ = the wind speed in mph

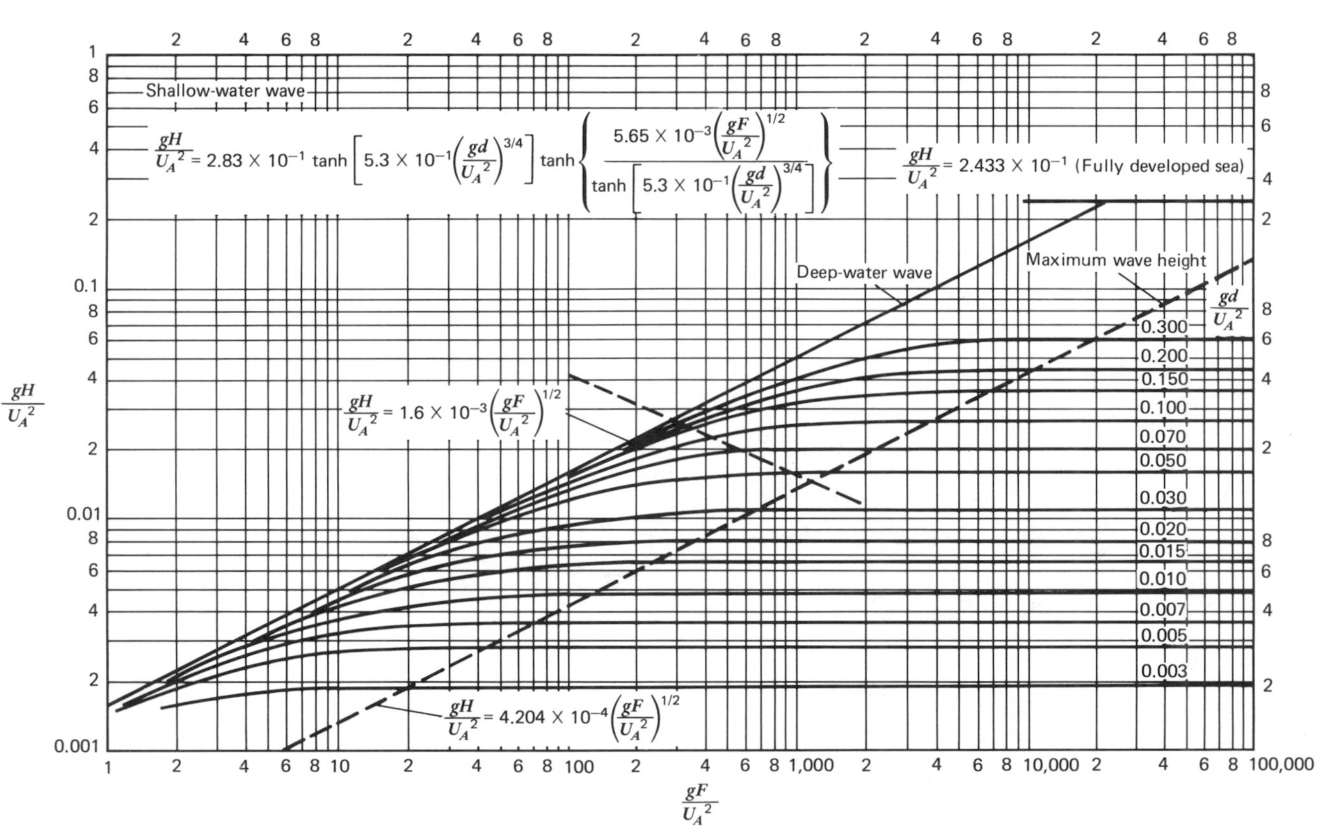


*Fig 20-4.* Forecasting curves for wave height. Constant water depth. (Source: Reference 7.)

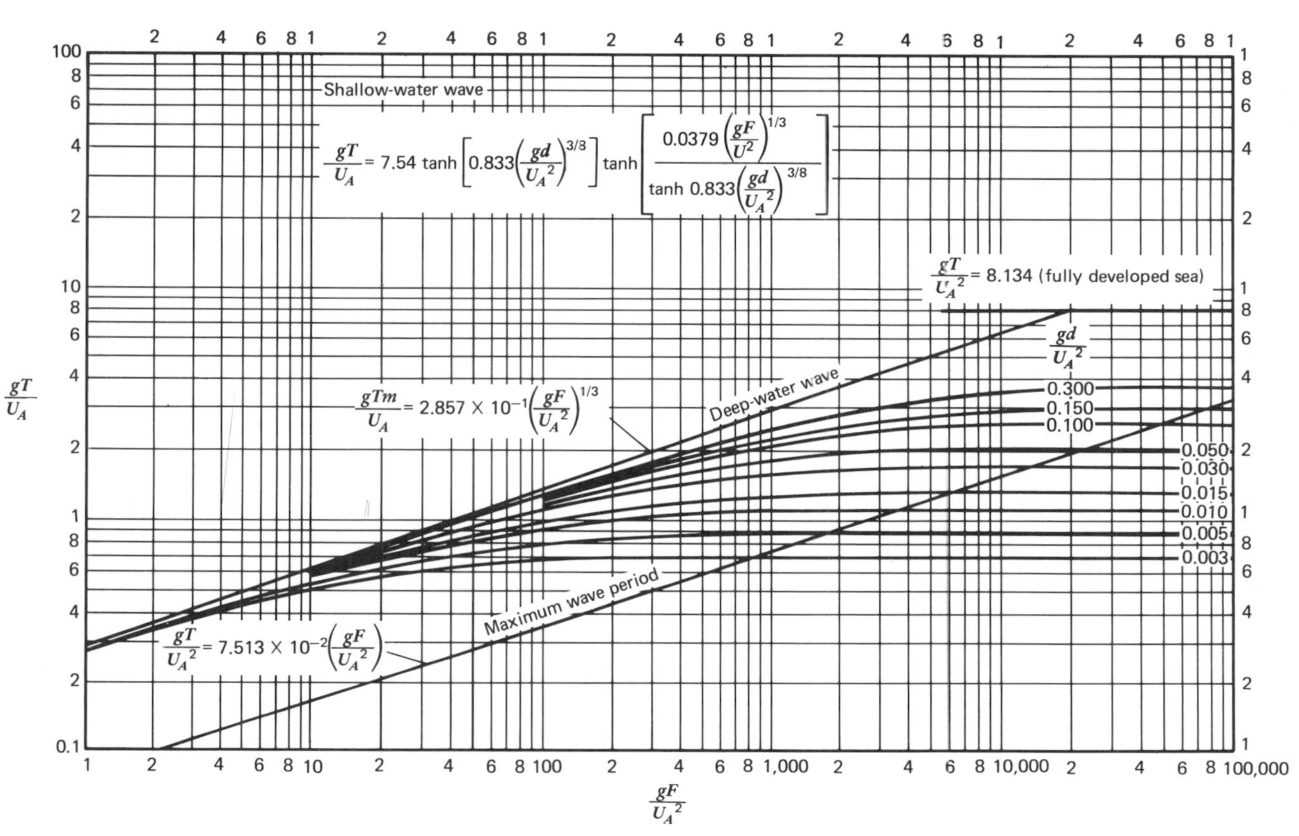


*Fig 20-5.* Forecasting curves for wave period. Constant water depth. (Source: Reference 7.)

The nomogram reproduced as Fig. 20-6 shows wave prediction curves of empirical values which can be used to check the reasonableness of the theoretical solutions of Fig. 20-4 and 20-5. The nomogram can be used to estimate the significant height of deep-water waves. These curves provide a direct method for estimating wave heights in simple wind fields in deep-water areas away from the coasts. However, near the coast, the significant wave methods may produce overestimates of wave heights due to failure to consider the directional spreading of the waves [8].

In shallow or transitional water, the water depth affects wave generation. Where waves are generated in such areas, smaller wave heights and shorter wave periods will be experienced. Several researchers [9, 10, 11, 12] have contributed to the development of forecasting methods for waves generated in shallow waters. The Corps of Engineers [7] has published a group of graphs for wave forecasting in relatively shallow waters. The graphs, exemplified by Fig. 20-7, were developed by successive approximations in which wave energy is added due to wind stress and subtracted due to bottom friction and percolation.

Consider the following examples which illustrate the use of the wave forecasting graphs.

## EXAMPLE 20-1

Given a fetch of 10 mi, a wind stress factor of 66 ft/sec and a mean water depth of 20 ft, determine the wave height and period.

$$\frac{gd}{U_A^2} = \frac{32.2(20)}{(66)^2} = 0.148$$

$$\frac{gF}{U_A^2} = \frac{32.2(52{,}800)}{(66)^2} = 390$$

From Figs. 20-4 and 20-5:

$$\frac{gH}{U_A^2} = 0.025 \quad \text{and} \quad \frac{gT}{U_A} = 1.8$$

$$H = \frac{(66)^2(0.025)}{32.2} = 3.4 \text{ ft}$$

$$T = \frac{(66)(1.8)}{32.2} = 3.7 \text{ sec}$$

Checking the results of Example 20-1 with Fig. 20-7, and entering the chart with fetch = 52,800 ft and wind stress factor = 45 mph,

$$H = 3.2 \text{ ft}$$

$$T = 3.6 \text{ sec}$$

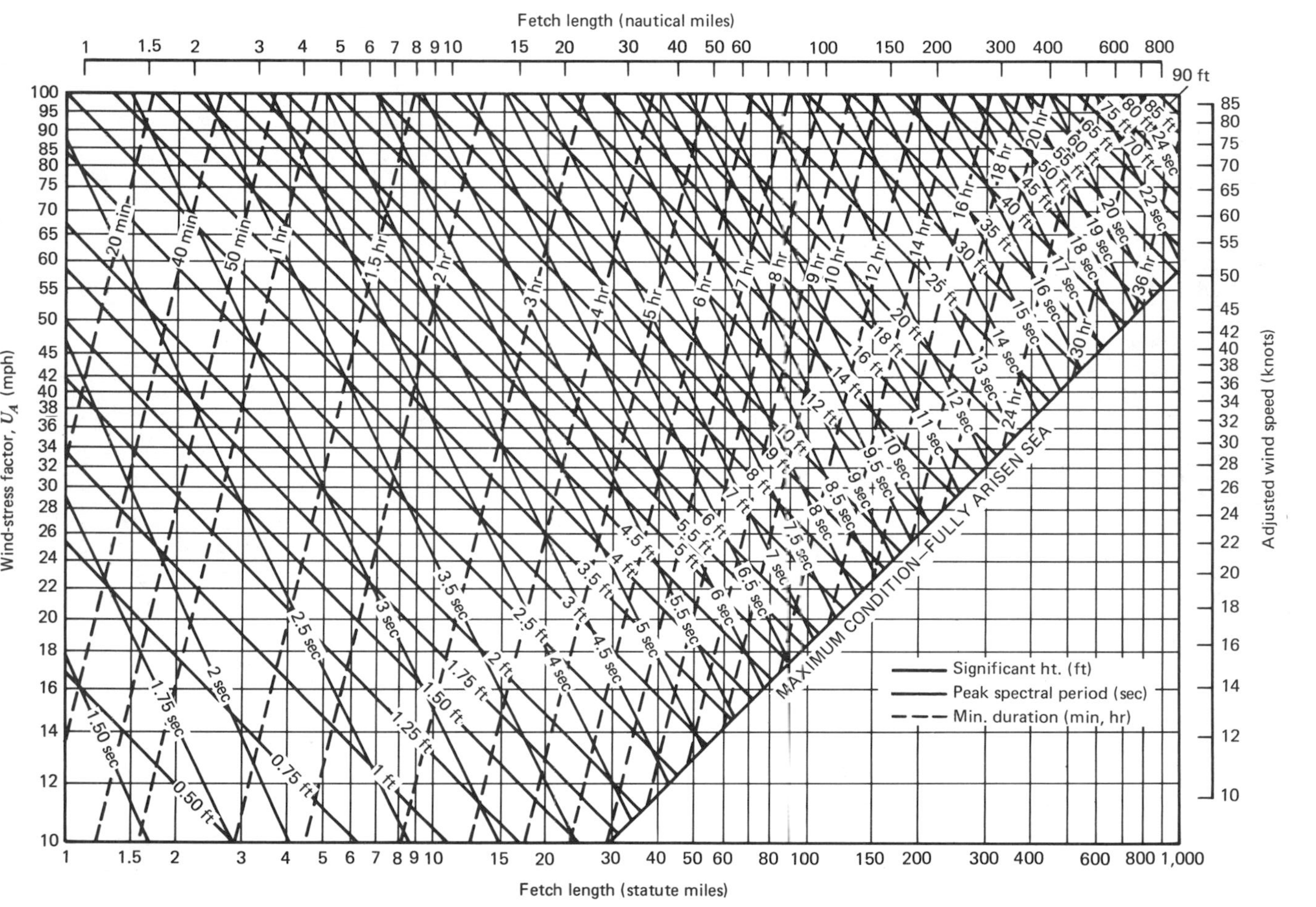


*Fig 20-6.* Nomograms of deep-water significant wave prediction curves as functions of windspeed, fetch length, and wind duration. (Source: Reference 7.)

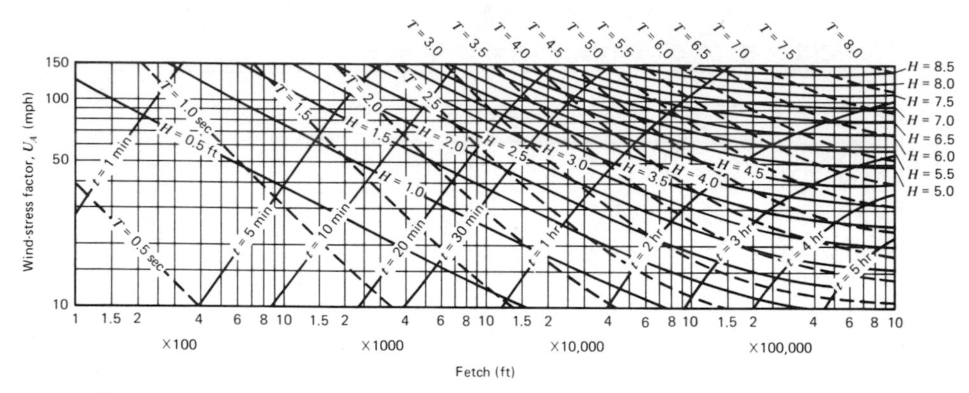


*Fig 20-7.* Forecasting curves for shallow-water waves with constant depths = 20 ft. (Source: Reference 7.)

## EXAMPLE 20-2

Given a fetch of 10 mi, a wind stress factor of 66 ft/sec, and a mean water depth of 60 ft, determine the wave height and period. If the storm that produced the winds lasted only 1.5 hr, how would this affect the answer?

$$\frac{gd}{U_A^2} = \frac{(32.2)(60)}{(66)^2} = 0.44$$

$$\frac{gF}{U_A^2} = 390 \text{ as in Example 20-1}$$

From Figs. 20-4 and 20-5:

$$\frac{gH}{U^2} = 0.032 \quad \text{and} \quad \frac{gT}{U_A} = 2.2$$

$$H = \frac{(66)^2(0.032)}{32.2} = 4.3 \text{ ft}$$

$$T = \frac{(66)(2.2)}{32.2} = 4.5 \text{ sec}$$

Checking the results with Fig. 20-6,

$$H = 4.25 \text{ ft}$$

$$T = 4.3 \text{ sec}$$

If the storm lasted only 1.5 hr, Fig. 20-6 indicates that $H = 3.3$ ft and $T = 3.2$ sec. This is known as a duration-limited wave.

## 20-8. CURRENTS

Much of water movement, as in waves, occurs in form rather than in the translation of water particles. Since water is viscous, however, the rotating particles that constitute waves do not return to their original position but rather drift in the direction of the wave movement. This flow, which occurs at a velocity smaller than the velocity of the wave itself, is called a current.

Offshore currents may result from hydraulic head when water is piled up along the coast because of mass translation associated with tides or waves. An example of this type of current is the Gulf Stream which flows northeasterly at a speed of about 4 mph between the southern tip of the Florida peninsula and the Bahama Islands.

The engineer's principal interest in currents lies in his or her efforts to stabilize erodible shoreline and to maintain navigable inlets.

## 20-9. TIDE

Some knowledge is required by the designer of port and harbor facilities of the nature and effects of tides. The tide is the alternate rising and falling of the surface of the oceans, gulfs, bays, and coastal rivers caused by the gravitational attraction of the moon and sun. In most places, such as on the Atlantic coast of the United States, the tide ebbs and flows twice in each lunar day (24 hr and 50 min). Larger-than-usual tides, called spring tides, occur when the sun and moon act in combination as when there is a new moon or full moon. Smaller-than-usual tides, called neap tides, occur when the moon is at first or third quarter.

In addition to the effects of the moon and sun, the magnitude and nature of a tide at a given location and time will be influenced by:

1. geographical location;
2. physical character of the coastlines;
3. atmospheric pressure;
4. currents.

At certain inland and landlocked seas, such as the Mediterranean Sea and the Gulf of Mexico, the tides are practically negligible. At other places, such as the Bay of Fundy, local topographical peculiarities contribute to tides as high as 100 ft.

Tidal charts and tables are published for the various ports of the world by the National Geodetic Survey, the British Admiralty, and other organizations.

## 20-10. PHYSICAL AND MATHEMATICAL MODELS

There are many problems in port planning and design that do not lend themselves to simple and straightforward analysis. Such problems are often addressed with physical or hydraulic models. Physical models may be used to facilitate the overall design of a harbor to determine the best locations and dimensions of structures to provide required protection against wave attack. Such models allow port planners to determine the proper types and alignments of coastal structures, the locations, shapes, and widths of navigation openings, and the effects of proposed dredging operations within the harbor. They are also used to evaluate the structural design of coastal facilities and to study inlet stability and hydraulics.

Figure 20-8 shows a physical model of Murrells Inlet in South Carolina. A spectral wave generator produces waves that simulate the actual waves that approach the inlet. The model, which was developed by the U.S. Army Engineer Waterways Experiment Station, has a vertical scale of 1:60 and a horizontal scale of 1:200.

An important recent development in port and harbor planning has been the development of mathematical modelling by means of advanced digital computers. Mathematical models can take into account virtually any physical

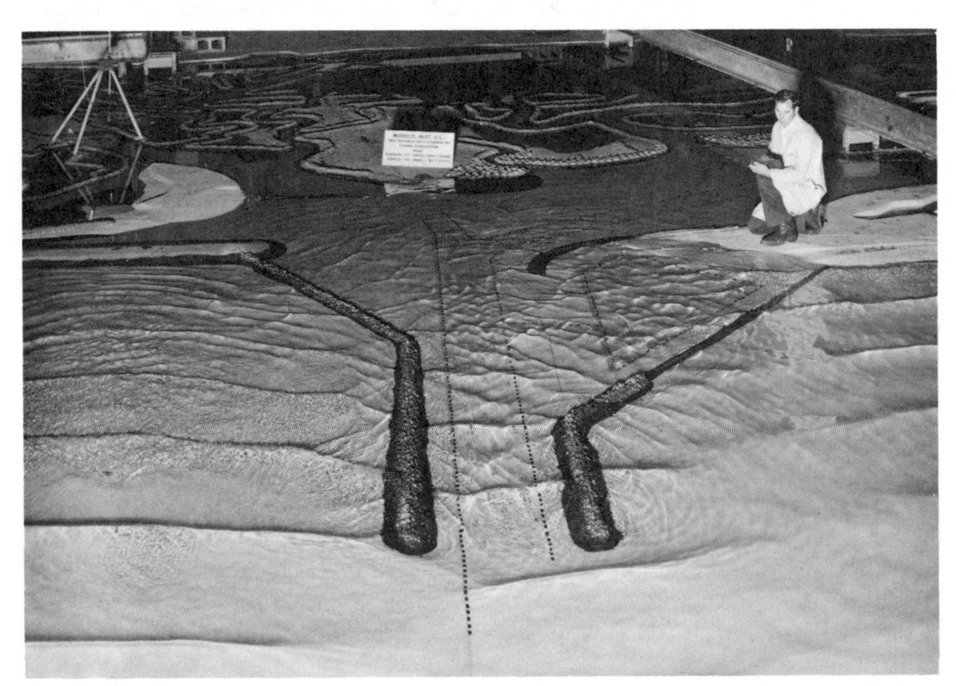

*Fig 20-8.* A physical model of Murrells Inlet in South Carolina. (Courtesy U.S. Army Corps of Engineers.)

phenomena that can be described in physical form. Thus, with mathematical models, it is possible to describe many phenomena that would be impossible or extremely costly to describe with a physical model. For example, mathematical models can describe the effects of winds on currents and water levels in shallow waters and on ships maneuvering at slow speeds [3]. Mathematical models are also less costly to maintain and to remobilize and are not subject to the errors of "scale effect" that may be present in physical models. Nevertheless, at least for the foreseeable future, mathematical models are likely to complement rather than replace physical models in providing solutions to coastal hydraulic problems.

Figure 20-9 shows a computer-plotted wave pattern for the port of Valencia, Spain.

## DETERIORATION AND TREATMENT OF MARINE STRUCTURES

Port and harbor structures must withstand some of the most destructive environmental conditions found in the world. This is especially true of piles

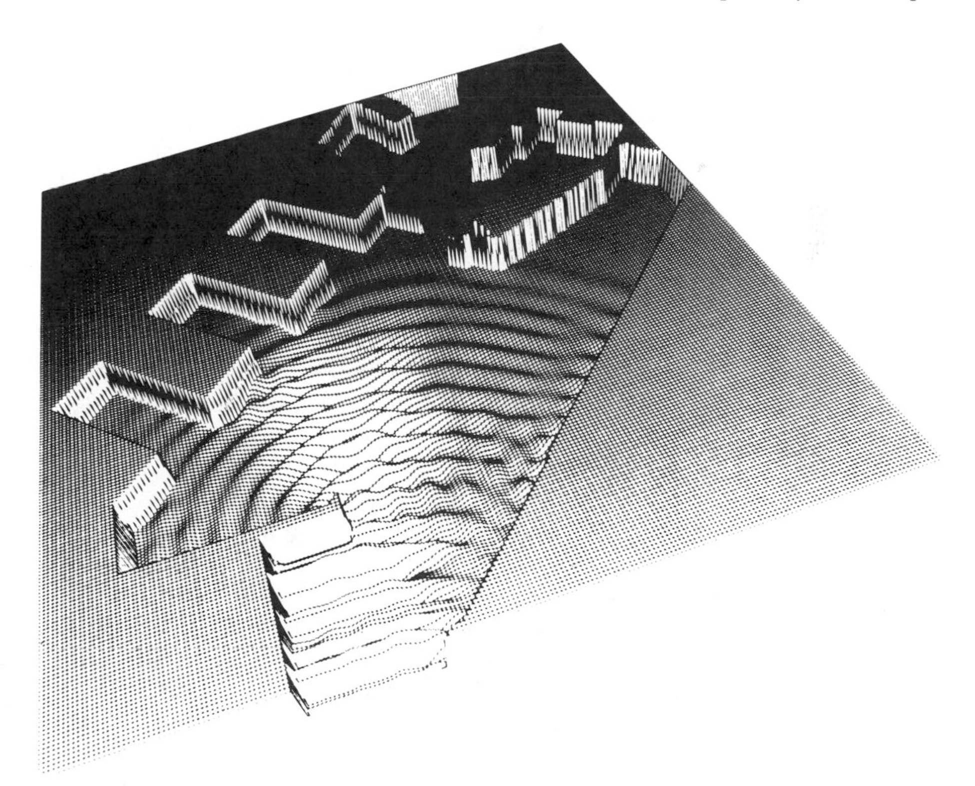

*Fig 20-9.* A computer plotted wave pattern for the port of Valencia, Spain. (Source: Reference 3.)

and other elements of substructures which are subject to attack by marine life as well as the corrosive effects of salt and sea. Generally these destructive effects are most pronounced in a seawater environment and some of the destructive agents, such as certain species of marine borers, are not found in fresh waters.

In Sections 20-11, 20-12, and 20-13 the causes of deterioration and methods of protection of wood, concrete, and steel waterfront structures will be discussed. Reference 13 gives a more extensive treatment of this important subject.

## 20-11. DETERIORATION AND TREATMENT OF WOOD STRUCTURES

Timber used in coastal structures may be damaged by three forms of attack: decay, insects (termites and wharf borers), and marine borers. The decay and insects tend to damage structures above the water level, while marine borers attack below the water level.

*Decay.* All forms of decay are caused by certain low forms of plant life called fungi. The fungi consume certain substances in the wood, causing it to disintegrate. The species of fungi that cause decay require food, air, adequate moisture, and a favorable temperature. Poisoning the food supply by impregnating the timber with a suitable preservative is the most common method of preventing decay.

*Termites.* Sometimes called "white ants," termites are not really ants, but resemble them somewhat in appearance and method of life. There are two types of termites: (1) the "subterranean" type that requires moisture and therefore must have access to the ground at all times, and (2) the "dry wood" type that flies and does not need contact with the ground. The subterranean termites are widely distributed throughout the world, but the dry wood termites tend to be found only in the warmer climates.

The chief food of termites is cellulose, which they obtain from dead wood. Most species of timber are subject to the attack of termites. It is therefore necessary, to prevent termite damage, to use commercial timbers that have been properly treated, or to insulate all wood from the ground so that the termites cannot attack it. However, insulation will not prevent attack by dry wood termites.

*Wharf Borers.* These animals can do extensive damage to wood structures, but they are not of as much concern as others of the organisms attacking timber. The damage is caused by the young of the winged beetle. The beetle lays its eggs in the cracks and crevices of the timber, and the larvae or worms hatched from these eggs are the cause of the damage to the timber. Wharf borers do not work below water level but more commonly in timber that is not far above high water, or timber that is wet by salt spray at times.

*Marine Borers.* There are two main divisions of these very destructive animals: (1) the Molluscan group, which is related to the oyster and clam; and (2) the Crustacean group, which is related to the lobster and crab. Their methods of attack on timber are completely different. The Molluscan group enters the timber through a tiny hole, and the animals grow as they destroy the interior of the timber. The Crustacean group destroys the timber from the outside. The attack of the Molluscan borers can only be detected by the most careful inspection of the surface, or by cutting into it. The attack of the Crustacean borers is easily seen and measured by surface inspection, and the rate of destruction in heavy attacks is less rapid by the Crustaceans than by the Molluscans.

Molluscan borers are classified biologically into several genera and many species. One group, the toredo or shipworm, has grayish, slimy, wormlike bodies with the shells on the head used for boring. The size of the mature animals of the common species may be as large as 1 in. in diameter and 4 to 5 ft long, but more commonly they are about $\frac{3}{8}$ in. in diameter and 5 to 6 in. in length. In areas of heavy attack, these borers may totally destroy an unprotected pile's bearing value in 6 to 8 months.

Another group, the pholad, also uses its shells for boring but the body of the animal is enclosed by the shells. Some species of the group bore in soft rock and mud and even concrete. The entrance holes made by this group tend to be larger than those made by the toredo group, but may still be hard to find by surface inspection. This group is most commonly active in tropical and semitropical waters.

Crustacean borers are classified into several genera and species. One of the species resembles a wood louse; another is related to the ordinary sand hopper. These borers have a body that is typically $\frac{1}{8}$ to $\frac{1}{2}$ in. in length and a width of about one-third to one-half of the length. Crustacean borers exist in a wide variety of climates and environments and may be present in either saltwater or freshwater. Hundreds of these animals per square inch have been counted on timber under heavy attack. They commonly destroy timber gnawing interlacing branching burrows on the surface of the timber.

Many methods have been tried to protect timber from marine borer attack including covering the timber with metal sheathings or encasing it in concrete, cast iron, or vitrified pipe. The most common and reliable form of protection has been impregnation of the timber with a toxin such as coal tar creosote. This type of treatment also protects the wood from decay, termites, and wharf borers.

## 20-12. DETERIORATION AND PROTECTION OF CONCRETE STRUCTURES

Concrete piles that are entirely embedded in earth generally are not subject to significant deterioration. However, in isolated instances, concrete piles may be damaged by the percolation of groundwater that contains destructive chemicals from industrial wastes, leaky sewers, or leaching from storage piles of coal or cinders containing acids or other destructive compounds.

Reinforced concrete piles above the ground surface may be damaged by weathering and destructive elements carried in the air. Moisture in the air may penetrate permeable concrete and reach the reinforcement, causing it to rust and spall the sides and edges of the pile. In addition, damage to reinforced piles in waterfront structures may be caused by:

1. Chemical action of polluted waters.
2. Attack by molluscan borers.
3. Abrasion by floating objects or scouring sands.
4. Frost action.

Because of the harshness of the coastal environment, special care is required to ensure that the concrete meets high standards of materials and workmanship. Experience has shown that the composition of the portland cement, the quality of the aggregates, the amount of cover over the steel, and the workmanship in mixing and placing the concrete are the most important factors that affect the durability of the concrete piles.

## 20-13. DETERIORATION AND PROTECTION OF STEEL STRUCTURES

Steel piles entirely embedded in relatively impervious earth receive little damage from corrosion. From a level of about 2 ft below the ground surface downward, atmospheric oxygen is blanketed off by the surrounding soil, inhibiting progressive corrosion. Occasionally, however, steel piles that are completely embedded may deteriorate if the surrounding earth or groundwater contains corrosive compounds.

Steel piles protruding from the ground into open water are subjected to severe deterioration in a saltwater environment. Steel piles in freshwater generally do not require protection.

Steel piles in seawater tend to experience less corrosion if in protected waters than if subjected to wave action in the open ocean. Deterioration may be worse because of the abrasive action of waterborne sand agitated by waves and currents. This condition usually exists only in shallow waters where wave and tidal action is most prevalent. Destructive organic substances consisting of decayed marine life deposited on the bottom may also cause severe deterioration in a narrow zone at the mud line. Where either of these conditions exist, it may be desirable to protect the piling by some form of encasement in the vicinity of the mud line.

Above the mud line, corrosion is usually more active near the water surface where the oxygen content of the water is greatest. It tends to be more severe in regions exposed to wetting and drying or to saltwater spray. In these areas, various coatings and paints, or in extreme cases, concrete casings are used to inhibit corrosion.

## PROBLEMS

**1.** Estimate the velocity of a 110-ft-long wave in water 40 ft deep using Eqs. 20-3 and 20-4.

**2.** Express Eq. 20-3 in SI (metric) units.

**3.** Estimate the maximum deep-water wave height for a fetch of 170 statute mi and a wind velocity of 18 statute mph:

   a. Using Eq. 20-7;

   b. Using Eq. 20-8;

   c. Using the forecasting curves.

**4.** Develop an equation similar to Eq. 20-8 in SI (metric) units.

**5.** Given a fetch of 40 nautical mi and a wind speed of 28 knots:

   a. Estimate the maximum wave height.

   b. What would be the significant wave height?

   c. Estimate the significant wave period.

   d. What minimum duration of a 28-knot wind would cause this wave?

   e. Estimate the speed of the significant wave.

**6.** Using the equation developed in Problem 4, estimate the wave height in meters given a wind speed of 70 km/hr and a fetch of 48 km.

**7.** Given a fetch of 16 mi, a wind speed of 74 ft/sec, and a mean water depth of 13 ft, determine the wave height and period.

**8.** Develop an equivalent formula in SI (metric) unit for the shallow-water wave equation shown in Fig. (20-5).

## REFERENCES

1. Carr, Braxton B., "Inland Water Transportation Resources," *U.S. Transportation Resources, Performance and Problems,* National Academy of Sciences, National Research Council, Publication 841-S, Washington, D. C., August 1960.
2. *1977 Inland Waterborne Commerce Statistics,* The American Waterways Operators, Inc., Arlington, Virginia, 1978.
3. Agerschou, Hans, Helge Lundgren, and Torben Sorensen, *Planning and Design of Ports and Marine Terminals,* John Wiley, New York, 1983.

4. *Breakers and Surf,* U. S. Navy Hydrographic Office, November 1944.
5. Molitar, D.A., "Wave Pressure on Sea-walls and Breakwaters," *Proceedings,* American Society of Civil Engineers, 1934.
6. Wilson, Basil W., *Transactions,* American Society of Civil Engineers, Vol. 128, Part IV, Paper No. 3416, 1963.
7. *Shore Protection Manual,* Vols. I and II, Department of The Army, Waterways Experiment Station, Corps of Engineers, Vicksburg, Mississippi, 1984.
8. Borgman, L.E., and D.T. Resio, "Extremal Prediction in Wave Climatology," *Ports '77,* American Society of Civil Engineers, 1977.
9. Sverdrup, H.U. and W.H. Munk, *Wind, Sea, and Swell: Theory of Relations for Forecasting,* TR-1, H. O. No. 601, U. S. Navy Hydrographic Office, Washington, D. C., 1947.
10. Bretschneider, C.L., "Revised Wave Forecasting Relationships," *Proceedings of the Second Conference on Coastal Engineering,* American Society of Civil Engineers, 1952.
11. Bretschneider, C.L., and R.O. Reid, *Modification of Wave Height Due to Bottom Friction, Percolation, and Refraction,* TM-45, U. S. Army Corps of Engineers, Washington, D. C., October 1954.
12. Thijsse, J. Th., and J.B. Schijf, "Penetration of Waves and Swells into Harbors," *Proceedings,* XVII International Navigation Congress, Lisbon, 1949.
13. Nucci, Louis R., "Inspection, Repair and Maintenance of Marine Structures," *Infrastructure—Repairs and Inspection,* Mahendra J. Shah, (ed.), American Society of Civil Engineers, New York, April 29, 1987.

## OTHER REFERENCES

*Symposium on Treated Wood for Marine Use,* Special Technical Publication No. 275, American Society for Testing Materials, Philadelphia, Pennsylvania, 1960.

Laque, Francis L., *Marine Corrosion: Causes and Prevention,* John Wiley, New York, 1975.

Cornick, Henry F., *Dock and Harbour Engineering,* Vol. 2, Charles Griffin and Company, Ltd., London, 1959.

Mitchell, C. Bradford, "Pride of the Seas," *American Heritage,* December 1967.

Quinn, Alonzo, *Design and Construction of Ports and Marine Structures,* Second Edition, McGraw-Hill, New York, 1972.

Bruun, Per, *Port Engineering,* Third Edition, Gulf Publishing, Houston, 1981.

# Chapter 21

# Planning and Design of Harbors

## 21-1. INTRODUCTION

A harbor is a partially enclosed area of water that serves as a place of refuge for ships. The term port refers to a portion of a harbor that serves as a base for commercial activities. Harbor and coastal structures are the means by which protection from waves and winds is provided and the erosion of beaches and coastlines is controlled. Port facilities make it possible for ships to obtain fuel and supplies, to be repaired, and to transfer passengers and cargo.

This chapter will deal with the planning and design of coastal and harbor structures; Chapter 22 will be concerned with the planning and design of port facilities.

## 21-2. CLASSES OF HARBORS

Harbors may be classified into one of several categories according to function and protective features. There are many natural harbors in the world where protection from storms is provided by natural topographical features. Natural harbors may be found in bays, inlets, and estuaries and may be shielded by offshore islands, peninsulas, or reefs. Other natural harbors are protected by virtue of the fact that they are located in river channels some distance from the sea. Several famous world harbors are located as much as 50 mi inland.

When sufficient protection from storms has not been provided by nature, it may be provided by the construction of breakwaters and jetties. Harbors thus formed are known as artificial harbors. Planning considerations and design criteria for breakwaters and jetties will be discussed in Section 21-4.

Some degree of protection from storms is provided by harbors of refuge and roadsteads. These areas are easily accessible but generally offer less protection than does a harbor. The term harbor of refuge refers to convenient protected anchorage areas that are usually found along established sea routes and dangerous coasts. A roadstead is a tract of water that is protected from heavy seas by a bank, shoal, or breakwater. Port facilities are not provided in harbors of refuge or roadsteads although these facilities may be provided within a nearby harbor.

Additional classifications of harbors include: commercial harbors, which provide protection for ports engaged in foreign or coastwise trade; and military harbors within which the dominant activity is the accommodation of naval vessels.

## 21-3. DESIRABLE FEATURES OF A HARBOR SITE

There are at least five desirable features of a harbor site:

1. Sufficient depth.
2. Secure anchorage.
3. Adequate anchorage area.
4. Narrow channel entrance in relation to harbor size.
5. Protection against wave action.

The depth of harbor and approach channel should be sufficient to permit fully loaded ships to navigate safely at the lowest water. Obviously, the harbor depth required depends principally on the draft[1] of the ships using the harbor.

The harbor and channel depth below the lowest low water should be at least the maximum draft anticipated plus an additional 5 ft, approximately. This additional depth is to allow for the tendency of a ship to "surge" when in motion and to provide a clearance of at least 3 ft below the ship's keel as a factor of safety.

A summary of the average drafts of various types of ships in the world's ocean fleet is given in Chapter 4, along with average drafts of typical inland waterway vessels. Generally, the average draft of oceangoing ships varies from 21 to 37 ft depending on ship type, while the draft of the largest commercial oceangoing ship (a tanker) is 94 ft [1]. Few if any ocean ports have channel depths to accommodate the largest tankers and special provisions must be made to unload these vessels offshore to smaller tankers or by means of pipelines.

Generally, ocean ports maintain harbor and channel depths of 35 to 40 ft; however, the trend to larger ships indicates that modern ocean harbors may require depths in excess of 40 ft.

On the inland waterway system, a 9 ft operating depth is considered standard. Channel depths on the inland system vary a great deal, however, and about 26 percent of the inland waterway channel miles has a depth of less than 6 ft while approximately 18 percent is 14 ft and over [2].

Conceivably, the selection of a harbor site may be influenced by the soil

[1] Draft is the vertical distance between the waterline and the keel. Unless otherwise noted, when the word draft is used here it will refer to the full load draft.

conditions along the bottom of the anchorage area. Generally, firm cohesive materials provide good anchorage, while light sandy bottoms are poor anchorage areas. Other factors being equal, more anchorage area will be required when poor bottom soil conditions prevail. This follows from the fact that maximum resistance to ship movement occurs when the anchor cable is as nearly horizontal as possible.

The shape and extent of anchorage area is dependent principally on five factors:

1. The maximum number of ships to be served.
2. The size of the ships.
3. The method of mooring.
4. Maneuverability requirements.
5. Topographic conditions at the proposed site.

Because objectionable waves will be generated in large harbors, artificial harbors should be built as small as possible and still be consistent with the needs for convenient safe maneuvering and mooring.

Space requirements for ships vary a great deal depending on the method of mooring. A ship that is secured by a single anchor will occupy a circular area with a radius of the ship length plus approximately three times the water depth. Thus, a 600-ft-long ship anchored by a single anchor in a harbor 50 ft deep will require about 40 acres or anchorage area. Considerably less area is required for a ship that is secured by two anchors. Such a ship will occupy a circular area with a diameter little more than the ship length, or about 6.5 acres for a 600-ft ship. However, when a clear area is maintained for the anchor cable, a rectangular area 1000 to 1200 ft long by about 500 to 600 ft wide is required for a typical merchant freighter. Thus, a merchant freighter secured by two anchors requires a total of about 16 acres of harbor space.

The minimum turning radius for a ship is equal to about twice the ship length. Thus a typical oceangoing freighter will require an additional 30 to 35 acres for maneuvering if a full-size turning basin is provided.

As one might expect, harbors of the world vary a great deal in area, the smallest fishing harbors being less than 10 acres and the largest harbors being more than 10 $mi^2$.

To minimize wave action within the harbor, the harbor entrance should be as narrow as possible provided it meets the requirements for safe and expeditious navigation and provided it does not cause excessive tidal currents. Currents in excess of 4 to 5 ft/sec will affect navigation adversely and may cause scour of breakwaters and other protective works.

The required entrance width naturally will be influenced by the size of the harbor and the ships that use it. As a rule of thumb, the width of entrance should be roughly equal to the length of the largest ship using it. While this guideline should be helpful for planning purposes, the entrance width and location used for design purposes should be determined by model tests.[2]

## PROTECTIVE COASTAL WORKS

The term protective coastal works used here will be used in the broad sense to include:

1. Offshore structures (breakwaters) to lessen wave heights and velocities.
2. Structures that are built at an angle to the shore, such as jetties and groins to control littoral drift.
3. Structures built at or near the shoreline to protect the shore from the erosive forces of waves. In this category are seawalls, bulkheads, and revetments.
4. Natural coastal features such as protective beaches and sand dunes that help to control and dissipate waves without creating adverse environmental effects.

### 21-4. BREAKWATERS AND JETTIES

Breakwaters are massive structures built generally parallel to the shoreline to protect a shore area or to develop an artificial harbor. A jetty is a structure built roughly perpendicular to the shore extending some distance seaward for the purpose of maintaining an entrance channel and protecting it from waves and excessive or otherwise undesirable currents. Jetties usually are built in pairs, one on each side of a channel or the mouth of a river. Structurally, breakwaters and jetties are similar; however, the design standards for jetties may be slightly lower than are those for breakwaters. This results from the fact that jetties are not subject to direct wave attack to as great an extent as are breakwaters.

To avoid redundance, the remaining discussion of this section will focus principally on the planning and design of breakwaters, and little else will be said about jetties. However, one especially important feature relating to the design of jetties should be mentioned, that of hydraulic design. Normally, when longshore transport occurs predominantly in one direction, jetties can cause accretion of the updrift shore and erosion of the downdrift shore. The stability of the shore in the vicinity of inlets can be improved by a process known as *sand bypassing,* which involves artificially nourishing sand-deprived

[2] Reference 3 provides an excellent case study illustrating the usefulness of hydraulic modeling to quantitatively investigate the effect of future harbor development plans.

areas by means of land-based dredging plants, floating dredges, or land-based vehicles [4].

Because of the possibility of erosion and accretion due to changes to the velocity and direction of channel currents, the determination of distance between jetties must be made carefully. In making this decision, the designer should study the magnitude and direction of existing tidal currents and the effect that the construction of jetties might have on these currents. Of course, important inputs to this decision will be existing topographical features of the area and the channel width needed for navigation. Unfortunately, few analytical tools are available to help the designer with this problem, and he or she must rely on engineering judgment and, whenever feasible, model studies.

While a wide variety of breakwaters have been built, those successfully employed generally fall into two classes: mound breakwaters and wall breakwaters.

By far the most popular type of breakwater is the rubble mound breakwater. As the name suggests, this type of breakwater consists of a mound of large stones extending in a line from the shore or lying parallel to and some distance from the shore. Typically, this type of breakwater is constructed of stones ranging in weight from 500 lb to more than 16 tons each. The smaller stones are used to construct the core, while the largest sizes, being most resistant to displacement, serve as armor stones that comprise the outer layer of the mound. Commonly, the largest armor stones are used on the seaward side of the breakwater.

Where adequate quantities of armor stone are not available in suitable sizes, precast concrete armor units may be used. Various shapes of these units have been used, including tetrapods, quadripods, and tribars. See Fig. 21-1 and Table 21-1. Some of these shapes are patented and a royalty charge must be paid for their use. A rubble mound breakwater utilizing an armor layer of tetrapods is shown as Fig. 21-2.

The dolos armor unit, illustrated by Fig. 21-3, was developed by E. M. Merrifield, Republic of South Africa, in 1963. This concrete shape is similar to a ship anchor or the letter "H" with one vertical perpendicular to the other [4].

In certain instances, a relatively impervious material is used in the core of the breakwater. When a sand-clay or shale is used for the base and core material, the breakwater is classed as a solid fill structure. The use of a fine-grained material in the core may be used because of economy or in order to prevent the effect of waves passing through the structure. The voids of the upper portion of the core of the breakwater may also be filled with the hot asphaltic concrete or portland cement concrete in order to improve the stability or imperviousness of the structure.

Vertical wall breakwaters constitute a second major class of breakwaters. This class of breakwater differs in concept and design from rubble mound breakwaters. The designer of a vertical wall breakwater must be concerned with the ability of the total structure to remain stable under the attack of

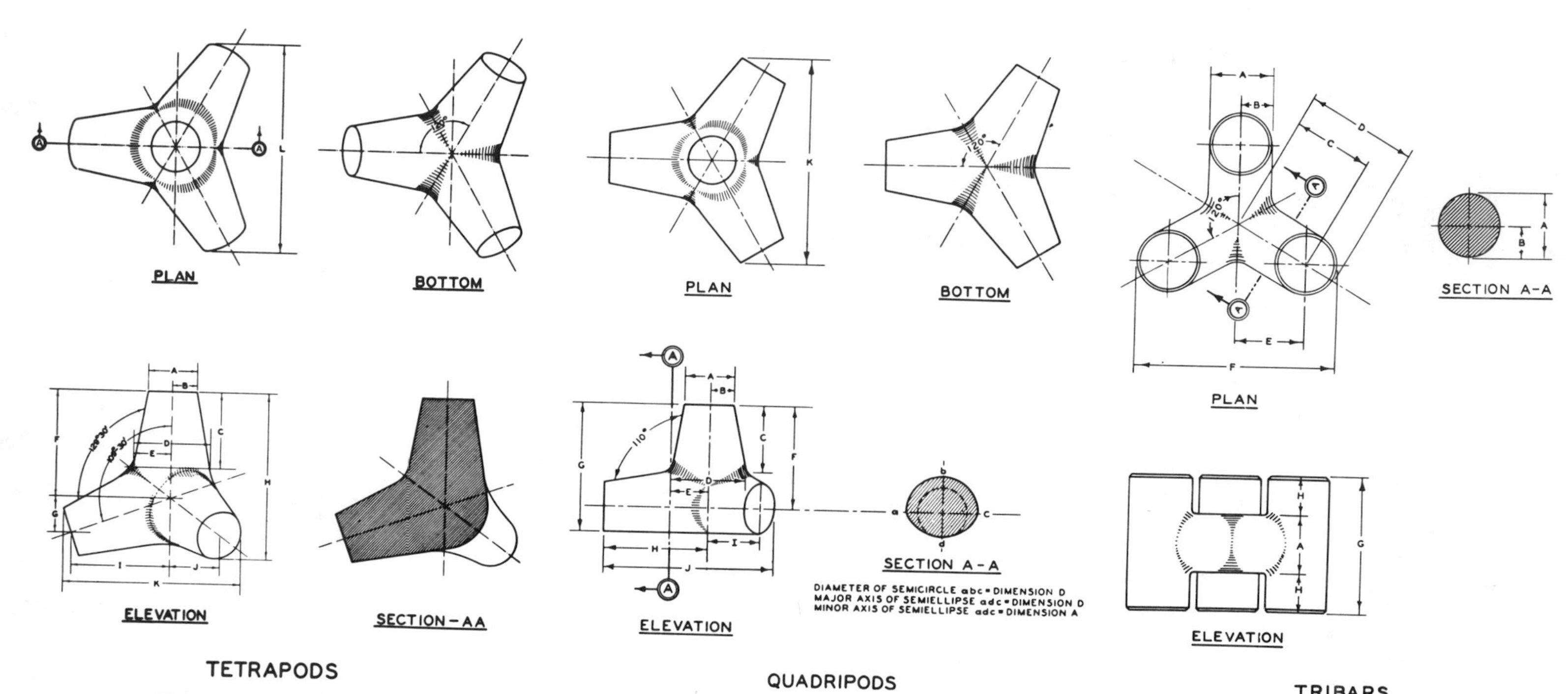


*Fig 21-1.* Typical concrete armor units. (Source: Shore Protection Manual, U.S. Army Corps of Engineers, 1977.)

*Table 21-1*

**Design Information for Concrete Armor Units**

| | *Tetrapods* | | | *Quadripods* | | | *Tribars* | | |
|---|---|---|---|---|---|---|---|---|---|
| | *Small* | *Medium* | *Large* | *Small* | *Medium* | *Large* | *Small* | *Medium* | *Large* |
| Volume of armor unit, $ft^3$ | 7.14 | 214.29 | 571.43 | 7.14 | 214.29 | 571.43 | 7.14 | 214.29 | 571.43 |
| Weight of armor unit, tons[a] | 0.53 | 16.02 | 42.71 | 0.53 | 16.02 | 42.71 | 0.53 | 16.02 | 42.71 |
| Average thickness of two layers placed pell-mell, ft | 4.01 | 12.45 | 17.26 | 3.66 | 11.37 | 15.77 | 3.85 | 11.97 | 16.60 |
| Number of armor units/1000 $ft^2$ (2 layers, pell-mell) | 280.18 | 29.02 | 15.18 | 261.05 | 27.04 | 14.15 | 161.34 | 16.71 | 8.74 |
| Dimension, ft | | | | | | | | | |
| A | 0.89 | 2.76 | 3.83 | 0.93 | 2.88 | 4.01 | 1.05 | 3.25 | 4.51 |
| B | 0.44 | 1.38 | 1.91 | 0.46 | 1.44 | 2.00 | 0.52 | 1.62 | 2.25 |
| C | 1.40 | 4.36 | 6.05 | 1.28 | 3.98 | 5.52 | 1.25 | 3.90 | 5.41 |
| D | 1.38 | 4.30 | 5.96 | 1.38 | 4.28 | 5.94 | 1.78 | 5.52 | 7.66 |
| E | 0.69 | 2.15 | 2.98 | 0.69 | 2.14 | 2.97 | 1.09 | 3.38 | 4.68 |
| F | 1.89 | 5.88 | 8.16 | 1.97 | 6.12 | 8.49 | 3.22 | 10.00 | 13.87 |
| G | 0.63 | 1.96 | 2.72 | 2.43 | 7.57 | 10.49 | 2.09 | 6.50 | 9.01 |
| H | 2.94 | 9.14 | 12.68 | 1.97 | 6.12 | 8.49 | 0.52 | 1.62 | 2.25 |
| I | 1.78 | 5.54 | 7.69 | 0.99 | 3.06 | 4.25 | | | |
| J | 0.89 | 2.77 | 3.84 | 3.36 | 10.43 | 14.47 | | | |
| K | 3.21 | 9.97 | 13.83 | 3.88 | 12.05 | 16.70 | | | |
| L | 3.54 | 10.98 | 15.23 | | | | | | |

[a] Assuming specific weight = 149.5 $lb/ft^3$.

SOURCE: *Shore Protection Manual*, U.S. Army Corps of Engineers, 1984.

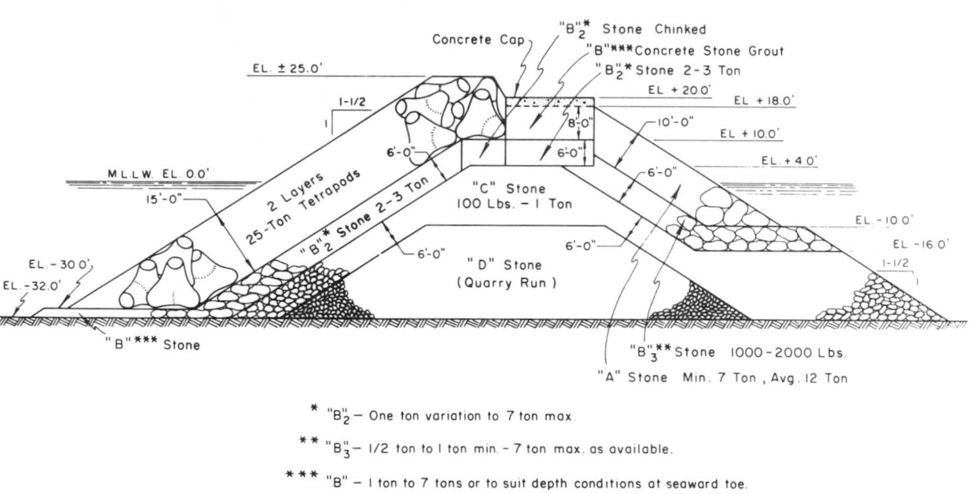


*Fig 21-2.* Tetrapod-rubble-mound breakwater at Crescent City, California. (Courtesy U.S. Army Corps of Engineers.)

waves, whereas in the case of rubble mound breakwaters, attention must be focused on the stability of the individual stones.

The principal advantage of vertical wall breakwaters is that less rock is required than is needed for rubble mound construction. These breakwaters, being less massive, provide more usable harbor area and make it possible to have a narrower harbor entrance. On the other hand, rubble mound breakwaters can be constructed on foundations that would be unsuitable for the

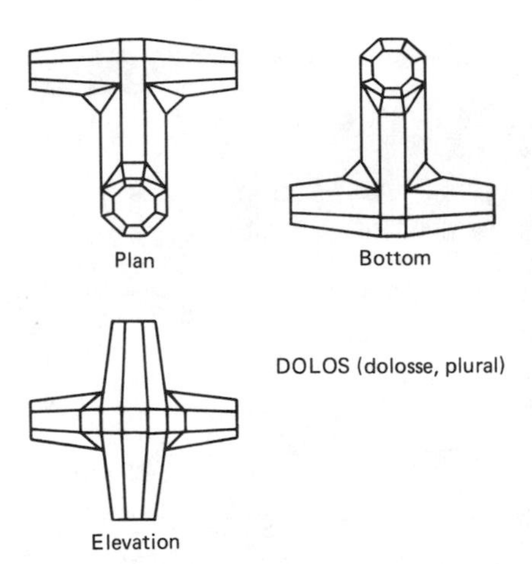


*Fig 21-3.* Views of dolos armor unit.
(Source: Reference 4.)

support of a vertical wall breakwater. Waves tend to break and be dissipated on the slopes of rubble mound breakwaters, and consequently these structures do not have to be constructed to heights as great as those required for vertical wall breakwaters.

Vertical wall breakwaters include the following types:

1. Timber or precast concrete cribs filled with large stones.
2. Concrete caissons filled with stone or sand.
3. Sheet piling breakwaters.

Timber or precast concrete cribs consist of large boxlike compartments of open construction which are placed on a prepared foundation and then filled with stone. Concrete caissons are massive watertight boxes which are floated into position, settled on a prepared foundation, and filled with stone, earth, or sand. Usually, these caissons are then covered with a concrete slab.

The simplest sheet piling breakwater consists of a single row of piling which may or may not be strengthened by vertical piles. Another type of sheet piling breakwater consists of double walls of sheet piling connected by tie bars with the space between the walls filled with stone or sand. Finally, vertical wall breakwaters may be built of cellular sheet pile structures which are filled with earth, stone, or sand to provide stability. A photograph of such a breakwater is shown in Fig. 21-4.

Of course, one can find breakwaters that are neither mound type nor vertical wall type structures, but that are composite structures containing features common to both broad classifications. Small vertical walls are often superimposed on top of rubble mound breakwaters, and it is not uncommon to find rubble mound foundations that support massive vertical wall breakwaters.

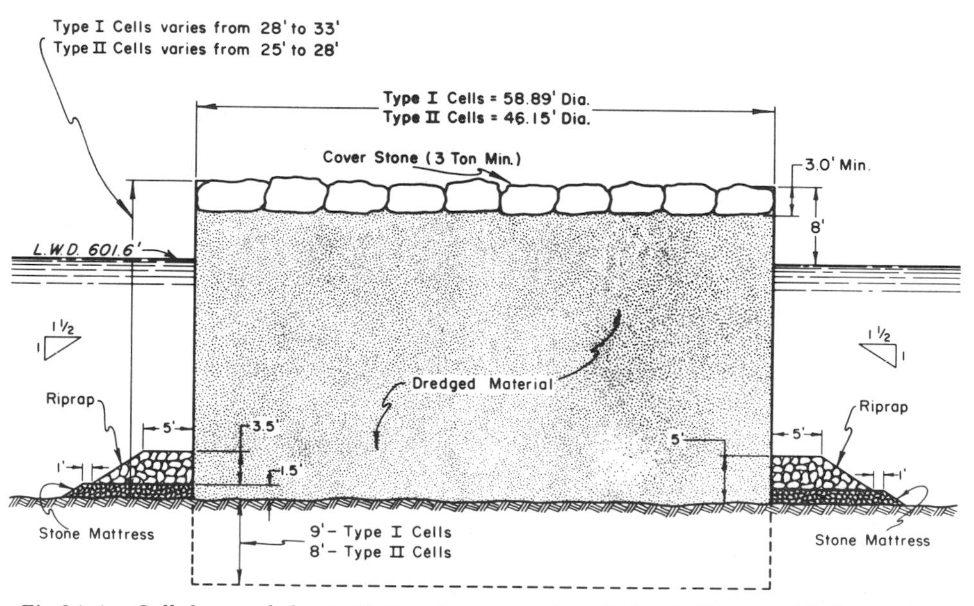


*Fig 21-4.* Cellular steel sheet-pile breakwater at Grand Marais Harbor, Michigan. (Courtesy U.S. Army Corps of Engineers.)

Several novel concepts in breakwater design have been studied, including floating breakwaters, hydraulic breakwaters, and bubble breakwaters. These proposals appear to have very limited usefulness. In the final analysis, the selection of breakwater type will depend on its purpose, foundation conditions, wave forces, availability of materials, and costs.

Breakwaters may not be connected to the shore and may be constructed as a single unit or as a series of relatively short structures. The latter approach may provide good protection from waves without the formation of undesirable sand shoals between the breakwater and the shore.

The height of a breakwater will depend principally on maximum tide elevation, wave height, and breakwater type. Waves tend to break on mound structures, and the required height will vary with the angle of breakwater slope, wave height and length, and the smoothness and permeability of the face of the structure. A wave will rise higher on a vertical wall breakwater, and these structures should be built to an elevation equal to or greater than about 1.5 times the maximum wave height plus the elevation of maximum tide.

As a rule of thumb, the minimum width of the top of a mound breakwater should be approximately equal to the height of the maximum wave [5]. The width of the top and bottom of a vertical wall breakwater should be determined by an analysis of the various forces on the structure. These dimensions should be sufficient to prevent the structure from overturning, allowing a suitable factor of safety.

## 21-5. GROINS

The erosion of beach areas represents one of the most complex and difficult problems facing the coastal engineer. This erosion results from the effects of breaking waves, especially when the waves approach the shoreline at an oblique angle. Waves approaching a shoreline obliquely create a current that generally parallels the shoreline. This current, which is called a longshore current, sweeps along the sandy particles that constitute the beach bottom. The material that moves along the shore under the influence of waves and the longshore current is called littoral drift.

It is not always understood that manmade structures may stabilize one beach area while creating additional problems of beach erosion in adjoining areas. Impermeable seawalls, bulkheads, or revetments that are constructed along the foreshore often upset the delicate natural regime and cause or increase erosion immediately in front of the structure and along the unprotected coastline downdrift from the structure.

The most common approach to the control of beach erosion is to build a groin or a system of groins. A groin is a structure that is constructed approximately perpendicular to the shore in order to retard erosion of an existing beach or to build up the beach by trapping littoral drift. The groin serves as a partial dam which causes material to accumulate on the updrift or windward side. The decrease in the supply of material on the downdrift side causes the downdrift shore to recede. The shoreline changes that follow the construction of a properly designed groin system are shown in Fig. 21-5.

There are two broad classes of groins: permeable and impermeable. Permeable groins permit the passage of appreciable quantities of littoral drift through the structure. Impermeable groins, the most common type, serve as a virtual barrier to the passage of littoral drift.

A wide variety of groins has been constructed, utilizing timber, steel, concrete, and stone. A typical timber-steel sheet-pile groin is shown in Fig. 21-6. The type shown provides an impermeable barrier. Cellular steel sheet-pile groins also have been employed successfully, as have prestressed concrete sheet-pile groins and rubble mound groins. The latter usually is constructed with a core of fine quarry run stone to prevent the passage of sand through the structure.

The selection of the type of groin will depend to a large degree on the following factors:

1. availability of materials;
2. foundation conditions;
3. presence or absence of marine borers;
4. topography of the beach and uplands.

The hydraulic behavior of a system of groins is exceedingly complex and its performance will be influenced by:

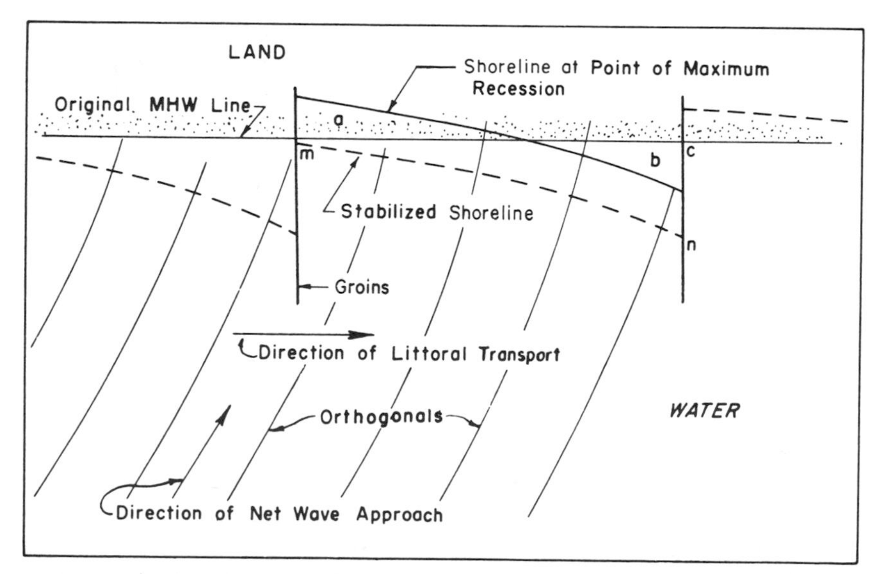


*Fig 21-5.* Groin system operation. (Courtesy U.S. Army Corps of Engineers.)

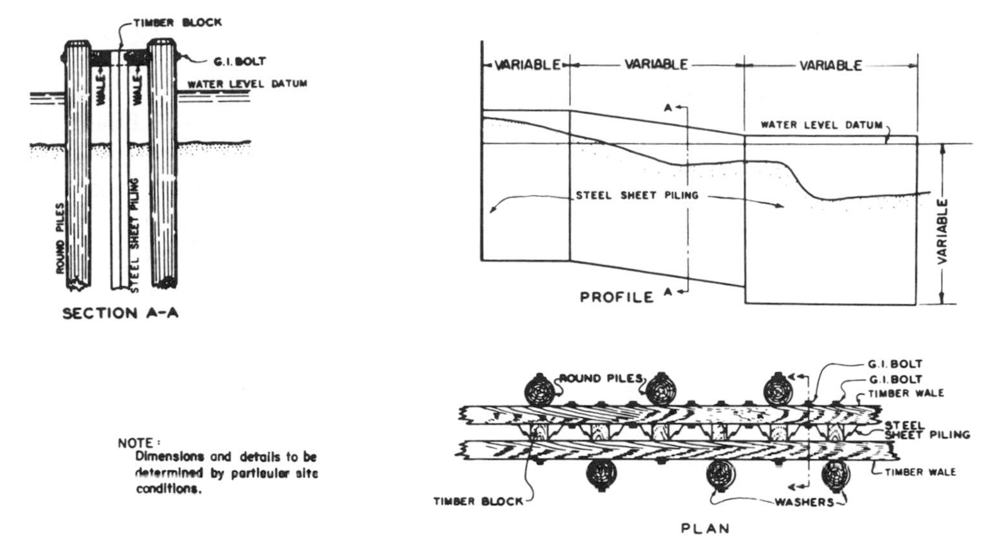


*Fig 21-6.* A typical timber-steel sheet-pile groin. (Courtesy U.S. Army Corps of Engineers.)

1. The specific weight, shape, and size of the particles that constitute the littoral drift.
2. The height, period, and angle of attack of approaching waves.
3. The range of the tide and the magnitude and direction of tidal currents.
4. The design features of the groin system, including the groin orientation, length, spacing, and crown elevation.

Groins usually are built in a straight line and are oriented approximately normal to the shoreline. There appears to be little advantage to the use of curved structures or groins of the T or L head types. These types tend to be more expensive to build and will normally experience more scour at the end of the structure than will be experienced with the straight groin.

There are no reliable analytical techniques for the determination of the desirable length of a groin. The length will depend on the nature and extent of the prevailing erosion and the desired shape and location of the stabilized shoreline. The total length,which typically is on the order of 100 to 150 ft, is comprised of three sections:

1. The horizontal shore section which extends from the landward end to the desired location of the crest of the berm,
2. The intermediate sloped section, which extends between the horizontal shore section and the outer section, roughly paralleling the slope of the foreshore,
3. The outer section, which is horizontal.

The spacing of groins in a groin system depends on the groin length and the desired alignment and location of the stabilized shoreline. As a rule of thumb, the U.S. Army Corps of Engineers [5] recommends that the spacing between groins should be equal to two or three times their length from the berm crest to the seaward end.

The elevation of the crest of a groin will determine to some extent the amount of sand trapped by the groin. In cases where it is desirable to maintain a supply of sand on the leeward side of the groin, it may be built to a low height, allowing certain waves to overtop the structure. If no passage of sand beyond the groin is desired, the elevation of the crest should be such that storm waves will not overtop the structure.

## 21-6. SEAWALLS, REVETMENTS, AND BULKHEADS

Seawalls, revetments, and bulkheads are structures that are placed parallel to the shoreline to separate the land from the water. While these structures have this general purpose in common, there are significant differences in specific function and design. Seawalls are massive structures that are usually placed along otherwise unprotected coasts to resist the force of waves. Seawalls are gravity structures that are subjected to the forces of waves on the seaward side and the active earth pressure on the shoreward side.

A revetment is also used to protect the shore from the erosive action of waves. It is essentially a protective pavement that is supported by an earth slope. A bulkhead is not intended to resist heavy waves but simply to serve as a retaining wall to prevent existing earth or fill from sliding into the sea.

Typical structural types of seawalls are shown in Fig. 21-7. These include a sloping wall, stepped-face wall, and a curved wall. The latter wall face may be either the non-reentrant type, which is essentially a vertical wall, or a reentrant type, which turns the wave back upon itself.

Sloping or vertical-faced seawalls offer the least resistance to wave overtopping. The amount of wave overtopping can be decreased substantially by the use of an armor block facing, such as tetrapods [6]. The stepped-face sea wall is used under moderate wave conditions. It, too, may experience objectionable wave overtopping when subjected to heavy seas and high winds.

Under the most severe wave conditions, massive curved sea walls are used most frequently. For this type of structure, the use of a sheet pile cutoff wall at the toe of the seawall is recommended to reduce scour and undermining of the base. As a further precautionary measure to prevent scour, large rocks may be piled at the toe of the structure. Both of these features may be seen in Fig. 21-8.

There are two broad classes of revetments, rigid and flexible. The rigid type of revetment consists of a series of cast-in-place concrete slabs. In essence,

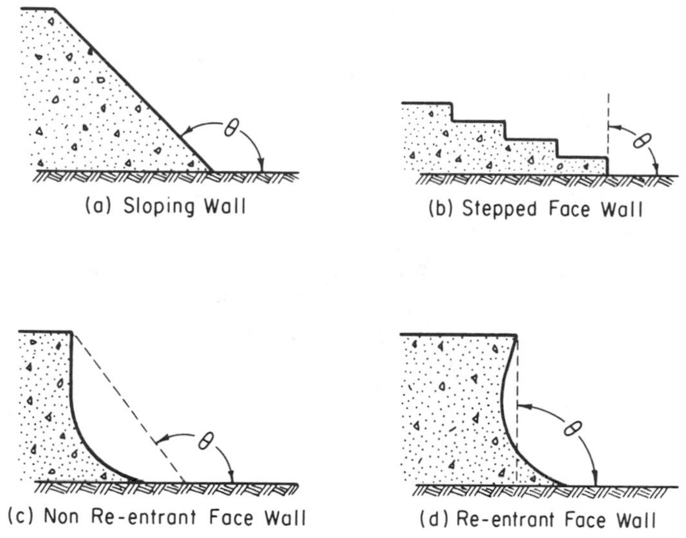


*Fig 21-7.* Typical sea walls.

this type of revetment is a small sloping sea wall.[3] Flexible or articulated armor-type revetments are constructed of riprap or interlocking concrete blocks which cover the shore slope. Typical riprap revetments have armor stones that weigh from ½ to 3 tons. Concrete blocks used for revetments are commonly 1.5 to 4.0 ft square and 2 to 12 in. thick. Sketches of typical revetments are shown in Fig. 21-9.

Two common types of bulkheads are illustrated by Fig. 21-10. A bulkhead is supported in a cantilever fashion by the soil into which it is driven. Additional support for a bulkhead may be provided by tie rods connected to vertical piles, which are driven some distance shoreward.

The choices of the location and length of seawalls, revetments, and bulkheads are usually straightforward depending as they do on local circumstances. The location of these structures with relation to the shoreline generally will coincide with the line of defense against further erosion and encroachment of the sea. The length of these structures depends on how much shoreline is to be separated from the water or protected from the sea.

A critical factor in the design of seawalls, revetments, and bulkheads is the determination of the height of the structure. This determination will hinge on a choice between two basic approaches to the problem. One approach is to design the structure so as to prevent wave overtopping which might damage the structure, flood facilities on the landward side, and possibly endanger

[3]The distinction between seawalls, revetments, and bulkheads is not sharp. A sloping faced seawall in one locality may be called a revetment in another. Similarly, a vertical wall structure may be termed a bulkhead by some observers and a seawall by others.

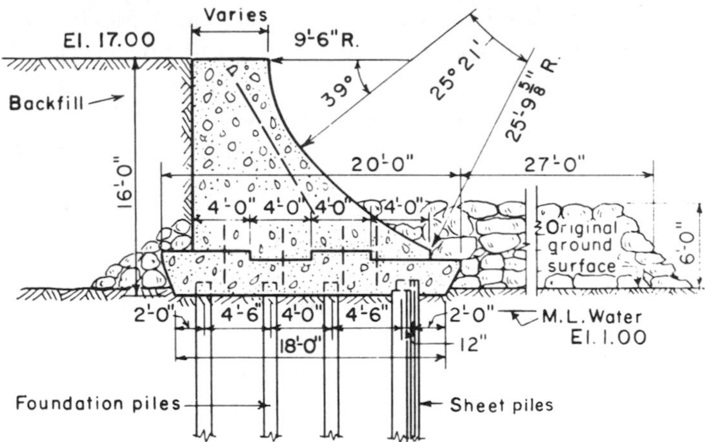


*Fig 21-8.* Concrete curved-face seawall at Galveston, Texas. (Courtesy U.S. Army Corps of Engineers.)

human lives. A second design approach is to recognize that it may not be feasible to construct the protective facility high enough to ensure that no wave overtopping will ever occur. This approach attempts to estimate the volume of water that will pass over the top of the structure under the most critical wave conditions and to attempt to provide appropriate facilities to expel the over-topped water.

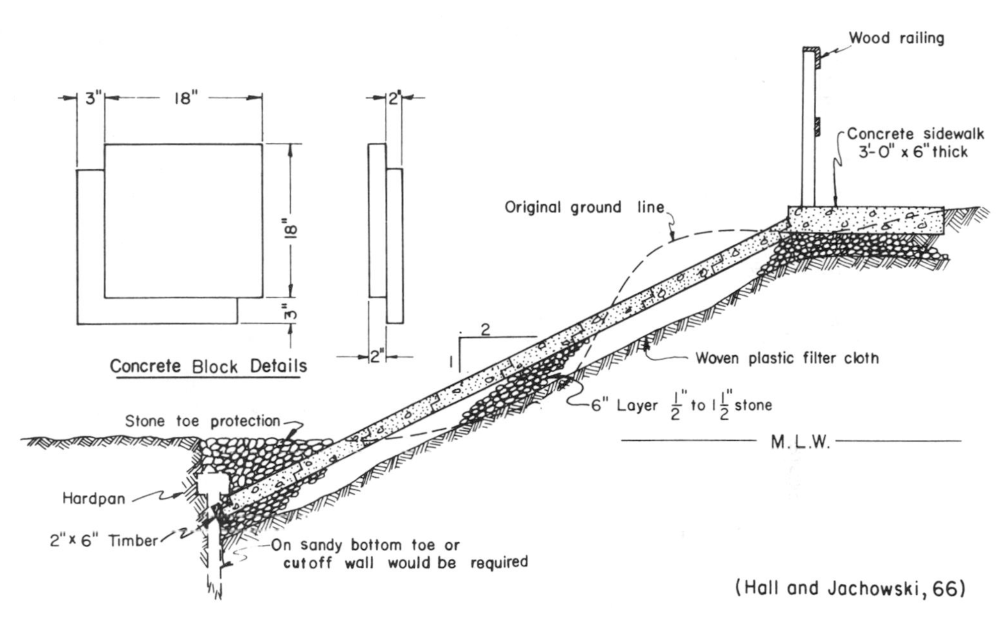


INTERLOCKING CONCRETE-BLOCK REVETMENT
( BENEDICT, MARYLAND)(MODIFIED)

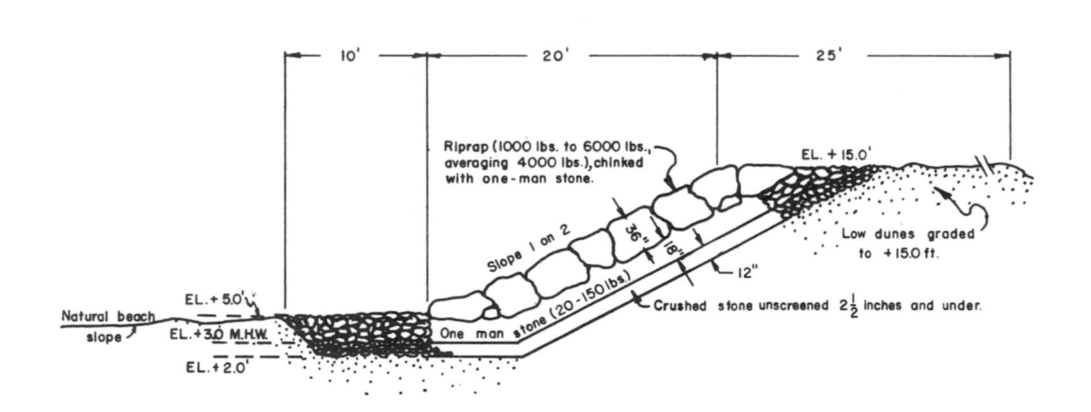


RIPRAP REVETMENT
( FORT STORY, VIRGINIA )

*Fig 21-9.* Typical flexible revetments. (Courtesy U.S. Army Corps of Engineers.)

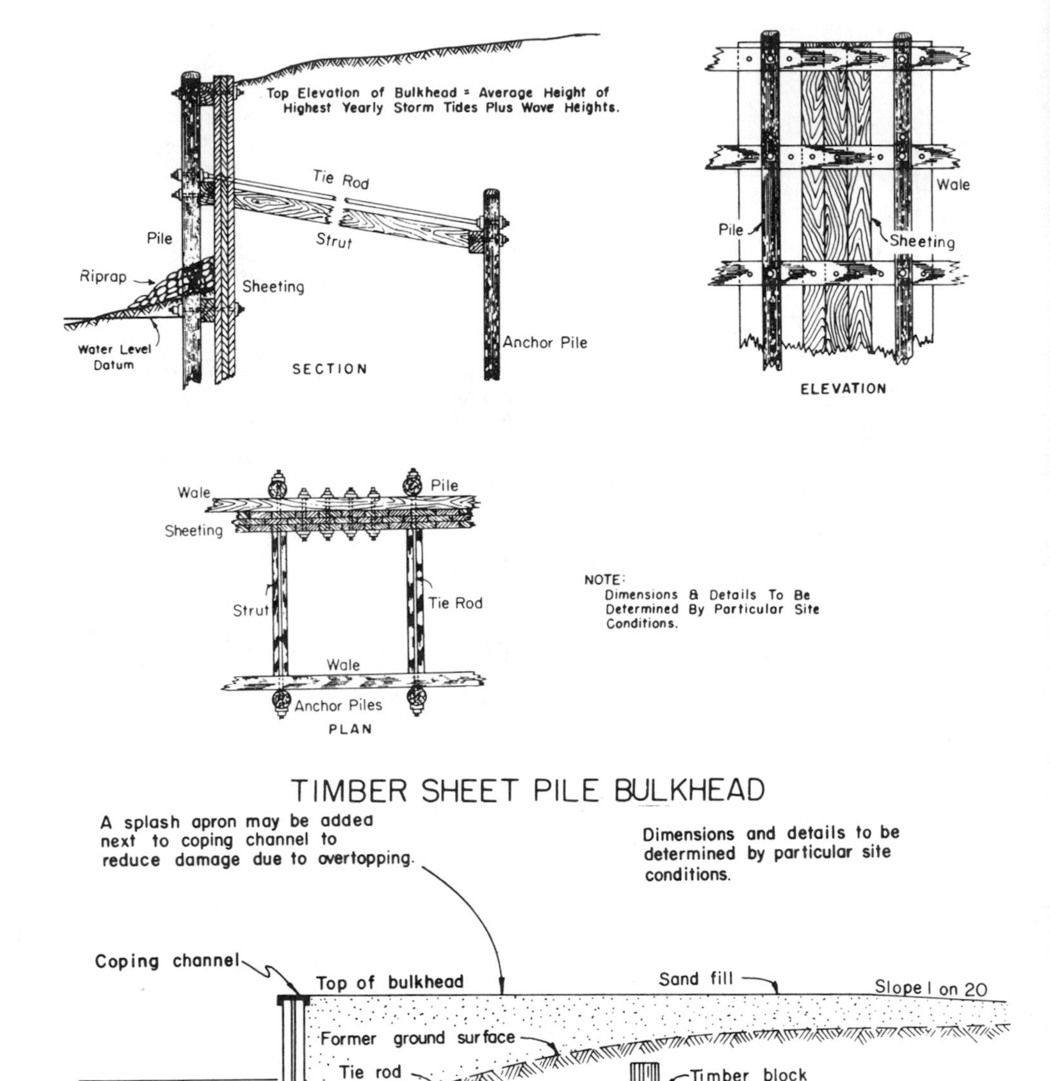


*Fig 21-10.* Typical bulkheads. (Courtesy U.S. Army Corps of Engineers.)

## 21-7. PROTECTIVE BEACHES

The *Shore Protection Manual* [4] describes the functions of protective beaches as follows:

> Beaches can effectively dissipate wave energy and are classified as shore protection structures of adjacent uplands when maintained at proper dimensions. Existing beaches are part of the natural coastal system and their wave dissipation usually occurs without creating adverse environmental effects. Since most beach erosion problems occur when there is a deficiency in the natural supply of sand, the placement of borrow material on the shore should be considered as one shore stabilization measure. It is advisable to investigate the feasibility of mechanically or hydraulically placing sand directly on an eroding shore, termed *beach restoration,* to restore or form, and subsequently maintain, an adequate protective beach, and to consider other remedial measures as auxiliary to this solution. Also, it is important to remember that the replenishment of sand eroded from the beach does not in itself solve an ongoing erosion problem and that periodic replenishment will be required at a rate equal to natural losses caused by the erosion. Replenishment along an eroding beach segment can be achieved by stockpiling suitable beach material at its updrift end and allowing longshore processes to redistribute the material along the remaining beach. The establishment and periodic replenishment of such a stockpile is termed *artificial beach nourishment.* Artificial nourishment then maintains the shoreline at its restored position. When conditions are suitable for artificial nourishment, long reaches of shore may be protected at a cost relatively low compared to costs of other alternative protective structures.

Reference 4 gives recommended procedures for:

1. Estimating the nourishment required to maintain stability of the shore.
2. Sampling and testing native beach sand to obtain a standard for comparing the suitability of potential borrow sediments.
3. Selecting borrow material.
4. Determining an appropriate elevation, width, and slope of a beach berm.
5. Treating the transition from the fill to the existing shoreline.
6. Planning the proper location of a stockpile of nourishment material.

## 21-8. SAND DUNES

Sand dunes are also an important protective coastal formation. The line of dunes nearest the coast, known as foredunes, can prevent the movement of storm tides and waves into the land areas behind the beach. Sand dunes near the beach may also serve as stockpiles to feed the beach. Dune ridges located farther inland also provide protection from waves but to a lesser degree than foredunes.

Reference 4 gives guidelines and suggestions for creation and stabilization of protective dunes by the use of slat-type snow fencing. Figure 21-11 illustrates sand accumulation by a series of four single-fence lifts along the Outer Banks, North Carolina.

Vegetation can also be employed to create stabilized dunes along the shore [4].

## WAVE RUNUP AND OVERTOPPING

Where it is feasible to design a coastal structure to prevent wave overtopping, it is necessary to estimate the magnitude of the wave runup. Wave runup is defined as the vertical height above stillwater level to which water will rise on the face of the structure. Thus the runup, when added to the stillwater elevation, establishes the minimum elevation of the crest of the structure.

In recent years numerous laboratory investigations have been conducted to quantify the effects of several variables on runup. The research has shown that runup depends primarily on the structure shape and roughness, the water depth at the structure toe, the bottom slope in front of the structure, and certain wave characteristics.

The Corps of Engineers [7] has published a series of empirical curves, exemplified by Fig. 21-12, by which wave runup on smooth slopes can be

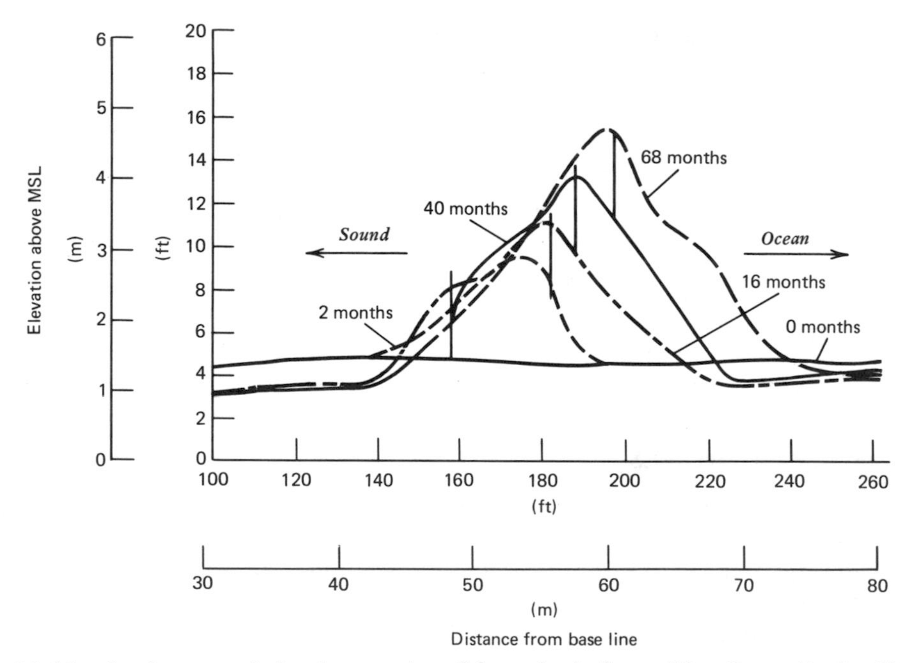


*Fig 21-11.* Sand accumulation by a series of four single-fence lifts, Outer Banks, North Carolina. (Source: Reference 4.)

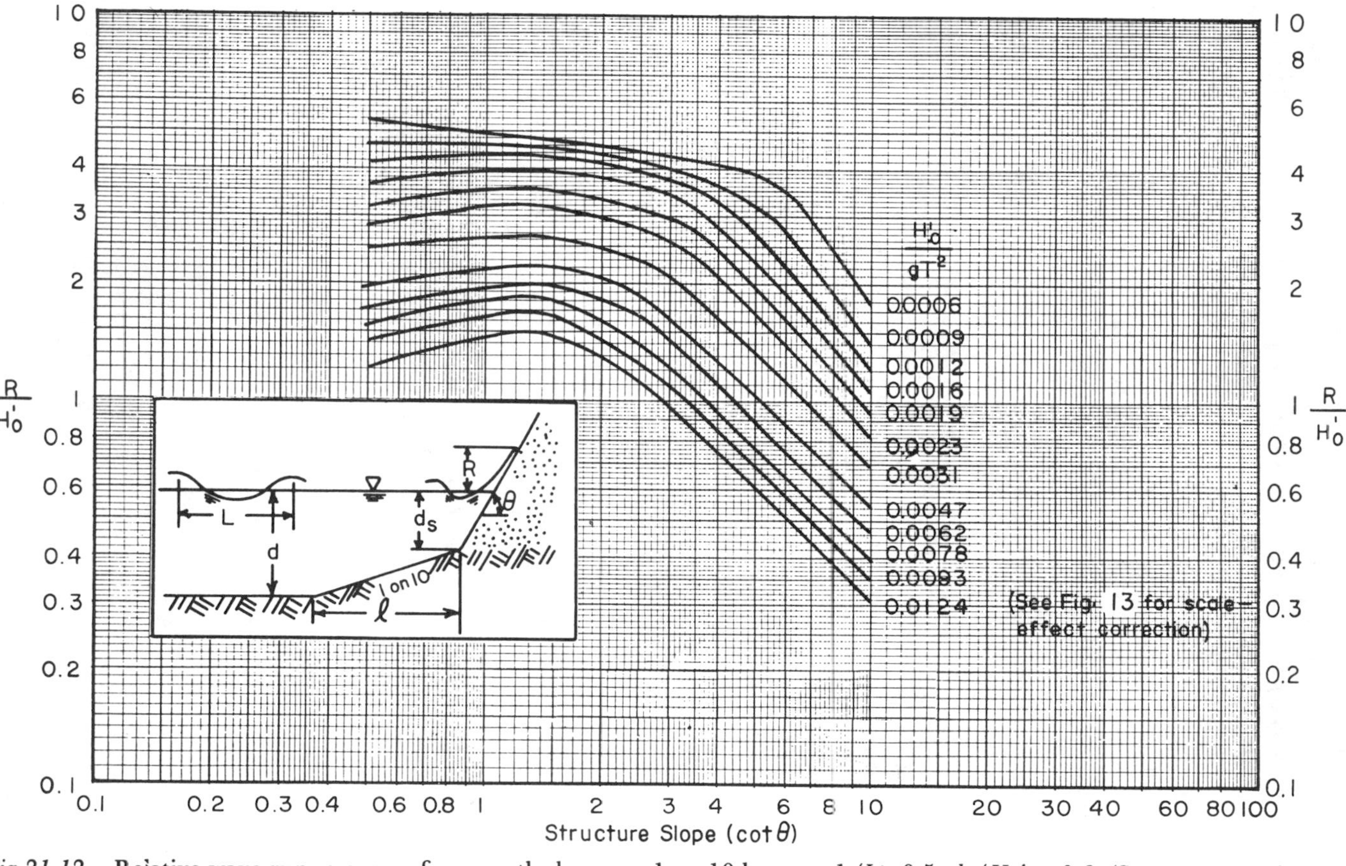


*Fig 21-12.* Relative wave runup curves for smooth slopes on 1 on 10 bottom; $1/L \geq 0.5$; $d_s/H_o' = 0.6$. (Source: *Revised Wave Runup Curves for Smooth Slopes*, U.S. Army Corps of Engineers, July 1978.)

estimated. Similar curves have been published for runup on slopes covered with riprap, rubble, and concrete armor units [8]. The curves are in dimensionless form giving the relative runup $R/H'_0$ as a function of the deep-water wave steepness, $H'_0/gT^2$, and structure slope, cot $\theta$. Here $R$ is the runup height measured vertically from the stillwater level; $H'_0$ is the unrefracted deep-water wave height; $T$ is the wave period; and $g$ is the acceleration of gravity.

Most of the studies of wave runup have been conducted using small-scale models. It has been found that there is a scale effect whereby the predicted values of wave runup from small-scale tests are smaller than those observed in practice. Table 21-2, which shows a scale correction factor as a function of the structure slope, indicates the magnitude of the scale effect.

The following example illustrates the use of Fig. 21-12 and similar figures.

EXAMPLE 21-1

Given: A design similar to that shown in Fig. 21-12 where a smooth, impermeable 1-on-3 structure is fronted by a 1-on-10 bottom slope. The depth at the toe of the structure $d_s = 10$ ft, and the bottom slope extends seaward to a depth, $d = 50$ ft. The deep-water wave height is 16 ft, and the wave period is 7 sec. Estimate the wave runup.

$$\text{The wave steepness, } \frac{H'_0}{gT^2} = \frac{16}{32.2(7)^2} = 0.0101$$

$$\text{From Fig. 21-11, } \frac{R}{H'_0} = 1.05$$

$$R = 16.8 \text{ ft}$$

From Table 21-2, the scale correction factor, $K = 1.12$ and corrected $R = 16.8$ $(1.12) = 18.8$ ft

Model studies have revealed that the volume rate of wave overtopping that occurs at a structure depends on the wave height and period, the water depth at the structure toe, and height, slope, type, and roughness of the

*Table 21-2*
**Runup Scale-Effect Correction Factors**

| *Structure Slope, cot θ* | *Correction Factor,* k | *Structure Slope, cot θ* | *Correction Factor,* k |
|---|---|---|---|
| 0.1 | 1.00 | 3.0 | 1.12 |
| 0.5 | 1.11 | 4.0 | 1.10 |
| 1.0 | 1.14 | 5.0 | 1.09 |
| 2.0 | 1.14 | 10.0 | 1.03 |

SOURCE: Adapted from *Revised Wave Runup Curves for Smooth Slopes,* U.S. Army Corps of Engineers, July 1978.

structure. An empirical procedure for estimating the overtopping rate for various types of structures, water depths, and wave conditions is given by Reference 5.

## WAVE FORCES

Attempts by engineers to control the powerful and relentless forces of waves have often met with disappointment and failure. Rarely is an engineer called upon to design a structure to withstand forces of the magnitude of those imposed by ocean waves during a storm. Furthermore, wave phenomena are exceedingly complex and reliable analytical equations for the estimate of wave forces generally are not available. Yet by means of empirical data collected over a period of many years, the design of coastal and harbor structures can be approached with confidence by the application of sound engineering principles and procedures.

The magnitude of wave forces on coastal and harbor structures will vary a great deal depending on whether or not the waves break, or the shape and slope of the face of the structure, and on its roughness and permeability.

The following material on wave forces is included for illustrative purposes only. For an in-depth treatment of this complex subject, the reader is referred to the *Shore Protection Manual* [4].

### 21-9. VERTICAL WALLS SUBJECTED TO NONBREAKING WAVES

Coastal structures located in protected areas or deep water may be subjected to nonbreaking waves. In the case of vertical walls subjected to the nonbreaking waves, a satisfactory estimate of wave forces may be obtained by the Sainflou method. This method, which was proposed by the Frenchman George Sainflou in 1928, is based on the assumption that pressures due to nonbreaking waves are essentially hydrostatic.

Observations have shown that when a nonbreaking wave strikes a vertical wall the reflected wave augments the next oncoming wave and results in the formation of a standing wave, or clapotis. The height of a clapotis is approximately twice the height of the original wave.

The orbit center of the standing wave lies a distance $h_o$ above the still-water level. This distance may be computed by the following equation:

$$h_o = \frac{\pi H^2}{L} \coth\left(\frac{2\pi d}{L}\right) \tag{21-1}$$

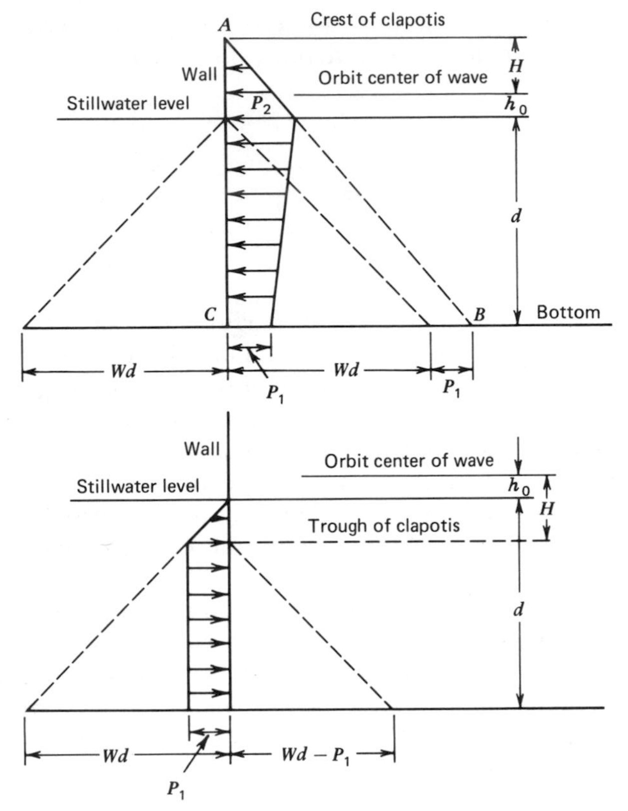


*Fig 21-13.* Wave pressures on vertical walls, according to Sainflou method.

where

$H$ = wave height, ft
$L$ = wave length, ft
$d$ = still-water depth, ft

Assuming the same still-water level exists on both sides of the wall, the pressures on the wall are shown by Fig. 21-12. When the wave is in the crest position the resultant pressures are landward as shown by Fig. 21-13*a*. When the trough of the clapotis is at the face of the wall, the net pressures are directed seaward as shown by Fig. 21-13*b*.

The pressure $P_1$ is given by the equation:

$$P_1 = \frac{wH}{\cosh\left(\frac{2\pi d}{L}\right)} \tag{21-2}$$

where

$w$ = specific weight of water, lb/ft$^3$

Graphical solutions for $h_0$ and $P_1$ may be determined, respectively, by Figs. 21-14 and 21-15.

By simple proportion, it can be seen that:

$$P_2 = (wd + P_1)\left(\frac{H + h_0}{H + h_0 + d}\right) \tag{21-3}$$

It is noted that if there is no water on the landward side of the wall, the resultant pressure with the crest at $A$ would be that shown by the triangle $ABC$, Fig. 21-13$a$.

Experimental observations have indicated that the Sainflou method tends to overestimate forces due to steep, nonbreaking waves [4]. The method appears to be most suitable for smooth vertical walls where the wave reflection is most complete. Where wales, tiebacks, or other structural elements increase the surface roughness of the wall and retard the vertical motion of the water, it is suggested that the pressure $P_1$ be decreased by 5 percent.

## 21-10. FORCES DUE TO BREAKING WAVES

Bulkheads, seawalls, and vertical wall breakwaters often are located so as to be exposed to the force of breaking waves. Research has shown that these waves are much more complex than are nonbreaking waves. Model studies have indicated that structures exposed to breaking waves must withstand both hydrostatic and dynamic pressures. The dynamic pressure typically is intense and of short duration.

The method commonly used to evaluate forces due to breaking waves was developed by R. R. Minikin [9]. The Minikin method was originally developed to analyze forces on a composite breakwater consisting of a concrete superstructure supported by a rubble mound substructure. According to this method, the dynamic pressure is assumed to be a maximum at still-water level, decreasing parabolically to zero at a distance of one-half the wave height above and below the still-water level. (See Fig. 21-16.)

The magnitude of the maximum dynamic pressure in lb/ft$^2$ is given by:

$$P_m = \pi g w \left(\frac{H}{L}\right) \frac{d}{D} (D + d) \tag{21-4}$$

where

$d$ = depth of water at the toe of the vertical wall, ft

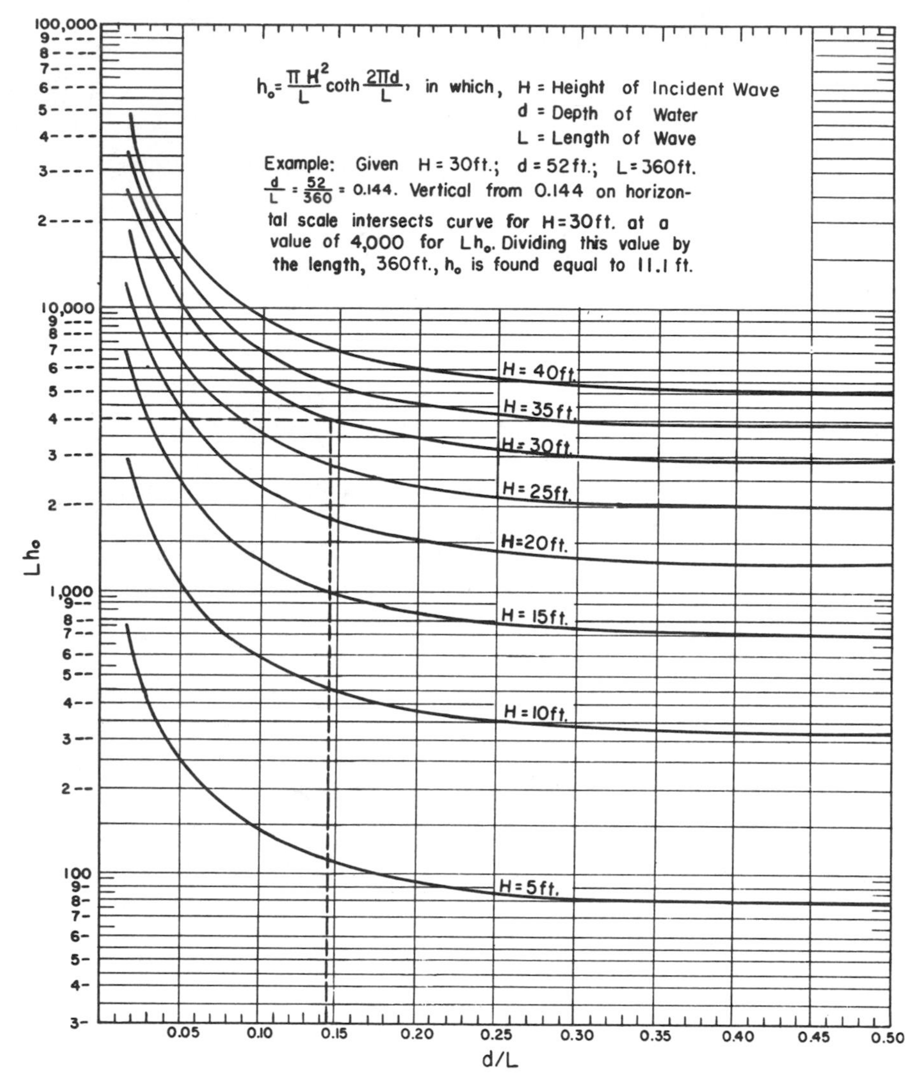


*Fig 21-14.* Determination of value of $h_0$ in Sainflou's formula. (Source: U.S. Army Corps of Engineers, *Shore Protecting, Planning, and Design,* Third Edition, 1966.)

Other values in Eq. 21-4 are as defined previously.

The dynamic force per linear foot of structure is obtained from the area of the force diagram in Fig. 21-16*a*:

$$\text{Dynamic force per linear foot} = \tfrac{1}{3} P_m H \qquad (21\text{-}5)$$

In addition to the dynamic pressure, the Minikin method recognizes that this is a hydrostatic pressure acting shoreward due to the height of the wave above still-water level. (See Fig. 21-16*b*.) The magnitude of this force at the still-water level is:

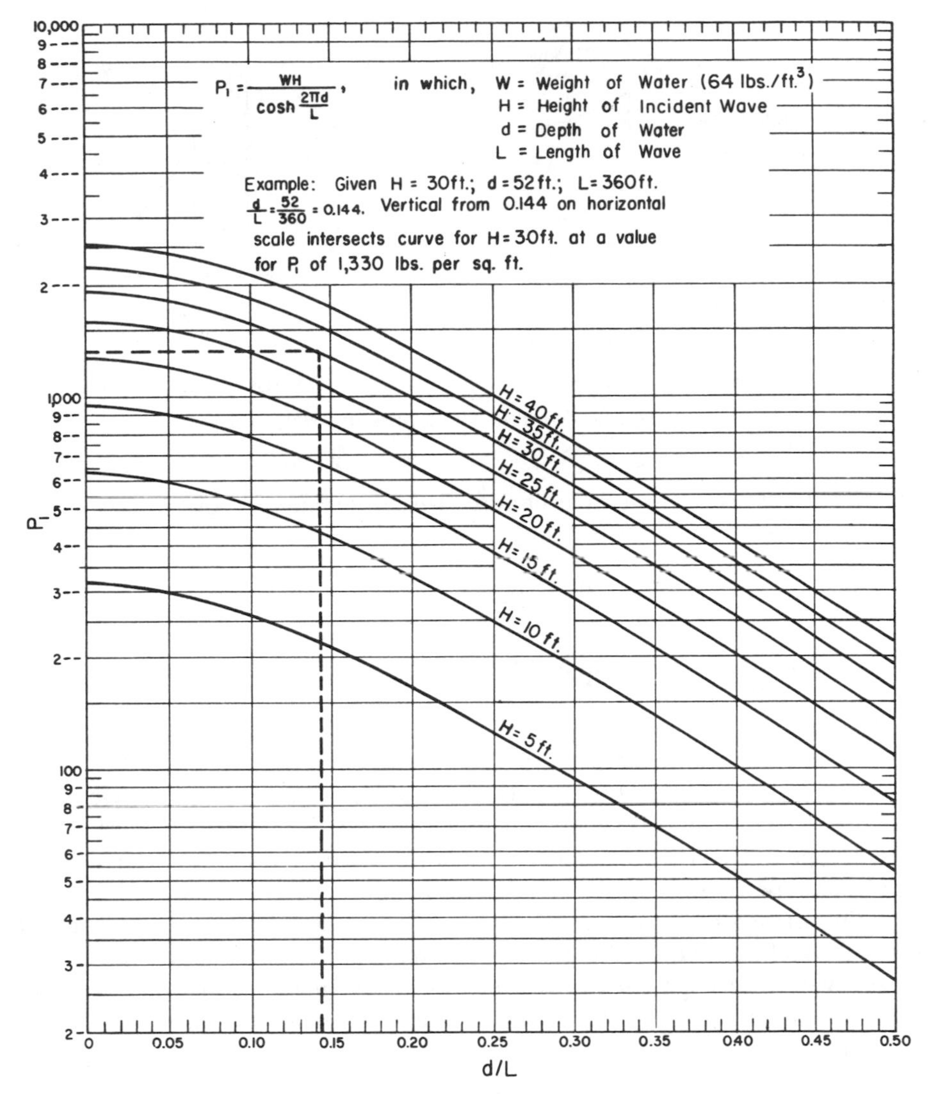


*Fig 21-15.* Determination of value of $P_1$ Sainflou's formula. (Source: U.S. Army Corps of Engineers, *Shore Protection, Planning, and Design,* Third Edition, 1966.)

$$P_s = \frac{wH}{2} \tag{21-6}$$

Assuming that hydrostatic pressures exist on both sides of the wall, the net hydrostatic force per linear foot of structure is:

$$\text{Hydrostatic force/linear ft} = P_s d + \frac{P_s(II/2)}{2} \tag{21-7}$$

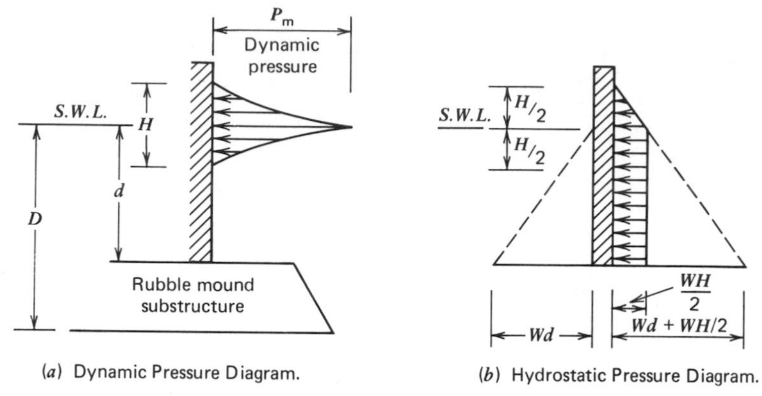


(*a*) Dynamic Pressure Diagram. (*b*) Hydrostatic Pressure Diagram.

*Fig 21-16.* Minikin wave pressure diagram.

It follows that the resultant unit wave force per linear foot of structure is:

$$R = \frac{P_m H}{3} + P_s d + \frac{P_s H}{4} \tag{21-8}$$

An adaption of the Minikin method has been used to calculate forces on caissons or walls that have no substructure. In this case, the values of $D$ and $L$ in Eq. 21-4 refer, respectively, to the water depth and wavelength measured one wavelength seaward from the structure. This is illustrated by Fig. 21-17. Reference 10 indicates that this adaption of the Minikin method yields reasonably reliable results provided the bottom slope is at least 1 : 15. When the actual slope is flatter than 1 : 15, it is recommended that pressure derived for a 1 : 15 slope be used.

The preceding equation for dynamic wave pressures are applicable to vertical wall faces and should give approximate values for stepped-faced structures. Walls that slope backward will experience smaller dynamic forces, and the Minikin gives the following equation for the horizontal dynamic wave pressure:

$$P'_m = P_m \sin^2 \theta \tag{21-9}$$

where

$\theta =$ the angle between the wall face and the horizontal

## 21-11. WAVE FORCES ON RUBBLEMOUND STRUCTURES

The stability of rubblemound structures depends on the ability of the individual armor units that comprise the armor layer to resist displacement. Thus,

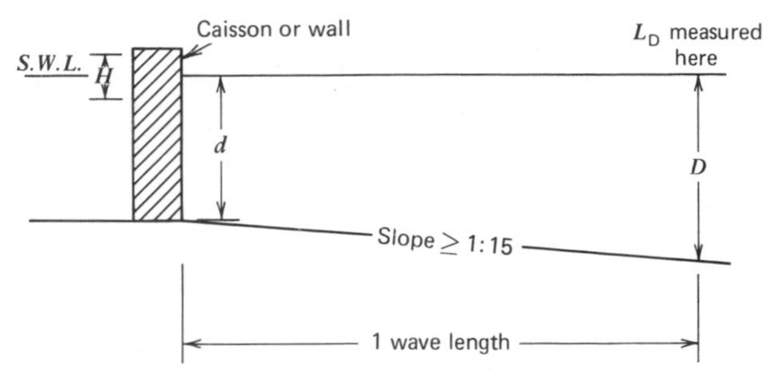


*Fig 21-17.* Parameters for adaptation of Minikin method.

the designer's task is to determine the size of individual armor units required to withstand the attack of storm waves. Because of the complexity of the problem it is necessary to rely on empirical equations that have been derived by means of extensive laboratory model tests as well as by observation of wall failures.

The following equation developed by the U.S. Army Corps of Engineers [5] is considered to be the most reliable guide for the estimation of the weight of each armor unit:

$$W = \frac{w_r H^3}{K_D(S_r - 1)^3 \cot \alpha} \tag{21-10}$$

where

$W$ = weight of armor unit in primary cover layer, lb
$w_r$ = unit weight (saturated surface dry) of armor unit, lb/ft$^3$
$H$ = design wave height at the structure site
$S_r$ = specific gravity of armor unit, relative to the water in which structure is situated $\left(S_r = \dfrac{w_r}{w_w}\right)$
$w_w$ = unit weight of water, freshwater = 62.4 lb/ft$^3$, seawater = 64.0 lb/ft$^3$
$\alpha$ = angle of breakwater slope measured from horizontal, degrees
$K_D$ = an empirical constant (see Table 21-3)

This equation directly accounts for variations due to the wave height, slope of the structure face, and the unit weights of the water and the armor unit.

The empirical coefficient, $K_D$, is used to allow for the effect of the following variables:

1. Shape and roughness of the armor units.
2. Number of units comprising the thickness of the armor layer.

*Table 21-3*

**$K_D$ Values for Use in Determining Armor Unit Weight, No-Damage Criteria**

| | | | Structure Trunk | | Structure Head | |
|---|---|---|---|---|---|---|
| *Armor Units* | $n^a$ | *Placement* | *Breaking Wave* | *Nonbreaking Wave* | *Breaking Wave* | *Nonbreaking Wave* |
| Smooth rounded quarrystone | 2 | Random | 1.2 | 2.4 | 1.1 | 1.9 |
| Smooth rounded quarrystone | >3 | Random | 1.6 | 3.2 | 1.4 | 2.3 |
| Rough angular quarrystone | 2 | Random | 2.0 | 4.0 | 1.3–1.9[b] | 2.3–3.2[b] |
| Rough angular quarrystone | >3 | Random | 2.2 | 4.5 | 2.1 | 4.2 |
| Tetrapod | 2 | Random | 7.0 | 8.0 | 3.5–5.0[b] | 4.0–6.0[b] |
| Quadripod | 2 | Random | 7.0 | 8.0 | 3.5–5.0[b] | 4.0–6.0[b] |
| Tribar | 2 | Random | 9.0 | 10.0 | 6.0–8.3[b] | 6.5–9.0[b] |
| Tribar | 1 | Uniform | 12.0 | 15.0 | 7.5 | 9.5 |
| Dolos, slope = 2.0 | 2 | Random | 15.8 | 31.8 | 8.0 | 16.0 |
| Dolos, slope = 3.0 | 2 | Random | 15.8 | 31.8 | 7.0 | 14.0 |

[a]$n$ is the number of units comprising the thickness of the armor layer.

[b]$K_D$ varies with the slope of the structure. Smaller values of $K_D$ is applicable to slope, cot $\alpha = 3.0$. Larger value applies to cot $\alpha = 1.5$.

SOURCE: U.S. Army Corps of Engineers, *Shore Protection Manual,* 1984.

3. Permeability of the structure as affected by the placement of the armor units.
4. Whether the structure is subjected to breaking or nonbreaking waves.
5. Whether the armor unit is to cover the structure's trunk or head.

Experience has shown that the head of a breakwater or jetty is more likely to sustain extensive damage than is the trunk. The provision of different values of $K_D$ for the head and trunk allows for this fact.

It is noted that the values given in Table 21-3 provide little or no factor of safety.

## PROBLEMS

**1.** Given: A smooth impermeable seawall has a 1-on-3 slope fronted by a 1-on-10 bottom slope similar to that shown in Fig. 21-12. The depth at the toe of the structure is 7 ft, and the bottom slope extends seaward to a depth of 63 ft. The deep-water wave height is 12 ft, and the wave period is 6 sec. Estimate the wave runup.

**2.** Develop an equivalent expression for Eq. 21-4 using SI (metric) units.

**3.** According to the Sainflou method, estimate the shoreward pressure on a vertical wall at the bottom due to a nonbreaking wave that is 25 ft in height and 250 ft in length. Compute these pressures using the appropriate equations, then check your results using Figs. 21-14 and 21-15. The still-water depth is 50 ft.

**4.** Assuming that the variables in Eq. 21-10 are expressed in SI (metric) units, develop $K_D$ values (as shown in Table 21-3) for tribars that incorporate appropriate unit conversion factors.

**5.** A composite breakwater is subjected to the following wave conditions:

Wave height, $H = 20$ ft
Wave length, $L = 270$ ft
Still-water depth to top of substructure, $d = 45$ ft
Depth to bottom, $D = 70$ ft

Determine: (a) maximum dynamic pressure and the dynamic force per linear foot on the breakwater; (b) hydrostatic force per linear foot.

**6.** Estimate the shoreward pressure on a vertical wall at the still-water level due to a nonbreaking wave that is 5 m in height and 65 m in length according to the Sainflou method. The still-water depth is 12 m. Compute the pressure in SI (metric) units.

## REFERENCES

1. *A Statistical Analysis of the World's Merchant Fleets,* U.S. Department of Commerce, Maritime Administration, Washington, D.C., January 1, 1984.
2. *1977 Inland Waterborne Commerce Statistics,* The American Waterways Operators, Inc., Arlington, Virginia.
3. *Ports '77,* Fourth Annual Symposium of the Waterway, Port, Coastal and Ocean Division of American Society of Civil Engineers, pp. 1-177, New York, 1977.
4. *Shore Protection Manual,* Vols. I and II, Coastal Engineering Research Center, U.S. Army Corps of Engineers, Waterways Experiment Station, Vicksburg, Mississippi, 1984.
5. Quinn, Alonzo, *Design and Construction of Ports and Marine Structures,* Second Edition, McGraw-Hill, New York, 1972.
6. Shiraishi, Naofumi, Atsushi Numat, and Taiji Endo, "On the Effect of Armor Block Facing on the Quantity of Wave Overtopping," *Proceedings of Eleventh Conference on Costal Engineering,* American Society of Civil Engineers, September 1968.
7. Stoa, Philip N., *Revised Wave Runup on Smooth Slopes,* Technical Aid No. 78-2, Coastal Engineering Research Center, U.S. Army Corps of Engineers, Fort Belvoir, Virginia, July 1978.
8. Stoa, Philip N., *Wave Runup on Rough Slopes,* Technical Aid No. 79-1, Coastal Engineering Research Center, U.S. Army Corps of Engineers, Fort Belvoir, Virginia, July 1979.
9. Minikin, R. R., *Wind, Waves, and Maritime Structures,* Second Edition, Charles Griffin and Co., Ltd., London, 1963.

# Chapter 22

# Planning and Design of Port Facilities

## 22-1. INTRODUCTION

Increasing world population, the industrialization of nations, and the reshaping of national boundaries have caused substantial increases in the volume of ocean commerce. The needs of increased world trade have resulted in increases in the number and size of the merchant fleet, which has created a growing need for the construction of new ports and the enlargement and modernization of existing port facilities.

In planning and designing port improvements, engineers must choose between alternate port layout schemes and substructure designs. They must obtain reliable forecasts of the number and time distribution of ship movements in order to anticipate the number of required berths. Adequate dimensions for channels and berths must be provided to allow safe and expeditious ship movements and berthing. Sufficient apron space also must be provided for the loading and unloading of ships and for the taking on of fuel and supplies.

General cargo terminals require properly designed transit sheds for sorting and temporary storage of packaged freight. Long-term storage facilities, both open and covered, usually are required and special facilities may be needed for the handling and storage of chemicals, grains, and other materials in bulk.

Every attempt should be made in planning and designing a port to anticipate innovations and improvements in cargo-handling technology in order that prohibitively expensive alterations will not be required at a later date.

## 22-2. GENERAL LAYOUT AND DESIGN CONSIDERATIONS

In the early planning stages of the development of a port facility, certain basic decisions must be made regarding the general arrangement and layout of the facility. One such decision regards the choice of type of wharf.

A wharf is a structure built on the shore of a river, canal, or bay so that vessels may lie alongside and receive and discharge cargo and passengers. A wharf built generally parallel to the shoreline is called a marginal wharf or quay. A wharf built at an angle to the shore is called a finger pier or simply a pier. The berthing and maneuvering space between adjacent piers is called a slip.

Marginal wharves provide for easier berthing of ships and, therefore, generally are preferred over finger piers by steamship and stevedoring companies. Other advantages claimed for marginal wharves are:

1. The cost per berth may be lower.
2. Operational needs of ship owners, truckers, and stevedoring companies are satisfied better.
3. The costs of channel maintenance are less.
4. There is less hazard from waterfront fires and explosions.
5. The continuous line of wharf apron facilitates emergency movements of cargo between adjacent buildings or ships along the waterfront.

Piers, which generally favor rail operations, are preferred by railroad companies. The pier-type layout usually provides more berths per unit length of waterfront than does the marginal wharf arrangement.

The finger pier or slip type of construction may be obtained by extending the piers outward from the shoreline or by reshaping the shoreline by dredging the area between adjacent piers.

The choice of type of wharf layout will depend to a large extent on local conditions. At a port that has ample harbor area for ship maneuvering, but that has a scarcity of waterfront land, constructing piers from the shore may prove advantageous. At a location where there is a scarcity of water area for ship movement and berthing, consideration should be given to dredging of slips. In the latter case, space must be provided for dumping the spoilage that is removed. Such excavation, of course, also removes from service land that otherwise could be devoted to on-shore port operations and storage.

## 22-3. PIER AND WHARF SUBSTRUCTURES

Basically, there are two broad classes of wharf substructures: (1) solid fill type, and (2) the open type. The solid fill type, illustrated by Fig. 22-1, consists of a vertical wall that is backfilled by earth that supports a paved deck. The wall is commonly a cantilevered, anchored steel sheet-pile bulkhead. Gravity structures such as cellular steel bulkheads and cribs of timber or concrete also have been employed successfully.

In the open-type construction, the wharf superstructure is supported by timber, concrete, or steel piles. In this type of substructure, transverse rows of bearing piles are driven and capped with concrete girders. Longitudinal beams may also be provided to support unusually heavy concentrated loads.

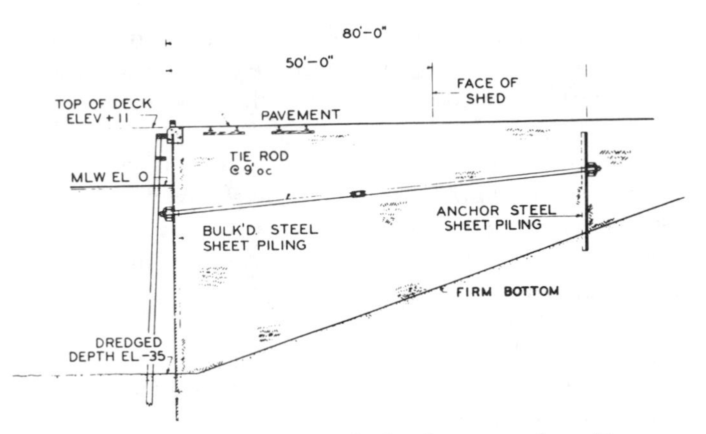


*Fig 22-1.* Solid fill type of wharf construction. (Courtesy The American Association of Port Authorities.)

Alternatively, where heavy concentrated loads are not expected, the piles are spaced closely and capped with a flat slab. A typical open type of wharf substructure is shown in Fig. 22-2.

The principal advantage of the solid fill type of wharf substructure is that its great mass provides adequate resistance to the impact of mooring ships. Solid fill substructures are inexpensive (except in deep water), stable, and require little maintenance. Because this type of substructure serves as a barrier to currents and tides, it is used principally to support marginal wharves.

The open type of wharf substructure is more economical in deep-water locations and where a high-level superstructure is required. Since this type of substructure offers little restriction to water movements, it can be used to support piers in rivers and coastal areas alike.

Open substructures supported by timber piles are subject to decay and may be attacked by marine borers. Considerable risk of wharf fires is asso-

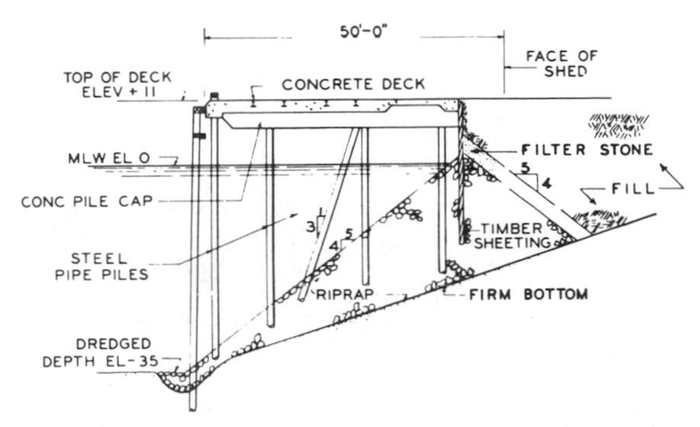


*Fig 22-2.* High-level open type of wharf construction. (Courtesy The American Association of Port Authorities.)

ciated with the use of timber piles. These objections to the open type of substructure, of course, may be largely overcome by using concrete piles. Steel piles usually are encased in concrete above the low water line to inhibit corrosion.

A variation of the open-type wharf substructure utilizes a relieving platform on which fill is superimposed, capped by a paved deck. This type of design offers the advantages of high resistance to impact and economy of construction. High load concentrations are spread by the fill, lessening or eliminating the need for massive longitudinal girders. The relieving platform type of design is less subject to deterioration and decay than is the conventional open-type design, especially if the platform is located below the elevation of mean low water. A typical open-type wharf substructure with a relieving platform is shown in Fig. 22-3.

## 22-4. FENDER SYSTEMS

During the mooring process, a great deal of damage can be done to the wharf and ship unless some sort of protective device is provided to absorb the energy of the moving vessel. Such a device also is needed to lessen the effect of the bumping and rubbing of the ship against the wharf while the ship is secured. Protective installations that meet these needs are called fenders. A wide variety of fenders has been employed, including:

1. pile fenders;
2. timber-hung systems;
3. rubber fenders;
4. gravity-type fender systems.

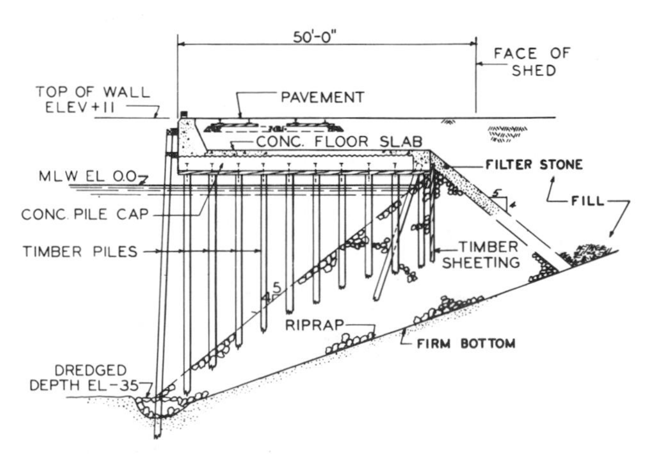


*Fig 22-3.* Open-type wharf substructure with concrete relieving platform. (Courtesy The American Association of Port Authorities.)

One of the simplest fender systems involves a row of vertical wood piles that are driven on a slight batter and secured to the top edge of the wharf. By this system, impact is absorbed by deflection and by compression of the wood. A timber pile fender may be seen in Fig. 22-3. A floating log, called a camel, often is placed between the ship and the pile fenders to distribute impact loads along the fender system and keep the ship away from the face of the dock. The energy absorption of timber piles depends on the pile diameter and length and the type of wood. (See Fig. 22-4.) It should be remembered that the energy absorption capabilities of wood piles, which are rather limited at best, decrease sharply with deterioration and wear.

Pile fenders constructed of steel and concrete also have been used, but generally these systems do not perform as satisfactorily as do timber piles.

In locations where the water is calm and the tidal range is small, a timber-hung fender system may be used. In this system, vertical wood members are secured to the face of the dock and terminate near the water surface. Typically, horizontal wood members are attached between the vertical members and the face of the dock. Timber-hung systems have a low energy-absorption capacity which depends entirely on the compression of the wood.

Rubber has been used extensively and effectively in fender systems and in a variety of ways. Cylindrical or rectangular rubber blocks sometimes are used in wood fender systems to improve energy-absorption capability. These blocks, which are placed behind horizontal members or vertical piles, absorb energy by compression.

Several patented rubber fender devices have been employed, including

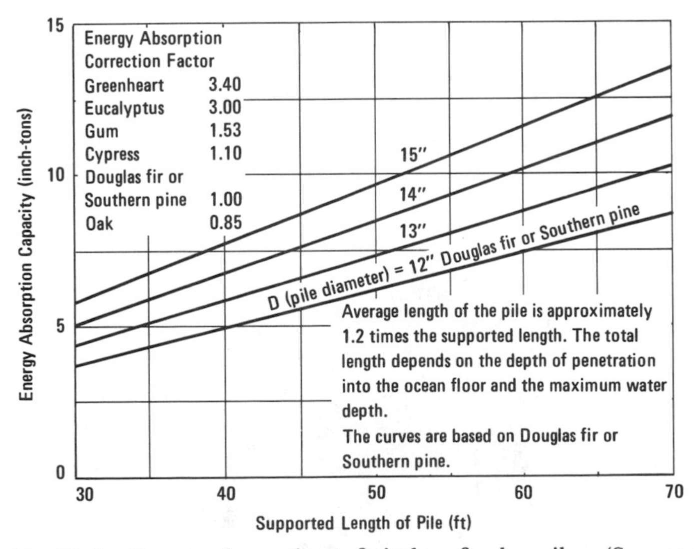


*Fig 22-4.* Energy-absorption of timber fender piles. (Source: Theodore T. Lee, "Design Criteria Recommended for Marine Fender Systems," *Proceedings of Eleventh Conference on Coastal Engineering,* September 1968.)

Raykin fender buffers and Lord fenders. Raykin fender buffers consist of layers of rubber cemented to steel plates and formed in a "V-shape." (See Fig. 22-5.) Energy is absorbed by this device as the rubber layers distort in response to the shearing forces imposed by a mooring vessel. Lord fenders consist of an arc-shaped rubber block bonded between two steel plates. Impact energy is absorbed by the bending and compression of an arc-shaped rubber column.

Hollow rubber cylinders have also been used as fenders. These vary in size from about 5 to 18 in. in outside diameter; the inside diameter is typically one-half the outside diameter. These cylinders are draped along the face of a wharf, suspended by a heavy chain. These fenders are suitable to protect a solid and deep wall such as the face of a relieving platform-type of substructure.

The energy-absorption characteristics of rubber fenders are described in graphs and tables available from the rubber companies that sell these products.

Gravity-type fenders are made of large concrete blocks or cylinders that are suspended from the edge of the wharf deck. When a ship impacts the fender system, these heavy objects are lifted a short distance absorbing the energy of impact. One such fender uses concrete-filled steel tubes which measure 2 to 3 ft in diameter and 20 to 25 ft in length. These cylinders, which weigh about 15 tons, are suspended vertically along the side of the wharf and are hinged in such a way so as to be lifted when a lateral impact force is applied. Wood rubbing strips usually are attached to the seaward side of these cylinders.

In the preceding paragraphs, a brief summary description of some of the more popular fender systems has been given. Table 22-1 lists the principal advantages and disadvantages of these systems.

In selecting a type of fender system the designer must consider a variety of factors, but the most important factors are:

1. Mass of ships to be berthed.
2. Speed of berthing (normal to the dock).
3. Environmental conditions at the port.

The speed of approach normal to the dock taken for design purposes varies from about 0.1 to 1.25 ft/sec, depending principally on the ship size,

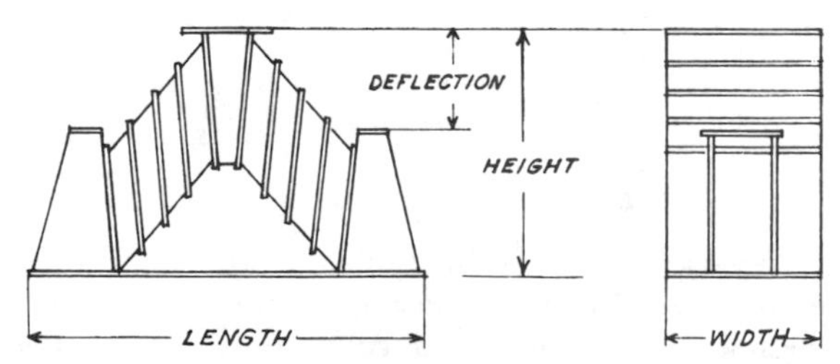


*Fig 22-5.* The Raykin fender buffer.

*Table 22-1*
**Comparison of Various Types of Fender Systems**

| *Fender System* | *Advantages* | *Disadvantages* |
|---|---|---|
| Standard pile, timber | Low initial cost<br>Timber piles are abundant in United States, and most world regions | Energy absorption capacity is limited, it declines as result of biodeterioration<br>Susceptible to mechanical damage and biological deterioration<br>High maintenance cost of damage and deterioration is significant |
| Standard pile, steel | High strength<br>Feasible for difficult seafloor conditions | Vulnerability to corrosion<br>High cost |
| Standard pile, reinforced concrete | Insignificant effects of biodeterioration | Energy-absorption capacity is very limited<br>Corrosion of steel reinforcement through cracks |
| Timber-hung system | Very low initial cost<br>Less biodeterioration hazard | Low energy-absorption capacity<br>Unsuitability for locations with significant tide and current effects |
| Rubber fender systems: Rubber-in-compression | Simplicity and adaptability<br>Effectiveness at reasonable cost | High concentrated loading may result; frictional force may be developed if rubber fenders contact ship hull directly<br>Higher initial cost than standard pile system without resilient units |
| Rubber fender systems: Rubber-in-shear | Capable of cushioning berthing impact from lateral, longitudinal, and vertical directions<br>Most suitable for dock-corner protection<br>High energy-absorbing capacity for serving large ships of relatively uniform size<br>Favorable initial cost for very heavy-duty piers | Raykin buffers tend to be too stiff for small vessels and for moored ships subject to wave and surge action<br>Steel plates are subject to corrosion<br>Bond between steel plate and rubber is a problem<br>High initial cost for general cargo berths |

*continued*

*Table 22-1*
**Comparison of Various Types of Fender Systems—Continued**

| *Fender System* | *Advantages* | *Disadvantages* |
|---|---|---|
| Gravity-type fender systems | Smooth resistance to impacts induced by moored ships under severe wave and swell action<br>High energy-absorption and low terminal load can be achieved through long travel for locations where excessive distance between ship and dock is not a problem. | Heavy berthing structure is required<br>Heavy equipment is required for installation and replacement<br>High initial and maintenance costs<br>Excessive distance between dock and ship caused by the gravity fender is undesirable for general cargo piers and wharves |

SOURCE: Theodore T. Lee, "Design Criteria Recommended for Marine Fender Systems," *Proceedings of Eleventh Conference on Coastal Engineering,* American Society of Civil Engineers, September 1968.

exposure of the wharf to wind, waves, tides and currents, and the availability and type of docking assistance. Some guidance on selection of docking speed has been published by Lee [1].

The kinetic energy of a docking ship is given by the equation:

$$E = \frac{1}{2} Mv^2 = \frac{1}{2} \frac{W}{g} v^2 \tag{22-1}$$

where

$M$ = mass of ship
$v$ = berthing velocity normal to face of the dock
$W$ = displacement of the ship
$g$ = acceleration due to gravity

For ships docking at moderate to high speeds, the kinetic energy is increased due to the mass of water moving alongside the ship. To allow for this effect, the mass value used in Eq. 22-1 should be increased by about 60 percent.

For planning purposes, it may be assumed that one-half of the kinetic energy is to be absorbed by the fender system.

The load-deflection characteristics of various fender systems is shown in Fig. 22-6.

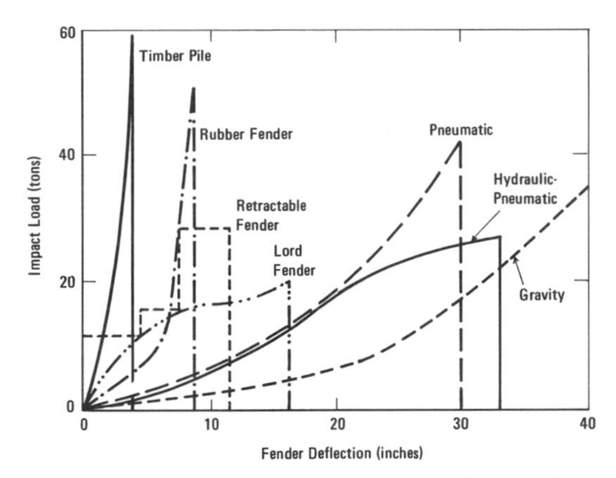


*Fig 22-6.* Load-deflection characteristics of various fender systems. (Source: Theodore T. Lee, "Design Criteria Recommended for Marine Fender Systems," *Proceedings of Eleventh Conference on Coastal Engineering,* September 1968.)

## 22-5. ESTIMATION OF REQUIRED NUMBER OF BERTHS

The planning of the size of a port begins with an appraisal of the volume of present and future commerce and types of shipping. This essential information allows the port planner to estimate the number, type, and sizes of the ships to be accommodated. The forecasting of the anticipated volume and type of freight is an area of concern of economists and planners, and engineers seldom are involved in this type of activity. While the detailed procedures for making such a study do not fall within the purview of this chapter, a brief general discussion of certain aspects of this problem will be given.

The commerce to be shipped through a proposed port will depend most of all on the nature and size of its tributary area, or hinterland. The hinterland generally is defined as that area within which the overall cost of freight movements through the port in question is equal to or less than corresponding costs via competing ports, based on existing rates and charges [2].

The forecasting of freight movements for a proposed port normally would include an evaluation of existing and future levels of activity in manufacturing, mining, oil drilling, agriculture, and forestry within the hinterland. Significant changes in berthing, cargo handling, and storage capabilities of competing ports should also be considered carefully. When expansion of existing port facilities is contemplated, the current level of port activity and its efficiency in handling current traffic are considerations of foremost concern.

Basically, a marine terminal must be able to perform three functions:

1. Unload and load a ship's cargo with efficiency and dispatch.

2. Provide adequate temporary and long-term storage for incoming and outgoing cargo.
3. Provide rail and/or highway connections for movement of freight into and out of the port area.

While the practical capacity of a port may be limited by any one of these functions, the first is generally the controlling factor. Thus, the operating capacity is essentially the product of the cargo-handling rate (tons/day/occupied berth) and the number and extent of utilization of berths.

The rate of loading and discharging of cargo depends on:

1. types of cargo;
2. vessel type and size (especially, number of hatches);
3. availability and size of stevedore gangs;
4. degree of mechanization and methods of cargo handling.

The hourly rate per hatch is especially sensitive to the types of cargo being handled and the cargo handling methods. For example, the rate could range from about 15 tons/hr/hatch for a bulky, low-density commodity such as baled cotton handled by ship's gear, to as much as 300 tons/hr/hatch for rolls of coiled steel handled by a shoreside crane.

At a typical general cargo berth, the average loading or unloading rate is about 25 to 30 tons/hr/hatch, or 200 to 240 tons/hatch/8-hr day. Obviously, the cargo-handling rate will depend on the number of hatches being worked simultaneously and the length of working day. For planning purposes, an average of 40 stevedore gang-hours per working day may be used [2]. This may consist of five gangs working 8 hr each, four gangs working 10 hr each, or any similar combination. On this basis, the maximum loading rate for an occupied berth is 1000 to 1200 tons/berth/day. In using these rates to estimate annual berth capacities, it should be remembered that a given berth will not be occupied 100 percent of the time during the year.

The estimation of the number of berths required must be made in the face of fluctuations in demand. In cold climates, allowance must be made for the closing of the port in winter months due to the formation of ice. Consideration may also have to be given to seasonal fluctuations in transportation of certain products.

Independent studies by Plumlee [3], Nicolaou [4], and Fratar, et al. [5], have shown that ships arrive at a public seaport in accordance with a random pattern and that the Poisson probability distribution may be used to predict the number of days on which a particular number of ships will be present.

According to the Poisson law:

$$F_n = \frac{T(\bar{n})^n e^{-\bar{n}}}{n!} \tag{22-2}$$

where

$F_n$ = the number of units of time that $n$ ships are present during $T$ time units

$\bar{n}$ = the average number of ships present

$e$ = the Naperian logarithmic base, 2.71828

The Poisson equation makes it possible to calculate the distribution of ship arrivals if only the average number of ships present during the period in question is known. For example, if it is known that an average number of ships present at a certain port is 8.0, the number of hours during a one-year period (8760 hr) that 10 ships will be present is:

$$F_{10} = \frac{8760(8)^{10}e^{-8}}{10!} = 872$$

This and similar values are plotted in Fig. 22-7 which shows the Poisson distribution for $\bar{n} = 8$ and a period of one year. The calculations indicate that at a nine-berth port, there are 872 hr during the year in which one ship will be waiting for a berth. Similarly, at a port with 11 berths, there will be 872 hr during the year in which one berth will be vacant.

Since there are costs involved in idle berths as well as in waiting ships, the optimum number of berths is a compromise, best determined by minimization

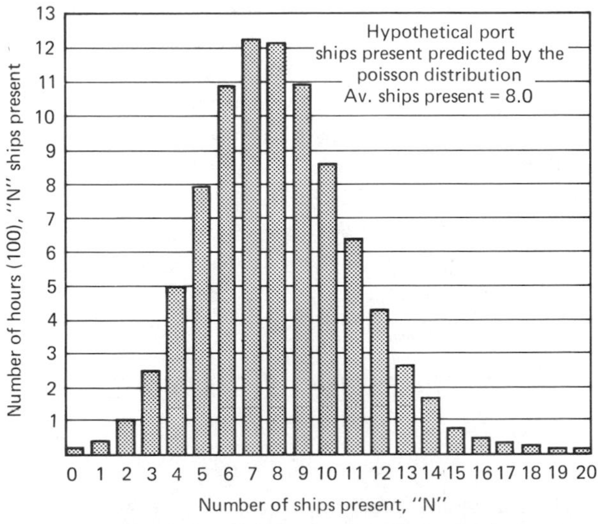


*Fig 22-7.* Ship distribution at a hypothetical port. (Courtesy American Society of Civil Engineers.)

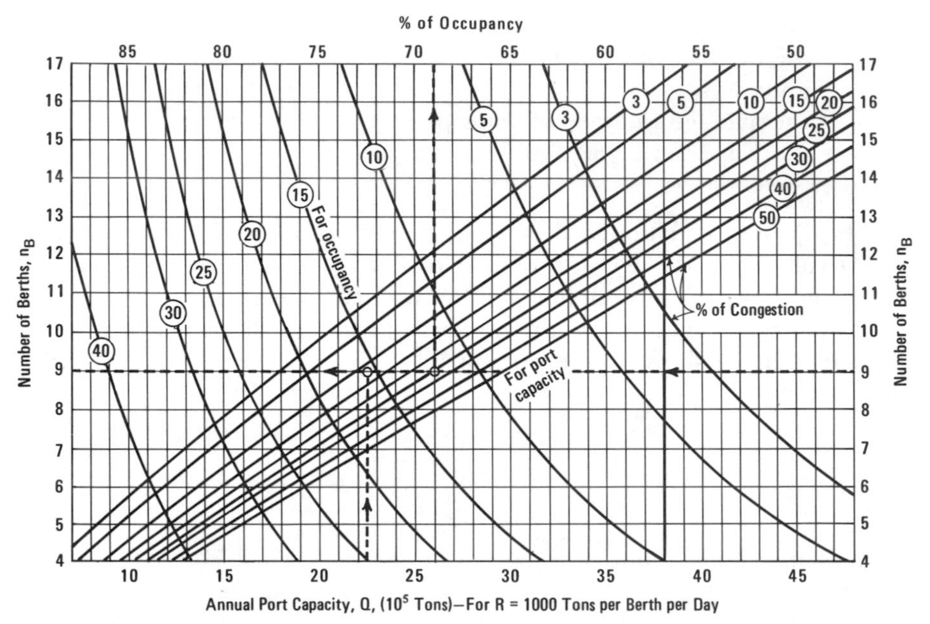


*Fig 22-8.* Relationship between annual port capacity, congestion, and berth occupancy. (Source: S. N. Nicolaou, "Berth Planning by Evaluation of Congestion and Cost," *Proceedings,* American Society of Civil Engineers, November 1967.)

of the sum of the annual costs of vacant berths and waiting ships.[1] Using this approach, Nicolaou [4] has developed a useful graph, shown in Fig. 22-8, which relates annual port capacity to the number of berths.

It will be noted that two parameters are shown on Fig. 22-8: percent of congestion and percent of occupancy. Percent of congestion is defined as the percent of time for which the number of ships in port exceeds the number of berths available. Percent of occupancy is defined as the percent of time that the total number of berths available at the port, $N_B$, is occupied.

Nicolaou [4] showed that percent of congestion is related to average cost of a ship waiting for a berth, $C_S$, and the average cost of an idle berth, $C_B$, by the following inequality:

$$\text{Percent congestion} < 100\left(1 - \frac{C_S}{C_B + C_S}\right) \tag{22-3}$$

[1] Wanhill [6] has described an iterative procedure based on a wider criterion: The minimization of the total port usage cost including the service time costs as well as the costs of the ship's waiting time and the costs of providing the berths.

The annual port capacity, $Q$, is given by the following equation:

$$Q = N_B R T \left(\frac{\text{percent occupancy}}{100}\right) \tag{22-4}$$

where

$N_B$ = number of berths
$R$ = annual average cargo handling rate per berth, tons/day
$T$ = time period, usually 365 days

The use of Fig. 22-8 is illustrated by a numerical example.

EXAMPLE 22-1

Given: $T = 365$ days; $Q = 1.8 \times 10^6$ tons; $R = 800$ tons/day; $C_B = \$250$; and $C_S = \$900$.

$$Q' = Q\left(\frac{1000}{R}\right) = (1.8 \times 10^6)\left(\frac{1000}{800}\right) = 2.25 \times 10^6 \text{ tons}$$

By Eq. 22-3,

$$\text{Percent congestion} < 100\left(1 - \frac{900}{250 + 900}\right)$$

$$\text{Percent congestion} < 21.7\%$$

By entering Fig. 22-8 with a value of $Q' = 2.25 \times 10^6$ tons and for a value of percent congestion $< 21.7$ (interpolated from the straight-sloped lines), we find that the chart gives a value of $N_B = 9$. Note that $N_B = 8$ gives a percent congestion = 24, which is too large. Corresponding to $N_B = 9$, the percent congestion = 12 percent. Reentering the chart at the right ordinate with a value of $N_B = 9$ and intersecting the percent of congestion curves as shown, we find that:

$$\text{Percent occupancy} = 69.0$$

from which

$$Q = 9(800)(365)\left(\frac{69.0}{100}\right) = 1.813 \times 10^6 \text{ tons}$$

Figure 22-8 should give suitable planning estimates of the number of required berths for the range of values shown. More complete analyses, which are beyond the scope of this text, may be made by the use of queuing theory [7] or by computer simulation [8, 9].

## 22-6. BERTH AND SLIP DIMENSIONS

The space required alongside a wharf for the berthing depends on:

1. size and type of ships served;
2. wharf configuration;
3. mooring procedures.

Terminals that accommodate tankers and other bulk cargo ships vary widely with regard to berth and mooring space requirements. There are no generally accepted standards for berth and slip dimensions for these facilities, as these space needs will depend not only on the factors listed above but also on the cargo-handling procedures and equipment. In contrast, terminals that serve oceangoing general cargo ships are conventional and typical dimensions may be useful in the planning of these facilities.

From the standpoint of ease of mooring, the choices of wharf configuration in order or desirability are:

1. marginal wharf;
2. two-berth pier;
3. four-berth pier.

The required berth length is equal to the length of a ship, plus a small clearance between adjacent ships and space for the ships' lines. Thus, a typical berth length for oceangoing general cargo ships is about 750 to 850 ft.

In planning a marginal wharf to serve oceangoing general cargo ships, about 750 ft should be allowed for a one-berth facility and a minimum of 650 ft per berth is required for multiple-berth facilities. A desirable pier length for a two-berth pier is 850 ft and for a four-berth pier, about 1500 ft.

For a wharf configuration utilizing two-berth piers, a slip width of at least 250 ft is needed. This dimension is roughly the sum of the beams of two ships and the length of a tugboat. When four-berth piers are used, a minimum slip width of about 325 ft is required. This dimension is based on the beams of two ships (moored) plus the beam of another ship (mooring) plus the length of a tugboat.

In summary, the minimum and desirable pier length and slip width for a two-berth pier and a four-berth pier are:

| | *Pier Length (ft)* | | *Slip Width (ft)* | |
|---|---|---|---|---|
| | *Minimum* | *Desirable* | *Minimum* | *Desirable* |
| Two-berth piers | 750 | 850 | 250 | 300 |
| Four-berth piers | 1375 | 1500 | 325 | 375 |

## 22-7. TRANSIT SHEDS AND APRON

At a typical general cargo terminal, temporary dry storage space is provided in a large building called a transit shed. In addition to providing short-term storage, the transit shed may also provide space for customs activities and for port administration and security. The transit shed should not be used for long-term storage, and when long-term storage facilities are required, warehouses and open storage areas usually are located shoreward of the transit shed.

Experience has shown that for a typical dry-cargo ship, a transit shed that has an area of 85,000 to 90,000 ft$^2$ is desired. This value provides adequate space to accommodate a single average-sized dry cargo vessel and includes an allowance of about 40 percent for aisles and other nonstorage areas. The 90,000 ft$^2$ requirement allows for the discharging and loading of the entire contents of an average cargo vessel. With larger vessels currently in service, transit sheds with areas as large as 120,000 ft$^2$ have been provided at some major ports [10]. Proportionately smaller transit sheds may be used at small terminals where ships are expected to discharge and take on only partial loads.

A common length of transit shed is 500 to 550 ft. The larger value provides a sheltered storage area adjacent to the entire length of a large merchant cargo vessel. A minimum transit shed width of 165 ft is recommended for marginal wharf terminals. (See, for example, Fig. 22-9.) Where a single shed serves two berths, one on each side of a pier, approximately twice

*Fig 22-9.* Port facilities at Port Everglades, Florida. (Courtesy The American Association of Port Authorities.)

this value would be required to provide the specified 90,000 ft$^2$ per berth. The choice of configuration of the transit shed may take many forms and will be governed by local conditions [10].

A 20-ft-wide covered platform along the rear (shoreward) side of the transit shed is recommended to facilitate the transfer of cargo between trucks and railroad cars and the transit shed. Door openings usually are provided in alternate bays along both the apron side and the platform side of the transit shed.

The portion of the wharf or pier that lies between the waterfront and the transit shed is called the apron. This uncovered space is needed for mooring, and for the loading, unloading, and movement of cargo into the transit shed. Along the waterfront edge of the apron, space must be reserved for bollards, cleats, and other mooring devices. Connections for electric power and for telephone and water service must be provided. These connections usually are housed in service boxes built into the deck of the apron along the waterfront edge. At least two service boxes per berth are recommended.

When railroad service is desired along the apron, the rails are constructed flush with the apron deck and the tracks are constructed on 13.5 ft center-to-center spacing. Rail-supported gantry cranes may be installed along the apron, in which case 3 to 5 ft of apron width must be allocated for each crane rail.

The required total apron width will vary from about 20 ft to more than 60 ft depending on the facilities provided. Table 22-2 gives principal dimensions for general cargo terminals recommended by the American Association of Port Authorities [10]. Figure 22-10 shows a cross section of a typical transit shed and apron.

*Table 22-2*

**Principal Dimensions for General Cargo Terminals**

| | |
|---|---|
| Berth lengths | |
| Wharves | 750 ft multiples[a] |
| Piers | 850 ft |
| Apron widths | |
| No railroad tracks | 30 ft |
| One railroad track | 30 ft |
| Two railroad tracks | 38.5 ft |
| Clear stacking height—sheds | 20 ft |
| Gross transit shed space per berth | 50,000 to 120,000 ft$^2$ |
| Interior column spacing | 40 ft minimum |

[a] 1 ft = 0.3048 m.

SOURCE: *Port Planning, Design and Construction,* The American Association of Port Authorities, Inc., Washington, D. C., 1973.

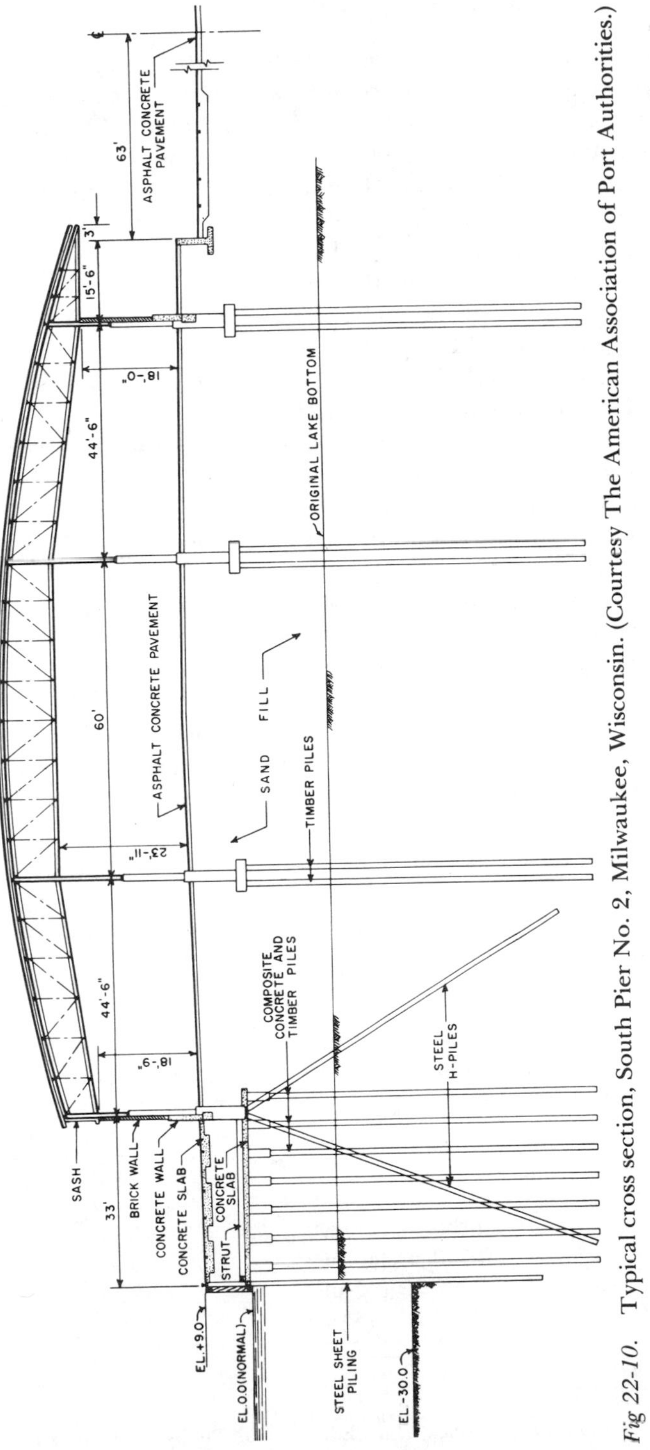


*Fig 22-10.* Typical cross section, South Pier No. 2, Milwaukee, Wisconsin. (Courtesy The American Association of Port Authorities.)

## 22-8. CONTAINER PORTS

Prior to 1956, cargo at ocean ports was handled in small units. This package-by-package cargo-handling procedure was time consuming and costly, accounting for a large percentage of the overall costs of ocean shipping.

In 1956 and 1957, Sea-Land Service, Inc. and Matson Navigation Company independently and almost simultaneously introduced a new concept in ocean shipping that came to be known as containerizaton [11]. It has been said that containerization has made perhaps the greatest impact on the shipping industry since the invention of the steam engine.

Containers are simply boxes, typically 8 by 8 ft in cross section and 12 to 40 ft in length. Table 22-3 gives the standard dimensions and weights of containers proposed by the International Standards Organization. Sizes of 12, 24, 27, and 35 ft have been used [12].

The usage of containers eliminates much of the handling of small units of cargo at a port facility. A container usually is loaded and sealed at the place of origin and remains sealed until it arrives at its destination. However, a fraction (up to 40 percent) of the containers are usually loaded (stuffed) and unloaded (stripped) at the port.

Containers may arrive at the port facility by rail, but more commonly arrive by trucks. Upon arrival the containers are weighed, logged in, and stored temporarily in an assigned location in a marshaling area. The containers are later moved from the marshaling area to wharfside and transferred to a ship by means of heavy container cranes. This procedure is reversed at the destination end of the trip. The principal elements of a large container port may be seen in Fig. 22-11.

At least five advantages may be listed for a container port:

1. The berth capacity is great, often being five times as high as the capacity of a traditional general cargo berth.

*Table 22-3*

**Standard Dimensions and Weights of Containers**

| *Length (ft)*[a] | *Volume* | *Weight (kg)*[b] | |
|---|---|---|---|
| | | *Empty* | *Loaded* |
| 20 | 1100 $ft^3$ (31 $m^3$) | 1850 | 20,320 |
| 30 | 1600 $ft^3$ (46 $m^3$) | 2600 | 25,400 |
| 40 | 2200 $ft^3$ (62 $m^3$) | 3200 | 30,580 |
| 40 (insulated) | 2050 $ft^3$ (58 $m^3$) | 4350 | 30,480 |

[a] 1 ft = 0.3048 m.

[b] 1 kg = 2.2046 lb.

SOURCE: Bruun, Per, *Port Engineering*, Third Edition, Gulf Publishing Company, Houston, 1981.

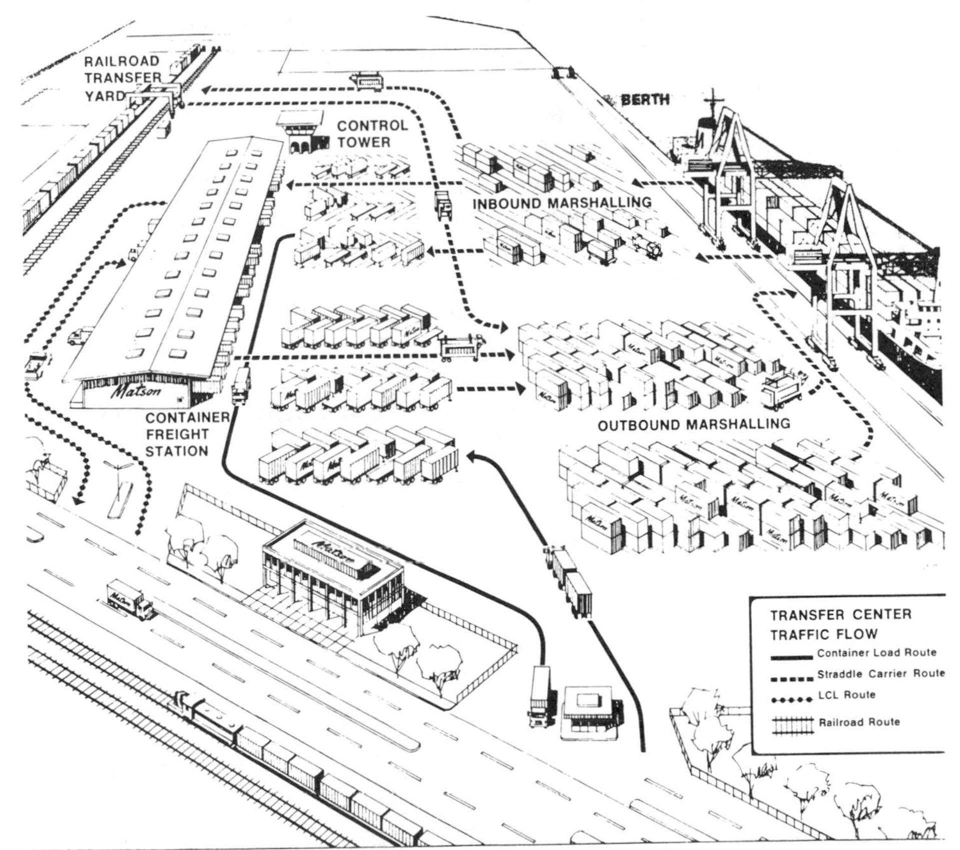


*Fig 22-11.* Example of a containership terminal and traffic-flow. The drawing illustrates how the flow of containers is organized in a Matson transfer center. The arrows show how less-than-container-load cargo moves to the container freight station first and comes out of the container freight station last. Full container shipments go direct to the container yard. The rest of the movement within the transfer center is handled by straddle trucks and gantry cranes. (Courtesy The American Association of Port Authorities.)

2. Overall transit time is less.
3. Container ports offer greater safety for waterfront employees.
4. There is less damage to cargo.
5. Less pilferage occurs at a container port.

The principal disadvantage of a container port is the large amount of land required for the marshaling area. The need to provide large expensive container cranes is also a disadvantage of container ports.

Planning considerations and design criteria for container ports, largely abstracted from Ref. 10, are described briefly in the following paragraphs.

Since containers are designed to be interchangeable between ship, truck, and rail transportation, it is important that container ports be easily accessible by highways and railroads. Preferably, container berths should be placed along a marginal wharf rather than a pier because of the need to provide a large supporting area for each berth.

The components that comprise a typical container terminal are: (1) the ship berth, (2) the container cranes, (3) the marshaling area, (4) the container packing shed, (5) the entry facilities, and (6) the garage and inspection building. Most of these components can be seen in Fig. 22-11.

*Ship Berths.* The berth lengths will depend upon the size of the vessel to be accommodated. The largest containerships in service have a length of about 950 ft, a width of about 110 ft, and a draft of approximately 35 ft. The American Association of Port Authorities [10] recommends a minimum berth length of 850 ft up to a maximum of 1000 ft to accommodate the largest containership in service. If the container berths are arranged along a marginal wharf, a certain amount of flexibility in berth length will be provided, and the choice of a design berth length is not critical.

*Fig 22-12.* A containership terminal. (Courtesy The Port of Long Beach.)

*Cranes.* Container ships usually do not have shipboard cranes, and container cranes on shore must be provided. Normally, two or more cranes working simultaneously will load and unload a containership. These cranes are rail-mounted on the dock and typically have a capacity of 20 to 30 long tons. (See Fig. 22-12.)

*Marshaling Areas.* Two general classes of container storage systems are used for the marshaling area: chassis storage and stacked storage. With chassis storage, a container unloaded from a ship is placed on a semitrailer chassis and then hauled by a yard tractor (Fig. 22-13*a*) to an assigned space in the terminal where it remains until retrieved by a highway tractor. Similarly, arriving export containers are assigned locations in the marshaling yard and remain there until being moved to the loading area by yard tractors.

The chassis storage system provides operational efficiency, as each container is immediately available to a tractor unit. However, chassis storage requires more yard space and more chassis than do other systems.

In chassis storage areas containers are commonly stored back-to-back with at least 4 ft between the backs of containers. The aisles between the rows should be at least 65 ft in width and preferably 70 to 75 ft to facilitate the parking maneuvers. The parking spaces are typically 10 ft wide, and the length is determined by the size of the container to be parked [13].

In the stacked storage system containers are stacked by straddle carriers or by a traveling bridge crane. Straddle carriers are special pieces of equipment that straddle the containers and transport them between shipside and storage areas or onto trucks or railroad cars (see Fig. 22-13*b*). Straddle carriers can stack containers two or three high in long rows that are one container in width. In these yards it is usually advantageous to orient the container stacks at a 45-degree angle to the traffic flow patterns. Traveling bridge cranes stack the containers three or four high. (See Fig. 22-13*c*). These yards usually are laid out at right angles to the traffic flow [13]. Stacked storage systems require considerably less area than do chassis storage systems but are not as efficient because nonproductive handlings are required to retrieve containers.

*Container Packing Shed.* A container packing shed normally is provided where less-than-container load (LTCL) shipments are handled. It need not be contiguous to the marshaling area, and definitely should not be placed at the normal location of a transit shed. There it would tend to encumber the steady movement of containers to and from the cranes during loading and unloading operations. The size of the packing shed varies widely. Its general configuration resembles a typical truck terminal, allowing delivery trucks to arrive at one side of the building and the cargo to be moved from these trucks directly into waiting containers on the opposite side. Thus, the building tends to be long and narrow with emphasis on the number of truck and container doors necessary. As an example, one of the major packing sheds at the Elizabeth, New Jersey, container port is 1000 ft long and 100 ft wide with a total of 162 truck doors [10].

*Entry Facilities.* The truck entrance to the terminal facility usually consists

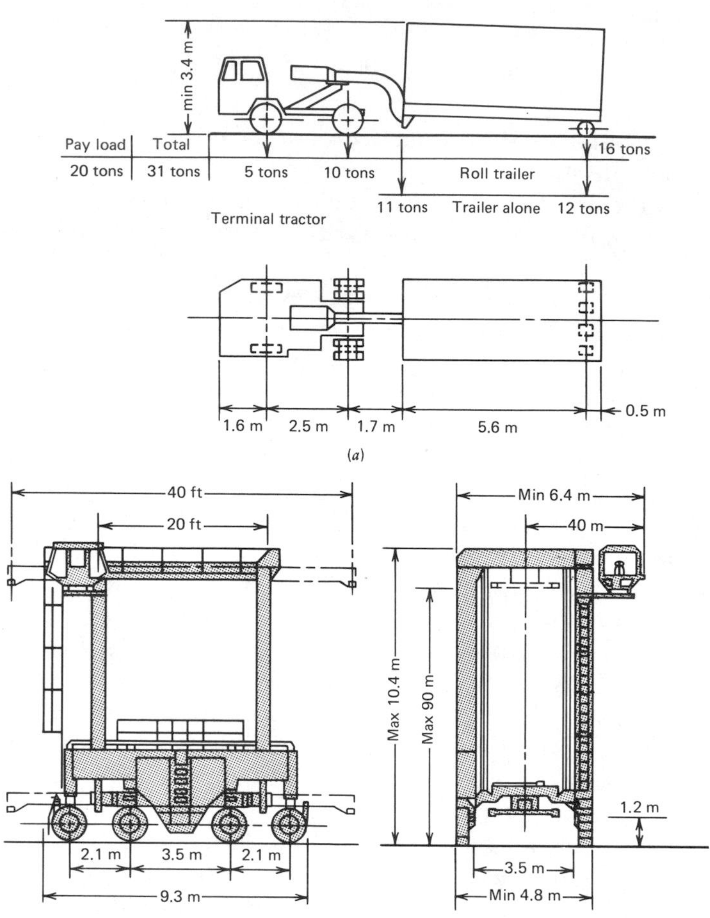


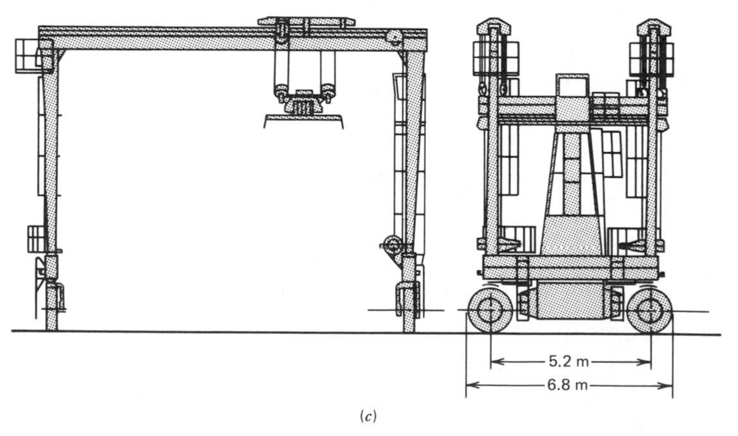


*Fig 22-13.* Equipment used to transport containers at a container terminal. (*a*) Typical yard tractor and 20-ft trailer. (*b*) Typical straddle carrier. (c) Typical gantry crane on rubber wheels. (Source: *Planning and Design of Ports and Marine Terminals,* by Agerschou, Hans, et al., John Wiley, New York, 1983.)

of two to six lanes in each direction. Each lane is normally provided with a truck scale to weigh the entering and departing containers. A receiving and delivery office usually is located at the entry-exit point to handle the necessary paperwork and to assign storage positions to incoming containers.

Container ports need to be served by major traffic arteries. Container operations generate substantial truck traffic, especially on days that ships are in port. This peaking necessitates truck queues waiting to enter the terminal, and unless generous approach roads are provided, enormous congestion will result.

*Garage and Inspection Building.* A small building for the physical inspection of arriving or departing containers normally is located in the vicinity of the entrance and adjacent to the marshaling area. In addition, a garage may be provided to maintain the stevedoring devices used to handle the containers in the marshaling yard.

Direct transfer of containers from rail to ship is not generally feasible. In such a transfer, the string of rail cars would have to be moved continuously during the loading and unloading operation to position the containers under the crane. Instead, it is usually better to unload the containers off-site and transport them to the marshaling area on rubber tires. Therefore, tracks will not be required at the stringpiece, although they should be provided to the packing shed.

## 22-9. ROLL-ON/ROLL-OFF FACILITIES

Because of the high investment costs of container ports with conventional lift-on/lift-off (LO/LO) facilities, many ports have developed facilities for roll-on/roll-off (RO/RO) cargo handling. These systems are often combined with conventional handling systems.

RO/RO service can be provided by three basic methods [12]:

1. The tractor and trailer unit drives on, remains on the ship during its voyage, and drives off at the destination port. This procedure is generally suitable only for short trips since the tractor unit utilizes space and remains idle during the sea journey.
2. The tractor unit tows the container from the storage area onto the ship and drives off, leaving only the trailer and container for the sea journey. Another tractor will be required to tow the trailer off the ship at the destination port. The tractor unit may be a normal highway unit or a smaller unit used only in the port area.
3. A tractor unit tows the trailer and container onto the ship where a ship crane lifts it off and stores it, or else a straddle carrier transports the container aboard and stacks it on the ship. Only the container remains on the ship during the sea trip. The process is reversed at the destination port.

RO/RO operations may also vary because of differences in ship design. Loading ramps may be provided in the bow, stern, or side of the ship.

Where RO/RO operations are used, special mooring devices may be required to hold the ship properly in place. Special adjustable ramps may be required on the wharf to accommodate variations in ship design and in the tide. Where cargo is handled through the side of a vessel, additional dock apron area will be needed to allow for ramp clearance. Since RO/RO loadings and unloadings can be performed in a short period of time, a large service area will be required for the processing and temporary storage of the freight.

## 22-10. LIGHTER-ABOARD-SHIP TERMINALS

A novel unitized cargo concept designed to maximize utilization of oceangoing ships and terminal facilities is the lighter-aboard-ship system. With this system, cargo is loaded onto large barges (lighters), then the barges and cargo are placed aboard a specially designed mother ship for the sea journey. The lighters are loaded aboard the mother ship either by a large crane on an overhang at the stern, or are lifted by an elevator onto the ship after being floated into a stern well. In the destination harbor the barges and cargo are unloaded and moved to a port facilities where the cargo is unloaded from the barges. (See Fig. 22-14.)

*Fig 22-14.* A seabee vessel. (Source: Lykes Lines.)

The mother ship may not need a land berth; however, some of the lighter-carrying ships are designed to transport both lighters and containers and therefore must have access to a container port.

Special requirements of lighter-aboard-ship terminals include facilities for lighter berthing and loading, berthing areas for tugboats, and a storage basin for the empty and loaded lighters.

More information on the functional planning of lighter-aboard-ship terminals is given in Reference 10.

## 22-11. DRY BULK CARGO TERMINALS

A wide variety of materials can best be transported unpackaged without the use of containers or pallets. Such commodities, termed dry bulk cargo, include grain, coal, sand, gravel, salt, sugar, scrap metal, and ores. Specialized port facilities are required to handle dry bulk cargo. These facilities may be placed on-shore or off-shore, but preferably are separated from general cargo and container ports.

Dry bulk cargo ships tend to be larger than general cargo ships and may require deeper channels and berths, more rugged fender systems, and more extensive storage areas. The storage areas may be open, enclosed, or a combination of both, depending on the nature of the material to be handled at the port.

Cargo-handling systems are varied, depending on the characteristics of the materials being handled. They include belt conveyor systems, cranes with clam shell buckets, pipelines that transport light weight materials pneumatically or in a hydraulic suspension, and many others.

At a dry bulk cargo terminal, special facilities may be required for weighing the cargo, for protecting against fires and explosions, and for controlling dust and pollution [10].

## 22-12. PETROLEUM TERMINALS

In recent years, worldwide increases in demand for petroleum have created a need for larger and more efficient tankers. The size of tankers joining the fleet has increased dramatically; the largest tanker now in service measures over 1300 ft in length and more than 200 ft in width. From 1967 to 1977, the average draft of the world's tankers increased from 31 to 37 ft. Many of the newest tankers have drafts approaching 100 ft [14].

The large draft of these ships has brought about a need for a number of new deep-water terminals. Because of the high operating costs of these vessels, it has become more important to develop terminal facilities capable of providing ship turnaround in the least possible time.

While a number of ports have dredged deeper channels, others have developed petroleum terminals thousands of feet off-shore, connected to the shore by a large submarine pipeline.

Two basic types of off-shore berthing facilities have been developed: (1) buoy moorings and (2) fixed platforms.

A conventional and simple means for the off-shore handling of petroleum, buoy moorings consist of a suitable anchorage facility adjacent to a system of submerged pipelines equipped with hoses lying on the bottom and marked by buoys. Generally, a ship using the facility is brought in and secured, facing in the direction of the wind and swells. The ship's hoisting tackle is used to bring the hoses to the surface and to facilitate their connection to the ship.

Buoy moorings tend to become impractical when more than 12-in. hoses are required. Experience has shown that hoses larger than 12 in. are difficult to handle with the ship's gear. Further, the system is unsuitable where high winds and unsteady seas prevail [10].

Typically, the fixed-platform facility consists of a tower on which a rotatable mooring boom is mounted. During loading and unloading operations the lines are attached to the mooring boom, with the ship heading into the wind and waves. A 500-to 600-ft loading arm extends along the tanker to facilitate the handling of loading hoses that connect the tanker and a submarine pipeline extending to the shore. When not in use the loading arm is submerged, during loading, it is swung alongside the ship. A terminal of this type is described in Reference 15.

Although high in capital cost, fixed platform terminals are reliable and efficient, providing a ship turnaround of approximately 18 hr [10].

## PROBLEMS

1. Ships arrive at a port in accordance with a Poisson probability distribution. On the average there are three ships in port.
   a. Determine the percentage of time during the year that there will be 0, 1, 2, 3, 4, 5, 6, 7, and 8 ships present.
   b. Plot a histogram of these data similar to Fig. 22-7.
   c. What percentage of time during the year will there be more than 5 ships present?
2. Develop a graph showing the energy absorption characteristics of a 14-in. timber pile using SI (metric) units based on Fig. 22-4.
3. Suppose that on the average it costs $2000/day for a ship to wait for a berth. Using the data given in Problem 1, what would be the annual costs of waiting at a port that has eight berths?
4. Develop a graph showing the load-deflection of characteristics for gravity fenders using SI (metric) units based on Fig. 22-6.
5. Suppose that the cost of an idle berth is $500/day. Using the arrival data

given in Problem 1, what would be the annual costs of idle berth time at a port that has eight berths?

6. Three million tons of cargo is to be handled at a port that has a daily cargo handling rate of 1100 tons/berth. The average cost of a waiting ship is $1500, and the average cost of an idle berth is $500/day.
   a. How many berths should be provided?
   b. What will be the percent of congestion?
   c. What will be the percent of occupancy?
   d. Determine the annual capacity of the port.

## REFERENCES


1. Lee, Theodore T., "Design Criteria Recommended for Marine Fender Systems," *Proceedings of Eleventh Conference on Coastal Engineering,* American Society of Civil Engineers, September 1968.
2. Brant, Austin E., Jr., "The Port of Chicago," *Proceedings,* American Society of Civil Engineers, Vol. 84, No. WW4, September 1958.
3. Plumlee, Carl H., "Optimum Size Seaport," *Proceedings,* American Society of Civil Engineers, Vol. 92, No. WW3, August 1966.
4. Nicolaou, S.N., "Berth Planning by Evaluation of Congestion and Cost," *Proceedings,* American Society of Civil Engineers, Vol. 93, No. WW4, November 1967.
5. Fratar, T.J., A.S. Goodman, and A.E. Brant, Jr., "Prediction of Maximum Practical Berth Occupancy," *Proceedings,* American Society of Civil Engineers, Vol. 86, No. WW2, June 1960.
6. Wanhill, Stephen R.C., "Further Analysis of Optimum Size Seaport," *Journal of The Waterways, Harbors and Coastal Engineering Division,* American Society of Civil Engineers, No. WW4, November 1974.
7. Miller, Alan J., "Queuing at Single-Berth Shipping Terminal," *Journal of the Waterways, Harbors and Coastal Engineering Division,* American Society of Civil Engineers, No. WW1, February 1971.
8. Nilsen, Kenneth O., and Usamah Abdus-Samad, "Simulation and Queuing Theory in Port Planning," *Ports '77,* American Society of Civil Engineers, 1977.
9. Soros, Paul, and Andrew T. Zador, "Port Planning and Computer Simulation," *Ports '77,* American Society of Civil Engineers, 1977.
10. *Port Planning, Design and Construction,* The American Association of Port Authorities, Washington, D.C., 1973.
11. Tozzoli, Anthony J., "Containerization and Its Impact on Port Development," *Journal of the Waterways, Harbors and Coastal Engineering Division,* American Society of Civil Engineers, August 1972.
12. Bruun, Per, *Port Engineering,* Third Edition, Gulf Publishing Company, Houston, 1981.
13. Mazorol, William J., "Design of Container Terminals," *Ports '77,* American Society of Civil Engineers, 1977.
14. *A Statistical Analysis of the World's Merchant Fleets,* U.S. Department of Transportation, Maritime Administration, January 1, 1985.

## *Appendix*

# EQUIVALENTS AMONG METRIC (SI) AND TRADITIONAL INCH-POUND-UNITS OF MEASUREMENT

| *Traditional to Metric* | *Metric to Traditional* |
|---|---|
| *Lengths* | |
| 1 in. = 2.54 cm | 1 cm = 0.3937 in. |
| 1 ft = 0.3048 m | 1 m = 3.2808 ft |
| 1 mi = 1.6093 km | 1 km = 0.6214 mi |
| *Areas* | |
| 1 in.$^2$ = 6.4516 cm$^2$ | 1 cm$^2$ = 0.1550 in$^2$ |
| 1 ft$^2$ = 0.0929 m$^2$ | 1 m$^2$ = 10.764 ft$^2$ |
| 1 mi$^2$ = 2.590 km$^2$ | 1 km$^2$ = 0.3861 mi$^2$ |
| 1 acre = 4047 m$^2$ | 1 m$^2$ = 0.0002471 acres |
| 1 acre = 0.4047 hectares | 1 hectare = 2.471 acres |
| *Volumes* | |
| 1 gal (liquid) = 3.7854 liter | 1 liter = 0.2642 gal (liquid) |
| 1 ft$^3$ = 28.3169 liters | 1 liter = 0.35313 ft$^3$ |
| 1 ft$^3$ = 0.02832 m$^3$ | 1 m$^3$ = 35.3147 ft$^3$ |
| 1 yd$^3$ = 0.7646 m$^3$ | 1 m$^3$ = 1.3079 yd$^3$ |
| *Velocities* | |
| 1 mph = 1.6093 km/hr | 1 km/hr = 0.6214 mph |
| *Mass* | |
| 1 lbm = 0.4536 kg | 1 kg = 2.2028 lbm |
| 1 ton (2000 lbm) = 907.18 kg | 1 kg = 0.0011 ton (2000 lbm) |
| *Force* | |
| 1 lbf = 4.4482 N | 1 N = 0.2248 lbf |
| *Pressure or Stress* | |
| 1 lbf/ft$^2$ = 47.880 N/m$^2$ | 1 N/m$^2$ = 0.02088 lbf/ft$^2$ |
| *Flow Rates* | |
| 1 ft$^3$/sec = 0.02832 m$^3$/sec | 1 m$^3$/sec = 35.3147 ft$^3$/sec |

# INDEX